Calculus with Analytic Geometry

Second Edition

Calculus with Analytic Geometry

NATHAN O. NILES

Associate Professor of Mathematics, Retired
U.S. Naval Academy

GEORGE E. HABORAK

Professor of Mathematics
and Vice President
for Student Affairs
College of Charleston

PRENTICE-HALL, INC. *Englewood Cliffs, New Jersey 07632*

Library of Congress Cataloging in Publication Data

NILES, NATHAN O.
Calculus with analytic geometry.

Includes index.
1. Calculus. 2. Geometry, Analytic. I. Haborak,
George E. II. Title.
QA303.N55 1982 515'.15 81-11971
ISBN 0-13-112011-5 AACR2

Editorial/Production Supervision: KAREN WINKLER WAGSTAFF
and LYNN S. FRANKEL
Art Production: JILL S. PACKER
Interior Design: KAREN WINKLER WAGSTAFF
Cover Design: MIRIAM RECIO
Manufacturing Buyer: GORDON OSBOURNE

ISBN 0-13-112011-5

Printed in the United States of America

10 9 8 7 6 5 4 3

Prentice-Hall International, Inc., *London*
Prentice-Hall of Australia Pty. Limited, *Sydney*
Prentice-Hall of Canada, Ltd, *Toronto*
Prentice-Hall of India Private Limited, *New Delhi*
Prentice-Hall of Japan, Inc., *Tokyo*
Prentice-Hall of Southeast Asia Pte. Ltd., *Singapore*
Whitehall Books Limited, *Wellington*, *New Zealand*

Contents

14 DIFFERENTIAL EQUATIONS 412

15 FUNCTIONS OF TWO VARIABLES AND PARTIAL DIFFERENTIATION 461

16 REPEATED INTEGRATION 481

17 INFINITE SERIES 503

Preface to Second Edition

OBJECTIVES

The aim of this second edition, as of the first, is to provide a good working knowledge of calculus which students can apply in other courses and later in their occupations. To achieve this goal, many practical problems are utilized, and skill in computational technique and problem solving is emphasized.

The material is suitable for use in technical colleges and institutes, ECPD programs, engineering and technical programs in community colleges, and anywhere that a knowledge of the techniques of calculus is desired. The material is presented in an informal conversational style using the intuitive approach. Formal proofs of theorems are given when they help clarify the discussion.

FEATURES

There are three major changes from the first edition.

1. The insertion of a problem similar to the illustrative example that it immediately follows. In effect, this says: "We did one and this is what you do. Now you do one." Thus students have the opportunity to do a problem immediately after reading an example.
2. **Warnings** are given to help students avoid common types of errors and/or when additional emphasis is needed. These warnings are indicated by the following symbol: ☛
3. The derivatives of the trigonometric functions are obtained by applying

the delta formulation to the tangent function and then making use of basic trigonometric identities and the power rule. (This approach eliminates the need for developing $\lim_{x \to 0} (\sin x)/x = 1$ or for trigonometric formulas which a student may have forgotten or never mastered.) Students really like and understand this procedure.

At the end of each chapter students are asked to express in their own words their understanding of the concepts discussed in that chapter.

Because many of the limits treated in calculus depend on manipulating the indeterminate form of 0/0, we have discussed division involving zero and have included many exercises relating to it. More emphasis has been placed on the fact that the derivative of a function with respect to a variable (or the derivative) is merely a new function derived from the original function according to a certain rule, and this new function is a "rate of change" which leads to many applications.

As in the first edition, we start with the calculus of polynomials. However, in Chapter 2, to help students in their work in concurrent courses, we state the rules for differentiation of products, quotients, powers, and sines and cosines. (We develop these rules later.) This adds flexibility to the book for teaching courses of varying lengths and topics. A look at the table of contents will show the number of varied topics that we discuss. The discussion in the first six chapters of basic applications of calculus using polynomial functions provides students with a solid base for their study in later chapters of the calculus of other functions.

The material on volumes of revolution has been rewritten to give a smoother and easier discussion of the "disk, washer, and cylindrical shell" methods. Several other topics have been rewritten where we feel it will benefit students. To help clarify the discussion of motion along a curve, we have introduced the position vector in **ij** form and also discuss the tangential and normal unit vectors. We have added more practical problems and have included some from the field of pharmacokinetics.

Since electronic calculators are now readily available, the answers to most problems and exercises are given in decimal form. Also, there is not now a need for tables of functions. Answers to all problems and odd-numbered exercises and/or parts (a), (c), (e), and so on, are included as the final section of the book.

We wish to express our sincere thanks and gratitude to the people of Prentice-Hall, Inc. who have helped in the publishing of this book. They have been most kind and patient with us. In particular a special 'thank you' to Karen Winkler Wagstaff, and Lynn Frankel, production editors, whose enthusiasm and keen senses of humor have made this edition a delight to work on.

N. O. Niles

G. E. Haborak

Preface to First Edition

This book is intended to give readers a clear understanding of the fundamentals of elementary calculus. Instead of dwelling on rigorous epsilonic proofs of theorems, we define the basic concepts and use the theorems necessary for a thorough development of the topics covered. We believe that elementary calculus can be explained intuitively, clearly, and correctly without having to prove all the facts that are used. We do this by giving many illustrative examples without going into all of the many exceptions or singularities of each topic. We utilize common examples in developing concepts—for example, because the concept of speed is far more familiar than the concept of slope of a curve, we develop the derivative as a rate of change, and do not discuss slope until we give a geometrical interpretation of the derivative. Similarly, we emphasize integration as a summation with the area under a curve being a by-product.

In developing the concepts of derivatives and integration of functions, we treat polynomial functions first, as these are the easiest functions to start with. Since exponential functions have many technical applications, they are introduced before the trigonometric functions. In discussing the latter subject, we emphasize the trigonometric functions of real numbers. This helps students in studying periodic motions, such as vibrations, to understand how the argument can be expressed in time and to realize that there are some vibration problems in which no angle is involved.

The basic applications of calculus are introduced in the discussion of polynomials, and they are repeated as the calculus of the other functions are studied.

The exercises at the end of most sections are divided into three groups:

A, B, and C. (Not all Exercise lists have Group B and/or C). Group A has straightforward computation exercises that follow closely the illustrative examples. Group B exercises involve more difficult algebraic manipulations, while those in Group C involve more difficult ideas and some theoretical developments. Some Group C exercises are extensions of topics in the text. Many of the exercises involve technical applications.

We have taken "poetic license" and have described many physical phenomena in terms of polynomials, such as electric current, charge on a capacitor, and magnetic flux. This is done to enable students to note that the derivative may be used in many ways. (This can be somewhat justified when one considers that the elementary functions can usually be expanded into Maclaurin series.)

By looking at the table of contents one can see that the arrangement of the material allows a great deal of flexibility in designing a calculus course. For example, a short course might use only the first six or seven chapters.

This text is written to be accessible to students. Thus some topics may be assigned as collateral reading, and it is hoped that some of the Group C problems might encourage students to pursue further their study of mathematics. As a further aid to students we have included at the end of each chapter a summary of the important words and concepts of that chapter.

We wish to thank John Wiley & Sons, Inc, for their "blanket permission" to use material from their texts *Plane Trigonometry*, Second Edition, and *Algebra and Trigonometry*, both by Nathan O. Niles. We also wish to thank McGraw-Hill, Inc., for their permission to use several problems from the text *Engineering Problems Illustrating Mathematics* by John W. Cell, published in 1943. These problems are designated by the word "Cell" placed after them. Our heartiest thanks for an excellent job to the typists of our manuscript, Miss Bertha Nowottnick and Mrs. V. N. Robinson, and to Professor J. H. White for his assistance in obtaining answers to many of the problems.

N. O. Niles

G. E. Haborak

Note to the Student

Calculus is a fascinating topic. It enables you to find solutions to many interesting problems. As you study this book, have pencil and paper at hand. Read carefully and make notes as you proceed. Be sure that you use the symbols of mathematics correctly. They are the "tools" of mathematicians just as a saw and a hammer are the tools of a carpenter. In reading the illustrative examples, read the equations as sentences, which they really are. For example, read the equation $y = 4x^2$ as "y equals four times x squared." The illustrative examples are for your benefit. Follow them carefully and rewrite them supplying any of the missing steps. To learn something you must get involved in it. To help you learn calculus we have stated a problem immediately after each illustrative example. You should make every effort to do these problems on your own. If you encounter difficulty, go back and study the example again. The answers to these problems are provided as the last section of the book.

The abbreviations for units of measurement are the standard symbols of the IEEE. For your help we list a few of them here.

A	Ampere	ft/s	feet per second	mi	mile
cm	centimeter	H	Henries	min	minute
C	Coulombs	h	hour	Ω	Ohms
F	Farads	in.	inch	s	seconds
ft	feet	m	meter	V	Volts
				W	Watts

Sometimes more than one reading is necessary to grasp the point of a new topic or concept. Do not becomes discouraged if you must reread a sentence or paragraph three or four times. Perhaps on the fifth reading it will make sense! You can only learn mathematics by doing mathematics; it is not a spectator sport. The problems in Group A follow very closely the illustrative examples. Be sure that you can do them. The more problems you do, the easier they become.

Good luck to you in your study of calculus.

N. O. Niles

G. E. Haborak

Calculus with Analytic Geometry

Chapter I

Some Preliminaries

1.1 INTRODUCTION

To some people *calculus* is a beautiful word. To others it is sometimes terrifying. In layman's language we might say that it means calculations. Once the terminology used in calculus is known, some of the terrifying aspects disappear. The words *function*, *limit*, *differential*, *derivative*, and *integral* are the backbone words of calculus. The subject of calculus is sometimes divided into two parts: *differential calculus* and *integral calculus*. We shall give an idea of the meaning of the words differential and integral shortly. Perhaps to the uninitiated the symbols used by mathematicians are frightening. They are used to shorten many statements and expressions. As you use the symbols you will become more familiar with them and begin to see their usefulness.

In differential calculus the main symbol used is dy. This is read as "dee-wye," is called the *differential of* y, and means a "little bit of y." It does *not* mean d *times* y as in algebra. Similarly, dx is "dee-ex" and represents a "little bit of x." The symbols du and dt then are read "dee-you" and "dee-tee" and mean a "little bit of u" and a "little bit of t," respectively. Thus dy, dx, du, and dt are all differentials. We shall give the mathematician's definition of a differential later.

Another extremely important symbol in differential calculus is dy/dx. This is read as "dee-wye dee-ex," is called the *derivative of* y *with respect to* x, and compares a "little bit of y" with a "little bit of x." It is also a symbol of the operation of differentiation. Similarly, dy/dt, ds/dt, and du/dx are the *derivative of* y *with respect to* t, the *derivative of* s *with respect to* t, and the *derivative of* u *with respect to* x, respectively.

The main symbol used in integral calculus is an elongated S written as $\int$. It is called an *integral* and means a *summation*. It is usually written with a differential symbol. Thus, since dx means a "little bit of x," the symbol $\int dx$ means the "sum of a lot of little bits of x," which is just x, and we write $\int dx = x$. (We shall give some refinements of this later.)

We shall study applications and uses of the symbols d and $\int$ and see how differential calculus and integral calculus are tied together by the *fundamental theorem of calculus.* To do this advantageously we make use of analytic geometry (which we develop as needed), which is merely an algebraic analysis of geometry. That is, we represent geometric figures by algebraic equations, and vice versa.

Calculus builds upon ideas from algebra and geometry. For example, with algebra we can find the average velocity of an object over an interval of time. Calculus enables us to find the instantaneous velocity of an object at a particular time. Using geometry we can find the area of regions bounded by straight lines and/or circles. Using calculus we can find the area of regions bounded by curves other than circles. Work done by a constant force can be determined using algebra. It takes calculus to find the work done by a varying force. As we proceed with calculus you will see how much more powerful it is than algebra.

The development of calculus depends on the concept of a *limit.* This is sometimes a difficult concept to grasp. Do not be discouraged if you do not understand it for a while. As a first step we might think of a "limit" as something we approach and get close to but never reach. For example, if we desired to approach a wall but not reach it, we might do so by going halfway to the wall each time we moved. In Fig. 1.1.1 if we started at A and then moved to B,

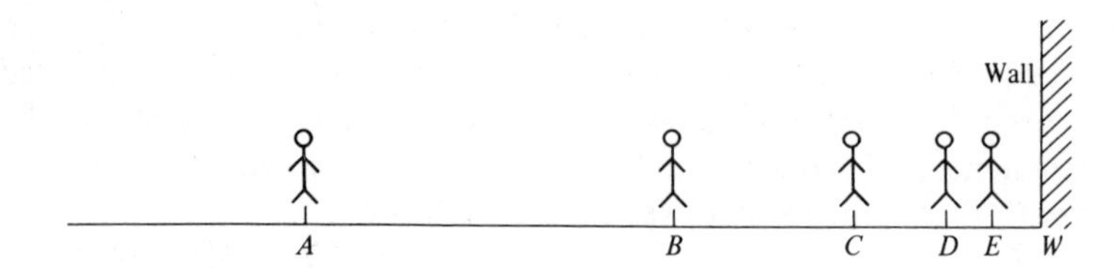

Figure 1.1.1 Concept of limit.

then to C, then to D, and so forth, where $AB = \frac{1}{2}AW$, $BC = \frac{1}{2}BW$, $CD = \frac{1}{2}CW$, $DE = \frac{1}{2}DW$, we could get as close to the wall, W, as we desire. Mathematicians have a more exact definition of *limit* which we shall give when needed.

The words *limit*, *differential*, *derivative*, and *integral* are used in conjunction with the word *function*. In giving a good definition of the concept of a function it is convenient to have at our disposal the concept of sets. In Sec. 1.2 we shall briefly discuss sets.

To get to the derivative of a function it is desirable to use the capital Greek letter *delta*, Δ, to indicate the *change* in a quantity. If the value of a quantity is 7 and some other time its value is 11, we say that the change in the quantity is 4. If, however, the second value had been 3, then the change would have been -4. The change in any quantity is always its *second value minus its first value.* To indicate the first value of a quantity, say y, we use the subscript 1 and write y_1 (read "y sub one"; note that the subscript is written to the right and below the letter). The second value is indicated by the subscript 2, and we write y_2. The *change in y* is then written as

$$\Delta y = y_2 - y_1$$

and is read "delta y."

Warning ***Be sure to subtract the first value from the second value to find the change,*** Δ.

EXAMPLE 1. If the electric current in a circuit is 3 A 5 s after the switch is closed and 20 A 15 s after the switch is closed, express the change in the current and the change in the time in terms of delta.

Solution. Using I to indicate current and t to indicate time, we have

$$I_1 = 3 \qquad t_1 = 5$$

$$I_2 = 20 \qquad t_2 = 15$$

Hence

$$\Delta I = I_2 - I_1 = 20 - 3 = 17$$

and

$$\Delta t = t_2 - t_1 = 15 - 5 = 10$$

Thus we can say that the current changed 17 A in 10 s.

PROBLEM 1. If the area of a circle is 13 m² when $t = 7$ s and is 5 m² at $t = 9$ s, express the change in the area and the change in the time using delta notation.

If in Example 1 we compare the change of I to the change of t, we may write

$$\frac{\Delta I}{\Delta t}$$

This ratio is called the *average rate of change of I over the time interval* Δt. Ratios of this type are basic for the definition of derivatives. Briefly, what we do is to study the limit of such ratios as the denominator and numerator get smaller and smaller or approach zero. If the limit of the ratio exists, then we have a derivative.

EXERCISES 1.1

Group A

1. For each of the following, find the change in the quantity.

(a) $t_1 = 22$, $t_2 = 29$.
(b) $s_1 = 9$, $s_2 = -7$.
(c) $HP_1 = 390$, $HP_2 = 315$.
(d) $I_1 = 6.1$, $I_2 = 6.15$.

2. (a) If $t_1 = 12$ and $\Delta t = 8$, find t_2.
   (b) If $s_1 = 96$ and $\Delta s = -20$, find s_2.
   (c) If $V_2 = 15$ and $\Delta V = -5$, find V_1.
   (d) If $I_2 = 26$ and $\Delta I = 0.5$, find I_1.

3. If $I_1 = 5$, $t_1 = 0$, $I_2 = 5.6$, and $t_2 = 0.1$, find the average rate of change of I over the interval Δt.

4. If $s = 5t$, complete the following table. What number is approached by the average rate of change of s over the intervals Δt as Δt gets smaller?

t_1	t_2	Δt	s_1	s_2	Δs	$\frac{\Delta s}{\Delta t}$
2.0	2.5	0.5	10.0	12.5	2.5	5
2.0	2.4					
2.0	2.3					
2.0	2.2					
2.0	2.1					

5. Consider the sequence of numbers $1, 1\frac{1}{2}, 1\frac{3}{4}, 1\frac{7}{8}, 1\frac{15}{16}, 1\frac{31}{32}, \ldots$. If we keep writing numbers in this sequence, will we ever get to the number 2? How close can we get to 2? Do you know the largest number that is less than 2? What do you think is the limit of this sequence of numbers?

6. Consider the numbers $1, \frac{1}{10}, \frac{1}{100}, \frac{1}{1000}, \ldots$. If the numbers are represented by $1/x$, what do you think is the limit of $1/x$ as x gets very large?

Group B

1. If a gauge pressure indicates 10 lb/in.^{2} of air in a tire and then more air is added until the gauge pressure indicates 32 lb/in.^{2}, express each pressure as p with a subscript. Write the change of pressure using delta notation.

2. If the volume of a balloon is given by $V = \frac{4}{3}\pi r^3$, find ΔV when the radius r is changed from (a) $r_1 = 2$ in. to $r_2 = 3$ in.; (b) $r_1 = 2$ in. to $r_2 = 1$ in.

3. If $s = 5t^2$, make a table as in Exercise 4 in Group A using the same values for t_1 and for t_2 with the addition of 2.01 and 2.001 for t_2. What number is approached by the average rate of change of s over the intervals Δt as t_2 gets closer to t_1?

4. Describe your idea of the concept of a limit.

5. What do you think is the smallest positive number? If I choose a small positive number, can you choose a smaller one?

1.2 INTRODUCTION TO SETS

In modern mathematics the function concept is one of the most basic concepts. We will soon define this concept in terms of sets of ordered pairs. Sets are so basic that some mathematicians believe that every mathematical concept may ultimately be defined in terms of them. For our purposes we like to think of a set as a "collection of elements." The elements may be anything. We will for the most part use real numbers. There are two ways to describe sets explicitly. One way is the *roster method*, in which the elements are listed and enclosed in braces. For example, the set of the letters of the word *Mississippi* is denoted as {i, M, p, s}. Note that an element of a set is not listed more than once, and the order of the listing is not important. Also note that the elements are separated by commas.

EXAMPLE 1. Let set $D = \{1, 2\}$ and $R = \{-3, 0, 4\}$. Write the set F whose elements are pairs of numbers with the first member of each pair an element of D and the second member an element of R.

Solution. $F = \{(1, -3), (1, 0), (1, 4), (2, -3), (2, 0), (2, 4)\}$.

Set F in Example 1 is an example of a set of ordered pairs of numbers. An *ordered pair* of numbers is a pair in which there is a first member and a second member. The ordered pair $(1, -3)$ is not the same as the ordered pair $(-3, 1)$. Why?

PROBLEM 1. Write the set F whose elements are ordered pairs of numbers with the first number an element of $D = \{-1, 0, 2\}$ and the second number an element of $R = \{5, -7, 0\}$.

A second way of describing sets is to use set-builder notation and a rule for inclusion in the set. We call this method the *rule method.* This method is convenient if one is describing a "large" set. Here a rule which determines those objects that are to be members of the set is enclosed in braces. Schematically, we use the symbol

{ ——————	——————	|	——————	}
start of set	member	"such that"	rule to be members	end of set

For example, if we desired the set of multiples of 10 such as 10, 20, 30, 40, and so forth, we might write $\{x \mid x = 10n, n \text{ a positive integer}\}$. This would be read "the set of all numbers x such that x equals $10n$, where n is a positive integer." The vertical line means "such that."

EXAMPLE 2. Use the roster method to describe the set of ordered pairs $\{(x, y) \mid y = x, x = -1, 0, 2, 5\}$.

Solution. $\{(-1, -1), (0, 0), (2, 2), (5, 5)\}$.

PROBLEM 2. Use the roster method to describe the set of ordered pairs $\{(x, y) \mid y = 3x^2, x = -2, -1, 0, 1, 2\}$.

A set that has no members is called a *null set* and is denoted by the symbol $\varnothing$.

EXAMPLE 3. The following set is an example of a null set:

$$\{x \mid x^2 = 1 \text{ and } x + 2 = 0\} = \varnothing$$

PROBLEM 3. Explain why the set in Example 3 is a null set. Give another example of a null set.

These few basic remarks about sets will be sufficient for this text.

EXERCISES 1.2

Group A

1. Write the set F that has for its elements pairs of ordered numbers with the first member an element of D and the second member an element of R.

(a) $D = \{-2, 3, 0\}$, $R = \{4, -1\}$.
(b) $D = \{-3, -5\}$, $R = \{9, 25, 0\}$.
(c) $D = \{a, i\}$, $R = \{s, t, n\}$.
(d) $D = \{9, 4\}$, $R = \{3, -3, 2, -2\}$.

2. Write out the set F of ordered pairs.

(a) $F = \{(x, x^2)\}$, $x = -2, -1, 0, 1, 2$.
(b) $F = \{(x, 2x)\}$, $x = -3, -1, 2, 4$.
(c) $F = \{(x, 3x + 7)\}$, $x = \frac{1}{3}, 1$.
(d) $F = \{(x, y)\}$, $x = 1, 2, 3, 4; y = x - 2$.
(e) $F = \{(x, y)\}$, $x = \frac{1}{2}, \frac{1}{3}, \frac{1}{4}; y = 12x$.
(f) $F = \{(x, y)\}$, $x = -2, -3, 0; y = x^2 - 2x$.

3. Given that $D = \{-2, -1, 0, 1, 2\}$, write out the set F where x is an element of D.

(a) $F = \{(x, y) \mid y = x\}$.
(b) $F = \{(x, y) \mid x^2 + y^2 = 4\}$.
(c) $F = \{(x, y) \mid y = 4x^2\}$.
(d) $F = \{(x, y) \mid x^2 + y^2 = -4\}$.

4. On a rectangular coordinate system (an XY-set of axes), plot the elements of each of the following sets, where each ordered pair represents the coordinates of a point.

(a) $A = \{(0, 0), (1, 1), (2, 2)\}$.
(b) $B = \{(0, 1), (1, 1), (2, 1)\}$.
(c) $C = \{(-3, 9), (-2, 4), (-1, 1), (0, 0), (1, 1), (2, 4), (3, 9)\}$.
(d) $D = \{(1, 1), (2, 2), (2, 3), (2, 4), (3, 5), (4, 5)\}$.

5. On a rectangular coordinate system, plot the ordered pairs of each set F in Exercise 3.

Group B

1. The axes of the rectangular coordinate system divide the XY-plane into four regions called *quadrants*. They are designated QI, QII, QIII, and QIV, as you should recall. (If not, see Sec. 1.4.) They may be defined in terms of set notation. For example, $\text{QI} = \{(x, y) \mid x > 0, y > 0\}$. Define in terms of sets QII, QIII, and QIV.

2. Shade the regions of the XY-plane indicated by each of the following sets.

(a) $\{(x, y) \mid y < 3\}$.
(b) $\{(x, y) \mid x > -2\}$.
(c) $\{(x, y) \mid -2 < x < 4, \ -1 < y < 3\}$.
(d) $\{(x, y) \mid x < -2, x > 4, y < -1, y > 3\}$.

3. Plot several points of the set $\{(x, y) \mid y = 2x - 3\}$. Can these be connected by a straight line?

1.3 FUNCTIONS

The concept of a function is very important. In fact, it is a central concept of modern mathematics. A function is similar to a vending machine in which you put some money, push a button, and a product comes out. Figure 1.3.1 shows

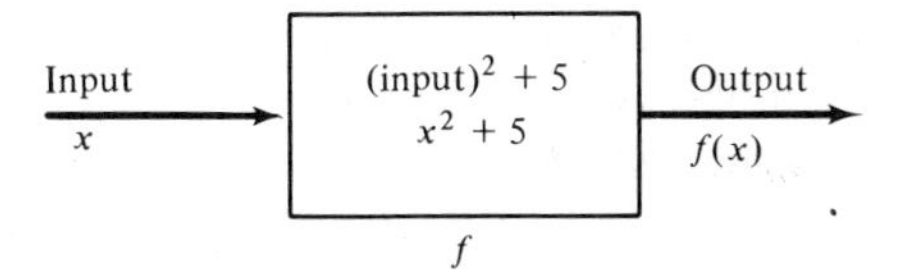

Figure 1.3.1

a function box where the output is obtained by adding 5 to the square of the input. If we denote the input by x, the function by f, and the output by the symbol $f(x)$, read "f at x," we may write the equation

$$f(x) = x^2 + 5$$

This equation, or formula, gives the rule for finding the output for a given input. The symbol $f(x)$ has been used in mathematics since the 1700s to denote the functional value of x. We must not confuse it with f times x as in algebra.

EXAMPLE 1. For the function represented in Fig. 1.3.1, find the output when the input is 3.

Solution. Here we denote the input by $x = 3$. Using this value for x in the formula, we obtain the output

$$f(3) = (3)^2 + 5 = 14$$

PROBLEM 1. Find the output for the function represented in Fig. 1.3.1 when the input is -2.

Different letters are used to denote functions. Some of them are f, g, s, G, F, u, v, and w.

EXAMPLE 2. The expression $g(7) = -11$ means that for an input of 7, the output of the g (read "gee") function is -11. The expression $F(x + h)$ is the output of the F (read "cap eff") function for an input of $x + h$.

PROBLEM 2. Tell what each of the following expressions means. (a) $f(-5) = 2$; (b) $G(0) = -6$; (c) $g(x + h) = x^2 + 2hx + h^2$; (d) $f(x + h)$.

The set of numbers that is used for the inputs is called the *input set* or the *domain*. The set of the corresponding outputs is called the *output set* or the *range*.

We are now ready to state the definition of a function.

Definition 1.3.1 *A function* $\boldsymbol{f}$ *with domain* $\mathbf{D}$ *(input set) and range* $\mathbf{R}$ *(output set) assigns to each element* $\boldsymbol{x}$ *of* $\mathbf{D}$ *exactly one element* $\boldsymbol{f}(\boldsymbol{x})$ *of* $\mathbf{R}$*. The element* $\boldsymbol{f}(\boldsymbol{x})$ *is said to be the value of the function at the argument* $\boldsymbol{x}$*.*

Upon a careful look at this definition we note that a function consists of three parts: (1) an input set called the domain; (2) an output set called the range; and (3) a rule of correspondence that assigns to every element of the input set one and only one element of the output set. This rule is usually given in the form of an equation or a formula.

The symbol used to denote an element of the domain (input set) is called the *independent variable*. The symbol used to denote an element of the range (output set) is called the *dependent variable*.

If we let y be an element of the range R, corresponding to an element x of the domain D, we may write

$$y = f(x)$$

Here x is the independent variable and y is the dependent variable.

Functions with a domain and range consisting only of real numbers are called *real functions*. These are the types of functions we shall study. In specifying a particular function, it is necessary to specify all three parts of the function either directly or indirectly. Noting that the rule of correspondence is the important part of a function when the domain and range are quite evident [such as we do not divide by zero (see Sec. 1.5) or take square roots of negative numbers], we loosely speak of

$$f(x) = \text{rule of correspondence}$$

as the function. This implies that the domain and range consist of numbers for which $f(x)$ makes sense.[1] For example, $f(x) = x^2$ implies that f is the *squaring* function with all real numbers as its domain and with all nonnegative numbers as its range.

EXAMPLE 3. For $f(x) = 3x^2 - x - 7$, find $f(0), f(-2), f(x + h)$, and $f(x + h) - f(x)$.

[1] This means that we reject even roots of negative numbers and division by zero. In some applications relevance is included in "making sense." For example, a payroll clerk notes that wages (W) are related to hours (t) and an hourly rate (r) through the function $W = rt$. Generally, however, $0 \leqq t \leqq 40$, and this is relevant for this function. For $t > 40$, a new function would apply.

Solution. The symbol $f(a)$ denotes the value of the function for an input of a. If we replace each x in the rule by a pair of parentheses we have for

$$f(x) = 3x^2 - x - 7$$
$$f(\ ) = 3(\ )^2 - (\ ) - 7$$

Now all we have to do is place the given input in each pair of parentheses and proceed with the evaluation. Thus

$$f(0) = 3(0)^2 - (0) - 7 = -7$$
$$f(-2) = 3(-2)^2 - (-2) - 7 = 7$$
$$f(x + h) = 3(x + h)^2 - (x + h) - 7 = 3x^2 + 6xh + 3h^2 - x - h - 7$$
$$f(x + h) - f(x) = 3x^2 + 6xh + 3h^2 - x - h - 7 - (3x^2 - x - 7) = 6xh + 3h^2 - h$$

The expression $f(x + h) - f(x)$ will be used a great deal in Chap. 2.

Warning ***Remember that $f(x + h)$ is the output of the f function for an input of $x + h$ and is not $fx + fh$.***

PROBLEM 3. For the squaring function $f(x) = x^2$, find $f(4), f(-1), f(h)$, $f(x + h)$, and $f(x + h) - f(x)$.

Sometimes it is useful in graphing functions to use the following equivalent definition of a function.

*A function **f** is a set of ordered pairs (x, y) with the first element from the input set and the second element the corresponding output, such that no two of the ordered pairs have the same first element.*

EXAMPLE 4. For the function $f(x) = x^2 - 3$, find $f(0)$, $f(2)$, and $f(-1)$. List the set of ordered pairs {*input*, *output*} for the indicated inputs.

Solution. Here the inputs, the values of x, are indicated to be 0, 2, and -1. (They are the numbers enclosed in parentheses after f.) Letting $y = f(x)$, we have

	Ordered pair (input, output) (x, y)
For input $x = 0$, $y = f(0) = (0)^2 - 3 = -3$.	$(0, -3)$
For input $x = 2$, $y = f(2) = (2)^2 - 3 = 1$.	$(2, 1)$
For input $x = -1$, $y = f(-1) = (-1)^2 - 3 = -2$.	$(-1, -2)$

The required set of ordered pairs is

$$\{(0, 3), (2, 1), (-1, -2)\}$$

PROBLEM 4. For the function $f(x) = 3x^2 - 7$, find $f(0), f(1)$, and $f(-2)$. List the set of ordered pairs {(*input*, *output*)} for the indicated inputs.

Any symbol or variable name can be used to denote the input. Thus the functions $f(x) = 3x^2 - x - 7$, $f(t) = 3t^2 - t - 7$, and $f(*) = 3(*)^2 - (*) - 7$ are the same real function.

There are many interesting functions. When the domain is the set of all real numbers and the range has only one number, the function is called the *constant* function. That is, $f(x) = c$, where c is a constant. When the output is the same as the input, $f(x) = x$, we have the *identity* function.

PROBLEM 5. For $f(x) = 7$, find $f(3), f(-11), f(a), f(0)$, and $f(x + h)$. What is the name of this function? What is the value of $f(x + h) - f(x)$?

PROBLEM 6. For $f(x) = x$, find $f(1), f(-7), f(h)$, and $f(x + h)$. What is the name of this function? Is $f(x + h) - f(x)$ equal to h or $f(h)$?

EXAMPLE 5. The absolute value function is denoted as $y = f(x) = |x|$, where $|x| = x$ for $x \geqq 0$, and $|x| = -x$ for $x < 0$. Here the domain is the set of all real numbers, and the range is the set of all nonnegative numbers. Thus for

$$f(x) = |x|$$

we have

$$f(3) = |3| = 3$$
$$f(-3) = |-3| = 3$$
$$f(0) = |0| = 0$$

PROBLEM 7. For $f(x) = |x - 3|$, find $f(0), f(3)$, and $f(-3)$.

EXAMPLE 6. The voltage E in a dc circuit is given by $E = IR$, where I is the current and R the resistance. If we state that $E = f(I) = IR$, then the voltage is a function of the current and R is considered to be constant. If, on the other hand, we wish to hold I fixed and vary R, then we would write $E = f(R) = IR$. At times both I and R must be considered as variables, and then we would have a function of two variables, $f(I, R) = IR$ (see Sec. 15.1). Note that changing the domain changes the function.

PROBLEM 8. The volume of a right circular cylinder is $V = \pi r^2 h$. Use functional notation to express the volume as a function (a) of r; (b) of h.

EXAMPLE 7. Consider the equation $y^2 - x = 0$. If we solve for y, to obtain $y = \pm\sqrt{x}$, we do not have a function as y would have two values for each value of x, the positive square root and the negative square root. This is an example of a *relation* which is merely a set of ordered pairs. If we solve for x to obtain $x = y^2$, then $x = f(y) = y^2$ is a function of y, because for each value of y we obtain only one value of x. The expression $y = \pm\sqrt{x}$ may be thought of as two different functions, $y_1 = f_1(x) = \sqrt{x}$ and $y_2 = f_2(x) = -\sqrt{x}$. The domain of f_1 is the set of all nonnegative numbers, as is the range. However, for f_2 the domain is the set of all nonnegative numbers, and the range is the set of all nonpositive numbers. Thus $f_1(4) = \sqrt{4} = 2$, while $f_2(4) = -\sqrt{4} = -2$.

PROBLEM 9. The circle of radius 5 whose center is at the origin of the rectangular coordinate system has the equation $x^2 + y^2 = 25$. Can this circle be expressed as a function of x? Why?

EXAMPLE 8. State the domain and range implied by

$$\text{(a)}\quad y = \sqrt{x - 4} \qquad \text{(b)}\quad y = \frac{x + 7}{x - 4}$$

Solution. In part (a) we must have $(x - 4) \geqq 0$ or $x \geqq 4$ in order not to have a square root of a negative number. Hence the domain is the set $\{x \mid x \geqq 4\}$. The range is the set of all nonnegative numbers.

In part (b), since division by zero is not allowed (see Sec. 1.5), we have to exclude $x = 4$. Hence the domain is the set of all numbers except 4, and the range is the set of all numbers except 1.

PROBLEM 10. State the domain and range implied by

$$\text{(a)}\quad f(x) = \sqrt{5 - x} \qquad \text{(b)}\quad f(x) = \frac{x - 3}{x + 5}$$

EXERCISES 1.3

Group A

1. State what each of the following expressions means.

(a) $f(-3) = 0$. (b) $F(x) = 3x$. (c) $f(x)$. (d) $f(x + h)$.

2. If $f(x) = 11$, find $f(2)$, $f(0)$, $f(-3)$, $f(x + h)$, and $f(x + h) - f(x)$. Why is the function $f(x) = c$ called the constant function?

3. If $f(x) = 3x$, find $f(-3)$, $f(3)$, $f(x + h)$, and $f(x + h) - f(x)$.

4. If $E = f(I) = IR$, find $f(15)$ and $f(0.1)$. If $E = f(R) = IR$, find $f(1000)$ and $f(500)$. Mathematically, the domain of $f(R)$ could include negative values of R. Does R, the resistance, usually have negative values in an electric circuit?

5. Functions can often be represented by an equation, but not every equation is a function. Which of the following equations are functions of x? [*Hint*: Solve for y and then replace y by $f(x)$.]

(a) $x - 4y = 7$.
(b) $x^2 - y + 13 = 0$.
(c) $x^2 + y^2 = 16$.
(d) $y^2 - 4x = 0$.
(e) $y = 3$.
(f) $x = 3$.

6. For each of the following functions, state, if possible, which of the variables are independent and which are dependent.

(a) $y + x^2 + 8x = 0$.
(b) $s - \frac{1}{2}gt^2 = 0$.
(c) $s - 10t = 16t^2$.
(d) $i + 6 + 15t = 0$.
(e) $y^2 + x = 0$.
(f) $x^2 + 4y = 0$.
(g) $x = 5 \cos t$.
(h) $s - 7 \sin t = 0$.

7. State the implied domain and range of each of the following functions.

(a) $f(x) = 9x^2$.
(b) $f(x) = \dfrac{3}{x + 5}$.
(c) $y = \sqrt{4 - x^2}$.
(d) $y = \sqrt{x^2 - 4}$.
(e) $y = |x - 1|$.
(f) $y = |x| + x$.
(g) $V = (\frac{4}{3})\pi r^3$, where V is the volume of a sphere of radius r.
(h) $A = \pi r^2$, where A is the area of a circle of radius r.
(i) $Q = 6t^3 + 2t$, where Q is the charge that passes a point in an electric circuit at time t.

Group B

1. The distance s a body falls due to the constant force of gravity g in time t is given by $s = \frac{1}{2}gt^2$.

(a) Express s as a function of t.
(b) Does the domain include negative values of t?
(c) Use the result of part (a) to find (i) $f(t + \Delta t)$; (ii) $f(t + \Delta t) - f(t)$.

2. If $f(x) = 4x^2 + 3$, show that (a) $f(2) = f(-2)$; (b) $f(-x) = f(x)$. When a function $f(x)$ satisfies the equality in part (b), $f(x)$ is said to be an *even* function.

3. If $f(x) = x^3 - 2x$, show that (a) $f(2) = -f(-2)$; (b) $f(-x) = -f(x)$. When a function $f(x)$ satisfies the equality in part (b), $f(x)$ is said to be an *odd* function.

4. Consider $f(x)$, $g(x)$, and $h(x)$. Let $h(x) = f(x)g(x)$. Make use of Exercises 2 and 3 to show that (a) $h(x)$ is even if $f(x)$ and $g(x)$ are both even; (b) $h(x)$ is odd if $f(x)$ is odd and $g(x)$ is even.

5. State whether the following equations represent odd or even functions or neither.

(a) $x + y = 0$; (b) $x^2 + y + 3 = 0$; (c) $x^2 + y + x = 3$.

6. If $f(x) = 3x^2$ and $x = g(t) = t - 3$, find (a) $f(2)$; (b) $g(2)$; (c) $f[g(2)]$.

7. The perimeter of a rectangle is 40 in. Denote one side of the rectangle by x and express the area A as a function of x; that is, $A = f(x)$.

8. An open box with a rectangular bottom is to be made from a piece of tin 10 by

20 in. by cutting out of each corner a square of side x in. and then folding up the sides of height x in. Find a function f such that the volume V is given by $V = f(x)$.

9. The cost C of fuel for running a train for an hour varies as the cube of the speed V. Express C as a function of V. If $C = \$25$ when $V = 40$, find C when $V = 80$.

10. The period T of a simple pendulum varies directly as the square root of its length L and inversely as the square root of the gravitational attraction g. Express T as a function of L. If for $L = 20$ and $g = 32$, $T = 4.95$, find T for $L = 30$ and $g = 32$.

11. The strength S of a rectangular beam varies jointly as the width w and the square of the depth d. Express S as a function of w and d. If for $S = 14{,}400$, $w = 4$, and $d = 12$, find S when $w = 12$ and $d = 4$.

12. The illumination I given by a light varies directly as the candlepower C and inversely as the square of the distance d from the light. Express I as a function of C and d.

13. The electrical resistance R of a wire varies directly as the length l and inversely as the square of its radius r. Express R as a function of l and r. If $R = 20\ \Omega$, $l = 100$ ft, and $r = 0.03$ in., find the resistance R if the length of the wire is doubled.

Group C

1. In Exercise 4 of Group B, what do you think is the domain of the function h; the range of h?

2. List some functions familiar to you which are even; odd.

3. What can be said about reciprocals of odd (even) functions; sums and differences of odd and even functions?

4. Let $f(x)$ be a function whose domain is all the real numbers. (a) Show that the function $e(x) = [f(x) + f(-x)]/2$ is even (see Exercise 2 of Group B). (b) Show that the function $o(x) = [f(x) - f(-x)]/2$ is odd (see Exercise 3 of Group B). (c) Show that any function whose domain is all the real numbers can be expressed as the sum of an odd function and an even function.

5. In Example 7 of Sec. 1.3, the concept of relation was introduced. Explain how *relation* and *function* are alike and how they are different.

1.4 GRAPHS OF FUNCTIONS

Graphs of functions give us a visual picture of the function and how changes in the independent variable affect the value of the function. In Exercises 1.2 it was assumed that you had some knowledge of plotting points. Here we briefly review the rectangular Cartesian coordinate system.

A directed line is a line to which a positive and a negative direction has been assigned. If A and B are any two distinct points on a directed line, then the directed distance from A to B is the negative of the directed distance from B

to A; that is, $AB = -BA$. Let two directed perpendicular lines in a plane (a flat sheet of paper) intersect at point O (Fig. 1.4.1). Call one of these lines the X-axis and the other line the Y-axis. The X-axis is usually drawn horizontally with the *positive* direction to the *right*. The Y-axis is usually drawn vertically with the *positive* direction *up*. These axes divide the plane into four regions called quadrants I, II, III, and IV. We now construct a coordinate system on each of these axes with the origin at O. We choose on each positive axis a unit distance from the origin and label it 1. Using this distance as a base, we can then assign every point on the line a number called its *coordinate*. In this way every number

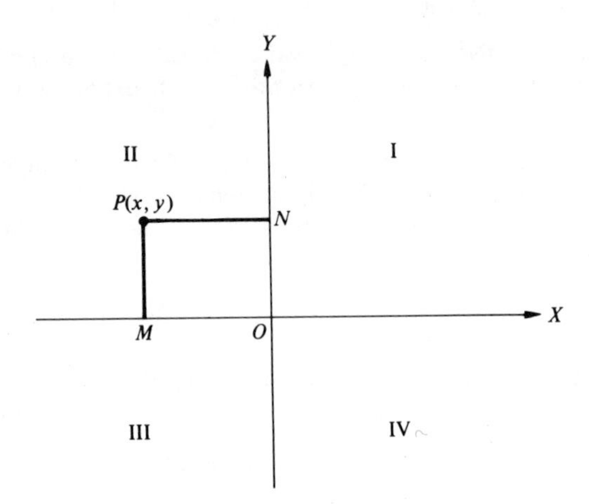

Figure 1.4.1 The Cartesian coordinate system of axes.

can be associated with a point on a coordinate axis, and we thus have set up a one-to-one correspondence between a point and a number.

Let P be any point in the plane. Through P draw a line perpendicular to the Y-axis intersecting it at N. Also draw a line through P perpendicular to the X-axis intersecting it at M.

Definition 1.4.1 *The directed distance* ***OM*** *is the* ***x****-coordinate or abscissa of point* ***P*** *and is denoted by* ***x****. The directed distance* ***ON*** *is the* ***y****-coordinate or ordinate of point* ***P*** *and is denoted by* ***y****.*

x and y are called the *rectangular coordinates* of point P and are written as the ordered pair (x, y). (Under no circumstances may you omit the parentheses surrounding the ordered pair when describing a point.) We are thus able to assign to each point in the plane an ordered pair of real numbers. To each ordered pair of real numbers there corresponds only one point in the plane. We thus have a one-to-one correspondence between a point and its rectangular coordinates. The symbol used to denote that *any* point P has coordinates x and y is $P(x, y)$.

The coordinates of a point may be negative since directed distances were used in the definition of rectangular coordinates.

EXAMPLE 1. Plot the points $A(3, 4)$, $B(-2, 5)$, $C(-1, -3)$, and $D(4, -2)$.

Solution. Draw two directed perpendicular lines and mark on them units of a suitable size (see Fig. 1.4.2). It is customary to let the unit distance be the same on both axes.

For point A, the abscissa equals 3 units and the ordinate equals 4 units.

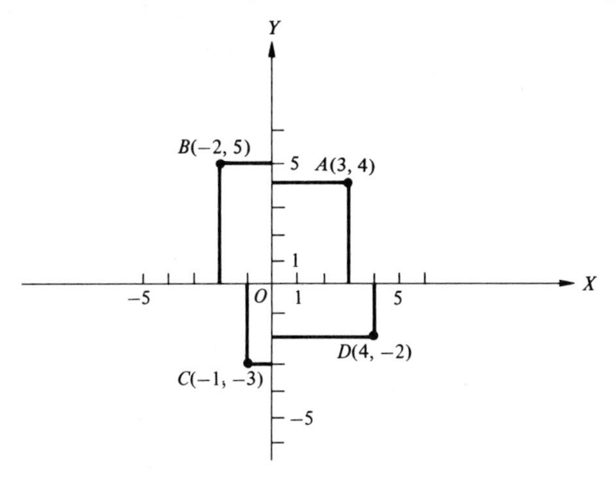

Figure 1.4.2 Plotting of points.

Mark them off and thus locate A. For point B, the abscissa equals -2 units and the ordinate equals 5 units, and so forth.

PROBLEM 1. Plot the points $A(2, 1)$, $B(-3, 0)$, $C(-2, -2)$, and $D(2, 0)$.

Now we are ready to define the graph of a function.

Definition 1.4.2 *The graph of a function in the* ***XY****-plane is the set of points in the plane whose coordinate pairs are the ordered pairs of the function.*

For example, suppose that the function is the set of ordered pairs $\{(0, 0), (1, 1), (2, 4), (3, 9)\}$. Then the points in the plane whose coordinate pairs are $(0, 0)$, $(1, 1)$, $(2, 4)$, and $(3, 9)$ comprise the graph of the function, which is shown in Fig. 1.4.3.

The graph of the function whose rule of correspondence is $y = f(x) = x^2$ for $0 \leqq x \leqq 3$ is shown in Fig. 1.4.4. It contains the points of the preceding example as well as an infinite number of points between them. As we cannot plot all the points of this function, we approximate them by drawing a curve through a sampling of plotted points. The points of the preceding example will do, as they indicate how the graph behaves. Note that values of inputs and outputs can be read from the graph. For an input of $x = 2.5$, we read from the graph that the output is $f(x) = 6.25$.

We now state three rules to follow in constructing a graph of a function in the XY-plane.

1. Obtain a table of ordered pairs of real numbers. Pick numbers, x, from the domain for the first entries and use the rule of correspondence, $f(x)$, to find the associated second entries.

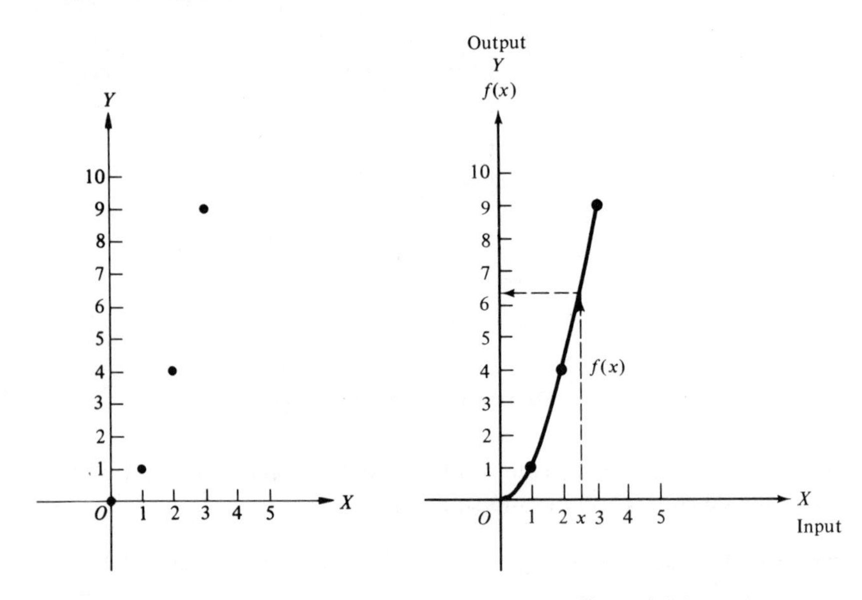

Figure 1.4.3 Graph of a finite set.

Figure 1.4.4

2. Construct a rectangular coordinate system and plot the ordered pairs. It is customary to let the unit distance on each axis be the same. Label each axis.
3. If the domain includes the line segments on the X-axis between the abscissas of the plotted points, connect adjacent plotted points by curves which approximate the function values in these intervals.

EXAMPLE 2. Draw the graph of $y = |x|$. [This means plot the graph of the function whose rule of correspondence is $f(x) = |x|$.]

Solution. First, since the rule of correspondence makes sense everywhere, we assume that the domain is the set of all real numbers. From Example 5 of Sec. 1.3 we know that

$$|x| = \begin{cases} x & \text{if} \quad x \geqq 0 \\ -x & \text{if} \quad x < 0 \end{cases}$$

So we see that $f(x)$ is built up from two functions—one whose domain is defined on 0 and the positive real axis, and the other whose domain is defined on the negative real axis. (These two functions do not fit together smoothly—thus the corner in Fig. 1.4.5.) We shall draw only a portion of the graph and use our imagination to extend it. First, we form a table of ordered pairs. Thus

x	-3	-2	-1	0	1	2	3
$y = \|x\|$	3	2	1	0	1	2	3

We plot these points, and since it is appropriate (the domain contains the intervals in between), we connect them by curves (see Fig. 1.4.5).

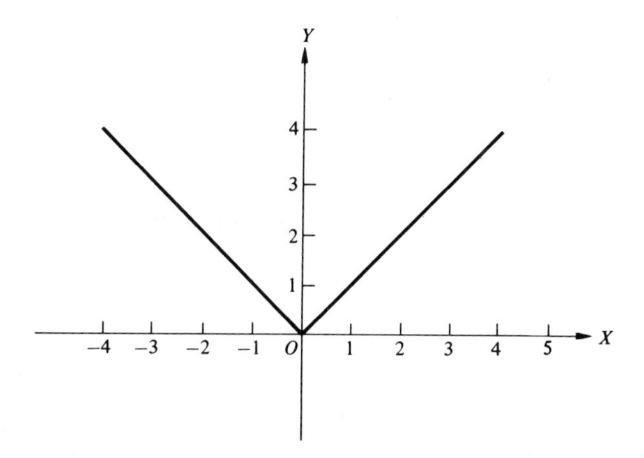

Figure 1.4.5 Graph of $y = |x|$.

PROBLEM 2. Draw the graph of $f(x) = |x - 2|$.

The next two examples show how to graph a function when the domain is given.

EXAMPLE 3. Draw the graph of the function whose rule of correspondence is

$$y = \begin{cases} x & 0 \leqq x \leqq 2 \\ 2 & x > 2 \end{cases}$$

Solution

x	0	1	2	3	4	5
y	0	1	2	2	2	2

The graph is shown in Fig 1.4.6.

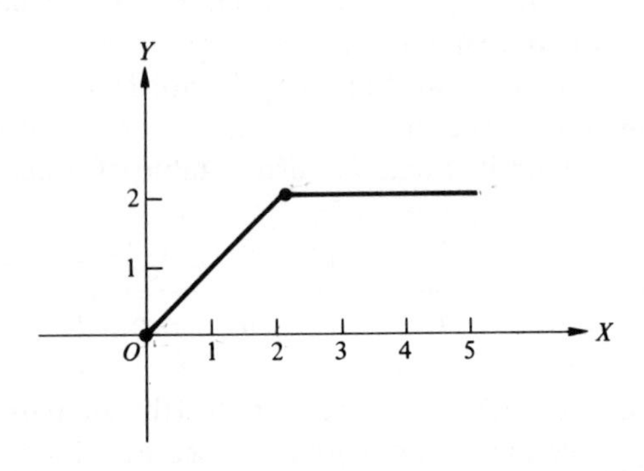

Figure 1.4.6 Graph of $y = \begin{cases} x, & 0 \leqq x \leqq 2, \\ 2, & x > 2. \end{cases}$

Warning ***You must use the correct rule. For x between 0 and 2 we had to use the rule y = x. For x > 2, we had to use the rule y = 2.***

PROBLEM 3. Draw the graph of

$$f(x) = \begin{cases} -1 & x \leqq 0 \\ x - 1 & x > 0 \end{cases}$$

EXAMPLE 4. Draw the graph of the function defined by

$$y = \begin{cases} 1 & x \geqq 1 \\ -1 & x < 1 \end{cases}$$

Solution

x	-1	0	1	2	3
y	-1	-1	1	1	1

The graph is shown in Fig. 1.4.7. Note that we use an "open" circle when a point is not included. A "solid" circle is used when the end point is included.

PROBLEM 4. Draw the graph of

$$f(x) = \begin{cases} 1 & x \leqq 2 \\ -1 & x > 2 \end{cases}$$

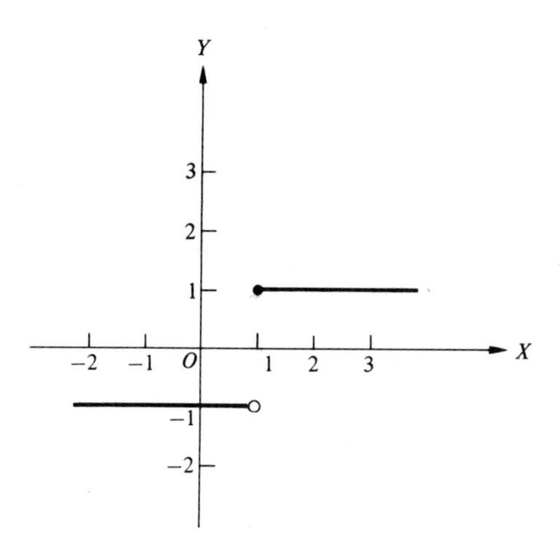

Figure 1.4.7 Graph of $y = \begin{cases} 1, & x \geqq 1, \\ -1, & x < 1. \end{cases}$

Note that the graph in Fig. 1.4.4 is smooth—it has no corners as do those in Figs. 1.4.5 and 1.4.6. Functions having smooth graphs are called *differentiable*. Functions whose graphs are not broken (functions that can be drawn without lifting the pencil from the paper) are called *continuous*. The graph in Fig. 1.4.7 is not continuous at $x = 1$. These properties are important in our later work and subsequently we shall give a precise meaning to these concepts.

Not all graphs of functions need to be drawn on an XY-coordinate system. It is often desirable to label the axis with the letters denoting the independent and dependent variables. Usually, the independent variable is used to label the horizontal axis. The axes may also be labeled input and output.

EXAMPLE 5. If the power of an electric circuit is given as $P = I^2R$, draw the graph of $P = f(I)$ for $R = 0.5\ \Omega$.

Solution. Since R is constant we have $P = 0.5I^2$, where the current I is the independent variable. We shall thus label the horizontal axis I (input) and the vertical axis P (output) (see Fig. 1.4.8).

I	0	1	2	3	4	5	6
P	0	0.5	2	4.5	8	12.5	18

PROBLEM 5. The resistance of a wire is given by $R = 5.2l/d^2$. Draw the graph of $R = f(d)$ for $l = 2$.

The *unit step function*

$$u(x - a) = \begin{cases} 0 & x < a \\ 1 & x > a \end{cases} \tag{1.4.1}$$

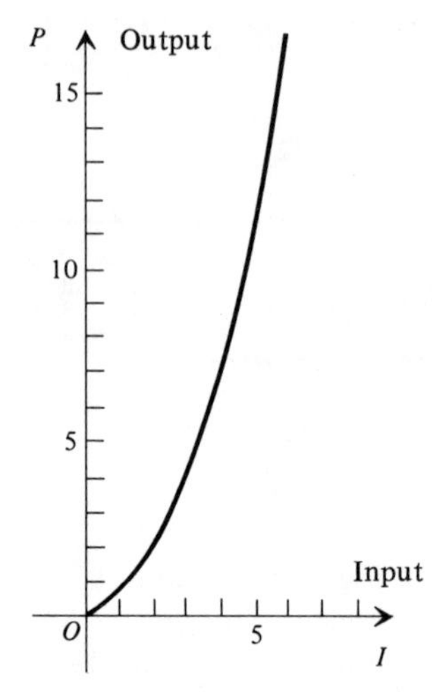

Figure 1.4.8 Graph of $P = 0.5I^2$.

is helpful in expressing functions, which have different rules of correspondence for different intervals, with only one rule of correspondence. Note that u is zero for a negative argument and is $+1$ for a positive argument. We may extend Eq. 1.4.1 to

$$\begin{aligned} U(x) &= [u(x-a) - u(x-b)] \\ &= \begin{cases} 0 & x < a \\ 1 & a < x < b \quad (a < b) \\ 0 & x > b \end{cases} \end{aligned} \tag{1.4.2}$$

The function $U(x)$ is sometimes called the *filter function*. You may graph $U(x)$ as an exercise. The product of a function f and U will be zero except for x between a and b, where it will be f.

EXAMPLE 6. Write

$$f(x) = \begin{cases} 0 & x < 2 \\ x^2 & 2 < x < 4 \\ 0 & x > 4 \end{cases}$$

using only one rule of correspondence.

Solution. $f(x) = x^2[u(x-2) - u(x-4)]$. Note that for $x < 2$, $u(x-2) = u(x-4) = 0$, so $f(x) = 0$. For $2 < x < 4$, $u(x-2) = 1$, $u(x-4) = 0$, so $f(x) = x^2$. For $x > 4$, $u(x-2) = u(x-4) = 1$, so $f(x) = 0$. The graph of $f(x)$ is shown in Fig. 1.4.9.

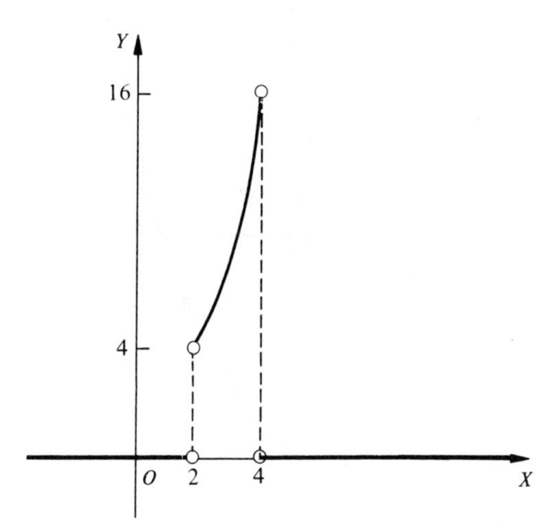

Figure 1.4.9 Graph showing use of filter function.

PROBLEM 6. Write

$$f(x) = \begin{cases} 0 & x < 1 \\ x + 2 & 1 < x < 3 \\ 0 & x > 3 \end{cases}$$

using only one rule of correspondence. Sketch the graph of $f(x)$.

EXERCISES 1.4

Group A

1. Let A, B, and C, be three points on the X-axis with coordinates as given.

(a) $A(3, 0)$, $B(5, 0)$, $C(9, 0)$. Find AB, AC, and CB. Show that $AC = AB + BC$.
(b) $A(-2, 0)$, $B(7, 0)$, $C(3, 0)$. Find AB, CA, and BC. Show that $AC = AB + BC$.

2. Plot the points $A(-3, 4)$, $B(5, -12)$, $C(12, 5)$, $D(-4, -3)$, $E(0, 7)$, and $F(0, -2)$.

3. Draw the graph of $y = x^2$ for $-3 \leqq x \leqq 0$.

4. Draw the graph of $y = x^2$. Is this graph the same as the graph in Exercise 3? Why?

5. In Fig. 1.4.10 the curve OPB is the graph of $y^2 = x^3$. Express (a) DP in terms of y; (b) GP in terms of x; (c) PE in terms of y; (d) PF in terms of x.

6. Sketch the graph of $y = 2x - 3$. Where does this graph intersect the X-axis; the Y-axis?

7. Draw the graph of the function

$$y = \begin{cases} 3x & 0 \leqq x \leqq 2 \\ 6 & x > 2 \end{cases}$$

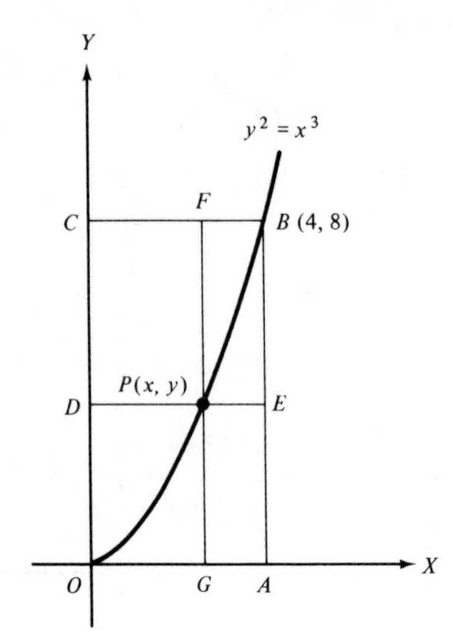

Figure 1.4.10

8. Draw the graph of the function

$$y = \begin{cases} x & 0 \leqq x \leqq 2 \\ 1 & x > 2 \end{cases}$$

9. Draw the graph of the unit pulse function

$$f(t) = \begin{cases} 0 & t < 0 \\ 1 & 0 \leqq t \leqq 1 \\ 0 & t > 1 \end{cases}$$

10. The current in an electric circuit is given as

$$I = \begin{cases} 1 & 0 \leqq t \leqq 1 \\ 0 & 1 < t < 2 \\ 1 & 2 \leqq t \leqq 3 \\ 0 & t > 3 \end{cases}$$

Draw the graph of $I = f(t)$.

11. Sketch the graph of each of the following.

(a) $u(x - 2)$.

(b) $U(x) = u(x - 2) - u(x - 4)$.

(c) $f(x) = 3[u(x - 2) - u(x - 4)]$.

(d) $f(x) = x[u(x - 1) - u(x - 3)]$.

(e) $f(x) = x[u(x + 1) - u(x - 1)]$.

(f) $f(x) = x^2[u(x - 1) - u(x - 2)]$.

(g) $f(x) = 2[u(x - 1) - u(x - 2)] - 2[u(x - 2) - u(x - 3)]$.

(h) $f(x) = x[u(x) - u(x - 1)] + 2[u(x - 1) - u(x - 2)]$.

12. Write each of the following using only one rule of correspondence.

(a) $f(x) = \begin{cases} 0 & x < 2 \\ 1 & 2 < x < 3 \\ 0 & x > 3. \end{cases}$

(b) $f(x) = \begin{cases} -1 & x < 1 \\ 0 & x > 1. \end{cases}$

(c) $f(x) = \begin{cases} 3 & 1 < x < 2 \\ 0 & \text{elsewhere.} \end{cases}$ (d) $f(x) = \begin{cases} x & 1 < x < 3 \\ 0 & \text{elsewhere.} \end{cases}$

13. Sketch the graph of each of the following for $0 < t < 4$.

(a) $I(t) = t - t[u(t-1)] + (t-2)[u(t-2)] - (t-2)[u(t-3)]$.
(b) $\varphi(t) = t^2 - t^2[u(t-1)] + (t^2-4)[u(t-2)] - (t^2-4)[u(t-3)]$.

Group B

1. If points A, B, and C are any three points on a directed line, show that $AC = AB + BC$ for the arrangement shown.

(a) $A \quad B \quad C \quad +.$ (b) $B \quad A \quad C \quad +.$

(c) $B \quad C \quad A \quad +.$ (d) $C \quad A \quad B \quad +.$

2. Sketch the graph of $y^2 = x$. Is this the graph of a function? In how many points may a line parallel to the Y-axis intersect the function $y = f(x)$?

3. In a dc circuit, the voltage, E, is given as $E = IR$, where I is the current in amperes and R is the resistance in ohms. For a constant voltage of $E = 12$ V, plot the graph of $I = f(R)$.

4. An automobile with mass of 1000 slugs experiences a force in the direction of its motion according to the formula $F = ma$. Plot the graph of $F = f(a)$.

5. The deflection, y, in a horizontal beam 2 ft long is given by

$$y = x^4 - 4x^3 + 8x$$

Plot the graph of the deflection curve.

6. Shade that region of the plane for the set A, where $A = \{(x, y) \mid x < 3\}$.

7. Shade that region of the plane for the set B, where $B = \{(x, y) \mid y > 2\}$.

8. Shade that region of the plane for the set C, where C is the region common to A and B, of Exercises 6 and 7.

9. With only one rule of correspondence, write

$$f(x) = \begin{cases} 2 & 1 < x < 3 \\ -2 & 3 < x < 5 \\ 0 & \text{elsewhere.} \end{cases}$$

Group C

1. If point P_1 has an abscissa of x_1 and point P_2 has an abscissa of x_2, show that for any arrangement of points P_1 and P_2 on the X-axis, $P_1P_2 = x_2 - x_1$. (*Hint:* Make use of Definition 1.4.1.)

1.5 DIVISION INVOLVING ZERO

As we proceed in our development of calculus we shall see that zero plays an important role, especially in the concept of a limit. The fraction

$$\frac{a}{b} = c$$

implies that

$$a = b \cdot c$$

To investigate division involving zero, we have three cases to consider.

Case 1. If $a = 0$ and $b \neq 0$, then $0/b = c$ implies that

$$0 = b \cdot c$$

Since $b \neq 0$, this can be true only if $c = 0$. Hence

$$\boxed{\frac{0}{b} = 0 \qquad \text{for} \qquad b \neq 0}$$

Case 2. If $a \neq 0$ and $b = 0$, then $a/0 = c$ implies that

$$a = c \cdot 0$$

But this says that $a = 0$, which is contrary to the hypothesis that $a \neq 0$. Thus $a/0$ has no meaning, and we say that

$$\boxed{\frac{a}{0} \text{ does not exist}}$$

Case 3. If $a = 0$ and $b = 0$, then $0/0 = c$ implies that

$$0 = 0 \cdot c$$

But this is true for any and all values of c, and we say that the fraction a/b is *indeterminate* for $a = 0$ and $b = 0$. That is,

$$\boxed{\frac{0}{0} \text{ is indeterminate}}$$

Thus division by zero is not allowed.

Warning ***Whenever a mathematical expression contains a fraction with a variable in the denominator, be on guard.***

We must not confuse $0/0$ with a/a, which equals 1 for $a \neq 0$. The rule to remember is

NEVER DIVIDE BY ZERO.

Use a calculator to perform the following operations: (a) 0/3; (b) 5/0; (c) 0/0. What does the calculator show for parts (b) and (c)?

We will make extensive use of Case 3, 0/0, in developing the concept of the derivative.

EXERCISES 1.5

Group A

1. Perform the indicated operation.

(a) $5 + 0$. (b) $(\frac{1}{7})(0)$. (c) $0/3$.

(d) $13/0$. (e) $-9/0$. (f) $\frac{6-2}{2-2}$.

(g) $\frac{5-5}{3-3}$. (h) $\frac{(8)(-2)+(4)(4)}{-2-(-2)}$.

2. In each of the following, state the value of x for which the fraction has no meaning (is undefined).

(a) $\frac{x+2}{x-3}$. (b) $\frac{x+2}{x+3}$. (c) $\frac{(x-1)(x+2)}{(x-4)(x-1)}$.

3. In each of the following, state the value of x for which the fraction is indeterminate.

(a) $\frac{xy}{3x}$. (b) $\frac{x-3}{(x-3)(x+3)}$. (c) $\frac{3x+2}{6x+4}$.

Group B

1. As we proceed in the study of calculus a variable x is allowed to take on an infinite set of values, causing numerators or denominators to *approach* but *not reach* certain values. The approach toward zero is critical. Give your judgment of the effect on values of the given expression as x changes as indicated.

(a) $5x$ as x becomes very large. (b) $5x$ as x approaches zero.

(c) $5 - x$ as x approaches 5. (d) $\frac{5}{x}$ as x becomes very large.

(e) $\frac{5}{x}$ as x approaches zero. (f) $\frac{1}{5-x}$ as x approaches zero.

(g) $\frac{1}{5-x}$ as x becomes very large. (h) $\frac{1}{5-x}$ as x approaches 5.

(i) $\frac{2x}{3x}$ as x becomes very large. (j) $\frac{2x}{3x}$ as x approaches zero.

2. Read the following discussion and find the error(s) in the reasoning.

Let $a = b$; then $a^2 - b^2 = 0$. Factoring gives $(a - b)(a + b) = 0$. Dividing by $(a - b)$ yields $a + b = 0$ or $a = -b$. Since $a = b$ we have $a = -a$, which upon dividing by a yields $1 = -1$.

1.6 LIMITS

The concept of a limit is fundamental to the development of calculus. Mathematicians have a rigorous definition of a limit which they use to work in analysis. At this stage we shall give an intuitive meaning of the concept of a limit and leave the rigor to more advanced books.

As a start let us consider that we hold two pieces of chalk. We shall then place one-half of the chalk which we hold in a box. We then hold one piece of chalk and the box has one piece of chalk. The difference between the amount in the box and the total amount 2 is 1 (see line 2 of Table 1.6.1). Let us repeat this process. We then hold $\frac{1}{2}$ piece of chalk and the box now has $1\frac{1}{2}$ pieces. The

TABLE 1.6.1

	Amount We Hold	Amount in Box	Difference Between Amount in Box and Total Amount
1.	2	0	$2 - 0 = 2$
2.	1	1	$2 - 1 = 1$
3.	$\frac{1}{2}$	$1 + \frac{1}{2} = 1\frac{1}{2}$	$2 - 1\frac{1}{2} = \frac{1}{2}$
4.	$\frac{1}{4}$	$1\frac{1}{2} + \frac{1}{4} = 1\frac{3}{4}$	$2 - 1\frac{3}{4} = \frac{1}{4}$
5.	$\frac{1}{8}$	$1\frac{3}{4} + \frac{1}{8} = 1\frac{7}{8}$	$2 - 1\frac{7}{8} = \frac{1}{8}$
6.	$\frac{1}{16}$	$1\frac{7}{8} + \frac{1}{16} = 1\frac{15}{16}$	$2 - 1\frac{15}{16} = \frac{1}{16}$
7.	$\frac{1}{32}$	$1\frac{15}{16} + \frac{1}{32} = 1\frac{31}{32}$	$2 - 1\frac{31}{32} = \frac{1}{32}$
.	.	.	.
.	.	.	.
.	.	.	.
8.	$\frac{1}{1024}$	$1\frac{511}{512} + \frac{1}{1024} = 1\frac{1023}{1024}$	$2 - 1\frac{1023}{1024} = \frac{1}{1024}$

difference between the amount in the box and the total amount 2 is $\frac{1}{2}$ (see line 3 of Table 1.6.1). Continuing to repeat this process, as in Table 1.6.1, we see that as long as we hold some chalk the amount in the box will approach 2 but never equal 2. The difference between 2 and the amount in the box becomes very small and gets smaller, or approaches 0, as we add more chalk to the box. We then say that the *limit* of the amount of chalk in the box is 2, although the amount is never exactly 2 nor does it ever exceed 2. This example may be expressed as

$$1 + \tfrac{1}{2} + \tfrac{1}{4} + \tfrac{1}{8} + \tfrac{1}{16} + \tfrac{1}{32} + \cdots \longrightarrow 2$$

where the symbol $\longrightarrow$ signifies *approach*, or we can say that

$$\text{limit}\,(1 + \tfrac{1}{2} + \tfrac{1}{4} + \tfrac{1}{8} + \tfrac{1}{16} + \tfrac{1}{32} + \cdots) = 2$$

We summarize this concept as follows.

Definition 1.6.1 *If a variable **x** approaches nearer and nearer a fixed number **L** such that the difference between **L** and **x** becomes and remains numerically smaller than any preassigned small positive number, however* *small, then the constant* **L** *is said to be the limit of* **x**, *or* **x** *is said to approach the limit* **L.**

In the preceding example the variable x is the amount of chalk in the box and the limit L is 2. The important thing to notice here is that the absolute value of the difference between the variable and its limit becomes and remains smaller than any small positive number you may previously choose. This small positive number is usually denoted by the Greek letter epsilon, ϵ. In the chalk example if we choose $\epsilon = \frac{1}{4096}$, then we can make the difference between 2 and the amount in the box less than ϵ by repeating the process 13 times. (You may show this.)

As another example consider the area of an n-sided regular polygon inscribed in a circle. The area of the polygon differs from the area of the circle by a certain amount. As the number of sides, n, increases, this difference becomes less and less. If we choose a small ϵ, we can always find an inscribed polygon whose area differs from that of the circle by less than ϵ, and once this polygon has been found the area of any polygon with a greater number of sides will differ from the area of the circle by less than ϵ. We then say that the limit of the areas of inscribed polygons of n sides is the area of the circle.

Notice that Definition 1.6.1 does not say that a variable cannot reach its limit, because sometimes it does, For example, a swinging pendulum may eventually come to rest, and the limit of the swing is zero.

Also notice that the word *remains* is important in the definition of a limit. If a car approaches an intersection, the difference between the position of the car and the intersection becomes less than any preassigned small number ϵ; however, if the car continues past the intersection, the difference does not remain small and no limit is approached.

In the preceding examples we observe that the phenomena of the sums of pieces of chalk and the areas of the polygons depend on how many steps one has taken in the process of summation or increasing sides. That is, the sum at any step is a function of that step, and the area of the polygon is a function of the number of sides of the polygon. Hence the sums and areas are really variables dependent on other variables. We reformulate Definition 1.6.1 to give the definition of the *limit of a function.*

Definition 1.6.2 *If the values of function* **f** *approach nearer and nearer a fixed number* **L** *(as its argument approaches a fixed number* **a**) *in such a way that the difference between* **L** *and* **f(x)** *becomes and remains numerically smaller than any preassigned small positive number, however small, whenever the difference between the argument* **x** *and the number* **a** *is sufficiently small, then the constant* **L** *is said to be the limit of the function* **f** *as* **x** *approaches* **a**. *The symbol used to denote the limit* **L** *of the function* **f** *as its argument* **x** *approaches* **a** *is*

$$\lim_{x \to a} f(x) = L$$

A working rule for considering the limit of a function is to replace x by a and see what result is obtained. For the present we will have to resort to algebra when we obtain 0/0 (see Example 4).

EXAMPLE 1. Find $\lim_{x \to 3} x^2$.

Solution

$$\lim_{x \to 3} x^2 = (3)^2 = 9$$

PROBLEM 1. Find $\lim_{x \to 2} x^3$.

EXAMPLE 2. Find $\lim_{x \to 0} (x/3)$.

Solution

$$\lim_{x \to 0} \frac{x}{3} = \frac{0}{3} = 0$$

PROBLEM 2. Find $\lim_{x \to 3} (3 - x)/5$.

EXAMPLE 3. Find $\lim_{x \to 2} [5/(x - 2)]$.

Solution

$$\lim_{x \to 2} \frac{5}{x - 2} = \frac{5}{2 - 2} = \frac{5}{0} \qquad \text{does not exist}$$

PROBLEM 3. Find $\lim_{x \to -2} [7/(x + 2)]$.

EXAMPLE 4. Find $\lim_{x \to 2} [(x^2 - 4)/(x - 2)]$.

Solution

$$\lim_{x \to 2} \frac{x^2 - 4}{x - 2} = \frac{4 - 4}{2 - 2} = \frac{0}{0} \qquad \text{is indeterminate}$$

One way to remove this indeterminancy is to apply some algebra to the function to see if it can be simplified.

Here we can factor the numerator and cancel the common factors. Thus

$$\lim_{x \to 2} \frac{x^2 - 4}{x - 2} = \lim_{x \to 2} \frac{(x + 2)\cancel{(x - 2)}}{\cancel{(x - 2)}}$$

$$= \lim_{x \to 2} (x + 2) = (2 + 2) = 4$$

Notice carefully that when we canceled the terms $x - 2$ we did not violate the rule of "never divide by zero," since at this point $x \neq 2$, as we have not yet gone to the limit. Also note that we keep the symbol $\lim_{x \to 2}$ intact until we replace x by 2.

PROBLEM 4. Find $\lim_{x \to 4} [(x^2 - 16)/(x - 4)]$.

In Exercise 2 of Group C in Exercises 7.4, we discuss another method of finding the limit of an indeterminate form.

EXAMPLE 5. Find $\lim_{x \to 2} [(x^3 - 8)/(x - 2)]$.

Solution

$$\lim_{x \to 2} \frac{x^3 - 8}{x - 2} = \frac{(2)^3 - 8}{2 - 2} = \frac{8 - 8}{2 - 2} = \frac{0}{0}$$

Therefore, we apply some algebra. (Why?) Here factoring and canceling, we obtain

$$\begin{aligned} \lim_{x \to 2} \frac{x^3 - 8}{x - 2} &= \lim_{x \to 2} \frac{\cancel{(x - 2)}(x^2 + 2x + 4)}{\cancel{(x - 2)}} \\ &= \lim_{x \to 2} (x^2 + 2x + 4) = (2)^2 + (2)(2) + 4 \\ &= 4 + 4 + 4 = 12 \end{aligned}$$

Thus

$$\lim_{x \to 2} \frac{x^3 - 8}{x - 2} = 12$$

PROBLEM 5. Find $\lim_{x \to 3} [(x - 3)/(x^3 - 27)]$.

It might be of interest to see some numerical calculations of a limit. In Table 1.6.2 are the results of a program that computed $\lim_{x \to 2} [(x^3 - 8)/(x - 2)]$

TABLE 1.6.2

x	$\dfrac{x^3 - 8}{x - 2}$	$\dfrac{1}{(x - 2)^2}$
1.0	7.0	1
1.75	10.56	16
1.9375	11.6289	256
1.984375	11.906496	4,096
1.9960939	11.9765796	65,536
1.9990236	11.9941501	1,048,576
1.9997561	11.9985676	16,777,220
1.9999392	11.9997577	268,435,520
$x \longrightarrow 2$	$\lim_{x \to 2} \dfrac{x^3 - 8}{x - 2} = 12$	$\lim_{x \to 2} \dfrac{1}{(x - 2)^2}$ does not exist
2.0000615	12.0000019	268,435,520
2.0002446	12.0014667	16,777,220
2.0009770	12.0058613	1,048,576
2.0039067	12.0234546	65,536
2.015625	12.093996	4,096
2.0625	12.3789	256
2.25	13.56	16
3.0	19.0	1

and $\lim_{x \to 2} [1/(x-2)^2]$ on a computer. The value of x approaches 2. Read the values from the top and the bottom toward the middle of the table.

PROBLEM 6. Use a hand calculator to complete the table.

x	x^2	$\dfrac{x^2-9}{x-3}$	$\dfrac{1}{(x-3)^2}$
2.0			
2.5			
2.9			
2.99			
2.999			
3.001			
3.01			
3.1			
3.5			
4.0			

Use the results obtained in the table for the following. As $x \longrightarrow 3$ from the left, denoted as $x \longrightarrow 3^-$, and from the right, denoted as $x \longrightarrow 3^+$, what values are approached for (a) x^2; (b) $(x^2-9)/(x-3)$; (c) $1/(x-3)^2$?

The following three rules are helpful when one confronts limits in problems that mix algebra and the computation of limits.

Let $f(x)$ and $g(x)$ be functions defined on a domain D and let $\lim_{x \to a} f(x) = L$ and $\lim_{x \to a} g(x) = M$; then

1. $$\lim_{x \to a} [f(x) + g(x)] = \lim_{x \to a} f(x) + \lim_{x \to a} g(x) = L + M$$

This says that *the limit of a sum is the sum of the limits.*

2. $$\lim_{x \to a} [f(x)g(x)] = \lim_{x \to a} f(x) \cdot \lim_{x \to a} g(x) = L \cdot M$$

This says that *the limit of a product is the product of the limits.*

3. $$\lim_{x \to a} \left[\frac{f(x)}{g(x)}\right] = \frac{\lim_{x \to a} f(x)}{\lim_{x \to a} g(x)} = \frac{L}{M} \qquad M \neq 0$$

This says that *the limit of a quotient is the quotient of the limits provided that the limit of the denominator is not zero.*

Some of these rules were used in some of the previous examples.

EXERCISES 1.6

Group A

1. In the chalk example (see Table 1.6.1), what would be the limit of the amount of the chalk in the box if you put one-third of what you hold in the box each time?

2. Complete the following table. Use a calculator.

x	x^2	$\dfrac{x^2-4}{x-2}$	$\dfrac{1}{(x-2)}$
1.0			
1.5			
1.9			
1.99			
1.999			
2.001			
2.01			
2.1			
2.5			
3.0			

As $x \longrightarrow 2$ from below, denoted as $x \longrightarrow 2^-$, and from above, denoted as $x \longrightarrow 2^+$, what values are approached for (a) x^2; (b) $(x^2 - 4)/(x - 2)$; (c) $1/(x - 2)$? Compare the answer to part (b) with Example 4.

3. Find the required limit in each of the following.

(a) $\lim_{x \to 0} x^2 + 4$.

(b) $\lim_{x \to 0} \dfrac{x-2}{x+2}$.

(c) $\lim_{x \to 0} \dfrac{x^2-3}{x}$.

(d) $\lim_{x \to -2} \dfrac{x^2-2}{x+2}$.

(e) $\lim_{x \to 3} \dfrac{x^2-9}{x-3}$. (See Problem 6)

(f) $\lim_{x \to 6} \dfrac{x^2-36}{x-6}$.

(g) $\lim_{x \to 1} \dfrac{x^3-1}{x-1}$.

(h) $\lim_{x \to -1} \dfrac{x^3+1}{x+1}$.

4. Find the required limit in each of the following. (**Warning:** $x\,\Delta x$ is not Δx^2.)

(a) $\lim_{\Delta x \to 0} [2x\,\Delta x + (\Delta x)^2]$.

(b) $\lim_{\Delta x \to 0} \dfrac{[5x\,\Delta x + (\Delta x)^2]}{\Delta x}$

(c) $\lim_{\Delta x \to 0} \dfrac{(x + \Delta x)^2 - x^2}{\Delta x}$.

(d) $\lim_{\Delta t \to 0} \dfrac{(t + \Delta t) - t}{\Delta t}$.

(e) $\lim_{\Delta t \to 0} \dfrac{3(t + \Delta t)^2 - 3t^2}{\Delta t}$.

(f) $\lim_{\Delta x \to 0} \left\{\dfrac{[1/(x + \Delta x)] - (1/x)}{\Delta x}\right\}$.

(g) $\lim_{\Delta x \to 0} \left\{\dfrac{[3/(x + \Delta x)] - (3/x)}{\Delta x}\right\}$.

Group B

1. Find

$$\lim_{\Delta u \to 0} \frac{(u + \Delta u)^n - u^n}{\Delta u}.$$

[*Hint:* Make use of the binomial expansion, which is

$$(a + b)^n = a^n + \frac{na^{n-1}}{1} b + \frac{n(n-1)a^{n-2}}{1 \cdot 2} b^2 + \cdots + b^n.\Big]$$

2. Find $\lim_{\Delta x \to 0} [(\sqrt{x + \Delta x} - \sqrt{x})/\Delta x]$. (*Hint:* Multiply both numerator and denominator by $\sqrt{x + \Delta x} + \sqrt{x}$. That is, rationalize the numerator.)

3. The average rate of change of distance over a time interval Δt is given as $\Delta s/\Delta t$; if $s = 3t$, Δs may be expressed as $\Delta s = 3(t + \Delta t) - 3t$. As $\Delta t \longrightarrow 0$ we see that $\Delta s \longrightarrow 0$. Find $\lim_{\Delta t \to 0} (\Delta s/\Delta t)$.

4. The charge Q in a circuit is given as $Q = t^2 - t$. The current I in the circuit is defined to be $I = \lim_{\Delta t \to 0} (\Delta Q/\Delta t)$. If $\Delta Q = (t + \Delta t)^2 - (t + \Delta t) - (t^2 - t)$, find I.

Group C

1. Let $f(x) = x^2$. Draw the graph of f for $-3 \leqq x \leqq 3$. Using horizontal and vertical cuts on the graph, show how one may find the small differences mentioned in Definition 1.6.2. For convenience work around $x = 2$.

2. Draw the graph of

$$f(x) = \begin{cases} -2 & x < 1 \\ 2 & x \geqq 1 \end{cases}$$

What is the value of (a) $\lim_{x \to 1^-} f(x)$; (b) $\lim_{x \to 1^+} f(x)$? (c) Does $\lim_{x \to 1} f(x)$ exist? If so, what is it?

3. Give a physical interpretation to the limit in (a) Exercise 3 of Group B; (b) Exercise 4 of Group B.

1.7 CONTINUITY

Functions are either continuous or discontinuous at a point. If a pan of water at a temperature of 20°C is heated to the boiling point, the temperature of the water gradually increases to 100°C. We say that the temperature is a continuous function of time. A blinking light represents a discontinuous function of the voltage as a function of time.

In our discussion of graphs we noted that the graph of a continuous function could be drawn without lifting the pencil from the paper. This is not the definition of continuity. Rather, it is a consequence.

Definition 1.7.1 *A function **f** defined on a domain **D** is said to be continuous at **a** belonging to **D** provided that*

1. $\lim_{x \to a} f(x)$ exists.
2. $\lim_{x \to a} f(x) = f(a)$.

This means that no matter how x approaches the value of a, if the limit of $f(x)$ equals $f(a)$, then f is continuous at a. When we prescribe that a belongs to D we are saying that f is defined at a. A function not continuous at a point is said to be *discontinuous* there. The tangent function $y = \tan x$ is discontinuous at $x = \pi/2$, as you may recall.

Functions are continuous on an interval according to

Definition 1.7.2 *A function **f** defined on a domain **D** is said to be continuous on an interval **I** contained in **D** if **f** is continuous at each point of **I**.*

EXAMPLE 1. Is $f(x) = 1/(x - 2)$ everywhere continuous? Is it continuous on the interval $2.5 \leqq x \leqq 3$?

Solution. Since the denominator is zero at $x = 2, f(2)$ does not exist. Hence $f(x)$ is not everywhere continuous. It is discontinuous at $x = 2$. Yes, since $x = 2$ is not in the interval.

PROBLEM 1. Are the following functions everywhere continuous? If not, state the reason why, and where they are discontinuous. (a) $3/(x + 7)$; (b) $1/(x^2 + 1)$. (c) Are they continuous on the interval $-9 \leqq x \leqq 0$?

EXAMPLE 2. A flashing light operating on a 12-V battery is on for 1 s and off for 2 s. Express the voltage as a function of time for the first 3 s. Is this function discontinuous?

Solution. We have

$$E = \begin{cases} 12 & 0 < t \leqq 1 \\ 0 & 1 < t \leqq 3 \end{cases}$$

Graph this function (see Fig. 1.7.1). Note that as $t \longrightarrow 1$ from the left, denoted as $t \longrightarrow 1^-$, $E = 12$, and as $t \longrightarrow 1$ from the right, denoted as $t \longrightarrow 1^+$, $E = 0$. Since these limits are different, E is discontinuous at $t = 1$.

PROBLEM 2. A signal is on for 2 s and off for 3 s. The current is 4 A. Express the current as a function of time for the first 5 s. Is this function discontinuous? If so, state when and why.

Most of the functions we are going to work with will be continuous.

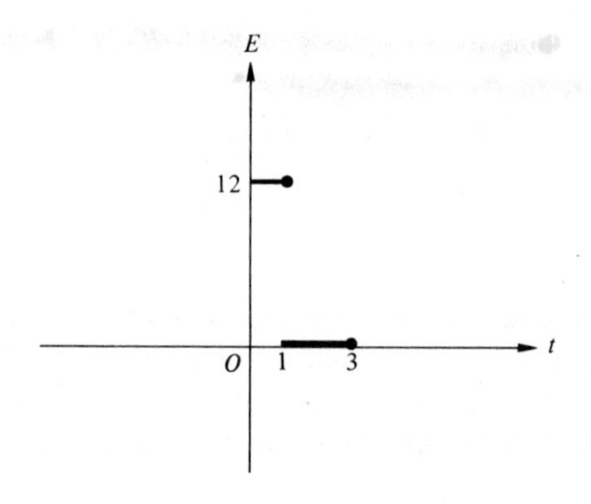

Figure 1.7.1 Graph of $E = \begin{cases} 12, & 0 < t \leqq 1, \\ 0, & 1 < t \leqq 3. \end{cases}$

EXERCISES 1.7

Group A

1. State whether each of the following functions is continuous or discontinuous over the specified interval. If discontinuous, state the points of discontinuity and give a reason for the discontinuity.

(a) $y = \dfrac{2}{x-3}$, $0 \leqq x \leqq 6$.

(b) $y = \sqrt{x}$, $x \geqq 0$.

(c) $f(x) = x^2$, $-3 \leqq x \leqq 3$.

(d) $f(x) = \dfrac{3}{x^2}$, $-2 \leqq x \leqq 2$.

(e) $f(t) = \begin{cases} 3 & 0 < t < 2 \\ 0 & 2 < t < 4. \end{cases}$

(f) $f(t) = \begin{cases} t & 0 < t < 2 \\ 2 & t > 2. \end{cases}$

(g) $I = \begin{cases} 3 & 0 < t < 3 \\ t & t \geqq 3. \end{cases}$

(h) $I = \begin{cases} t & 0 < t < 1 \\ 2 & t > 1. \end{cases}$

2. Determine where each of the following is continuous and where discontinuous. Give the reason for discontinuity.

(a) $y = x^2$, $-1 \leqq x \leqq 1$.

(b) $y = \begin{cases} x^2 & -1 \leqq x < 0 \\ -x^2 & 0 < x \leqq 1. \end{cases}$

(c) $f(x) = \begin{cases} 1 & 0 \leqq x \leqq 1 \\ -1 & -1 \leqq x < 0. \end{cases}$

(d) $f(x) = \begin{cases} 1 & 0 < x \leqq 1 \\ 0 & x = 0. \end{cases}$

(e) $g(t) = \begin{cases} \dfrac{1}{t} & 0 < t \leqq 1 \\ 0 & t = 0. \end{cases}$

(f) $h(t) = \dfrac{t^3 - 3}{t^2 - 1}$, $-3 \leqq t \leqq 3$.

(g) $q(t) = \dfrac{t^2 - 2t + 1}{t - 1}$, $-2 \leqq t \leqq 2$.

(h) $I(t) = \dfrac{t^3 - 3t + 1}{t - 1}$, $-1 \leqq t < 1$.

3. Sketch a graph of a function (a) continuous at $x = a$; (b) discontinuous at $x = a$.

Group B

1. Compare the concept of "limit of a function at a point" with the concept of "continuity of a function at a point." How are they alike and how are they different?

SUMMARY OF IMPORTANT WORDS AND CONCEPTS

For each of the following, state in your own words your understanding of the statement or word.

1. (a) Change in a quantity. (b) Average rate of change.

2. (a) Sets. (b) Ordered pairs of numbers. (c) Element of a set.

3. (a) Function. (b) Domain. (c) Range. (d) Input. (e) Output. (f) Rule of correspondence. (g) Argument. (h) Independent variable. (i) Dependent variable.

4. (a) Cartesian coordinate system. (b) Quadrants. (c) Rectangular coordinates. (d) Abscissa. (e) Ordinate. (f) Graphs.

5. (a) Division by zero. (b) Exist. (c) Indeterminate.

6. (a) Limit of a variable. (b) Limit of a function at a point. (c) Rules for computation with limits: sum, product, quotient.

7. (a) Continuity. (b) Continuous functions. (c) Continuity of a function at a point. (d) Continuity over an interval.

Chapter 2

The Derivative

2.1 AVERAGE RATE OF CHANGE

In this chapter we discuss the fundamental concept of differential calculus—the derivative of a function. But first we need to understand what is meant by the expression "the average rate of change."

The *ratio* of the change of the dependent variable of a function to the change of the independent variable is called the *average rate of change of the dependent variable with respect to the independent variable.* For example, if Q, the charge in an electric circuit, is a function of time t, then $\Delta Q/\Delta t$ is the average rate of change of the charge with respect to time. Many average rates of change have physical or geometric interpretations and have special names.

Warning ***The word "rate" as used here does not imply "speed" as it did in algebra.***

EXAMPLE 1. If $s = f(t)$, where s is the displacement of an object at time t, then $\Delta s/\Delta t$ is the *average velocity* during the time interval Δt.

EXAMPLE 2. If $v = f(t)$, where v is the velocity of an object at time t, then $\Delta v/\Delta t$ is the *average acceleration* during the time interval Δt.

EXAMPLE 3. If $Q = f(t)$, where Q is the charge in an electric circuit at time t, then $\Delta Q/\Delta t$ is the *average current* during the time interval Δt.

EXAMPLE 4. If $y = f(x)$, a geometric interpretation of $\Delta y/\Delta x$ is the *average slope* of the curve over the interval Δx. (We discuss slopes in Sec. 2.4.)

EXAMPLE 5. If $\theta = f(s)$, where θ is the angle of inclination of a tangent line, then $\Delta\theta/\Delta s$ is the *average curvature* over the interval Δs, where s is the distance along the curve.

EXAMPLE 6. If $W = f(t)$, where W is the amount of work done at time t, then $\Delta W/\Delta t$ is the *average power* during the interval Δt.

EXAMPLE 7. If $\theta = f(t)$, where θ is the angle, measured in radians, through which a wheel turns in t sec, then $\Delta\theta/\Delta t$ is the *average angular velocity* during the interval Δt.

EXAMPLE 8. If $\varphi = f(t)$, where φ is the magnetic flux passing through a circuit at time t, then $\Delta\varphi/\Delta t$ is the *average induced electromagnetic force* during the time interval Δt.

For some of our future work we shall need to know the distinction between *distance* and *displacement* as an object moves from a first position to a second position. The *distance* is the *length* of the path traversed. The *displacement* is the *length* and *direction* of a straight-line segment from the first position to the second position. Displacement is a vector quantity since both length and direction are required. As an example, consider points A and B at the opposite ends of a diameter of a circle of radius r (see Fig. 2.1.1). If an object is moved from A

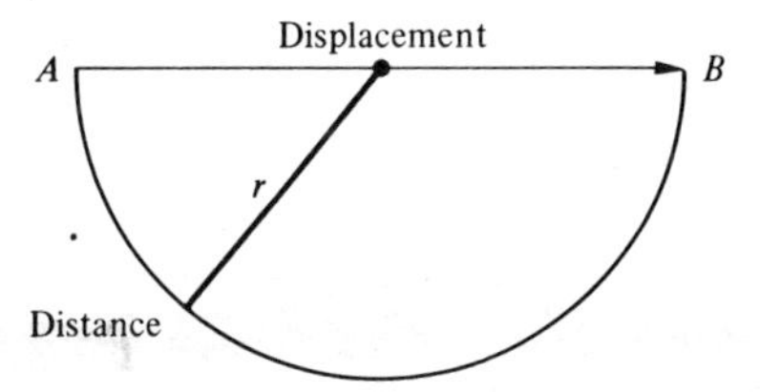

Figure 2.1.1 Displacement and distance.

to B along the semicircle, then the distance traveled is πr, and the displacement of B from A is $2r$ to the right. When we use the formula $s = f(t)$, to indicate the position of a particle at time t, s is the displacement from the origin, because s can be positive or negative, thereby indicating direction.

EXAMPLE 9. A particle moves along a straight line according to $s = t^2 - 4t$. Plot the position of the particle at times $t = 0, t = 1, t = 2, t = 3, t = 4, t = 5$, and $t = 6$ s. State the displacement, s feet, of the particle at each of these times. Assuming the particle reverses direction at $t = 2$, find the distance from the origin that the particle has traveled at each of these times.

Solution. Make a table of values and plot them (see Fig. 2.1.2). (All three horizontal lines should coincide but they are separated for convenience.)

t	0	1	2	3	4	5	6
s	0	-3	-4	-3	0	5	12

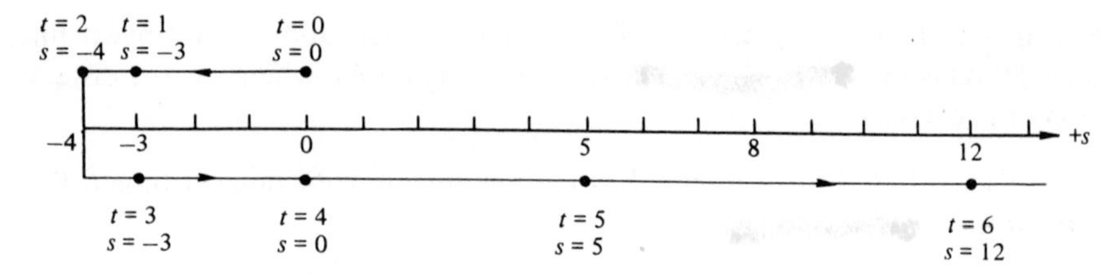

Figure 2.1.2 Displacement for $s = t^2 - 4t$ (Example 9).

The values of s in the table indicate the displacement at the corresponding time t. Note that the minus and plus signs indicate the direction. The distance traveled is found from the length of the path. Figure 2.1.2 helps in finding the distance. Thus

t	Displacement, s	Distance Traveled
0	0	0
1	−3	3
2	−4	4
3	−3	5
4	0	8
5	5	13
6	12	20

PROBLEM 1. A particle moves along a straight line according to $s = t^2 - 6t$. Plot the position of the particle at times $t = 0, 1, 2, 3, 4, 5$, and 6 s. State the displacement, s feet, of the particle at each of these times. Assuming that the particle reverses direction at $t = 3$, find the distance from the origin that the particle has traveled at each of these times.

The ratio $\Delta s/\Delta t$ is the *average velocity* (av. vel.) during the time interval Δt.

EXAMPLE 10. Using the table in Example 9, find the *average velocity* of the particle (a) during the first 3 s; (b) during the third second; (c) during the fifth and sixth seconds. (Time is stated at the end of an interval. That is, $t = 3$ indicates at the end of 3 sec.)

Solution

(a)
$$\Delta t = t_2 - t_1 = 3 - 0 = 3$$
$$\Delta s = s_2 - s_1 = -3 - 0 = -3$$
$$\therefore \quad \text{av. vel.} = \frac{\Delta s}{\Delta t} = \frac{-3}{3} = -1 \text{ ft/s (to the left)}$$

(b) Since the third second is that interval of time between $t = 2$ and $t =$

3, we have

$$\Delta t = t_2 - t_1 = 3 - 2 = 1$$

$$\Delta s = s_2 - s_1 = -3 - (-4) = 1$$

$$\therefore \quad \text{av. vel.} = \frac{\Delta s}{\Delta t} = \frac{1}{1} = 1 \text{ ft/s (to the right)}$$

(c)

$$\Delta t = t_2 - t_1 = 6 - 4 = 2$$

$$\Delta s = s_2 - s_1 = 12 - 0 = 12$$

$$\therefore \quad \text{av. vel.} = \frac{\Delta s}{\Delta t} = \frac{12}{2} = 6 \text{ ft/s (to the right)}$$

PROBLEM 2. Using the results of Prob. 1, find the *average velocity* of the particle (a) during the first 4 s; (b) during the fourth second; (c) during the fifth and sixth seconds.

The *average speed* during a time interval is given by

$$\text{av. speed} = \frac{\Delta(\text{distance})}{\Delta t}$$

EXAMPLE 11. Find the *average speed* of the particle of Example 9 during (a) the first 3 s; (b) during the third second; (c) during the fifth and sixth seconds.

Solution

(a)

$$\Delta t = t_2 - t_1 = 3 - 0 = 3$$

$$\Delta(\text{distance}) = \text{dist}_2 - \text{dist}_1 = 5 - 0 = 5$$

$$\therefore \quad \text{av. speed} = \tfrac{5}{3} = 1.67 \text{ ft/s}$$

(b)

$$\Delta t = t_2 - t_1 = 3 - 2 = 1$$

$$\Delta(\text{distance}) = \text{dist}_2 - \text{dist}_1 = 5 - 4 = 1$$

$$\therefore \quad \text{av. speed} = \tfrac{1}{1} = 1 \text{ ft/s}$$

(c)

$$\Delta t = t_2 - t_1 = 6 - 4 = 2$$

$$\Delta(\text{distance}) = \text{dist}_2 - \text{dist}_1 = 20 - 8 = 12$$

$$\therefore \quad \text{av. speed} = \tfrac{12}{2} = 6 \text{ ft/s}$$

PROBLEM 3. Find the *average speed* of the particle of Prob. 1 during (a) the first 4 s; (b) during the fourth second; (c) during the fifth and sixth seconds.

For any function $s = f(t)$, a general expression for the average velocity during any time interval may be obtained. All that is required are expressions for Δs and Δt. We know that $\Delta t = t_2 - t_1$ and $\Delta s = s_2 - s_1$. Then for any t and s a subsequent (or second) value can be indicated by

$$t_2 = t + \Delta t$$

and

$$s_2 = s + \Delta s$$

Thus for

$$s = f(t)$$

we have

$$s_2 = f(t_2) = f(t + \Delta t)$$

and

$$\Delta s = f(t + \Delta t) - f(t)$$

We then have

$$\boxed{\text{av. vel.} = \frac{\Delta s}{\Delta t} = \frac{f(t + \Delta t) - f(t)}{\Delta t}} \tag{2.1.1}$$

EXAMPLE 12. Find a general expression for the average velocity when $s = t^2 - 4t$.

Solution. Since $t_2 = t + \Delta t$ and here $s = t^2 - 4t$, we have

$$s_2 = (t + \Delta t)^2 - 4(t + \Delta t)$$

and

$$\Delta s = (t + \Delta t)^2 - 4(t + \Delta t) - (t^2 - 4t)$$

Simplifying, we obtain (note carefully that $t\,\Delta t$ is not Δt^2)

$$\begin{aligned} \Delta s &= t^2 + 2t\,\Delta t + (\Delta t)^2 - 4t - 4\Delta t - t^2 + 4t \\ &= 2t\,\Delta t + (\Delta t)^2 - 4\Delta t \end{aligned}$$

Thus

$$\text{av. vel.} = \frac{\Delta s}{\Delta t} = \frac{2t\,\Delta t + (\Delta t)^2 - 4\Delta t}{\Delta t}$$

or

$$\text{av. vel.} = 2t - 4 + \Delta t$$

It is interesting to check the answers to Example 10 using this result. For (a) we have $t = 0$ and $\Delta t = 3$, or

$$\text{av. vel.} = 2(0) - 4 + 3 = -1 \text{ ft/s}$$

For (b) we have $t = 2$ and $\Delta t = 1$, or

$$\text{av. vel.} = 2(2) - 4 + 1 = 1 \text{ ft/s}$$

For (c) we have $t = 4$ and $\Delta t = 2$, or

$$\text{av. vel.} = 2(4) - 4 + 2 = 6 \text{ ft/s}$$

All agree with Example 10.

PROBLEM 4. Find a general expression for the average velocity when $s = t^2 - 6t$. Use this result to check the answers of Prob. 2.

In previous examples we have used $s = f(t)$ quite extensively. There is no reason that other functions could not have been used to illustrate average rates.

Warning ***An average rate is during a specified interval and is given by the formula***

$$\textit{average rate} = \frac{\Delta(\textit{dependent variable})}{\Delta(\textit{independent variable})}$$

EXERCISES 2.1

Group A

1. A particle moves completely around a circle of radius 3 ft in 2 s. (a) What is its displacement at $t = 2$ s? (b) What distance has it traveled in 2 s? (c) What is its average velocity during the 2 s? (d) What is its average speed during the 2 s?

2. For each of the following, find a general expression for the average rate of change of y with respect to x. Then compute this value for the indicated values of x and Δx.

(a) $y = 3x^2$, $x = 4$, $\Delta x = 1$.
(b) $y = 3x$, $x = 4$, $\Delta x = 1$.
(c) $y = x^2 + 2x - 1$, $x = 2$, $\Delta x = 0.1$.
(d) $y = 3x^2 - 5x$, $x = 0$, $\Delta x = 0.1$.
(e) $y = x^3$, $x = 2$, $\Delta x = 1$.
(f) $y = x^3 - x$, $x = 1$, $\Delta x = 0.1$.
(g) $y = -5x$, $x = -2$, $\Delta x = 1$.
(h) $y = -x^2$, $x = -1$, $\Delta x = 1$.
(i) $y = (x + 2)^2$, $x = 0$, $\Delta x = 1$.
(j) $y = (x - 3)^3$, $x = -1$, $\Delta x = 2$.

3. The charge Q at time t in an electric circuit is given by $Q = 2t^2$. Complete the following table.

t_1	Q_1	t_2	Q_2	Δt	ΔQ	$\frac{\Delta Q}{\Delta t}$
3.0	18.0	3.2	20.48	0.2	2.48	12.4
3.0		3.1				
3.0		3.01				
3.0		3.001				
3.0		3.0001				

As $\Delta t \longrightarrow 0$, what do you think the current, I, will approach at $t = 3$? (See Example 3.)

Group B

1. If a particle moves in a straight line according to $s = 2t$, find its average velocity during (a) the first 3 s; (b) the fifth second. (c) Find a general expression for the average velocity of the particle.

2. A particle moves in a straight line according to $s = t^2$. (a) Find its average velocity during the fourth second. (b) Find its average speed during the fourth second. (c) Find a general expression for its average velocity. (d) Use the answer to part (c) to check the answer to part (a).

3. A particle moves in a straight line according to $s = 2t^2 - 12t$. It reverses direction at $t = 3$ s. (a) Find its displacement at $t = 0, 1, 2, 3, 4, 5, 6, 7$, and 8 s. (b) Find its average velocity during the first 3 s. (c) Find its average velocity during the first 6 s. (d) Find its average velocity during the eighth second. (e) Find its average speed during the first 6 s. (f) Find a general expression for its average velocity. (g) Use the answer to part (f) to check the answers to parts (b), (c), and (d).

4. If $v = 32t$, find a general expression for the average acceleration (see Example 2).

5. If $v = -16t^2$, find a general expression for the average acceleration.

6. The work, W, done by a certain engine is given by $W = t^2 - 4t + 3$. Find the average power during the second second (see Example 6).

7. If the magnetic flux,[1] φ, passing through an electric circuit at time t is $\varphi = 5t^2 + 7$, find the average induced electromagnetic force during the third second (see Example 8).

8. If the charge Q in an electric circuit at time t is given by $Q = 3t^2 - t$, find the average current during the fourth second.

9. If $Q = 3t^2 - t$, find a general expression for the average current.

10. The area of a circle is $A = \pi r^2$. Find a general expression for the average rate of change of A with respect to r.

11. The volume of a sphere is $V = \frac{4}{3}\pi r^3$. Find a general expression for the average rate of change of V with respect to r. How does this compare with the surface area, which is $S = 4\pi r^2$?

12. A car goes the first mile at 30 mi/h; it then goes 60 mi/h the second mile. How fast must it go the third mile if it is to average 60 mi/h for the 3-mi trip?

Group C

1. If $s = f(t)$ describes the motion of a particle moving in a straight line and it does not reverse direction, would the average speed equal the magnitude of the average velocity?

2. How are displacement and distance alike, and how are they different?

3. What do you think is instantaneous velocity? How is it similar to average velocity? How is it different from average velocity? Have you any idea of how one might compute instantaneous velocity using average velocity?

[1]Usually, the magnetic flux is expressed in terms of sines and cosines (Chap. 9). By virtue of infinite series (Chap. 17), we take liberty in expressing it as a polynomial.

2.2 DEFINITION OF THE DERIVATIVE

We are now ready to define the most important concept in differential calculus, the *derivative of a function.* All that we have previously discussed leads to this definition. In Sec. 2.1 we saw that the average rate of change of a function was defined over an interval. We now ask what happens when the interval becomes smaller and smaller or approaches zero. Consider $s = f(t)$; the average velocity is $\Delta s/\Delta t$. Now if the time interval $\Delta t \longrightarrow 0$, we shall get average velocities over smaller and smaller intervals of time. But note that as $\Delta t \longrightarrow 0$ so also does Δs, and if $\Delta t = 0$, so also does Δs, and we would have the indeterminate expression 0/0. This is where the limit concept comes to our rescue. If the limit of $\Delta s/\Delta t$ as $\Delta t \longrightarrow 0$ exists, we call this limit the *derivative of s with respect to t.* (We are interested in the ratio and not Δs and Δt independently.) It gives the velocity, v, at time t. Let us look at a numerical example to see how this works.

EXAMPLE 1. Find the velocity of a freely falling body at time $t = 3$ s.

Solution. For a freely falling body $s = \frac{1}{2}gt^2$, where s is the displacement (feet) of fall at time t (seconds) and g is the acceleration due to gravity. Usually, for convenience[2] in computation $g = 32$ ft/s². Let us make a table of values showing the average velocities over smaller and smaller time intervals as $t \longrightarrow 3$. Using $s = 16t^2$, we obtain for $t_1 = 3$ and $s_1 = 144$ the results shown in the table.

t_2	s_2	$\Delta t = t_2 - t_1$ $\Delta t = t_2 - 3$	$\Delta s = s_2 - s_1$ $\Delta s = s_2 - 144$	$\frac{\Delta s}{\Delta t}$ = av. vel.
3.1	153.76	0.1	9.76	97.6
3.01	144.9616	0.01	0.9616	96.16
3.001	144.096016	0.001	0.096016	96.016
3.0001	144.00960016	0.0001	0.00960016	96.0016

This table shows that as $t \longrightarrow 3$, the average velocity over a small interval of time is nearly 96 ft/s. We suspect that the $\lim_{\Delta t \to 0} (\Delta s/\Delta t) = 96$ ft/s, and we say that *at* $t = 3$, the velocity $v = 96$ ft/s.

PROBLEM 1. Find the velocity of a freely falling body at $t = 0.5$ s. (*Hint:* Start with $t_2 = 0.51$.)

The numerical work in Example 1 and Prob. 1 is tedious. The work is simplified by using the analytical concept of derivative and computing at the point of interest.

[2]For sea level at 40° latitude $g = 32.1570$ ft/s².

Definition 2.2.1 *If* $s = f(t)$, *where* s *is displacement at time* t *and if* t_0 *is in the domain of* f, *then the derivative of* s *with respect to* t *at* t_0 *is*

$$\lim_{t \to t_0} \frac{f(t) - f(t_0)}{t - t_0} = \lim_{\Delta t \to 0} \frac{\Delta s}{\Delta t} \tag{2.2.1}$$

provided that this limit exists. Various notations are used to denote this limit; some of them are

$$\left.\frac{ds}{dt}\right|_{t=t_0}, \quad f'(t_0), \quad v(t_0), \quad s'(t_0), \quad \dot{s}(t_0)$$

This derivative is called the velocity, v, *at* t_0.

We now redo Example 1 using Definition 2.2.1.

EXAMPLE 2. Find $v(3)$ for $s = 16t^2$.

Solution. From Definition 2.2.1 we have $f(t) = 16t^2$ and $t_0 = 3$. Then $f(t_0) = f(3) = 16(3)^2 = 144$. Substituting in formula (2.2.1), we have

$$\begin{aligned} v(3) &= \lim_{t \to 3} \frac{16t^2 - 144}{t - 3} = \lim_{t \to 3} \frac{16(t + 3)(t - 3)}{(t - 3)} \\ &= \lim_{t \to 3} 16(t + 3) = 16(3 + 3) \end{aligned}$$

and

$$v(3) = 96 \text{ ft/s}$$

This confirms our suspicion in Example 1.

PROBLEM 2. Use Definition 2.2.1 to find $v(0.5)$ for a freely falling body. Compare this answer with that of Prob. 1.

We can extend Definition 2.2.1 to include functions whose independent variable is not t. When $y = f(x)$ we would use

Definition 2.2.2 *If* $y = f(x)$ *is a function defined on a domain* D *and* x_0 *is an element of* D, *then the derivative of* f *with respect to* x *at* x_0 *is*

$$\lim_{x \to x_0} \frac{f(x) - f(x_0)}{x - x_0} = \lim_{\Delta x \to 0} \frac{\Delta y}{\Delta x} \tag{2.2.2}$$

provided that this limit exists. Various notations are used to denote this limit; some of them are

$$\left.\frac{dy}{dx}\right|_{x=x_0}, \quad f'(x_0), \quad y'(x_0)$$

Definition 2.2.3 *A function* f *is said to be differentiable on an interval of its domain if it is differentiable at each point in the interval.*

When the limit of the quotient, called the *difference quotient*, in Definitions 2.2.1 and 2.2.2 exists without specifying t_0 or x_0, we usually compute a general derivative formula and then substitute the point of interest in the formula (see Example 3).

For the general derivative of a function, we use

Definition 2.2.4 *If $y = f(x)$ is a function defined on a domain D, then the derivative of f with respect to x at any x in the domain is*

$$\lim_{\Delta x \to 0} \frac{f(x + \Delta x) - f(x)}{\Delta x} = \lim_{\Delta x \to 0} \frac{\Delta y}{\Delta x} \tag{2.2.3}$$

provided that this limit exists. Various notations are used to denote this limit; some of them are

$$\frac{dy}{dx}, \quad f'(x), \quad y'(x), \quad y'$$

Warning ***When the "prime" notation is used it always means the "derivative of the function with respect to the independent variable."***

Formula 2.2.3 may be written with symbols other than x, y, and f. The important thing to note is that the derivative is the limit of a difference quotient as the change in the independent variable approaches zero.

We now redo Example 1 using formula 2.2.3. This is called finding the velocity by the "delta" process.

EXAMPLE 3. Use the delta process to find the velocity of a freely falling body at time $t = 3$ s.

Solution. We will use formula 2.2.3 rewritten in terms of s and t, where $s = f(t)$. We need to find Δs and then divide that expression by Δt. For $s = 16t^2$ we have

$$s_2 = 16t_2^2$$

and since

$$s_2 = s + \Delta s \qquad \text{and} \qquad t_2 = t + \Delta t$$

we find

$$\Delta s = 16t_2^2 - 16t^2$$

or

$$\Delta s = 16(t + \Delta t)^2 - 16t^2$$

which upon expanding becomes

$$\Delta s = 16t^2 + 32t\,\Delta t + 16(\Delta t)^2 - 16t^2$$

and

$$\Delta s = 32t\,\Delta t + 16(\Delta t)^2$$

Then

$$\frac{\Delta s}{\Delta t} = \frac{32t\,\Delta t + 16(\Delta t)^2}{\Delta t}$$

Note that if we now let $\Delta t = 0$, we would have 0/0. However, in using limits we are interested in what happens when $\Delta t \neq 0$ but is very small. Therefore, we may use some algebra, namely divide out the Δt and obtain

$$\frac{\Delta s}{\Delta t} = 32t + 16\Delta t$$

Now take the limit of both sides as $\Delta t \longrightarrow 0$ and obtain

$$\lim_{\Delta t \to 0} \frac{\Delta s}{\Delta t} = \lim_{\Delta t \to 0} (32t + 16\Delta t)$$

$$\therefore \quad \frac{ds}{dt} = 32t$$

This is the general derivative of $s = 16t^2$ and is true for all values of t in the domain. We next substitute for t its value in which we are interested. For $t = 3$, we have $v = 32(3) = 96$ ft/s, which tells us that our suspicion in Example 1 was correct.

PROBLEM 3. Use the delta process to find the velocity at $t = 2$ s when $s = 32t^2$.

You probably noticed in Example 3 that ds/dt is a new function which was derived from the given function. This fact is emphasized by using the "prime" notation for the derivative. The function f' is derived from f: hence the name *derivative*.

Note carefully that Definitions 2.2.1, 2.2.2, and 2.2.4 tell us that *a derivative is a rate of change*. When using these formulas we are finding the derivative by the delta process. (Later we will use formulas for finding derivatives.)

Various derivatives have special names reserved for them. Some of these are given in the following example.

EXAMPLE 4. (a) If $v = f(t)$, then $\lim_{\Delta t \to 0} (\Delta v/\Delta t) = dv/dt = a$, where a is the acceleration *at* time t.

(b) If $Q = f(t)$, then $\lim_{\Delta t \to 0} (\Delta Q/\Delta t) = dQ/dt = I$, where I is the current *at* time t.

(c) If $y = f(x)$, then $\lim_{\Delta x \to 0} (\Delta y/\Delta x) = dy/dx$ is the slope (see Sec. 2.7) of the curve *at* the point (x, y).

PROBLEM 4. For each part of Example 4, write the formula for the derivative in functional notation, that is, in the form found in formula 2.2.3.

For the present we shall consider the symbol dy/dx as one symbol and not as the fraction of two separate things. Later (Sec. 4.1) we give the meaning of dy and dx.

EXAMPLE 5. Let $y = x^3 - 2x$. Find dy/dx by the delta process.

Solution. We will use formula 2.2.3. As in Example 12 of Sec. 2.1, we have $x_2 =$

$x + \Delta x$ and $y_2 = y + \Delta y$; thus

$$y_2 = (x + \Delta x)^3 - 2(x + \Delta x)$$

and

$$\Delta y = [(x + \Delta x)^3 - 2(x + \Delta x)] - (x^3 - 2x)$$
$$= x^3 + 3x^2 \Delta x + 3x(\Delta x)^2 + (\Delta x)^3 - 2x - 2\Delta x - x^3 + 2x$$
$$= 3x^2 \Delta x + 3x(\Delta x)^2 + (\Delta x)^3 - 2\Delta x$$
$$\frac{\Delta y}{\Delta x} = \frac{3x^2 \Delta x + 3x(\Delta x)^2 + (\Delta x)^3 - 2\Delta x}{\Delta x}$$
$$= 3x^2 + 3x(\Delta x) + (\Delta x)^2 - 2$$
$$\lim_{\Delta x \to 0} \frac{\Delta y}{\Delta x} = \lim_{\Delta x \to 0} [3x^2 + 3x(\Delta x) + (\Delta x)^2 - 2]$$

Thus

$$\frac{dy}{dx} = 3x^2 - 2$$

PROBLEM 5. Let $y = x^3 + 5x$. Use the delta process to find dy/dx.

To find the value of the derivative at a point, we usually use formula 2.2.3 to find the derivative at a general point and then evaluate it at the specified point. Sometimes it is convenient to use the f notation.

EXAMPLE 6. Find $f'(-2)$ for $f(x) = 3x^2$.

Solution. First we shall find $f'(x)$ using formula 2.2.3 and then evaluate the derivative at $x = -2$.

$$f(x) = 3x^2$$
$$f(x + \Delta x) = 3(x + \Delta x)^2$$
$$= 3x^2 + 6x\,\Delta x + 3(\Delta x)^2$$
$$f(x + \Delta x) - f(x) = [3x^2 + 6x\,\Delta x + 3(\Delta x)^2] - 3x^2$$
$$= 6x\,\Delta x + 3(\Delta x)^2$$
$$\frac{f(x + \Delta x) - f(x)}{\Delta x} = \frac{6x\,\Delta x + 3(\Delta x)^2}{\Delta x}$$
$$= 6x + 3\Delta x$$
$$\lim_{\Delta x \to 0} \frac{f(x + \Delta x) - f(x)}{\Delta x} = \lim_{\Delta x \to 0} (6x + 3\Delta x)$$
$$\therefore \quad f'(x) = 6x$$

Hence

$$f'(-2) = 6(-2) = -12$$

PROBLEM 6. Find $f'(3)$ for $f(x) = 4x^2$.

EXAMPLE 7. If $Q = t^2 - t + 3$, find I at $t = 1$ and $t = 4$.

Solution. Since $I = dQ/dt$ we must find the derivative of Q and then evaluate it at the specified times.

$$Q = t^2 - t + 3$$

$$Q + \Delta Q = (t + \Delta t)^2 - (t + \Delta t) + 3$$

$$\Delta Q = t^2 + 2t\,\Delta t + (\Delta t)^2 - t - \Delta t + 3 - (t^2 - t + 3)$$

$$= 2t\,\Delta t + (\Delta t)^2 - \Delta t$$

$$\frac{\Delta Q}{\Delta t} = 2t + \Delta t - 1$$

$$\lim_{\Delta t \to 0} \frac{\Delta Q}{\Delta t} = \lim_{\Delta t \to 0} (2t + \Delta t - 1)$$

$$\therefore \quad \frac{dQ}{dt} = 2t - 1$$

and

$$I = 2t - 1$$

At $t = 1$, we have

$$I|_{t=1} = 2(1) - 1 = 2 - 1 = 1$$

At $t = 4$, we have

$$I|_{t=4} = 2(4) - 1 = 7$$

Note carefully that at $t = 4$, $dQ/dt = 7$. This means that the charge Q is changing 7 times as fast as t at this instant. Here we used the symbol $I|_{t=t_0}$ to indicate that I is evaluated at $t = t_0$.

PROBLEM 7. If $Q = 3t^2 - t + 2$, find $I|_{t=1}$ and $I|_{t=5}$. Interpret the answer.

Warning ***The derivative is a rate of change at a value of the independent variable. The derivative is sometimes called the instantaneous rate of change. Recall that the average rate of change is during an interval of time or over an interval of the independent variable.***

PROBLEM 8. (a) If $f(x) = 10$, find $f'(x)$. (b) If $f(x) =$ constant, find $f'(x)$.

Note that Prob. 8 tells us that the derivative of a constant is zero. (There is no rate of change for a constant.)

Warning ***When we use the word "derivative" we mean the derivative of the function with respect to its independent variable, unless otherwise noted.***

EXERCISES 2.2

Group A

1. Use the delta process to find the derivative function of each of the following functions.

(a) $y = x^2$. (b) $y = 3x^2 - x$. (c) $s = 5t^2 - 4$.
(d) $s = t^2 + 3t$. (e) $Q = 6t + 7$. (f) $Q = 3t^2 + 2t - 3$.

2. Use the delta process to find the derivative function of each of the following functions.

(a) $y = x^3$. (b) $y = 2x^3 - 4$.
(c) $s = 2t^3 - t^2$. (d) $s = -t^3 + 2t$.
(e) $Q = t^3 - t^2 + 3t$. (f) $Q = 2t^3 - t^2 + 3t + 5$.

3. Use the delta process to find the derivative function of each of the following functions.

(a) $f(x) = 3x - 2$. (b) $f(x) = x^2 + 3x$.
(c) $f(x) = \frac{1}{x}$. (d) $f(x) = \frac{2}{x - 3}$.
(e) $y = \sqrt{x}$.

[*Hint* for part (e): Rationalize the numerator of $\Delta y/\Delta x$; that is, multiply both numerator and denominator by $(\sqrt{x + \Delta x} + x)/(\sqrt{x + \Delta x} + x)$.]

(f) $f(x) = x - \sqrt{x}$. (g) $f(x) = \frac{1}{\sqrt{x}}$.

4. If $f(x) = 3x^2 + x - 4$, find $f'(-1)$, $f'(0)$, $f'(1)$, and $f'(3)$.

5. If $Q = 3t^2 - 2t$, find the current I at $t = 1$, $t = 3$, and $t = 10$ s. Interpret the answers.

6. In each of the following, find the value of the derivative at the indicated value of the independent variable.

(a) $y = 5x^2 - x$; $x = 0, 2, 5$. (b) $Q = t^3 - 2t$; $t = 1, 2, 3.5$.
(c) $s = 4t^2 - 3t + 7$; $t = 0, 2, 5$. (d) $u = v - 2v^2$; $v = 1, 3$.

7. Find the velocity of a body whose displacement s centimeters is given by $s = 5t - t^2$ at $t = 3$ s.

8. For $s = f(t)$, $v = ds/dt$, and $a = dv/dt$. Find a at $t = 2$ if $s = t^3$.

Group B

1. Construct a table of values to find the average velocities of a freely falling body ($s = 16t^2$) over smaller and smaller time intervals as $t \longrightarrow 2^-$.

2. For $s = 16t^2$, find the velocity v at $t = 2$. How does this compare with the average velocities found in Exercise 1?

3. Use the delta process to find the derivative of each of the following functions: (a) $y = 4$; (b) $y = 0$. (c) What can you say about the derivative of a constant?

4. The area of a circle is given by $A = \pi r^2$. Find the rate of change of the area of the circle with respect to its radius when $r = 3$. At this instant how much faster is the area changing compared to the change in the radius?

5. Use Exercise 4 to show that the rate of change of the area of a circle is equal to its circumference.

6. The volume of a sphere is given by $V = \frac{4}{3}\pi r^3$ and its surface area by $S = 4\pi r^2$. Show that the rate of change of the volume of a sphere equals its surface area.

7. A right circular cylinder has volume of $V = \pi r^2 h$. If h is kept constant at 5 cm, find the rate of change of the volume with respect to its radius when $r = 3$ cm.

8. Each of the following defines the derivative of a function. In each case tell what that function is.

(a) $\lim\limits_{\Delta x \to 0} \dfrac{3(x + \Delta x)^2 - 3x^2}{\Delta x}$.

(b) $\lim\limits_{\Delta x \to 0} \dfrac{\sqrt{x + \Delta x} - \sqrt{x}}{\Delta x}$.

(c) $\lim\limits_{\Delta x \to 0} \dfrac{\sin(x + \Delta x) - \sin x}{\Delta x}$.

(d) $\lim\limits_{\Delta x \to 0} \dfrac{\log(x + \Delta x) - \log x}{\Delta x}$.

9. Each of the following defines the derivative of a function. State the function and its derivative.

(a) $\lim\limits_{\Delta \theta \to 0} \dfrac{\tan(\theta + \Delta\theta) - \tan\theta}{\Delta\theta} = \sec^2\theta$.

(b) $\lim\limits_{\Delta x \to 0} \dfrac{(e^{x+\Delta x} - e^x)}{\Delta x} = e^x$.

(c) $\lim\limits_{\Delta x \to 0} \dfrac{(x + \Delta x)^n - x^n}{\Delta x} = nx^{n-1}$.

(d) $\lim\limits_{\Delta x \to 0} \dfrac{\ln(x + \Delta x) - \ln x}{\Delta x} = \dfrac{1}{x}$.

Group C

1. Using the idea that speed is $\lim\limits_{\Delta t \to 0} [\Delta(\text{distance})/\Delta t]$, explain how a "speed trap" might be constructed to determine the speed of a car at a certain point.

2. Show that if y is a constant, then $dy/dx = 0$.

3. Tell in your words what a derivative is.

4. Give an example of a function that does not have a derivative at some point in its domain.

5. If the function $y = f(x)$ has domain D and its derivative $y' = f'(x)$ has domain D', what can you say about the relative sizes of D and D'?

6. In what ways (geometric) can a function fail to have a derivative at a point in its domain?

7. Are continuity at a point and differentiability at a point related? How?

8. Tell how Definitions 2.2.2 and 2.2.4 are alike and how they are different.

2.3 THE DERIVATIVE OF $y = ax^n$

The delta process of finding the derivative is at times quite tedious. We shall apply it to the general form of some functions and thereby arrive at some formulas for differentiation. For the present we shall work mostly with polynomial functions and certain simple power functions. Later, we shall extend the ideas to such functions as exponential, logarithmic, trigonometric, and inverse trigonometric.

The functions $y = x^3$, $y = 7x^5 - 2x^3 + 3$, and $y = -7$ are examples of a general class of functions called *polynomials*. In general, a polynomial function is defined as

$$y = f(x) = a_0x^n + a_1x^{n-1} + a_2x^{n-2} + \ldots + a_{n-1}x + a_n$$

where $a_0, a_1, \ldots, a_n$ are constants and n is a *nonnegative integer*. ($y = x^{1/3} + x$ is not a polynomial because $\frac{1}{3}$ is not an integer. Why is $y = x^{-1} = 1/x$ not a polynomial?) If the constant n is not a nonnegative integer, then $f(x)$ is called a *power function*.

> ***Theorem 2.3.1*** *If* $y = ax^n$, *where* a *is a constant and* n *is a positive integer, then* $dy/dx = nax^{n-1}$.

(We give the proof of this theorem after Prob. 1.) Note carefully that this theorem tells us that to find dy/dx for ax^n, we multiply the coefficient by the power of x and then we decrease the power by 1.

EXAMPLE 1. (a) If $y = x^3$, then $dy/dx = 3x^{3-1} = 3x^2$. (b) If $y = 2x^{15}$, then $dy/dx = 2(15)x^{15-1} = 30x^{14}$. We usually do the multiplying and the subtraction mentally and just write down the answer.
(c) If $f(x) = 3x^7$, $f'(x) = 21x^6$.
(d) If $s = 16t^2$, $v = 32t$.

PROBLEM 1. (a) If $y = x^5$, find dy/dx. (b) If $y = -3x^4$, find dy/dx. (c) If $f(x) = 6x^7$, find $f'(x)$. (d) Find v for $s = 3.5t^2$.

We now give a proof for Theorem 2.3.1.

Proof of Theorem 2.3.1 Using the Delta Process

$$y = ax^n$$
$$y + \Delta y = a(x + \Delta x)^n$$
$$\Delta y = a(x + \Delta x)^n - ax^n$$

Expanding by the binomial theorem we have

$$\Delta y = a\left[x^n + \frac{nx^{n-1}}{1}\Delta x + \frac{n(n-1)x^{n-2}}{1 \cdot 2}(\Delta x)^2 + \ldots + (\Delta x)^n\right] - ax^n$$

Then dividing by Δx, we obtain

$$\frac{\Delta y}{\Delta x} = \frac{\dfrac{anx^{n-1}}{1}\Delta x + \dfrac{an(n-1)x^{n-2}}{1\cdot 2}(\Delta x)^2 + \ldots + a(\Delta x)^n}{\Delta x}$$

or

$$\frac{\Delta y}{\Delta x} = anx^{n-1} + \frac{an(n-1)x^{n-2}(\Delta x)}{1\cdot 2} + \ldots + a(\Delta x)^{n-1}$$

$$\lim_{\Delta x\to 0}\frac{\Delta y}{\Delta x} = \lim_{\Delta x\to 0}\left[anx^{n-1} + \frac{an(n-1)x^{n-2}}{1\cdot 2}(\Delta x) + \ldots + a(\Delta x)^{n-1}\right]$$

Since in the expression on the right all terms except the first contain Δx, we have for

$$\boxed{\begin{aligned} y &= ax^n \\ \frac{dy}{dx} &= anx^{n-1} \end{aligned}} \qquad (2.3.1)$$

This concludes the proof for n, a positive integer. (Refer to Prob. 8 of Sec. 2.2 for $n = 0$.)

Warning ***Formula 2.3.1 can be used only when finding the derivative with respect to the variable which is raised to the power. That is, for $y = ax^n$,***

$$\frac{dy}{dx} = anx^{n-1}$$

(arrows under dx and x, labelled: ***same variable***)

For $y = t^3$, we cannot use formula 2.3.1 to find dy/dx because x and t are different variables. (However, we can find $dy/dt = 3t^2$.)

It can be shown that formula 2.3.1 is true when n is any real constant. All radical terms must be changed to fractional exponents.

EXAMPLE 2. (a) If $y = 5x^{2/3}$, then $dy/dx = \frac{10}{3}x^{-1/3}$.

(b) If $s = 3/t^2$, we write $s = 3t^{-2}$, and $ds/dt = -6t^{-3} = -6/t^3$.

(c) If $f(x) = \sqrt[3]{x}$, we write $f(x) = x^{1/3}$ and $f'(x) = \frac{1}{3}x^{-2/3} = 1/3\sqrt[3]{x^2}$.

PROBLEM 2. (a) If $y = 3x^{2/3}$, find dy/dx. (b) If $v = 2/t^3$, find dv/dt. (c) If $W = \sqrt{t}$, find dW/dt.

To find the derivative of the sum of functions we consider the following. Suppose that f and g are functions which are both differentiable at x_0; then we have

$$\lim_{x \to x_0} \frac{[f(x) + g(x)] - [f(x_0) + g(x_0)]}{x - x_0}$$
$$= \lim_{x \to x_0} \frac{[f(x) - f(x_0)] + [g(x) - g(x_0)]}{x - x_0}$$
$$= \lim_{x \to x_0} \frac{f(x) - f(x_0)}{x - x_0} + \lim_{x \to x_0} \frac{g(x) - g(x_0)}{x - x_0} = f'(x_0) + g'(x_0)$$

This means that *the derivative of a sum of terms is the sum of the derivatives of each of its terms.*

EXAMPLE 3. (a) If $y = 5x^4 - 2x^3 + x - 7$, then $dy/dx = 20x^3 - 6x^2 + 1$.
(b) If $Q = 3t^3 - t^2 + 5$, then $I = dQ/dt = 9t^2 - 2t$.
(c) If $f(x) = 3x^2 - 2x^{-1}$, find $f'(2)$, $f'(-2)$, and $f'(0)$.

Solution. First we find $f'(x)$, then we evaluate it at the required values of x,

$$f'(x) = 6x + 2x^{-2} = 6x + \frac{2}{x^2}$$

$$f'(2) = 6(2) + \frac{2}{(2)^2} = 12 + 0.5 = 12.5$$

$$f'(-2) = 6(-2) + \frac{2}{(-2)^2} = -12 + 0.5 = -11.5$$

$$f'(0) = 6(0) + \frac{2}{(0)^2} \qquad \text{does not exist}$$

PROBLEM 3. (a) If $y = 3x^6 - 5x^2 + 2x - 9$, find dy/dx.
(b) If $Q = 2t^5 - 3t^2 + t - 1$, find $I|_{t=1}$.
(c) If $f(x) = 2x^3 - 2/x^3$, find $f'(1)$, $f'(0)$, and $f'(-1)$.
(d) If $f(t) = 5t^3 - 3t + 2\sqrt{t} - 7/t$, find $f'(t)$.

EXERCISES 2.3

Group A

1. Find the derivative with respect to the independent variable of each of the following functions.

(a) $y = 3x$.
(b) $y = 2x + 5$.
(c) $f(x) = 3x^2$.
(d) $f(x) = 3x^2 + 5x - 4$.
(e) $Q = 5t^2 - 4t$.
(f) $Q = \frac{1}{3}t^3 - t^2 + 5t$.
(g) $s = 16t^2 + 32t - 64$
(h) $v = 32t - 16$.

2. Find the derivative with respect to x of each of the following functions.

(a) $y = x^3 - 2x^{-2}$.
(b) $y = \dfrac{1}{x}$.
(c) $y = \sqrt[4]{x}$.
(d) $y = x - \dfrac{1}{3\sqrt{x}}$.

(e) $f(x) = 5x + \frac{5}{x^2}$.   (f) $f(x) = x^{1/3} - 3x^{-1/6}$.

3. Perform the indicated operations on each of the following.

(a) $f(t) = t^2 - 3t$; find $f'(-2)$, $f'(0)$, and $f'(1)$.
(b) $f(t) = 3t$; find $f'(-2)$, $f'(0)$, and $f'(3)$.
(c) $f(t) = 1$; find $f'(-2)$, $f'(0)$, and $f'(2)$.
(d) $f(t) = \sqrt{t}$; find $f(0)$ and $f'(0)$.
(e) $f(t) = 1/t$; find $f'(-2)$, $f'(0)$, and $f'(2)$.

Group B

1. (a) If $s = t^3 + 3t^2 - 6t + 4$, find v at $t = 10$.
(b) If $v = 200 - 32t$, find a at $t = 10$.

2. (a) If $Q = 6t^2 - 3t$, find I at $t = 3$ s.
(b) If $Q = 3t$, find I at $t = 10$ s.

3. If a stone is thrown up from the surface of the earth with a velocity of 128 ft/s, its displacement at t seconds is given by $s = 128t - 16t^2$. Find t when its velocity is zero. Also find s at this time. If $t = 5$ s, is the stone still going up or is it falling back toward the earth?

4. The force of attraction between two bodies is given by $F = k(m_1m_2/r^2)$. If $F = 50/r^2$, find the rate of change of F with respect to r when $r = 10$.

5. For $W = t^2 - 3t$, find the power at $t = 3$.

6. For $\theta = -3t^2 + 15$, find the angular velocity at $t = 2$.

7. For $\phi = 0.7t^3 - t$, find the induced electromagnetic force at $t = 5$.

8. The velocity of a particle is given by the formula $v = 6t^2 + 2t + 1$. Find its acceleration (a) at time $t = 1$; (b) at any time $t \geqq 0$.

9. If work is expressed by the formula $W(t) = \frac{1}{2}t^3 + 2t$, find the power (a) at time $t = 3$; (b) at any time $t \geqq 0$.

10. The magnetic flux passing through a circuit at time t is given by the formula $\phi(t) = t^4 + 2t^{1/2}$. Find the induced electromagnetic force (a) at time $t = 4$; (b) at any time $t \geqq 0$.

11. The angle of inclination of a tangent line is given by the formula $\theta = s^4 + 2s^2 + 1$. Find the curvature for any distance $s \geqq 0$.

12. The angle θ through which a wheel turns in t seconds is given by the formula $\theta = 3t^3 + 2t^2 + 1$. Find the angular velocity at any time $t \geqq 0$.

13. The formula used to change degrees Celsius, C, to degrees Fahrenheit, F, is $F = \frac{9}{5}C + 32$. Find (a) dF/dC; (b) dC/dF.

14. The voltage E in a circuit is given by $E = 3t^2 - t$. (a) Find $(dE/dt)|_{t=2}$. (b) Find the value of $C(dE/dt)$ when $C = 2 \times 10^{-4}$ and $t = 1$.

15. Find $(dI/dt)|_{t=5}$ when (a) $I = 3t^2 - 7t$; (b) $5I + 2t - 25 = 0$.

16. $f(x) = \begin{cases} 2 & 0 \leqq x < 4 \\ x^2 - x & 4 \leqq x < 8. \end{cases}$ Find $f'(2)$ and $f'(5)$.

17. $f(t) = \begin{cases} 3t & 0 \leqq t < 2 \\ 3t - 6 & 2 \leqq t < 4. \end{cases}$ Find $f'(1)$, $f'(2)$, and $f'(3)$.

18. For $f(x) = \sqrt{x}$, do $f(x)$ and $f'(x)$ have the same input set? Why?

19. For $y = 3x^2 - 2x + 7$, find the value of x when the rate of change of y with respect to x will be 10.

20. Find the time when the current in a circuit will be 0 for $Q = t^3 - 3t^2$.

21. Write the value of each of the following limits without going through the delta process.

(a) $\lim_{\Delta x \to 0} \frac{5(x + \Delta x)^3 - 5x^3}{\Delta x}$. (b) $\lim_{\Delta t \to 0} \frac{16(t + \Delta t)^2 - 16t^2}{\Delta t}$.

22. Let

$$f(x) = \begin{cases} x^2 - 2x & x \leqq 0 \text{ and } x \geqq 2 \\ -x^2 + 2x & 0 < x < 2 \end{cases}$$

(a) Draw the graph of $f(x)$. (b) Discuss the derivative of $f(x)$, giving special attention to the points (0, 0) and (2, 0).

23. (a) We noted earlier that the limit of a sum is the sum of the limits. Also, the derivative of a sum is the sum of the derivatives. Since the limit of a product is the product of the limits, do you think that the derivative of a product is the product of the derivatives? (b) Consider a product of the functions f and g and compute the difference quotient for the product $f(x)g(x)$. Can you compute the limit of this quotient as $x \longrightarrow x_0$? What is wrong with applying the limit of a product rule to find the derivative of a product?

2.4 SLOPE OF A STRAIGHT LINE

Before we can discuss the geometric interpretation of the derivative, we need to know something about a quantity called the *slope*. It is used to measure the steepness of a graph. The best place to start is with a straight line.

The straight line L in Fig. 2.4.1 is any arbitrary line intersecting the X-axis at $A(a, 0)$. The angle θ (read "theta"), measured counterclockwise from the positive X-axis to the line L, is called the *angle of inclination* of line L. If L intersects the X-axis, θ is between 0° and 180°. The angle of inclination of a line parallel to the X-axis is 0°, and the angle of inclination of a line perpendicular to the X-axis is 90°. Thus we have $0° \leqq \theta < 180°$.

Definition 2.4.1 *The slope of a line, denoted by* **m**, *is given by*

$$\boxed{m = \tan \theta} \tag{2.4.1}$$

where θ is the angle of inclination.

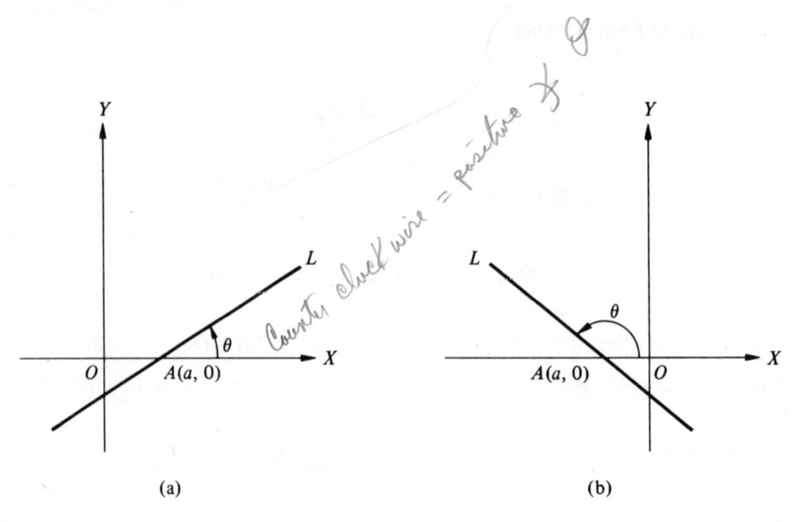

Figure 2.4.1 Slope of a straight line.

Since the tangent of an angle between 0° and 90° is positive, we say that a line running *up* to the right has *positive* slope. Similarly, since the tangent of an angle between 90° and 180° is negative, we say that a line running *down* to the right has a *negative* slope. Since tan 90° does not exist, we say that a line parallel to the Y-axis has an indeterminate slope. A line parallel to the X-axis has zero slope. In Fig. 2.4.2 several lines are shown with their slope indicated by the value of m. You should become familiar with the steepness of the line as indicated by its slope.

We now will develop a formula for finding the slope of a line joining any two points. In Fig. 2.4.3 let $P_1(x_1, y_1)$ and $P_2(x_2, y_2)$ be any two points on a

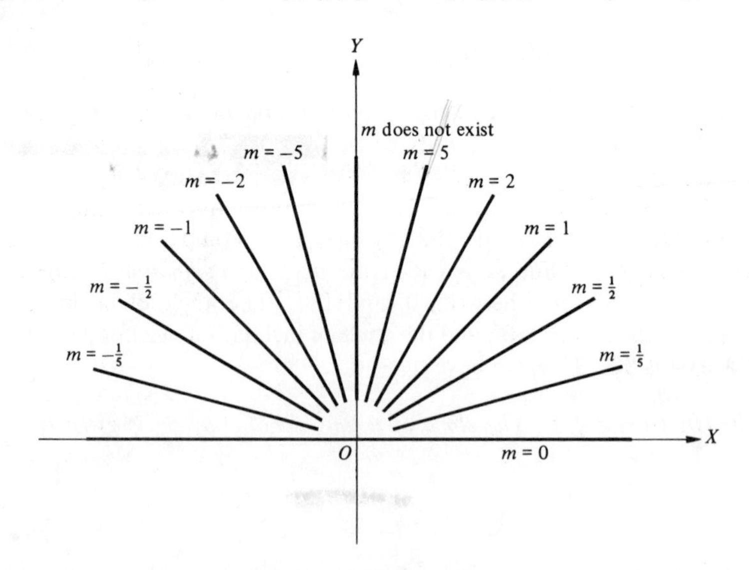

Figure 2.4.2 Slopes of lines

straight line L. In Fig. 2.4.3(a) L has a positive slope, while in Fig. 2.4.3(b) L has a negative slope.

If we imagine that a point moves from P_1 along the line L to P_2, we see that the X-coordinate changes from x_1 to x_2, that is, $\Delta x = x_2 - x_1$, and that the Y-coordinate changes from y_1 to y_2 or $\Delta y = y_2 - y_1$. In Fig. 2.4.3(a) we have that

$$\tan \theta = \frac{\Delta y}{\Delta x}$$

Hence

$$\boxed{m = \frac{\Delta y}{\Delta x} = \frac{y_2 - y_1}{x_2 - x_1}} \tag{2.4.2}$$

In Fig. 2.4.3(b) note that $\tan \theta = -\tan(180° - \theta)$, but since $y_2 < y_1$, Δy is negative, so Eq. 2.4.2, $m = \Delta y/\Delta x$, holds in this case also. In fact, Eq. 2.4.2 is true for all arrangements of points P_1 and P_2 on line L. The slope is sometimes said to be the *rise over the run*. (Can you tell why?)

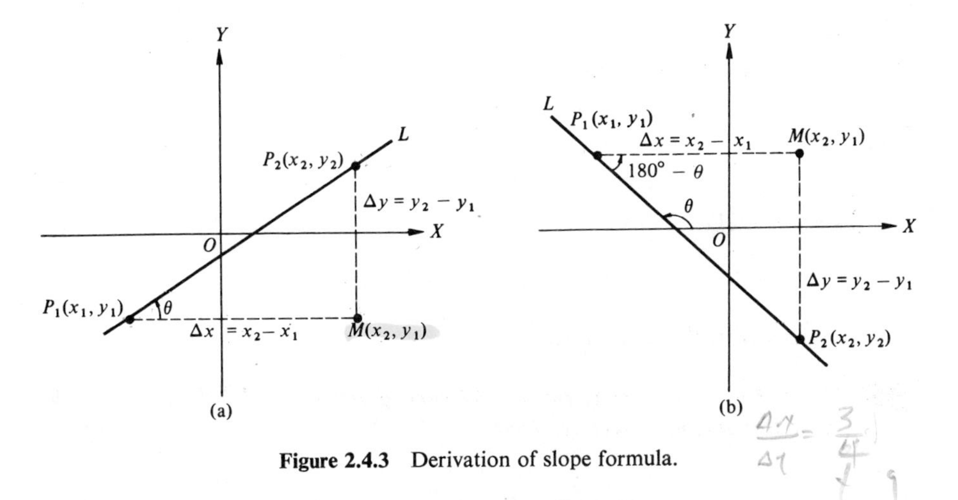

Figure 2.4.3 Derivation of slope formula.

EXAMPLE 1. Find the slope of a line that passes through the points (2, −3) and (5, 1).

Solution. Plot the points and draw line L through them (see Fig. 2.4.4). Either of the points could be chosen as P_1. For convenience, to keep Δx positive, let us choose as P_1, that point having the smallest X-coordinate (that is, the point farthest to the left). Thus here we have $P_1(2, -3)$ and $P_2(5, 1)$. Then

$$\Delta x = x_2 - x_1 = 5 - 2 = 3$$

and

$$\Delta y = y_2 - y_1 = 1 - (-3) = 4$$

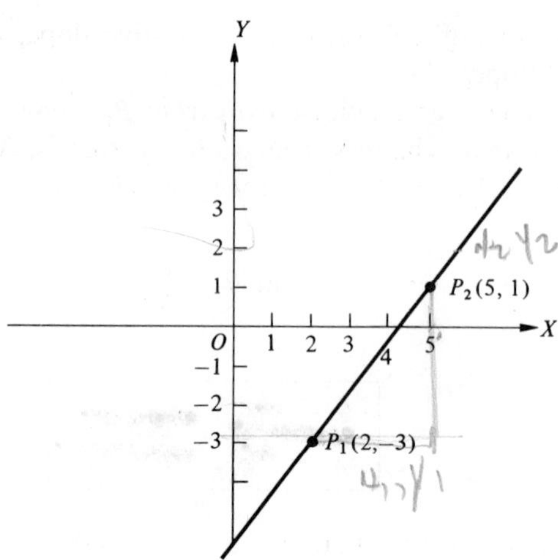

Figure 2.4.4 Example 1.

Therefore,

$$m = \frac{\Delta y}{\Delta x} = \frac{4}{3} = 1.3 \qquad \text{(from Eq. 2.4.2)}$$

If we had chosen the points in the other order, that is, had we chosen $P_1(5, 1)$ and $P_2(2, -3)$, we would have

$$\Delta x = x_2 - x_1 = 2 - 5 = -3$$

and

$$\Delta y = y_2 - y_1 = -3 - 1 = -4$$

and using formula 2.4.2, we have

$$m = \frac{\Delta y}{\Delta x} = \frac{-4}{-3} = \frac{4}{3} = 1.3$$

which agrees with the first choice.

Warning ***When using the second part of formula 2.4.2, be sure that the subscripts line up correctly. That is,***

$$m = \frac{y_2 - y_1}{x_2 - x_1}$$

PROBLEM 1. Find the slope of a line that passes through the points (3, −2) and (1, 5).

EXAMPLE 2. The slope of a line is $\frac{2}{5}$. If a point on the line is at (−3, −4), find the coordinates of another point on the line.

Solution. Let P_1 be the given point, that is, $P_1(-3, -4)$. Since $m = \Delta y/\Delta x = \frac{2}{5}$, we can find one solution by letting $\Delta y = 2$ and $\Delta x = 5$. This gives us

$$\Delta y = y_2 - y_1 = y_2 - (-4) = y_2 + 4 = 2$$

or

$$y_2 = -2$$

Similarly,

$$\Delta x = x_2 - x_1 = x_2 - (-3) = x_2 + 3 = 5$$

or

$$x_2 = 2$$

Thus P_2 is at (2, −2). There are, of course, many solutions to this problem, as any values of Δy and Δx that give a ratio of $\frac{2}{5}$ will do. Thus we could choose $\Delta y = 4$ and $\Delta x = 10$ and then

$$\Delta y = y_2 - y_1 = y_2 - (-4) = 4$$

or

$$y_2 = 0$$

Similarly,

$$\Delta x = x_2 - x_1 = x_2 - (-3) = 10$$

or

$$x_2 = 7$$

Hence another point on line L is $P_3(7, 0)$.

PROBLEM 2. The slope of a line is $-\frac{5}{2}$. One point on the line is at (−4, 7). Find another point on the line.

If two lines have the same slope, their angles of inclination are equal and the lines are parallel. If the slopes are not equal, then the lines must intersect. The angle of intersection can be obtained by using the slopes. In Fig. 2.4.5 the two lines L_1 and L_2 intersect as shown. Let $m_1 = \tan \theta_1$ and $m_2 = \tan \theta_2$. Let angle β (read "beta") be the angle measured from L_1 to L_2 in a counterclockwise direction. From geometry we have that $\beta = \theta_2 - \theta_1$. In trigonometry we saw

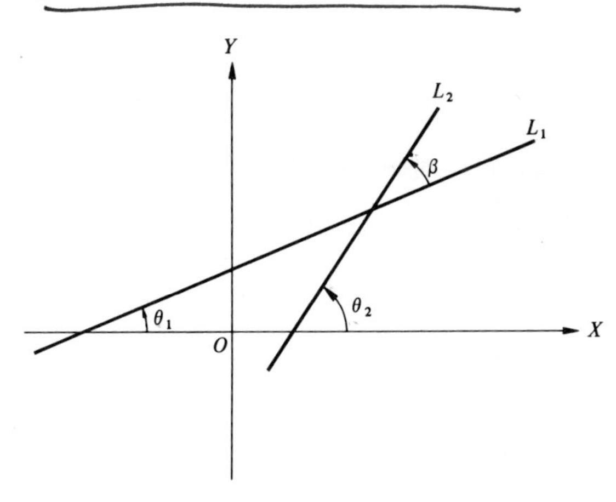

Figure 2.4.5 Angle of intersection.

that

$$\tan \beta = \tan (\theta_2 - \theta_1) = \frac{\tan \theta_2 - \tan \theta_1}{1 + \tan \theta_1 \tan \theta_2}$$

Hence we have that

$$\boxed{\tan \beta = \frac{m_2 - m_1}{1 + m_1 m_2}} \tag{2.4.3}$$

Note carefully that Eq. 2.4.3 gives the tangent of the angle β, which has for its initial side the line L_1 with slope m_1 and for its terminal side the line L_2 with slope m_2 when β is measured in a counterclockwise direction. If $\tan \beta$ is known, then β can be found from a table of tangents of angles, or a calculator.

EXAMPLE 3. Line L_1 passes through $P_1(3, 0)$ and $P_2(0, 3)$. Line L_2 passes through $P_3(-3, 0)$ and $P_4(3, 3)$. Find their angle of intersection.

Solution. Plot the points and draw the lines (see Fig. 2.4.6).

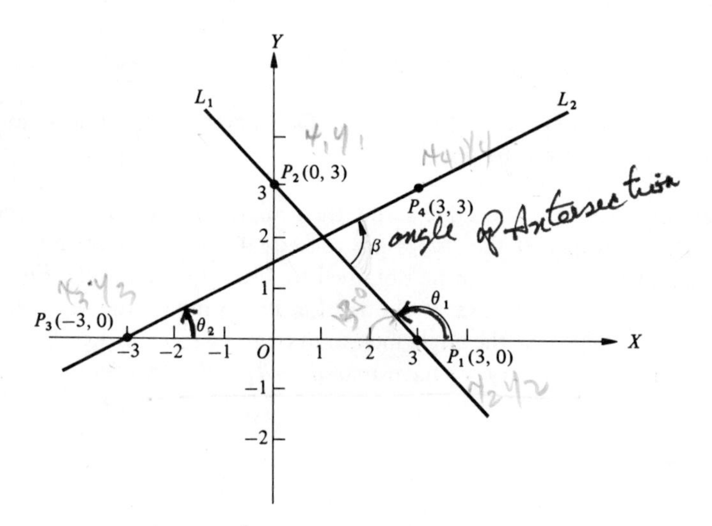

Figure 2.4.6 Example 3.

$$m_1 = \frac{3 - 0}{0 - 3} = -1 \quad \text{and} \quad m_2 = \frac{3 - 0}{3 - (-3)} = \frac{1}{2}$$

Using Eq. (2.4.3), we have

$$\tan \beta = \frac{m_2 - m_1}{1 + m_1 m_2} = \frac{(\frac{1}{2}) - (-1)}{1 + (\frac{1}{2})(-1)} = \frac{\frac{3}{2}}{\frac{1}{2}}$$

and

$$\tan \beta = 3$$

From a calculator, we find that $\beta = 71.6°$. (Your calculator should have trigonometric functions.)

PROBLEM 3. Line L_1 passes through $P_1(0, -1)$ and $P_2(4, 2)$. Line L_2 passes through $P_3(-1, 4)$ and $P_4(5, 0)$. Find their angle of intersection.

Two cases are of particular interest.

Case 1. If L_1 and L_2 are *perpendicular*, then $\beta = 90°$, and $\tan 90°$ does not exist. This implies that the denominator in Eq. 2.4.3 for $\tan \beta$ is zero; otherwise, $\tan \beta$ would be a real number and β would not be 90°. Thus we have $1 + m_1 m_2 = 0$, or

$$m_1 m_2 = -1$$

and

$$\boxed{m_2 = -\frac{1}{m_1}} \tag{2.4.4}$$

That is, *if two lines are perpendicular, their slopes are negative reciprocals.*

Case 2. If L_1 and L_2 are parallel, then $\beta = 0°$, and $\tan 0° = 0$. This implies that the numerator in Eq. 2.4.3 for $\tan \beta$ is zero. Thus we have

$$m_1 - m_2 = 0$$

or

$$\boxed{m_2 = m_1} \tag{2.4.5}$$

That is, *if two lines are parallel, their slopes are equal.*

EXAMPLE 4. Line L_1 has slope of $\frac{3}{2}$. Find another point on L_2 if L_2 passes through $P_1(1, 3)$ and is perpendicular to L_1.

Solution. Since $L_1 \perp L_2$ we have from Eq. 2.4.4

$$m_2 = -\frac{1}{m_1} = -\frac{1}{\frac{3}{2}} = -\frac{2}{3} \quad \text{or} \quad -\frac{2k}{3k} \quad k \neq 0$$

Hence

$$m_2 = \frac{y_2 - y_1}{x_2 - x_1} = \frac{y_2 - 3}{x_2 - 1} = -\frac{2k}{3k}$$

and we have

$$y_2 - 3 = -2k \quad \text{and} \quad x_2 - 1 = 3k$$

Therefore, for $k = 1$

$$y_2 = 1 \quad \text{and} \quad x_2 = 4$$

Thus a point is $P_2(4, 1)$. (Another point could be found using a different k.)

PROBLEM 4. Line L_2 has slope of $\frac{2}{5}$. Find another point on line L_1 if L_1 passes through $P_1(-2, 1)$ and is perpendicular to L_2.

EXERCISES 2.4

Group A

1. In each of the following, plot the designated points and draw line L through them. Find the slope of L, and label the value of m on each line.

(a) $P_1(2, 4)$, $P_2(4, 6)$.
(b) $P_1(2, 1)$, $P_2(-4, 7)$.
(c) $P_1(-2, 0)$, $P_2(0, 2)$.
(d) $P_1(0, -4)$, $P_2(-4, 3)$.
(e) $P_1(-2, -2)$, $P_2(-2, 8)$.
(f) $P_1(-3, -4)$, $P_2(1, 4)$.
(g) $P_1(0, -3)$, $P_2(0, 3)$.
(h) $P_1(-1, -3)$, $P_2(8, -3)$.

2. In each of the following, draw line L with the given slope through the designated point. Find at least one other point on line L.

(a) $m = 1$, $P_1(2, 1)$.
(b) $m = -1$, $P_1(-2, 3)$.
(c) $m = \frac{1}{2}$, $P_1(1, 0)$.
(d) $m = \frac{2}{3}$, $P_1(-3, -2)$.
(e) $m = -3$, $P_1(3, -1)$.
(f) $m = 0$, $P_1(-2, -4)$.
(g) $m = -5$, $P_1(-4, 0)$.
(h) $m = \frac{1}{10}$, $P_1(-10, 5)$.

3. In each of the following, draw line L through the designated point with the indicated angle of inclination and label each line with the value of the slope.

(a) $P_1(0, 0)$, $\theta = 45°$.
(b) $P_1(1, 2)$, $\theta = 30°$.
(c) $P_1(1, 2)$, $\theta = 135°$.
(d) $P_1(-1, 3)$, $\theta = 0°$.
(e) $P_1(-2, 0)$, $\theta = 90°$.
(f) $P_1(0, 3)$, $\theta = 90°$.
(g) $P_1(5, -1)$, $\theta = 120°$.
(h) $P_1(-2, 2)$, $\theta = 0°$.

4. In each of the following, m_1 is the slope of line L_1. Find the slopes of lines L_2 and L_3.

(a) $m_1 = \frac{1}{3}$, $L_2 \parallel L_1$, $L_3 \perp L_1$.
(b) $m_1 = -2$, $L_2 \parallel L_1$, $L_3 \perp L_1$.

5. In each of the following, find the slope of a line perpendicular to the line segment joining the points.

(a) $(-2, 4)$ and $(4, 8)$.
(b) $(0, -1)$ and $(-3, 5)$.
(c) $(4, -4)$ and $(0, -4)$.
(d) $(3, 2)$ and $(3, -7)$.

6. In each of the following, find the angle of intersection of line L_1 through P_1 and P_2 and line L_2 through P_3 and P_4.

(a) $P_1(2, 3)$, $P_2(-1, 4)$, $P_3(0, 2)$, $P_4(-1, 1)$.
(b) $P_1(-1, 1)$, $P_2(1, 3)$, $P_3(1, 1)$, $P_4(0, 3)$.

7. Find the angle of intersection between two lines whose slopes are 2 and $\frac{1}{3}$, respectively.

Group B

1. If three points are collinear (that is, on the same straight line), then the slopes of the line segments P_1P_2 and P_1P_3 are equal. Show that each set of points is collinear.

(a) $\{(1, 6), (-3, 10), (2, 5)\}$.
(b) $\{(0, 0), (1, -2), (-3, 6)\}$.

2. Find x such that $(x, 3)$, $(0, -2)$, and $(2, -1)$ are collinear.

3. Find x such that the line through P_1 and P_2 is perpendicular to the line through P_3 and P_4 for the points $P_1(x, 2)$, $P_2(0, 4)$, $P_3(0, -4)$, and $P_4(x, -3)$.

4. If two lines have slopes m and $[(m - 1)/(m + 1)]$, respectively, find their angle of intersection.

2.5 EQUATIONS OF A STRAIGHT LINE

A straight line may be characterized by its slope and a point through which it passes. Consider a straight line L with slope m passing through point $P_1(x_1, y_1)$. Let $P(x, y)$ be any other point on L. If we consider this variable point as point P_2 in the slope formula, Eq. 2.4.2,

$$m = \frac{y_2 - y_1}{x_2 - x_1}$$

we have

$$m = \frac{y - y_1}{x - x_1}$$

Clearing fractions, we obtain

$$\boxed{y - y_1 = m(x - x_1)} \tag{2.5.1}$$

Equation 2.5.1 is *the equation of a line through point* (x_1, y_1) *with a constant slope of* m, since it is satisfied by the coordinates of any point on the line and no other point. It is called the *point-slope* form of the straight line.

EXAMPLE 1. Find an equation of a line through $(-3, 2)$ with slope of -2.

Solution. Since we have a point and a slope given, we shall use the point-slope formula. Here $x_1 = -3$, $y_1 = 2$, and $m = -2$. Substitution in Eq. 2.5.1 yields

$$y - 2 = -2[x - (-3)]$$

or

$$y - 2 = -2(x + 3)$$

This can be written in simpler form as

$$y = -2x - 4$$

You can verify that the coordinates of the given point $(-3, 2)$ satisfy this equation by direct substitution.

PROBLEM 1. Find an equation of a line through $(2, -5)$ with slope of 7.

☛ **Warning** ***In the point-slope formula m must be a constant and x_1 and y_1 must be the coordinates of a fixed point on the line.***

$$\overbrace{y - y_1 = \underset{\substack{\uparrow \\ \textit{constant} \\ \textit{number}}}{m}(x - x_1)}^{\textit{numbers}}$$

(The arrows under "numbers" point to y_1 and x_1.)

If the line is not parallel to the Y-axis, we may denote the point where it intersects the Y-axis as $(0, b)$. Then Eq. 2.5.1 reduces to

$$y - b = m(x - 0)$$

or

$$\boxed{y = \underset{\substack{\uparrow \\ \text{slope}}}{m}x + \underset{\substack{\uparrow \\ Y\text{-intercept}}}{b}} \tag{2.5.2}$$

Equation 2.5.2 is called the *slope-intercept* form of the straight line, where m is the slope and b is the Y-intercept. [The point of intersection is $(0, b)$.] If the line is parallel to the X-axis, $m = 0$ and Eq. 2.5.2 becomes

$$y = b \tag{2.5.3}$$

If the line is parallel to the Y-axis, then m does not exist, and

$$x = x_1 \tag{2.5.4}$$

EXAMPLE 2. Find an equation of the line with slope of -2 intersecting the Y-axis at $(0, -4)$.

Solution. Since we are given the slope and the Y-intercept, we use the slope-intercept form with $m = -2$ and $b = -4$. Thus

$$y = -2x - 4$$

PROBLEM 2. Find an equation of the line with slope of 3 intersecting the Y-axis at $(0, 2)$.

EXAMPLE 3. What is the slope and Y-intercept of the straight line whose equation is $2x + 3y - 9 = 0$?

Solution. Solving for y, we have

$$y = -\tfrac{2}{3}x + 3$$

Comparing with Eq. 2.5.2, we can say that

$$m = -\tfrac{2}{3} \qquad \text{and} \qquad b = 3$$

PROBLEM 3. What is the slope and Y-intercept of the straight line whose equation is $3x - 2y + 7 = 0$?

EXAMPLE 4. The equation $y = 4$ is a straight line parallel to the X-axis and 4 units above it (see Eq. 2.5.3).

PROBLEM 4. Describe the graph of the equation $y = -2$. Sketch the graph. What is its slope?

EXAMPLE 5. The equation $x + 3 = 0$ is a straight line parallel to the Y-axis and 3 units to the left of it since $x = -3$ (see Eq. 2.5.4).

PROBLEM 5. Describe the graph of the equation $2x - 5 = 0$. Sketch the graph. What is its slope?

Any equation of the form $Ax + By + C = 0$, where A, B, and C are constants, *represents a straight line* since it can be written as

$$y = -\frac{A}{B}x - \frac{C}{B}$$

We say that the equation $Ax + By + C = 0$ is a *linear equation* and the function $f(x) = mx + b$ is a *linear function*, since its graph is a straight line.

There are several different forms for the equation of a straight line. They are all useful at times. We now list some of them.

$y - y_1 = m(x - x_1)$	point-slope
$y - y_1 = \frac{y_2 - y_1}{x_2 - x_1}(x - x_1)$	two-point
$y = mx + b$	slope-intercept
$Ax + By + C = 0$	general
$\frac{x}{a} + \frac{y}{b} = 1$	intercept

EXAMPLE 6. Sketch the graph of the equation $2x + 3y - 6 = 0$.

Solution. We recognize this equation as the general form of a linear equation; hence its graph is a straight line. If we can find two points on the line, the straight line can be drawn readily. Probably the easiest points are the X- and the Y-intercepts. If $x = 0$, we have $y = 2$. Thus the Y-intercept is at (0, 2). If $y = 0$, we have $x = 3$, so the X-intercept is at (3, 0). Plot these two points and draw the line (see Fig. 2.5.1).

If we had written the equation as

$$2x + 3y = 6$$

and then divided by 6 to obtain

$$\frac{x}{3} + \frac{y}{2} = 1$$

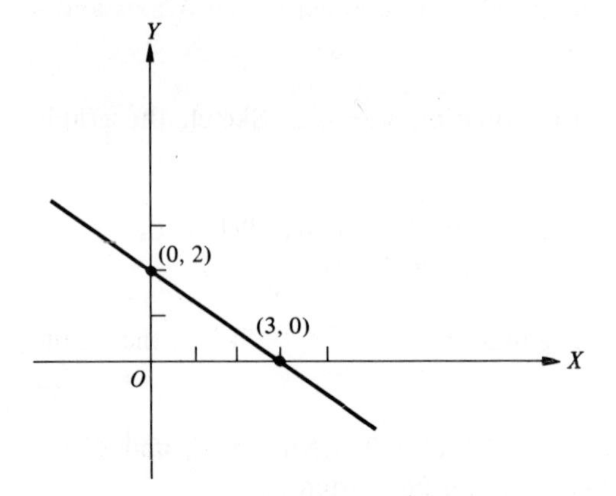

Figure 2.5.1 Graph of $2x + 3y - 6 = 0$.

we would have the intercept form. Note carefully that the denominators give the X- and Y-intercepts, respectively.

PROBLEM 6. Sketch the graph of $3x - 4y + 12 = 0$.

When the linear equation is written in the slope-intercept form, $y = mx + b$, we have that

$$\frac{dy}{dx} = m$$

Thus we see that *the derivative of the linear function gives the slope of the line.* (We will extend this idea in Sec. 2.7.)

Many applied problems may be solved graphically.

EXAMPLE 7. In an electric circuit the voltage–time (E–t) graph is the triangular wave shown in Fig. 2.5.2. Find dE/dt (a) during the first 5 s; (b) at $t = 10$ s.

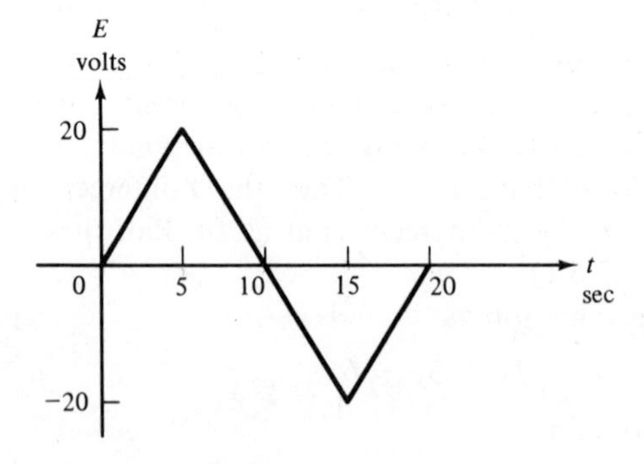

Figure 2.5.2 Example 7.

Solution. The graph consists of straight-line segments. We can find the slope $\Delta E/\Delta t$ and hence dE/dt for each of the segments.

(a) During the first 5 s,

$$m = \frac{\Delta E}{\Delta t} = \frac{20}{5} = 4$$

Thus

$$\frac{dE}{dt} = 4 \qquad \text{for} \qquad 0 < t < 5$$

(b) Since $t = 10$ is between $t = 5$ and $t = 15$, we have

$$m = \frac{\Delta E}{\Delta t} = \frac{-20 - 20}{15 - 5} = -4$$

Thus at $t = 10$ s.

$$\frac{dE}{dt} = -4$$

As the derivative is a rate of change, we can say that for any time between 0 and 5 s, since $dE/dt = 4$, the voltage is increasing four times as fast as the time. For example, if t changes 0.01 s, the voltage will change 0.04 s. At $t = 10$ s, if t changes 0.01 s, the voltage will decrease by 0.04 V.

PROBLEM 7. The graph of current–time (I–t) for an electric circuit is shown in Fig. 2.5.3. Find dI/dt (a) during the first 2 s; (b) at $t = 4$ s; (c) at $t = 7$ s. (d) Interpret the answers in parts (a), (b), and (c).

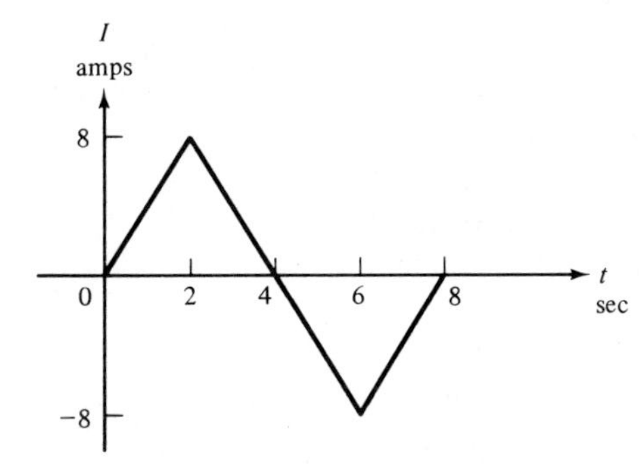

Figure 2.5.3 Problem 7.

EXERCISES 2.5

Group A

1. In each of the following, find an equation of the line through the given point having the given slope. Sketch the graph.

(a) $(3, 1)$, $m = 3$.
(b) $(-2, 0)$, $m = -2$.
(c) $(4, -9)$, $m = -\frac{1}{2}$.
(d) $(-2, -3)$, $m = \frac{2}{3}$.

2. In each of the following, state the slope and the Y-intercept of the line. Sketch the graph.

(a) $y = 3x + 2$.
(b) $y = -2x + 5$.
(c) $y = -\frac{2}{3}x$.
(d) $y = \frac{4}{3}x - 2$.

3. Find the slope of each of the following lines.

(a) $x + y = 0$.
(b) $x - y = 3$.
(c) $2x - y = 7$.
(d) $x + 2y = 11$.
(e) $3x + 2y - 7 = 0$.
(f) $2x - 5y + 4 = 0$.

4. Which of the following lines are parallel or perpendicular to $2x - 3y + 6 = 0$?

(a) $2x - 3y + 7 = 0$.
(b) $3x + 2y + 6 = 0$.
(c) $3x - 2y + 6 = 0$.
(d) $2x + 3y = -6$.
(e) $4x - 6y = 0$.
(f) $6x + 4y = 0$.

5. Sketch the graph of each of the following.

(a) $y = 2x - 3$.
(b) $y = -3x$.
(c) $3x + 4y - 12 = 0$.
(d) $3x - 2y + 6 = 0$.
(e) $x + y = 0$.
(f) $4x + 4y = 0$.
(g) $2x - 3y + 9 = 0$.
(h) $5x + 6y = 30$.
(i) $x = -5$.
(j) $2y + 7 = 0$.

Group B

1. Write an equation of the line and graph it.

(a) Parallel to the X-axis and 3 units below it.
(b) Parallel to the X-axis and 2 units above it.
(c) Parallel to the Y-axis and 4 units to the left of it.
(d) Perpendicular to the X-axis and through $(-2, -1)$.

2. In each of the following, write an equation of the straight line using the specified conditions. Sketch the graph.

(a) Passing through $(3, -2)$ and $(2, 3)$.
(b) Passing through $(-4, 5)$ and $(1, -6)$.
(c) Passing through $(0, 2)$ and $(5, 0)$.
(d) Passing through $(0, -5)$ perpendicular to $3x - y + 5 = 0$.
(e) Passing through $(0, 0)$ parallel to $2x + 3y = 4$.

3. Find the angle of intersection between the lines $2x - y = 0$ and $5x - y - 3 = 0$.

4. Find the angle of intersection between the line $x + 2y + 7 = 0$ and $2x - y + 11 = 0$.

5. Find an equation of the line whose angle of inclination is $\theta = 30°$ and which passes through $(-2, 5)$.

6. Find an equation of the line passing through $(3, -1)$ whose angle of inclination is $\theta = 120°$.

7. A voltage–time graph is shown in Fig. 2.5.4. Find dE/dt (a) at $t = 1$; (b) for $4 < t < 8$; (c) at $t = 9$. (d) Find I, where $I = C(dE/dt)$, at $t = 11$, for $C = 10^{-6}$ F.

8. A current–time graph is shown in Fig. 2.5.5. Find dI/dt (a) at $t = 1$; (b) for $2 < t < 4$. (c) Find E, where $E = L(dI/dt)$, at $t = 5$, for $L = 10^{-2}$ H.

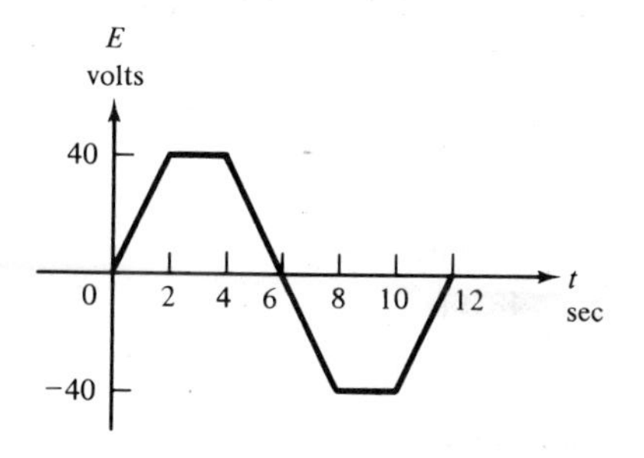

Figure 2.5.4 Exercise 7.

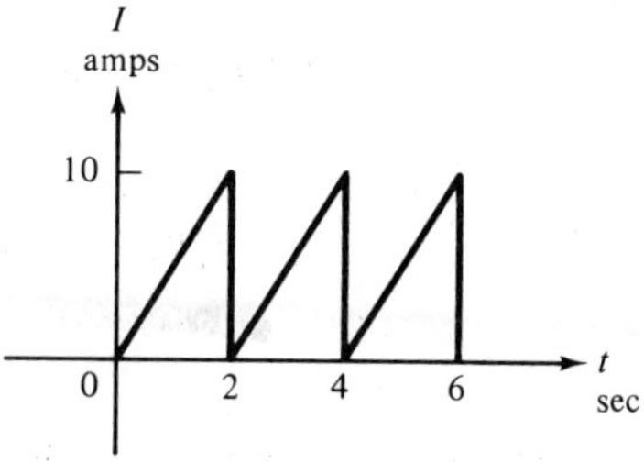

Figure 2.5.5 Exercise 8.

2.6 SOME MORE ABOUT POINTS AND LINES

One of the greatest conveniences of the Cartesian coordinate system is that of being able to put geometric concepts into algebraic formulas. (This branch of mathematics is called *analytic geometry.*) In this section we treat three such topics, the first of which is the distance between two points.

(1) Distance Between Two Points

Let P_1 and P_2 be two points in the plane. We wish to determine the distance between these two points. We define this distance to be the length of the straight-line segment connecting points P_1 and P_2 and denote it as $|P_1P_2|$. Denoting the coordinates of the points P_1 and P_2 by (x_1, y_1) and (x_2, y_2), respectively, in Fig. 2.6.1, we see that the solution to our problem is an application of the Pythagorean theorem to the right triangle P_1MP_2. Thus $(P_1P_2)^2 = (P_1M)^2 + (MP_2)^2$. Since P_1M is the change in x and MP_2 is the change in y, we have $P_1M =$

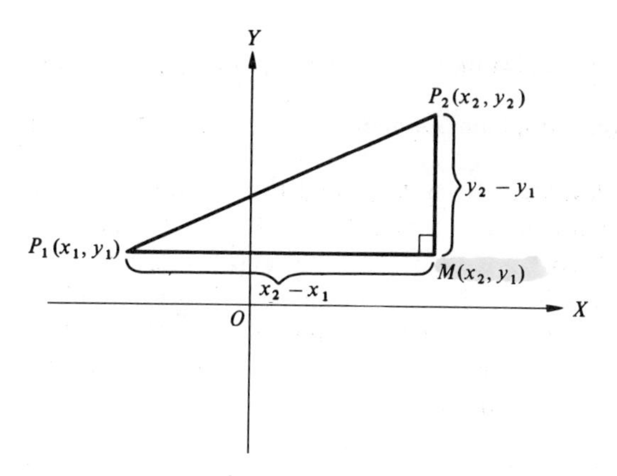

Figure 2.6.1 Distance formula.

$x_2 - x_1$ and $MP_2 = y_2 - y_1$. (See also Exercise 1 of Group C in Exercises 1.4.) Thus

$$(P_1P_2)^2 = (x_2 - x_1)^2 + (y_2 - y_1)^2$$

or

$$|P_1P_2| = \sqrt{(x_2 - x_1)^2 + (y_2 - y_1)^2}$$

$$\text{or} \quad |P_1P_2| = \sqrt{(\Delta x)^2 + (\Delta y)^2} \tag{2.6.1}$$

Equation 2.6.1 is known as the *distance formula.*[3]

EXAMPLE 1. Find the distance between the points $(-1, 2)$ and $(3, 4)$.

Solution. Let P_1 be $(-1, 2)$ and P_2 be $(3, 4)$. Then substituting these values into Eq. 2.6.1, we have

$$|P_1P_2| = \sqrt{[3 - (-1)]^2 + (4 - 2)^2} = \sqrt{4^2 + 2^2}$$
$$= \sqrt{16 + 4} = \sqrt{20}$$

and

$$|P_1P_2| = 2\sqrt{5} = 4.47$$

If we had chosen P_1 to be $(3, 4)$ and P_2 to be $(-1, 2)$, we would have arrived at the same result. Thus

$$|P_1P_2| = \sqrt{(-1 - 3)^2 + (2 - 4)^2}$$
$$= \sqrt{16 + 4} = 2\sqrt{5} = 4.47$$

This is true because $(x_2 - x_1)^2 = (x_1 - x_2)^2$, as you may show.

Warning ***In using the distance formula, remember that*** $\sqrt{a^2 + b^2}$ ***is not equal to*** $a + b$. $\sqrt{16 + 4} = \sqrt{20} = 4.47$, ***whereas*** $\sqrt{16} + \sqrt{4} = 4 + 2 = 6$.

PROBLEM 1. Find the distance between the points $(1, -2)$ and $(4, 7)$.

(2) Midpoint of a Line Segment

Next, we develop a formula to find the coordinates of the midpoint of a line segment joining any two points P_1 and P_2. Let P_1 and P_2 be the two points in the plane with coordinates (x_1, y_1) and (x_2, y_2), respectively (see Fig. 2.6.2). Let M be the midpoint of P_1P_2, and let (x, y) be its coordinates. We denote the point with coordinates (x_2, y_1) by Q and complete the right triangle as shown in Fig. 2.6.2. Construct the line MN from M perpendicular to P_1Q meeting it at $N(x, y_1)$ and the line MR perpendicular to QP_2 meeting it at $R(x_2, y)$. From

[3]This formula gives what is called the *Euclidean distance.* Other distance formulas may occur in the study of mathematics.

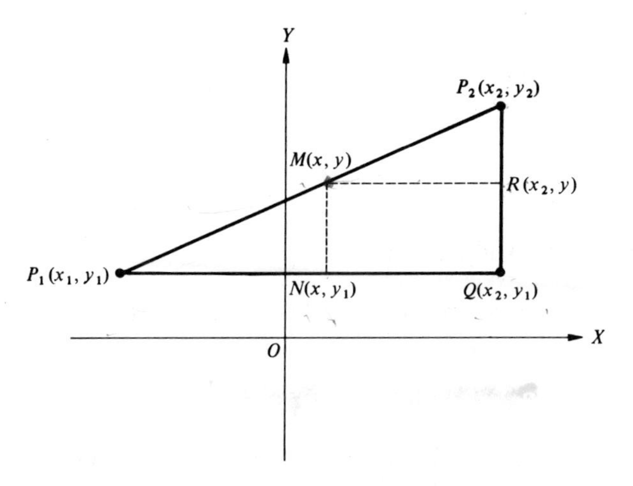

Figure 2.6.2 Midpoint of a line segment.

the similar triangles MRP_2 and P_1QP_2, we have $MR/P_1Q = P_1M/P_1P_2$. Since $P_1M = \frac{1}{2}P_1P_2$, we obtain

$$\frac{x_2 - x}{x_2 - x_1} = \frac{1}{2}$$

or

$$x = \tfrac{1}{2}(x_1 + x_2)$$

Similarly, we have

$$\frac{NM}{QP_2} = \frac{P_1M}{P_1P_2}$$

or

$$\frac{y - y_1}{y_2 - y_1} = \frac{1}{2}$$

and

$$y = \tfrac{1}{2}(y_1 + y_2)$$

Thus the coordinates of the midpoint M are

$$\boxed{\begin{aligned} x &= \tfrac{1}{2}(x_1 + x_2) \\ y &= \tfrac{1}{2}(y_1 + y_2) \end{aligned}} \qquad (2.6.2)$$

Equations 2.6.2 are called the *midpoint formulas*.

EXAMPLE 2. Find the midpoint of the line segment joining the points $(-1, 2)$ and $(3, 4)$.

Solution. Let P_1 be $(-1, 2)$ and P_2 be $(3, 4)$. Then substituting in Eqs. 2.6.2 we obtain

$$x = \frac{-1 + 3}{2} = \frac{2}{2} = 1$$

$$y = \frac{2 + 4}{2} = \frac{6}{2} = 3$$

Thus the midpoint M is at (1, 3).

PROBLEM 2. Find the midpoint of the line segment joining the points (1, −2) and (4, 7).

(3) Distance Between a Point and a Line

The development of the formula for the distance between a point and a line involves concepts and algebra that we will like to leave to you. Some of the Group C exercises at the end of this section will lead you through the development. We shall now state the formula.

The distance d between a point P with coordinates (x_0, y_0) and a straight line with the general equation $Ax + By + C = 0$ is given by

$$\boxed{d = \frac{|Ax_0 + By_0 + C|}{\sqrt{A^2 + B^2}}} \tag{2.6.3}$$

When the general equation of the straight line $Ax + By + C = 0$ is written as

$$\frac{A}{\sqrt{A^2 + B^2}}x + \frac{B}{\sqrt{A^2 + B^2}}y + \frac{C}{\sqrt{A^2 + B^2}} = 0 \tag{2.6.4}$$

we say that we have written the equation in *normal form*. Note the relationship between Eqs. 2.6.3 and 2.6.4. Thus we may obtain the distance between the point $P(x_0, y_0)$ and the straight line $Ax + By + C = 0$ by writing the equation of the line in normal form and substituting the values of x_0 and y_0 for x and y, respectively. The absolute value of the resulting fraction is the required distance. (The distance between two points is always nonnegative.)

EXAMPLE 3. Find the distance between the point (2, 1) and the line $4x - 3y + 5 = 0$.

Solution. Write the equation of the given line in the normal form. Thus

$$\frac{4x - 3y + 5}{\sqrt{4^2 + (-3)^2}} = 0$$

or

$$\frac{4x - 3y + 5}{5} = 0$$

Now substitute the coordinates of the given point for x and y. Thus

$$d = \left|\frac{4(2) - 3(1) + 5}{5}\right| = \frac{10}{5} = 2$$

PROBLEM 3. Find the distance between the point $(3, -2)$ and the line $5x + 12y - 15 = 0$.

EXERCISES 2.6

Group A

1. In each of the following, find the distance between the given points.

(a) $(1, 2)$ and $(4, 6)$.
(b) $(16, 3)$ and $(12, 6)$.
(c) $(-6, 2)$ and $(2, 1)$.
(d) $(-5, -3)$ and $(-3, -1)$.
(e) $(-5, -3)$ and $(3, 1)$.
(f) $(18, 6)$ and $(30, 5)$.

2. In each of the following, find the coordinates of the midpoint of the line segment between the points given.

(a) $(1, 2)$ and $(3, 4)$.
(b) $(6, 19)$ and $(-3, 1)$.
(c) $(18, 25)$ and $(\sqrt{2}, 1)$.
(d) $(-5, -9)$ and $(-2, -9)$.
(e) $(-4, 1)$ and $(-4, 2)$.
(f) $(6, 36)$ and $(-18, -9)$.

3. In each of the following, find the distance between the given points and lines.

(a) $3x + 4y + 1 = 0$, $(1, 2)$.
(b) $5x - 12y + 4 = 0$, $(-2, -1)$.
(c) $-3x + 12y - 3 = 0$, $(2, -1)$.
(d) $6x + 9y - 14 = 0$, $(-3, 2)$.
(e) $y - 12 = 0$, $(2, 12)$.
(f) $x + 13 = 0$, $(13, 5)$.
(g) $x + y - 1 = 0$, $(0, 0)$.

Group B

1. Given the triangle ABC with the vertices at the points $A(-3, 6)$, $B(-9, -2)$, and $C(6, 2)$.

(a) Find the slopes of the sides of the triangle.
(b) Find the slopes of the altitudes of the triangle.
(c) Find the slopes of the medians of the triangle.

2. Determine the coordinates of the midpoint of the line segment joining the points $(a + b, a)$ and $(-b, b - a)$.

3. Find the distance between the two parallel lines $2x + 4y + 6 = 0$ and $2x + 4y - 5 = 0$. (*Hint:* Find the coordinates of a point on one of the lines.)

4. Find the distance between the two parallel lines $Ax + By + C = 0$ and $Ax + By + C' = 0$.

5. Show that the points $(1, 1)$, $(-1, 1)$, $(-1, -1)$, and $(1, -1)$ are the vertices of a square.

6. Show that the points (2, 3), (−2, 1), (0, −3) and (4, −1) are the vertices of a rectangle.

7. Show that the point (−1, 1) is on the perpendicular bisector of the line segment between (−1, −1), and (1, 1).

8. Show that the point (−3, 2) is on the line that is the perpendicular bisector of the line segment between the points (2, 3) and (−2, −3).

Group C

1. Let (x_0, y_0) be a given point and let $Ax + By + C = 0$ be a given line.

(a) Find the equation of the line through (x_0, y_0) and perpendicular to $Ax + By + C = 0$.
(b) Find the point of intersection of the solution to part (a) with $Ax + By + C = 0$.
(c) Find the distance between the point (x_0, y_0) and the point that is the solution to part (b). Does this agree with Eq. 2.6.3?

2. (a) Let P_1 and P_2 be two points with coordinates (x_1, y_1) and (x_2, y_2), respectively. Show that the coordinates of the point one-third of the way from P_1 to P_2 are given by $x = (2x_1 + x_2)/3$ and $y = (2y_1 + y_2)/3$. (b) Find the coordinates of the point that is one-third of the way from $(-3, -\frac{1}{2})$ and (3, 4).

3. Show that $(x_2 - x_1)^2 = (x_1 - x_2)^2$.

2.7 SLOPE OF A CURVE

Now we are ready to discuss the geometric interpretation of the derivative. Consider the graph of $y = f(x)$ as in Fig. 2.7.1. The point $P(x, y)$ is any point on the graph of $y = f(x)$ and point $Q(x + \Delta x, y + \Delta y)$ is any nearby point on

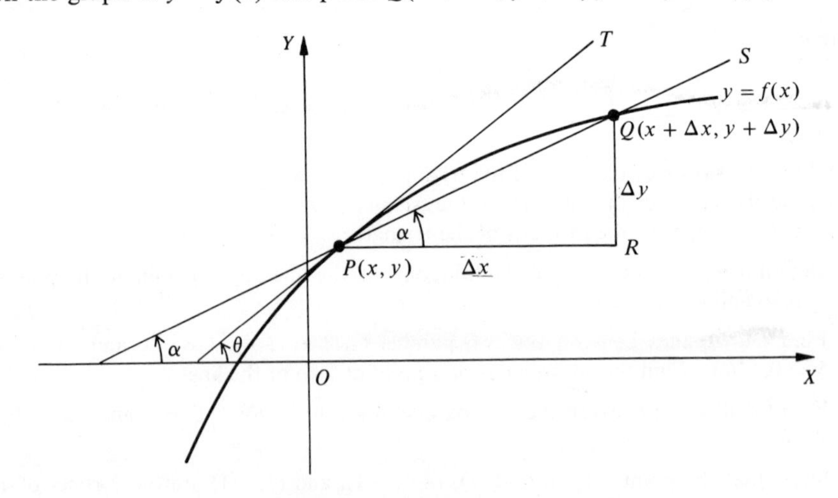

Figure 2.7.1 Slope of a curve.

the graph. The line PS, which passes through and beyond points P and Q, is called the *secant line* of the curve. It has an angle of inclination α, and its slope, $\Delta y/\Delta x$, is the average slope of the curve over the interval Δx. Now if we hold P fixed and let $\Delta x \longrightarrow 0$, point Q approaches point P along the curve and in the limit line PS approaches line PT. We define line PT to be the *tangent line* of the curve at point P. (This definition of the tangent line of a curve is consistent with the definition of the tangent to a circle which you learned in plane geometry. For our purposes this definition is more useful.) We now state the definition of the slope of a curve.

> **Definition 2.7.1** *The slope of a curve at point* **P** *is defined to be the slope of the tangent line at* **P**.

To find a value that may be used for the slope of the curve at point P, we use the limit process as $Q \longrightarrow P$, or as $\Delta x \longrightarrow 0$. Thus for Fig. 2.7.1 we have

$$m_{PQ} = \tan \alpha = \frac{\Delta y}{\Delta x}$$

As $\Delta x \longrightarrow 0$, $\alpha \longrightarrow \theta$, and $\tan \alpha \longrightarrow \tan \theta$. In the limit as $\Delta x \longrightarrow 0$ we have

$$\lim_{\Delta x \to 0} \frac{\Delta y}{\Delta x} = \lim_{\Delta x \to 0} (\tan \alpha)$$

and

$$\boxed{\frac{dy}{dx} = \tan \theta = m} \tag{2.7.1}$$

This is summarized by saying that *the slope of a curve $y = f(x)$ at any point is given by the value of the derivative, dy/dx, at that point.*

Warning ***dy/dx is the slope of the curve at a point, while $\Delta y/\Delta x$ is the average slope of the curve over the interval Δx.***

EXAMPLE 1. Find the slope of $y = x^2$ at $(1, 1)$ and $(-1, 1)$. Draw the tangent lines at these points.

Solution. Find dy/dx in general, and then evaluate dy/dx at the required points. Thus for $y = x^2$,

$$\frac{dy}{dx} = 2x$$

At $(1, 1)$ we find that

$$\left.\frac{dy}{dx}\right|_{x=1} = 2(1) = 2$$

At $(-1, 1)$ we find that

$$\left.\frac{dy}{dx}\right|_{x=-1} = 2(-1) = -2$$

Thus the slope at (1, 1) is 2, and at (−1, 1) it is −2. The tangents are shown in Fig. 2.7.2.

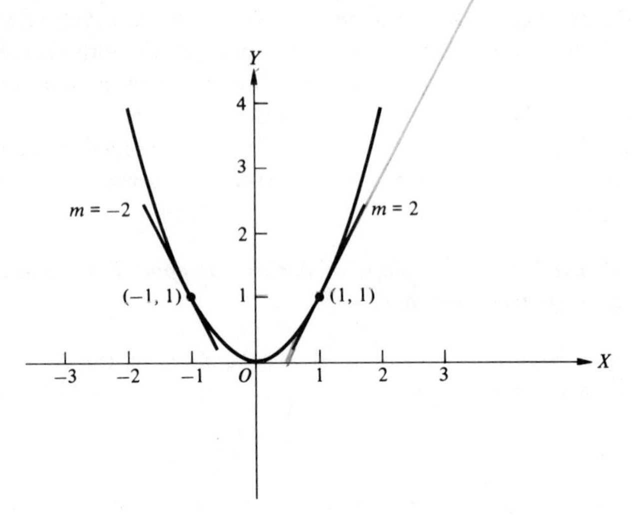

Figure 2.7.2 Tangent lines.

PROBLEM 1. Find the slope of $y = \frac{1}{4}x^2$ at (2, 1) and (−2, 1). Draw the tangent lines at these points.

EXAMPLE 2. Find the point at which the slope of $f(x) = -x^2 + 4x$ is zero.

Solution. For $f(x) = -x^2 + 4x$,

$$f'(x) = -2x + 4$$

Now let $-2x + 4 = 0$ and we find $x = 2$. To find the point we must use $f(x)$. Thus $f(2) = -(2)^2 + 4(2) = 4$, and the required point is (2, 4).

Warning ***To find the point at $x = a$ we have to evaluate $f(x)$ at $x = a$. The slope is found from $f'(x)$. Be sure to distinguish between $f(x)$ and $f'(x)$.***

PROBLEM 2. Find the point at which the slope of $f(x) = x^2 - 6x$ is zero.

EXAMPLE 3. Find an equation of the tangent line to the curve $y = x^2 + 5x$ at the point (−1, −2).

Solution. We know the point; now we need to find the slope. For $y = x^2 + 5x$,

$$\frac{dy}{dx} = 2x + 5$$

Thus at (−1, −2) the slope is

$$m = \frac{dy}{dx} = 2(-1) + 5 = 3$$

For an equation of the tangent line we use the point-slope formula:

$$y - y_1 = m(x - x_1)$$

Thus

$$y - (-2) = 3[x - (-1)]$$

or

$$y = 3x + 1$$

is an equation of the tangent line.

PROBLEM 3. Find an equation of the tangent line to the curve $y = x^3 - 4x$ at the point (2, 0).

Warning ***Remember that the slope m at a point is a constant. It is wrong to say that m is 2x + 5.***

It is not always necessary to use x and y as the variables of a function. However, if the graph of the function is drawn on a rectangular coordinate system in which the horizontal axis represents the independent variable and the vertical axis represents the dependent variable, then the derivative with respect to the independent variable gives the slope of the graph of the function.

EXAMPLE 4. In a dc circuit $P = I^2R$. For a fixed resistance R, $P = f(I)$. Find the slope of the PI curve when $I = 3$ and $R = 10$.

Solution. Since $P = f(I)$, R is constant and we have $P = 10I^2$. Then

$$\frac{dP}{dt} = 20I$$

and

$$\left.\frac{dP}{dI}\right|_{I=3} = 20(3) = 60$$

Thus the slope of the PI curve is 60 when $I = 3$.

PROBLEM 4. The voltage across a capacitor is given by $E = 5t^2 + 3t$. Find the slope of the E–t curve when $t = 0.1$.

The *normal* to the curve at point P is the line perpendicular to the tangent at P.

EXAMPLE 5. Find an equation of the normal to the curve $y = x^3 - x$ at $(-2, -6)$.

Solution. For $y = x^3 - x$,

$$\frac{dy}{dx} = 3x^2 - 1$$

The slope of the tangent at $(-2, -6)$ is

$$m_T = 3(-2)^2 - 1 = 11$$

Thus the slope of the normal at $(-2, -6)$ is

$$m_N = -\frac{1}{11}$$

We then find an equation of the normal by using the point-slope formula:

$$y - (-6) = -\frac{1}{11}[x - (-2)] \qquad \text{or} \qquad x + 11y + 68 = 0$$

is an equation of the normal.

PROBLEM 5. Find an equation of the normal to the curve $f(x) = 2x^3 - 4x$ at the point where $x = 1$.

The *angle of intersection* of two curves is defined to be the angle between their tangent lines at the point of intersection.

EXAMPLE 6. Find the angle of intersection of the curves $y = x^2$ and $y = x^3$.

Solution. We must first find the points of intersection. This is done by equating the values for y. Thus $x^2 = x^3$. To solve this equation rewrite it as

$$x^3 - x^2 = 0 \qquad \text{or} \qquad x^2(x - 1) = 0$$

This yields $x^2 = 0$ and $x - 1 = 0$. Thus $x = 0$ and $x = 1$, and the points of intersection are $(0, 0)$ and $(1, 1)$. Call $y_1 = x^2$ curve 1 and $y_2 = x^3$ curve 2 for convenience; then

$$\frac{dy_1}{dx} = 2x \qquad \text{and} \qquad \frac{dy_2}{dx} = 3x^2$$

At $(0, 0)$ we have $m_1 = 0$ and $m_2 = 0$. At $(1, 1)$ we have $m_1 = 2$ and $m_2 = 3$. At $(0, 0)$, since the slopes are equal, we have the angle of intersection $\theta = 0$. At $(1, 1)$ we have

$$\tan\theta = \frac{m_2 - m_1}{1 + m_1 m_2} = \frac{3 - 2}{1 + (2)(3)} = \frac{1}{7} = 0\ 143$$

or

$$\theta = 8.1^\circ \text{ approximately}$$

PROBLEM 6. Find the angle of intersection of the curves $y = -x^2 + 6x$ and $y = 2x + 3$.

EXERCISES 2.7

Group A

1. In each of the following, find the slope of $y = f(x)$ at the indicated points. Draw the tangent line at each of these points.

(a) $y = 3x^2$, $(1, 3)$, $(-1, 3)$.
(b) $y = x^2 + 2x$, $(0, 0)$, $(1, 3)$, $(-2, 0)$.
(c) $f(x) = x^3$, $(-2, -8)$, $(0, 0)$, $(1, 1)$, $(2, 8)$.
(d) $f(x) = 2x$, $(1, 2)$, $(-1, -2)$.

2. In each of the following, find the point at which the slope is zero.

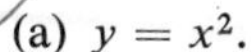

(a) $y = x^2$.
(b) $y = x^2 - 2x$.
(c) $y = -3x^2 - 2x$.
(d) $y = x^3 - 3x + 1$.

3. In each of the following, find an equation of the tangent line to the curve at the indicated point.

(a) $y = x^2 - 3x$, $(1, -2)$.
(b) $y = -x^2 + 2$, $(0, 2)$.
(c) $f(x) = x^3$, $(1, 1)$.
(d) $f(x) = x^3 - x^2$, $(1, 0)$.

4. Find an equation of the normal line in each part of Exercise 3.

5. In each of the following, find the angle of intersection between y_1 and y_2.

(a) $y_1 = x^2$, $y_2 = x$.
(b) $y_1 = -x^2$, $y_2 = 2x$.
(c) $y_1 = x^2$, $y_2 = -x^2$.
(d) $y_1 = x^3$, $y_2 = x + 6$.

6. In each of the following, tell whether the curve runs up to the right or down to the right at the indicated point. (*Hint*: Consider the slope.)

(a) $y = x^2$, $(-2, 4)$, $(2, 4)$.
(b) $y = x^2 - 2x$, $(0, 0)$, $(2, 0)$.
(c) $y = -x^2$, $(-1, -1)$, $(1, -1)$.
(d) $y = x^3$, $(-1, -1)$, $(1, 1)$.

Group B

1. Show that for $P = I^2R$ in a dc electric circuit that the slope of the PI curve is twice the voltage.

2. Find the area of the triangle formed by the coordinate axes and the line tangent to the graph of $y = x^2 - 4$ at the point $(1, -3)$.

3. The curves $y = x^2$ and $y = 8 - x^2$ intersect at the points $(-2, 4)$ and $(2, 4)$. Find equations of the lines tangent to the curves at the point $(2, 4)$ and the angle of intersection of the curves at this point.

4. The velocity of a particle is given by the formula $v(t) = 36t^2 + t$. Is the velocity increasing (running up to the right) or decreasing (running down to the right) for $t \geqq 0$?

5. The magnetic flux passing through one circuit is given by $\phi_1(t) = 7t^3 + 2t$ and the magnetic flux passing through a second circuit is given by $\phi_2(t) = 3t^7 + t^2$. Find the time $0 \leqq t \leqq 1$ for which ϕ_1 and ϕ_2 are increasing (running up to the right) at the same rate.

Group C

1. Given $f(x) = \sqrt{x}$, show that the slope does not exist at $x = 0$.

2. Given

$$f(t) = \begin{cases} t & 0 < t \leqq 2 \\ 2 & t > 2 \end{cases}$$

show that the derivative is discontinuous at $t = 2$.

3. Given $y = a^{2/3} - x^{2/3}$. (a) Is the function continuous at $x = 0$? (b) Is the slope continuous at $x = 0$?

4. Given $f(x) = |x|$. (a) Find the slope. (b) Is the slope for $x < 0$ the same as for $x > 0$? (c) Is $f(x)$ continuous for all x? (d) Is $f'(x)$ everywhere continuous? Why?

2.8 IMPLICATIONS OF THE SIGN OF THE DERIVATIVE

The algebraic sign of the derivative can tell us whether the function is increasing or decreasing. Consider the function $f(x) = x^3$. Its derivative is $f'(x) = 3x^2$. If $x < 0$, $f'(x) > 0$, and since the slope is positive, we say that the curve is running up to the right or that if we increase x, $f(x)$ will increase. Also, for $x > 0$, $f'(x) > 0$, and $f(x)$ again will increase. Note that for $x = 0$, $f'(0) = 0$. We could use this information to help graph $f(x)$ (see Fig. 2.8.1). (We discuss graphing in more detail in Sec. 3.1.) That is, we can say that the graph of $f(x)$ rises sharply for negative x. As x increases, it flattens out as $x \longrightarrow 0$ and then rises sharply again. [To see this, evaluate $f'(x)$ at several values of x.]

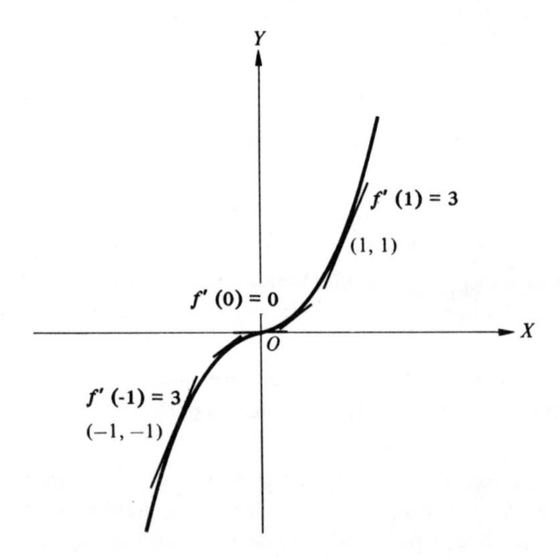

Figure 2.8.1 Graph of $y = x^3$.

We say that a function is increasing on a closed interval $[a, b]$ (a and b are included) if $f(x_1) < f(x_2)$ whenever $x_1 < x_2$. We say that a function is decreasing on a closed interval $[a, b]$ if $f(x_1) > f(x_2)$ whenever $x_1 < x_2$.

We could guess that if a function is increasing at a point x_0, then $f'(x_0) \geqq 0$. This guess is based on the consideration of the quotient

$$\frac{f(x) - f(x_0)}{x - x_0} = \frac{+}{+} = + \qquad x > x_0$$

$$\frac{f(x) - f(x_0)}{x - x_0} = \frac{-}{-} = + \qquad x < x_0$$

so

$$\lim_{x \to x_0} \frac{f(x) - f(x_0)}{x - x_0} \geqq 0$$

But an even more important result is the following. If $f'(x) > 0$ on $[a, b]$, then f is increasing on the open interval (a, b) (a and b not included). This conclusion is guaranteed by a very famous theorem in calculus called the mean-value theorem, which we discuss in Chap. 3. The same theorem guarantees the conclusion that if $f'(x) < 0$ on an interval, then f is decreasing on this interval.

EXAMPLE 1. Find the intervals of x for which y increases and for which y decreases given that $y = 2x^3 + 3x^2 - 36x + 7$.

Solution. Since we want to find out how y changes as x changes, we shall investigate the sign of the derivative, that is, find values of x for which $dy/dx > 0$ and for which $dy/dx < 0$. For $y = 2x^3 + 3x^2 - 36x + 7$,

$$\frac{dy}{dx} = 6x^2 + 6x - 36$$

or

$$\frac{dy}{dx} = 6(x^2 + x - 6)$$

Since zero separates positive from negative, set $dy/dx = 0$ and find the roots of this equation. First we factor and write $dy/dx = 6(x + 3)(x - 2)$. Then setting $dy/dx = 0$, we have $6(x + 3)(x - 2) = 0$, from which we find that $x = -3$ and $x = 2$. Mark the roots on the X-axis and consider the regions into which the roots divide the X-axis. (See the tabular form at the end of this example.) They are $x < -3$, $-3 < x < 2$, and $x > 2$. For each of these regions choose a value of x and investigate the algebraic signs of the factors of dy/dx and finally by their product the sign of dy/dx. This determines the sign of dy/dx throughout the region. Thus for the region $-3 < x < 2$, choose $x = 0$ for convenience. Then we have the factors $(x + 3) > 0$ and $(x - 2) < 0$, and hence $dy/dx < 0$. The following tabular form helps simplify the process.

Regions:	$x < -3$	$-3 < x < 2$	$x > 2$
		-3	2 $\longrightarrow X$
Factors of dy/dx	$6(x + 3)(x - 2)$	$6(x + 3)(x - 2)$	$6(x + 3)(x - 2)$
Signs of factors	$(-)(-)$	$(+)(-)$	$(+)(+)$
Thus	$dy/dx > 0$	$dy/dx < 0$	$dy/dx > 0$
and	y increases	y decreases	y increases

To summarize,

$$\frac{dy}{dx} > 0 \qquad \text{for } x < -3 \text{ and } x > 2, \text{ and } y \text{ increases}$$

$$\frac{dy}{dx} < 0 \qquad \text{for } -3 < x < 2, \text{ and } y \text{ decreases}$$

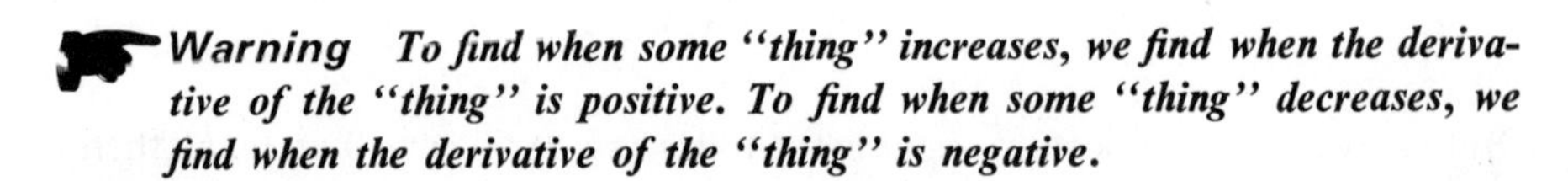

Warning ***To find when some "thing" increases, we find when the derivative of the "thing" is positive. To find when some "thing" decreases, we find when the derivative of the "thing" is negative.***

PROBLEM 1. Find the intervals of t for which the voltage, E, increases and when E decreases given that $E = 4t^3 - 24t^2 + 36t + 5$.

EXERCISES 2.8

Group A

1. In each of the following, find the intervals of x for which y increases and for which y decreases.

(a) $y = \frac{1}{2}x^2 - x + 1$.
(b) $y = \frac{1}{3}x^3 - x^2$.
(c) $y = 2x^3 + 3x^2 - 12x + 4$.
(d) $y = 2x^3 - 9x^2 + 12x$.

2. If $s = -16t^2 + 64t$, find the interval of t for which s is decreasing. When is the velocity, v, equal to zero?

3. If $v = -32t + 64$, when does v increase?

4. The curvature of a certain curve is given by the formula $k(s) = 2s^3 - 24s$. Find the intervals $s \geqq 0$ for which k is increasing and for which k is decreasing.

5. The work of a system is given by the formula $W(t) = 6t^2 - 9t$. Is the work increasing more rapidly at $t = 1$ or at $t = 2$?

6. The magnetic flux passing through a circuit is given by $\phi(t) = t^3 - 7.5t^2 + 18t$. Find the intervals $t \geqq 0$ for which ϕ is increasing and for which ϕ is decreasing.

7. The electromagnetic force in a circuit is given by emf $(t) = t^2 - 2t$. Where is the emf increasing and where is it decreasing for $t \geqq 0$?

8. The angular velocity for a particle is given by $\omega(t) = 8t^3 + 12t^2 - 144t + 28$. Find the intervals of $t > 0$ for which ω is increasing and for which ω is decreasing.

9. Let $f(x) = x[u(x) - u(x-1)] + (x-1)[u(x-1) - u(x-2)] + (x-2)[u(x-2) - u(x-3)]$ (see formula 1.4.2). (a) Graph $f(x)$. (b) On what intervals is f increasing? (c) What is the behavior of f at $x = 1$; $x = 2$?

2.9 HIGHER-ORDER DERIVATIVES

When we take the derivative of a function we obtain a new function. This new function may also have a derivative. We may continue this process and obtain many new functions by taking derivatives of derivatives. Such derivatives are called *higher-order derivatives*. If we consider that dy/dx is the derivative with respect to x of y, we can write dy/dx as $(d/dx)(y)$. Then the second derivative, or the derivative of the derivative, can be symbolized by $(d/dx)(dy/dx)$ or d^2y/dx^2. (This is read "dee-two-wye dee-ex-squared.") Continuing in this way, we can write any order derivative. The following list shows how higher-order derivatives are indicated. If

		The derivatives may also be written as:
$y = f(x)$		
$\dfrac{dy}{dx} = f'(x)$	(first-order derivative)	$\dfrac{d}{dx}(y)$
$\dfrac{d^2y}{dx^2} = f''(x)$	(second-order derivative)	$\dfrac{d}{dx}\left(\dfrac{dy}{dx}\right)$
$\dfrac{d^3y}{dx^3} = f'''(x)$	(third-order derivative)	$\dfrac{d}{dx}\left(\dfrac{d^2y}{dx^2}\right)$
$\vdots \quad \vdots$		$\vdots$
$\dfrac{d^ny}{dx^n} = f^n(x)$	(nth-order derivative)	$\dfrac{d}{dx}\left(\dfrac{d^{n-1}y}{dx^{n-1}}\right)$

Frequently when discussing higher-order derivatives, we do not use the word *order*. For example, we might say "Find the second derivative" instead of saying "Find the second-order derivative."

EXAMPLE 1. If $y = x^4 - 2x^3 + x$, find dy/dx, d^2y/dx^2, and d^3y/dx^3.

Solution

$$y = x^4 - 2x^3 + x$$

$$\frac{dy}{dx} = 4x^3 - 6x^2 + 1$$

$$\frac{d^2y}{dx^2} = 12x^2 - 12x$$

$$\frac{d^3y}{dx^3} = 24x - 12$$

(Note that d^5y/dx^5 and higher-order derivatives would be zero for this function.)

PROBLEM 1. For $f(x) = 3x^5 - 2x^4 + x^2 - 7$, find $f'(x), f''(x), f'''(0)$, and $f^4(-1)$.

The second derivative is useful in drawing graphs and is used to find when the slope is increasing or decreasing. We discuss this in detail in Sec. 3.2.

EXAMPLE 2. Given that $y = x^2$, find when the slope is increasing and when decreasing.

Solution. Since we are interested in the slope, we must find when the derivative of the slope, that is, $(d/dx)(dy/dx)$ or d^2y/dx^2, is positive and when negative. Thus for $y = x^2$,

$$\frac{dy}{dx} = 2x$$

and

$$\frac{d^2y}{dx^2} = 2$$

Since d^2y/dx^2 is always positive, the slope is always increasing. (Recall that going from a slope of -6 to -1 is an increase.)

PROBLEM 2. Find the intervals of x for which the slope is increasing and when decreasing for $f(x) = \frac{1}{3}x^3 - x$.

Sometimes higher-order derivatives are given special names. Thus for

$$s = f(t) \qquad \text{(displacement)}$$

we have

$$\frac{ds}{dt} = v = f'(t) \qquad \text{(velocity)}$$

$$\frac{d^2s}{dt^2} = \frac{dv}{dt} = a = f''(t) \qquad \text{(acceleration)}$$

$$\frac{d^3s}{dt^3} = \frac{d^2v}{dt^2} = \frac{da}{dt} = f'''(t) \qquad \text{(jerk)}$$

A little thought will show that "jerk" is an appropriate name for da/dt, since da/dt is the rate of change of the acceleration.

A circle that has the same curvature as a curve at a point is called the *circle of curvature* at that point. The radius of this circle is called the *radius of curvature.* It is denoted by the Greek letter ρ (rho) and is given by

$$\rho = \left| \frac{\left[1 + \left(\frac{dy}{dx}\right)^2\right]^{3/2}}{\frac{d^2y}{dx^2}} \right| \tag{2.9.1}$$

EXERCISES 2.9

Group A

1. In each of the following, find the indicated derivative.

(a) $y = x^2 + 3x; \dfrac{d^2y}{dx^2}$.

(b) $f(x) = -x^3; f''(x)$.

(c) $f(x) = 3x^4 - 2x^3; f''(x)$.

(d) $y = -x^5 + 7; \dfrac{d^2y}{dx^2}$.

2. In each of the following, find the indicated derivative.

(a) $y = \sqrt{x}$; $\dfrac{d^2y}{dx^2}$.

(b) $y = \sqrt{x} - x$; $\dfrac{d^3y}{dx^3}$.

(c) $y = x^2 - 3x + 5$; $\dfrac{d^4y}{dx^4}$.

(d) $f(x) = x^{2/3}$; $f''(x)$.

3. In each of the following, find the required quantity.

(a) $f(x) = 3x^4 - x^2 + 7$; $f''(2)$.
(b) $y = x^{1/2} - x$; d^2y/dx^2 at $x = 9$.
(c) $s = 16t^2 - 32t + 8$; $v(3)$, $a(3)$.
(d) $f(x) = x^{5/3}$; $f'(0)$, $f''(0)$, $f'''(0)$.

4. In each of the following, find the intervals of x for which the slope increases and for which the slope decreases.

(a) $f(x) = x^3 - 3x^2$.
(b) $y = -x^3 + 7$.
(c) $y = x^4 + 2x^3 + 3x - 11$.
(d) $f(x) = x^4 - 6x^2 + 2x$.

5. For each of the following, find the radius of curvature at the indicated point.

(a) $y = x^2$, $(1, 1)$.
(b) $y = x^3$, $(1, 1)$.
(c) $y = x$, $(1, 1)$.
(d) $y = x^{1/2}$, $(1, 1)$.
(e) $y = x^2 + 3$, $(1, 4)$.
(f) $y = x^2 + 2x + 1$, $(-1, 1)$.

Group B

1. If $s = 16t^2 - 64t$, find the velocity and the acceleration at $t = 2$ s. Does s ever decrease for positive t? Does v ever decrease?

2. If you want an elevator to have a smooth motion, what should be the value of the jerk? Which of the following equations of motion for an elevator will give a smooth ride to the passengers?

(a) $s = 4t^2 - 3t$.
(b) $s = \frac{1}{3}t^3 + t$.

3. The flux in an electric circuit is given by $\phi(t) = 0.7t^3 - t^2$. Where is the emf increasing and where is it decreasing?

4. The equation of a plane curve is given by $y = x^3 + 2x + 1$. Where is the slope of the curve increasing and where is it decreasing?

5. The equations of two plane curves are given by $y = x^4 + 2x^2 + 1$ and $y = 8x^2 + 3x + 1$. Where do the slopes of the curves increase at the same rate?

6. If work is expressed by the formula $W(t) = t^4 - 6t^2 + 2t$, where is the power increasing and where is it decreasing?

7. The charge on a capacitor in a simple electric circuit is given by $Q(t) = 4t^3 - 3t^2 + 1$. Does the current ever decrease?

8. (a) How fast is the slope of the curve $f(x) = 2x^3 - x^2 + 1$ changing at the point where $x = 1$? (b) What are the coordinates of that point? (c) What is the slope of the curve at that point?

9. $f(x) = ax^2 + bx + c$ is a second-degree polynomial. If $f(0) = 5$, $f'(1) = -3$, and $f''(2) = 4$, find $f(1)$.

10. For an electric circuit where

$$E = L\frac{dI}{dt} \quad \text{and} \quad I = C\frac{dE}{dt}$$

with L and C constants, show that

$$E = L\frac{d^2E}{dt^2} - \frac{E}{C} = 0$$

2.10 SOME DERIVATIVE RULES

At this point, before we discuss many applications of the derivative using mostly polynomials, we state some rules on derivatives that might be useful in your concurrent courses. We give them now without proof. They will be proved when discussed in later sections. This section may be omitted if your instructor wishes. In these rules u and v are differentiable functions.

The Chain Rule

If $y = f(u)$ and $u = g(x)$, then $y = f[g(x)]$. (y is called a *composite* function.) If there is a change in x, this will change u, which in turn will change y. It turns out that

$$\boxed{\frac{dy}{dx} = \frac{dy}{du}\frac{du}{dx}} \tag{2.10.1}$$

Formula 2.10.1 is called the *chain rule.*

EXAMPLE 1. Given $y = 9(3x^5 - 2x + 4)^7$. Find dy/dx.

Solution. The given function does not fit the form of formula 2.3.1, so we cannot use it. But if we let

$$u = 3x^5 - 2x + 4$$

then we have

$$y = 9u^7$$

Now we use formula 2.3.1 to find

$$\frac{dy}{du} = 63u^6 \quad \text{and} \quad \frac{du}{dx} = 15x^4 - 2$$

Using the chain rule, formula 2.10.1, we write

$$\frac{dy}{dx} = \underbrace{63u^6}_{dy/du}\underbrace{(15x^4 - 2)}_{du/dx}$$

or

$$\frac{dy}{dx} = 63(3x^5 - 2x + 4)^6(15x^4 - 2)$$

PROBLEM 1. Find dy/dx for $y = 2(4x^3 - 5x^2 + 7)^{13}$.

A Function Raised to a Power

If $y = au^n$, where a and n are constants and u is a differentiable function of x, then

$$\boxed{\text{for } y = au^n, \quad \frac{dy}{dx} = anu^{n-1}\frac{du}{dx}} \tag{2.10.2}$$

This follows by combining the chain rule and formula 2.3.1. In words, formula 2.10.2 says that the derivative of a function raised to a power, n, is the product of the power, n, the function raised to the $(n - 1)$th power, and the derivative of the function.

EXAMPLE 2. Find dy/dx if $y = 9\sqrt{x^2 - 25}$.

Solution. First write y in fractional exponent form.

$$y = 9(x^2 - 25)^{1/2}$$

Now use formula 2.10.2 to obtain

$$\frac{dy}{dx} = \underbrace{9}_{a}\underbrace{\left(\frac{1}{2}\right)}_{n}\underbrace{(x^2 - 25)^{-1/2}}_{u^{n-1}}\underbrace{(2x)}_{du/dx}$$

$$= \frac{9x}{\sqrt{x^2 - 25}}$$

Warning ***When using formula 2.10.2, always end with the "derivative of the function."***

PROBLEM 2. Find ds/dt for $s = 3(5t^2 - 7t)^{3/2}$.

Derivative of a Constant

The derivative of a constant is zero.

$$\boxed{\frac{d(c)}{dx} = 0} \tag{2.10.3}$$

Derivative of a Constant Times a Function

The derivative of a constant times a function is the product of the constant and the derivative of the function.

$$\frac{d(cu)}{dx} = c\frac{du}{dx} \tag{2.10.4}$$

EXAMPLE 3. (a) If $y = 29$, $dy/dx = 0$.
(b) If $y = 7(3x^2 - 5x + 2)$, $dy/dx = 7(d/dx)(3x^2 - 5x + 2) = 7(6x - 5)$.

PROBLEM 3. (a) Find ds/dt for $s = 16$.
(b) Find dQ/dt for $Q = 13(2t^3 - 9t + 14)$.

Derivative of a Sum

The derivative of a sum of a finite number of functions is equal to the sum of the derivatives of the functions.

$$\frac{d}{dx}(u + v) = \frac{du}{dx} + \frac{dv}{dx} \tag{2.10.5}$$

EXAMPLE 4. For $E = (t^2 - 5)^3 + (7t + 4)^{1/3}$,

$$\frac{dE}{dt} = 3(t^2 - 5)^2(2t) + (\tfrac{1}{3})(7t + 4)^{-2/3}(7)$$

PROBLEM 4. For $y = \sqrt{x^4 + 16} - 2(x^3 - x)^5$, find dy/dx.

Derivative of a Product

The derivative of a product of two functions is the sum of the product of the first function and the derivative of the second function, and the product of the second function and the derivative of the first function.

$$\frac{d}{dx}(uv) = u\frac{dv}{dx} + v\frac{du}{dx} \tag{2.10.6}$$

EXAMPLE 5. Find $(ds/dt)|_{t=2}$ for $s = (3t^2 - t)(t^4 + 8)^{1/2}$.

Solution. Since this is the product of two functions, we use formula 2.10.6 to find ds/dt. Then we will evaluate the derivative at $t = 2$.

$$\frac{ds}{dt} = \underbrace{(3t^2 - t)}_{\text{(first)}}\underbrace{\tfrac{1}{2}(t^4 + 8)^{-1/2}(4t^3 + 0)}_{\text{(derivative of second)}} + \underbrace{(t^4 + 8)^{1/2}}_{\text{(second)}}\underbrace{(6t - 1)}_{\text{(derivative of first)}}.$$

$$= \frac{(3t^2 - t)(2t^3)}{\sqrt{t^4 + 8}} + \sqrt{(t^4 + 8)}(6t - 1)$$

Substituting $t = 2$, we find that $(ds/dt)_{t=2} = 86.5$. (It is not necessary to simplify algebraically when a numerical answer is required.)

PROBLEM 5. Find $(dy/dx)|_{x=1}$ for $y = (3x^3 + 5x)^{-1/3}(7x^2 - 9)$.

Derivative of a Quotient

The derivative of a quotient of two functions is equal to the denominator times the derivative of the numerator minus the numerator times the derivative of the denominator, all divided by the square of the denominator.

$$\frac{d}{dx}\left(\frac{u}{v}\right) = \frac{v\dfrac{du}{dx} - u\dfrac{dv}{dx}}{v^2} \tag{2.10.7}$$

Warning ***Do not get the numerator and denominator reversed.***

EXAMPLE 6. Find dQ/dt for

$$Q = \frac{3t^2 - 7t}{5t^9 + 4}$$

Solution. Since this is a fraction, we use formula 2.10.7.

$$\frac{dQ}{dt} = \frac{\overbrace{(5t^9 + 4)(6t - 7)}^{\text{(denom.)(deriv. of num.)}} - \overbrace{(3t^2 - 7t)(45t^8 + 0)}^{\text{(num.)(deriv. of denom.)}}}{\underbrace{(5t^9 + 4)^2}_{\text{(square of denom.)}}}$$

(You may simplify this if you wish.)

PROBLEM 6. Find $f'(-2)$ for

$$f(x) = \frac{x^2 + 1}{5x - 3}$$

The next two formulas we state are important in courses in electricity and other applied subjects. We discuss them in more detail in Chap. 9. In calculus the argument used for the trigonometric functions is a real number.[4] The trigonometric function of a real number u is equivalent to the same trigonometric function of an angle θ measured in radians where $u = \theta$. That is,

$$\underbrace{\cos 2.5}_{\substack{\text{real}\\\text{number}}} = \underbrace{\cos 2.5}_{\substack{\text{angle}\\\text{in radians}}} = -0.801$$

[4]See *Modern Technical Mathematics* by N. O. Niles, Reston Publishing Co., Reston, Va., 1978.

Derivative of sin u and of cos u

When u is a differentiable function of x, then

$$\frac{d}{dx}(\sin u) = \cos u \frac{du}{dx} \tag{2.10.8}$$

$$\frac{d}{dx}(\cos u) = -\sin u \frac{du}{dx} \tag{2.10.9}$$

EXAMPLE 7. For $y = 2 \sin 3x$,

$$\frac{dy}{dx} = 2\underbrace{(\cos 3x)}_{\cos u}\underbrace{(3)}_{du/dx}$$

or

$$\frac{dy}{dx} = 6 \cos 3x$$

EXAMPLE 8. If $E = 110 \cos \pi t^2$,

$$\frac{dE}{dt} = 110\underbrace{(-\sin \pi t^2)}_{(-\sin u)}\underbrace{(\pi 2t)}_{(du/dt)}$$

or

$$\frac{dE}{dt} = -220\pi t \sin \pi t^2$$

PROBLEM 7. Find I if $Q = 50 \sin 60t$.

PROBLEM 8. Find v if $s = 6 \cos 2t^3$.

EXAMPLE 9. Find $f'(1)$ for $f(x) = 5x^2 \sin x^3$.

Solution. Here we have a product. We first apply formula 2.10.6 to obtain

$$\begin{aligned} f'(x) &= (5x^2)\frac{d(\sin x^3)}{dx} + (\sin x^3)\frac{d(5x^2)}{dx} \\ &= (5x^2)(\cos x^3)(3x^2) + (\sin x^3)(10x) \\ &= 15x^4 \cos x^3 + 10x \sin x^3 \end{aligned}$$

Now let $x = 1$ to obtain

$$\begin{aligned} f'(1) &= 15(1)^4 \cos (1^3) + 10(1) \sin (1^3) \\ &= 15 \cos 1 + 10 \sin 1 \\ &= 15(0.540) + 10(0.841) \end{aligned}$$

and

$$f'(1) = 16.51$$

(Note that $x = 1$ is a real number. Hence we had to find the cosine and the sine of the real number 1.)

PROBLEM 9. Find the velocity of a particle at $t = 2.5$ if $s = t^3 \cos t^2$.

EXAMPLE 10. Find the slope of the curve $y = 2\cos^3 5x$ at $x = 0.1$.

Solution. We first rewrite $y = 2(\cos 5x)^3$, then apply the power rule, formula 2.10.2, to obtain

$$\frac{dy}{dx} = 2(3)(\cos 5x)^2(-\sin 5x)(5)$$

or

$$\frac{dy}{dx} = -30\cos^2 5x \sin 5x$$

At $x = 0.1$ the slope is

$$\left.\frac{dy}{dx}\right|_{x=0.1} = -30(\cos 0.5)^2 \sin 0.5$$

$$= -11.1$$

PROBLEM 10. Find the slope of the curve $y = 3\sin^2 4x$ at $x = 0.2$.

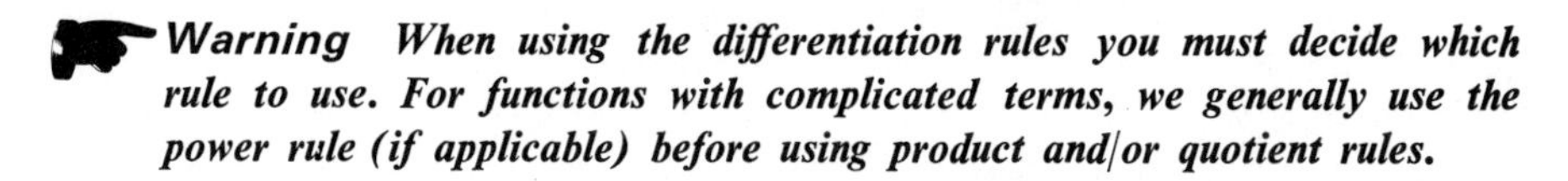

Warning ***When using the differentiation rules you must decide which rule to use. For functions with complicated terms, we generally use the power rule (if applicable) before using product and/or quotient rules.***

EXERCISES 2.10

Group A

1. Find dy/dx for each of the following. (Leave the answer in terms of u and x.)

(a) $y = u^3 - u$, $u = 5x - 3$.
(b) $y = u^{-1}$, $u = x^{1/2} - x^{-1/2}$.
(c) $y = 3u$, $u = 5x^2 - 7x + 1$.
(d) $y = \sqrt{u}$, $u = \sqrt{x^2 - 25}$.

2. Find dy/dx for each of the following.

(a) $y = (x^2 + 25)^{1/3}$.
(b) $y = 3x^{-2} + 5x^2$.
(c) $y = (2x^3 - x^{-1})^{-2}$.
(d) $y = \sqrt{4x^4 - 16}$.

3. In each of the following, find the required value.

(a) $f(r) = (r^2 + 5r - 7)^5$; find $f'(-1)$.
(b) $f(t) = (t^2 + 16)^2 - \sqrt{4t^2 + 9}$; find $f'(2)$.
(c) $v(t) = (3t^{-3} + 2t^2)^{1/3}$; find $v'(1)$.
(d) $f(x) = (4x^2 - 16)^{-1/2}$; find $f'(3)$.

4. Find the derivative with respect to the independent variable.

(a) $y = 3$.
(b) $y = 0$.
(c) $s = 7t^0$.
(d) $Q = \pi$.

(e) $E = \pi t^{\pi}$.
(f) $U = -2(t^3 - t^{-1})^2$.
(g) $f(v) = (2t - 3)^5 - 4(3t^2 - t)^{1/2}$.
(h) $y = \sqrt{4t^2 - 25} - \sqrt{25 - 4t^2}$.

5. Find the derivative with respect to t of each of the following. (Do not simplify the answer.)

(a) $E = (3t - 1)(t^2 - 3t + 5)$.
(b) $A = \sqrt{3t}\,(t^3 - 2t)$.
(c) $I = (t^2 - 1)(t^2 - 2t + 1)$.
(d) $Q = (3t + 5)^2\sqrt{t^2 + 4}$.

6. In each of the following, find the required value.

(a) $f(x) = (x^2 - 1)(\sqrt{3x} + 1)$; find $f'(2)$.
(b) $Q = \sqrt{2t + 1}(3t^2 - 2)$; find $(dQ/dt)|_{t=0.5}$.
(c) $A = (r^2 - 3)^3\sqrt{2r - 1}$; find $(dA/dr)|_{r=1}$.
(d) $f(x) = \sqrt{x^2 + 16} - \sqrt[3]{8 - x^3}$; find $f'(0)$.

7. In each of the following, find dy/dx.

(a) $y = \dfrac{3x + 1}{2x - 1}$.
(b) $y = \dfrac{5x}{x^2 - 1}$.
(c) $y = \dfrac{3}{x}$.
(d) $y = \dfrac{5}{3x^2}$.
(e) $y = \dfrac{1}{(x + 1)^2}$.
(f) $y = \dfrac{\sqrt{x} - 1}{\sqrt{x} + 1}$.
(g) $y = \dfrac{(3x - 1)^2}{(5x + 7)}$.
(h) $y = \dfrac{\sqrt{x} - 1}{(x - 1)^2}$.

8. In each of the following, find the required value.

(a) $Q = t^2 - \dfrac{3}{t^2 + 1}$; find $(dQ/dt)|_{t=1}$.

(b) $s = \dfrac{t + 2}{\sqrt{t} - 2}$; find $v|_{t=9}$.

(c) $f(x) = \dfrac{x}{(x - 1)^2}$; find $f'(1.5)$.

(d) $v = \dfrac{\sqrt{t}}{\sqrt{t} + 1}$; find $a|_{t=3}$.

9. Find dy/dx for each of the following.

(a) $y = 3 \sin 5x$.
(b) $y = 2 \sin \left(\dfrac{x}{2}\right)$.
(c) $y = 5 \cos 7x$.
(d) $y = -4 \cos 2x$.
(e) $y = 2 \sin x^3$.
(f) $y = 2 \sin^3 x$.
(g) $y = \frac{1}{2} \cos (x + 1)^2$.
(h) $y = 3 \cos^2 (x + 1)$.
(i) $y = x \sin^3 2x$.
(j) $y = x^2 \cos 3x$.
(k) $y = \dfrac{x}{\sin 2x}$.
(l) $y = \dfrac{\cos 3x}{x}$.

10. In each of the following, find the required value.

(a) $f(t) = 2 \sin 3t$; find $f'(0.1)$.
(b) $Q = 5 \sin 3t^2$; find $Q'(0.5)$.
(c) $s = t \sin 5t$; find $v|_{t=1}$.
(d) $y = \cos^2 x$; find $(dy/dx)|_{x=\pi/2}$.
(e) $E = t^2 \sin t^2$; find $(dE/dt)|_{t=1}$.
(f) $f(x) = \dfrac{\sin x}{x}$; find $f'(0.5)$.

Group B

1. Find $f'(x)$ for $f(x) = 1/x$ using (a) the quotient rule; (b) the power rule ($1/x = x^{-1}$). (c) Compare the answers of parts (a) and (b).

2. Given $y = \dfrac{x}{(x+1)^2}$. Find dy/dx using (a) the quotient rule; (b) the product rule. (c) Compare the answers of parts (a) and (b).

3. Given $y = (3x + 1)(x - 2)^2$. Find dy/dx using (a) the product rule; (b) the quotient rule. (c) Compare the answers of parts (a) and (b).

4. Find $f'(x)$ for

$$f(x) = \sqrt{\frac{x^2 - 1}{x^2 + 1}}$$

5. Find $f'(1)$ for $f(x) = [(2x - 1)^2(x^2 + 5)]^{3/2}$.

6. Given $y = uv = u/v^{-1}$, use the quotient rule to find dy/dx. Is the answer like the product rule for uv?

7. Given $y = u/v = uv^{-1}$, use the product rule to find dy/dx. Is the answer like the quotient rule for u/v?

8. For $f(x) = (x^3 - 2)/(2x + 1)$, is f increasing or decreasing at the point where $x = 1$? Find an equation of the line tangent to the curve of f at the point where $x = 1$.

9. The displacement, s, of a weight vibrating on a vertical spring where resonance is involved is given by $s = t \sin t$. Find the velocity of the weight when $t = 2$ s. Explain, in your own words, why we need trigonometric functions of numbers. (Is an angle involved in this vibration? What two quantities are angles usually measured in?)

10. The equation

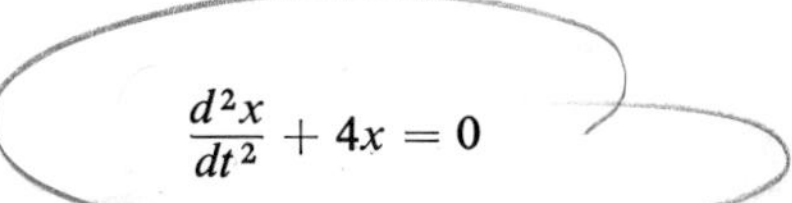

$$\frac{d^2x}{dt^2} + 4x = 0$$

is a differential equation. (a) Show that $x = 4 \sin 2t$ is a solution of this equation. (A solution of an equation makes the equation true.) (b) Show that $x = 3 \cos 2t$ is also a solution. (c) Show that $x = 4 \sin 2t + 3 \cos 2t$ is also a solution. (d) Is $x = 5 \sin (2t + 0.64)$ also a solution?

11. Use the quotient rule to find $f'(x)$ for $f(x) = \tan x$. ($\tan x = \sin x/\cos x$.)

12. It is true that $\sin^2 x + \cos^2 x = 1$. Take the derivative with respect to x of $\sqrt{1 - \sin^2 x}$ and show that $d(\cos x)/dx = -\sin x$.

SUMMARY OF IMPORTANT WORDS AND CONCEPTS

For each of the following, state in your own words your understanding of the statement or word.

1. (a) Average rate of change of a quantity. (b) Average current. (c) Average power.

2. (a) Distance. (b) Displacement. (c) Average speed. (d) Average velocity. (e) Average acceleration.

3. (a) Derivative. (b) Delta process. (c) Instantaneous rate of change. (d) Velocity. (e) Acceleration. (f) Derivative of a polynomial.

4. (a) Slope of a straight line. (b) Angle of inclination. (c) Angle between two lines. (d) Perpendicular lines. (e) Parallel lines.

5. (a) Equations of a straight line. (b) Point-slope. (c) Slope-intercept. (d) General, intercept, normal form.

6. (a) Linear equation. (b) Linear function.

7. (a) Distance between two points. (b) Midpoint of a line segment. (c) Distance between a point and a line.

8. (a) Slope of a curve at a point P on the curve. (b) Tangent line at a point P on the curve. (c) Normal line at a point P on the curve.

9. (a) Algebraic sign of the derivative. (b) Increasing and decreasing functions.

10. (a) Higher-order derivatives. (b) Radius of curvature.

11. Derivative rules: (a) Power. (b) Product. (c) Quotient. (d) Sine. (e) Cosine.

Chapter 3

Some Applications of the Derivative

3.1 SMOOTH CURVES AND CONTINUITY

In Sec. 2.8 we saw that the algebraic sign of the derivative enables us to tell when a function increases and when it decreases. A question we might ask is: If a function changes from an increasing to a decreasing function, or vice versa, is the change smooth or is it abrupt? Consider the curves in Fig. 3.1.1. The curve in Fig. 3.1.1(a) is said to be *smooth*. At the point (x_1, y_1) the slope, dy/dx, is zero as the function changes from an increasing function to a decreasing function. In Fig. 3.1.1(b) the curve is not smooth at point (x_2, y_2). There is an abrupt change from an increasing to a decreasing function. This is because the slope, dy/dx, is not defined at (x_2, y_2).

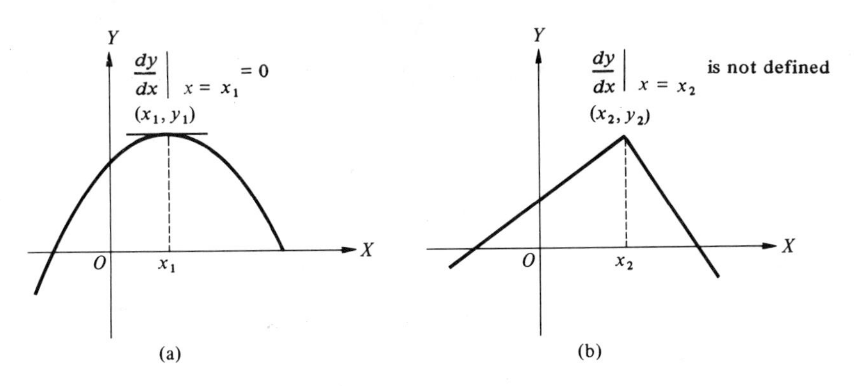

Figure 3.1.1 Smoothness of curves.

Definition 3.1.1 *If* dy/dx *exists at all points in an interval, then the graph of* $y = f(x)$ *is smooth over that interval.*

Definition 3.1.1 tells us when there are no "kinks" or sharp corners in the graph.

EXAMPLE 1. For $y = x^2$, $dy/dx = 2x$, and the derivative exists for all values of x. Hence the graph of $y = x^2$ is a smooth curve (see Fig. 3.1.2).

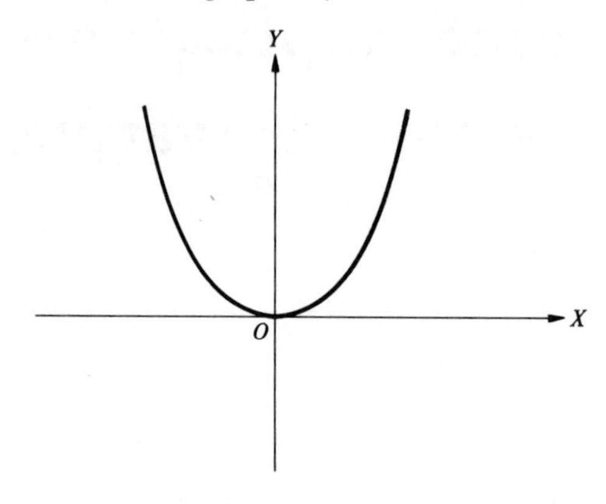

Figure 3.1.2 Graph of $y = x^2$.

EXAMPLE 2. For $y = x^{2/3}$, $dy/dx = \frac{2}{3}x^{-1/3} = 2/3x^{1/3}$. Here dy/dx does not exist at $x = 0$; hence the graph of $y = x^{2/3}$ is not smooth at $x = 0$ (see Fig. 3.1.3).

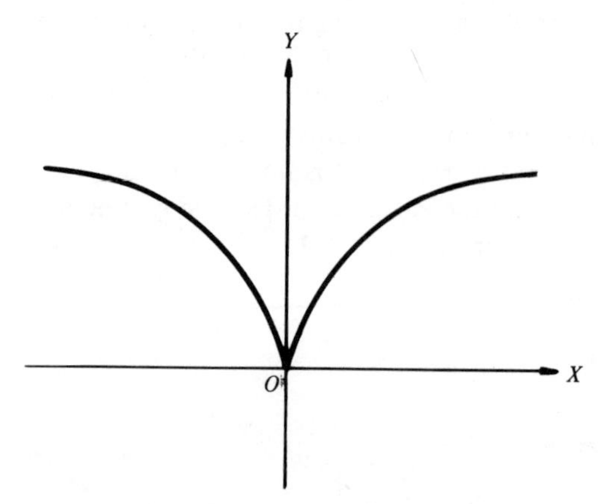

Figure 3.1.3 Graph of $y = x^{2/3}$.

PROBLEM 1. Is the graph of $y = x^3 - 2x$ smooth everywhere? Why?

PROBLEM 2. (a) Is the graph of $y = 3x^{1/3}$ smooth everywhere? If not, where not and why not?

(b) Is the graph of $y = \sqrt{x}$ smooth everywhere? (Be careful.)

We notice in Example 2 that even though the function is continuous at the origin it is not differentiable there. However, differentiability at a point does imply continuity at that point, as stated in Theorem 3.1.1.

Theorem 3.1.1 *If $f(x)$ is differentiable at $x = a$, it is continuous there.*

Theorem 3.1.1 means that wherever dy/dx exists, $y = f(x)$ is continuous. An important property of continuous functions is given in Theorem 3.1.2.

Theorem 3.1.2 *Let $f(x)$ be a continuous function over the interval $a \leqq x \leqq b$. If $f(a) \neq f(b)$, then if c is any number between $f(a)$ and $f(b)$ there is at least one value of x, say X, between a and b for which $f(X) = c$.*

Theorem 3.1.2 is called the *intermediate value theorem* for continuous functions and is proved in advanced calculus books. It says that the graph of a continuous function $y = f(x)$, satisfying the hypothesis of Theorem 3.1.2, must necessarily cross the line $y = c$ at some point between $x = a$ and $x = b$. This allows us to plot a few points and connect them by a curve, as illustrated in Fig. 3.1.4. It also guarantees a solution to an equation whose graph is continuous and crosses the X-axis (see Exercise 1 of Group C in Exercises 3.1).

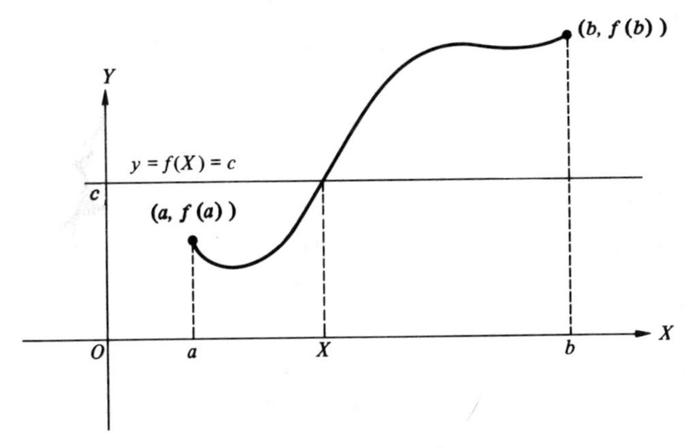

Figure 3.1.4 Intermediate value theorem.

EXERCISES 3.1

Group A

1. For each of the following, tell whether the graph of the function will be smooth or not smooth. If not smooth, at what points does the curve change abruptly?

(a) $y = x$. (b) $y = x^3$. (c) $y = 3x^2 + 5x + 7$.
(d) $y = (x - 1)^{1/3}$. (e) $y = x^{4/3}$. (f) $y = \sqrt{x + 2}$.
(g) $y = |x|$. (h) $y = |x| + x$.

2. For each of the following, tell whether the graph of the function is (i) continuous and (ii) smooth. If not continuous or smooth, tell at what points these properties do not hold.

(a) $f(x) = \begin{cases} x^2 & 0 \leqq x < 2 \\ 4 & x \geqq 2. \end{cases}$

(b) $g(t) = \begin{cases} t^3 & 0 \leqq t < 1 \\ 2 & t \geqq 1. \end{cases}$

(c) $I(t) = \begin{cases} t^2 + 2t + 1 & 0 \leqq t \leqq 5 \\ 3t + 21 & t > 5. \end{cases}$

(d) $\theta(s) = \begin{cases} s^4 + 2 & -1 \leqq s \leqq 1 \\ 3 & |s| > 1. \end{cases}$

(e) $f(x) = \begin{cases} x^2 & x \text{ not an integer} \\ 2x & x \text{ an integer.} \end{cases}$

Group B

1. Given that a particle moves according to the equation $s = 4t^3 - 2t^2$, discuss the smoothness of the graphs of s, v, and a.

2. Given that the charge in a circuit is $Q = t^2 + t^{3/2}$, discuss the smoothness of the graphs of Q and I.

Group C

1. For $f(x)$ continuous and satisfying $f(a) < 0$ and $f(b) > 0$, why does $f(x) = 0$ for some x satisfying $a < x < b$?

2. Using the function $f(x) = x^2 - 5x + 6$ and the points $a = \frac{5}{2}$ and $b = \frac{7}{2}$, find an x satisfying $\frac{5}{2} < x < \frac{7}{2}$ and $f(x) = 0$.

3.2 RELATIVE MAXIMA AND MINIMA AND INFLECTION POINTS

(1) Relative Maxima and Minima

Let us consider the function $y = f(x)$. If $f(a)$ is greater than $f(x)$ at any nearby point, then the point $(a, f(a))$ is said to be a *relative maximum*. Similarly, if $f(a)$ is less than $f(x)$ at any nearby point, then the point $(a, f(a))$ is said to be a *relative minimum*. A relative maximum occurs when $f(x)$ changes from an increasing function to a decreasing function as x increases. This happens only where $f'(x) = 0$ or where $f'(x)$ does not exist. [We may have a corner where $f'(x)$ does not exist.] A relative minimum occurs when $f(x)$ changes from a decreasing function to an increasing function as x increases. This also happens only where $f'(x) = 0$ or where $f'(x)$ does not exist. Figure 3.2.1(a) and (b) illustrate relative maximum and minimum points, respectively.

The values of x for which $f'(x) = 0$ or for which $f'(x)$ does not exist are called *critical values*. Their corresponding points $(x, f(x))$ are called *critical points*. All relative maxima and minima points are critical points. The converse is not true, as we shall see in Example 2.

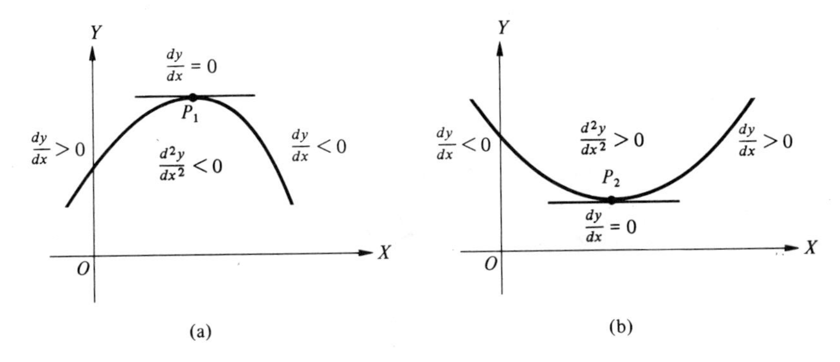

Figure 3.2.1 (a) Point P_1 is a relative maximum. (b) Point P_2 is a relative minimum.

EXAMPLE 1. Find the critical points of

$$y = \tfrac{1}{3}x^3 - 2x^2 + 3x + 1$$

Then find the relative maxima and minima points.

Solution. The critical values occur when $dy/dx = 0$. Since $dy/dx = x^2 - 4x + 3$, we set $dy/dx = 0$ and obtain

$$x^2 - 4x + 3 = 0$$

Factoring and solving for x, we find from

$$(x - 1)(x - 3) = 0$$

that the critical values are $x = 1$ and $x = 3$. Using these values to solve for y, we find that the critical points are

$$(1, \tfrac{7}{3}) \qquad \text{and} \qquad (3, 1)$$

To find the relative maxima and minima points we apply the techniques of Sec. 2.8 to find the values of x for which y is an increasing or a decreasing function:

Regions:	$x < 1$	$1 < x < 3$	$x > 3$
		1	3 → X
Factors of dy/dx	$(x - 1)(x - 3)$	$(x - 1)(x - 3)$	$(x - 1)(x - 3)$
Signs of factors	$(-)(-)$	$(+)(-)$	$(+)(+)$
Thus	$dy/dx > 0$	$dy/dx < 0$	$dy/dx > 0$
And	y increases	y decreases	y increases

At $x = 1$, y changes from an increasing to a decreasing function. Hence $(1, \tfrac{7}{3})$ is a relative maximum point. At $x = 3$, y changes from a decreasing to an increasing function. Hence $(3, 1)$ is a relative minimum point (see Fig. 3.2.2).

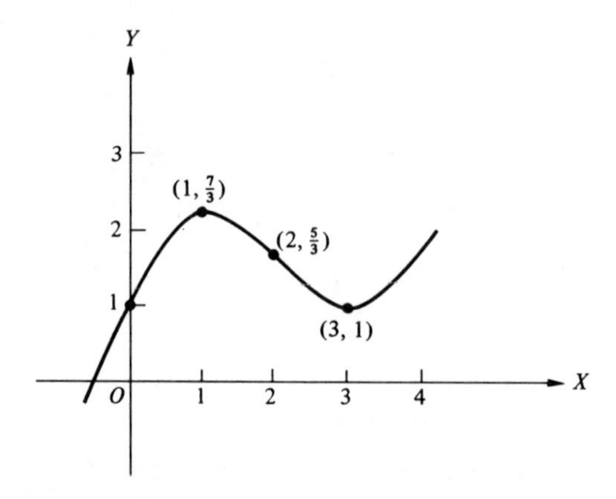

Figure 3.2.2 Graph of $y = \frac{1}{3}x^3 - 2x^2 + 3x + 1$.

PROBLEM 1. Find the critical points of $y = x^3 - 3x + 1$. Then find the relative maxima and minima points.

EXAMPLE 2. Investigate $y = x^3$ for relative maxima or minima points.

Solution. For $y = x^3$, $dy/dx = 3x^2$. Setting $dy/dx = 0$, we have $3x^2 = 0$, and thus $x = 0$ is a critical value. Since any nonzero number squared is positive, dy/dx does not change sign and $y = x^3$ has no relative maxima or minima points.

PROBLEM 2. Investigate $y = x^3 - 3x^2 + 3x - 1$ for relative maxima or minima points.

EXAMPLE 3. Investigate $y = \sin 4x$ for relative maxima or minima points in the interval $0 \leqq x \leqq \pi$.

Solution. For $y = \sin 4x$, $dy/dx = 4 \cos 4x$ (see Sec. 2.10). Upon setting $dy/dx = 0$, we have $\cos 4x = 0$. Recalling that the cosine function is zero when its argument is an odd multiple of $\pi/2$, we write $4x = (2n + 1)\pi/2$, $n = 0, \pm 1, \pm 2, \ldots$, and

$$x = \frac{2n + 1}{8}\pi$$

In the interval of interest, $\cos 4x = 0$ at $x = \pi/8$, $3\pi/8$, $5\pi/8$, and $7\pi/8$.

Regions:	$0 < x < \pi/8$	$\pi/8 < x < 3\pi/8$	$3\pi/8 < x < 5\pi/8$	$5\pi/8 < x < 7\pi/8$	$x > 7\pi/8$
X	0	$\pi/8$	$3\pi/8$	$5\pi/8$	$7\pi/8$
$y' = 4 \cos 4x$	+ y inc.	− y dec.	+ y inc.	− y dec.	+ y inc.

We conclude that $(\pi/8, 4)$ and $(5\pi/8, 4)$ are relative maximum points and $(3\pi/8, -4)$ and $(7\pi/8, -4)$ are relative minimum points.

PROBLEM 3. Investigate the function $y = \cos 3x$ for relative maxima and minima points in the interval $0 \leqq x \leqq 2\pi$. Then sketch its graph.

See Exercise 7 of Group C in Exercises 3.2 for a case where a relative maximum occurs at a point where $f'(x) \neq 0$.

(2) Inflection Points

Referring to Fig. 3.2.1(a), we note that as x increases, the slope, dy/dx, decreases, and in Fig. 3.2.1(b) we note that as x increases, the slope, dy/dx, increases. This property of the slope, increasing or decreasing, is determined by the algebraic sign of d^2y/dx^2, since

$$\frac{d^2y}{dx^2} = \frac{d}{dx}\left(\frac{dy}{dx}\right) = \frac{d}{dx}(\text{slope})$$

1. If $d^2y/dx^2 > 0$, the slope increases with an increase of x, and the graph of $y = f(x)$ is said to be *concave up* [see Fig. 3.2.1(b)].
2. If $d^2y/dx^2 < 0$, the slope decreases with an increase of x, and the graph of $y = f(x)$ is said to be *concave down* [see Fig. 3.2.1(a)].

The point at which a curve changes from concave up to concave down (or vice versa) is called an *inflection point*, and it occurs at values of x for which $f''(x) = 0$. We may not have an inflection point at every x for which $f''(x) = 0$. (The algebraic sign of d^2y/dx^2 must change.)

In Fig. 3.2.2 we see that the curve is concave down to the left of the point $(2, \frac{5}{3})$ and is concave up to the right of that point. Thus $(2, \frac{5}{3})$ is an inflection point. We shall prove this analytically in Example 4.

2.6 − 8 + 6 + 1

EXAMPLE 4. Find the inflection point for

$$y = \tfrac{1}{3}x^3 - 2x^2 + 3x + 1$$

(This is the same function as in Example 1.)

Solution. For $y = \frac{1}{3}x^3 - 2x^2 + 3x + 1$,

$$\frac{dy}{dx} = x^2 - 4x + 3$$

$$\frac{d^2y}{dx^2} = 2x - 4$$

To find the value of x at which an inflection point might occur we set $d^2y/dx^2 = 0$ and solve for x. Thus

$$\frac{d^2y}{dx^2} = 2(x - 2) = 0$$

and

$$x = 2$$

We now test the sign of d^2y/dx^2:

Regions:	$x < 2$	$x > 2$
		2 → X
Factors of d^2y/dx^2	$2(x-2)$	$2(x-2)$
Signs of factors	$(+)(-)$	$(+)(+)$
Thus	$d^2y/dx^2 < 0$	$d^2y/dx^2 > 0$

Since the sign of d^2y/dx^2 changes at $x = 2$, the point $(2, \frac{5}{3})$ is an inflection point (see Fig. 3.2.2).

PROBLEM 4. Find the inflection point for $y = x^3 - 3x + 1$.

EXAMPLE 5. Does $y = x^4$ have an inflection point?

Solution. For $y = x^4$,

$$\frac{dy}{dx} = 4x^3$$

$$\frac{d^2y}{dx^2} = 12x^2$$

At $x = 0$, $d^2y/dx^2 = 0$. The sign of d^2y/dx^2 is positive for all x (not equal to zero); hence there is no point of inflection and the graph is concave up everywhere.

PROBLEM 5. Find the points of inflection for $y = x^4 + 4x^3 + 6x^2 + 4x + 1$ (if there are any).

The following test may be used in finding inflection points. (It usually is used only when the third derivative of a function is easy to find.) If for $f(x)$, $f''(a) = 0$ and $f'''(a) \neq 0$, then the point $(a, f(a))$ is an inflection point. If $f'''(a) = 0$, no conclusion can be drawn. Use this test for Prob. 4 and compare your results.

(3) Tests for Relative Maxima and Minima

There are two standard tests for determining relative maximum and minimum points of a function $y = f(x)$ for which $f'(x)$ exists.

Test I *For the function* $\boldsymbol{y = f(x)}$, *if* $\boldsymbol{f'(a)} = 0$ *and if the sign of* $\boldsymbol{dy/dx}$ *changes from plus to minus as* $\boldsymbol{x}$ *increases through* $\boldsymbol{a}$, *then the point* $\boldsymbol{(a, f(a))}$ *is a relative maximum point. If the sign of* $\boldsymbol{dy/dx}$ *changes from minus to* *plus as* x *increases through* a, *then the point* $(a, f(a))$ *is a relative minimum point.*

Test I was illustrated in Example 1.

Test II *If* $f'(a) = 0$ *and if* $f''(a) > 0$, *then* $(a, f(a))$ *is a relative minimum point. If* $f'(a) = 0$ *and if* $f''(a) < 0$, *then* $(a, f(a))$ *is a relative maximum point. If* $f'(a) = 0$ *and if* $f''(a) = 0$, *this test fails and test I must be used.*

Test II is illustrated in Fig. 3.2.1.

EXAMPLE 6. Use Test II to find the relative maximum and minimum points of $f(x) = 4x^3 - 3x + 4$.

Solution. For $f(x) = 4x^3 - 3x + 4$,

$$f'(x) = 12x^2 - 3$$

$$f''(x) = 24x$$

Setting $f'(x) = 12x^2 - 3 = 3(4x^2 - 1) = 0$, the critical values of $f(x)$ are found to be $x = \frac{1}{2}$ and $x = -\frac{1}{2}$.

Testing $f''(x)$, we find that

$$f''(\tfrac{1}{2}) = 24(\tfrac{1}{2}) = 12$$

Hence $(\frac{1}{2}, 3)$ is a relative minimum point. From

$$f''(-\tfrac{1}{2}) = 24(-\tfrac{1}{2}) = -12$$

we find that $(-\frac{1}{2}, 5)$ is a relative maximum point.

PROBLEM 6. Use Tests I and II to find the relative maxima and minima and inflection points for $y = x^4 - 2x^2 + 1$. Use this information to sketch the graph of this function.

Warning ***The given function is used to determine points; the first derivative is used to determine critical values; the second derivative is used in determining inflection points.***

You might wonder if relative maxima and minima points always occur. A theorem called the *mean-value theorem* guarantees the existence of relative maxima or minima points for certain functions. It is given as

Theorem 3.2.1 *If* $f(x)$ *is continuous at all* x *in the interval* $a \leqq x \leqq b$ *and if its derivative exists at each* x *within the interval, then there is at least one value of* X, *where* $a < X < b$, *such that*

$$\frac{f(b) - f(a)}{b - a} = f'(X)$$

A geometric interpretation of Theorem 3.2.1 is illustrated in Fig. 3.2.3. It implies that the slope of the line QR which joins the end points of the curve over the interval $a \leqq x \leqq b$ is the same as the slope of the curve at some intermediate point P, where $x = X$. (See Exercise 4 of Group C in Exercises 3.2 for a discussion of the existence of a relative maximum or minimum point.)

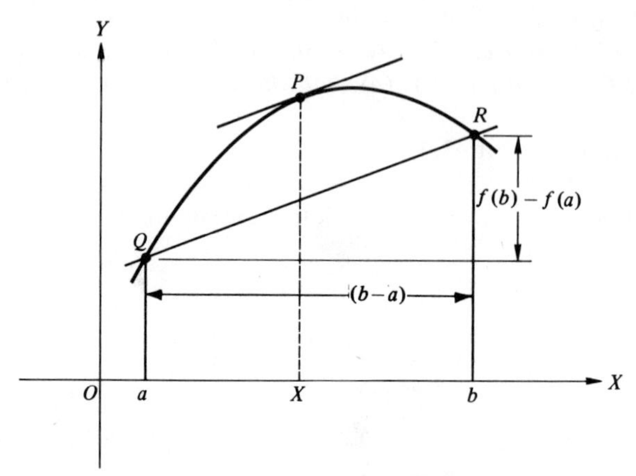

Figure 3.2.3 Illustration of Theorem 3.2.1.

The mean-value theorem is one of the most important and famous theorems of the calculus. It has many important theoretical and applied consequences. We saw one such consequence when we studied increasing and decreasing functions in Chap. 2. The Mathematical Association of America has a very good film that explores the consequences of the mean-value theorem as applied to a driver on a turnpike. The driver got caught for speeding. Can you guess how the mean-value theorem caught him?

EXERCISES 3.2

Group A

1. For each of the following functions, find the critical points and the relative maxima and minima points.

(a) $y = x^3 - 3x$.
(b) $y = \frac{1}{4}x^4 - \frac{1}{3}x^3 - \frac{1}{2}x^2 + x + 10$.
(c) $y = x^2 + 2x + 1$.
(d) $y = x^3 + 1$.
(e) $y = x^4 + 2$.
(f) $y = x^4 + 2x^2 + 2$.
(g) $y = |x|$.

2. For each of the following functions, find the critical points, the relative maxima and minima points, and the inflection points.

(a) $y = x^3 + 3x^2 + 2$.
(b) $y = x^4 + 3x^2 + 2$.
(c) $y = x^3 + 4$.
(d) $y = x^6 + 6$.

(e) $y = x^4 + 4x^3 + 12x^2 + 16$. (f) $y = x^4 - 4x^3 + 6x^2 - 8x + 1$.
(g) $y = 3x^5 - 5x^3 + 1$.

3. (This exercise is to be worked if you have studied Sec. 2.10.) For each of the following functions, find the critical points, the relative maxima and minima points, and the points of inflection.

(a) $f(t) = (t - 1)^5$. (b) $f(x) = \dfrac{\sqrt{x-1}}{x}$. (c) $f(x) = \dfrac{1}{1 + x^2}$.
(d) $y = \sin x$ (one cycle only). (e) $y = \cos x$ (one cycle only).
(f) $E = 110 \sin 60\pi t$ (one cycle only). (g) $y = x \sin x$ (one cycle only).

Group B

1. Tell how you would find the relative maxima, relative minima, and inflection points for a function.

2. How are critical points and relative maxima and minima points alike and how are they different?

3. Figure 3.2.4 is the graph of a function $y = f(x)$. Classify the points (x_i, y_i), $i = 1, 2, \ldots, 10$, as a relative maximum, relative minimum, or inflection point.

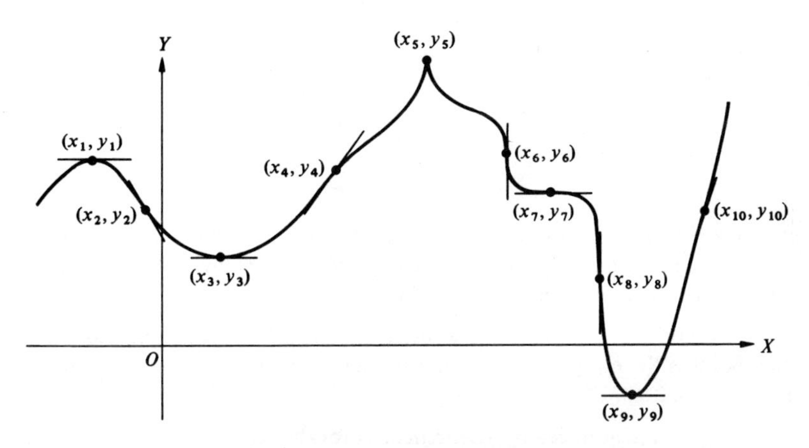

Figure 3.2.4 Exercise 3.

4. The charge in an electric circuit is given by $Q = t^3 - 3t^2 + 3t + 1$. Determine whether the charge achieves a relative maximum or relative minimum as time t goes from 0 to 5 s.

5. The displacement s of a particle from its starting point is given by $s = t^4 - 2t^2 + 3$. Determine whether the graph of this particle has any critical points or inflection points during the first 10 s of travel.

6. The work of a system is given by the formula

$$W(t) = \tfrac{5}{4}t^4 - \tfrac{20}{3}t^3 + 10t^2$$

Determine whether the power associated with this work has any maximum or minimum points.

7. In Exercise 6, determine whether the work has any maximum or minimum points.

8. The angle of inclination of a tangent line is given by the formula $\theta(s) = \frac{1}{12}s^4 - \frac{5}{6}s^3 + 3s^2 + 2$. Determine whether the curvature has any maximum or minimum points.

9. The magnetic flux passing through a circuit is given by $\phi(t) = t^4 - 2.6t^3 + 2.4t^2 + 0.7$. Determine whether the electromagnetic force has any maximum or minimum points.

10. The equation of a plane curve is given by

$$y = x^6 + 3x^5 - 20x^4 + 16x$$

Determine whether the slope of this curve has any maximum or minimum points.

11. The current in an electric circuit is given by

$$I(t) = t - t[u(t - 1)] + (t - 2)[u(t - 2)] - (t - 2)[u(t - 3)]$$

for $0 \leqq t \leqq 4$. Does the current have any relative maximum or minimum points in this interval of time? (See Exercise 13 of Group A in Exercises 1.4.)

12. The magnetic flux in an electric circuit is given by

$$\phi(t) = t^2 - t^2[u(t - 1)] + (t^2 - 4)[u(t - 2)] - (t^2 - 4)[u(t - 3)]$$

for $0 \leqq t \leqq 4$. Does the flux have any relative maximum or minimum points in this time interval?

Group C

1. Let $y = ax^3 + bx^2 + cx + d$, where $a \neq 0$. (a) Show that if $b^2 - 3ac > 0$, y has one relative maximum and one relative minimum. (b) Show that if $b^2 - 3ac = 0$, y has neither a relative maximum nor a relative minimum.

2. For $f(x) = ax^3 + bx^2 + cx + d$, where $a \neq 0$ and $b^2 - 3ac > 0$, show that the inflection point is the midpoint of the line segment joining the relative maximum and minimum points.

3. Find the value X guaranteed by Theorem 3.2.1 for the function $f(x) = x^2 - 2x + 1$ in the interval $-1 \leqq x \leqq 1$.

4. (a) Show that if $f(x)$ is a continuous function on an interval $a < x < b$ and differentiable on the interval $a < x < b$, then $f(a) = f(b) = 0$ implies that there is at least one critical point in the interval $a \leqq x \leqq b$. (b) Let $f(x) = x^2 + 5x + 6$. Find the critical point in the interval $-3 \leqq x \leqq -2$ and determine whether it is a relative maximum point or a relative minimum point. (c) Is it necessary for $f(a) = f(b) = 0$ to have Theorem 3.2.1 guarantee the existence of a relative maximum or minimum point?

5. Consider the graph in Fig. 3.2.5. At no point of the graph where $a \leqq x \leqq b$ is the tangent parallel to line PQ. Is this a contradiction to Theorem 3.2.1? Explain.

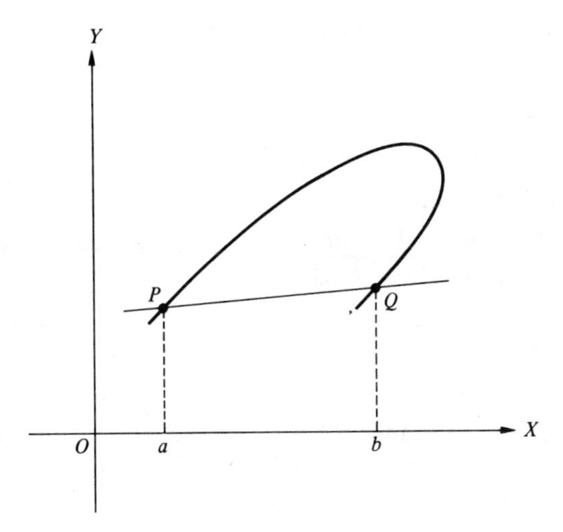

Figure 3.2.5 Exercise 5.

6. Use the conclusion of the mean-value theorem to verify the criteria used in Chap. 2 to find where a function is increasing or decreasing.

7. Consider the function

$$f(x) = \begin{cases} x & 0 \leqq x \leqq 2 \\ -x + 4 & 2 < x \leqq 4 \end{cases}$$

(a) Find $f'(x)$ for the closed interval $0 \leqq x \leqq 2$. (b) Find $f'(x)$ for the half-open interval $2 < x \leqq 4$. (c) Sketch the graph of $f(x)$. (d) Does $f(x)$ have a relative maximum? Where does it occur? Is $f'(x) = 0$ there? Does $f'(x)$ change sign there? (e) Must $f'(x) = 0$ at all relative maxima or minima points? Explain.

3.3 GRAPHS OF POLYNOMIALS

The derivative of the polynomial

$$y = a_0x^n + a_1x^{n-1} + \cdots + a_{n-1}x + a_n$$

is

$$\frac{dy}{dx} = a_0nx^{n-1} + a_1(n - 1)x^{n-2} + \cdots + a_{n-1}$$

and this exists for all values of x. Thus the graph of a polynomial is smooth and continuous. (Why?) We need to find only a few points on the graph and then make use of Theorem 3.1.1 to complete the graph. The easiest points to use are the relative maxima and minima points, the inflection points, and the Y-intercept. The Y-intercept is that point for which $x = 0$. It is helpful to draw the tangent line at these points. Its slope is found from dy/dx.

The steps to follow in graphing a polynomial $y = f(x)$ are:

1. Find dy/dx.
2. Find d^2y/dx^2.
3. Set $dy/dx = 0$. Find the critical values of x and the critical points (y is from the original function).
4. Test each of the critical points for relative maxima or minima.
5. Set $d^2y/dx^2 = 0$. Find the possible points of inflection (sometimes denoted as PPI).
6. Test the PPI to see if they are points of inflection.
7. Find the Y-intercept.
8. Find the slope of the tangent line at each of the relative maxima, relative minima, and inflection points and the Y-intercept.
9. Draw each of the tangent lines found in step 8.
10. Connect the points of step 8 by a smooth continuous curve. (Note that the graph crosses the tangent line at points of inflection but nowhere else.)

If the X-intercepts (when $y = 0$) are easily found, these points may be used also. However, they are not usually easily found. One of the big problems in algebra is finding of the roots of a polynomial equation.

EXAMPLE 1. Sketch the graph of $y = x^3 - 3x + 4$.

Solution. $y = x^3 - 3x + 4$.

Step 1. $dy/dx = 3x^2 - 3$.

Step 2. $d^2y/dx^2 = 6x$.

Step 3. $(dy/dx = 0)$; $3(x^2 - 1) = 0$

$$3(x + 1)(x - 1) = 0$$

Thus $x = -1$ and $x = 1$ are critical values. The critical points are $(-1, 6)$ and $(1, 2)$.

Step 4. At $x = -1$, $d^2y/dx^2 = 6(-1) = -6$. Since d^2y/dx^2 is < 0, $(-1, 6)$ is a relative maximum. At $x = 1$, $d^2y/dx^2 = 6(1) = 6$. Since $d^2y/dx^2 > 0$, $(1, 2)$ is a relative minimum.

Step 5. $(d^2y/dx^2 = 0)$; $6x = 0$ and $x = 0$. Thus a possible point of inflection is $(0, 4)$.

Step 6.

Regions:	$x < 0$	$x > 0$
	0	$\to X$
Factors of d^2y/dx^2	$(6)(x)$	$(6)(x)$
Signs of factors	$(+)(-)$	$(+)(+)$
Thus	$d^2y/dx^2 < 0$	$d^2y/dx^2 > 0$

Therefore, (0, 4) is a point of inflection. (d^2y/dx^2 changes sign.)

Step 7. The Y-intercept is (0, 4).

Step 8. It is helpful to construct a table at this step:

x	y	dy/dx	d^2y/dx^2	Classification of Points	Shape of Curve
-1	6	0	-6	Relative maxima	Concave down
0	4	-3	0	Inflection and Y-intercept	
1	2	0	6	Relative minima	Concave up

Step 9. On an XY-coordinate system of axes, plot the special points and draw the tangent lines [see Fig. 3.3.1(a)].

Step 10. Draw a smooth continuous curve through these points as in Fig. 3.3.1(b).

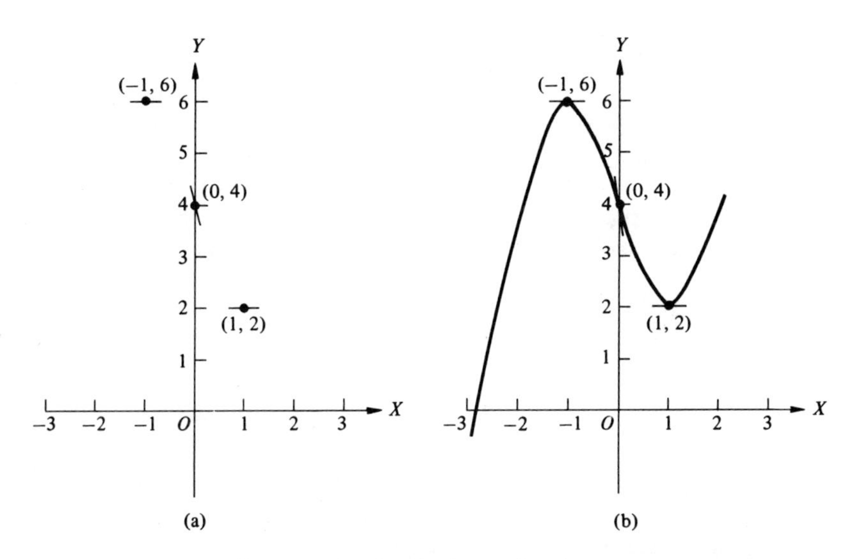

Figure 3.3.1 Graph of $y = x^3 - 3x + 4$.

Note that there is only one zero of the polynomial and it is not readily determined. However, we may find that $f(-2) = 2$ and $f(-3) = -14$; hence the graph crosses the X-axis between -2 and -3. (In Sec. 3.6 we shall discuss a method of finding zeros of a polynomial.)

PROBLEM 1. Sketch the graph of $y = x^3 - 3x^2 + 2$.

Sometimes the functions are simple enough so that only a table need be constructed before sketching the graph.

EXAMPLE 2. Sketch the graph of the charge $Q = t^3 + 1$.

Solution

$$Q = t^3 + 1$$

$$\frac{dQ}{dt} = 3t^2$$

$$\frac{d^2Q}{dt^2} = 6t$$

We see that both dQ/dt and d^2Q/dt^2 are zero only for the critical value $t = 0$. We also note that dQ/dt is always positive. Hence no relative maxima or minima exist. The following table will suffice to show the information we need:

t	Q	dQ/dt	d^2Q/dt^2	Classification of Points	Shape of Curve
−1	0	3	−6		Concave down
0	1	0	0	Inflection	
1	2	3	6		Concave up

The values of $t = -1$ and $t = 1$ were chosen for convenience as they are on different sides of the critical value. This problem shows that even though $dQ/dt = 0$ at $t = 0$, there is no relative maxima or minima. The graph is drawn in Fig. 3.3.2.

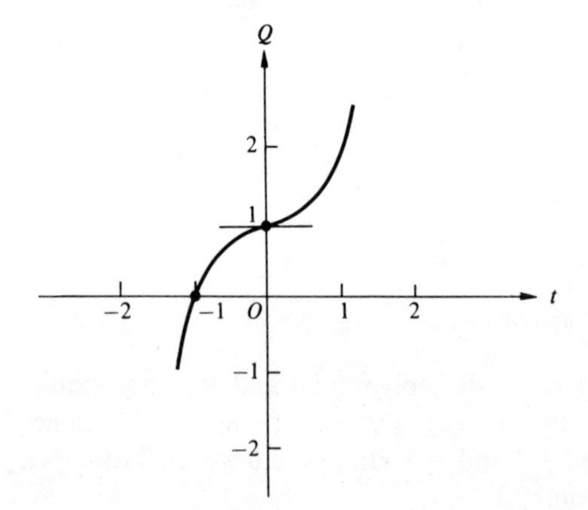

Figure 3.3.2 Graph of $Q = t^3 + 1$.

PROBLEM 2. Sketch the graph of $E = t^3 - 3t^2 + 3t$.

EXERCISES 3.3

Group A

1. Sketch the graph of each of the following functions.

(a) $y = x^2$. (b) $y = x^3$.
(c) $y = x^2 + 2x + 1$. (d) $y = x^3 + x^2$.
(e) $y = \frac{1}{3}x^3 + \frac{3}{2}x^2 + 2x + 1$. (f) $y = -x^3 + 3x + 1$.

2. Sketch the graph of each of the following functions.

(a) $s = t^4 + 2t^3 - 12t^2 + 14t$. (b) $Q = t^4 + 2t^3 + 6t^2 + 10t - 4$.
(c) $y = 2x^3 - 3x^2 - 12x - 5$. (d) $E = -3t^4 + 8t^3$.

3. Sketch the graph of each of the following functions.

(a) $\theta(s) = s^3 - 9s^2 + 14s - 24$. (b) $\phi(t) = t^3 - 3t^2 + 2.75t - 0.75$.
(c) $W(t) = t^3 - 6t^2 + 11t - 6$.

4. For each of the following, sketch a small portion of the graph using the given information.

(a) $f(3) = 2, f'(3) = 0, f''(3) = 6$. [*Hint*: At the point (3, 2) the slope is zero.]
(b) $f(-2) = -1, f'(-2) = 0, f''(-2) = 1$.
(c) $f(0) = 0, f'(0) = 0, f''(0) = -3$.
(d) $f(3) = 2, f'(3) = 1, f''(3) = 0, f''(x) > 0$ for $x > 3$, $f''(x) < 0$ for $x < 3$.
(e) $f(0) = 0$, $f'(0)$ does not exist, $f''(0) > 0$ for $x < 0$, $f''(0) < 0$ for $x > 0$.

Group B

1. A thrown object moves vertically according to the equation $s = 128t - 16t^2$. Sketch the graph of its motion.

2. The deflection, y, of a beam is given by $y = x^4 - 8x^3 + 24x^2$. Sketch the graph of the deflection over the interval $0 \leqq x \leqq 2$.

3. Given a square piece of cardboard with 8-in. sides. A person makes a box by cutting squares of length x from the corners and folding up the edges (see Fig. 3.3.3). Write a function to represent the volume of the resultant box and sketch the graph of the function.

4. Work is given by the formula $W(t) = 2t^3 - 15t^2 + 36t + 5$. Sketch the graph for $t \geqq 0$.

5. The curvature of a curve is given by $k(s) = 2s^3 - 15s^2 + 24s$. Sketch the graph of k.

6. The magnetic flux in a circuit is given by $\phi(t) = 0.1t^3 - 0.2t^2$. Sketch the graph of the flux.

7. The speed of a particle is given by $v(t) = |t^2 - 4|$. Sketch the graph of the speed for $t \geqq 0$.

8. The equation of a plane curve is given by $y = \frac{1}{3}x^3 + x^2 + 2x + 3$. Sketch the graph of the slope of this curve.

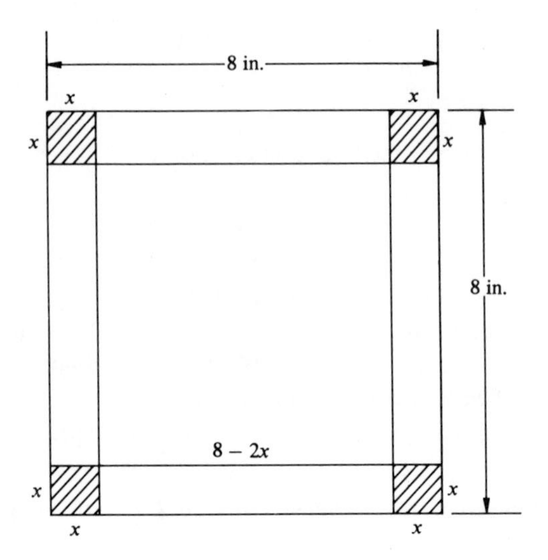

Figure 3.3.3 Exercise 3.

9. How are the graphs of $y = x^2$ and $y = x^4$ alike and how are they different?

10. How are the graphs of $y = x^3$ and $y = x^5$ alike and how are they different?

11. Compare the graphs of $y = x^4$ and $y = (x - 1)^4$.

Group C

1. Let $y = ax^3 + bx^2 + cx + d$. Find a, b, c, and d such that the graph of y has a relative maximum at $x = -1$ and a relative minimum at $x = 1$, a point of inflection at $x = 0$, and a Y-intercept of 4.

2. (a) Sketch the graph of $f(x) = x^2$. (b) Sketch the graphs of $h(x) = x^2 + 3$ and $h(x) = x^2 - 3$. (c) Sketch the graphs of $g(x) = (x + 3)^2$ and $g(x) = (x - 3)^2$. (d) Notice that $h(x) = f(x) + C$ and $g(x) = f(x + C)$, where C is a constant. (e) What is the relation of the graph of $y = f(x) + C$ to the graph of $y = f(x)$? (f) What is the relation of the graph of $y = f(x + C)$ to the graph of $y = f(x)$?

3.4 APPLIED MAXIMA AND MINIMA

We can utilize some of the techniques of Sec. 3.3 in answering such questions as: What is the maximum rectangular area that may be enclosed by a given length of fence?

Before proceeding with problems of this type, we need some more ideas.

The largest value a function assumes over its domain is called its *absolute maximum*. Correspondingly, the smallest value a function assumes over its domain is called its *absolute minimum*. The possibilities most frequently encountered are described below.

If $a \leqq x \leqq b$, then x is said to be in a *closed interval*. The absolute maximum or minimum points of a function, $f(x)$, in a closed interval, $a \leqq x \leqq b$, occur at the end points $(a, f(a))$ or $(b, f(b))$, or at the points where $f'(x) = 0$, or at the points where $f'(x)$ does not exist [see Fig. 3.4.1(a) and (b)]. In Fig. 3.4.1(a) the minimum is the same as the relative minimum and the maximum is $f(b)$. In Fig. 3.4.1(b) the maximum occurs where the derivative does not exist.

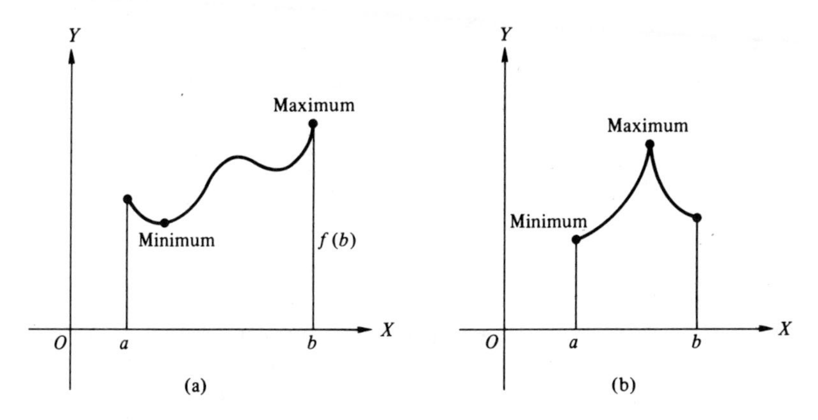

Figure 3.4.1 Absolute maximum or minimum points in a closed interval.

If $a < x < b$, then x is said to be in an *open interval*. If the absolute maximum or minimum points of a function, $f(x)$, in an open interval, $a < x < b$, exist, they will occur at the points where $f'(x) = 0$ or at the points where $f'(x)$ does not exist (see Fig. 3.4.2). Since the function is not defined at $x = a$ and $x = b$, we cannot say what happens there. Note that $y = 1/x$, for $0 < x \leqq 1$ has no absolute maximum (see Fig. 3.4.3).

We usually refer to the absolute maximum and minimum values as just maximum and minimum. Frequently, these values are the same as the relative

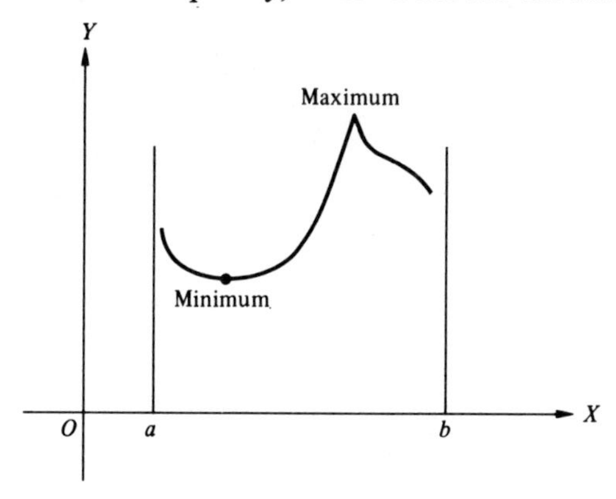

Figure 3.4.2 Maximum and minimum points in an open interval.

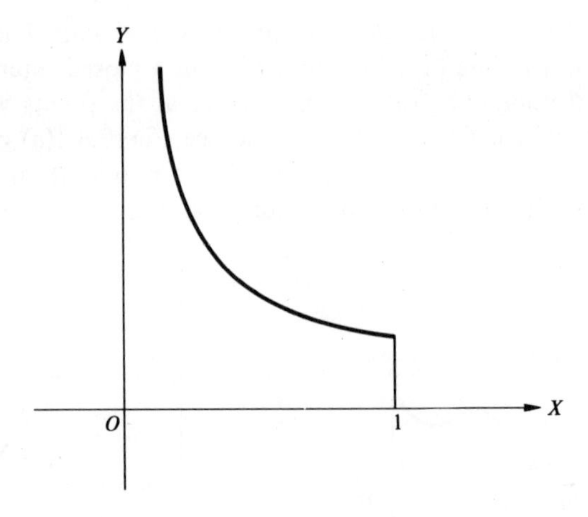

Figure 3.4.3 Graph of $y = 1/x$, $0 < x \leqq 1$.

maximum and minimum. In many applied problems we deal with closed intervals in which the derivative exists at all points in the interval.

These problems are usually "word problems." Thus we need to introduce variables and express the dependent variable as a function of the independent variable, keeping in mind that the dependent variable is limited to values in a specific range.

EXAMPLE 1. A 500-ft fence is to be erected in the shape of a rectangle. What is the maximum area that may be enclosed?

Solution. Draw a figure representing the problem and introduce variables (see Fig. 3.4.4). The area of the rectangle is given by $A = xy$. But here we are in trouble as we have not yet dealt with a function of two variables. However, the perimeter of this rectangle is

$$2x + 2y = 500$$

so

$$y = 250 - x$$

and now

$$A = x(250 - x)$$

or

$$A = 250x - x^2 \qquad 0 \leqq x \leqq 250$$

A
y
x

Figure 3.4.4 Example 1.

The limitations on x are due to the length of the fence. Since we want to maximize A, we set $dA/dx = 0$ and solve for x. Thus

$$\frac{dA}{dx} = 250 - 2x$$

Therefore,

$$250 - 2x = 0 \qquad \text{and} \qquad x = 125 \text{ ft}$$

Since the maximum area is required, we must test for it. We test this value of x for relative maximum by using the second derivative. Since $d^2A/dx^2 = -2$, we have the relative maximum of A at $x = 125$. This value of A is

$$A = 125(250 - 125) = 15{,}625 \text{ ft}^2$$

Since x is in the closed interval $0 \leq x \leq 250$, we must test the end points for maximum A. At $x = 0$, $A = 0(250 - 0) = 0$. At $x = 250$, $A = 250(250 - 250) = 0$. Thus the maximum area is $A = 15{,}625 \text{ ft}^2$. We might note that since $y = 250 - x$, that $y = 125$ ft and the area enclosed is a square.

We now summarize the steps to follow in working an applied maximum-minimum problem.

Step 1. Sketch a figure (if appropriate) and introduce the necessary variables.

Step 2. Express the quantity to be maximized or minimized in terms of the other variables.

Step 3. If there is more than one independent variable, use other relationships between the variables to write the quantity of step 2 as a function of one variable.

Step 4. Differentiate the function of step 3 and set the derivative equal to zero.

Step 5. Solve the equation of step 4 and test these values as to relative maximum or minimum.

Step 6. Find the values of the function at the end points of the interval.

Step 7. Compare steps 5 and 6 to choose the maximum or minimum value.

Warning ***Be sure that the sought-after maximum or minimum is the dependent variable in your equation relating the variables.***

PROBLEM 1. A box with an open top and a square base is to have a capacity of 108 cm³. Find the dimensions so that the total surface of the box is a minimum.

EXAMPLE 2. A right circular cylinder is inscribed in a sphere of radius a. Find the dimensions of the cylinder such that its volume is a maximum.

Solution. Sketch the figure (see Figure 3.4.5; step 1). The volume of a right circular cylinder is

$$V = \pi r^2 h \qquad \text{(step 2)}$$

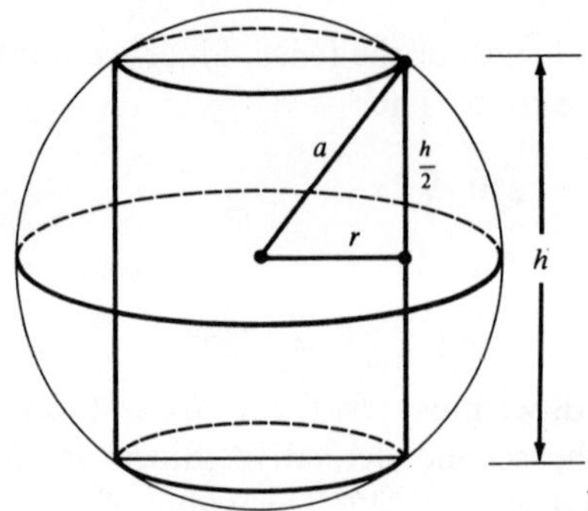

Figure 3.4.5 Example 2.

We must obtain a relation between r and h where $0 < h < 2a$ and $0 < r < a$. Why? Referring to Fig. 3.4.5, we see that

$$r^2 = a^2 - \frac{h^2}{4} \qquad \text{(step 3)}$$

Thus

$$V = \pi\left(a^2 - \frac{h^2}{4}\right)h \qquad \text{(step 3)}$$

or

$$V = \pi a^2 h - \pi\frac{h^3}{4} \qquad \text{(step 4)}$$

and

$$\frac{dV}{dh} = \pi a^2 - \frac{3\pi h^2}{4} \qquad \text{(step 4)}$$

To find h that gives a maximum V, we set $dV/dh = 0$ and obtain

$$\pi a^2 - \frac{3\pi h^2}{4} = 0 \qquad \text{or} \qquad h = \pm\frac{2a}{\sqrt{3}} \qquad \text{(steps 4 and 5)}$$

We choose $h = 2a/\sqrt{3}$, since a negative h has no physical meaning, and also the positive h makes d^2V/dh^2 negative. (You should check this.) From $r^2 = a^2 - (h^2/4)$ and $h = 2a/\sqrt{3}$, we have

$$r^2 = a^2 - \frac{1}{4}\left(\frac{4a^2}{3}\right) = \frac{2}{3}a^2$$

and

$$r = \tfrac{1}{3}\sqrt{6}\,a$$

Checking the end points of the intervals for r and h (step 6), we see that if $r = a$, $h = 0$ and $V = 0$. For $h = 2a$, $r = 0$ and $V = 0$. Thus the dimensions for maximum volume are

$$r = \left(\frac{\sqrt{6}}{3}\right)a = 0.82a \qquad \text{and} \qquad h = \left(\frac{2\sqrt{3}}{3}\right)a = 1.15a$$

PROBLEM 2. A cone is generated by revolving a right triangle about one leg. If the hypotenuse is always 12 cm, what is the maximum volume the cone may have? ($V = \frac{1}{3}\pi r^2 h$.)

EXERCISES 3.4

Group A

1. Find the relative and absolute maxima and minima for each of the following functions. If an absolute maximum or minimum does not exist, give a reason for the nonexistence. Sketch the curve of each function.

(a) $y = x^2,\ -1 \leqq x \leqq 1.$

(b) $y = x^2,\ -1 < x < 1.$

(c) $y = 3x + 2,\ 0 \leqq x < 2.$

(d) $y = 3x + 2,\ 0 < x \leqq 2.$

(e) $y = 3x + 2,\ 0 < x < 2.$

(f) $y = x^3 + 2x^2 + x + 1,\ -3 \leqq x \leqq 3.$

(g) $y = x^3 + 2x^2 + x + 1,\ -3 < x < 3.$

(h) $y = x^3 + 2x^2 + x + 1,\ -3 \leqq x < 3.$

(i) $y = 1/x^2,\ 0 < x \leqq 1.$

(j) $y = \begin{cases} x^2 & 0 \leqq x < 1 \\ 3 & x = 1 \\ x & 1 < x \leqq 2. \end{cases}$

(k) $y = \begin{cases} \dfrac{1}{x-1} & 0 \leqq x < 1 \\ 0 & x = 1 \\ \dfrac{1}{x-1} & 1 < x \leqq 2. \end{cases}$

Group B

1. A 700-ft fence is to be erected in the shape of a rectangle. What is the maximum area that may be enclosed?

2. One side of a rectangular cattle corral is to lie along part or all of the base of a cliff 40 ft long. The other three sides are to be fenced in with 36 ft of fencing. Find the dimensions of the largest corral that can be so constructed.

3. A square of cardboard is 36 in. on a side. If a box with no cover is made by cutting small squares from each corner and folding the flaps upward, find the size of cut which enables the box to hold the most.

4. A horizontal, simply supported, uniform beam of length L bends under its own weight, which is w lb/ft. The equation of its elastic curve is given by

$$y = \frac{w}{24EI}(x^4 - 2Lx^3 + L^3x)$$

Find the maximum deflection, y, in the beam.

5. The strength of a rectangular beam varies as the product of the square of the breadth and the cube of the depth. Find the dimensions of the strongest beam that can be cut from a cylindrical log whose radius is 12 in.

6. A rectangle is to have a perimeter of 20 in. and its diagonal is to be a minimum. Find its dimensions. (*Hint*: d is minimum when d^2 is minimum.)

7. Find two positive numbers whose sum is 24 such that the sum of the cube of one number plus 3 times the square of the other number is a minimum.

8. Find the dimensions of the largest rectangle that can be inscribed in a triangle of base 10 and height 8.

9. A rectangle has an area of 64 in.^{2}. Find its dimensions if its diagonal is to be a minimum.

10. A ladder is to reach over an 8-ft-high fence to a wall 1 ft behind the fence. What is the length of the shortest ladder that can be used?

11. A cabinet company finds that it makes a net profit of \$6 per cabinet for 150 cabinets or less in a week. If the profit decreases 5 cents per cabinet for every cabinet over 150, what number of cabinets per week would yield the greatest profit?

12. A box factory wishes to minimize the cost of making a certain size box. The box has a base whose sides are in proportion 3 to 2, and is to contain 144 ft^3. If the bottom and sides cost 8 cents/ft^2, and the top 12 cents/ft^2, what dimensions would make the cost a minimum?

13. The power delivered to an electric circuit by a generator is given by $P = 40I - 5I^2$. Find the current I for which the generator will deliver maximum power.

14. A cylindrical tank with an open top is to be constructed out of 20π ft^2 of material. What are its dimensions if the volume is to be a maximum?

15. Show that for all rectangles with a given area, the square has the smallest perimeter.

16. Show that for all rectangles with a given perimeter, the square has the maximum area.

17. The specific weight of water at temperature t degrees Celsius is given by $s = 1 + 5.3 \times 10^{-5}t - 6.53 \times 10^{-6}t^2 + 1.4 \times 10^{-8}t^3$. Find the temperature at which water has a maximum specific weight.

Group C

1. (a) Write a formula for finding the distance from the point (x_1, y_1) to any point on the line $Ax + By + C = 0$. (b) Show that the minimum distance is given by

$$\frac{Ax_1 + By_1 + C}{\sqrt{A^2 + B^2}}$$

2. How are relative and absolute maxima (minima) alike and how are they different?

3. Does a straight-line segment ever have a relative maximum (minimum)? Why? Does a straight-line segment ever have an absolute maximum (minimum)? Where?

4. Of all rectangles with a given area, A, find (a) the one with the smallest perimeter; (b) the one with the shortest diagonal.

5. Explain why the maximum or minimum values of a function cannot always be found by the derivative method.

3.5 RECTILINEAR MOTION

When a particle, or body, moves along a straight line it is said to have *rectilinear motion*. If s, the displacement from the origin, is a function of time, t, then $s = f(t)$, and $v = ds/dt$ is the velocity and $a = dv/dt = d^2s/dt^2$ is the acceleration.

The *speed* of the body is always positive and is the *absolute value of the velocity*, which may be positive or negative. Making use of Sec. 2.8, we have

Theorem 3.5.1 *If the velocity is positive, the body moves so as to increase* **s**. *If the velocity is negative, the body moves so as to decrease* **s**.

EXAMPLE 1. If a stone is thrown straight up from the ground with a speed of 32 ft/s, its displacement from the ground at time t is $s = 32t - 16t^2$. Find the interval of time for which s increases and for which s decreases. Also find the height to which it goes.

Solution. To find the interval of time for which s increases or decreases we must investigate the sign of ds/dt or v. For $s = 32t - 16t^2$,

$$v = \frac{ds}{dt} = 32 - 32t$$

Setting $v = 0$, we have

$$32 - 32t = 0$$

$$32(1 - t) = 0$$

and the critical value of $t = 1$:

$t < 1$	$t > 1$
v is plus $\therefore s$ increases	v is minus $\therefore s$ decreases

(axis t, critical point 1)

The physical interpretation of the answer is that the stone goes upward for 1 s and then starts down. At $t = 1$, since $v = 0$, the stone is at its highest point, which is found to be $s = 32(1) - 16(1)^2 = 16$ ft.

PROBLEM 1. A ball is projected straight up from the ground such that its displacement from the ground at time t is $s = 96t - 16t^2$. Find its initial velocity, the interval of time for which s increases and for which s decreases. Also find how high it goes, and its displacement, velocity, and acceleration at the end of 4 s.

The sign of the acceleration $a = dv/dt$ indicates an increasing or decreasing velocity.

Theorem 3.5.2 *If the acceleration is positive, the body moves with an increase in its velocity. If the acceleration is negative, the body moves with a decrease in its velocity.*

Since the speed of a body is the absolute value of the velocity, we must be extremely careful in discussing increasing or decreasing speeds. For example, if the velocity is negative, an increase in the velocity means the speed decreases. Thus if $v = -10$ and increases to $v = -7$, the speed changes from $|-10| = 10$ to $|-7| = 7$, which is a decrease in the speed. Similarly, for a negative velocity a decrease in the velocity means an increase in the speed. Thus if $v = -10$ and decreases to $v = -15$, the speed changes from $|-10| = 10$ to $|-15| = 15$, which is an increase in the speed.

Table 3.5.1 summarizes these ideas. Note that the speed increases only when v and a have the same sign.

TABLE 3.5.1

Velocity v	Acceleration a	Displacement s	Velocity v	Speed
+	+	Increasing	Increasing	Increasing
+	−	Increasing	Decreasing	Decreasing
−	+	Decreasing	Increasing	Decreasing
−	−	Decreasing	Decreasing	Increasing

EXAMPLE 2. A body moves in a straight line according to $s = t^3 - 3t^2 + 2$. Discuss its motion for positive t and sketch the graphs of $s = f(t)$, $v = g(t)$, and $a = h(t)$ on the same set of axes.

Solution. For $s = t^3 - 3t^2 + 2$,

$$v = 3t^2 - 6t$$

$$a = 6t - 6$$

The critical values of t for s are determined by setting $v = 0$. Thus

$$v = 3t(t - 2) = 0$$

$$t = 0 \qquad \text{and} \qquad t = 2$$

The critical values of t for v are determined by setting $a = 0$. Thus

$$a = 6(t - 1) = 0$$

$$t = 1$$

Regions:	$0 < t < 1$	$1 < t < 2$	$t > 2$
(t-axis)	0	1	2
Factors of v	$3(t)(t-2)$	$3(t)(t-2)$	$3(t)(t-2)$
Sign of v	$(+)(+)(-) = (-)$	$(+)(+)(-) = (-)$	$(+)(+)(+) = (+)$
	s decreases	s decreases	s increases
Factors of a	$6(t-1)$	$6(t-1)$	$6(t-1)$
Sign of a	$(+)(-) = (-)$	$(+)(+) = (+)$	$(+)(+) = (+)$
	v decreases	v increases	v increases
	Speed increases	Speed decreases	Speed increases

The graphs are shown in Fig. 3.5.1.

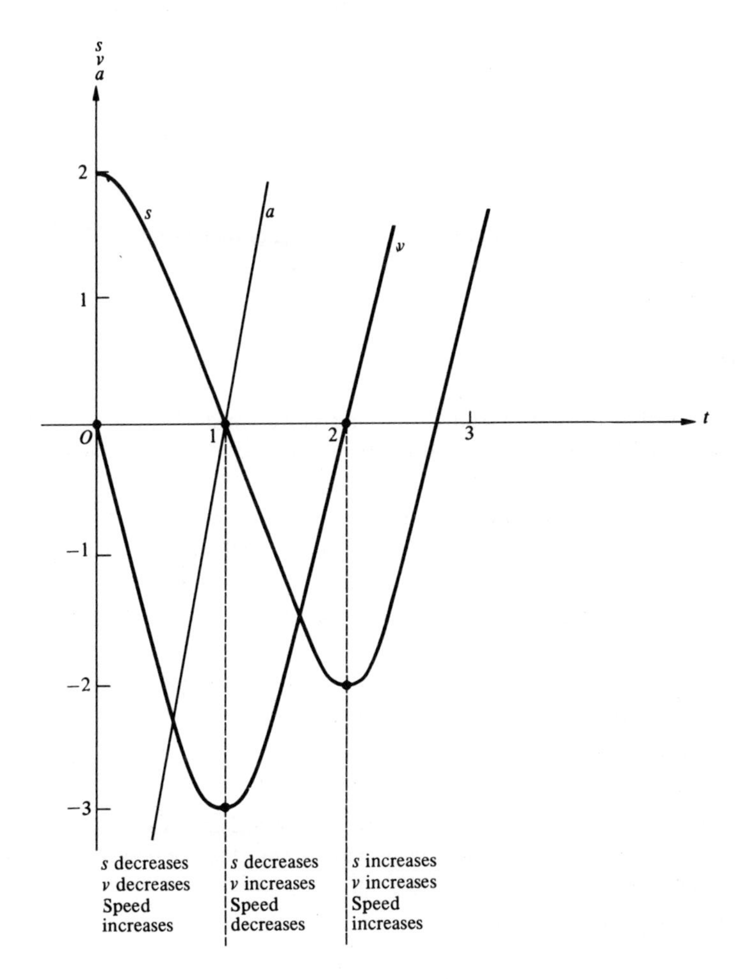

Figure 3.5.1 Example 2.

PROBLEM 2. A body moves in a straight line according to $s = t^3 - 6t^2 + 4$. Discuss its motion for positive t and sketch the graphs of $s = f(t)$, $v = g(t)$, and $a = h(t)$ on the same set of axes.

EXERCISES 3.5

Group A

1. A particle moves in a straight line according to $s = t^3 - 6t^2 + 9t$. (a) Find v at $t = 0$ and at $t = \frac{1}{2}$. Has the speed increased or decreased, and by how much? (b) Find v at $t = 2$ and at $t = 3$. Has the speed increased or decreased, and by how much?

2. In each of the following, find the interval of time ($t > 0$) (i) for which s increases and for which s decreases, (ii) for which v increases and for which v decreases, and (iii) for which the speed increases and for which the speed decreases.

(a) $s = 608t - 16t^2$.
(b) $s = -5t^2 + 50t$.
(c) $s = t^2 - 3t + 1$.
(d) $s = 1 + 3t^2 - t^3$.
(e) $s = t^3 - 2t^2 + 5t + 1$.
(f) $s = 8 - 4t - 2t^2 + t^3$.

3. For each motion described in Exercise 2, draw the graphs of s, v, and a on the same set of coordinate axes.

Group B

1. If a stone is thrown upward from the ground with a speed of 64 ft/s, its displacement from the ground at time t is $s = 64t - 16t^2$. Find the interval of time for which s increases and for which s decreases.

2. A body moves in a straight line according to $s = 2t^3 - 6t^2 + 4$. Discuss its motion for positive t and sketch the graphs of $s = f(t)$, $v = g(t)$ and $a = h(t)$.

3. An errant punt is kicked straight up in the air with a speed of 96 ft/s. Its displacement from the ground at time t is $s = 96t - 16t^2$. Find the interval of time for which s increases and for which s decreases.

4. A body moves in a straight line according to $s = t^4 + 2t^3 + t + 1$. Discuss its motion for positive t and sketch the graph of $s = f(t)$, $v = g(t)$, and $a = h(t)$.

5. Discuss the motion of a particle that moves in a straight line according to $s = t^3 + 3t^2 - t + 2$.

6. A certain weight on a vibrating spring has its motion characterized by $y = 3 \sin t$. (a) Find the first time for maximum displacement. (b) Find the first time and position for maximum velocity. (c) Find the first time and position for maximum acceleration.

7. Tell how velocity and speed are alike and how they are different.

3.6 NEWTON'S METHOD OF FINDING ROOTS

The use of the derivative in finding the roots of an equation is a very powerful method. One such method that gives good approximations to the root is Newton's method. It is based on a rather simple geometric idea. Suppose that we want to find the zero of $y = f(x)$, that is, the root of $f(x) = 0$. Sketch the graph of $y = f(x)$ as in Fig. 3.6.1. The point $(r, 0)$ where the graph crosses the X-axis gives us the value of the root r. Let x_1 be an approximation to r. At the point (x_1, y_1) draw the tangent to the curve $y = f(x)$. Let x_2 be the abscissa of the point where this tangent line crosses the X-axis. This value of x_2 is a second approximation to the required root r. We can find a formula for x_2 as follows. The slope of the tangent line is $f'(x_1)$, and using the point-slope form of a straight line its equation is

$$y - y_1 = f'(x_1)(x - x_1)$$

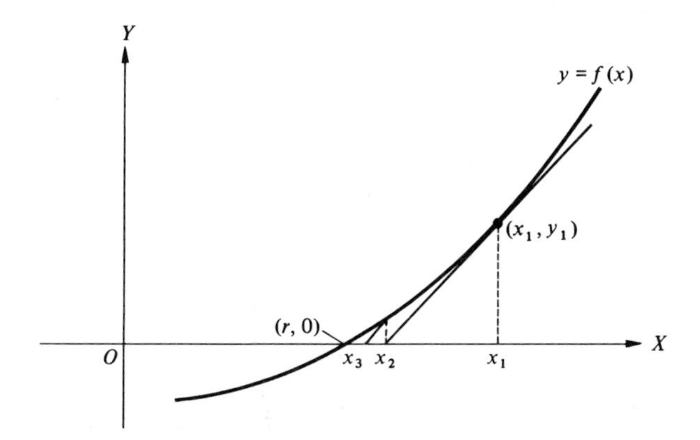

Figure 3.6.1 Newton's method.

Where this line crosses the X-axis, $y = 0$ and $x = x_2$; hence we have

$$0 - y_1 = f'(x_1)(x_2 - x_1)$$

and

$$x_2 = x_1 - \frac{y_1}{f'(x)} = x_1 - \frac{f(x_1)}{f'(x_1)} \qquad (3.6.1)$$

At the point $(x_2, f(x_2))$ this process may be repeated and a third approximation may be found with

$$x_3 = x_2 - \frac{f(x_2)}{f'(x_2)}$$

In general, we have

$$\boxed{x_{n+1} = x_n - \frac{f(x_n)}{f'(x_n)}} \qquad (3.6.2)$$

This method could lead to trouble under certain conditions. However, if there are no relative maxima or minima and no inflection points between x_1 and r, we usually encounter no difficulty.

EXAMPLE 1. Use Newton's method to find the real solution of $x^3 - 2x - 3 = 0$ correct to three decimal places.

Solution. Let $y = f(x) = x^3 - 2x - 3$. Then $f'(x) = 3x^2 - 2$. The graph is shown in Fig. 3.6.2. It is seen that the real root lies between $x = 1$ and $x = 2$. (The other two roots are imaginary.) To find this root chose $x_1 = 2$ and apply Eq. 3.6.1. Here $x_1 = 2$, $f(x_1) = f(2) = 1$, and $f'(x_1) = f'(2) = 10$. Thus

$$x_2 = 2 - \frac{f(2)}{f'(2)} = 2 - \frac{1}{10} = 1.9$$

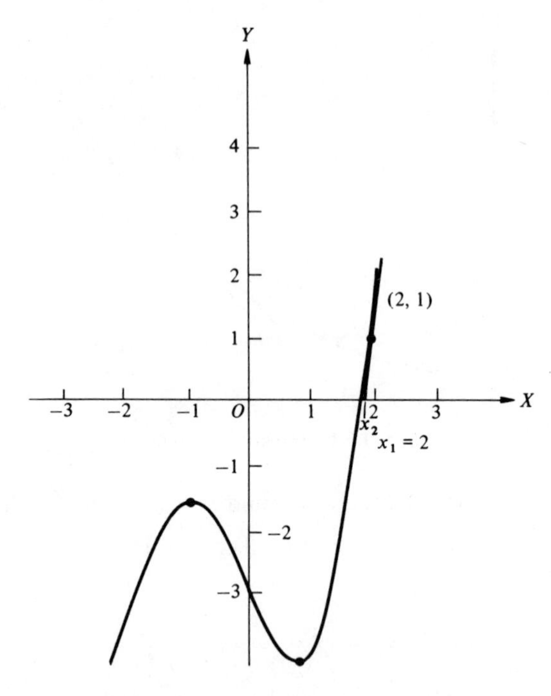

Figure 3.6.2 Root of $x^3 - 2x - 3 = 0$.

Now use $x_2 = 1.9$ and obtain a new approximation, x_3. Thus

$$f(1.9) = (1.9)^3 - 2(1.9) - 3 = 0.059$$

and

$$f'(1.9) = 3(1.9)^2 - 2 = 8.83$$

Using Eq. 3.6.2, we have

$$x_3 = 1.9 - \frac{f(1.9)}{f'(1.9)} = 1.9 - \frac{0.059}{8.83}$$

and

$$x_3 = 1.8933$$

Now use $x_3 = 1.8933$ and obtain a new approximation, x_4. Thus

$$f(1.8933) = (1.8933)^3 - 2(1.8933) - 3 = 0.0001$$

and

$$f'(1.8933) = 3(1.8933)^2 - 2 = 1.5846$$

Since

$$\frac{f(1.8933)}{f'(1.8933)} = \frac{0.0001}{1.5846} = 0.00006$$

this will not affect the third decimal place and we have $x = 1.893$ as our desired answer.

PROBLEM 1. Use Newton's method to find the real root of $x^3 - 3x - 3 = 0$ correct to three decimal places.

EXERCISES 3.6

Group A

1. Using Newton's method, find the root of $3x^3 - 7x^2 - 9x + 21 = 0$ between 2 and 3 correct to two decimal places.

2. Using Newton's method, find the root of $16x^3 - 16x^2 - 25x + 7 = 0$ between 1 and 2 correct to two decimal places.

3. Using Newton's method, find the real root of $x^3 - 0.33x^2 + x - 0.33 = 0$ correct to two decimal places.

4. Using Newton's method, find the real solution of $x^3 - 3x - 4 = 0$ correct to three decimal places.

5. Using Newton's method, find the real solution of $x^3 - 0.4412x^2 + x - 0.4412 = 0$ correct to three decimal places.

6. In each of the following, use Newton's method to find the answer correct to three decimal places. Use your calculator to check your answer.

(a) $\sqrt{5}$. (*Hint*: $\sqrt{5}$ is the positive root of $x^2 - 5 = 0$.)
(b) $\sqrt{27}$. (c) $\sqrt[3]{2}$. (d) $\sqrt[3]{-41}$.

SUMMARY OF IMPORTANT WORDS AND CONCEPTS

For each of the following, state in your own words your understanding of the statement or word.

1. (a) Smooth curves. (b) Continuity. (c) Continuous function.

2. Intermediate-value theorem.

3. (a) Relative maxima and minima. (b) Critical values. (c) Critical points. (d) Tests for relative maxima and minima.

4. (a) Inflection points. (b) Concave up. (c) Concave down. (d) Test for inflection point.

5. Theorem of the mean for differentiable functions.

6. Graphs of polynomials (10 steps).

7. (a) Absolute maximum and minimum. (b) Closed interval. (c) Open interval. (d) Seven steps used in solving applied maximum–minimum problems.

8. (a) Rectilinear motion. (b) Increasing and decreasing s. (c) Increasing and decreasing v (Table 3.5.1).

9. Newton's method of approximating roots.

Chapter 4

Differentials and Antiderivatives

4.1 THE DIFFERENTIAL

The derivative of the function $y = f(x)$ was defined using the limit of the fraction $\Delta y/\Delta x$ at points in the domain of y. However, its symbol dy/dx was not treated as a fraction; it was treated as a function. In some of our future work we shall need to consider the symbols dy and dx separately. These symbols are called the *differential of* y and the *differential of* x, respectively.

Definition 4.1.1 *For the function $y = f(x)$, the differential of the independent variable, x, is defined to be Δx; that is,*

$$dx = \Delta x \tag{4.1.1}$$

The differential of y is defined in Definition 4.1.2.

Definition 4.1.2 *For the function $y = f(x)$, the differential of the dependent variable, y, is defined to be the product of the derivative of the function and the differential of the independent variable; that is,*

$$dy = f'(x)dx \tag{4.1.2}$$

Definition 4.1.2 is convenient as it allows us to express the derivative as the quotient of two differentials; that is,

$$f'(x) = \frac{dy}{dx}$$

In Example 1 we illustrate that dy is not always equal to Δy.

EXAMPLE 1. For $y = x^2$, find Δy and dy.

Solution. We find Δy by use of the delta process of Sec. 2.1. Thus for $y = x^2$,

$$y + \Delta y = (x + \Delta x)^2 = x^2 + 2x\,\Delta x + (\Delta x)^2$$
$$\Delta y = 2x\,\Delta x + (\Delta x)^2$$

From Definition 4.1.1, $dx = \Delta x$, and we may write $\Delta y = 2x\,dx + (dx)^2$. Since from Definition 4.1.2 $dy = f'(x)\,dx$, and here $f'(x) = 2x$, we have $dy = 2x\,dx$. Comparing

$$\Delta y = 2x\,dx + (dx)^2 \qquad \text{and} \qquad dy = 2x\,dx$$

we see that for this function dy and Δy differ by $(dx)^2$. Note that if dx is very small, $(dx)^2$ is very very small. This means that if Δy is the actual change in y, the dependent variable, we can guess that dy is a reasonably good approximation to Δy, especially when dx is small.

We may interpret Example 1 as the area of a square of side x with an increase dx of the side x (see Fig. 4.1.1). That is, for $A = x^2$, we may find that

$$\Delta A = 2x\,dx + (dx)^2$$

where the change in the area, ΔA, is made up of the sum of areas (1), (2), and (3). The differential of the area, $dA = 2x\,dx$, is made up of the sum of areas (1) and (2). Thus ΔA and dA differ by the area of (3), which is $(dx)^2$. If dx is quite small, then ΔA and dA are nearly the same; that is, if $dx = 0.001$, then ΔA and dA differ by the amount of $(dx)^2 = 0.000001$.

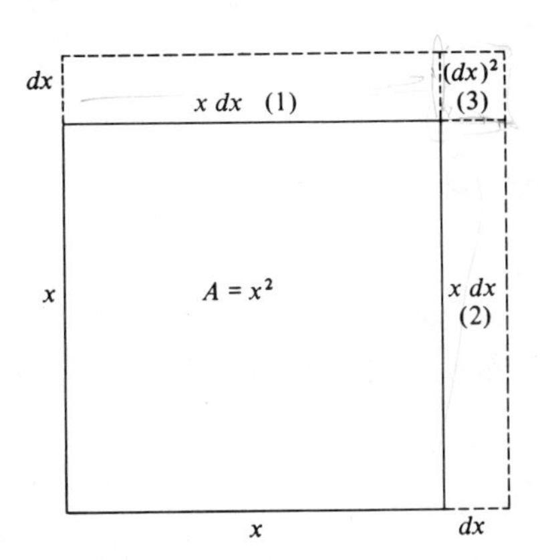

Figure 4.1.1 Differential of area.

PROBLEM 1. Compare Δy and dy for $y = 3x^2 - x$.

For a geometric interpretation of the differential consider the graph of $y = f(x)$ as in Fig. 4.1.2(a) and (b). In both cases $PR = \Delta x = dx$, the change in x, and $QR = \Delta y$, the exact change in y due to x changing as we move from P to Q along the curve of $y = f(x)$. If PT is the tangent to the curve at P, then by our previous work we know that $dy/dx = f'(x) = \tan\theta$ or $dy = (\tan\theta)\,dx$.

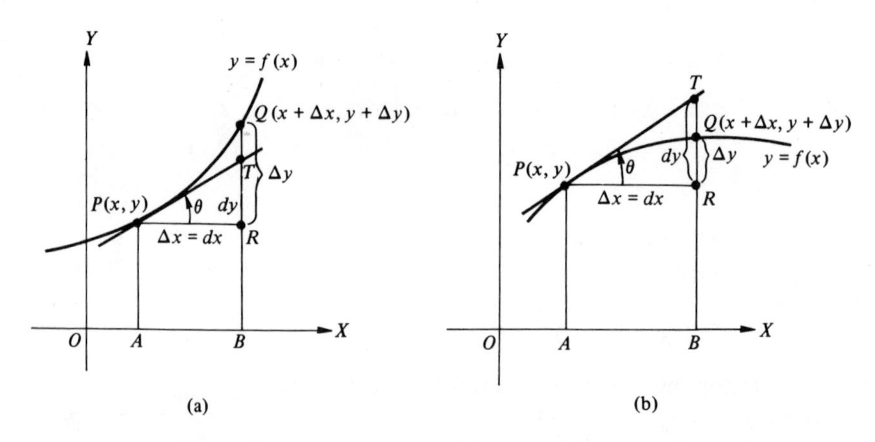

Figure 4.1.2 Geometric interpretation of the differential.

Then, since $\tan\theta = RT/PR = RT/dx$ or $RT = (\tan\theta)dx$, we have $dy = RT$. That is, if instead of moving along the curve we move along the tangent line, the length of RT represents the *approximate* change in y. The magnitude of the difference between Δy and dy is the line segment QT, and if dx is very small, QT is going to be very small.

Warning ***You must remember that Δx and dx are the same but Δy and dy are not the same; that is, $dx = \Delta x$, but $dy = f'(x)\,dx$.***

EXERCISES 4.1

Group A

1. Write the expression for dy for each of the following functions y.

(a) $y = x$.
(b) $y = 3x^2$.
(c) $y = x^2 - 2x + 1$.
(d) $y = x^3 - 2x^2 + x + 1$.
(e) $y = x^4 + 3x^2 + x$.
(f) $y = x^{3/2} + x + 2$.

2. Write expressions for dy and Δy for each of the following functions y.

(a) $y = 10$.
(b) $y = 5x^2$.
(c) $y = x^3$.
(d) $y = x^3 - x^2 + x + 1$.
(e) $y = x^4 - 3x^3 + 2x^2 + 1$.
(f) $y = x^2 - 3x + 2$.

3. In each of the following, find the values of dy and Δy for the given values of x and Δx.

(a) $y = x$, $x = 2$, $\Delta x = 0.2$.
(b) $y = x^2$, $x = -1$, $\Delta x = 0.02$.
(c) $y = x^2 - 2x + 1$, $x = 2$, $\Delta x = 0.04$.
(d) $y = x^4 + 3x^2 - 1$, $x = 1$, $\Delta x = 0.01$.
(e) $y = 10$, $x = 19$, $\Delta x = 0.0001$.
(f) $y = x^3 + x + 2$, $x = 4$, $\Delta x = 0.02$.

4. If you have studied Sec. 2.10, find the differential of the dependent variable in each of the following.

(a) $y = (3x^2 - 5x)^3$. (b) $s = t^{3/2}(5t - 1)^2$. (c) $v = (3t + 2)/(2t - 5)$.
(d) $y = 5 \sin^2 3x$. (e) $y = 2 \cos 5x$. (f) $E = 120 \sin 60t$.

Group B

1. If the motion of a particle is given by $s = 16t^2$, find Δs and ds for $t = 3$ and $\Delta t = 0.01$.

2. If $P = I^2R$ for a constant resistance of $R = 1.0\ \Omega$, find the change in the power ΔP as the current changes from $I = 20$ to $I = 21$. Find dP at $I = 20$ for $\Delta I = 1$.

3. Given a cubic box 3 ft on a side, find the change in volume, ΔV, of the box if each side is increased to 3.3 ft. Find dV at $x = 3$ for $\Delta x = 0.3$.

4. A circle has its radius increased from r_1 to r_2. If A is the area of the circle, find ΔA and dA at r_1.

5. A sphere has radius 2. How much should the radius of the sphere be increased to give a 5% increase in its volume? Find dV.

6. The current in an electric circuit is given by $I = 4t^2 + 2t + 2$. Find ΔI and dI for $t = 1$ and $\Delta t = 0.01$.

7. The equation of a plane curve is given by $y = \frac{1}{3}x^3 + x^2 + x$. If $y'(x)$ denotes the slope of y at x, find $\Delta y'$ and dy' at $x = 2$ for $\Delta x = 0.02$.

8. The magnetic flux in a circuit is given by

$$\phi(t) = t - t[u(t - 1)] + (t - 2)[u(t - 2)] - (t - 2)[u(t - 3)]$$

for $0 \leq t \leq 4$. Find (a) $d\phi$ and $\Delta\phi$ at $t = 0.75$ for $\Delta t = 0.35$; (b) $d\phi$ and $\Delta\phi$ at $t = 1.75$ for $\Delta t = 0.2$.

9. The work in a system is given by the equation

$$W(t) = t^3 + 3t^2 + 3$$

If P denotes the power associated with this work, find dP and ΔP at $t = 3$ for $\Delta t = 0.001$.

10. The angular velocity of a fly sitting on a spinning wheel is given by $\omega(t) = 2t^2 + 6t + 2$. Find $\Delta\omega$ and $d\omega$ at $t = 5$ for $t = 0.005$.

Group C

1. For the function $y = f(x)$, how are Δy and dy alike and how are they different?

2. Find a straight-line approximation (equation of the tangent line) to the curve $y = x^2$ in the neighborhood at the point where $x = 1$. How good is the approximation?

3. Is it ever true that $\Delta y = dy$? Give an example.

4. Let $y = x^3$ and interpret x as the side of a cube. By drawing this cube and a cube with sides $x + \Delta x$, show diagrammatically the quantities Δy and dy. Show the geometrical representation of $\Delta y - dy$ as a volume.

4.2 APPROXIMATIONS

The differential is useful in approximating certain calculations when one desires to find the effect on a function caused by a slight change in the value of the independent variable. This is due to the fact that dy, the differential of y, and Δy differ by a very small amount, which becomes even less as Δx is made smaller. The value of dy is called the *approximate change* in y. The value of Δy is called the *actual change* in y.

EXAMPLE 1. Find approximately the change in the area of a square 5 in. on a side if 0.01 in. is cut off two adjacent sides.

Solution. Let x be the side of the square. Then its area, A, is $A = x^2$ and

$$dA = 2x\,dx$$

For $x = 5$ and $dx = -0.01$ (we use the minus sign because we are decreasing x), we have

$$dA = 2(5)(-0.01) = -0.1 \text{ in.}^2$$

This is the approximate change in the area. The actual change in the area, ΔA, may be found by computing the area for $x = 5$ and also for $x = 4.99$ (or one may use the delta process). For $x = 5$, $A_1 = (5)^2 = 25$. For $x = 4.99$, $A_2 = (4.99)^2 = 24.9001$, and the actual change in the area is

$$\Delta A = A_2 - A_1 = 24.9001 - 25 = -0.0999 \text{ in.}^2$$

Hence our approximation is in error by 0.0001 in.^{2}.

PROBLEM 1. The voltage is given by $E = 2t^3 - t + 5$. Find the voltage when $t = 2$ and how much it changes approximately when t increases to 2.0005.

Recalling that $\Delta y = f(x + \Delta x) - f(x)$, $\Delta x = dx$, and $dy = f'(x)\,dx$, we can use differentials to approximate the value of $f(x + \Delta x)$ by

$$f(x + \Delta x) = f(x) + \Delta y \doteq f(x) + dy$$

or

$$f(x + dx) \doteq f(x) + f'(x)\,dx \tag{4.2.1}$$

where the symbol $\doteq$ means "appropriately equal to." In using formula 4.2.1 we must choose dx to be small and choose x such that $f(x)$ and $f'(x)$ are easily obtained.

EXAMPLE 2. Approximate the value of $\sqrt{34}$.

Solution. Let $f(x) = \sqrt{x} = x^{1/2}$; then

$$f'(x) = \frac{1}{2x^{1/2}} = \frac{1}{2\sqrt{x}}$$

The perfect square nearest to 34 is 36, so we choose $x = 36$ and $dx = -2$. Using formula 4.2.1, we have

$$f(x + dx) = f(x) + f'(x)\,dx$$

or

$$f(36 - 2) = \sqrt{36} + \frac{1}{2\sqrt{36}}(-2)$$

Then

$$\sqrt{34} \doteq 6 + \frac{1}{2(6)}(-2) = 6 - \frac{1}{6}$$

and
$$\sqrt{34} \doteq 5.833.$$

Warning ***The word "approximate" implies that we must use differentials.***

PROBLEM 2. Approximate the value of $\sqrt[3]{27.5}$.

Sometimes we are interested in the relative error or the percentage error incurred by using differentials in making approximations. The ratio dy/y is called the *relative error* and $100dy/y$ (%) is called the *percentage error.*

EXAMPLE 3. Find the percentage error in the volume of a sphere of radius 8 in. if the error in measuring its radius is 0.1 in.

Solution. To find percentage error we need to find the relative error first. The volume of a sphere is $V = \frac{4}{3}\pi r^3$. Therefore,

$$dV = 4\pi r^2\,dr$$

and

$$dV = 4\pi(8)^2(0.1) = 25.6\pi$$

is the approximate change in volume. For $r = 8$, $V = \frac{4}{3}\pi(8)^3 = (2048/3)\pi = 682.67\pi$. The relative error is

$$\frac{dV}{V} = \frac{25.6\pi}{682.67\pi} = 0.0375$$

The percentage error is

$$100\frac{dV}{V}\,\% = 3.75\,\%$$

PROBLEM 3. A cube has its sides measured as 5 ± 0.01 cm. What is the percentage error in measuring (a) its sides; (b) its volume; (c) its total surface area?

The previous examples may help you see why in Sec. 1.1 we remarked that dy may be thought of as a "little bit of y."

EXERCISES 4.2

Group A

1. Find the approximate change for each of the given functions with the associated data.

(a) $V = x^3$; $x = 2$, $dx = 0.3$.
(b) $V = \frac{4}{3}\pi r^3$; $r = 3$, $dr = 0.4$.
(c) $V = \pi r^2 h$; $h = 4$, $r = 1$, $dr = 0.5$.
(d) $A = 4\pi r^2$; $r = 10$, $dr = 0.01$.
(e) $A = bh$; $b = 3$, $h = \sqrt{2}$, $dh = 0.005$.

2. Find the relative and percentage errors for each of the exercises in Exercise 1.

3. Use differentials to approximate each of the following.

(a) $\sqrt{17}$. (b) $\sqrt{22}$. (c) $\sqrt[3]{12}$.
(d) $\sqrt[5]{33}$. (e) $\sqrt{143}$. (f) $\sqrt[3]{122}$.

4. Write a formula for the approximate change of each of the following. (a) The volume of a cube. (b) The volume of a sphere. (c) The volume of a right circular cylinder (1) with constant height and (2) with constant radius. (d) The area of the surface of a sphere.

5. Write a formula for the relative error for each of the parts of Exercise 4.

Group B

1. A cylinder in an automobile is 3 in. in diameter and 6 in. high. Its diameter is increased by 0.01 in. Approximately how much metal is cut out?

2. A cubical block of wood has edges 12 in. long. If a layer $\frac{1}{4}$ in. thick is cut off of each face, find (a) the approximate decrease in volume of the block; (b) the approximate decrease in surface area of the block; (c) the percentage decrease in edge length; (d) the approximate percentage decrease in surface area.

3. If a weather balloon is inflated to a 6-ft diameter and then has its radius increased by $\frac{1}{4}$ in., what is the approximate percentage increase in surface area of the balloon?

4. The voltage in an electrical circuit is given by $E = IR$. For a constant resistance of 10 Ω, find the actual change and the approximate change in the voltage as I increases from 10 to 12 A.

5. A cubical box is to be made to hold 27 ft^3. Find approximately how accurately the dimension of the inner edge must be if the volume is to be accurate to within 0.5 ft^3.

6. If the work in a system is given by $W(t) = 4t^2 + 2t + 2$, find the approximate change in the work and power between $t = 1$ and $t = 1.2$.

7. If the charge on a capacitor in an electric circuit is given by $Q(t) = t^{3/2} + t + 1$, find the approximate change in the charge and the current between $t = 4$ and $t = 4.5$.

8. The angle through which a particle on a wheel moves in t seconds is given by

$\theta(t) = t^2 - 2t + 1$. Find the approximate change in θ and in the angular velocity of the particle between $t = 3$ and $t = 3.003$.

9. State in your own words what is meant by the approximate change in a quantity. Compare it with the actual change.

4.3 ANTIDERIVATIVES

Many problems involve the reverse process of differentiation. That is, we desire to find the function whose derivative is known. This process is known as *finding the antiderivative.* The first question to ask is: What function must be differentiated to obtain the given derivative? Recall that for $y = ax^n$, $dy/dx = nax^{n-1}$. Notice that here the exponent of x has been reduced by 1 and the coefficient has been multiplied by n. To find the antiderivative of ax^n we must increase the exponent by 1 and divide the coefficient by the new exponent.

For example, if $dy/dx = 3x^2$, one value of the antiderivative is $y = [3/(2+1)]x^{2+1} = x^3$. We say "one value" because if $y = x^3 + 7$, $dy/dx = 3x^2$; also, if $y = x^3 - 11$, $dy/dx = 3x^2$. These values differ only by a constant, and since the derivative of a constant is zero, the most general antiderivative of $3x^2$ is $x^3 + C$, where C is an arbitrary constant.

One nice thing about finding antiderivatives is that you can always check your answer by differentiation. You may thus prove the following theorem.

Theorem 4.3.1 *If* $\boldsymbol{dy/dx = ax^n}$, *where* $\boldsymbol{a}$ *and* $\boldsymbol{n}$ *are constants and* $\boldsymbol{n \neq -1}$, *then the antiderivative* $\boldsymbol{y}$ *is given by*

$$y = \frac{a}{n+1}x^{n+1} + C \tag{4.3.1}$$

where $\boldsymbol{C}$ *is an arbitrary constant.*

The restriction that $n \neq -1$ avoids division by zero if we try to find the antiderivative of $1/x$. The antiderivative of this function will be found after we study differentiation of the logarithmic function (Sec. 8.6).

EXAMPLE 1. Find y for each of the following.

$$\text{(a)}\quad \frac{dy}{dx} = 5x^{1/2} \qquad \text{(b)}\quad \frac{dy}{dx} = 3x^5 - 2x + 7$$

Solution. (a) Apply Theorem 4.3.1. Thus

$$y = \frac{5}{\frac{1}{2}+1}x^{1/2+1} + C$$

or

$$y = \frac{5}{\frac{3}{2}}x^{3/2} + C = \frac{10}{3}x^{3/2} + C$$

(b) Apply Theorem 4.3.1 to each term separately. Thus

$$y = \tfrac{1}{2}x^6 - x^2 + 7x + C$$

These answers may be checked by differentiation.

Warning ***An arbitrary constant is always part of an antiderivative.***

PROBLEM 1. Find y (the antiderivative) for each of the following. Check your answer by differentiation.

$$\text{(a)}\quad \frac{dy}{dx} = 3x^2 \qquad \text{(b)}\quad \frac{dy}{dx} = x^{2/3} + 3x + 5$$

In Sec. 2.10 we stated formulas for derivatives of the sine and cosine functions. Reversing them, we can state formulas for their antiderivatives.

$$\text{For } \frac{dy}{dx} = a \sin bx \qquad y = -\frac{a}{b} \cos bx + C \tag{4.3.2}$$

$$\text{For } \frac{dy}{dx} = a \cos bx \qquad y = \frac{a}{b} \sin bx + C \tag{4.3.3}$$

EXAMPLE 2. Find the antiderivative of each of the following.

$$\text{(a)}\quad \frac{ds}{dt} = 3 \sin 5t \qquad \text{(b)}\quad \frac{dQ}{dt} = 10 \cos \pi t$$

Solution. (a) Apply formula 4.3.2 to obtain

$$s = -\tfrac{3}{5} \cos 5t + C$$

(b) Apply formula 4.3.3 to obtain

$$Q = \frac{10}{\pi} \sin \pi t + C$$

(You should check these answers using differentiation.)

PROBLEM 2. Find the antiderivative of

$$\text{(a)}\quad \frac{dy}{dx} = 5 \sin 3x \qquad \text{(b)}\quad \frac{dE}{dt} = \pi \cos 3\pi t$$

Warning ***Remember the minus sign when finding the antiderivative of the sine function.***

EXERCISES 4.3

Group A

1. Find and check the antiderivative of each of the following.

(a) $2x$. (b) $3x^2$. (c) $4x^3$.

(d) nx^{n-1}. (e) $\frac{1}{5}(x^{-4/5} + 5)$. (f) $2x^2 - x$.

(g) $\frac{x^5}{6} + \frac{x^4}{5} - \frac{x^3}{4}$. (h) $7x^{1/7}$. (i) $3x^{0.7}$.

(j) $-2.7x^{1.3}$.

2. Find and check the antiderivative of each of the following.

(a) $\frac{1}{x^2}$. (b) $\frac{4}{x^2}$. (c) $\frac{1}{x^3}$.

(d) $\frac{-5}{x^3}$. (e) $\frac{1}{x^3} + \frac{1}{x^4}$. (f) $x^3 - \frac{1}{x^2}$.

(g) $\frac{1}{x^{1/2}} + x^4$. (h) $\frac{1}{3x^{3/2}}$.

3. Find and check the antiderivative of each of the following.

(a) $\frac{dy}{dx} = 2x^2 + 3x^{-3}$. (b) $\frac{ds}{dt} = 16t^2 - 32t$.

(c) $\frac{dQ}{dt} = t^{-2} - t^2$. (d) $\frac{du}{dv} = -5v^3 + 3v^{-2}$.

4. Find and check the antiderivative of each of the following.

(a) $\frac{ds}{dt} = 2 \sin 2t$. (b) $\frac{dv}{dt} = -3 \sin 6t$.

(c) $\frac{dy}{dx} = 5 \cos 3x$. (d) $\frac{da}{dt} = -\frac{1}{2} \cos 4t$.

(e) $\frac{dQ}{dt} = 110 \sin 60t$. (f) $\frac{dE}{dt} = \frac{3}{2} \cos \left(\frac{2}{3}t\right)$.

(g) $\frac{dy}{dx} = 3 \cos 2x - 5 \sin 2x$. (h) $\frac{ds}{dt} = \sin \left(\frac{x}{2}\right) - \cos \left(\frac{x}{2}\right)$.

Group B

1. Find and check the antiderivative of each of the following.

(a) $x^2 + 3x + 7$. (b) $\frac{1}{5}x^3 + \frac{1}{9}x^2$.

(c) $\frac{1}{8}x^{9/10} + 2$. (d) $x^{100} + x^{-1/2}$.

(e) a_0x^n. (f) $x^{1/16} + x^{-1/32}$.

(g) $a_0x^n + a_1x^{n-1} + \cdots + a_n$.

2. Find and check the antiderivative of each of the following.

(a) $4x^{16} + 3x^{15} + 6x^9 + 2x + 5$. (b) $\frac{1}{x^{1/15}} + x^{1/30}$.

(c) $4x^{100} + 3x^{75} + 2x^{50} + x^{25} + 1$. (d) $\frac{1}{x^{25}} + 2x^3$.

(e) $\frac{1}{x^4} + 6x^5$. (f) $\frac{2}{3x^{3/2}} + \frac{2x^{3/2}}{3}$.

(g) $\frac{1}{x^4} + x^4$. (h) $\frac{1}{x^5} + x^5$.

Group C

1. Prove Theorem 4.3.1.

2. Show what happens when formula 4.3.1 is used in finding the antiderivative of $dy/dx = 1/x$. Can you conclude from your answer that $1/x$ has no antiderivative?

4.4 SOME SIMPLE DIFFERENTIAL EQUATIONS

An equation containing differentials or derivatives is called a *differential equation*. The process of finding antiderivatives is one method of solving one type of differential equation. This process yields the *general solution*. A solution is an expression free of derivatives or differentials which makes the equation true.

EXAMPLE 1. Find the general solution of the differential equation $dy/dx = x^2$.

Solution. For $dy/dx = x^2$, the general solution is $y = \frac{1}{3}x^3 + C$.

PROBLEM 1. Find the general solution of the differential equation $dQ/dt = 4t^3 - t$.

In practical problems a value for the constant C is determined by specifying a value of y for some value of x.

EXAMPLE 2. Find the equation of the curve whose slope is three times the square of the abscissa at any point and which passes through the point (2, 4).

Solution. We must first write an equation representing the statement of the problem. Since dy/dx is the slope of a curve and x is the abscissa at any point, we have

$$\frac{dy}{dx} = 3x^2 \tag{4.4.1}$$

Solving Eq. 4.4.1 by taking the antiderivative, we have

$$y = x^3 + C \tag{4.4.2}$$

Equation 4.4.2 represents a *family* of curves all satisfying Eq. 4.4.1 (see Fig. 4.4.1).

To complete the problem we want the member of the family that passes through the point (2, 4). In Eq. 4.4.2 we substitute $x = 2$ and $y = 4$ to find the value of C. Thus

$$4 = (2)^3 + C \qquad \text{and} \qquad C = -4$$

Hence the required equation is

$$y = x^3 - 4$$

PROBLEM 2. Find the equation of the curve whose slope is two times the cube of the abscissa at any point and which passes through the point $(-2, 10)$.

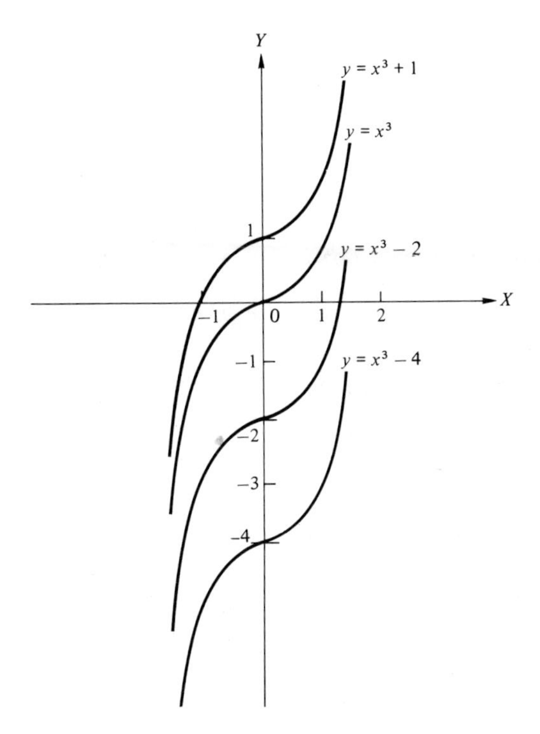

Figure 4.4.1 Family of curves $y = x^3 + C$.

EXAMPLE 3. A body falls from rest with a constant acceleration of g ft/s². Find its velocity and its displacement at any subsequent time t.

Solution. Since $a = dv/dt$, we have

$$\frac{dv}{dt} = g \qquad \text{and} \qquad v = gt + C_1 \tag{4.4.3}$$

Since $v = ds/dt$, we have

$$\frac{ds}{dt} = gt + C_1$$

and

$$s = \frac{g}{2}t^2 + C_1 t + C_2 \tag{4.4.4}$$

To find the arbitrary constants C_1 and C_2, we need some conditions. Since the body falls from rest, we have, at $t = 0$, $v = 0$ ft/s. If we measure its displacement from its position at $t = 0$, we have $t = 0$ and $s = 0$. To find C_1 we use $t = 0$ and $v = 0$ in Eq. 4.4.3. Thus

$$0 = g(0) + C_1 \qquad \text{and} \qquad C_1 = 0$$

Therefore,

$$v = gt \tag{4.4.5}$$

To find C_2 we use $t = 0$ and $s = 0$ in Eq. 4.4.4. Thus

$$0 = \frac{g}{2}(0)^2 + 0 + C_2 \qquad \text{and} \qquad C_2 = 0$$

Therefore,

$$s = \tfrac{1}{2}gt^2 \tag{4.4.6}$$

PROBLEM 3. A body is thrown downward with an initial velocity of 5 ft/s. It has a constant acceleration of g ft/s². If its displacement is $s = 0$ when $t = 0$ and the positive displacement is down, find formulas for its velocity v, and displacement s at any subsequent time t. If $g = 32$ ft/s², what is its velocity and how far has it dropped at the end of 3 s?

EXERCISES 4.4

Group A

1. Find the general solution of each of the following differential equations.

(a) $\dfrac{dy}{dx} = x^3$. (b) $\dfrac{dQ}{dt} = t^5 + 2t^4 + t$.

(c) $\dfrac{ds}{dt} = \dfrac{1}{5}t^4 + \dfrac{1}{4}t^{1/5} + 2$. (d) $\dfrac{dy}{dx} = x^{-4} + x^{-2} + x + x^{1/4} + 9$.

2. Find the solution of each of the following differential equations with the given conditions.

(a) $\dfrac{dy}{dx} = 4x^3$; $y = 16$ when $x = 0$. (b) $\dfrac{ds}{dt} = 4t^3$; $s = 16$ when $t = 2$.

(c) $\dfrac{dy}{dx} = \dfrac{x^{-2/3}}{3}$; $y = 2$ when $x = 8$. (d) $\dfrac{dE}{dt} - 4t^5 = 0$; $E = \frac{2}{3}$ when $t = 1$.

(e) $\dfrac{dy}{dx} - 3x^{-4/5} = 0$; $y = 6$ when $x = 32$.

3. If you have studied Sec. 2.10, find the solution of each of the following differential equations with the given condition.

(a) $\dfrac{dy}{dx} = 2\cos x$; $y = 3$, $x = \pi/2$. (b) $\dfrac{dE}{dt} = 110 \sin 60t$; $E = 0$, $t = \pi/120$.

(c) $\dfrac{dQ}{dt} = 3 \sin 2t$; $Q = 0$, $t = 0$. (d) $\dfrac{dx}{dt} = 5 \cos 3t$; $x = 9$, $t = \pi$.

Group B

1. Find the equation of the curve whose slope is twice the abscissa at any point and which passes through the point $(2, -1)$.

2. A ball is thrown up from the ground with an initial velocity of 70 ft/s. Find formulas for its velocity and its height at any later time t. When will it hit the ground?

3. An automobile is traveling at 60 mi/h and the driver slows the car at a constant rate of 22 ft/s^2. How far will the car travel before it stops?

4. From what height must an object be dropped if it is to reach the ground in 3 s?

5. The rate of change of resistance per unit temperature is given by the formula $dR/dT = \frac{1}{300}$. If $R = 1$ when $T = 0$ and $T = 150P$, where P is the power in watts, find the resistance when P is 5 W.

6. Let EI be the flexural rigidity of a beam which we assume to be a constant and suppose that

$$EI\frac{d^2y}{dx^2} = \frac{wx^2}{2} - \frac{wLx}{2}$$

where L is the length of the beam and w is its weight. Find the equation for y, the deflection of the beam, if $y = 0$ when $x = 0$ and $x = L$.

7. If power is expressed by the formula $dW/dt = 3t + 4$, find the work done at any time t if $W = 7$ when $t = 1$.

8. The emf in an electric circuit is given by $d\phi/dt = 5t^2 - 6t$. Find the magnetic flux ϕ if $\phi = 20$ when $t = 2$.

9. The current in an electric circuit is given by $dQ/dt = t^2 + 2t + 1$. Find the charge Q on the capacitor in the circuit if $Q = 5$ when $t = 0$.

10. Find s when $a = 3t^2 - t$ with $v = 1$, $s = 0$ at $t = 0$.

11. In an electric circuit it was found that $L(d^2Q/dt^2) = 12t - 5$. Find Q at $t = 4$ if $L = 2$, and at $t = 0$, $dQ/dt = 3$, $Q = 0$.

Group C

1. Sketch several members of the family of curves that are solutions to the equation $dy/dx = x$.

2. Sketch several members of the family of curves that are solutions to the equation $dy/dx = 5x^4$.

3. Sketch several members of the family of curves that are solutions to the equation $dy/dx = x^2 + 5x^4$.

4. (a) Sketch several members of the family of curves that are solutions to the equation $dy/dx = 2x + 3x^3$. (b) What member of the family passes through the point (0, 1)?

5. (a) Sketch several members of the family of curves that are solutions to the equation $dx/dy = -2y$. (b) What member of the family passes through (1, 0)?

4.5 AREAS BY ANTIDERIVATIVES

The important problem of finding the area bounded by the curve $y = f(x)$, the X-axis, and two different ordinates may be done by solving a rather simple differential equation.

For the following discussion refer to Fig. 4.5.1. For now we impose the restrictions $f(x) \geqq 0$ and $a \geqq 0$. However, the end result may be used without these restrictions. Let it be required to find the area bounded by the curve $y = f(x)$, the X-axis, and the ordinates at $x = a$ and $x = b$: that is, the area $EBCDE$. Let A be the area of the region bounded above by the curve $y = f(x)$, below by the X-axis, to the left by the ordinate ED at $x = a$, and on the right by the ordinate $MP = l$ at an arbitrary point x such that $a < x < b$. As x changes by an amount Δx, A will change by an amount ΔA. Thus A is a function of x. Let this function be denoted by $F(x)$; it is this function that we shall determine. To do this note that

$$\Delta A = \text{area } MNQPM$$

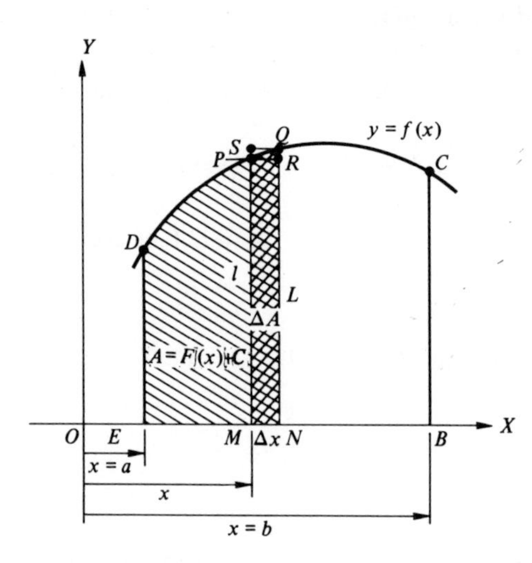

Figure 4.5.1 Area by antiderivative.

Let $MN = \Delta x$, let $l = MP$ be the shortest ordinate of the curve between MP and NQ, and let $L = NQ$ be the longest ordinate of the curve between MP and NQ. Then

$$l\,\Delta x < \Delta A < L\,\Delta x \tag{4.5.1}$$

where $l\,\Delta x$ is the area of the rectangle $MPRNM$ and $L\,\Delta x$ is the area of the rectangle $MSQNM$. Divide Eq. 4.5.1 by Δx to obtain

$$l < \frac{\Delta A}{\Delta x} < L \tag{4.5.2}$$

As $\Delta x \longrightarrow 0$, NQ approaches MP, or $L \longrightarrow l$, $\Delta A/\Delta x$ becomes dA/dx, and since $l = f(x)$, we have in the limit the differential equation

$$\frac{dA}{dx} = f(x) \tag{4.5.3}$$

We may solve Eq. 4.5.3 by taking the antiderivative and letting $F(x)$ be any function whose derivative is $f(x)$. Thus

$$A = F(x) + C \tag{4.5.4}$$

We can find C by noting that at $x = a$, $A = 0$. Substituting in Eq. 4.5.4, we have

$$0 = F(a) + C \qquad \text{and} \qquad C = -F(a)$$

Thus Eq. 4.5.4 becomes

$$A = F(x) - F(a) \tag{4.5.5}$$

Since we want the area to be bounded on the right by $x = b$, we place $x = b$ in Eq. 4.5.5 and obtain

$$A = F(b) - F(a) \tag{4.5.6}$$

as a formula for the required area.

We will see Eq. 4.5.6 again in Sec. 5.3 as Eq. 5.3.4 when we find areas by a summation process. It will also occur in the statement of the fundamental theorem of calculus in Sec. 5.4.

In doing problems we start with Eq. 4.5.3 rather than Eq. 4.5.6.

EXAMPLE 1. Find the area bounded by the curve $y = x^3$, the X-axis, and the ordinates $x = 1$ and $x = 2$.

Solution. Sketch the curve and the boundaries of the required area (see Fig. 4.5.2. Use Eq. 4.5.3, where $f(x) = x^3$; that is,

$$\frac{dA}{dx} = x^3$$

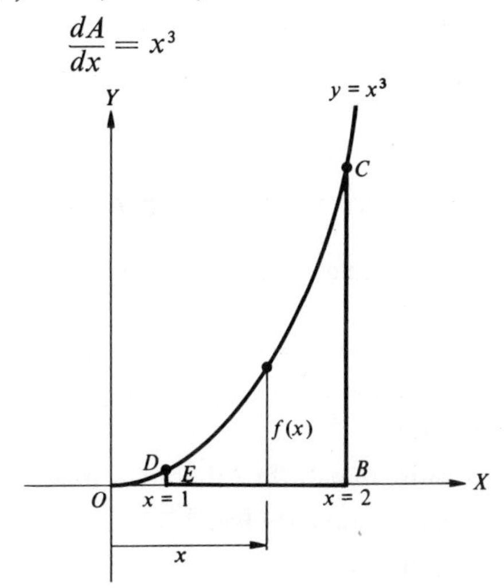

Figure 4.5.2 Area under $y = x^3$.

Finding the antiderivative, we have

$$A = \tfrac{1}{4}x^4 + C$$

When $x = 1$, $A = 0$; hence

$$0 = \tfrac{1}{4}(1)^4 + C \qquad \text{and} \qquad C = -\tfrac{1}{4}$$

Thus $A = \frac{1}{4}x^4 - \frac{1}{4}$, and when $x = 2$,

$$A = \tfrac{1}{4}(2)^4 - \tfrac{1}{4} = 3\tfrac{3}{4} = 3.75$$

which is the required area.

PROBLEM 1. Find the area bounded by the curve $y = 3x^2 + 2$, the X-axis, and the ordinates $x = 1$ and $x = 3$.

EXERCISES 4.5

Group A

1. In each of the following, find the area bounded above by the given curve, below by the X-axis, and the given vertical boundaries. Sketch the area to be found.

(a) $y = 2x + 1$; $x = 0$, $x = 2$.
(b) $y = x^2$; $x = 0$, $x = 5$.
(c) $y = x^4 + 2$; $x = -1$, $x = 1$.
(d) $y = x^{1/2}$; $x = 0$, $x = 1$.
(e) $y = -x^2 - x + 6$; $x = -3$, $x = 2$.

2. In each of the following, find the area bounded below by the given curve, above by the X-axis, and the given vertical boundaries.

(a) $y = -x^2$; $x = 0$, $x = 2$.
(b) $y = x^3$; $x = -1$, $x = 0$.
(c) $y = x^2 - 5x + 6$; $x = 2$, $x = 3$.
(d) $y = -3x + 2$; $x = \frac{2}{3}$, $x = 5$.
(e) $y = x^3 + 5$; $x = -3$, $x = -2$.

3. Find the area bounded by $y = \sin x$, the X-axis, $x = 0$, and $x = \pi$ (see Sec. 2.10).

4. Find the area bounded by $y = \cos x$, the X-axis, $x = 0$, and $x = \pi/2$ (see Sec. 2.10).

5. Explain why Eq. 4.5.3 gives an answer of zero when used to find the area bounded by the curve $y = x^3$, the X-axis, $x = -1$, and $x = 1$.

SUMMARY OF IMPORTANT WORDS AND CONCEPTS

For each of the following, state in your own words your understanding of the statement or word.

1. (a) Differential of independent variable. (b) Differential of dependent variable.

2. (a) Actual change. (b) Approximate change in dependent variable. (c) Relative error. (d) Percentage error.

3. (a) Antiderivative of ax^n. (b) Arbitrary constant.

4. (a) Differential equation. (b) Family of curves.

5. Area by antiderivative.

Chapter 5

The Integral

5.1 INTRODUCTION

Calculus is based on two fundamental concepts, that of the derivative and that of the integral. We have defined the derivative and have seen that it was the limit of a rate of change and that it had many interpretations, such as velocity, acceleration, current, slope, and so forth. Now we shall study the concept of the integral. We shall see that the integral tells us of the cumulative effect of change. It is defined in terms of a limit of sums as they are determined by the function being analyzed. For example, one of the simplest interpretations is that of finding the area of a region bounded by the graph of a positive function, two vertical lines, and the X-axis. After we have defined the integral and find out how its value is determined, we shall give it different interpretations, all of which will be related to the cumulative effect of change.

In Sec. 4.5 we found the area of a plane figure by use of the antiderivative in solving a differential equation. In Secs. 5.2 and 5.3 the same general area is found by other methods and a comparison is made which indicates that antiderivatives and integrals are related.

5.2 TRAPEZOIDAL RULE

Let it be required to find the area bounded above by the curve $y = f(x)$, below by the X-axis, on the left by $x = a$, and on the right by $x = b$ (see Fig. 5.2.1). Let the interval $AB = b - a$ be divided into n equal parts by the points x_0, x_1, $x_2, \ldots, x_n$. where $x_0 = a$, $x_1 = x_0 + \Delta x$, $x_2 = x_1 + \Delta x, \ldots,$ and $x_n = b$. At

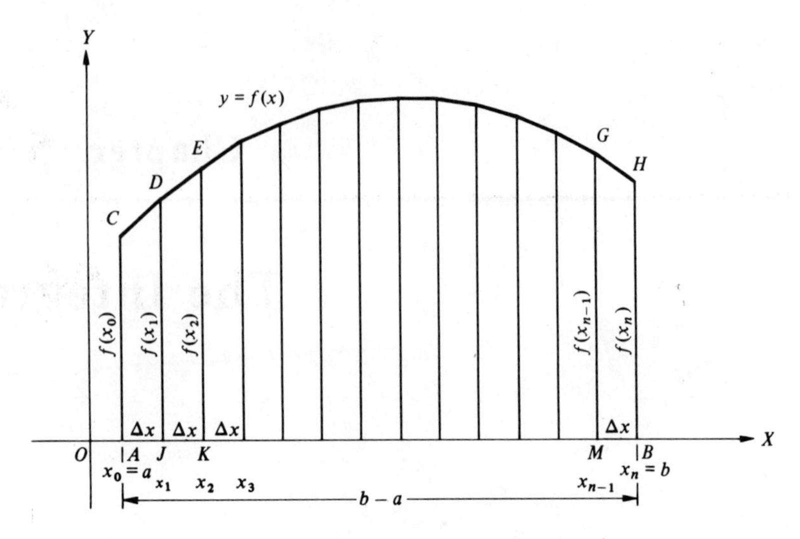

Figure 5.2.1 Trapezoidal rule.

each of these points draw a vertical line to the curve $y = f(x)$. Connect the points where these lines intersect the curve by straight lines such as CD, DE, . . . , GH, thus forming trapezoids. If n is large, it seems reasonable that the sum of the areas of these trapezoids will closely approximate the area under the curve. From geometry we know that the area of a trapezoid is one-half its height times the sum of its bases. The height of each trapezoid in Fig. 5.2.1 is

$$\Delta x = \frac{b - a}{n}$$

The bases of the trapezoids are the ordinates

$$f(x_0), f(x_1), f(x_2), \ldots, f(x_{n-1}), f(x_n)$$

We now find the area of each of the trapezoids. This is most clearly indicated by the following table:

Trapezoid	Area
$ACDJA$	$\frac{1}{2}(\Delta x)[f(x_0) + f(x_1)]$
$JDEKJ$	$\frac{1}{2}(\Delta x)[f(x_1) + f(x_2)]$
.	.
.	.
.	.
$MGHBM$	$\frac{1}{2}(\Delta x)[f(x_{n-1}) + f(x_n)]$

The sum of the areas of all of these trapezoids is

$$A = \frac{\Delta x}{2}[f(x_0) + 2f(x_1) + 2f(x_2) + \ldots + 2f(x_{n-1}) + f(x_n)] \quad (5.2.1)$$

Equation 5.2.1 is called the *trapezoidal rule*. It is a formula for a numerical method of computing the approximate area under a curve.

EXAMPLE 1. Use the trapezoidal rule to find the approximate area under the curve $y = x^2$, above the X-axis, and between $x = 0$ and $x = 4$. Use eight trapezoids.

Solution. The required area is shown in Fig. 5.2.2. Here $a = 0$, $b = 4$, and $n = 8$, so

$$\Delta x = \frac{b - a}{n} = \frac{4 - 0}{8} = 0.5$$

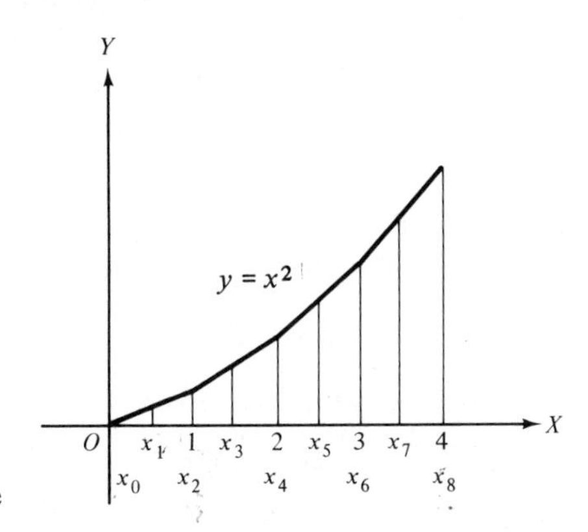

Figure 5.2.2 Area by trapezoidal rule

To use Eq. 5.2.1 it is best to exhibit the work in tabular form; thus

$x_0 = 0$	$f(x_0) = (0)^2 = 0$	$f(x_0) = 0.0$
$x_1 = 0.5$	$f(x_1) = (0.5)^2 = 0.25$	$2f(x_1) = 0.5$
$x_2 = 1.0$	$f(x_2) = (1.0)^2 = 1.0$	$2f(x_2) = 2.0$
$x_3 = 1.5$	$f(x_3) = (1.5)^2 = 2.25$	$2f(x_3) = 4.5$
$x_4 = 2.0$	$f(x_4) = (2.0)^2 = 4.0$	$2f(x_4) = 8.0$
$x_5 = 2.5$	$f(x_5) = (2.5)^2 = 6.25$	$2f(x_5) = 12.5$
$x_6 = 3.0$	$f(x_6) = (3.0)^2 = 9.0$	$2f(x_6) = 18.0$
$x_7 = 3.5$	$f(x_7) = (3.5)^2 = 12.25$	$2f(x_7) = 24.5$
$x_8 = 4.0$	$f(x_8) = (4.0)^2 = 16.0$	$f(x_8) = 16.0$
		Sum $= 86.0$

Now using Eq. 5.2.1, we find that

$$A = \frac{0.5}{2}[86] = 21.5$$

(In this example we chose $n = 8$ for convenience in obtaining a nice value for Δx.)

PROBLEM 1. Use the trapezoidal rule to find the approximate area under the curve $y = x^2$, above the X-axis, and between $x = -1$ and $x = 5$. Use six trapezoids.

EXERCISES 5.2

Group A

1. Use the trapezoidal rule to approximate the area bounded above by the given curve, below by the X-axis, and by the vertical lines at the given values of x. Use n trapezoids.

(a) $y = x^2$, $x = 0$, $x = 5$, $n = 10$.
(b) $y = 4x - x^2$, $x = 0$, $x = 4$, $n = 4$.
(c) $y = x$, $x = 0$, $x = 1$, $n = 5$.
(d) $y = x^3 + 2$, $x = 1$, $x = 3$, $n = 6$.
(e) $y = x^4$, $x = -1$, $x = 1$, $n = 10$.

2. Use antidifferentiation to compute the area of each of the regions described in Exercise 1 and compare with the results found by using the trapezoidal rule.

Group B

1. In each of the following, use the trapezoidal rule to approximate the area bounded by the given curves and the X-axis.

(a) $y = \dfrac{1}{x}$, $x = 1$, $x = 3$, $n = 5$.

(b) $y = \dfrac{1}{1 + x^2}$, $x = -1$, $x = 0$, $n = 5$.

(c) $y = x\sqrt{1 + x^2}$, $x = 2$, $x = 4$, $n = 4$.

(d) $y = \dfrac{1}{x^4}$, $x = -1$, $x = 1$, $n = 10$.

(e) $y = \sqrt{1 - x}$, $x = 0$, $x = 1$, $n = 3$.

(f) $y = \sin x$, $x = 0$, $x = \pi$, $n = 6$.

(g) $y = \cos x$, $x = 0$, $x = \pi/2$, $n = 4$.

Group C

1. When does the trapezoidal rule, Eq. 5.2.1, give an exact area?

2. How are the trapezoidal rule and antidifferentiation alike and how are they different in evaluating areas?

3. Why does a larger n in the trapezoidal rule give a better approximation to the actual area?

4. Explain how one could use the trapezoidal rule to compute an area above the X-axis bounded by a curve drawn using empirical data points.

5. Could trapezoids be used to approximate an area if the height (Δx) of each trapezoid is not the same? Explain.

6. The table of data gives the "specific heat" s of water at θ°C. Use the trapezoidal rule to approximate the total heat, area under the s–θ curve, required to raise the temperature of 1 g of water from 10°C to 50°C.

θ	10°C	20°C	30°C	40°C	50°C
s	1.0097	1.0062	1.0044	1.0037	1.0041

7. The table of data gives the "specific heat" s of mercury at θ°C. Use the trapezoidal rule to approximate the total heat, area under the s–θ curve, required to raise the temperature of 1 g of mercury from 0°C to 80°C.

θ	0°C	20°C	40°C	60°C	80°C
s	0.1402	0.1394	0.1385	0.1377	0.1373

8. When a ferromagnetic material is taken through a cycle of magnetization, the energy dissipated as heat per cm³ is $1/4\pi$ times the area under the H–B curve, where H is the magnetizing force and B is the magnetic inductance. Use trapezoids to approximate the total energy dissipated for the following data:

B	0	13.0	14.0	15.4	16.3	17.2
H	0	8	10	15	20	30

(*Caution*: The intervals are not uniform.)

5.3 AREA BY SUMMATION

Let us again consider finding the area under a continuous curve which is above the X-axis. But this time we shall use rectangles instead of trapezoids as elements of area. Let it be required to find the area bounded above by $y = f(x)$, below by the X-axis, on the left by $x = a$, and on the right by $x = b$ (see Fig. 5.3.1). For convenience let $a < b$ and let $f(x)$ be positive for all x between a and b.

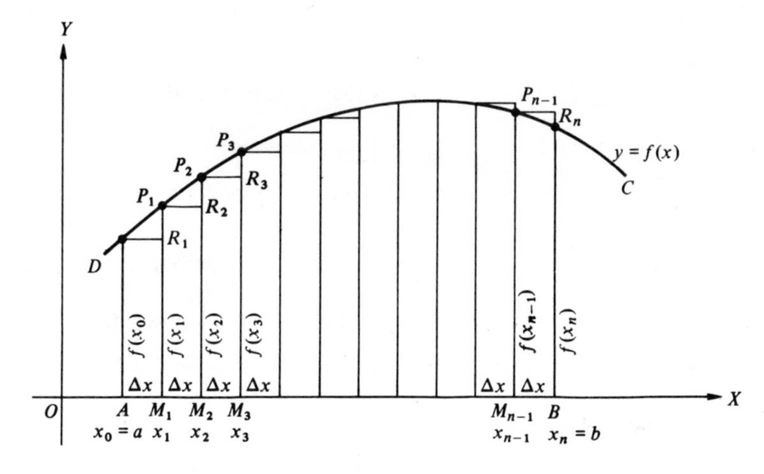

Figure 5.3.1

Let the interval $AB = b - a$ be divided into n equal parts, Δx, by the points $x_0, x_1, x_2, \ldots, x_{n-1}, x_n$, where $x_0 = a$, $x_1 = x_0 + \Delta x$, $x_2 = x_1 + \Delta x, \ldots$, and $x_n = b$. At each of these points erect a vertical line to or above the curve $y = f(x)$. Now construct rectangles by drawing lines $DR_1, P_1R_2, P_2R_3, \ldots, P_{n-1}R_n$ parallel to the X-axis. Note that these lines are drawn from the left ordinate and may intersect the vertical line at the right below or above the curve. Thus some of the rectangles so formed are within the required area and some are outside. We now obtain the area of each of these rectangles:

Rectangle	Area
ADR_1M_1A	$f(x_0)\,\Delta x$
$M_1P_1R_2M_2M_1$	$f(x_1)\,\Delta x$
$M_2P_2R_2M_3M_2$	$f(x_2)\,\Delta x$
$\vdots$	$\vdots$
$M_{n-1}P_{n-1}R_nBM_{n-1}$	$f(x_{n-1})\,\Delta x$

The sum of the areas of these rectangles is

$$f(x_0)\,\Delta x + f(x_1)\,\Delta x + f(x_2)\,\Delta x + \ldots + f(x_{n-1})\,\Delta x \tag{5.3.1}$$

This sum approximates the required area. As the number of rectangles is increased, that is, as n increases without bound, the sum expressed by Eq. 5.3.1 will give the exact area under the curve.

We need a symbol to express the sum 5.3.1 concisely. Such a symbol is the Greek capital letter sigma, denoted $\sum$. The notation

$$\sum_{i=1}^{i=5} i^2$$

means let i take on all integers from 1 to 5 inclusive, square them, and then add the squares. Thus

$$\sum_{i=1}^{i=5} i^2 = (1)^2 + (2)^2 + (3)^2 + (4)^2 + (5)^2$$

Using the sigma notation, Eq. 5.3.1 becomes

$$\sum_{i=0}^{i=n-1} f(x_i)\,\Delta x \tag{5.3.2}$$

The limit of this sum as $n \longrightarrow \infty$ is expressed by the symbol $\int_a^b f(x)\,dx$, where $\Delta x = dx$ and $\int$ is called the *integral sign*. The symbol $\int_a^b f(x)\,dx$ is read "the integral from a to b of $f(x)\,dx$." We now make Definition 5.3.1.

Definition 5.3.1

$$\lim_{n\to\infty} \sum_{i=0}^{i=n-1} f(x_i)\,\Delta x = \int_a^b f(x)\,dx$$

(It can be shown that the continuity of the curve guarantees the existence of this limit.)

We also define that the required area A is

$$A = \int_a^b f(x)\,dx \tag{5.3.3}$$

Our next question is: Can we easily determine the value of $\int_a^b f(x)\,dx$? In Sec. 4.5 we found that the area under the curve was given by $A = F(b) - F(a)$, Eq. 4.5.6, where $F(x)$ is an antiderivative of $f(x)$; that is, $F'(x) = f(x)$. The expression $F(b) - F(a)$ is usually denoted by $[F(x)]_a^b$. Thus we have from Eq. 5.3.3,

$$A = \int_a^b f(x)\,dx = [F(x)]_a^b = F(b) - F(a) \tag{5.3.4}$$

[Formula 5.3.4 is an easy solution to our problem provided that $F(x)$ is easily found.]

From Eq. 4.5.5, where A is expressed as a function of x, we may obtain

$$dA = f(x)\,dx$$

since $F'(x) = f(x)$ and $F'(a) = 0$ because $F(a)$ is a constant. Hence the expression $f(x)\,dx$ which appears in Eq. 5.3.4 is called an *element of area* and we see that to find the total area all we have to do is add a lot of little areas. The simplest little area is a rectangle. We may also write $dA = y\,dx$.

EXAMPLE 1. Find the area bounded by the curve $y = x^2$, above the X-axis, and between $x = 0$ and $x = 4$.

Solution. Sketch the required area (see Fig. 5.3.2). At any point x between 0 and 4, draw a vertical line to the curve and construct a *rectangle* of width dx. Then the little bit of area dA is given by

$$dA = y\,dx = x^2\,dx$$

and using Eq. 5.3.3, where $a = 0$ and $b = 4$, we have

$$A = \int_0^4 x^2\,dx$$

Taking an antiderivative on the right-hand side and using Eq. 5.3.4 we find that

$$A = \left[\frac{x^3}{3}\right]_0^4 = \frac{(4)^3}{3} - \frac{(0)^3}{3} = \frac{64}{3} = 21\frac{1}{3} = 21.3$$

(Compare this value with the answer to Example 1 of Sec. 5.2.)

The arbitrary constant C for the antiderivative is not included here since in Sec. 4.5 it was found to be equal to $F(a)$. We will also see in Sec. 5.4 that it cancels out in finding the value of a definite integral.

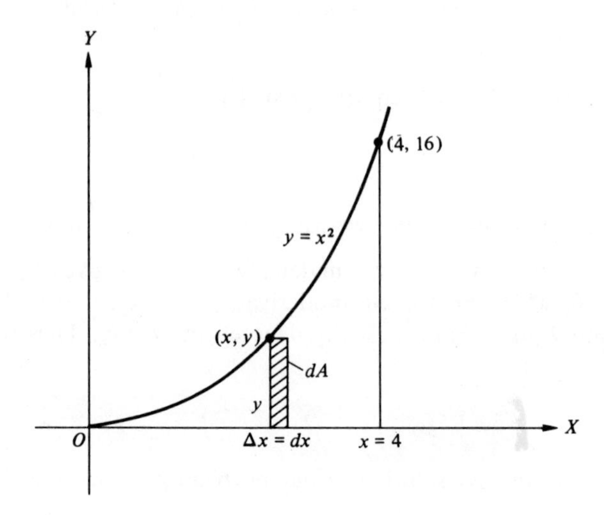

Figure 5.3.2 Area by integration.

PROBLEM 1. Find the area bounded by the curve $y = x^2$, above the X-axis, and between $x = -1$ and $x = 5$. Compare the answer with that of Prob. 1 of Sec. 5.2.

In applying Eq. 5.3.4 it is best to make dx positive. This may be done by taking a as the smaller of the two quantities a and b. If the curve $y = f(x)$ lies above the X-axis, then $f(x_i)\,dx$ is positive throughout the entire region and A is positive. If the curve lies entirely below the X-axis, then $f(x_i)\,dx$ is negative and A is negative. If the curve lies partly above and partly below the X-axis, then Eq. 5.3.4 gives the algebraic sum of the areas. In this case to find the entire area the positive and negative regions must be found separately and their absolute values must be added.

EXAMPLE 2. Find the area bounded by the curve $f(x) = x^3 - x^2 - 2x$ and the X-axis.

Solution. Sketch the curve (see Fig. 5.3.3). Setting $f(x) = 0$, we have $x^3 - x^2 - 2x = x(x^2 - x - 2) = x(x - 2)(x + 1) = 0$ and we find that the curve crosses the X-axis at $x = -1$, $x = 0$, and $x = 2$.

Since part of the curve is above the X-axis and part below, we find the areas of the positive and negative regions separately.

For region A_1 at any point x between -1 and 0, construct a rectangle of width dx. Then

$$dA_1 = y\,dx = (x^3 - x^2 - 2x)\,dx$$

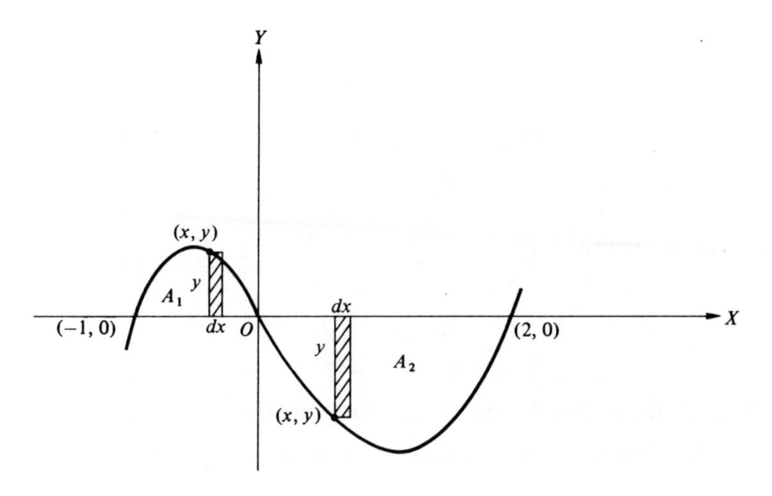

Figure 5.3.3 Area of positive and negative regions.

and

$$A_1 = \int_{-1}^{0} (x^3 - x^2 - 2x)\,dx$$
$$= \left[\frac{x^4}{4} - \frac{x^3}{3} - x^2\right]_{-1}^{0}$$
$$= \left[\frac{(0)^4}{4} - \frac{(0)^3}{3} - (0)^2\right] - \left[\frac{(-1)^4}{4} - \frac{(-1)^3}{3} - (-1)^2\right]$$
$$= 0 - [\tfrac{1}{4} + \tfrac{1}{3} - 1] = \tfrac{5}{12}$$

For region A_2 at any point x between $x = 0$ and $x = 2$, construct a rectangle of width dx. Here y has a negative value so we expect A_2 to be negative.

$$dA_2 = y\,dx = (x^3 - x^2 - 2x)\,dx$$

and

$$A_2 = \int_{0}^{2} (x^3 - x^2 - 2x)\,dx$$
$$= \left[\frac{x^4}{4} - \frac{x^3}{3} - x^2\right]_{0}^{2}$$
$$= \left[\frac{(2)^4}{4} - \frac{(2)^3}{3} - (2)^2\right] - \left[\frac{(0)^4}{4} - \frac{(0)^2}{3} - (0)^2\right]$$
$$= 4 - \tfrac{8}{3} - 4 = -\tfrac{8}{3}$$

The required area A is given by

$$A = |A_1| + |A_2|$$
$$= |\tfrac{5}{12}| + |-\tfrac{8}{3}|$$
$$= \tfrac{5}{12} + \tfrac{8}{3} = \tfrac{37}{12} = 3.08$$

If we had blindly written

$$A = \int_{-1}^{2} (x^3 - x^2 - 2x)\, dx$$

(not considering the positive and negative regions), we would find that

$$\begin{aligned} A &= \left[\frac{x^4}{4} - \frac{x^3}{3} - x^2\right]_{-1}^{2} \\ &= \left[\frac{(2)^4}{4} - \frac{(2)^3}{3} - (2)^2\right] - \left[\frac{(-1)^4}{4} - \frac{(-1)^3}{3} - (-1)^2\right] \\ &= (4 - \tfrac{8}{3} - 4) - (\tfrac{1}{4} + \tfrac{1}{3} - 1) \\ &= -\tfrac{8}{3} + \tfrac{5}{12} = -\tfrac{27}{12} = -2.25 \end{aligned}$$

which is not the required area. (What does this number represent?)

PROBLEM 2. Find the area bounded by the curve $f(x) = -x^3 + x^2 + 6x$ and the X-axis.

EXERCISES 5.3

Group A

1. In each of the following, use Eq. 5.3.4 to find the area bounded by the curve, the X-axis, and the vertical lines at the given values of x. Sketch the required area.

(a) $y = x^2$, $x = 0$, $x = 2$.
(b) $y = x^3$, $x = -1$, $x = 1$.
(c) $y = x^2 + 1$, $x = -1$, $x = 2$.
(d) $y = \sqrt{x}$, $x = 0$, $x = 4$.
(e) $y = x^3 + 1$, $x = -1$, $x = 1$.

2. For each part of Exercise 1, find the approximate area bounded by the given curves by using Eq. 5.3.1. Use elements of area obtained from the rectangles whose height is given by the left ordinate of each rectangle and whose base is dx, where $dx = \Delta x = (b - a)/n$. Compare the results with the answers to Exercise 1.

(a) Use $n = 4$.
(b) Use $n = 4$.
(c) Use $n = 6$.
(d) Use $n = 4$.
(e) Use $n = 5$.

3. In each of the following, find the area bounded by the given curves and the X-axis.

(a) $f(x) = x^3 + x^2 - 2x$.
(b) $f(x) = -x^3 + x$.
(c) $f(x) = x^3 - 3x^2 - x + 3 = (x^2 - 1)(x - 3)$.
(d) $f(x) = (x + 1)(x)(x - 1)(x - 2)$.

4. In each of the following, use Eq. 5.3.4 to find the area bounded by the curve, the X-axis, and the vertical lines at the given values of x. Sketch the required area.

(a) $y = \sin x$, $x = -\pi/2$, $x = \pi$.
(b) $y = \cos x$, $x = 0$, $x = 3\pi/2$.
(c) $y = \sin 2x$, $x = 0$, $x = \pi$.
(d) $y = \cos 2x$, $x = -\pi/2$, $x = \pi$.

Group B

1. In each of the following, find the area between the straight line, the X-axis, and the vertical lines at the given values of x by using Eq. 5.3.4. Sketch each area and check your answer by using elementary geometry.

(a) $y = x + 1$, $x = 0$, $x = 2$.
(b) $y = -x + 1$, $x = -1$, $x = 1$.
(c) $2x + 3y = 9$, $x = -\frac{3}{2}$, $x = \frac{9}{2}$.
(d) $6x + y = 4$, $x = -\frac{1}{6}$, $x = 0$.
(e) $y = -\frac{2}{5}x + 20$, $x = 5$, $x = 20$.

2. In each of the following, find the area bounded by the curve and the X-axis. Sketch each area.

(a) $y = -x^2 + 4$.
(b) $x^2 + 9y = 36$.
(c) $y = \frac{1}{4}(-x^2 + 10x - 9)$.
(d) $x^2 + 6x + 3y = 0$.

3. In each of the following, find the area between the arch above the X-axis and the X-axis. Sketch each area.

(a) $y = -x^3 + 16x$.
(b) $y = x^4 - 17x^2 + 16$.
(c) $y = x^3 + x^2 - 17x + 15$.
(d) $y = -x^4 + 3x^3 + x - 3$.

Group C

1. Compare the approximate areas obtained by summing the areas of the rectangles whose height is the left ordinate and by the trapezoidal rule for $y = x^2$, $x = -1$ to $x = 1$, and $n = 10$. Compare with the exact area.

2. Formulate a method for finding the area between two horizontal lines, the Y-axis, and the curve of a function which has y as the independent variable and x as the dependent variable.

3. Find the area bounded by the Y-axis, the given curve, and horizontal lines.

(a) $x = y^2$, $y = 0$, $y = 2$.
(b) $x = \sqrt{y}$, $y = 0$, $y = 4$.
(c) $x = y - 2$, $y = 3$, $y = 5$.
(d) $y = x^3$, $y = 1$, $y = 2$.

5.4 THE DEFINITE INTEGRAL

In Sec. 5.3 we found the area under a curve by summing an infinite number of elements of area. These elements were rectangles whose height was determined by the left ordinates of each rectangle and whose base was dx. Mathematicians have proved that the height may also be the right-hand ordinate or any ordinate within the rectangle and that the base of each rectangle need not be the same width as long as the widest rectangle approaches zero as the number of rectangles is increased. Then the limit of the sum of elements $f(x)\,dx$ is the same. In this book we consider only the case where the Δx's are equal.

Equation 5.3.4,

$$\int_a^b f(x)\,dx = F(b) - F(a)$$

which we developed in determining the area of a plane region, has far wider applications than simply area. In fact, any problem that requires the limit of the sum of products of the type $f(x)\,dx$ may be solved by Eq. 5.3.4. For example, if $f(x)$ happened to be a function representing an area, then $f(x)\,dx$ might be an element of volume. More about this later.

We now define the definite integral.

Definition 5.4.1 *If $f(x)$ is continuous over the interval $a \leqq x \leqq b$ and $\Delta x = dx = (b - a)/n$, n an integer, then*

$$\lim_{n\to\infty} \sum_{i=0}^{i=n-1} f(x_i)\,\Delta x = \int_a^b f(x)\,dx$$

The symbol $\int_a^b f(x)\,dx$ is called the definite integral. The numbers a and b are called the *lower limit* and the *upper limit*, respectively, of the definite integral. (The use of the word *limit* here is not the same as when a variable approaches a limit.) The quantity $f(x)\,dx$ is called the *element of integration*, $f(x)$ is called the *integrand*, and dx is called the *differential*.

Mathematicians have proved the fundamental theorem of calculus, which relates the definite integral and the antiderivative. It ties together integration and differentiation. We have hinted at it in Sec. 5.3, especially in Eq. 5.3.4. Using it, we can evalute the definite integral. We now state this important theorem, but do not prove it.

The Fundamental Theorem of Calculus *If $f(x)$ is continuous on the closed interval $a \leqq x \leqq b$ and if $F(x)$ is any function whose derivative is $f(x)$ at each x in the interval, then*

$$\int_a^b f(x)\,dx = F(b) - F(a)$$

In Sec. 4.3 we saw that the antiderivative contained an arbitrary constant C. Yet in the fundamental theorem of calculus the C did not appear in the evaluation of $\int_a^b f(x)\,dx$, because if it is added, it will become eliminated in the subtraction process. Thus

$$\int_a^b f(x)\,dx = [F(x) + C]_a^b = [F(b) + C] - [F(a) + C]$$
$$= F(b) - F(a)$$

The steps to follow in evaluating the definite integral $\int_a^b f(x)\,dx$ are:

1. Find $F(x)$, the antiderivative of $f(x)$ (omit the C).
2. Substitute b for x.
3. Substitute a for x.
4. Subtract the result of step 3 from step 2.

EXAMPLE 1. Evaluate $\int_2^5 3x^2\,dx$.

Solution

$$\begin{aligned}\int_2^5 3x^2\,dx &= [x^3]_2^5 \qquad \text{(step 1)}\\ &= \underbrace{[(5)^3]}_{\text{step 2}} - \underbrace{[(2)^3]}_{\text{step 3}}\\ &= 125 - 8\\ &= 117 \qquad \text{(step 4)}\end{aligned}$$

PROBLEM 1. Evaluate $\int_1^3 4x^3\,dx$.

Warning ***Be sure that you subtract the value at the lower limit from the value at the upper limit.***

Because of the close relation between antiderivatives and the definite integral, we use the symbol $\int f(x)\,dx$ to denote the antiderivative of $f(x)$. The symbol $\int f(x)\,dx$ is called an *indefinite integral* (there are no upper or lower limits on the integral sign). Henceforth, we shall probably not use the word *antiderivative* but will denote it as an indefinite integral. Thus the indefinite integral

$$\int x^2\,dx = \frac{x^3}{3} + C$$

while the definite integral

$$\int_{-1}^5 x^2\,dx = \left[\frac{x^3}{3}\right]_{-1}^5 = \frac{1}{3}[(5)^3 - (-1)^3] = 42$$

In general,

$$\boxed{\int f(x)\,dx = F(x) + C}$$

For $f(x) = ax^n$, we have

$$\boxed{\int ax^n\,dx = \frac{ax^{n+1}}{n+1} + C \qquad n \neq -1} \tag{5.4.1}$$

Warning ***For an indefinite integral the arbitrary constant C is needed; for the definite integral it is not.***

When we find the indefinite integral of a function we find a function or whole family of functions. However, the definite integral of a function is a number.

The process of evaluating an integral is called *integration*. Before an expression can be integrated the integrand and differential must be of the *same variable*. For example, if y is not a constant, $\int y\,dx$ cannot be determined. But if we know that $y = x^3$, then $\int y\,dx$ becomes

$$\int x^3\,dx = \frac{x^4}{4} + C$$

PROBLEM 2. Find $\int 3x^5\,dx$.

It should be pointed out that the value of a definite integral does not depend on the letter used to denote the variable in the element of integration. That is,

$$\int_a^b f(x)\,dx = \int_a^b f(u)\,du = \int_a^b f(*)\,d(*)$$

It is well to think of the symbol $\int_a^b f(x)\,dx$ as a summation and of the symbol $\int f(x)\,dx$ as an antiderivative.

EXAMPLE 2. Evaluate $\int_{-2}^{1} (t^2 - t)\,dt$.

Solution

$$\begin{aligned}\int_{-2}^{1} (t^2 - t)\,dt &= [\tfrac{1}{3}t^3 - \tfrac{1}{2}t^2]_{-2}^{1} \\ &= [\tfrac{1}{3}(1)^3 - \tfrac{1}{2}(1)^2] - [\tfrac{1}{3}(-2)^3 - \tfrac{1}{2}(-2)^2] \\ &= (\tfrac{1}{3} - \tfrac{1}{2}) - (-\tfrac{8}{3} - 2)\end{aligned}$$

$$\therefore \quad \int_{-2}^{1} (t^2 - t)\,dt = \tfrac{9}{2} = 4.5$$

PROBLEM 3. Evaluate $\int_{-3}^{2} (t^3 - 3t^2 + 4)\,dt$.

EXERCISES 5.4

Group A

1. Evaluate each of the following definite integrals.

(a) $\int_0^3 x^2\,dx$.

(b) $\int_{-1}^{2} (x^2 + 2x + 1)\,dx$.

(c) $\int_1^4 t^{1/2}\,dt.$ (d) $\int_a^b (4u^3 + u^2 + 3u + 1)\,du.$

(e) $\int_1^8 t^{1/3}\,dt.$ (f) $\int_0^5 (y^4 + 2y^2 + 1)\,dy.$

(g) $\int_{-1}^1 (x^4 + 2)\,dx.$ (h) $\int_{-2}^0 (y^5 + y^3)\,dy.$

(i) $\int_0^{16} (t^{3/2} - t^{1/2})\,dt.$ (j) $\int_{-1/2}^{1/4} (u^2 + 2u + 5)\,du.$

2. Find each of the following indefinite integrals.

(a) $\int x^2\,dx.$ (b) $\int (x^{5/2} + x + 1)\,dx.$

(c) $\int (t^4 + 2t^3 + t + \frac{1}{2})\,dt.$ (d) $\int u^{15/16}\,du.$

(e) $\int (y^{2/3} + y + y^{-1/2})\,dy.$ (f) $\int (x^3 + x^2 + 2x + 1)\,dx.$

(g) $\int (\sqrt{x} + \sqrt{2}x + 1)\,dx.$ (h) $\int (t^5 + 4t - t^{1/2})\,dt.$

(i) $\int (u^4 + u^3 - u^2 - u + 1)\,du.$ (j) $\int (y^{3/2} + y^{-2})\,dy.$

3. Evaluate each of the following definite integrals.

(a) $\int_0^{\pi/2} 3 \sin x\,dx.$ (b) $\int_{\pi/2}^{\pi} \sin 2x\,dx.$

(c) $\int_{-\pi}^0 2 \cos x\,dx.$ (d) $\int_{\pi/2}^{\pi} \cos 2x\,dx.$

4. Find each of the following definite integrals.

(a) $\int 5 \sin 3x\,dx.$ (b) $\int -2 \sin 5x\,dx.$

(c) $\int 2 \cos \frac{x}{2}\,dx.$ (d) $\int -3 \cos 2x\,dx.$

Group B

1. In each of the following, find the approximate area under the curve between the X-axis and the two given values of x. Use as elements of area the rectangles whose height is given by the right ordinate and whose base is dx, where dx is determined by the given n, and sum these elements of area. Sketch the required area in each problem. Compare your results with those of Exercises 1 and 2 of Group A in Exercises 5.3.

(a) $y = x^2, x = 0, x = 2, n = 4.$ (b) $y = x^3, x = -1, x = 1, n = 4.$

(c) $y = x^2 + 1, x = -1, x = 2, n = 6.$ (d) $y = \sqrt{x}, x = 0, x = 4, n = 4.$

(e) $y = x^3 + 1, x = -1, x = 1, n = 5.$

2. Evaluate the following definite integrals.

(a) $\int_0^1 (\sqrt{x} + 2)^2\,dx.$ (b) $\int_{-2}^{-1} \frac{3 - x + x^2}{x^4}\,dx.$

(c) $\int_3^4 \frac{\sqrt{t}-1}{t^2}\,dt.$ (d) $\int_1^2 \frac{u^2-3u+4}{\sqrt{u}}\,du.$

(e) $\int_0^1 t\sqrt{t}\,dt.$ (f) $\int_0^4 u\left(\sqrt{u}+\frac{1}{u}\right)du.$

(g) $\int_{-2}^2 \frac{t^2-4}{t+2}\,dt.$ (h) $\int_{-5}^{-1} x^3\left(x^{-3}+\frac{1}{x^2}\right)dx.$

Group C

1. How are $\int_a^b f(x)\,dx$ and $\int f(x)\,dx$ alike and how are they different?

2. Does the function $F(x) = \int_a^x f(t)\,dt$ have a derivative?

3. Make a list of some physical phenomena that can be interpreted as the cumulative effect of a change process.

5.5 PROPERTIES OF THE DEFINITE INTEGRAL

We now state some properties of definite integrals.

$$\int_a^b f(x)\,dx = -\int_b^a f(x)\,dx \tag{5.5.1}$$

$$\int_a^a f(x)\,dx = 0 \tag{5.5.2}$$

If $a < b < c$,

$$\int_a^c f(x)\,dx = \int_a^b f(x)\,dx + \int_b^c f(x)\,dx \tag{5.5.3}$$

If c is a constant,

$$\int_a^b cf(x)\,dx = c\int_a^b f(x)\,dx \tag{5.5.4}$$

$$\int_a^b [f(x)+g(x)]\,dx = \int_a^b f(x)\,dx + \int_a^b g(x)\,dx \tag{5.5.5}$$

The proof of these properties follow almost directly from the definition of a definite integral.

Property 5.5.1 states that reversing the upper and lower limits changes the sign of the integral. Property 5.5.2 should be clear. Property 5.5.3 may be thought of as the "whole is equal to the sum of its parts." Property 5.5.4 says that a constant can be factored out from under the integral sign. Property 5.5.5 states that an integral of a sum is equal to the sum of the integrals.

EXAMPLE 1. Verify property 5.5.3 for $\int_{-1}^3 2x\,dx = \int_{-1}^0 2x\,dx + \int_0^3 2x\,dx.$

Solution

$$\int_{-1}^{3} 2x\,dx = [x^2]_{-1}^{3} = 9 - (-1)^2 = 8$$

$$\int_{-1}^{0} 2x\,dx = [x^2]_{-1}^{0} = 0 - (-1)^2 = -1$$

$$\int_{0}^{3} 2x\,dx = [x^2]_{0}^{3} = 9 - 0 = 9$$

$$8 = -1 + 9$$

$$= 8$$

PROBLEM 1. Verify property 5.5.2 for $\int_{2}^{2} 3x^2\,dx$.

EXERCISES 5.5

Group A

1. Verify Eq. 5.5.1 using each of the following.

(a) $\int_{0}^{1} x^4\,dx$. (b) $\int_{-1}^{1} x^2\,dx$.

(c) $\int_{0}^{2} x^3\,dx$. (d) $\int_{1}^{4} x^{1/2}\,dx$.

2. Verify Eq. 5.5.3 using each of the following.

(a) $\int_{-1}^{1} x^2\,dx$ with $b = 0$. (b) $\int_{1}^{3} x^3\,dx$ with $b = 2$.

(c) $\int_{-2}^{0} x^2\,dx$ with $b = -1$. (d) $\int_{-3}^{-1} x^4\,dx$ with $b = -2$.

3. Verify Eq. 5.5.4 using each of the following.

(a) $\int_{2}^{3} 4x^3\,dx$. (b) $\int_{-2}^{-1} 3x^2\,dx$.

(c) $\int_{1}^{5} (3x^2 + 6x)\,dx$. (d) $\int_{-3}^{0} (4x^3 - 16)\,dx$.

4. Verify Eq. 5.5.5 using each of the following.

(a) $\int_{0}^{1} (x^{3/2} + 2x)\,dx$. (b) $\int_{-1}^{0} (x^4 + 2x^2)\,dx$.

(c) $\int_{-1}^{2} (x^3 + 2x)\,dx$. (d) $\int_{0}^{8} (x^2 + x^{1/3})\,dx$.

Group B

1. Prove that Eq. 5.5.3 is true. (Make use of Definition 5.4.1.)

2. Prove that Eq. 5.5.5 is true.

SUMMARY OF IMPORTANT WORDS AND CONCEPTS

For each of the following, state in your own words your understanding of the statement or word.

1. Trapezoidal rule.

2. (a) Area by summation. (b) Summation symbol $\sum$. (c) Integral sign. (d) Element of area.

3. (a) Definite integral (four steps for evaluating). (b) Lower limit. (c) Upper limit. (d) Element of integration. (e) Integrand. (f) Fundamental theorem of calculus.

4. Indefinite integral.

5. Integration.

6. Properties of the definite integral (Eqs. 5.5.1 through 5.5.5).

Chapter 6

Some Applications of the Definite Integral

6.1 AREA BETWEEN TWO CURVES

The area between two curves may be found by considering a typical small element of area and summing an infinite number of them. That is, the total area is the sum of a lot of little areas. Consider the area bounded by $y = f(x)$ and $y = g(x)$ as in Fig. 6.1.1. The points of intersection may be found by solving the equations simultaneously. At any x between a and b draw a line from the lower curve to the upper curve parallel to the Y-axis. From the end points of this line, U and L, draw a rectangle of width dx such that dx is positive. The length of the line LU is given by

$$LU = y_u - y_l$$

where y_u is the y of the upper curve and y_l is the y of the lower curve. Then the element of area dA is

$$dA = (y_u - y_l)\, dx$$

or

$$dA = [f(x) - g(x)]\, dx$$

In taking the sum of these elements of area we go in a positive direction; hence we go from $x = a$ to $x = b$. Thus

$$A = \int_a^b [f(x) - g(x)]\, dx \qquad (6.1.1)$$

In doing problems of this type it is best not to work directly from Eq. 6.1.1 but instead to start with the element of area.

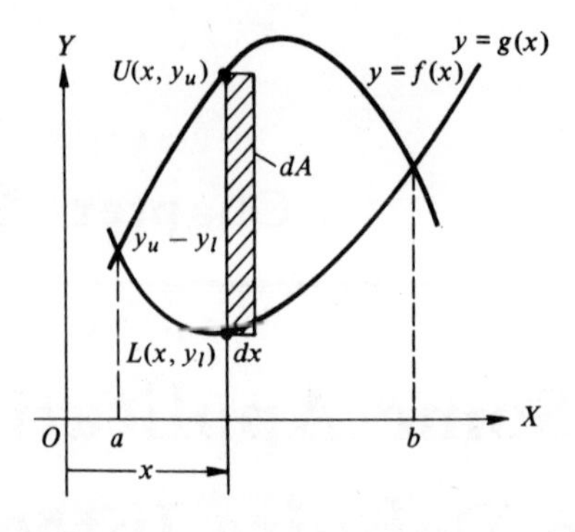

Figure 6.1.1 Area between two curves.

***Warning** The element of area, dA, is "parallel" to the Y-axis and for all $a \leqq x \leqq b$, the top edge of the small rectangle is on y_u and the bottom edge is on y_l.*

EXAMPLE 1. Find the area bounded by $y = x^2 - 4$ and $y = 3x$.

Solution. Sketch the curves and find the points of intersection (see Fig. 6.1.2). Solving $y = x^2 - 4$ and $y = 3x$ simultaneously, we find

$$x^2 - 4 = 3x$$

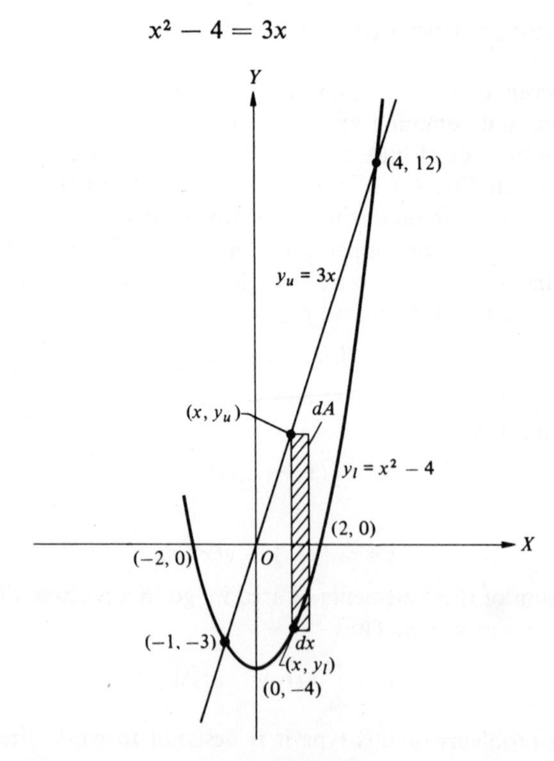

Figure 6.1.2 Area bounded by $y = x^2 - 4$ and $y = 3x$.

or

$$x^2 - 3x - 4 = (x - 4)(x + 1) = 0$$

hence $(-1, -3)$ and $(4, 12)$ are the points of intersection. At any x such that $-1 < x < 4$, draw a line from $y_l = x^2 - 4$ to $y_u = 3x$ and construct a rectangle with positive dx. We see that for the entire region the top edge of the rectangle is on y_u and the lower edge is on y_l. Thus

$$dA = (y_u - y_l)\,dx$$

and

$$dA = [3x - (x^2 - 4)]\,dx$$

or

$$dA = (3x - x^2 + 4)\,dx$$

(Note carefully that dA has its upper end on $y = 3x$ and its lower end on $y = x^2 - 4$ throughout the entire region.)

Since the total area is the sum of all its little areas or $A = \int dA$, and taking the sum in the positive direction of x, we have

$$\begin{aligned} A &= \int_{-1}^{4} (3x - x^2 + 4)\,dx \\ &= \left[\frac{3}{2}x^2 - \frac{x^3}{3} + 4x\right]_{-1}^{4} \\ &= [\tfrac{3}{2}(4)^2 - \tfrac{1}{3}(4)^3 + 4(4)] - [\tfrac{3}{2}(-1)^2 - \tfrac{1}{3}(-1)^3 + 4(-1)] \\ \therefore\quad A &= \tfrac{125}{6} = 20.8 \text{ square units} \end{aligned}$$

PROBLEM 1. Find the area bounded by $y = x^2$ and $x + y - 2 = 0$.

When the roles of x and y are interchanged, we rewrite formula 6.1.1 as

$$A = \int_{c}^{d} [f(y) - g(y)]\,dy \tag{6.1.2}$$

Formula 6.1.2 gives the area bounded by the curves $x_R = f(y)$ on the right and $x_L = g(y)$ on the left, and the lines $y = c$ and $y = d$. To use formula 6.1.2 the element of area dA is drawn parallel to the X-axis with its right edge always on $x_R = f(y)$ and its left edge always on $x_L = g(y)$ throughout the entire region. For a positive area $c \leqq y \leqq d$, and $dA = (x_R - x_L)\,dy$.

EXAMPLE 2. Find the area bounded by the parabola $x - y^2 + 1 = 0$ and the line $x - y - 1 = 0$.

Solution. Sketch the curves (see Fig. 6.1.3). Find their points of intersection by solving the equations simultaneously. Thus from $x = y + 1$ and $x = y^2 - 1$ we have

$$y^2 - 1 = y + 1 \qquad \text{or} \qquad y^2 - y - 2 = (y - 2)(y + 1) = 0$$

and $y = -1$ and 2. The points of intersection then are $(0, -1)$ and $(3, 2)$. From Fig. 6.1.3 we see that an element of area parallel to the Y-axis would have

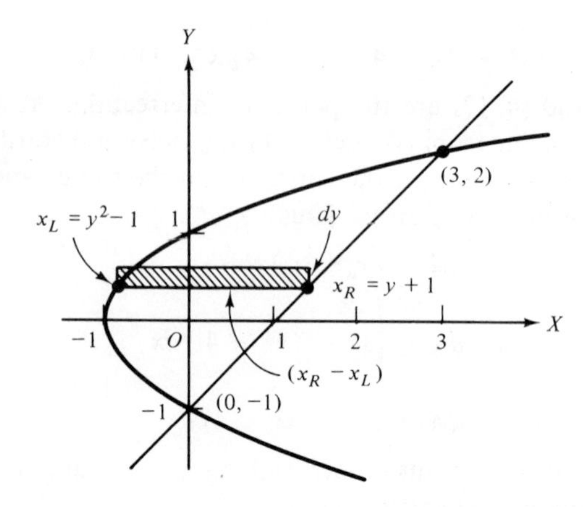

Figure 6.1.3 Area bounded by $x = y^2 - 1$ and $x = y + 1$.

its lower edge on the parabola for part of the region and on the straight line for the rest of the region. This is not good. However, an element of area parallel to the X-axis has its right edge on the straight line and its left edge on the parabola for the entire region. This is good. At any y for $-1 \leqq y \leqq 2$, draw a horizontal line from $x_L = y^2 - 1$ to $x_R = y + 1$ and construct a rectangle with positive dy. Then

$$\begin{aligned} dA &= (x_R - x_L)\, dy \\ &= [(y + 1) - (y^2 - 1)]\, dy \\ &= (-y^2 + y + 2)\, dy \end{aligned}$$

Upon summing all of these little areas in the positive y direction, we have

$$\begin{aligned} A &= \int_{-1}^{2} (-y^2 + y + 2)\, dy \\ &= \left[-\frac{y^3}{3} + \frac{y^2}{2} + 2y\right]_{-1}^{2} \\ &= \left[-\frac{(2)^3}{3} + \frac{(2)^2}{2} + 2(2)\right] - \left[-\frac{(-1)^3}{3} + \frac{(-1)^2}{2} + 2(-1)\right] \\ &= (-\tfrac{8}{3} + 2 + 4) - (\tfrac{1}{3} + \tfrac{1}{2} - 2) \\ &= \tfrac{9}{2} = 4.5 \text{ square units} \end{aligned}$$

PROBLEM 2. Find the area bounded by the parabola $x - y^2 + 1 = 0$ and the line $x + y - 1 = 0$.

Warning ***When the element of area, dA, is parallel to the X-axis, for all $c \leqq y \leqq d$, the right edge is on x_R and the left edge is on x_L.***

EXERCISES 6.1

Group A

1. Find each of the following areas.

(a) Between $y = 8 - x^2$ and $y = x^2$ [see Fig. 6.1.4(a)].
(b) Between $y = x^2 - 4$ and $y = 4 - x^2$ [see Fig. 6.1.4(b)].
(c) Between $y = x^2 - 4x + 1$ and $y = -x + 5$ [see Fig. 6.1.4(c)].
(d) Between $4y = x^3$ and $y = x$, $x \leqq 0$ [see Fig. 6.1.4(d)].

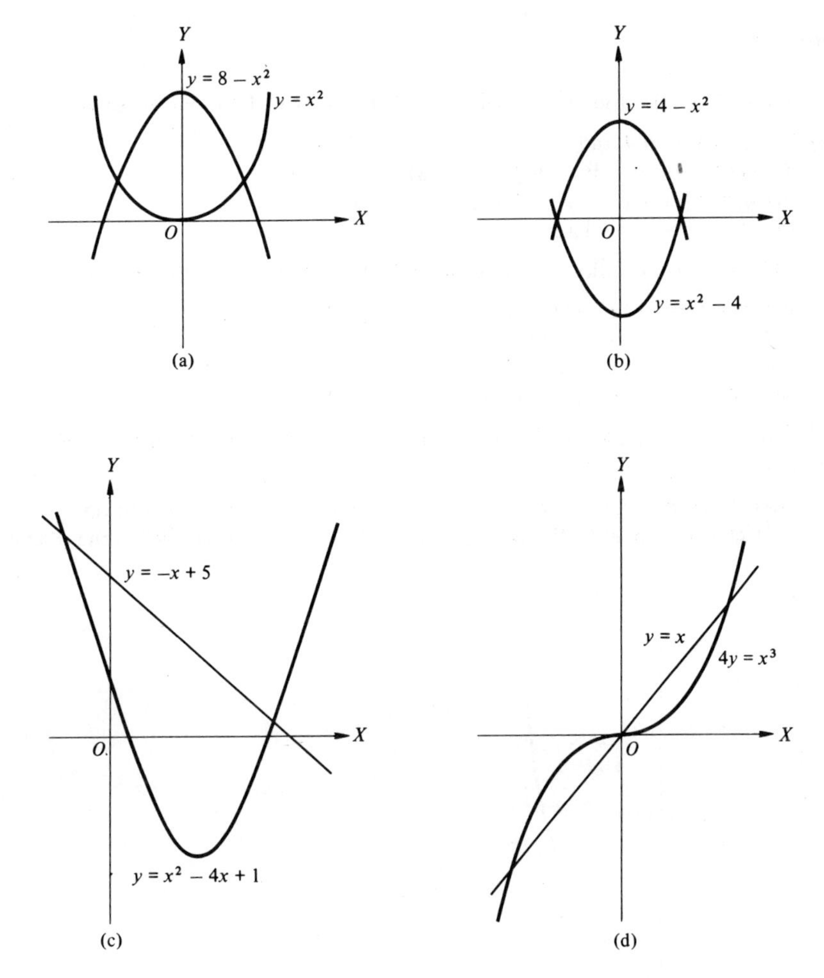

Figure 6.1.4 Exercise 1.

2. Find each of the following areas. Sketch the area to be found in each case.

(a) Between $y = 2x^2$ and $y = 12 - x^2$.
(b) Between $y = x^2 - 16$ and $y = 16 - x^2$.

(c) Between $y = 3 - 2x - x^2$ and $y = -x + 1$.
(d) Between $4y = x^3$ and $y = x$, $x \geqq 0$.

3. Find each of the following areas. Sketch the area to be found in each case.

(a) Between $x + y^2 - 1 = 0$ and $x - y + 1 = 0$.
(b) Between $x = y^2 + 2y$ and $x = 3$.
(c) Between $x = 4y - y^2$ and $x = 3$.
(d) Between $y^2 = 2x + 3$ and $x = y$.

Group B

1. Find each of the following areas. Sketch the area to be found in each case.

(a) Between $x^2y = 4$ and $3x + y = 7$.
(b) Between $x^2 + y - 9 = 0$ and $(x + 3)^2 + 4y = 0$.
(c) Between $x^4 - 4x^2 - y = 0$ and $y + 4 = 0$.
(d) Between $x^2 - y = 0$ and $y - 1 = 0$.

2. Find each of the following areas. Sketch the area to be found in each case.

(a) Between $y = -2x^2 - 4x$ and $y = 2x$.
(b) Between $y = 3x - x^3$ and $y = 2$.
(c) Between $y = x^2 + 2x$ and $y = \frac{1}{2}(x - 1)$.
(d) Between $y = x^3 + 3x^2 - 4$ and $y = 2x^3$.

3. Find the area of the indicated region in Fig. 6.1.5. Draw the element of area as directed.

(a) $OABO$, dA parallel to the Y-axis.
(b) $OABO$, dA parallel to the X-axis.
(c) $OBCO$, dA parallel to the X-axis.
(d) $OBCO$, dA parallel to the Y-axis.

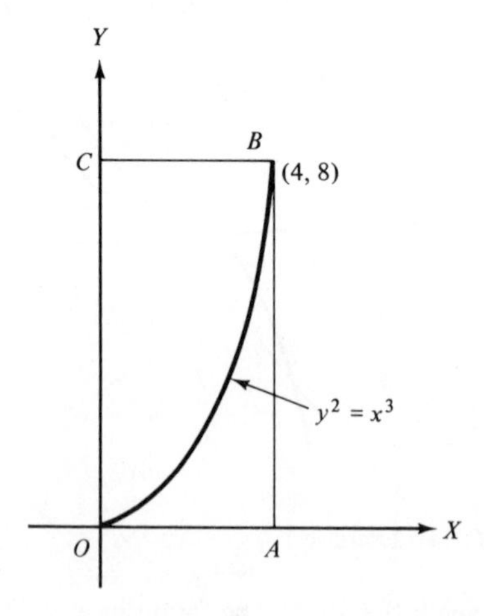

Figure 6.1.5 Exercise 3.

4. Sketch the curves $y^2 = x$ and $x - 1 = 0$. Find the area between these two curves by integrating with respect to y.

5. Find the area in each of the following problems in two ways.

(a) Between $4y = x^3$ and $y = x$, $x \leqq 0$. (b) Between $4y = x^3$ and $y = 3x + 4$.

6. In finding the area between two curves, explain when you would integrate with respect to x and when you would integrate with respect to y.

7. Find the area bounded by the curves $y = \sin x$, $y = \cos x$, $x = 0$, and $x = \pi/4$.

Group C

1. Consider the triangle ABC whose vertices are at $A(1, 2)$, $B(4, 6)$, and $C(3, 8)$. (a) Find the equations of the sides of the triangle. (b) Explain why the area cannot be found by using only one integral. (c) Draw a line through C parallel to the Y-axis dividing the area into two regions, A_1 to the left of the line and A_2 to the right. Use integration to find A_1 and A_2. Now find the area of the triangle.

6.2 FIRST MOMENT; CENTROID

Consider a system of n particles of mass $m_1, m_2, \ldots, m_n$ all lying in a plane at points whose coordinates are $(x_1, y_1), (x_2, y_2), \ldots, (x_n, y_n)$, respectively. Figure 6.2.1 shows a system of three particles.

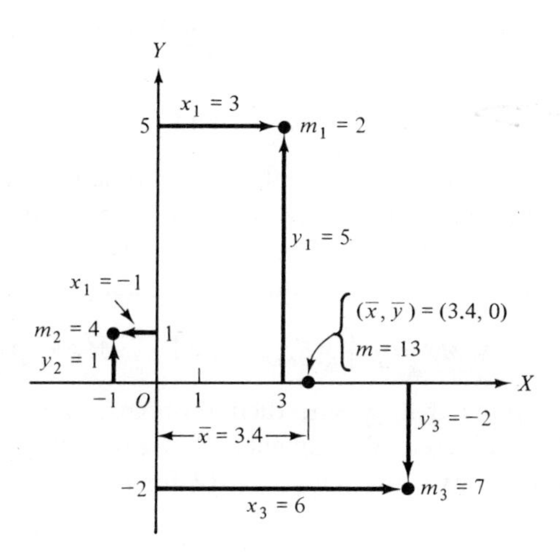

Figure 6.2.1 First moment.

The moment (or first moment) of each particle about the Y-axis, M_Y, is defined to be the product

$$M_Y = x_j m_j \qquad j = 1, \ldots, n$$

Similarly, the moment of each particle about the X-axis, M_X, is the product

$$M_X = y_j m_j \qquad j = 1, \ldots, n$$

The moments of the entire system of particles about the Y- and X-axes, respectively, are given by

$$M_Y = \sum_{j=1}^{j=n} x_j m_j \qquad \text{and} \qquad M_X = \sum_{j=1}^{j=n} y_j m_j \tag{6.2.1}$$

EXAMPLE 1. Consider the system of three particles with $m_1 = 2$ at $(3, 5)$, $m_2 = 4$ at $(-1, 1)$, and $m_3 = 7$ at $(6, -2)$ (see Fig. 6.2.1). Find the moment of the system (a) about the Y-axis; (b) about the X-axis.

Solution. Using Eqs. 6.2.1, we have

$$M_Y = \underset{x_1\, m_1}{(3)(2)} + \underset{x_2\ m_2}{(-1)(4)} + \underset{x_2\, m_2}{(6)(7)} = 44$$

and

$$M_X = \underset{y_1\, m_1}{(5)(2)} + \underset{y_2\, m_2}{(1)(4)} + \underset{y_3\ m_3}{(-2)(7)} = 0$$

PROBLEM 1. Consider the system of four particles with $m_1 = 3$ at $(1, -1)$, $m_2 = 2$ at $(4, 1)$, $m_3 = 1$ at $(5, 0)$, and $m_4 = 4$ at $(-1, 2)$. Find the moment of the system (a) about the Y-axis; (b) about the X-axis.

Warning ***Remember that a moment must be about a line or a point and is the product of a distance and a mass.***

A rigid body may be considered as a system of an infinite number of particles. If we extend Eqs. 6.2.1 to a rigid body by taking the sum as $n \longrightarrow \infty$, we have

$$M_Y = \int x\, dm \qquad \text{and} \qquad M_X = \int y\, dm \tag{6.2.2}$$

We may extend this concept to the moment of an area about an axis by noting that the mass of a body is the product of its mass density and area. That is, $m = \rho A$, where the Greek letter ρ (rho) is the mass density. For the present we assume that the density, ρ, is constant. Then Eqs. 6.2.2 become

$$M_Y = \rho \int x\, dA \qquad \text{and} \qquad M_X = \rho \int y\, dA \tag{6.2.3}$$

When we assume that $\rho = 1$, then the expressions $x\, dA$ and $y\, dA$ may be considered as differential moments of areas. That is, $dM_Y = x\, dA$ for dA parallel to the Y-axis and $dM_X = y\, dA$ for dA parallel to the X-axis.

EXAMPLE 2. Find the moment of the area of a rectangle of length a, width b, and density ρ about its edges.

Solution. For convenience, let one corner of the rectangle be placed at the origin of a set of axes and two adjacent sides be placed on the X- and the Y-axes, respectively (see Fig. 6.2.2).

Interpreting Eqs. 6.2.3 as the "total moment is equal to the sum of all the little moments" we need to find dM_Y and dM_X. If we draw dA parallel to

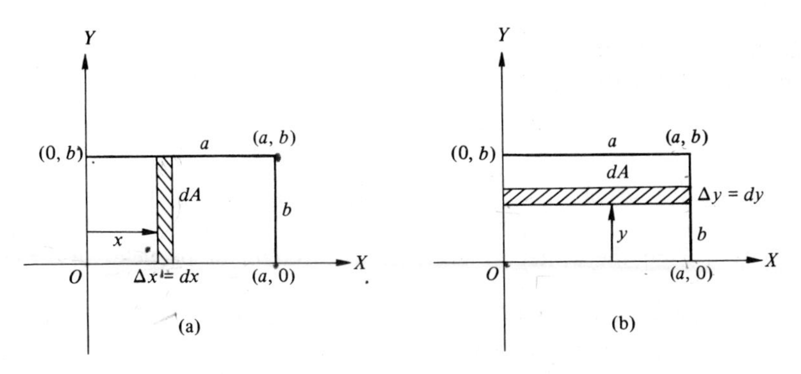

Figure 6.2.2 Moment of a rectangle with respect to its edge.

the Y-axis as in Fig. 6.2.2(a), we note that in the limit as $\Delta x \longrightarrow 0$ all the particles of dA would be at a distance x from the Y-axis. Hence

$$dM_Y = \rho x\, dA = \rho xy\, dx = \rho xb\, dx$$

and

$$M_Y = \int dM_Y = \rho b \int_0^a x\, dx = \frac{\rho a^2 b}{2}$$

We cannot use this same dA to find dM_X because every particle of dA is at a different distance from the X-axis. So we draw a new dA parallel to the X-axis as in Fig. 6.2.2(b). Then

$$dM_X = \rho y\, dA = \rho yx\, dy = \rho ay\, dy$$

and

$$M_X = \int dM_X = \rho a \int_0^b y\, dy = \rho \frac{ab^2}{2}$$

PROBLEM 2. Use integration to find the moment of the area of a rectangle of length 5 and width 3 about its edges. Assume that $\rho = 2$.

There is a point called the *center of mass* which is closely associated with moments about the axes. It is denoted by the symbol $(\bar{x}, \bar{y})$ (read "x bar, y bar"). If the total mass, m, of a system could be replaced by a single particle of mass m, we then define

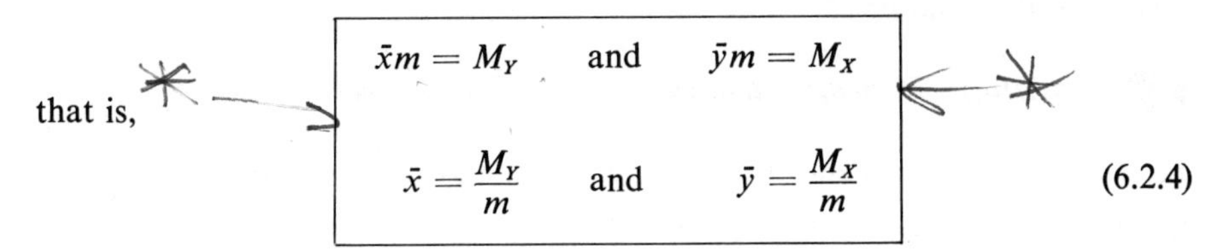

$$\bar{x}m = M_Y \quad \text{and} \quad \bar{y}m = M_X$$

that is,

$$\bar{x} = \frac{M_Y}{m} \quad \text{and} \quad \bar{y} = \frac{M_X}{m} \tag{6.2.4}$$

EXAMPLE 3. Find the center of mass for the system of particles of Example 1.

Solution. From Example 1, $m = m_1 + m_2 + m_3 = 2 + 4 + 7 = 13$. $M_Y = 44$ and $M_X = 0$. Using formulas 6.2.4, we have

$$\bar{x} = \frac{M_Y}{m} = \frac{44}{13} = 3.4 \qquad \text{and} \qquad \bar{y} = \frac{M_X}{m} = \frac{0}{13} = 0$$

Thus the center of mass is at $(\bar{x}, \bar{y}) = (3.4, 0)$ (see Fig. 6.2.1).

PROBLEM 3. Find the center of mass for the system of particles of Prob. 1.

Sometimes the center of mass is called the *center of gravity*. In the case of plane areas with uniform density ρ, the point $(\bar{x}, \bar{y})$ coincides with the geometric center of the plane area and is called the *centroid of the plane area*. We define

$$\bar{x}\rho A = M_Y \qquad \text{and} \qquad \bar{y}\rho A = M_X$$

or

$$\bar{x} = \frac{M_Y}{\rho A} \qquad \text{and} \qquad \bar{y} = \frac{M_X}{\rho A} \tag{6.2.5}$$

EXAMPLE 4. Find the centroid of the area of a rectangle of length a and width b.

Solution. From the results of Example 2, we we have

$$M_Y = \rho\frac{a^2b}{2} \qquad \text{and} \qquad M_X = \rho\frac{ab^2}{2}$$

Since the total mass of the rectangle is $\rho A = \rho ab$, we use Eqs. 6.2.5 to find

$$\bar{x} = \frac{M_Y}{\rho A} = \frac{\rho a^2b/2}{\rho ab} = \frac{a}{2}$$

and

$$\bar{y} = \frac{M_X}{\rho A} = \frac{\rho ab^2/2}{\rho ab} = \frac{b}{2}$$

$$\therefore \quad (\bar{x}, \bar{y}) = \left(\frac{a}{2}, \frac{b}{2}\right)$$

(which is where we should have expected it to be).

PROBLEM 4. Find the centroid of the area of the rectangle of Prob. 2. Is it where you think it should be?

***Warning** Remember that for $\bar{x}$ we use M_Y and for $\bar{y}$ we use M_X.*

EXAMPLE 5. Find the centroid of the region bounded by $y = 4x - x^2$ and the X-axis.

Solution. Sketch the area (see Fig. 6.2.3). At any x such that $0 < x < 4$, draw an ordinate and construct a rectangle of width dx. Then use Eqs. 6.2.5 to find

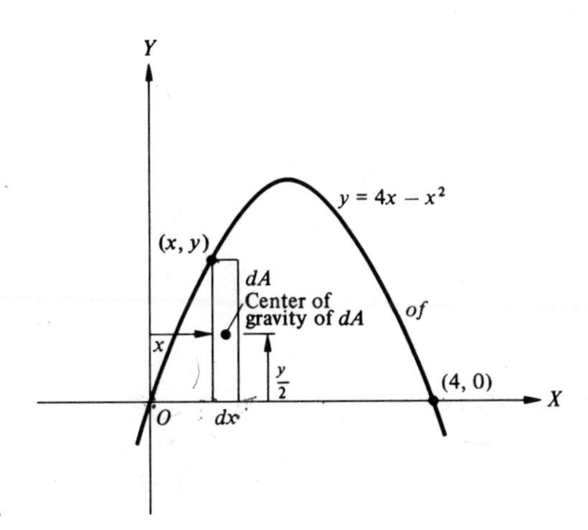

Figure 6.2.3 Centroid of region.

$\bar{x}$ and $\bar{y}$. Thus

$$dA = y\,dx = (4x - x^2)\,dx$$

and

$$\rho A = \rho \int dA = \rho \int_0^4 (4x - x^2)\,dx = \rho\left[2x^2 - \frac{x^3}{3}\right]_0^4$$

or

$$\rho A = \rho \frac{32}{3}\ .$$

To find M_Y we note that dA is parallel to the Y-axis, so that

$$dM_Y = \rho x\,dA = \rho x(y\,dx) = \rho x(4x - x^2)\,dx$$

and

$$M_Y = \int dM_Y = \int_0^4 \rho x(4x - x^2)\,dx = \rho \int_0^4 (4x^2 - x^3)\,dx$$

Upon integrating, we find that

$$M_Y = \rho \frac{64}{3}$$

Hence from Eqs. 6.2.5 we have $\bar{x} = M_Y/\rho A = \rho\frac{64}{3}/\rho\frac{32}{3} = 2$.

To find M_X we note that dA is not parallel to the X-axis and that if it were parallel to the X-axis, we would have trouble finding its length. So we use Eqs. 6.2.5 and obtain $dM_X = \rho(y/2)\,dA$, where $y/2$ is the $\bar{y}$ of the rectangle dA as found in Example 4. Thus

$$dM_X = \rho\tfrac{1}{2}\underbrace{(4x - x^2)}_{y}\underbrace{(4x - x^2)\,dx}_{dA}$$

or

$$dM_X = \rho\tfrac{1}{2}(16x^2 - 8x^3 + x^4)\,dx$$

Then

$$M_X = \int dM_X = \tfrac{1}{2}\rho \int_0^4 (16x^2 - 8x^3 + x^4)\,dx$$
$$= \tfrac{1}{2}\rho[\tfrac{16}{3}x^3 - 2x^4 + \tfrac{1}{5}x^5]_0^4$$
$$= \tfrac{1}{2}\rho[x^3(\tfrac{16}{3} - 2x + \tfrac{1}{5}x^2)]_0^4$$
$$= \tfrac{1}{2}\rho\{[4^3(\tfrac{16}{3} - 8 + \tfrac{16}{5})] - \tfrac{1}{2}[0]\}$$

and

$$M_X = \rho\tfrac{256}{15}$$

So from Eqs. 6.2.5 we have

$$\bar{y} = \frac{M_X}{\rho A} = \frac{\rho(256/15)}{\rho(32/3)} = \frac{8}{5} = 1.6$$

and the centroid is at

$$(\bar{x}, \bar{y}) = (2, \tfrac{8}{5}) = (2, 1.6)$$

PROBLEM 5. Find the centroid of the region bounded by $y = 6x - x^2$ and the X-axis.

EXAMPLE 6. Find the centroid of the area bounded by $y = x^2 - 4$ and $y = 3x$.

Solution. The area of this region was found to be $A = \frac{125}{6}$ in Example 1 of Sec. 6.1. The sketch of the region is shown in Fig. 6.1.2.

$$dM_Y = \rho x\,dA = \rho x(3x - x^2 + 4)\,dx$$

$$M_Y = \int dM_Y = \rho \int_{-1}^4 (3x^2 - x^3 + 4x)\,dx$$
$$= \rho\left[x^3 - \frac{x^4}{4} + 2x^2\right]_{-1}^4 = \rho\frac{125}{4}$$

So

$$\bar{x} = \frac{M_Y}{\rho A} = \frac{\rho\frac{125}{4}}{\rho\frac{125}{6}} = \frac{3}{2}$$

To find dM_X we need $\bar{y}$ of the centroid of dA. It is the ordinate of the midpoint of the line from y_l to y_u, which is $\frac{1}{2}(y_l + y_u)$ (see Eq. 2.6.2). Thus

$$dM_X = \rho\tfrac{1}{2}(y_l + y_u)\,dA$$
$$= \rho\tfrac{1}{2}(y_u + y_l)(y_u - y_l)\,dx = \rho\tfrac{1}{2}(y_u^2 - y_l^2)\,dx$$

Since $y_u = 3x$ and $y_l = x^2 - 4$, we have

$$dM_X = \tfrac{1}{2}(9x^2 - x^4 + 8x^2 - 16)\,dx$$

Then

$$M_X = \int dM_X = \tfrac{1}{2}\int_{-1}^4 (-x^4 + 17x^2 - 16)\,dx$$
$$= \rho\frac{1}{2}\left[\frac{-x^5}{5} + \frac{17x^3}{3} - 16x\right]_{-1}^4 = \rho\left(-\frac{175}{3}\right)$$

So

$$\bar{y} = \frac{M_X}{\rho A} = \frac{\rho(-\frac{175}{3})}{\rho\frac{125}{6}} = -\frac{14}{5}$$

and the centroid is at

$$(\bar{x}, \bar{y}) = (\tfrac{3}{2}, -\tfrac{14}{5}) = (1.5, -2.8)$$

PROBLEM 6. Find the centroid of the area bounded by $y = -x^2 + 4$ and $y = -x + 2$.

EXERCISES 6.2

Group A

1. In each of the following parts, we use the notation $m(x, y)$, where m is the mass of a particle and (x, y) the coordinates of its location. Find M_X and M_Y for each system of particles.

(a) $3(2, 2)$, $4(2, -2)$, $5(-2, 2)$, $2(-2, -2)$.
(b) $3(0, 0)$, $5(4, 0)$, $7(8, 0)$, $2(-10, 0)$, $8(-5, 0)$.
(c) $6(0, 0)$, $6(4, 0)$, $6(0, 4)$, $6(4, 4)$, $3(2, 2)$.
(d) $4(-1, -1)$, $5(-1, 1)$, $7(1, 1)$, $8(1, -1)$.

2. Find M_X and M_Y for each of the following bodies whose shape is bounded by the curves given and having constant density ρ.

(a) $y = -x + 2$, $x = 0$, and the X-axis.
(b) $y = x^3$, $x \geqq 0$, $x = 1$, and the X-axis.
(c) $y = -x^2 + 1$ and the X-axis.
(d) $y = 5x - x^2$ and the X-axis.

3. Find the center of mass for each part of Exercise 1.

4. Find the centroid of each of the areas determined by the curves given in Exercise 2.

5. Find the centroid of the area enclosed by a right triangle of base b and height h. (*Hint*: Let the right angle be at the origin with b on the X-axis and h on the Y-axis.)

6. Find the centroid of the area bounded by the given curves.

(a) $x = 4y - y^2$ and the Y-axis.
(b) $y^2 = x + 4$ and the Y-axis.

Group B

1. Find the centroid of the areas that have boundaries determined by the curves given in each of the following.

(a) $y = 6x - x^2$ and $y = x$.
(b) $y = x^2$ and $y = x + 2$.
(c) $y = x^2 - 4$ and $y = -x^2 + 4x - 4$.
(d) $y = x^3$ and $y = x$ for $x \geqq 0$.

2. Find the centroid of the areas that have boundaries determined by the curves given in each of the following.

(a) $y = 4 - x^2$, $x = 0$, $x = 2$, and $y = 0$.
(b) $y = \frac{1}{8}x^2$, $x = 4$ and $y = 0$.
(c) $y = 2(1 + x^3)$, $x = 0$ and $y = 0$.
(d) $y = \frac{1}{2}x^2$ and $y = x^3$.

3. Find the center of mass for the areas bounded by the given curves and whose density varies as given.

(a) $y = 4x - x^2$ and the X-axis; $\rho = x$.
(b) $y = x^2 - 4$ and $y = 3x$; $\rho = x/2$.

4. Set up the integrals necessary to find the centroid of the area bounded by $y = \sin x$, $y = 0$, $x = 0$, and $x = \pi$.

5. Set up the integrals necessary to find the centroid of the area bounded by $y = \cos x$, $y = 0$, $x = 0$, and $x = \pi/2$.

6. Explain how the center of mass and the centroid of plane areas are alike and how they are different.

7. Show that when dA is parallel to the X-axis and is bounded on the right by x_R and on the left by x_L, $dM_Y = \rho\frac{1}{2}(x_R^2 - x_L^2)\,dy$ (see Example 6).

8. Find the centroid of the area bounded by the given curves. (Make use of Exercise 7.)

(a) $y = x$, $y = -x + 4$, and $y = 4$. (b) $x = y^2 - 1$ and $x + y = 1$.

9. For each part of Exercise 3 of Group B in Exercises 6.1, set up the integrals necessary to find the centroid.

Group C

1. Use integration to find the centroid of the rectangular region positioned as in Fig. 6.2.4.

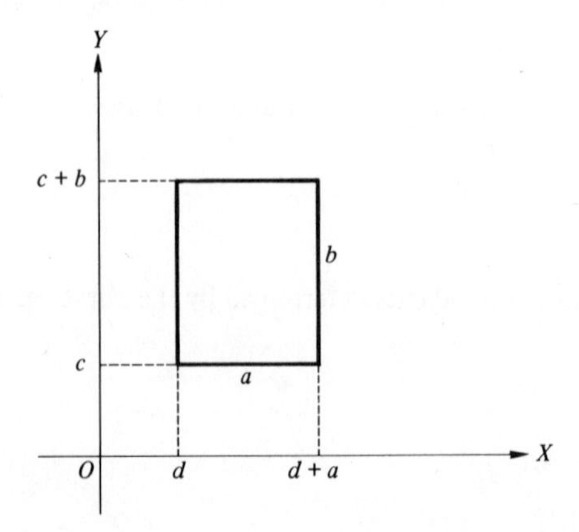

Figure 6.2.4 Exercise 1.

2. Use the following theorem to determine the centroids of each of the areas in Fig. 6.2.5.

> ***Theorem*** *If a body of mass* m *consists of* n *distinct parts of masses* m_1, $m_2, \ldots m_n$ *and if an auxiliary system of* n *particles is formed by concentrating the mass of each of the parts at its own center of mass, then the center of mass of the entire body coincides with the center of mass of the auxiliary system; that is*

$$\bar{x} = \frac{\bar{x}_1 m_1 + \bar{x}_2 m_2 + \ldots + \bar{x}_n m_n}{m_1 + m_2 + \ldots + m_n} \qquad \bar{y} = \frac{\bar{y}_1 m_1 + \bar{y}_2 m_2 + \ldots + \bar{y}_n m_n}{m_1 + m_2 + \ldots + m_n}$$

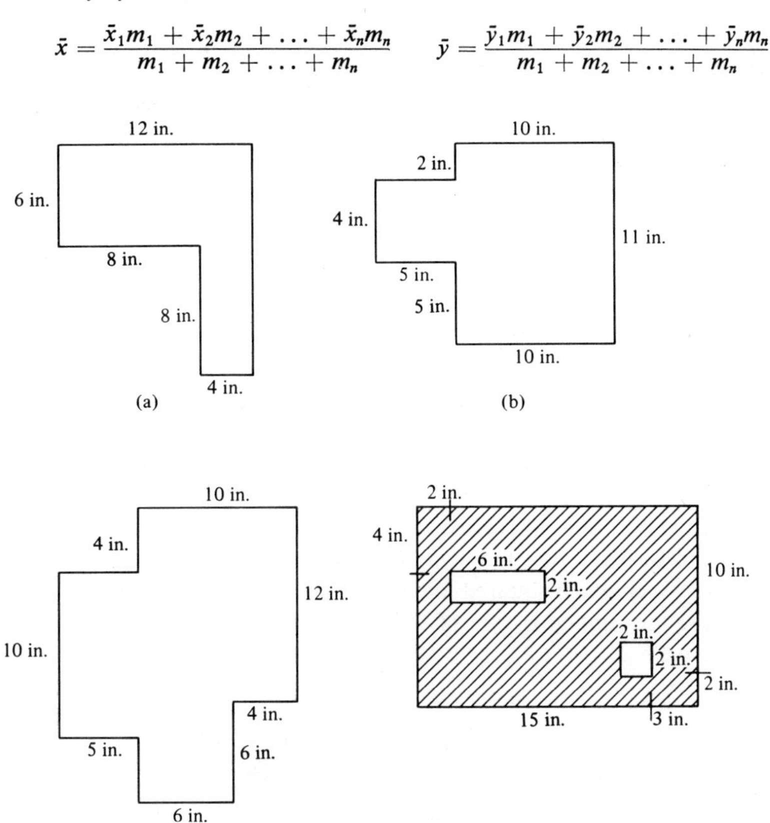

Figure 6.2.5 Exercise 2.

6.3 SECOND MOMENT; MOMENT OF INERTIA

Consider a system of n particles of masses $m_1, m_2, \ldots, m_n$ all lying in a plane at points whose coordinates are $(x_1, y_1), (x_2, y_2), \ldots, (x_n, y_n)$, respectively. The *second moment* of each particle about the Y-axis is defined to be the product

$$x_j^2 m_j \qquad j = 1, 2, \ldots, n$$

This product is also called the *moment of inertia* of the particle about the Y-axis. The total moment of inertia of the entire system about the Y-axis is denoted as i_Y and is given by

$$i_Y = \sum_{j=1}^{j=n} x_j^2 m_j \tag{6.3.1}$$

Similarly, the moment of inertia of the system about the X-axis is given by

$$i_X = \sum_{j=1}^{j=n} y_j^2 m_j \tag{6.3.2}$$

EXAMPLE 1. Consider the system of three particles with $m_1 = 2$ at (3, 5), $m_2 = 4$ at (−1, 1), and $m_3 = 7$ at (6, −2). Find the moment of inertia of the system (a) about the Y-axis; (b) about the X-axis.

Solution. Using Eqs. 6.3.1 and 6.3.2, we have

$$i_Y = \underset{(x_1)^2 m_1}{(3)^2(2)} + \underset{(x_2)^2\ m_2}{(-1)^2(4)} + \underset{(x_3)^2 m_3}{(6)^2(7)} = 274$$

and

$$i_X = \underset{(y_1)^2 m_1}{(5)^2(2)} + \underset{(y_2)^2 m_2}{(1)^2(4)} + \underset{(y_3)^2\ m_3}{(-2)^2(7)} = 82$$

PROBLEM 1. Consider the system of four particles with $m_1 = 3$ at (1, −1), $m_2 = 2$ at (4, 1), $m_3 = 1$ at (5, 0), and $m_4 = 4$ at (−1, 2). Find the moment of inertia of the system (a) about the Y-axis; (b) about the X-axis.

There is no point associated with moment of inertia such as the centroid is associated with the first moment. However, there is a distance called *radius of gyration* which when squared and multiplied by the total mass of the system will give the moment of inertia of the system. Denoting the radius of gyration by k, we have

$$i_Y = k_Y^2 m \quad \text{and} \quad i_X = k_X^2 m$$

or

$$k_Y = \sqrt{\frac{i_Y}{m}} \quad \text{and} \quad k_X = \sqrt{\frac{i_X}{m}} \tag{6.3.3}$$

where m is the total mass of the system.

EXAMPLE 2. Find the radius of gyration with respect to the Y-axis, k_Y, and with respect to the X-axis, k_X, for the system of particles in Example 1.

Solution. From Example 1 we have $i_Y = 274$ and $i_X = 82$. We find that $m = m_1 + m_2 + m_3 = 2 + 4 + 7 = 13$. Using Eqs. 6.3.3, we have

$$k_Y = \sqrt{\tfrac{274}{13}} = 4.59 \quad \text{and} \quad k_X = \sqrt{\tfrac{82}{13}} = 2.51$$

PROBLEM 2. Find the radius of gyration with respect to the Y-axis and with respect to the X-axis for the system of particles in Prob. 1.

The moment of inertia of an object has an important role when rotation of bodies is studied. It is analogous to the role of mass in Newton's law that states "force equals mass times acceleration."

Extending Eqs. 6.3.1 and 6.3.2 to a plane area with uniform density ρ as was done for the centroid, we have

$$i_Y = \rho \int x^2\, dA \qquad \text{and} \qquad i_X = \rho \int y^2\, dA \tag{6.3.4}$$

Equations 6.3.4 are good as long as dA is parallel to the axis we are working with. If it is not, we must wait until we study double integrals in Chap. 16 or use the result of Exercise 4 in Group C of Exercises 6.3.

EXAMPLE 3. Find the moment of inertia of the area of a rectangle of base b and height h about its base. Also find the radius of gyration with respect to the base.

Solution. For convenience, let the base be on the X-axis as in Fig. 6.3.1. Draw dA parallel to the X-axis, as then in the limit for small dy all particles of dA will be distance y from the X-axis. From Eqs. 6.3.4 we have

$$di_X = \rho y^2\, dA = \rho y^2(x\, dy) = \rho b y^2\, dy$$

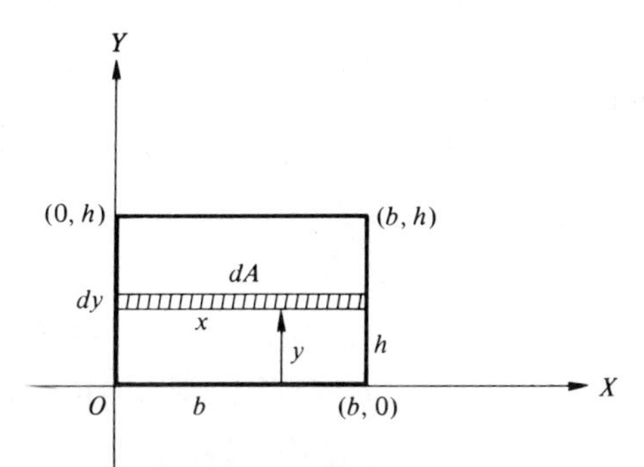

Figure 6.3.1 Moment of inertia of a rectangle.

and since "the total moment of inertia is the sum of all the little moment of inertias," we have

$$i_X = \int di_X = \rho b \int_0^h y^2\, dy = \rho \frac{bh^3}{3}$$

or

$$i_X = \tfrac{1}{3}\rho h^2 A$$

where $A = bh$ is the area of the rectangle. If $\rho = 1$, we may say that for a rectangle $i_{\text{base}} = \frac{1}{3}(\text{base})(\text{height})^3$.

The radius of gyration is found by using Eq. 6.3.3. For a rectangle $A = bh$ and $m = \rho\, bh$. Thus

$$k_X = \sqrt{\frac{\frac{1}{3}\rho bh^3}{\rho bh}} = \sqrt{\frac{h^2}{3}}$$

or

$$k_{\text{base}} = \frac{h}{\sqrt{3}} = 0.577h$$

(Note that the radius of gyration is not equal to the distance to the centroid, which for a rectangle is $0.5h$.)

PROBLEM 3. Use integration to find the moment of inertia about the X-axis of the area of the rectangle with $\rho = 1$ and whose corners are at (1, 1), (5, 1), (5, 3), and (1, 3). Also find k_x.

***Warning** Remember that a moment of inertia must be about a line or a point and is the product of the square of a distance and a mass.*

EXAMPLE 4. Find i_Y and k_Y for the area bounded by $y = x^2 - 4$ and $y = 3x$.

Solution. The region is sketched in Fig. 6.1.2. From Eqs. 6.3.4 we have

$$\begin{aligned} di_Y &= \rho x^2\, dA = \rho x^2(y_u - y_l)\, dx \\ &= \rho x^2(3x - x^2 + 4)\, dx \end{aligned}$$

and

$$\begin{aligned} i_Y &= \rho \int_{-1}^{4} (3x^3 - x^4 + 4x^2)\, dx \\ &= \rho\left[\frac{3}{4}x^4 - \frac{x^5}{5} + \frac{4}{3}x^3\right]_{-1}^{4} \\ &= \rho[(x^3)(\tfrac{3}{4}x - \tfrac{1}{5}x^2 + \tfrac{4}{3})]_{-1}^{4} \\ &= \rho\{[(4)^3(3 - \tfrac{16}{5} + \tfrac{4}{3})] - [(-1)^3(-\tfrac{3}{4} - \tfrac{1}{5} + \tfrac{4}{3})]\} \\ \therefore\quad i_Y &= \rho\tfrac{4375}{60} = 72.9\rho \end{aligned}$$

From Example 1, Sec. 6.1, we have $m = 20.8\rho$. Thus

$$k_Y = \sqrt{\frac{72.9\rho}{20.8\rho}} = 1.87$$

'ROBLEM 4. Find i_Y for the region bounded by $y = x^2$ and $x + y - 2$

For a discussion of the polar moment of inertia, see Sec. 16.3.

EXERCISES 6.3

Group A

1. Use Eqs. 6.3.1 and 6.3.2 to find i_Y and i_X for each of the systems of particles of Exercise 1 of Group A in Exercises 6.2. Also find k_Y and k_X.

2. Find i_Y for the area bounded by $y = -x^2 + 1$ and the X-axis.

3. Find i_Y for the area bounded by $y = 5x - x^2$ and the X-axis.

4. Find i_X and i_Y for the area bounded by $x = 0$, $y = -x + 2$, and the X-axis. Also find k_X and k_Y. (*Hint*: For i_X draw the element of area parallel to the X-axis.)

5. Find i_X and i_Y for the area bounded by $y = x^3$, $x = -1$, and the X-axis.

6. Find i_{base} and k_{base} of the area of a right triangle of base b and height h. (*Hint*: Let the right angle be at the origin with b on the X-axis and h on the Y-axis.)

7. Find i_Y for the area bounded by $y = 16 - x^2$ and the X-axis.

8. Find i_X for the area bounded by $y^2 = 4x$, $y = 2$, and the Y-axis.

9. Find i_X for the area bounded by $-y^2 + 9 = x$ and the Y-axis.

10. Find i_X for the area bounded by $y^2 - 1 = x$ and the Y-axis.

11. Find the radius of gyration corresponding to the required moment of inertia for (a) Exercise 7; (b) Exercise 8; (c) Exercise 9; (d) Exercise 10.

Group B

1. Find i_Y for the area bounded by $y = 3x - x^2$ and $y = x$.

2. Find i_Y for the area bounded by $y = x^2$ and $y = x + 2$.

3. Find i_Y for the area bounded by $y = x^2 + 2$ and $y = -x^2 + 4x + 8$.

4. Find i_X and i_Y for the area bounded by $y = x^3$ and $y = x$ for $x \leqq 0$.

5. Find the moment of inertia and the radius of gyration of a slender rod of length l about one end. Assume a constant density. (*Hint*: Let the rod lie on the X-axis. Then $dm = \rho dx$.) Compare your answer with that of Example 3 of Sec. 6.3.

6. Tell how the *moment* of a mass about a line and the *moment of inertia* of a mass about a line are alike and how they are different.

7. Tell how the centroid and radius of gyration are alike and how they are different.

8. Third and fourth moments are sometimes used in statistics. How would you define (a) the third moment of a mass about a line; (b) the fourth moment of a mass about a line?

Group C

1. One may define the moment of inertia of a point mass or system of mass particles about a given line L in the plane. If $m_1, m_2, \ldots, m_n$ are the masses of particles at distance $d_1, d_2, \ldots, d_n$ from a given line L, the moment of inertia of the system about L is defined to be

$$i_L = m_1 d_1^2 + m_2 d_2^2 + \ldots + m_n d_n^2 \tag{6.3.5}$$

(a) Let L be the line $y = x$. Find the moment of inertia with respect to L of the following system of mass particles [$m(x, y)$ denotes a particle of mass m at the point (x, y)]: $2(-1, 2)$, $4(1, 2)$, $2(-1, -2)$, $4(1, -2)$. (b) Let L be the line $y = -2x + 1$. Find the moment of inertia with respect to L of the following system of point masses: $3(1, -3)$, $4(2, 1)$, $5(3, 4)$ (see Eq. 2.6.3).

2. Find the radius of gyration k_L of the systems in Exercise 1.

3. Find the radius of gyration with respect to the Y-axis for the plane area bounded by $y = 3x - x^2$ and $y = x$, where $\rho = x$.

4. In using Eqs. 6.3.4, dA must be parallel to the axis in question. Use the result of Example 3 to show that for the area and dA shown in Fig. 6.3.2,

$$i_X = \tfrac{1}{3}\rho \int_a^b y^3\, dx \tag{6.3.6}$$

(Note that dA is perpendicular to the X-axis.)

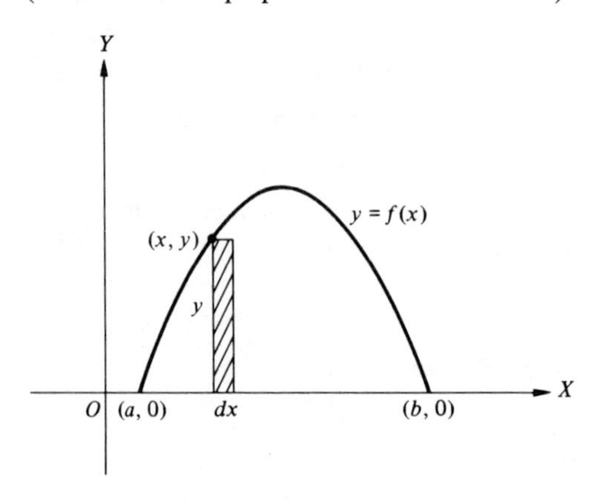

Figure 6.3.2 Exercise 4.

5. Use Eq. 6.3.6 to find i_X for each of the following.

(a) Exercise 2 of Group A.
(b) Exercise 3 of Group A.
(c) Exercise 1 of Group B. [*Hint*: di_X for $dA = (di_X \text{ for } dA_u) - (di_X \text{ for } dA_l)$.]
(d) Exercise 2 of Group B.
(e) Exercise 3 of Group B.

6.4 VOLUMES

We may find the volume of a solid by finding a typical differential element of volume, dV, and summing an infinite number of these as the magnitude of dV approaches zero. That is, if V is the volume to be found, then

$$V = \int dV$$

We might say that "the total volume is the sum of a lot of little volumes." Here we shall consider

$$dV = A\, dh$$

where A is the area of a typical cross section of the solid and dh is the differential thickness of the cross section.

When a solid is formed by revolving an area about an axis, its volume is called a *volume of revolution*. There are three methods used to find dV for a volume of revolution: (1) the *solid disk*, (2) the *washer*, and (3) the *cylindrical shell.*

(1) Solid Disk Method

dV is a solid disk when the element of area, a rectangle dA, is perpendicular to the axis of revolution and one end of dA touches the axis. The formula for dV for a solid disk is

$$dV = \pi(radius)^2(thickness) \tag{6.4.1}$$

This is so because as the rectangle dA is revolved about the axis it will generate a cylinder whose "height" is equal to the "thickness" of the disk. This method is used in Example 1.

EXAMPLE 1. Find the volume of a cone of base radius r and height h.

Solution. A cone may be generated by revolving the area of a right triangle about one of its legs. For convenience let the vertex of the cone be at the origin of axes and one leg of the right triangle be on the X-axis, as shown in Fig. 6.4.1(a). At any x such that $0 < x < h$, draw an ordinate and construct a rectangle perpendicular to the X-axis of height y, thickness dx, and area dA. As this rectangle is revolved about the X-axis it will generate a solid disk of radius y and thickness dx. This element of volume [Fig. 6.4.1(b)] so generated is a typical cross section of the cone. Using formula 6.4.1, we have

$$dV = \underbrace{\pi y^2}_{\text{radius}} \underbrace{dx}_{\text{thickness}} \tag{6.4.2}$$

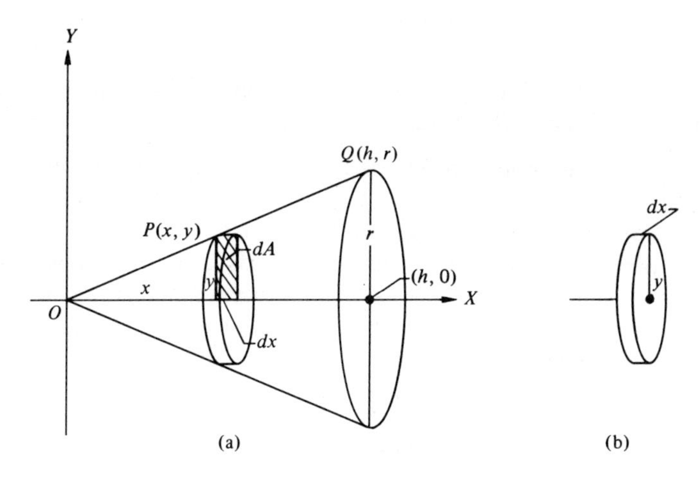

Figure 6.4.1 Volume of revolution (disk).

Before integrating we must have only one variable on the right-hand side of Eq. 6.4.2. The equation of the straight line through points O and Q is $y = (r/h)x$. (The slope is r/h and it passes through the origin.) Then we may write

$$dV = \pi\left(\frac{r}{h}x\right)^2 dx = \frac{\pi r^2}{h^2}x^2\,dx$$

and

$$V = \int dV = \int_0^h \frac{\pi r^2}{h^2}x^2\,dx = \frac{\pi r^2}{h^2}\int_0^h x^2\,dx$$

Integrating, we have

$$V = \frac{\pi r^2}{h^2}\left[\frac{x^3}{3}\right]_0^h$$

and

$$V = \tfrac{1}{3}\pi r^2 h$$

Warning ***Always draw a clear figure of dV outside the volume as we did in Fig. 6.4.1(b).***

PROBLEM 1. Use integration to find the volume of revolution generated by revolving about the X-axis the area bounded by $y = x/3$, $y = 0$, $x = 0$, and $x = 6$. (Check your answer by using the result of Example 1.)

(2) Washer Method

dV is a washer when the element of area, a rectangle dA, is perpendicular to the axis of revolution and does not touch the axis. The formula for dV for a washer is

$$dV = \pi[(outer\ radius)^2 - (inner\ radius)^2](thickness) \qquad (6.4.3)$$

This method is used in Example 2.

EXAMPLE 2. Find the volume generated by revolving the region bounded by $y = x$ and $y = x^2$ about the X-axis.

Solution. To find the points of intersection we set $x^2 = x$ and find that the curves intersect at (0, 0) and (1, 1). They are shown in Fig. 6.4.2(a). Draw a rectangle perpendicular to the X-axis in the region at any x such that $0 < x < 1$. Upon revolving this rectangle about the X-axis a washer is generated of thickness dx [see Fig. 6.4.2(b)]. The area of this cross-sectional washer is $\pi y_u^2 - \pi y_l^2 = \pi(y_u^2 - y_l^2)$; where $y_u = x$ is the outer radius and $y_l = x^2$ is the inner radius. Using formula 6.4.3, we have

$$\begin{aligned} dV &= \pi(y_u^2 - y_l^2)\,dx \\ &= \pi[(x)^2 - (x^2)^2]\,dx \\ &= \pi(x^2 - x^4)\,dx \end{aligned}$$

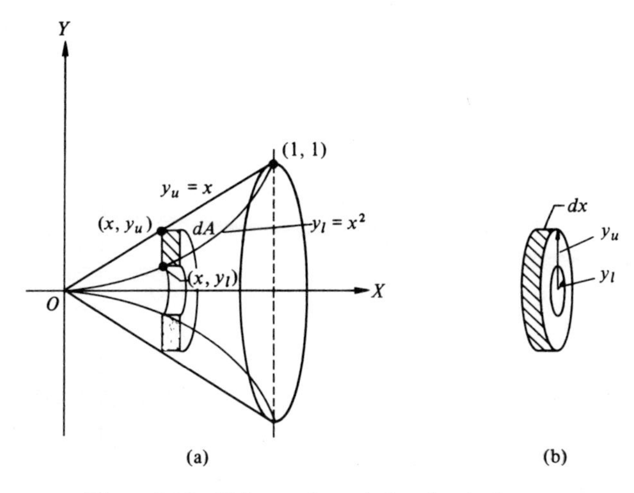

Figure 6.4.2 Volume of revolution (washer).

So

$$V = \pi \int_0^1 (x^2 - x^4)\, dx$$

$$= \pi \left[\frac{x^3}{3} - \frac{x^5}{5}\right]_0^1$$

$$= \frac{2\pi}{15} = 0.419$$

PROBLEM 2. Use the washer method to find the volume of revolution formed by revolving the area bounded by $y^2 = x$, and $y = x$ about the X-axis.

(3) Cylindrical Shell

dV is a cylindrical shell when the element of area, a rectangle dA, is parallel to the axis of revolution. The formula for dV for a cylindrical shell is

$$dV = 2\pi(radius)(height)(thickness) \tag{6.4.4}$$

This method is used in Example 3.

EXAMPLE 3. Use the cylindrical shell method to find the volume of revolution formed by revolving the area bounded by $y = x^2$ and $y^2 = x$ about the Y-axis.

Solution. The curves intersect at (0, 0) and (1, 1) (see Fig. 6.4.3.) Draw a rectangle dA parallel to the Y-axis in the region at any x such that $0 < x < 1$. Upon revolving this rectangle about the axis of revolution a cylindrical shell is generated [see Fig. 6.4.3(b)]. It has radius $= x$, height $= (y_u - y_l)$, and thickness $= dx$. Using formula 6.4.4, we have

$$dV = 2\pi(x)(y_u - y_l)\, dx$$

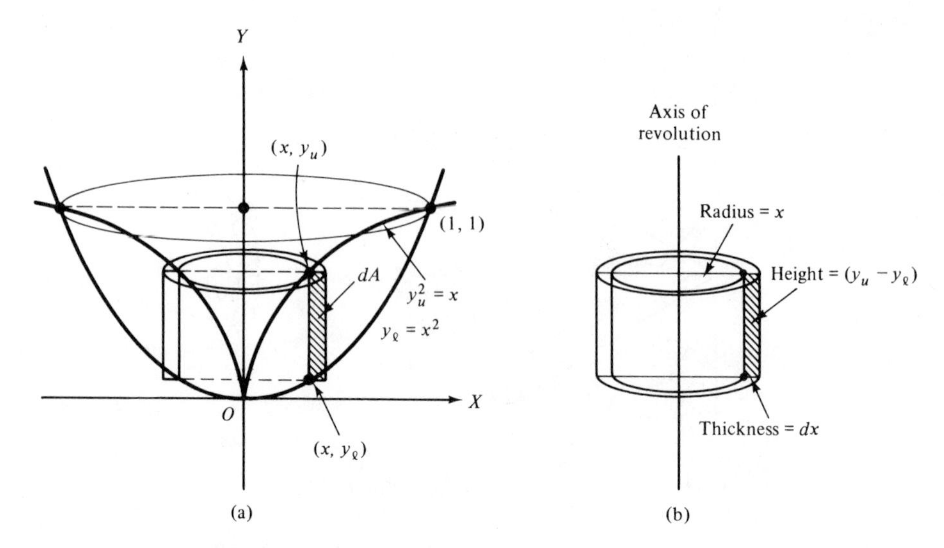

Figure 6.4.3 Volume of revolution (cylindrical shell).

Since $y_u = x^{1/2}$ and $y_l = x^2$, we have

$$dV = 2\pi(x)(x^{1/2} - x^2)\,dx$$
$$= 2\pi(x^{3/2} - x^3)\,dx$$

Thus

$$V = 2\pi \int_0^1 (x^{3/2} - x^3)\,dx$$
$$= 2\pi\left[\frac{2}{5}x^{5/2} - \frac{x^4}{4}\right]_0^1$$
$$= \frac{3\pi}{10} = 0.942$$

PROBLEM 3. Use the cylindrical shell method to find the volume of revolution generated by revolving about the Y-axis the area bounded by $y = x$ and $y = x^2$.

Moments of inertia about the axis of solids of revolution are very important. The cylindrical shell is extremely helpful in finding them since every particle of its mass may be considered to be the same distance from the axis for a small differential thickness.

EXAMPLE 4. Find i_Y for the volume of revolution described in Example 3.

Solution. From Sec. 6.3 we know that

$$i_Y = \int di_Y = \int (\text{distance})^2\,dm$$

If ρ is the volume density, then $dm = \rho\,dV$. Looking at Fig. 6.4.3, we see that for the cylindrical shell

$$dV = 2\pi x(y_u - y_l)\,dx$$

and the distance from the axis of revolution to the shell is x. For a constant volume density, we have

$$di_Y = \underbrace{(x)^2}_{\text{(dist.)}}\underbrace{\rho 2\pi x(y_u - y_l)\,dx}_{dm}$$
$$= 2\pi\rho x^3(x^{1/2} - x^2)\,dx$$
$$= 2\pi\rho(x^{7/2} - x^5)\,dx$$

Hence

$$i_Y = 2\pi\rho \int_0^1 (x^{7/2} - x^5)\,dx$$

Evaluating the integral, we find that

$$i_Y = \frac{\pi\rho}{9} = 0.349\rho$$

(From Example 3 the mass of this volume is 0.942ρ. Thus the radius of gyration is $k_Y = \sqrt{0.349\rho/0.942\rho} = 0.609$.)

PROBLEM 4. Find the moment of inertia of a cone of base radius r and height h about its axis of revolution. Assume constant density. [Refer to Example 1. Is the "height" of the cylindrical shell $(h - x)$?]

When a volume is not one of revolution, the slab (or slice) method may be useful. In this method we obtain an expression of a typical cross-sectional area as a function of the variable denoting the thickness of the slab. The formula for dV is

$$dV = (\textit{area of cross section})(\textit{thickness}) \tag{6.4.5}$$

This method is used in Example 5.

EXAMPLE 5. The base of a solid lies in the XY-plane and is bounded by the curve $y = x^2$. Cross sections perpendicular to the base are squares with two adjacent corners on the curve. Find its volume when $x = 2$.

Solution. See Fig. 6.4.4(a). Note the choice of axes. Since a typical cross section [Fig. 6.4.4(b)] is a square of side $2x$, its area is $A = (2x)^2$. Using Eq. 6.4.5, the element of volume of thickness dy is

$$dV = (2x)^2\,dy = 4x^2\,dy$$

Before integrating we must here express x in terms of y. Since $y = x^2$, we have

$$dV = 4y\,dy$$

and

$$V = 4\int_0^4 y\,dy$$

Hence

$$V = 4\left[\frac{y^2}{2}\right]_0^4 = 32$$

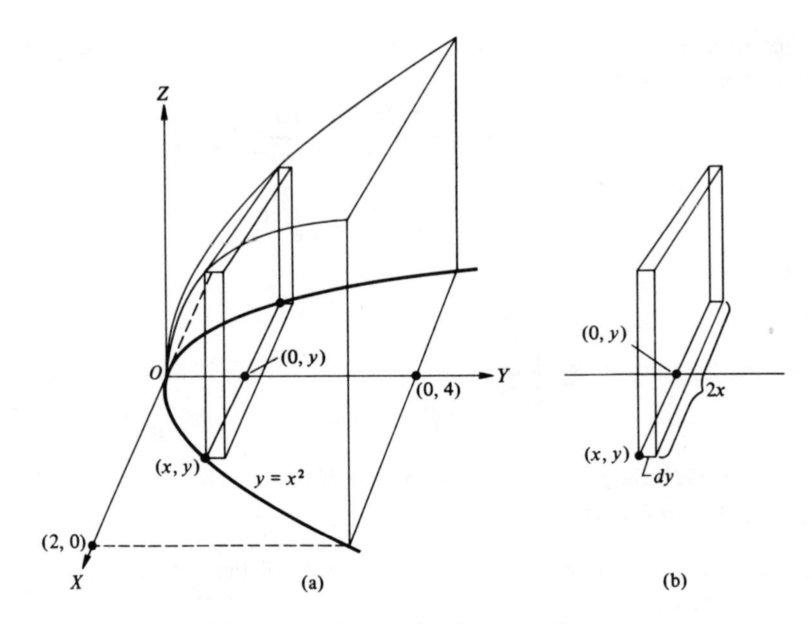

Figure 6.4.4 Volume by slice method.

PROBLEM 5. The base of a solid lies in the XY-plane. It is bounded by $y = 3x$, the Y-axis, and $y = 3$. Find the volume of the solid if every cross section perpendicular to the Y-axis is a square.

EXERCISES 6.4

Group A

1. Find the volume of the solid generated by revolving the region bounded by $y = 2\sqrt{x}$, $x = 3$, and the X-axis about the X-axis.

2. Find the volume of the solid generated by revolving the region bounded by $y = 2 - \frac{1}{2}x^2$, $x = 0$, $x = 2$, and the X-axis about the X-axis.

3. Find the volume of the solid generated by revolving the region bounded by $y = (x - 3)^2$, $x = 0$, $x = 3$, and the X-axis about the X-axis.

4. Find the volume of the solid generated by revolving the region bounded by $y = x$, $y = x^3$, $0 \leqq x \leqq 1$, about the X-axis.

5. Find the volume of the solid generated by revolving the region bounded by $y = x + 3$, $x = -3$, $x = 1$, and the X-axis about the X-axis.

6. Find the volume of the solid generated by revolving the region bounded by $y = 2x$ and $y = x^2$ about the X-axis.

7. Find the volume of the solid generated by revolving the region bounded by $y = -x$ and $y = x^2$ about the X-axis.

8. Let dA be perpendicular to the Y-axis and find the volume of the solid generated by revolving about the Y-axis the region bounded by the given curves.

(a) $y = x + 1$, $y = 3$, and $x = 0$.
(b) $y = x^2$, $y = 1$, $y = 9$, and $x = 0$.
(c) $y = x^2$, $x = 1$, and $y = 0$.
(d) $y = 2x + 1$, $x = 1$, and $y = 9$.

In Exercises 9 through 12, use the cylindrical shell method to find the volume generated by revolving the specified area about the specified axis.

9. Area bounded by $y = x^2$, $x = 2$, and the X-axis about the Y-axis.

10. Area bounded by $y = x^3$, $x = 3$, and the X-axis about the Y-axis.

11. Area bounded by $y^2 = x$ and $y = x^2$ about the X-axis.

12. Area bounded by $y = x$ and $y = x^3$ about the X-axis (first quadrant area only).

13. The base of a solid lies in the XY-plane. It is bounded by $y = x$, the X-axis, and $x = 2$. Find the volume of the solid if every cross section perpendicular to the X-axis is (a) a square; (b) an isosceles right triangle with the right angle on the X-axis.

14. The base of a solid lies in the XY-plane. It is a semicircle with its boundaries the X-axis and $x^2 + y^2 = 9$. Find the volume of the solid if every cross section perpendicular to the Y-axis is (a) a square; (b) a semicircle.

Group B

1. Derive the formula for the volume of a right circular cylinder of base radius r and height h. (*Hint*: Revolve a rectangle about one edge.)

2. Derive the formula for the volume of a sphere of radius r. (*Hint*: Revolve the circle $x^2 + y^2 = r^2$ about the X-axis.)

3. Explain in your own words the formulas for dV when dV is (a) a solid disk; (b) a washer; (c) a cylindrical shell. For part (c) sketch a figure showing the geometric object obtained when the cylindrical shell is sliced along its "height" and rolled out flat.

4. Find the volume of the solid generated by revolving the region bounded by $y = 2x + 1$, $x = 0$, $x = 4$, and the X-axis (a) about the X-axis; (b) about the Y-axis.

5. Find the volume of the solid generated by revolving the region bounded by $y = x^2 + 1$, $x = 0$, and $y = 3$ (a) about the X-axis; (b) about the Y-axis.

6. Find the volume of the solid generated by revolving the region bounded by $y^3 = x$, $y = 0$, $y = 2$, and the Y-axis (a) about the X-axis; (b) about the Y-axis.

7. Find the volume of the solid generated by revolving the region bounded by $y = x^3 - 2x^2 - 4x + 8$ and the X-axis about the X-axis.

8. For each of the following, refer to Fig. 6.4.5. Set up the integral ready to integrate to find the volume when the region A or B is revolved about the designated line. Do each part in two ways: (i) with dA perpendicular to the axis of revolution and (ii) with dA parallel to the axis of revolution.

(a) Region B about the X-axis.
(b) Region B about $x = 2$.
(c) Region A about the Y-axis.
(d) Region A about $y = 8$.
(e) Region A about $x = 2$.
(f) Region B about $y = 8$.

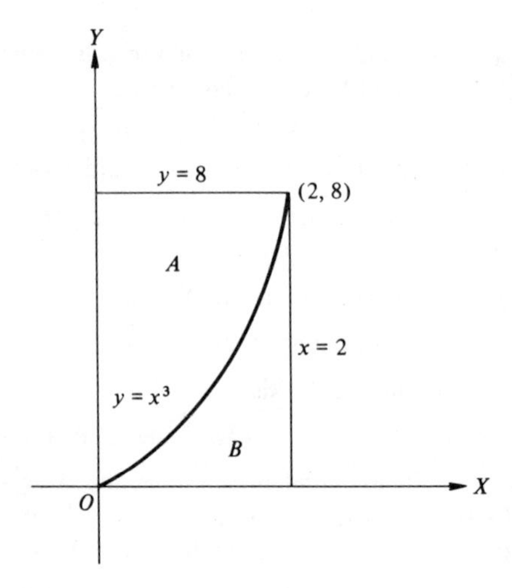

Figure 6.4.5 Exercise 8.

6.5 WORK

The amount of work done by a force in moving a body in a straight line from one position to another is defined to be the *product of the force and the distance the body moves*. The force must be in the direction of motion.

If F is a constant force, then the work, W, done by F lb in moving a body s ft is

$$W = F \cdot s \text{ ft lb}$$

If F is a variable force, then we may use integration methods to find the work done. That is, we "sum a lot of little works to obtain the total work."

EXAMPLE 1. Find the amount of work done in stretching a spring 4 in. from its natural length if it requires 10 lb to stretch the spring 2 in.

Solution. From physics we have a law called Hooke's law that says that the force is proportional to the stretch. That is, $F = kx$, where F is the force, k is the proportionality constant, called the *spring constant*, and x is the stretch. Here we have

$$10 = k(2) \qquad \text{or} \qquad k = 5 \text{ lb/in.}$$

as the spring constant. (A sketch of the spring is shown in Fig. 6.5.1.)

The differential amount of work done over a distance dx is

$$dW = F\,dx = 5x\,dx$$

and

$$W = \int_0^4 5x\,dx = \left[\tfrac{5}{2}x^2\right]_0^4$$

$$= 40 \text{ in. lb}$$

or

$$W = 3\tfrac{1}{3} = 3.3 \text{ ft lb}$$

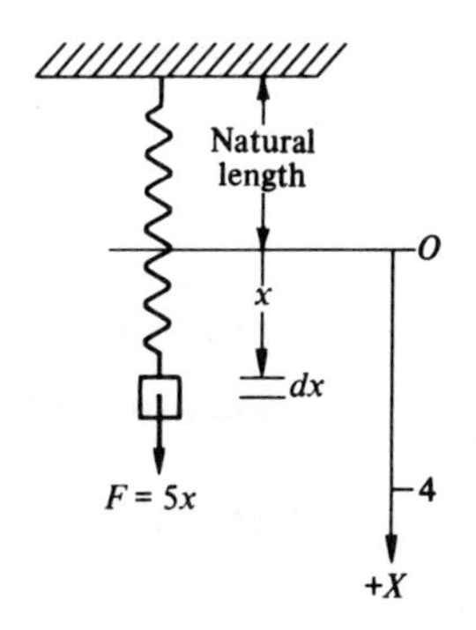

Figure 6.5.1 Work in stretching a spring.

PROBLEM 1. Find the work done in compressing a spring 6 in. from its natural length if it requires 12 lb to compress it 3 in.

EXAMPLE 2. A 60-ft chain hanging from the top of a shaft weighs 15 lb/ft. How much work is required to lift it to the top of the shaft?

Solution. In Fig. 6.5.2 we have chosen the origin of axes at the top and have designated $+X$ as down. Then the weight of x ft of chain is $15x$ lb and

$$dW = 15x\,dx$$

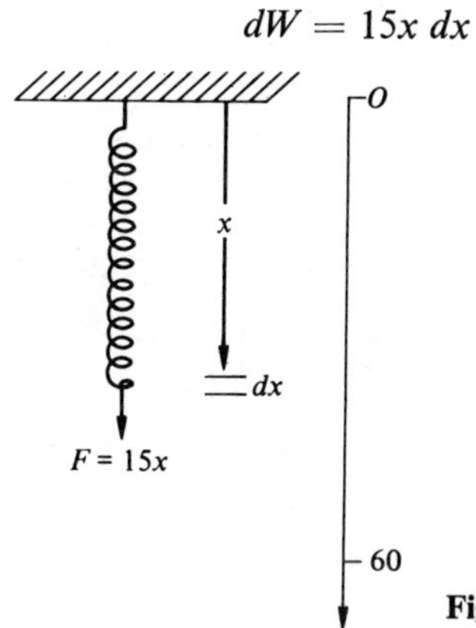

Figure 6.5.2 Work done in lifting a chain.

Since the total work is the sum of all the little works, we have

$$W = \int_{60}^{0} 15x\,dx$$

Note carefully the limits. (When we start to lift the chain $x = 60$; when it is at the top $x = 0$.) Integrating, we have

$$\begin{aligned} W &= [\tfrac{15}{2}x^2]_{60}^{0} \\ &= -27{,}000 \text{ ft lb} \end{aligned}$$

The negative sign occurs because of our choice of axes and we have done work against gravity.

PROBLEM 2. A 100-ft chain weighing 2 lb/ft supports a 200-lb weight. How much work is done in winding up the chain and weight?

EXAMPLE 3. A conical water tank, with vertex down, is filled with water. If the diameter of the top is 10 ft and the height is 15 ft, how much work is required to pump all the water out over the top?

Solution. We shall find the amount of work required to lift a small element of water to the top and then sum for the total work. Referring to Fig. 6.5.3, we note that a small volume element of water dV is

$$dV = \pi r^2\,dx$$

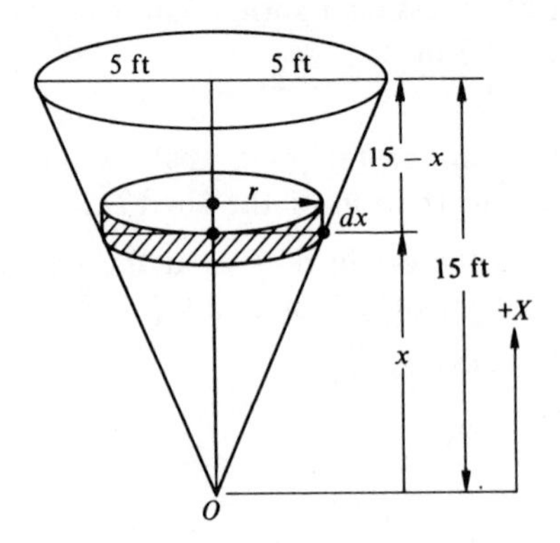

Figure 6.5.3 Work in pumping water out of a cone.

Since water weighs 62.4 lb/ft³, we have for the differential work

$$dW = \underbrace{(62.4\pi r^2\,dx)}_{\text{force}}\underbrace{(15 - x)}_{\text{distance}}$$

We must now obtain r in terms of x. From similar triangles we note that

$$\frac{r}{5} = \frac{x}{15}$$

or

$$r = \tfrac{1}{3}x$$

Then

$$dW = 62.4\pi(\tfrac{1}{3}x)^2(15 - x)\,dx$$

and

$$\begin{aligned} W &= \frac{62.4}{9}\pi \int_{15}^{0} (15x^2 - x^3)\,dx \\ &= -\frac{62.4}{9}\pi\left[(x^3)\left(5 - \frac{x}{4}\right)\right]_0^{15} \\ &= -\frac{62.4}{9}\pi\left[(15)^3\left(5 - \frac{15}{4}\right)\right] \\ &= -92{,}000 \text{ ft lb} \end{aligned}$$

PROBLEM 3. A hemispherical water tank (concave up) of radius 5 ft is filled with water. How much work is required to pump all the water out over the top? (*Hint*: Sketch the tank and show that $dV = \pi r^2\, dx$ and $r = \sqrt{25 - x^2}$.)

Warning ***Be sure that you always include a coordinate system for each figure. Use differentials in writing a general formula for a small amount of work. Express dW in only one variable before integrating.***

EXERCISES 6.5

Group A

1. In each of the following, a particle moves along the X-axis from point a to point b according to the force law, $F(x)$, which is given. Find the work done.

(a) $F(x) = x^2 + 2x + 1$, $a = 1$, $b = 5$.
(b) $F(x) = x^3 + x$, $a = 0$, $b = 4$.
(c) $F(x) = (x^3 + 3x + 1)(x + 2)$, $a = 1$, $b = 2$.
(d) $F(x) = 1/x^2$, $a = 1$, $b = 5$.

2. An 80-ft chain hanging from the top of a shaft weighs 20 lb/ft. How much work is required to lift it to the top of the shaft?

3. A spring with natural length 8 in. requires a force of 20 lb to stretch it $\frac{1}{2}$ in. Find the work done in stretching the spring from 8 to 15 in.

4. A swimming pool has dimensions of $50 \times 25 \times 10$ ft. How much work is done in pumping the water out of the top of the pool?

5. Two electrically unlike charged particles attract each other according to the inverse square law; that is, $F(x) = Q_1Q_2/x^2$, where Q_1 and Q_2 are the constant electrostatic charges of the two particles and x is the distance between them. If charge Q_1 is held at $x = 0$ and Q_2 is moved along the X-axis from $x = 1$ to $x = 4$, how much work is done?

Group B

1. A pail at the bottom of a well weighs 2 lb and contains 8 lb of water. While the pail is raised up the 10-ft shaft of the well, the water leaks out at the rate of $\frac{1}{4}$ lb/ft of lift. How much work is done in raising the pail to the top of the well?

2. In Exercise 4 of Group A, find the work done if the water is pumped out just to the top of the swimming pool when it is one-half full, that is, the water is 5 ft deep. Should this be one-half of the work of Exercise 4?

3. Find the work done in pumping the water out of the top of a full cylindrical storage tank that has a radius of 10 ft and is 50 ft high.

4. The natural length of a spring is 7 in., and 800 lb compresses the spring $\frac{1}{4}$ in. Find the work done in compressing the spring from 7 to 5 in.

5. Two like charged particles repel each other with force $F(r) = 14/r^2$, where r is the distance between the two particles. If one particle is on the Y-axis at point (0, 1) and the other particle is on the Y-axis at point (0, 5), find the work done in moving the second particle along the Y-axis from (0, 5) to (0, 2).

6. A 200-ft chain weighing 2 lb/ft supports a 100-lb load. How much work is done in winding up the chain and load?

7. The force exerted on a certain piston by an expanding gas is given by $F = (20)(10)^3 x^{-1.5}$. Find the work done on the piston as x increases from 6 to 12 in.

8. A rocket of initial weight 100 lb, including fuel, is thrust directly upward. The amount of fuel burned is given by $2x^{1/2}$ lb, where x is the distance above the ground. How much work has been done when the rocket has reached an altitude of 2000 ft?

6.6 AVERAGE VALUE; ROOT MEAN SQUARE

The average value of a function is used frequently. To find an average we usually think of adding the quantities and dividing by the number of them. Since the definite integral represents a sum, we can extend this idea to functions. The *average value* of a function $y = f(x)$ with respect to x over the interval $a \leqq x \leqq b$ is given by

$$\boxed{y_{\text{av}} = \frac{1}{b-a}\int_a^b f(x)\,dx} \tag{6.6.1}$$

EXAMPLE 1. Find the average height of y with respect to x for $y = x^2$ from (1, 1) to (3, 9).

Solution. Using formula 6.6.1, we have

$$y_{\text{av}} = \frac{1}{3-1}\int_1^3 x^2\,dx$$

$$= \frac{1}{2}\left[\frac{x^3}{3}\right]_1^3 = \frac{13}{3} = 4.3$$

PROBLEM 1. Find the average slope of $y = x^2$ with respect to x from (1, 1) to (3, 9). (*Hint:* Remember that the slope is given by dy/dx.)

EXAMPLE 2. Find the average voltage with respect to time during the first 2s if $E = 4t - t^2$.

Solution. Using formula 6.6.1, we have

$$E_{\text{av}} = \frac{1}{2-0}\int_0^2 (4t - t^2)\,dt = \frac{1}{2}\left[2t^2 - \frac{t^3}{3}\right]_0^2 = \frac{8}{3} = 2.67$$

This value would be the reading on a dc voltmeter.

PROBLEM 2. The power, P watts, developed in a resistor of R ohms for a dc current of I amperes is given by $P = RI^2$. Find the average power with respect to the current for a resistor of 5 Ω as the current increases from 3 A to 6 A.

Warning ***The average value is always taken with respect to some variable.***

The average value of a function is closely associated with the mean-value theorem for integrals (see Exercise 1 of Group C in Exercises 6.6).

In electricity the *root-mean-square* current (or effective current) is the value of the direct current that would develop the same quantity of heat in the same time as an alternating current. It is used mainly for periodic functions and the integral is evaluated over one period.

The root-mean-square value of a function $y = f(x)$ with respect to x for the interval $a \leqq x \leqq b$ is given by

$$y_{\text{rms}} = \sqrt{\frac{1}{b-a}\int_a^b [f(x)]^2\,dx} \tag{6.6.2}$$

EXAMPLE 3. Find I_{rms} for $I = 4t - t^2$ with period $T = 4$.

Solution. First we will evaluate the integral, then divide that answer by the value of T and take the square root.

$$\int_0^4 (4t - t^2)^2\,dt = \left[\frac{16}{3}t^3 - 2t^4 + \frac{1}{5}t^5\right]_0^4 = 33.9$$

Thus

$$I_{\text{rms}} = \sqrt{\frac{33.9}{4-0}} = 2.9$$

PROBLEM 3. Find E_{rms} for $E = 3t$ over the interval $0 \leqq t \leqq 2$.

EXAMPLE 4. Find E_{rms} for $E = 10 \sin 3t$, over the interval $0 \leqq t \leqq 2\pi/3$. (Recall that the period of this sine function is $2\pi/3$.)

Solution. Using formula 6.6.2, we find that

$$E_{\text{rms}} = \sqrt{\frac{1}{(2\pi/3) - 0}\int_0^{2\pi/3} 100 \sin^2 3t\,dt}$$

In Sec. 4.3 we studied antiderivatives of the sine function. Using the trigonometric identity $\sin^2\theta = (1 + \cos 2\theta)/2$, we can rewrite the integral and evaluate it as

$$\frac{100}{2}\int_0^{2\pi/3} (1 + \cos 6t)\,dt = 50\left[t + \frac{\sin 6t}{6}\right]_0^{2\pi/3} = \frac{100\pi}{3}$$

Thus

$$E_{\text{rms}} = \sqrt{\frac{100\pi/3}{2\pi/3}} = \sqrt{50} = 7.07$$

PROBLEM 4. Find E_{rms} for $E = 110 \sin 60\pi t$ over one period. (Here the period is $2\pi/60\pi$.)

EXERCISES 6.6

Group A

1. Find the average height of y with respect to x for each of the following.

(a) $y = 9 - x^2$ from $(0, 9)$ to $(3, 0)$.
(b) $y = x^2 + 2$ from $(1, 3)$ to $(5, 27)$.
(c) $y = x^3$ from $(0, 0)$ to $(3, 27)$.
(d) $y = x^2$ from $(-4, 0)$ to $(4, 0)$.

2. Find the average slope with respect to x of the graphs of the functions in Exercise 1.

3. Find the average velocity with respect to time during the first 4 s for a freely falling body where $v = 32t$.

4. Find the average velocity of a body with respect to time during the third second where $v = 96t - 16t^2$.

5. Find the average velocity with respect to time of a particle moving in a straight line such that $s = 2t^3 - 3t$ for the interval $1 \leqq t \leqq 4$.

6. Find the average voltage with respect to time for each of the following.

(a) $E = 6t - t^2$ for $t = 0$ to $t = 6$.
(b) $E = t(t - 7)$ for the interval $1 \leqq t \leqq 4$.
(c) $E = 12 \sin t$ for the interval $0 \leqq t \leqq \pi$.
(d) $E = 6 \cos 2t$ for the interval $0 \leqq t \leqq \pi$.

7. Find the root-mean-square value of each of the following.

(a) $y = 2x$ for $x = 1$ to $x = 3$.
(b) $y = x^2 - 3$ for $x = 0$ to $x = 2$.
(c) $E = t(1 - t)$ for $t = 0$ to $t = 1$.
(d) $I = t^2 - 2t + 3$ for $t = 0$ to $t = 2$.

Group B

1. Show that the average height of $f(x)$ with respect to x is the altitude of a rectangle of base $b - a$ whose area is equal to the area under the curve $y = f(x)$ above the X-axis from $x = a$ to $x = b$.

2. Show that for the current $I = I_0 \sin \omega t$ that I_{rms} for one period $(0 \leqq t \leqq 2\pi/\omega)$ is 70.7% of I_0.

3. For a freely falling body the displacement $s = 16t^2$. Find the average velocity with respect to s as the body falls from $s = 1$ to $s = 25$. (*Hint*: Use $v = ds/dt$, and then express v in terms of s.)

4. Compare the average height of y with respect to x and the root-mean-square value for each of the following. Do you expect them to be the same or different?

(a) $y = 4$ for $1 \leqq x \leqq 4$.
(b) $y = x$ for $1 \leqq x \leqq 4$.
(c) $y = x^2$ for $1 \leqq x \leqq 4$.
(d) $y = x^3$ for $-1 \leqq x \leqq 1$.

5. (a) Find the average slope with respect to x of $y = x^2$ from $A(-1, 1)$ to $B(2, 4)$. (b) Find the slope of line AB. (c) Compare the answers of parts (a) and (b). (d) Find the point on $y = x^2$ where the slope is equal to the average slope of part (a).

Group C

1. In Sec. 3.2 we stated the mean-value theorem for derivatives. There is also a mean-value theorem for integrals. It is important in theoretical discussions. It states that for f continuous on $a \leqq x \leqq b$, there is at least one number c where $a \leqq c \leqq b$ such that $\int_a^b f(x)\,dx = (b - a)f(c)$. Give a geometrical discussion of this theorem and compare it with Eq. 6.6.1.

SUMMARY OF IMPORTANT WORDS AND CONCEPTS

For each of the following, state in your own words your understanding of the statement or word.

1. Area between two curves.

2. (a) Moment. (b) Center of gravity. (c) Centroid. (d) Moment of inertia. (e) Radius of gyration.

3. (a) Volume of revolution. (b) Volume by plane slices. (c) Work.

4. (a) Average value. (b) Root mean square.

Chapter 7

Algebraic Functions

7.1 INTRODUCTION

In preceding chapters we have introduced and illustrated the fundamental ideas of differentiation and integration primarily by using polynomial functions. Now we extend these ideas to other functions.

Algebraic functions are functions that may be formed by any one of or any combination of the algebraic operations performed on power functions (see Sec. 2.3). The algebraic operations are addition, subtraction, multiplication, division, raising to a power, and taking roots. For example, $f(x) = [x(x^{1/3} - 7)]/(x + 5)$ and $f(x) = \sqrt{x^2 - 5}$ are algebraic functions. Polynomials are algebraic functions. $f(x) = 3^x$ is *not* an algebraic function. It is an exponential function and discussed in Sec. 8.2.

Before discussing derivatives of algebraic functions, we state four theorems on limits. In these we assume that we are taking the limits at the same point in the domain of each function.

Theorem 7.1.1 *The limit of the sum of a finite number of functions is equal to the sum of the limits of the functions; that is,* $\lim (u + v + w) = \lim u + \lim v + \lim w$.

Theorem 7.1.2 *The limit of the product of a finite number of functions is equal to the product of the limits of the functions; that is,* $\lim (uvw) = (\lim u)(\lim v)(\lim w)$.

Theorem 7.1.3 *The limit of a constant times a function is the product of the constant and the limit of the function; that is,* $\lim cu = c \lim u$.

Theorem 7.1.4 *The limit of the quotient of two functions is equal to the quotient of the limits of the functions only if the limit of the denominator is not zero; that is,* $\lim (u/v) = (\lim u)/(\lim v)$, $\lim v \neq 0$.

EXERCISES 7.1

Group A

1. Determine whether the following functions are algebraic or not algebraic.

(a) $f(x) = \dfrac{x^3 - 3x^2 + 4}{(x^2 + 2)^{1/2}}$. (b) $f(x) = x^{3/4} + 10$.

(c) $f(x) = (\frac{3}{4})^x + 2$. (d) $f(x) = \sqrt{1 - x^2}$.

(e) $f(x) = \dfrac{x^3 - 3x^2 + 4}{(x^2 + 2)^{1/2}} + \sqrt{1 - x^2} + 10^x$.

2. Compute the limits of each of the following functions. Make use of the theorems in Sec. 7.1 and indicate which theorem you have used.

(a) $\lim\limits_{x \to 2} (x^2 + 2)(1 + x^2)^{1/2}$. (b) $\lim\limits_{x \to 0} [(25 + 10x) + (1 - x^2)^{3/2}]$.

(c) $\lim\limits_{x \to 1} \dfrac{x^2 + 10x + 16}{x^2 - 2x + 3}$. (d) $\lim\limits_{x \to \sqrt{3}} (x^2 + 3)(1 + x^4)^{1/2}$.

(e) $\lim\limits_{x \to \pi} \dfrac{2x + 4\pi}{\sqrt{\pi^2 + 3x^2}}$.

7.2 DERIVATIVE OF A CONSTANT, CONSTANT TIMES A FUNCTION, AND DERIVATIVE OF THE SUM OF TWO FUNCTIONS

In Chap. 2 we stated and made use of some theorems on derivatives. In this and the next several sections we prove these theorems.

In the following discussions y, u, v, and w are all differentiable functions of x.

Theorem 7.2.1 *The derivative of a constant is zero; that is,*

$$\boxed{\frac{d(c)}{dx} = 0} \tag{7.2.1}$$

Proof: If $y = c$, $\Delta y = 0$, and $\lim\limits_{\Delta x \to 0} (\Delta y/\Delta x) = 0$.

$$\therefore \quad \frac{dy}{dx} = \frac{d(c)}{dx} = 0$$

EXAMPLE 1. For $y = \pi$, $dy/dx = 0$.

PROBLEM 1. What is ds/dt for $s = \pi^2$? Why?

Theorem 7.2.2 *The derivative of a constant times a function is the product of the constant and the derivative of the function; that is,*

$$\boxed{\frac{d(cu)}{dx} = c\frac{du}{dx}} \tag{7.2.2}$$

Proof: Let $y = cu$:

$$y + \Delta y = c(u + \Delta u)$$

$$\Delta y = c(u + \Delta u) - cu = c\,\Delta u$$

$$\frac{\Delta y}{\Delta x} = \frac{c\Delta u}{\Delta x}$$

$$\lim_{\Delta x \to 0} \frac{\Delta y}{\Delta x} = \lim_{\Delta x \to 0} c\frac{\Delta u}{\Delta x} = c \lim_{\Delta x \to 0} \frac{\Delta u}{\Delta x} \qquad \text{(by Theorem 7.1.3)}$$

$$\therefore \quad \frac{dy}{dx} = c\frac{du}{dx}$$

EXAMPLE 2. For $y = 7(x^2 - 3x + 5)$, $dy/dx = 7(d/dx)(x^2 - 3x + 5) = 7(2x - 3)$.

PROBLEM 2. Find dy/dx for $y = 13(x^3 - 5x^2 + 23)$.

Theorem 7.2.3 *The derivative of the sum of a finite number of functions is equal to the sum of the derivatives of the functions; that is, for two functions*

$$\boxed{\frac{d}{dx}(u + v) = \frac{du}{dx} + \frac{dv}{dx}} \tag{7.2.3}$$

Proof: Let $y = u + v$:

$$y + \Delta y = (u + \Delta u) + (v + \Delta v)$$

$$\Delta y = u + \Delta u + v + \Delta v - (u + v)$$

$$\Delta y = \Delta u + \Delta v$$

$$\frac{\Delta y}{\Delta x} = \frac{\Delta u}{\Delta x} + \frac{\Delta v}{\Delta x}$$

$$\lim_{\Delta x \to 0} \frac{\Delta y}{\Delta x} = \lim_{\Delta x \to 0} \frac{\Delta u}{\Delta x} + \lim_{\Delta x \to 0} \frac{\Delta v}{\Delta x} \qquad \text{(by Theorem 7.1.1)}$$

$$\frac{dy}{dx} = \frac{du}{dx} + \frac{dv}{dx}$$

$$\therefore \quad \frac{d(u + v)}{dx} = \frac{du}{dx} + \frac{dv}{dx}$$

EXAMPLE 3. For $y = (x^3 - 5x) + (x^{1/2} - 7)$, $dy/dx = (3x^2 - 5) + (\frac{1}{2})x^{-1/2}$.

PROBLEM 3. Find dy/dx for $y = 2(x^2 - 6) - (x^{1/3} - 4x)$.

The proof of Theorem 7.2.3 may be extended to the sum of any number of differentiable functions (see Exercise 1 of Group C in Exercises 7.2).

Warning ***Be sure that you have learned and understand the word statement of the theorems.***

EXERCISES 7.2

Group A

1. Find the derivative with respect to the independent variable of each of the following functions.

(a) $f(x) = \sqrt{95}$.
(b) $f(x) = 10(x^2 + 3x + 2)$.
(c) $f(x) = (x^3 + 13x + 5) + x^{1/2}$.
(d) $f(x) = x^{4/5} + 10x^2 + 35x^{1/8}$.
(e) $f(x) = 5[(x^4 + 2x^2 + 10) + x^{3/2}]$.
(f) $f(x) = \pi x^2 - 5x - \sqrt{x}$.
(g) $f(t) = 3/t^{1/2} + 3t^{1/2}$.
(h) $f(z) = \pi^2 + 3^2$.

Group B

1. An object moves straight up from the ground according to the rule

$$s = 3(-490t^2 + 2450t)$$

where s is measured in centimeters and t is measured in seconds. Compute the maximum height attained and the time required to attain it.

2. Find the critical points and inflection points of the function $f(x) = 13(x^4 - 5x^2 + 5)$.

3. If work $W = 12(t^2 - 3t^{1/2} + 5)$, find the power at $t = 4$.

4. How fast is the velocity changing at $t = 3$ if $v = -7(t^2 - 3) + 11(\sqrt{t} - t^{3/2})$?

5. When is the slope negative for $y = \sqrt{2}(x^{1/3} + x^{-1/3})$?

Group C

1. (a) Extend Theorem 7.2.3 to a finite number of differentiable functions. (b) Prove your theorem.

2. Write a formula for the derivative with respect to x of (a) $y = c(u + v)$; (b) $y = c(u - v - w)$.

7.3 DERIVATIVE OF A PRODUCT

The theorems of Sec. 7.2 seemed to follow logically from some of the rules of algebra. But now we get something different. Consider $f(x) = x^9$; we know that $f'(x) = 9x^8$. What would happen if we considered $f(x)$ as a product, say $f(x) = x^5 \cdot x^4$ and let $g(x) = x^5$ and $h(x) = x^4$? Here $g'(x) = 5x^4$ and $h'(x) = 4x^3$, so $g'(x) \cdot h'(x) = 20x^7$, which certainly is not $f'(x)$. Thus we see that the derivative of a product is *not* the product of the derivatives. Fortunately, Theorem 7.3.1 comes to our rescue.

Theorem 7.3.1 *The derivative of a product of two functions is the sum of the product of the first function and the derivative of the second function and the product of the second function and the derivative of the first function; that is,*

$$\boxed{\frac{d}{dx}(uv) = u\frac{dv}{dx} + v\frac{du}{dx}} \tag{7.3.1}$$

Proof: Let $y = uv$, where u and v are differentiable functions of x.

$$y + \Delta y = (u + \Delta u)(v + \Delta v)$$

$$\Delta y = uv + u\,\Delta v + v\,\Delta u + \Delta u\,\Delta v - uv$$

$$= u\,\Delta v + v\,\Delta u + \Delta u\,\Delta v$$

$$\frac{\Delta y}{\Delta x} = u\frac{\Delta v}{\Delta x} + v\frac{\Delta u}{\Delta x} + \frac{\Delta u}{\Delta x}\Delta v$$

$$\lim_{\Delta x \to 0} \frac{\Delta y}{\Delta x} = \lim_{\Delta x \to 0} u\frac{\Delta v}{\Delta x} + \lim_{\Delta x \to 0} v\frac{\Delta u}{\Delta x} + \lim_{\Delta x \to 0}\left(\frac{\Delta u}{\Delta x}\Delta v\right)$$

Because u and v are not affected as $\Delta x \longrightarrow 0$, and using theorems in Sec. 7.1, we may write

$$\lim_{\Delta x \to 0} \frac{\Delta y}{\Delta x} = u \lim_{\Delta x \to 0} \frac{\Delta v}{\Delta x} + v \lim_{\Delta x \to 0} \frac{\Delta u}{\Delta x} + \lim_{\Delta x \to 0} \frac{\Delta u}{\Delta x} \lim_{\Delta x \to 0} \Delta v$$

Since v is a function of x, $\Delta v \longrightarrow 0$ as $\Delta x \longrightarrow 0$; thus we have

$$\frac{dy}{dx} = u\frac{dv}{dx} + v\frac{du}{dx} + \frac{du}{dx} \cdot 0$$

and

$$\frac{d(uv)}{dx} = u\frac{dv}{dx} + v\frac{du}{dx}$$

EXAMPLE 1. For $y = x^5 \cdot x^4$, we have

$$\frac{dy}{dx} = x^5\frac{d(x^4)}{dx} + x^4\frac{d(x^5)}{dx} = x^5 \cdot 4x^3 + x^4 \cdot 5x^4$$

or

$$\frac{dy}{dx} = 4x^8 + 5x^8 = 9x^8$$

PROBLEM 1. Find dy/dx for $y = (3x^2 - x)(4x^3 + 5x)$ (a) by the product rule; (b) by expanding and then differentiating. Are the answers the same?

This theorem may be extended to products of more than two functions. Consider $y = uvw$. Then $y = u(vw)$ and

$$\frac{dy}{dx} = u\frac{d(vw)}{dx} + vw\frac{du}{dx}$$

$$= u\left(v\frac{dw}{dx} + w\frac{dv}{dx}\right) + vw\frac{du}{dx}$$

and

$$\frac{dy}{dx} = uv\frac{dw}{dx} + uw\frac{dv}{dx} + vw\frac{du}{dx}$$

EXAMPLE 2. For $y = (x^2 + 3x)(x^{1/2})(2x^3 - 7)$, $dy/dx = (x^2 + 3x)(x^{1/2})(6x^2) + (x^2 + 3x)(2x^3 - 7)(\frac{1}{2}x^{-1/2}) + (x^{1/2})(2x^3 - 7)(2x + 3)$.

PROBLEM 2. Find ds/dt for $s = (t^3 - 2)(t^{2/3})(5t^2 - 9t)$.

Warning ***Remember that the derivative of a product is always a sum.***

EXERCISES 7.3

Group A

1. Write the derivatives with respect to the independent variable of each of the following functions.

(a) $f(x) = x^4 \cdot x^{10}$. [Compare the result with $f'(x)$ for $f(x) = x^{14}$.]
(b) $f(x) = (x^3 + 2)(x^2 + 1)$.
(c) $f(x) = (x^{3/2} + 2x)(x^2 + \sqrt{2})$.
(d) $s(t) = t^3(t^2 + 2)$.
(e) $g(y) = y^{1/16}(16y^2)$.
(f) $I = t^{1/3}(t^3 - 3t)$.
(g) $Q = (3t^2 + 4)(\sqrt{t} - \pi)$.
(h) $y = (x^2 - 3x)(x^{1/3})(x^3 + 7)$.
(i) $I = (t - 2)(t^2 + 3)(2t - 1)$.
(j) $z = (t - 1)(t + 1)(t + 2)(t - 3)$.

Group B

1. Write each of the following as a product and then compute the derivative with respect to the independent variable.

(a) $y = \dfrac{x^2 + 2}{x^3}$. (b) $I = \dfrac{t^{1/3} + 2}{3t}$.

(c) $Q = \dfrac{3 + t}{3t}$. (d) $y = \dfrac{t^3 - t}{t^2}$.

(e) $y = (x^2 + 3)^2$. (f) $y = (x^3 - 2x)^3$.

2. Let $f(x) = (x^2 - 9)(x^3 + 2x^2 - x - 2)$. Find $f'(x)$ and $f''(x)$.

3. A particle moves horizontally with motion described by $s(t) = (t^2 + 2)(t + 1)$. Find the speed of the particle at the end of 5 s.

4. A missile is fired directly upward and its motion is described by $s(t) = 16(t - 500)t$, where s is measured in feet and t is measured in seconds. At what time, to the nearest second, will the missile start to fall to earth?

5. If the work $W = 3(t^2 + 5)(3t^{1/2} - 5)$, find the power at $t = 4$.

Group C

1. State a theorem in words for the derivative of the product of three or more functions.

2. (a) Write $y = (5x + 3)^2$ as $y = (5x + 3)(5x + 3)$ and find dy/dx.
(b) Write $y = (5x + 3)^3$ as $y = (5x + 3)(5x + 3)(5x + 3)$ and find dy/dx.
(c) Could this method be extended to $y = (ax + b)^n$? What would dy/dx be?

7.4 DERIVATIVE OF A QUOTIENT

Similar to the derivative of a product, the derivative of a fraction is not what one would suspect.

> **Theorem 7.4.1** *The derivative of a quotient of two functions is equal to the denominator times the derivative of the numerator minus the numerator times the derivative of the denominator, all divided by the square of the denominator; that is,*

$$\boxed{\frac{d}{dx}\left(\frac{u}{v}\right) = \frac{v\dfrac{du}{dx} - u\dfrac{dv}{dx}}{v^2}} \tag{7.4.1}$$

Proof: Let $y = u/v$, where u and v are differentiable functions of x:

$$y + \Delta y = \frac{u + \Delta u}{v + \Delta v}$$

$$\Delta y = \frac{u + \Delta u}{v + \Delta v} - \frac{u}{v} = \frac{uv + v\,\Delta u - uv - u\,\Delta v}{v(v + \Delta v)}$$

$$\frac{\Delta y}{\Delta x} = \frac{v\,\Delta u - u\,\Delta v}{v(v + \Delta v)\,\Delta x} = \frac{v(\Delta u/\Delta x) - u(\Delta v/\Delta x)}{v(v + \Delta v)}$$

$$\lim_{\Delta x \to 0} \frac{\Delta y}{\Delta x} = \frac{\lim_{\Delta x \to 0} v(\Delta u/\Delta x) - \lim_{\Delta x \to 0} u(\Delta v/\Delta x)}{\lim_{\Delta x \to 0} v(v + \Delta v)}$$

By using theorems of Sec. 7.1 and the fact that $\Delta v \longrightarrow 0$ as $\Delta x \longrightarrow 0$, we have

$$\frac{dy}{dx} = \frac{v(du/dx) - u(dv/dx)}{v^2}$$

or

$$\frac{d}{dx}\left(\frac{u}{v}\right) = \frac{v(du/dx) - u(dv/dx)}{v^2}$$

EXAMPLE 1. For $y = (x^2 - 2)/3x$,

$$\frac{dy}{dx} = \frac{\overbrace{(3x)}^{v}\overbrace{(2x)}^{du/dx} - \overbrace{(x^2 - 2)}^{u}\overbrace{(3)}^{dv/dx}}{\underbrace{(3x)^2}_{v^2}}$$

$$= \frac{6x^2 - 3x^2 + 6}{9x^2} = \frac{3x^2 + 6}{9x^2} = \frac{x^2 + 2}{3x^2}$$

Warning ***Do not get the numerator and denominator interchanged.***

PROBLEM 1. Find ds/dt for $s = (t^3 - 2t)/5t^2$.

If we let $u = 1$ in Theorem 7.4.1 we get a useful result:

$$\boxed{\frac{d}{dx}\left(\frac{1}{v}\right) = -\frac{1}{v^2}\frac{dv}{dx}} \qquad (7.4.2)$$

You may prove this by applying Theorem 7.4.1.

EXAMPLE 2. For $y = 3/(x^2 - 7)$,

$$\frac{dy}{dx} = -\frac{3}{(x^2 - 7)^2}(2x) = \frac{-6x}{(x^2 - 7)^2}$$

PROBLEM 2. Find dW/dt for $W = 5/(3t - 2)$.

PROBLEM 3. Write out formula 7.4.2 in your own words.

EXERCISES 7.4

Group A

1. Find the derivative with respect to the independent variable of each of the following functions.

(a) $f(x) = \dfrac{x-1}{x+1}$.

(b) $f(x) = \dfrac{3x+5}{2x}$.

(c) $y = \dfrac{2x-7}{x^3}$.

(d) $y = \dfrac{x^2}{3x-1}$.

(e) $I = \dfrac{3}{t^2}$.

(f) $I = \dfrac{3}{t^2-7}$.

(g) $Q = \dfrac{5}{t^{1/3}-4}$.

(h) $Q = \dfrac{-3}{2t^3-5t}$.

(i) $f(x) = \dfrac{x^2}{x^3+2}$.

(j) $f(x) = \dfrac{x^2+1}{4x+2}$.

(k) $f(x) = \dfrac{x^{3/4}}{2x^2+3x+1}$.

(l) $f(x) = \dfrac{x^4+3x^3+4x^2+1}{x}$.

(m) $f(x) = \dfrac{4x^5+2x+2}{x^{1/2}}$.

Group B

1. Find the derivative with respect to the independent variable of each of the following.

(a) $y = \dfrac{(x-3)(x+1)}{2x-7}$.

(b) $y = \dfrac{3x-2}{(2x)(x-1)}$.

(c) $y = \dfrac{(x^2-1)(3x^2)}{(x^3)(2x-1)}$.

(d) $y = \dfrac{(x+3)(x^4+2x)}{(3x-7)(x+9)}$.

(e) $y = \dfrac{\pi}{(3x-2)(x^2)}$.

(f) $y = \dfrac{-7}{(x^3-x)(x^{2/3})}$.

(g) $I = \dfrac{-3}{(t^2-1)(4t+3)}$.

(h) $I = \dfrac{\sqrt{\pi}}{(t-t^3)(t^{1/2}-1)}$.

2. Find an equation of the line tangent to $y = 3/(x^2+4)$ where $x = 2$.

3. Discuss the slope of the curve $xy = 1$. Sketch the curve.

4. For $E = IR$, find the instantaneous rate of change of I with respect to R when $R = 3$ and E is constantly 12 V.

Group C

1. Prove Eq. 7.4.2.

2. The ratio $f(x)/g(x)$ is called an *indeterminate form* if both $f(x)$ and $g(x)$ approach zero as x approaches a (that is the ratio is of the form 0/0). L'Hôpital's rule is useful in finding the limit of this ratio. In brief it states that

$$\lim_{x\to a}\frac{f(x)}{g(x)} = \lim_{x\to a}\frac{f'(x)}{g'(x)}$$

if $f(x)$, $g(x)$, $f'(x)$, and $g'(x)$ all exist at $x = a$ and $f(a) = g(a) = 0$. Note carefully that the right side of the formula is the *ratio of the derivatives* of the given numerator and denominator. (a) State how L'Hôpital's rule differs from the quotient rule for derivatives. Use L'Hôpital's rule for each of the following.

(b) $\lim_{x\to 3}\frac{x^2-9}{x-3}$. (c) $\lim_{x\to -2}\frac{x^3+8}{x^2-4}$.

(d) $\lim_{x\to 0}\frac{\sin x}{x}$. (e) $\lim_{x\to \pi/2}\frac{\cos x}{x-\pi/2}$.

(f) $\lim_{x\to 0}\frac{1-\cos x}{x}$. (g) $\lim_{x\to 2}\frac{x^3-6x^2+12x-8}{x^2-4x+4}$.

3. The form $\infty - \infty$ is also an indeterminate form. It can usually be rewritten in the form 0/0. Show that

$$\lim_{x\to 1}\left[\frac{2}{x^2-1} - \frac{1}{x-1}\right]$$

which is $\infty - \infty$, can be rewritten as

$$\lim_{x\to 1}\left[\frac{1-x}{x^2-1}\right]$$

Use L'Hôpital's rule to find the limit.

7.5 THE CHAIN RULE AND DERIVATIVE OF $y = u^n$

Sometimes functions are constructed out of simpler functions. Functions that are formed by the substitution of one function for the independent variable in another function are called *composite functions*. For example, if in $V = \frac{4}{3}\pi r^3$ we substitute $r = 3t$ to obtain $V = \frac{4}{3}\pi(3t)^3 = 36\pi t^3$, we say that V is a composite function.

If $y = f(u)$ and $u = g(x)$, then $y = f[g(x)]$. If there is a change in x, this will change u, which in turn will change y.

Theorem 7.5.1 *If y is a differentiable function of u and u is a differentiable function of x, then*

$$\boxed{\frac{dy}{dx} = \frac{dy}{du}\,\frac{du}{dx}} \qquad (7.5.1)$$

Proof: When $du/dx \neq 0$: Consider $\Delta y/\Delta x$; this may be written as

$$\frac{\Delta y}{\Delta x} = \frac{\Delta y}{\Delta u}\frac{\Delta u}{\Delta x}$$

$$\lim_{\Delta x \to 0} \frac{\Delta y}{\Delta x} = \lim_{\Delta x \to 0} \frac{\Delta y}{\Delta u} \lim_{\Delta x \to 0} \frac{\Delta u}{\Delta x} \qquad \text{(by Theorem 7.1.2)}$$

As $\Delta x \longrightarrow 0$, $\Delta u \longrightarrow 0$, and $\Delta y \longrightarrow 0$, also

$$\lim_{\Delta x \to 0} \frac{\Delta y}{\Delta u} = \lim_{\Delta u \to 0} \frac{\Delta y}{\Delta u}$$

$$\therefore \quad \frac{dy}{dx} = \frac{dy}{du}\frac{du}{dx}$$

Note. This proof is not valid if $du/dx = 0$, but mathematicians have proved Theorem 7.5.1 for this case also. Equation 7.5.1 is sometimes called the *chain rule*.

EXAMPLE 1. For $y = u^3$ and $u = 2x^3 - 3x$, find dy/dx.

Solution. Applying the chain rule (Eq. 7.5.1), we have

$$\frac{dy}{du} = 3u^2 \qquad \text{and} \qquad \frac{du}{dx} = 6x^2 - 3$$

Therefore,

$$\frac{dy}{dx} = \underbrace{3u^2}_{dy/du}\underbrace{(6x^2 - 3)}_{du/dx} = 3(2x^3 - 3x)^2(6x^2 - 3)$$

Of course dy/dx could have been found by first substituting the value for u in y. Thus $y = (2x^3 - 3x)^3$, then expanding by multiplication and differentiating the resulting polynomial. You will see the beauty of the chain rule as we proceed.

PROBLEM 1. Use the chain rule to find dV/dt for $V = \frac{4}{3}\pi r^3$ and $r = 3t^4 - 2t^2$.

We now apply the chain rule, Theorem 7.5.1, to find a formula for the derivative of a function raised to a power.

The derivative with respect to x of u^n where u is a differentiable function of x and n is a constant is given by

$$\boxed{\frac{d(u^n)}{dx} = nu^{n-1}\frac{du}{dx}} \qquad (7.5.2)$$

Proof: Consider $y = u^n$, where u is a differentiable function of x. Then $dy/du = nu^{n-1}$ (from Eq. 2.3.1), and using Theorem 7.5.1, we have

$$\frac{dy}{dx} = nu^{n-1}\frac{du}{dx}$$

We may state formula 7.5.2 in words as:

*The derivative of a function raised to a power, **n**, is the product of the power, **n**, the function raised to the (**n** − **1**)th power, and the derivative of the function.*

Formula 7.5.2 is sometimes referred to as the *power rule*.

EXAMPLE 2. For $y = (3x^2 - 5x)^7$, find dy/dx.

Solution. Here $n = 7$ and $u = 3x^2 - 5x$; therefore,

$$\frac{dy}{dx} = \underbrace{7}_{n}\underbrace{(3x^2 - 5x)^6}_{u^{n-1}}\underbrace{(6x - 5)}_{du/dx}$$

PROBLEM 2. Find dy/dx for $y = (5x^3 - 3x)^9$.

Note carefully that formula 7.5.2 is also true for $y = x^n$. For we then have $dy/dx = nx^{n-1}[d(x)/dx]$, but since $d(x)/dx = 1$, $dy/dx = nx^{n-1}$.

EXAMPLE 3. If $y = \sqrt{x^2 - 25}$, find dy/dx.

Solution. $y = (x^2 - 25)^{1/2}$.

$$\frac{dy}{dx} = \underbrace{\tfrac{1}{2}}_{n}\underbrace{(x^2 - 25)^{-1/2}}_{u^{n-1}}\underbrace{(2x)}_{du/dx}$$

$$= \frac{x}{\sqrt{x^2 - 25}}$$

PROBLEM 3. Find ds/dt for $s = \sqrt[3]{x^3 - 8}$.

Warning ***When using formula 7.5.2, always end with the "derivative of the function."***

In formula 7.5.2 u may be any differentiable function of x as shown in Example 4.

EXAMPLE 4. Find dy/dx for $y = 2 \sin^5 7x$.

Solution. Since $y = 2 \sin^5 7x = 2(\sin 7x)^5$, here $u = \sin 7x$ and $n = 5$. Thus we write

$$\frac{dy}{dx} = 2(5)(\sin 7x)^4 \frac{d}{dx}(\sin 7x)$$

$$= 10(\sin^4 7x)(7 \cos 7x) \qquad \text{(from Eq. 2.10.8)}$$

and

$$\frac{dy}{dx} = 70 \sin^4 7x \cos 7x$$

PROBLEM 4. Find dy/dx for $y = 3\cos^7 5x$.

Another theorem, which we state without proof, is

Theorem 7.5.2 *If $y = f(u)$ and $x = g(u)$,*

$$\frac{dy}{dx} = \frac{dy/du}{dx/du}, \quad \frac{dx}{du} \neq 0 \tag{7.5.3}$$

EXAMPLE 5. If $y = u^7$ and $x = u^3$, $dy/du = 7u^6$, $dx/du = 3u^2$, and $dy/dx = 7u^6/3u^2 = \frac{7}{3}u^4$.

PROBLEM 5. Find dy/dx if $y = u^3 - 5u$ and $x = (2u - 1)^4$.

EXERCISES 7.5

Group A

1. Find dy/dx for each of the following.

(a) $y = u^2$, $u = 3x^2 - 7x$.
(b) $y = 3u^5$, $u = x^3 + 2x^2 - 5$.
(c) $y = u^3 - u^2$, $u = 5x^2 + 4x + 7$.
(d) $y = u^2 + 3u - 7$, $u = x^3 - 7x + 11$.

2. Differentiate each of the following with respect to the independent variable.

(a) $f(x) = (x^2 + 2x + 5)^3$.
(b) $f(x) = (x^8 + 10x^4 + 2x^2)^{16}$.
(c) $f(x) = (x^2 + 2)^{1/2}$.
(d) $f(x) = \sqrt[3]{(x^4 + 2x^3 + 3)}$.
(e) $f(x) = (x^{3/4} + x^{1/2} + x^{1/4})^2$.
(f) $I = \sqrt[3]{(t - 2)^2}$.
(g) $I = \sqrt{3t} - t\sqrt{t}$.
(h) $y = \sqrt{1 - 4x^2}$.

3. In each of the following, find dy/dx.

(a) $y = u^3$, $x = u^2$.
(b) $y = u^5$, $x = u^3 - 2u$.
(c) $y = u^2 - 3u$, $x = u^7$.
(d) $y = 3u^2 + u$, $x = u^3 - u$.

Group B

1. Find dy/dx for $y = (u^2 - 3u)^{1/2}$, $u = 2x - 1$.

2. Find dy/dx for $y = (u^2 + 2u)^{1/3}$, $u = (x^2 - 3)^{1/2}$.

3. If $V = \frac{4}{3}\pi r^3$ and $r = 2t - 1$, find dV/dt at $t = 3$.

4. The volume of a sphere of radius r is $V = \frac{4}{3}\pi r^3$. Its surface area is $S = 4\pi r^2$. Find the rate of change of the volume with respect to the surface.

5. If the radius of a circle is given as $r = -3t^2 + 5$, how fast is the area changing when $t = 2$?

6. In a dc circuit, $E = IR$, the voltage is constantly 12. How fast is I changing at $t = 3$ if $R = 3(t^2 - t)$?

7. In each of the following, write the fraction as a product [see part (a)]; then find dy/dx. Check your answer by using the quotient rule.

(a) $y = \dfrac{3x}{2x - 5} = 3x(2x - 5)^{-1}$.
(b) $y = \dfrac{7x^2 - 3x}{x^2 - 1}$.
(c) $y = \dfrac{5x - 2}{(3x - 1)^2}$.
(d) $y = \dfrac{3x}{\sqrt{x^2 - 25}}$.

8. Find dy/dx for each of the following.

(a) $y = \sin^3 5x$.
(b) $y = \cos^2 3x$.
(c) $y = 5 \sin^2 x^3$.
(d) $y = 7 \cos^3 2x^2$.

9. At the point where $x = 3$ for the curve $y = \sqrt{25 - x^2}$, find (a) the slope; (b) an equation of the tangent line.

Group C

1. Give a verbal summary of the chain rule.

2. (a) Extend the chain rule to find dy/dx when $y = f(u)$, $u = g(v)$, and $v = h(x)$. (b) Find dy/dx for $y = u^2$, $u = 3v - 1$, $v = (2x + 7)^3$.

3. Prove Theorem 7.5.2.

4. Write a summary of all the rules of differentiation you have studied thus far.

5. If there is a one-to-one correspondence between the elements of the domain and range of a function, the function has an inverse (see Sec. 8.4). For such functions it may be shown that

$$\frac{dx}{dy} = \frac{1}{dy/dx} \tag{7.5.4}$$

Thus for $y = x^3$, $dy/dx = 3x^2$, and $dx/dy = 1/(3x^2)$. For each of the following, find dx/dy in two ways: (i) use Eq. 7.5.4; and (ii) write an equation expressing x in terms of y.

(a) $y = x^5$.
(b) $y = \sqrt{1 - x^2}$.
(c) $y = 6x - 7$.

6. Obtain Eq. 7.4.1 by considering $y = u/v$ as $y = uv^{-1}$.

7.6 SUMMARY OF DIFFERENTIATION FORMULAS

The general formulas of differentiation must be memorized. We now summarize them.

$$\frac{d(c)}{dx} = 0 \tag{7.2.1}$$

$$\frac{d(cu)}{dx} = c\frac{du}{dx} \tag{7.2.2}$$

$$\frac{d(u + v)}{dx} = \frac{du}{dx} + \frac{dv}{dx} \tag{7.2.3}$$

$$\frac{d(uv)}{dx} = u\frac{dv}{dx} + v\frac{du}{dx} \tag{7.3.1}$$

$$\frac{d(u/v)}{dx} = \frac{v(du/dx) - u(dv/dx)}{v^2} \tag{7.4.1}$$

$$\frac{d(1/v)}{dx} = -\frac{1}{v^2}\frac{dv}{dx} \tag{7.4.2}$$

$$\frac{d(u^n)}{dx} = nu^{n-1}\frac{du}{dx} \tag{7.5.2}$$

$$\frac{dy}{dx} = \frac{dy}{du}\frac{du}{dx} \tag{7.5.1}$$

$$\frac{dy}{dx} = \frac{dy/du}{dx/du} \tag{7.5.3}$$

Warning ***You must be able to express each of these formulas clearly in words as well as in symbols.***

Functions may consist of sums, products, quotients, and powers all mixed together. The differentiation of such functions requires the determination of the basic type of function (that is, is it a product, a quotient, or a power?) to get started with the correct differentiation formula.

EXAMPLE 1. For $y = (x^3 - 2x)^4/\sqrt{3x - 2}$, find dy/dx.

Solution. We note that basically we have a fraction with quantities raised to a power. We therefore use formulas 7.4.1 and 7.5.2. Thus

$$\frac{dy}{dx} = \frac{\sqrt{(3x-2)}4(x^3-2x)^3(3x^2-2) - (x^3-2x)^4\frac{1}{2}(3x-2)^{-1/2}(3)}{(\sqrt{3x-2})^2}$$

$$= \frac{\sqrt{(3x-2)}(4)(x^3-2x)^3(3x^2-2) - [3(x^3-2x)^4/2\sqrt{3x-2}]}{3x-2}$$

$$= \frac{8(3x-2)(x^3-2x)^3(3x^2-2) - 3(x^3-2x)^4}{2(3x-2)^{3/2}}$$

If you are clever in algebraic manipulations, you might reduce this further. All we desired to show was a technique of differentiating.

PROBLEM 1. Find ds/dt for $s = \sqrt{3t^2 - 16}/(2t + 5)^3$. Do not simplify.

EXAMPLE 2. For $y = [(2x^3 - x)(6x)]^7$, find dy/dx.

Solution. We note that basically we have a quantity raised to a power. Thus

$$\frac{dy}{dx} = \underbrace{7}_{n}\underbrace{[(2x^3 - x)(6x)]^6}_{u^{n-1}}\underbrace{[(2x^3 - x)(6) + (6x)(6x^2 - 1)]}_{du/dx}$$

You might try to simplify this.

PROBLEM 2. Find du/dv for $u = [(5v)(3v^2 - v)]^4$.

EXAMPLE 3. For $y = [(x^2 - 3)/(5x - 7)]^{1/2}$, find dy/dx.

Solution. Here y is a fraction raised to a power. Therefore,

$$\frac{dy}{dx} = \underbrace{\frac{1}{2}}_{n}\underbrace{\left(\frac{x^2 - 3}{5x - 7}\right)^{-1/2}}_{u^{n-1}}\underbrace{\left[\frac{(5x - 7)(2x) - (x^2 - 3)(5)}{(5x - 7)^2}\right]}_{du/dx}$$

You might try to simplify this.

PROBLEM 3. Find dV/dt for $V = [(2t + 3)/(3t^2 - 7)]^{1/2}$.

We may obtain formulas for differentials from some of the differentiation formulas.

$$d(c) = 0 \tag{7.6.1}$$

$$d(cu) = c\,du \tag{7.6.2}$$

$$d(u + v) = du + dv \tag{7.6.3}$$

$$d(uv) = u\,dv + v\,du \tag{7.6.4}$$

$$d\left(\frac{u}{v}\right) = \frac{v\,du - u\,dv}{v^2} \tag{7.6.5}$$

$$d(u^n) = nu^{n-1}\,du \tag{7.6.6}$$

EXAMPLE 4. Find dy if $y = (x^3 - 2x)^{1/3}$.

Solution. Use Eq. 7.6.6 (why?) to obtain

$$dy = \tfrac{1}{3}(x^3 - 2x)^{-2/3}(3x^2 - 2)\,dx$$

or

$$dy = \frac{3x^2 - 2}{3(x^3 - 2x)^{2/3}}\,dx$$

PROBLEM 4. Find dE if $E = \sqrt{16 - 4t^2}$.

EXERCISES 7.6

Group A

1. Differentiate with respect to x each of the following functions. State which formula you used.

(a) $(x^2 + 2)(x^{1/2})$.

(b) $(x^3 + 3x^2 + 2) + (4x^2 + 1)$.

(c) $\dfrac{x^{1/2}}{x^3 + 1}$.

(d) $\sqrt{x^2 + 2x + 1}$.

(e) $10(x^4 + 1)^2$.

(f) $\sqrt[3]{3 - x^3}$.

2. Differentiate with respect to x each of following functions. State which formula or combination of formulas you used.

(a) $[(x^5 + 2x^3 + 3) + (8x^4 + 2x^2 + 1)]^9$.

(b) $\dfrac{\sqrt{1 - x^2}}{x}$.

(c) $x^{3/2}(x^3 + 3x + 1)^{2/3}$.

(d) $\dfrac{x^2(\sqrt{x} - x^3)}{1 + x}$.

(e) $\dfrac{x^4}{\sqrt{1 - x^2}}$.

(f) $x^{1/2}(x^2 + 2x + 1)^2(x^3 + 1)$.

(g) $[(3x^2 - x)(2x)]^9$.

(h) $\left(\dfrac{x^2 + 1}{2x - 1}\right)^{2/3}$

3. Write the differential of each of the following functions. State which formula you used.

(a) $y = (x^2 + 1)^{1/3}$.

(b) $y = (2x + 1)^{1/2} + x^3$.

(c) $y = \dfrac{x^{1/4}}{x^2 + 1}$.

(d) $y = 21(x^2 + 3x + 5)^3$.

(e) $y = \sqrt{1 + 2\sqrt{2}}$.

(f) $y = 4(x^2 + 3x)^2(x^{1/2} - x^3)$.

4. Find an approximate value of (a) $x^2\sqrt{x^2 + 9}$ when $x = 3.99$; (b) $(x - 1)/\sqrt{26 - x^2}$ when $x = 1.1$.

5. Find d^2y/dx^2 for each of the following.

(a) $y = (3x^2 + 2)^5$.

(b) $y = \dfrac{2x - 3}{x + 1}$.

(c) $y = (x^3 - 2x)^2(2x - 1)$.

(d) $y = \sqrt{2x + 1}$.

Group B

1. Find the maximum and minimum points of the function $f(x) = x^{1/3}(x - 5)^2$.

2. Find the maximum and minimum points of the function $f(x) = (x - 1)/(x^2 + 3)$.

3. Find the rectangle of maximum area that can be inscribed in a semicircle of radius 5.

4. Find the dimensions of the right circular cylinder of maximum volume inscribed in a sphere of radius 4.

5. If the hypotenuse of a right triangle is constant, how long are the legs if the area is a maximum?

6. Find the angle of intersection of the curves $y = (x - 2)^2$ and $y = 2 - (x - 2)^2$.

7. A closed right circular cylinder is to have a volume of 54π in.^{3}. What are its dimensions if the amount of material used is to be a minimum?

8. A nuclear power plant is situated on one shore of a straight river that is 2 mi wide. Ten miles upstream on the opposite shore there is a city. It costs a dollars per mile to construct the transmission line on land and b dollars per mile over water. What is the most economical way to build the transmission line?

9. A cone is formed by revolving an isosceles triangle about its altitude. If the perimeter of the triangle is 20 in., what are the dimensions of the triangle for a cone of maximum volume? ($V = \frac{1}{3}\pi r^2 h$.)

10. The cross section of a ditch is an isosceles trapezoid. If the base is b units wide and the length of one leg is a units, how wide should the top be for maximum cross-sectional area?

11. The velocity of air passing through a flexible tube is given by $V = r^2(r_0 - r)/\pi ak$, where r_0 is the radius when no air is flowing and a and k are constants. What value of r will give the maximum air velocity?

12. The rate of a certain reaction is given by $r = kx(a - x)$. At what concentration x is the rate a maximum?

13. The reaction R of the body to a dose D of a drug is given by

$$R = D^2\left(\frac{k}{2} - \frac{D}{3}\right)$$

where the constant k is the maximum amount that may be given. Find D for the maximum change in R (that is, R' is max.).

Group C

1. Some differential equations may be solved by using the formulas for differentials. Thus $x\,dy + y\,dx = 0$ may be written as $d(xy) = 0$ by Eq. 7.6.4 and then from Eq. 7.6.1 we have $xy = c$. Make use of Eqs. 7.6.1, 7.6.4, and 7.6.5 to solve each of the following differential equations.

(a) $3x\,dy + 3y\,dx = 0$.

(b) $x^2\,dy + 2xy\,dx = 0$.

(c) $2xy^2\,dx + 2yx^2\,dy = 0$.

(d) $\dfrac{x\,dy - y\,dx}{x^2} = 0$.

(e) $\dfrac{y\,dx - x\,dy}{y^2} = 0$.

(f) $\dfrac{2xy\,dx - x^2\,dy}{y^2} = 0$.

7.7 GRAPHS OF RATIONAL FUNCTIONS

Definition 7.7.1 *A rational function, $f(x)$, is defined to be the quotient of two polynomials $P(x)$ and $Q(x)$. Symbolically, $f(x) = P(x)/Q(x)$.*

Examples of rational functions are $f(x) = (x^3 - 3x + 7)/(2x^5 - 1)$, $f(x) = (x^7 - 2x)/(x^2 + 1)$, and $f(x) = 1/x$.

In the following disscussion we assume that $P(x)$ and $Q(x)$ have no common factors. If they do, then we should divide both numerator and denominator by the common factor. For example, if $f(x) = (x^2 + 2x - 3)/(x^3 + 3x^2 + x + 3)$, we should factor and write $f(x) = [(x - 1)(x + 3)]/(x^2 + 1)(x + 3)] = (x - 1)/(x^2 + 1)$.

We are mainly interested in sketching the graph of a rational function. To help us in this endeavor we need a few preliminary remarks. Consider

$f(x) = 1/x$ and construct a table of values for x and $f(x)$. Since $f(0)$ does not exist (why?), we choose values for x around 0. Thus,

x	-1	$-\frac{1}{2}$	$-\frac{1}{5}$	$-\frac{1}{100}$	$-\frac{1}{10,000}$	0	$\frac{1}{10,000}$	$\frac{1}{100}$	$\frac{1}{5}$	$\frac{1}{2}$	1
$f(x) = \frac{1}{x}$	-1	-2	-5	-100	$-10,000$	does not exist	10,000	100	5	2	1

We see that as x approaches 0 through positive values, $f(x)$ becomes larger and larger in a positive sense. We then say that $f(x)$ *increases without bound* or $f(x)$ *approaches positive infinity* as x approaches 0 through positive values. Symbolically, we write $f(x) \longrightarrow \infty$, as $x \longrightarrow 0^+$. (Note the positive sign written as an exponent on the zero.) We also see that as x approaches 0 through negative values, $f(x)$ becomes larger and larger in a negative sense. We then say that $f(x)$ *approaches negative infinity*. Symbolically, we write $f(x) \longrightarrow -\infty$, as $x \longrightarrow 0^-$. (Note the minus sign written as an exponent on the zero.) The symbols ∞ and $-\infty$ are used only to indicate that a value increases or decreases without bound. They do not represent real numbers and they must not be treated as such.

In graphing rational functions we use the first and second derivatives and plot a few points and then investigate the trend of the graph as x approaches the zeros of the denominator.

EXAMPLE 1. Graph $f(x) = 1/(x - 3)$.

Solution. For $f(x) = 1/(x - 3)$, we have

$$f'(x) = -\frac{1}{(x-3)^2}$$

and

$$f''(x) = \frac{+(1)(2)(x-3)}{(x-3)^4} = \frac{2}{(x-3)^3}$$

We see that there are no relative maximum or minimum points and no points of inflection. (Why?) We note that $x = 3$ is a zero of the denominator of $f(x)$ and then construct a table of values for x around 3.

x	-2	-1	0	1	2	3	4	5	6
$f(x)$	$-\frac{1}{5}$	$-\frac{1}{4}$	$-\frac{1}{3}$	$-\frac{1}{2}$	-1	does not exist	1	$\frac{1}{2}$	$\frac{1}{3}$

Now plot these points (see Fig. 7.7.1).

Next we investigate the trend of $f(x)$ as $x \longrightarrow 3$ from values both above and below 3. As $x \longrightarrow 3^+$ (from the right of 3), we see from $f(x) = 1/(x - 3)$

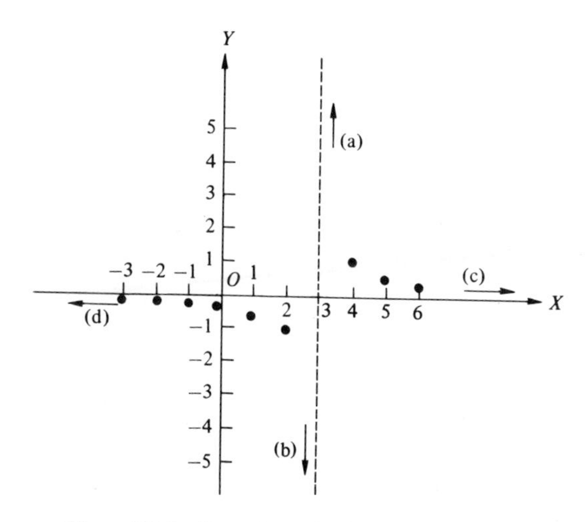

Figure 7.7.1 Partial graph of $f(x) = 1/(x - 3)$.

that the denominator approaches zero through positive values, and thus as $x \longrightarrow 3^+, f(x) \longrightarrow \infty$. As $x \longrightarrow 3^-$ (from the left of 3), we see that the denominator approaches zero through negative values. Hence as $x \longrightarrow 3^-$, $f(x) \longrightarrow -\infty$. At $x = 3$ draw a vertical line. This vertical line which the graph of x approaches, as the graph recedes from the origin, but never reaches is called an *asymptote*. Near the asymptote we indicate by arrows the trend of $f(x)$ as just discussed [see Fig. 7.7.1, arrows (a) and (b)].

Before completing the graph we investigate the trend of $f(x)$ as $x \longrightarrow \infty$ and as $x \longrightarrow -\infty$. Note from the table that as $|x|$ increases, $|f(x)|$ decreases. So here as $x \longrightarrow \infty, f(x) \longrightarrow 0^+$ (from above zero), and as $x \longrightarrow -\infty, f(x) \longrightarrow 0^-$ (from below zero). We also indicate this trend on the graph by arrows [see Fig. 7.7.1, arrows (c) and (d)].

Rational functions are continuous, except at the zeros of the denominator, never intersect a vertical asymptote, and have no sharp corners as seen from the derivative. Since $f'(x)$ is always negative, the curve continuously runs down to the right. For $x < 3$, $f''(x) < 0$; hence the curve is concave down. For $x > 3$, $f''(x) > 0$; hence the curve is concave up. Now we can complete the graph by connecting the arrows and the plotted points in each region of the plane determined by the vertical asymptotes (see Fig. 7.7.2).

PROBLEM 1. Graph $f(x) = 2/(x + 1)$.

Note that in Fig. 7.7.2, $y = 0$ is a *horizontal asymptote*. Horizontal asymptotes are found by letting $x \longrightarrow \infty$ and $x \longrightarrow -\infty$. The graph of a rational function may intersect a horizontal asymptote.

EXAMPLE 2. Find the horizontal asymptote for (a) $f(x) = 2x/(5x^2 - 1)$; (b) $f(x) = (3x^2 - x + 7)/(x^2 + x + 1)$.

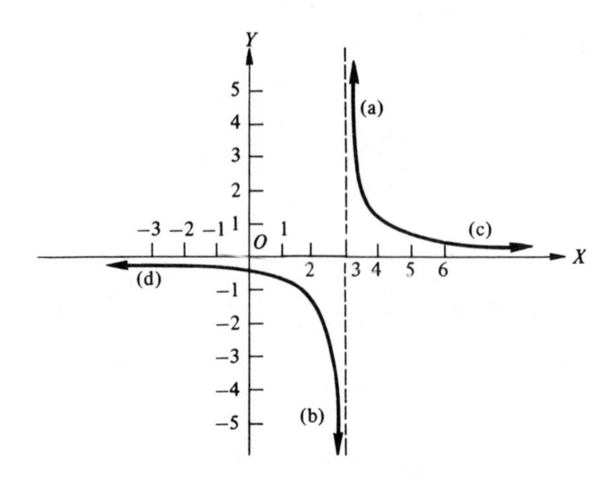

Figure 7.7.2 Graph of $f(x) = 1/(x - 3)$.

Solution. As x becomes large the leading term (highest power of x) dominates the polynomial. Thus we have for large x in (a) $f(x) \doteq 2x/5x^2 = 2/5x$, and as $x \longrightarrow \infty$, $f(x) \longrightarrow 0^+$. As $x \longrightarrow -\infty$, $f(x) \longrightarrow 0^-$. Thus $y = 0$ is the horizontal asymptote. (b) For large x, $f(x) \doteq 3x^2/x^2 = 3$, and as $x \longrightarrow \infty, f(x) \longrightarrow 3^+$, and as $x \longrightarrow -\infty$ $f(x) \longrightarrow 3^+$. Thus $y = 3$ is the horizontal asymptote.

PROBLEM 2. Find the horizontal asymptote for (a) $f(x) = (3x + 2)/(7x^2 - 1)$; (b) $f(x) = (x^2 - 5)/(2x^2 + 3x - 1)$.

In summary, to graph a rational function:

1. Reduce $f(x)$ until there are no common factors in the numerator and the denominator.
2. Find $f'(x)$ and $f''(x)$.
3. Investigate for relative maxima or minima and inflection points.
4. Find the vertical asymptotes (if any) and indicate the trend of $f(x)$ near them with arrows.
5. Find the horizontal asymptotes (if any) and indicate the trend of $f(x)$ near them with arrows.
6. Construct a table of convenient values of x and $f(x)$. Plot these points.
7. Sketch the graph, noting slope and concavity from $f'(x)$ and $f''(x)$.

EXAMPLE 3. Sketch the graph of $f(x) = x^2/(x^2 - 1)$.

Solution

$$f(x) = \frac{x^2}{x^2 - 1}$$

$$f'(x) = \frac{(x^2 - 1)(2x) - (x^2)(2x)}{(x^2 - 1)^2} = \frac{-2x}{(x^2 - 1)^2}$$

$$f''(x) = \frac{(x^2-1)^2(-2) - (-2x)(2)(x^2-1)(2x)}{(x^2-1)^4} = \frac{6x^2+2}{(x^2-1)^3}$$

At $x = 0$, $f'(0) = 0$ and $f''(0) = -2$; therefore, $(0, 0)$ is a relative maximum. There is no inflection point as $f''(x)$ does not have a real zero. The slope and concavity are listed in the following table:

Regions:	$x < -1$	$-1 < x < 0$	$0 < x < 1$	$x > 1$
		-1	0	1 → X
$f'(x)$	+	+	−	−
Slope	up	up	down	down
$f''(x)$	+	−	−	+
Concavity	up	down	down	up

Factor the denominator to obtain $f(x) = x^2/[(x+1)(x-1)]$. To find the vertical asymptotes set the denominator $(x+1)(x-1) = 0$ to obtain $x = -1$ and $x = 1$. Draw these as dashed vertical lines (see Fig. 7.7.3).

To investigate the sign of $f(x)$ as $x \longrightarrow -1^-$ (from the left) let x become very near to -1 but smaller than -1; for example, let $x = -1.1$. Then $f(-1.1) = (-1.1)^2/(-1.1+1)(-1.1-1)$ and the sign of $f(-1.1)$ is $+/(-)(-) = +$.

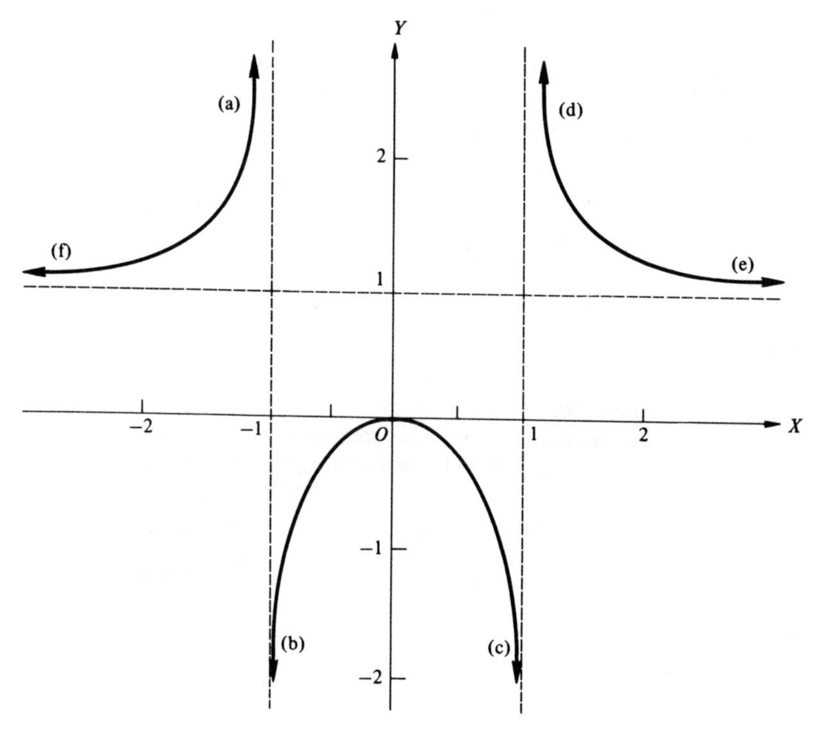

Figure 7.7.3 Graph of $f(x) = x^2/(x^2 - 1)$.

Hence as $x \longrightarrow -1^-$, $f(x) \longrightarrow +\infty$. Similarly, as $x \longrightarrow -1$ from the right, we might let $x = -0.9$, and we have $f(-0.9) = (-0.9)^2/[(-0.9 + 1)(-0.9 - 1)]$ and the sign of $f(-0.9)$ is $+/[(+)(-)] = -$. Hence as $x \longrightarrow -1^+$, $f(x) \longrightarrow -\infty$. Similarly, as $x \longrightarrow 1^-$, we let $x = 0.9$ for convenience in computation and find the sign of $f(0.9)$ to be minus, and as $x \longrightarrow 1^+$, we let $x = 1.1$ for convenience and find the sign of $f(1.1)$ to be plus. Summarizing, we have

$$
\begin{array}{rl}
\text{as } x \longrightarrow -1^- & f(x) \longrightarrow \infty \\
\text{as } x \longrightarrow -1^+ & f(x) \longrightarrow -\infty \\
\text{as } x \longrightarrow 1^- & f(x) \longrightarrow -\infty \\
\text{as } x \longrightarrow 1^+ & f(x) \longrightarrow \infty
\end{array}
$$

Using this information, draw the arrows labeled (a), (b), (c), and (d) in Fig. 7.7.3.

To find the horizontal asymptote let x become large positively and then $f(x) \doteq x^2/x^2 = 1$. Hence as $x \longrightarrow \infty$, $f(x) \longrightarrow 1^+$. Similarly, as $x \longrightarrow -\infty$, $f(x) \longrightarrow 1^+$. Draw the arrows labeled (e) and (f) in Fig. 7.7.3.

Next we construct a table of values and plot these points:

x	-2	0	2
$f(x) = y$	1.3	0	1.3

To complete the graph, connect the arrows within each region of the plane bounded by the vertical asymptotes by a smooth unbroken curve passing through the points just plotted making use of the slope and concavity.

PROBLEM 3. Sketch the graph of $f(x) = 2x^2/(x^2 - 1)$.

EXERCISES 7.7

Group A

1. For each of the following, if $x = a$ is a zero of the denominator, state clearly the trend of $f(x)$ as $x \longrightarrow a^-$ and as $x \longrightarrow a^+$. Also, state clearly the trend of $f(x)$ as $x \longrightarrow \infty$ and as $x \longrightarrow -\infty$. Find the vertical and horizontal asymptotes, if any.

(a) $f(x) = \dfrac{1}{x}$. (b) $f(x) = \dfrac{1}{x - 2}$. (c) $f(x) = \dfrac{1}{x + 5}$.

(d) $f(x) = \dfrac{3}{2x + 5}$. (e) $f(x) = \dfrac{1}{x^2}$. (f) $f(x) = \dfrac{1}{(x - 1)^2}$.

(g) $f(x) = \dfrac{1}{x^3}$. (h) $f(x) = \dfrac{1}{(x + 1)^4}$. (i) $f(x) = \dfrac{3}{x^2 - 4}$.

(j) $f(x) = \dfrac{-5}{x^2 - 16}$. (k) $f(x) = \dfrac{x^2 + 2}{3x^2 - 16}$. (l) $f(x) = \dfrac{5x^2}{x^2 + 4}$.

2. Graph each of the following.

(a) $f(x) = \dfrac{1}{x-1}$. (b) $f(x) = \dfrac{1}{(x-1)^2}$. (c) $f(x) = \dfrac{1}{x^2-1}$.

(d) $f(x) = \dfrac{x}{x^2-1}$. (e) $f(x) = \dfrac{2x^2}{x^2-9}$. (f) $f(x) = \dfrac{1}{x(x-6)}$.

(g) $f(x) = \dfrac{1}{x(x-6)^2}$. (h) $f(x) = \dfrac{1}{x^2(x-6)}$. (i) $f(x) = \dfrac{1}{x^2+1}$.

(j) $f(x) = \dfrac{x}{x^2+1}$. (k) $f(x) = \dfrac{x^2}{x^2+1}$. (l) $f(x) = \dfrac{x^2}{x^3+1}$.

Group B

1. Graph each of the following.

(a) $f(x) = \dfrac{x-3}{x(x-6)}$. (b) $f(x) = \dfrac{x(x-6)}{(x-2)(x-4)}$.

(c) $x^2y + x = 1$. (d) $s(t) = \dfrac{t}{(t-1)(t+2)}$.

(e) $f(x) = \dfrac{x^2-3x}{x^2-3x-4}$. (f) $f(x) = \dfrac{x+8}{(x-2)^3}$.

2. If the degree of $P(x)$ is greater than the degree of $Q(x)$, we perform long division before discussing the graph of $f(x) = P(x)/Q(x)$. In this case we might obtain an oblique asymptote. Graph each of the following.

(a) $f(x) = \dfrac{3x^3}{3x^2-2}$. (b) $f(x) = \dfrac{(x-4)^2}{x-2}$.

7.8 IMPLICIT DIFFERENTIATION

In the equation $x^2 + y^2 - 25 = 0$, we do not really know whether x or y is the independent variable. If we solve for y, we obtain the two functions $y_1 = \sqrt{25 - x^2}$ and $y_2 = -\sqrt{25 - x^2}$. (Recall that a function has only one output for an input.) If we solve for x, we obtain $x = \pm\sqrt{25 - y^2}$. In an equation containing both x and y which has not been solved for either x or y, y is called an *implicit function* of x, even though there may be more than one function of y. For example, $xy - 1 = 0$ and $3xy^2 + y^3x^2 = 3$ are called implicit functions.

An implicit function may be differentiated directly and this may be solved algebraically for the derivative.

EXAMPLE 1. For $x^2 + y^2 - 25 = 0$, find dy/dx.

Solution. Differentiate each term of the function with respect to x. Thus for $x^2 + y^2 - 25 = 0$ we have

$$\frac{d(x^2)}{dx} + \frac{d(y^2)}{dx} - \frac{d(25)}{dx} = \frac{d(0)}{dx}$$

$$2x\frac{dx}{dx} + 2y\frac{dy}{dx} - 0 = 0$$

$$2x + 2y\frac{dy}{dx} = 0$$

and

$$\frac{dy}{dx} = -\frac{x}{y}$$

In Example 1 we could have solved for $y = \pm\sqrt{25 - x^2}$ and then obtained $dy/dx = -x/\pm\sqrt{25 - x^2}$, which agrees with the result of Example 1.

PROBLEM 1. Find dy/dx for $x^3 - y^2 + 27 = 0$.

In some implicit functions it is impossible to solve for either x or y, but it is possible to find dy/dx. Note that we apply the chain rule to each term involving y to a power; that is,

$$\frac{d}{dx}(y^p) = py^{p-1}\frac{dy}{dx}$$

Warning ***Remember that dy/dx comes only from differentiating a y with respect to x. For Example 1 it is wrong to write***

$$\frac{dy}{dx} = 2x + 2y\frac{dy}{dx} = 0 \qquad \textit{(Why?)}$$

EXAMPLE 2. Find dy/dx for $x^3y^2 + xy^3 - 5 = 0$.

Solution. Note that we cannot solve for either x or y but that we can differentiate each term with respect to x. Thus

$$\frac{d(x^3y^2)}{dx} + \frac{d(xy^3)}{dx} - \frac{d(5)}{dx} = \frac{d(0)}{dx}$$

$$\frac{x^3\,d(y^2)}{dx} + y^2\frac{d(x^3)}{dx} + \frac{x\,d(y^3)}{dx} + y^3\frac{d(x)}{dx} - 0 = 0$$

$$x^3 2y\frac{dy}{dx} + y^2 3x^2 + x3y^2\frac{dy}{dx} + y^3 = 0$$

$$\frac{dy}{dx}(2x^3y + 3xy^2) = -(3x^2y^2 + y^3)$$

and

$$\frac{dy}{dx} = \frac{-(3x^2y^2 + y^3)}{2x^3y + 3xy^2}$$

PROBLEM 2. Find dy/dx for $x^2y^3 - x^3y + 7 = 0$.

EXAMPLE 3. Find an equation of the tangent line to the circle $x^2 + y^2 - 169 = 0$ at the point (5, 12).

Solution. To find an equation of the tangent at (5, 12) we need the slope of the curve at that point. Thus for $x^2 + y^2 - 169 = 0$, we have

$$2x + 2y\frac{dy}{dx} = 0$$

or

$$\frac{dy}{dx} = -\frac{x}{y}$$

At (5, 12),

$$\frac{dy}{dx} = -\frac{5}{12}$$

Then using the point-slope form of a straight line, we have

$$y - 12 = -\tfrac{5}{12}(x - 5)$$

or

$$5x + 12y - 169 = 0$$

as an equation of the required tangent line.

PROBLEM 3. Find an equation of the line tangent to the curve $x^2y + xy^2 - 6 = 0$ at the point (2, 1).

EXERCISES 7.8

Group A

1. Find dy/dx in each of the following.

(a) $y^3 = 4x$.

(b) $y^2x + 3xy = 25$.

(c) $x^2 + y^2 = 10$.

(d) $\dfrac{x}{y} - 4y = x$.

(e) $\sqrt{xy} + 8x^2 = \frac{1}{2}y$.

(f) $\dfrac{1}{xy} = 2$.

2. Find dy/dx in each of the following.

(a) $\dfrac{x^2 + y^2}{x + y} = x$.

(b) $\dfrac{x - y}{x + y} = x^2$.

(c) $y\sqrt{1 + x} + y^2 = x^2$.

(d) $\dfrac{y^3}{1 + y} = x$.

(e) $\sqrt[3]{xy} + \sqrt[3]{y^2} = x$.

(f) $4x^2 + 9y^2 = 36$.

3. In each of the following, find the slope of the curve at the given point. (*Hint*: It is best to substitute the values for x and y right after taking the derivative of each term. Then solve for dy/dx.)

(a) $y^3 - xy + x^3 = 1$; $(-1, 1)$.
(b) $x^2y - y^3x + 6 = 0$; $(1, 2)$.
(c) $3x^3 - x^2y^2 - 2y^3 - 2 = 0$; $(0, -1)$.
(d) $\sqrt{2xy} - \sqrt[3]{xy} = 2$; $(4, 2)$.

Group B

1. (a) Find two differentiable functions $y = f(x)$ defined by the equation $x^2 + y^2 = 4$. (b) Graph the equation $x^2 + y^2 = 4$ and indicate the two functions from part (a) on your graph. (c) Find dy/dx for each of the functions in part (a). (d) Find dy/dx using implicit differentiation. (e) Find equations of the lines tangent to the graph of the equation at $(1, \sqrt{3})$ and $(1/2, -\sqrt{15}/2)$.

2. (a) Find two differentiable functions $y = f(x)$ defined by the equation $y^2 = 9x$. (b) Graph the equation $y^2 = 9x$ and indicate each of the two functions from part (a) on your graph. (c) Find dy/dx for each function in part (a). (d) Find dy/dx using implicit differentiation. (e) Verify that the derivatives found in parts (c) and (d) are the same. (f) Find equations of the lines tangent to the graph of the equation at $(1, 3)$ and $(9, -9)$.

3. Find an equation of the tangent to the graph $4x^2 + 9y^2 = 36$ at the point where $x = 1$, and y is negative.

4. Find d^2y/dx^2 for (a) $xy^2 + x^2y = 7$; (b) $xy^3 + y = 1$.

5. Find the maximum and minimum points of $x^2 + xy - y^2 = 20$.

6. Find dy/dx for $xy + \sin y = 0$.

7. Find the slope of the curve $y = \sin(xy)$ at the point $(1, \pi/3)$.

8. A curve in the XY-plane is defined by the equation $x^4 - 10x^2y^2 + 9y^4 = 0$ and contains the points $(-2, 2)$ and $(3, -1)$. (a) Find a formula for dy/dx. (b) Find an equation of the line tangent to the curve at $(3, -1)$. (c) Find a point on the tangent line not on the curve. (d) Find a point on the curve not on the tangent line.

7.9 DIFFERENTIAL OF ARC; PARAMETRIC EQUATIONS

(1) Differential of Arc

Some applications of calculus deal with distance measured along a curve. These problems usually involve a differential of arc length. We shall now develop an expression for a differential of arc. In Fig. 7.9.1, $P(x, y)$ is a point on the curve $y = f(x)$ and $Q(x + \Delta x, y + \Delta y)$ is another point on the curve. Let the length of arc PQ be denoted by Δs and the chord PQ be denoted by Δc. In the right triangle PMQ, we have from the Pythagorean theorem

$$(\Delta c)^2 = (\Delta x)^2 + (\Delta y)^2$$

Dividing by $(\Delta x)^2$, we have

$$\left(\frac{\Delta c}{\Delta x}\right)^2 = 1 + \left(\frac{\Delta y}{\Delta x}\right)^2$$

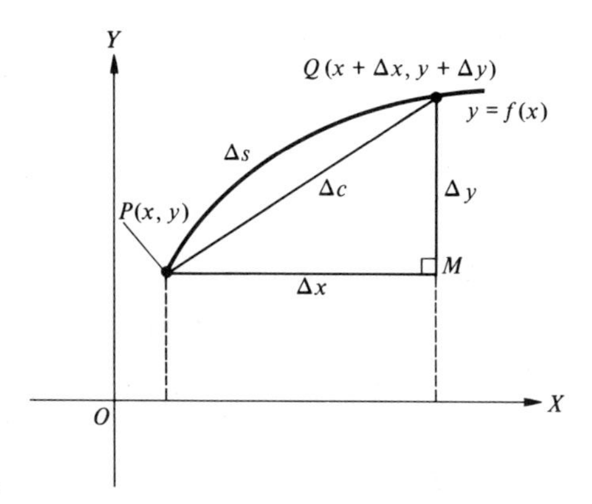

Figure 7.9.1 Differential of arc.

Now multiply both sides of the preceding equation by $(\Delta s/\Delta c)^2$ to obtain

$$\left(\frac{\Delta s}{\Delta c}\right)^2\left(\frac{\Delta c}{\Delta x}\right)^2 = \left(\frac{\Delta s}{\Delta c}\right)^2\left[1 + \left(\frac{\Delta y}{\Delta x}\right)^2\right]$$

or

$$\left(\frac{\Delta s}{\Delta x}\right)^2 = \left(\frac{\Delta s}{\Delta c}\right)^2\left[1 + \left(\frac{\Delta y}{\Delta x}\right)^2\right]$$

Our next step is to take the limit as $\Delta x \longrightarrow 0$, but before we do this we wonder about the expression $\Delta s/\Delta c$. Intuitively, it appears that as $\Delta x \longrightarrow 0$, that is, as Q gets close to P, the ratio $\Delta s/\Delta c \longrightarrow 1$, and mathematicians have proved that this is precisely what happens. Thus we have

$$\lim_{\Delta x \to 0}\left(\frac{\Delta s}{\Delta x}\right)^2 = \lim_{\Delta x \to 0}\left(\frac{\Delta s}{\Delta c}\right)^2\left[1 + \lim_{\Delta x \to 0}\left(\frac{\Delta y}{\Delta x}\right)^2\right]$$

and

$$\left(\frac{ds}{dx}\right)^2 = (1)^2\left[1 + \left(\frac{dy}{dx}\right)^2\right]$$

or

$$\left(\frac{ds}{dx}\right)^2 = 1 + \left(\frac{dy}{dx}\right)^2 \tag{7.9.1}$$

If we multiply Eq. 7.9.1 by $(dx)^2$, we obtain the desired result:

$$(ds)^2 = (dx)^2 + (dy)^2 \tag{7.9.2}$$

The differential of arc length is denoted as ds and can be expressed in several ways. For example,

$$ds = \sqrt{1 + \left(\frac{dy}{dx}\right)^2}\,dx \tag{7.9.3}$$

and

$$ds = \sqrt{1 + \left(\frac{dx}{dy}\right)^2}\,dy \tag{7.9.4}$$

The positive sign for the square root was chosen to indicate that in Eq. 7.9.3 s increases along the curve in the direction of increasing x and that in Eq. 7.9.4 s increases along the curve in the direction of increasing y.

A convenient device for memorizing Eq. 7.9.2, $(ds)^2 = (dx)^2 + (dy)^2$, is the "triangle" shown in Fig. 7.9.2. This does not constitute a proof as PMQ is not a right triangle.

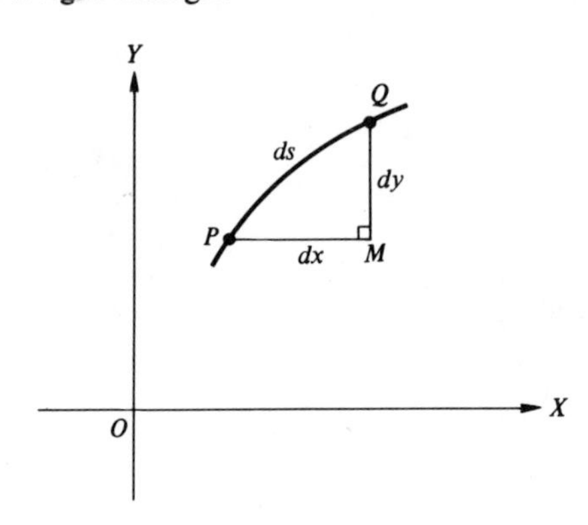

Figure 7.9.2 $(ds)^2 = (dx)^2 + (dy)^2$.

EXAMPLE 1. Find the differential of arc for the curve $y = x^2$.

Solution. For $y = x^2$

$$\frac{dy}{dx} = 2x$$

Using Eq. 7.9.3, we have

$$ds = \sqrt{1 + (2x)^2}\, dx$$

or

$$ds = \sqrt{1 + 4x^2}\, dx$$

(If we knew how to integrate this expression, we could find the length of arc of the given curve. We see how this may be done in Exercise 1(p) of Group B in Exercises 11.5.

PROBLEM 1. Find the differential of arc for the curve $x^2 + y^2 = 25$.

(2) Parametric Equations

If x is a function of t and y is also a function of t, then the pair of equations

$$x = g(t)$$
$$y = h(t)$$

are said to be *parametric equations*, where t is called the *parameter*. Parametric equations are especially useful in studying the motion of a body when its position is known as a function of time. We will make good use of them when we discuss motion along a curve (Sec. 7.11).

EXAMPLE 2. A body moves in the XY-plane in such a way that at any time t, $x = 3t$ and $y = 4t^2$. Find the differential of arc for its path.

Solution. In Eq. (7.9.2) if we divide by dt and take the positive square root, we have

$$ds = \sqrt{\left(\frac{dx}{dt}\right)^2 + \left(\frac{dy}{dt}\right)^2}\, dt \tag{7.9.5}$$

where s will increase for an increase in t. From $x = 3t$ and $y = 4t^2$, we have $dx/dt = 3$ and $dy/dt = 8t$;

$$\therefore \quad ds = \sqrt{9 + 64t^2}\, dt$$

PROBLEM 2. Find the differential of arc for the curve whose parametric equations are $x = 5t^2$, $y = 8t$.

In studying the graph of parametric equations it is sometimes useful to eliminate the parameter and obtain an equation in terms of x and y.

EXAMPLE 3. Find an xy-equation for the parametric equations of Example 2.

Solution. If we solve for t in the x-equation, we have $t = x/3$. Substituting this value in the y-equation, we have

$$y = 4\left(\frac{x}{3}\right)^2 = \frac{4}{9}x^2$$

(You may solve for t in either equation and substitute its value in the other equation.)

PROBLEM 3. Find an xy-equation for the parametric equations of Prob. 2.

In finding dy/dx for parametric equations, we make use of formula 7.5.3, that is,

$$\frac{dy}{dx} = \frac{dy/dt}{dx/dt}$$

EXAMPLE 4. For the parametric equations $x = 3t$, $y = 4t^3$, find at the point where $t = 3$, (a) the slope of the curve; (b) the value of d^2y/dx^2.

Solution. (a) From $x = 3t$, we have $dx/dt = 3$. From $y = 4t^3$, we have $dy/dt = 12t^2$. Then

$$\frac{dy}{dx} = \frac{dy/dt}{dx/dt} = \frac{12t^2}{3} = 4t^2$$

At $t = 3$ the slope is given by

$$\left.\frac{dy}{dx}\right|_{t=3} = 4(3)^2 = 36$$

(b) To find d^2y/dx^2 we differentiate $dy/dx = 4t^2$ with respect to x. Thus

$$\frac{d^2y}{dx^2} = \frac{d}{dx}(4t^2) = 8t\left(\frac{dt}{dx}\right) \qquad \text{(by the power and chain rules)}$$

$$= \frac{8t}{dx/dt} = \frac{8t}{3} = \frac{8}{3}t$$

At $t = 3$,

$$\left.\frac{d^2y}{dx^2}\right|_{t=3} = \frac{8}{3}(3) = 8$$

Warning ***In finding d^2y/dx^2, remember that we differentiate with respect to x.***

PROBLEM 4. For the parametric equations $x = 2t^3$, $y = 5t$, find at the point where $t = 2$, (a) the slope of the curve; (b) the value of d^2y/dx^2.

EXERCISES 7.9

Group A

1. Find the differential of arc, ds, for each of the following curves.

(a) $y = x^4$. (b) $x = y^3$. (c) $y = \sqrt{x}$.
(d) $y = x^2 - 4$. (e) $x = y^4 + y^2 + 1$. (f) $x^2 + 4y^2 = 16$.

2. Find the differential of arc, ds, for each of the following.

(a) $x = t$, $y = t^2$. (b) $x = 8t^2 + 5$, $y = 4t^4$.
(c) $x = t$, $y = \sqrt{t}$. (d) $x = \sqrt{t}$, $y = t$.
(e) $x = t^2 + 1$, $y = t + 1$. (f) $x = \sqrt{36 - 9t^2/2}$, $y = t$.
(g) $x = \cos t$, $y = \sin t$. (h) $x = 2 \cos t$, $y = 3 \sin t$.

3. In each of the following, find an xy-equation.

(a) $x = t$, $y = t^2$. (b) $x = \sqrt{t}$, $y = t$.
(c) $x = t^2$, $y = t^3$. (d) $x = t^2 + 1$, $y = t + 1$.
(e) $x = \cos t$, $y = \sin t$. (*Hint*: Square and add.)

4. In each of the following, find the slope and the value of d^2y/dx^2 at the point for the given value of t.

(a) $x = t^2$, $y = t$; $t = 2$. (b) $x = t^2$, $y = t^3$; $t = 3$.
(c) $x = t$, $y = \sqrt{t}$; $t = 2$. (d) $x = \sqrt{32 - t^2}$, $y = t$; $t = -4$.
(e) $x = \cos t$, $y = \sin t$; $t = \pi/4$.

Group B

1. In each of the following, write a set of parametric equations for the xy-equation. (*Hint*: Let x equal some function of t. Substitute that value for x and solve for y.)

(a) $x = y^2$. (b) $x^2 = y$. (c) $x^2 + y^2 = 25$. (d) $y^2 = x^3$.

2. (a) Find an xy-equation for $x = t$, $y = t^2$. (b) Find an xy-equation for $x = \sqrt{t}$, $y = t$. (c) Are the answers for parts (a) and (b) the same? What does this tell you about parametric equations for a given xy-equation? (d) Find ds for parts (a) and (b). Are they the same? Should they be?

Group C

1. When a differential arc length is revolved about an axis a circular band of width ds is formed. Consider the curve $y = f(x)$ and show from a sketch that the surface area of the band is $dS = 2\pi y\, ds$ when the curve is revolved about the X-axis. Find an expression for dS when an arc of $y = x^2$ is revolved about the X-axis.

2. Theorem 7.5.3 tells us how to find dy/dx for parametric equations, that is, if $x = g(t)$ and $y = h(t)$, then

$$\frac{dy}{dx} = \frac{dy/dt}{dx/dt} \qquad \text{for} \qquad dx/dt \neq 0$$

Let $y' = dy/dx$ and show that

$$\frac{d^2y}{dx^2} = \frac{dy'/dt}{dx/dt} \qquad \text{for} \qquad dx/dt \neq 0$$

3. Using the result of Exercise 2, find d^2y/dx^2 for each of the following parametric equations.

(a) $x = t^2$, $y = t$. (b) $x = 3t$, $y = 5t^2$.
(c) $x = t$, $y = \sqrt{8 - t^2}$. (d) $x = \sqrt{32 - t^2}$, $y = t$.

4. Find the radius of curvature of $x = 8t^2 + 5$, $y = 4t^4$ at the point where $t = 2$ (see Eq. 2.9.1).

7.10 INTRODUCTION TO VECTORS

At this point we stop the study of calculus and give a brief discussion of vectors, as they will be very useful in Sec. 7.11, where we discuss motion along a curve.

Quantities that need a direction as well as a magnitude to completely describe them are called *vector quantities.* Such quantities as force, displacement, velocity, acceleration, and momentum are vector quantities. Quantities such as distance, speed, time, mass, and temperature are called *scalar quantities*, as they need only a magnitude to describe them.

Geometrically, a vector quantity may be represented by a line segment that has both *magnitude* and *direction.* Its magnitude is represented by the length of the line segment and its direction is indicated by an arrow at the terminal end of the line segment. Such directed line segments are called *vectors.* (The words "vectors" and "vector quantities" are used interchangeably.) In this book vectors are usually denoted by capital *boldface* letters; thus **V** denotes the "vector V." Although boldface letters can be typeset, it is difficult to write them, so we always place an arrow over the letter when writing vectors: thus

$\vec{V}$. The magnitude of a vector is denoted by a lowercase letter and no arrow over the letter. It is always considered to be *positive*. Thus vector **V** has magnitude v. Symbolically, $v = |\mathbf{V}|$.

In Fig. 7.10.1, the vector **V** has initial point A and terminal point B. The vector **R** has initial point C and terminal point D. The length of **V** is v and the length of **R** is r.

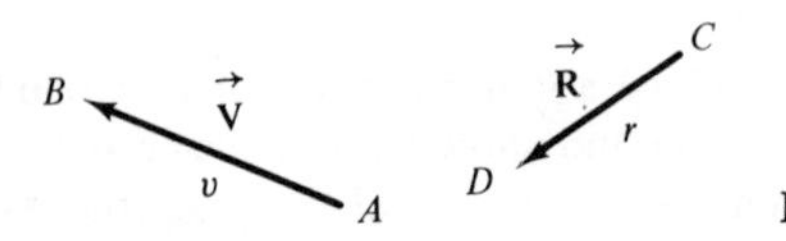

Figure 7.10.1 Vectors.

A Rule for Adding Two Vectors Geometrically

Two vectors **A** and **B** may be added geometrically as follows:

1. Draw vector **A**.
2. Draw vector **B** with its initial point at the terminal point of **A**.
3. Draw the sum, or resultant, vector **R** *from* the initial point of **A** *to* the terminal point of **B**. (**R** = **A** + **B**.)

EXAMPLE 1. Add geometrically the vectors **A** and **B** of Fig. 7.10.2(a).

Solution. See Fig. 7.10.2(b). Follow the rule given above.

PROBLEM 1. Add geometrically the vectors **C** and **D** of Fig. 7.10.2(c).

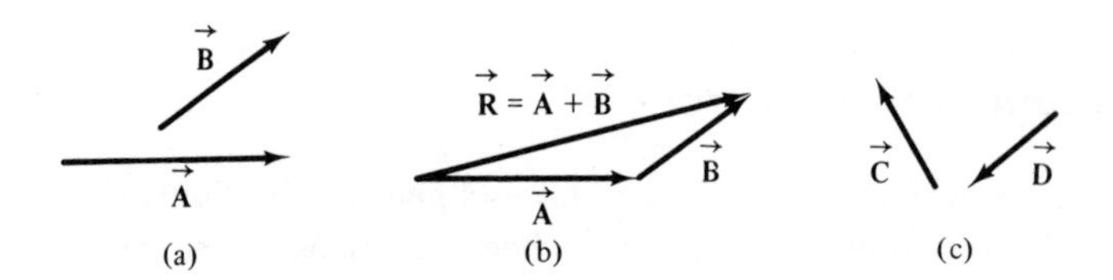

Figure 7.10.2 Sum of two vectors.

For algebraic operations with vectors we need to make some definitions.

Definition 7.10.1 *The product of a scalar, k, and a vector,* **V**, *is a vector whose magnitude is k times the magnitude of* **V**. *That is, if* **U** = k**V**, *then*

$$|\mathbf{U}| = u = kv = k|\mathbf{V}|$$

The vector $k\mathbf{V}$ is in the same direction as **V** for a positive k and is in the opposite direction for a negative k (see Fig. 7.10.3).

Definition 7.10.2 *A unit vector is a vector whose magnitude (or length) is 1 unit long.*

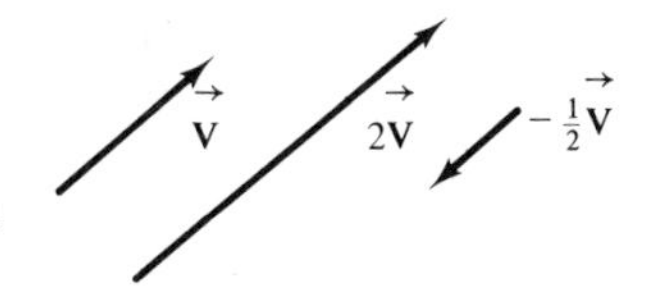

Figure 7.10.3 Product of a scalar and a vector.

So far the vector systems we have described are free of any coordinate system. It is frequently useful to imagine all vectors as described in terms of a coordinate basis. The usual coordinate basis is the Cartesian coordinate system with the two *unit* vectors **i** and **j**. (It is customary to use lowercase letters for **i** and **j**.) Their initial points are at the origin. The unit vector **i** points in the direction of the positive X-axis. The unit vector **j** points in the direction of the positive Y-axis (see Fig. 7.10.4). Any other vector with its initial point at the origin can be written as the sum of scalar multiples of **i** and **j**. Two vectors with the same magnitude and same direction are said to be *equivalent* vectors. Thus any vector in the XY-plane is equivalent to a vector written as the sum of scalar multiples of **i** and **j**.

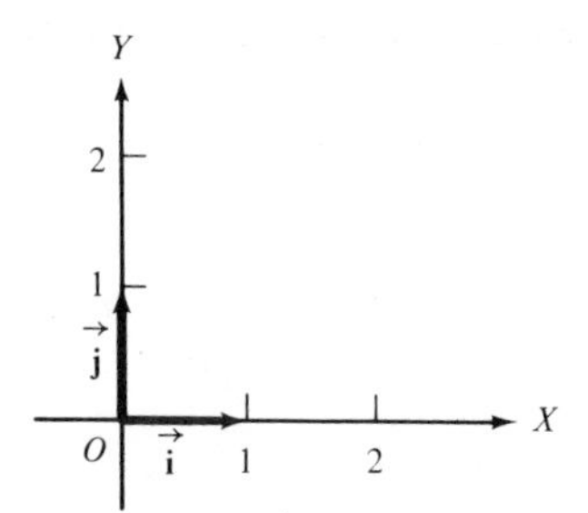

Figure 7.10.4 Unit vectors **i** and **j**.

EXAMPLE 2. Sketch the vector $\mathbf{R} = -4\mathbf{i} + 3\mathbf{j}$ with the initial point at the origin.

Solution. See Fig. 7.10.5. We use the rule for adding two vectors geometrically. $-4\mathbf{i}$ is a vector 4 units long in the direction of the negative X-axis. $3\mathbf{j}$ is a vector

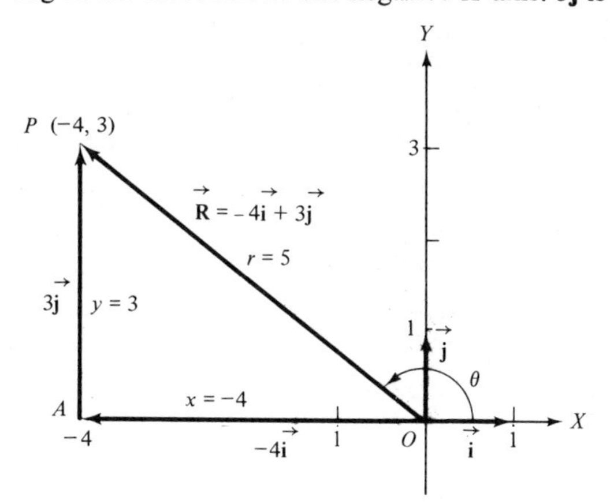

Figure 7.10.5 Example 2.

3 units long in the direction of the positive Y-axis. Note carefully that the terminal point of **R** is $(-4, 3)$. The magnitude of **R** is obtained by use of the Pythagorean theorem applied to right triangle OAP.

$$|\mathbf{R}| = r = \sqrt{(-4)^2 + (3)^2} = 5$$

The direction of **R** is the positive angle θ measured from the positive X-axis. $(0° \leqq \theta < 360°.)$ It is obtained from the right triangle OAP, where $\tan\theta = -\frac{3}{4} = -0.75$ and $\theta = 143.1°$.

PROBLEM 2. Sketch the vector $\mathbf{R} = 3\mathbf{i} + 5\mathbf{j}$ with its initial point at the origin. What are the coordinates of the terminal point of **R**? What is the magnitude and direction of **R**?

The vector, **R**, drawn from the origin to point $P(x, y)$, is called the *position vector* of point P. It is given by

$$\mathbf{R} = x\mathbf{i} + y\mathbf{j} \tag{7.10.1}$$

The coefficient of **i** (here the number x) is called the *X-component* or the *i-component* of **R**. The coefficient of **j** (here the number y) is called the *Y-component or the j-component* of **R**

From Fig. 7.10.6 and trigonometry the magnitude of **R** is given by

$$|\mathbf{R}| = r = \sqrt{x^2 + y^2} \tag{7.10.2}$$

and the direction of **R** is found from

$$\tan\theta = \frac{y}{x} \tag{7.10.3}$$

When the point $P(x, y)$ moves along a curve, the position vector rotates and its tip traces out the curve. Usually, x and y are expressed as functions of time and are given by the parametric equations $x = g(t)$, $y = h(t)$.

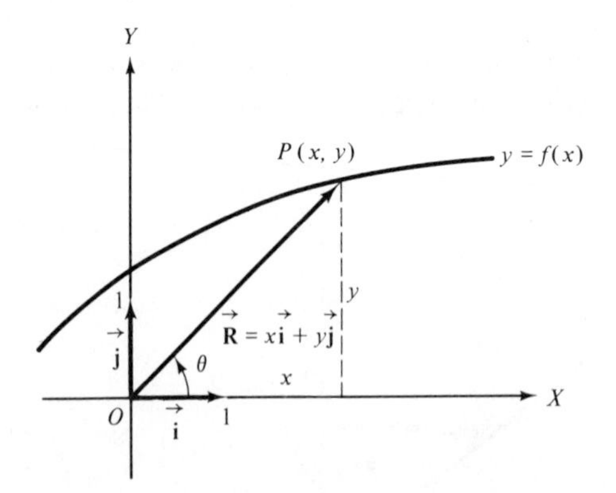

Figure 7.10.6 The position vector **R**.

EXAMPLE 3. Write the position vector for the parametric equations $x = t^2$, $y = t$.

Solution $\mathbf{R} = t^2\mathbf{i} + t\mathbf{j}$.

PROBLEM 3. For Example 3 let t have the values 0, 1, 2, 3 and draw **R** for each value. Then sketch the curve traced out by **R**.

EXERCISES 7.10

Group A

1. In each of the following, draw the vector with its initial point at the origin. Find the coordinates of the terminal point, and the magnitude and direction of the vector.

(a) $\mathbf{R} = 2\mathbf{i} + 2\mathbf{j}$. (b) $\mathbf{R} = -3\mathbf{i} + 5\mathbf{j}$. (c) $\mathbf{R} = -2\mathbf{i} - 3\mathbf{j}$.
(d) $\mathbf{R} = 3\mathbf{i} - 4\mathbf{j}$. (e) $\mathbf{R} = 4\mathbf{i}$. (f) $\mathbf{R} = -3\mathbf{j}$.

2. In each of the following, write the position vector for the given point. Find its magnitude and direction.

(a) (3, 2). (b) (−1.5, 3). (c) (−0.5, −0.4).
(d) (2.7, −0.5). (e) (−2, 0). (f) (0, 3).

3. In each of the following, write the position vector for the given parametric equations. Draw the position vector **R** for each of the given values of t and sketch the curve traced out by **R**.

(a) $x = t,\ y = t^2;\ t = 0, 1, 2, 3$. (b) $x = t,\ y = t^2;\ t = -1, 0, 1, 2$.
(c) $x = t^3,\ y = t;\ t = -1, 0, 1, 2$.
(d) $x = t^2 - 2t + 1,\ y = t - 1;\ t = 0, 1, 2$.

4. Show that each of the following are unit vectors.

(a) $\mathbf{U} = \frac{3}{5}\mathbf{i} + \frac{4}{5}\mathbf{j}$. (b) $\mathbf{A} = \frac{5}{13}\mathbf{i} - \frac{12}{13}\mathbf{j}$.
(c) $\mathbf{R} = (\cos t)\mathbf{i} + (\sin t)\mathbf{j}$. (d) $\mathbf{V} = -\sqrt{t}\,\mathbf{i} + \sqrt{1 - t}\,\mathbf{j}$.

Group B

1. Sometimes vectors are said to be added according to the parallelogram law. Verify this law for $\mathbf{A} + \mathbf{B} = \mathbf{C}$, where **C** is the diagonal of a parallelogram with **A** and **B** having the same initial point and are adjacent sides of the parallelogram.

2. When vectors are written in the **ij**-form they may be added algebraically by adding the **i**-components and adding the **j**-components. That is, for $\mathbf{A} = a_1\mathbf{i} + a_2\mathbf{j}$ and $\mathbf{B} = b_1\mathbf{i} + b_2\mathbf{j}$, $\mathbf{A} + \mathbf{B} = (a_1 + a_2)\mathbf{i} + (b_1 + b_2)\mathbf{j}$. For each of the following add the given vectors algebraically.

(a) $\mathbf{A} = 2\mathbf{i} + \mathbf{j},\ \mathbf{B} = \mathbf{i} + 4\mathbf{j}$. (b) $\mathbf{A} = -\mathbf{i} + 3\mathbf{j},\ \mathbf{B} = -2\mathbf{i} - 5\mathbf{j}$.
(c) $\mathbf{A} = -3\mathbf{i},\ \mathbf{B} = 2\mathbf{j},\ \mathbf{C} = \mathbf{i} + 4\mathbf{j}$.

3. Add geometrically $\mathbf{A} = a_1\mathbf{i} + a_2\mathbf{j}$ and $\mathbf{B} = b_1\mathbf{i} + b_2\mathbf{j}$ and show that $\mathbf{A} + \mathbf{B} = (a_1 + b_1)\mathbf{i} + (a_2 + b_2)\mathbf{j}$.

4. Vectors may be written in *polar form* as $\mathbf{V} = v\underline{/\theta}$ (read "v angle theta"), where v is the magnitude and θ is the direction measured from the positive X-axis. Show that $v\underline{/\theta} = (v \cos \theta)\mathbf{i} + (v \sin \theta)\mathbf{j}$. Write each of the following vectors in **ij**-form.

(a) $\mathbf{V} = 6\underline{/30°}$. (b) $\mathbf{R} = 2\underline{/120°}$. (c) $\mathbf{U} = 4\underline{/270°}$. (d) $\mathbf{A} = 3.5\underline{/225°}$.

5. Show that if $\mathbf{A} = \mathbf{i} + \mathbf{j}$, then $2\mathbf{A} = 2\mathbf{i} + 2\mathbf{j}$. (*Hint*: Compare the magnitudes and directions and use Definition 7.10.1.) In general, is $k(x\mathbf{i} + y\mathbf{j}) = kx\mathbf{i} + ky\mathbf{j}$?

6. Make use of Exercise 5 to perform the indicated operations for vectors $\mathbf{A} = 3\mathbf{i} - 2\mathbf{j}$, $\mathbf{B} = -\mathbf{i} + \mathbf{j}$, and $\mathbf{C} = -2\mathbf{i} - \mathbf{j}$.

(a) $\mathbf{A} + 2\mathbf{B}$. (b) $3\mathbf{A} - \mathbf{C}$. (c) $\mathbf{A} + \mathbf{B} + 2\mathbf{C}$. (d) $-3\mathbf{B} + \frac{1}{2}\mathbf{C}$.

7.11 MOTION ALONG A CURVE

When a body (idealized as a point) moves along a curve, the discussion of its velocity and acceleration is more complicated than for straight-line motion. The use of vectors helps clarify the matter.

Let us assume that the point $P(x, y)$ moves along the curve whose parametric equations are $x = g(t)$, $y = h(t)$, where the parameter t is time. Its position vector $\mathbf{R}$ is then a function of the time t. We define the vector $\mathbf{V} = d\mathbf{R}/dt$ to be the *velocity vector* of P [see Fig. 7.11.1(a)]. Since

$$\mathbf{R} = x\mathbf{i} + y\mathbf{j}$$

$$\frac{d\mathbf{R}}{dt} = \mathbf{V} = \left(\frac{dx}{dt}\right)\mathbf{i} + \left(\frac{dy}{dt}\right)\mathbf{j} \tag{7.11.1}$$

The magnitude (or length) of $\mathbf{V}$ is given by

$$v = |\mathbf{V}| = \sqrt{\left(\frac{dx}{dt}\right)^2 + \left(\frac{dy}{dt}\right)^2} \tag{7.11.2}$$

Comparing formula 7.11.2 with formula 7.9.5, we see that

$$v = \left|\frac{ds}{dt}\right|$$

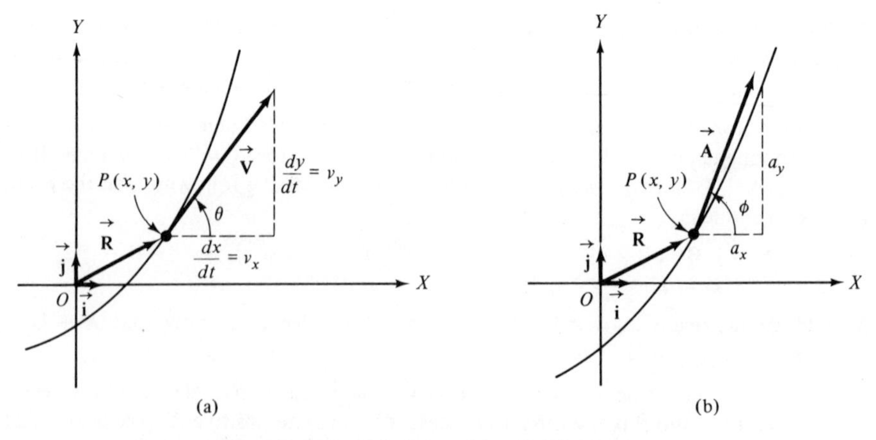

Figure 7.11.1 Velocity and acceleration vectors.

and indicates the rate at which the curve is being traversed. We call v the *speed* of the point P.

It is customary to use v_x and v_y for the X- and Y-components of the velocity. Then we write

$$v_x = \frac{dx}{dt} \qquad v_y = \frac{dy}{dt} \qquad v = \sqrt{v_x^2 + v_y^2}$$

The direction of $\mathbf{V}$ is found from

$$\tan\theta = \frac{dy/dt}{dx/dt} = \frac{dy}{dx} = \frac{v_y}{v_x} \tag{7.11.3}$$

where θ is the angle measured from the positive X-axis. Since dy/dx is the slope of the curve, we see that the velocity vector $\mathbf{V}$ has the *direction along the tangent* to the curve at point P.

We define the vector $\mathbf{A} = d\mathbf{V}/dt = d^2\mathbf{R}/dt^2$ to be the *acceleration vector* of point P. Thus

$$\mathbf{A} = \frac{d\mathbf{V}}{dt} = \frac{d^2x}{dt^2}\mathbf{i} + \frac{d^2y}{dt^2}\mathbf{j} = a_x\mathbf{i} + a_y\mathbf{j} \tag{7.11.4}$$

where $a_x = d^2x/dt^2$ and $a_y = d^2y/dt^2$ are the X- and Y-components of the acceleration. Its magnitude is

$$a = |\mathbf{A}| = \sqrt{\left(\frac{d^2x}{dt^2}\right)^2 + \left(\frac{d^2y}{dt^2}\right)^2} = \sqrt{a_x^2 + a_y^2}$$

and its direction is found from

$$\tan\varphi = \frac{d^2y/dt^2}{d^2x/dt^2} = \frac{a_y}{a_x}$$

where φ is the angle measured from the positive X-axis [see Figure 7.11.1(b)].

Usually, $\mathbf{V}$ and $\mathbf{A}$ have different directions.

Warning ***Remember that the velocity is a vector in the direction of the tangent to the curve. The speed is the magnitude of the velocity.***

EXAMPLE 1. A body moves along a curve whose parametric equations are $x = t^2$, $y = t^3 + 25$. Find its velocity and acceleration at $t = 3$s if x and y are measured in meters.

Solution. We first write the position vector and then differentiate with respect to t to find $\mathbf{V}$ and $\mathbf{A}$. Since

$$\mathbf{R} = x\mathbf{i} + y\mathbf{j}$$

we have

$$\mathbf{R} = t^2\mathbf{i} + (t^3 + 25)\mathbf{j}$$

Then

$$\mathbf{V} = \frac{d\mathbf{R}}{dt} = 2t\mathbf{i} + 3t^2\mathbf{j}$$

and

$$\mathbf{A} = \frac{d\mathbf{V}}{dt} = 2\mathbf{i} + 6t\mathbf{j}$$

At $t = 3$, $\mathbf{V} = 6\mathbf{i} + 27\mathbf{j}$ and $\mathbf{A} = 2\mathbf{i} + 18\mathbf{j}$.

Sometimes the velocity and acceleration are more meaningful if we state their magnitude and direction. For **V** we have

$$v = |\mathbf{V}| = \sqrt{(6)^2 + (27)^2} = 27.7 \text{ m/s}$$

with $\tan\theta = \frac{27}{6}$ and $\theta = 77.5°$. Thus the velocity of the body at $t = 3$ s is 27.7 m/s in a direction 77.5° above the X-axis.

For **A** we have

$$a = |\mathbf{A}| = \sqrt{(2)^2 + (18)^2} = 18.1 \text{ m/s}^2$$

with $\tan\varphi = \frac{18}{2}$ and $\varphi = 83.7°$. Thus the acceleration of the body at $t = 3$ s is 18.1 m/s² in a direction 83.7° above the X-axis.

PROBLEM 1. Find the velocity and acceleration of a particle at $t =$ 1s as it moves along a curve whose parametric equations are $x = t^2$, $y = t^3$, where x and y are measured in feet. What is the speed of the body when $t = 3$ s?

Quite often in applying Newton's law, $f = ma$, it is more useful to know the components of the acceleration vector, **A**, in terms of the local coordinate system formed by the tangential and normal (perpendicular to the tangent) directions rather than the **ij** components. We know that the direction of **V** is tangent to the curve. If **T** is a *unit tangent vector*, then $\mathbf{V} = v\mathbf{T}$. The acceleration of the particle, **A**, is

$$\mathbf{A} = \frac{d\mathbf{V}}{dt} = \frac{dv}{dt}\mathbf{T} + v\frac{d\mathbf{T}}{dt}$$

(*Note:* The product rule is valid when differentiating in this situation.) If we multiply the last term by ds/ds and $d\theta/d\theta$, we may write

$$\mathbf{A} = \frac{dv}{dt}\mathbf{T} + v\frac{ds}{dt}\frac{d\theta}{ds}\frac{d\mathbf{T}}{d\theta}$$

We already know that $v = ds/dt$. If θ is the angle between the positive direction of **T** and the positive X-axis, $d\theta/ds$ is the curvature of the path and its reciprocal is the *radius of curvature* ρ (the Greek letter rho). It can be shown that $d\mathbf{T}/d\theta$ is a unit vector perpendicular to **T** which is denoted by **N**. Its direction is usually toward the concave side of the path. Thus we have

$$\mathbf{A} = \frac{dv}{dt}\mathbf{T} + \frac{v^2}{\rho}\mathbf{N}$$

The component of **A** in the direction of **T** is called the *tangential acceleration* and is denoted as $a_\mathbf{T} = dv/dt$. The component of **A** in the direction of **N** is called the *normal acceleration* and is denoted as $a_\mathbf{N} = v^2/\rho$. Thus $\mathbf{A} = a_\mathbf{T}\mathbf{T} + a_\mathbf{N}\mathbf{N}$ (see Fig. 7.11.2).

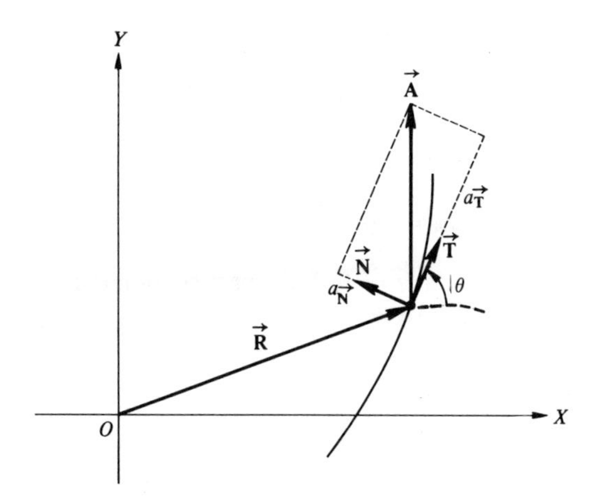

Figure 7.11.2 Acceleration vector.

In finding the acceleration components a_T and a_N on a curved path, $x = g(t)$, $y = h(t)$, we make use of formula 7.11.2 to obtain v; then $a_T = dv/dt$. The radius of curvature is found from the formula

$$\rho = \frac{[1 + (dy/dx)^2]^{3/2}}{d^2y/dx^2} \tag{7.11.5}$$

Then $a_N = v^2/\rho$.

EXAMPLE 2. Find a_T and a_N at $t = 2$s for a particle moving on the curve $x = t^2$, $y = t^3$ for x and y measured in feet.

Solution. We find v from

$$v = \sqrt{\left(\frac{dx}{dt}\right)^2 + \left(\frac{dy}{dt}\right)^2} \tag{7.11.2}$$

where

$$\frac{dx}{dt} = 2t \qquad \text{and} \qquad \frac{dy}{dt} = 3t^2$$

Thus

$$v = (4t^2 + 9t^4)^{1/2}$$

and

$$a_T = \frac{dv}{dt} = \frac{1}{2}(4t^2 + 9t^4)^{-1/2}(8t + 36t^3)$$

At $t = 2$,

$$a_T = \frac{8(2) + 36(2)^3}{2\sqrt{4(2)^2 + 9(2)^4}} = 12 \text{ ft/s}^2$$

To find ρ we need dy/dx and d^2y/dx^2.

$$\frac{dy}{dx} = \frac{dy/dt}{dx/dt} = \frac{3t^2}{2t} = \frac{3}{2}t \qquad \frac{d^2y}{dx^2} = \frac{d}{dx}\left(\frac{3}{2}t\right) = \frac{3}{2}\frac{dt}{dx} = \frac{3}{2}\cdot\frac{1}{2t} = \frac{3}{4t}$$

At $t = 2$, $dy/dx = 3$ and $d^2y/dx^2 = \frac{3}{8}$. Using formula 7.11.5, we obtain

$$\rho = \frac{[1 + (3)^2]^{3/2}}{(3/8)} = 72$$

At $t = 2$, we find that $v^2 = 4(2)^2 + 9(2)^4 = 160$. Thus

$$a_N = \frac{v^2}{\rho} = \frac{160}{72} = 2.2 \text{ ft/s}^2$$

PROBLEM 2. Find a_T and a_N at $t = 1$s for a particle moving on the curve $x = t^3$, $y = t^2$ for x and y measured in meters.

EXERCISES 7.11

Group A

1. In each of the following, the motion of an object along a curvilinear path is given. Find the speed of each object.

(a) $x = t$, $y = t^2$, at $t = 2$.
(b) $x = t^2$, $y = t^3$, at $t = 1$.
(c) $x = t^2 + 1$, $y = t + 1$, at $t = 5$.
(d) $x = t^3 + 2t^2 + 1$, $y = t$, at $t = 2$.
(e) $x = t$, $y = \sqrt{2 - t^2}$, at $t = 1$.
(f) $x = \dfrac{\sqrt{36 - 9t^2}}{2}$, $y = t$, at $t = 1$.

2. In each of the following, the motion of an object along a curvilinear path is given. Find the velocity of each object at the time indicated.

(a) $x = t^3$, $y = t^2$, at $t = 1$.
(b) $x = t$, $y = \sqrt{8 - t^2}$, at $t = 2$.
(c) $x = \sqrt{32 - t^2}$, $y = t$, at $t = -4$.
(d) $x = t^2$, $y = 4t^4$, at $t = 1$.

3. A particle moves on a circle of radius 2 ft with a constant speed of 4 ft/s. Find its acceleration.

4. A particle moves on a circle of radius 2 ft with a variable speed of $v = 3t$ ft/s. Find its acceleration when $t = 2$ s.

5. A particle moves upward on the parabola $y = x^2$ at a speed of 2 ft/s. Find its velocity and acceleration at the point (1, 1).

Group B

1. An object moves according to the parametric equations $x = t^2 - 1$, $y = 1/t^2$. Find the speed and velocity of the object at $t = 2$.

2. A particle moves according to the parametric equations $x = 4/(4 + t^2)$, $y = 4t/(4 + t^2)$. Find the speed and velocity of the object at $t = 4$.

3. A particle moves along the top branch of the parabola $y^2 = 4x$ with constant x-component of velocity 16 ft/s. Find the speed and velocity of the particle at $x = 4$.

4. A particle has an orbit described by $x = t$, $y = \sqrt{36 - 4t^2}/3$. Find the y-component of the velocity at $t = 2$.

5. A particle moves such that $x = 2t$, $y = 2\sqrt{-t^2 + 4t - 3}$. For what values of t is the motion defined? What is its velocity when $t = 1$?

6. The path of a projectile with an initial speed of v_0 is given as $x = v_0 t$, $y = v_0 t - \frac{1}{2}gt^2$. Show that the speed of the projectile is least when it is at its highest point.

7. A particle moves around a circle with a constant speed. Show that its acceleration is always directed toward the center of the circle. [*Hint*: $\mathbf{R} = (\cos t)\mathbf{i} + (\sin t)\mathbf{j}$.]

8. Explain why a particle moving on a curve with constant speed has a nonzero acceleration.

9. Explain how velocity and speed are alike and how they are different.

7.12 RELATED RATES

When we first developed the concept of the derivative we saw that dy/dx is the *rate of change of* y *with respect to* x. If in the function $y = f(x)$, x and y both vary with time t, we may differentiate the equation with respect to t. This result will contain expressions dy/dt and dx/dt, which are the *rates of change of* y *with respect to* t and *of* x *with respect to* t, respectively. If either of these rates are known, the other may be computed.

EXAMPLE 1. A stone is thrown into a smooth body of water. The radius of a circle, formed by the ripple, is increasing at the rate of 3 ft/s. How fast is the area of the circle increasing when the radius is 5 ft?

Solution. For a circle of radius r, its area, A, is given by $A = \pi r^2$. Since both A and r vary with t, we differentiate with respect to t to obtain

$$\frac{dA}{dt} = \pi 2r \frac{dr}{dt}$$

Since the radius is increasing at the rate of 3 ft/s, we have

$$\frac{dr}{dt} = 3$$

and the general formula for the rate of change of A with respect to t is

$$\frac{dA}{dt} = \pi 2(r)(3) = 6\pi r$$

Thus when $r = 5$, we have

$$\frac{dA}{dt} = 6\pi(5) = 30\pi = 94.2 \text{ ft}^2/\text{s}$$

and the area A is then increasing at the rate of 94.2 ft²/s.

PROBLEM 1. The volume of a cone is given by $V = (\pi/3)r^2h$. If the height, h, is always four times the radius, r, find how fast the volume is changing when $r = 3$ ft and is increasing at the rate of 2 ft/s.

The following five steps may help in solving related rate problems.

1. Draw a figure if needed. Denote by x, y, z, and so on, the quantities that vary with time.
2. Make a list of given and required quantities.
3. Obtain an equation between the variables involved which is true at any instant.
4. Differentiate the equation of step 3 with respect to time.
5. Substitute the known quantities in the result of step 4 and solve for the unknown.

EXAMPLE 2. In an electric circuit the current decreases at a rate of 0.5 A/s, while the resistance increases at a rate of 2 Ω/s. Find how fast the power changes when the current I is 15 A and the resistance R is 30 Ω.

Solution. Since the current decreases at 0.5 A/s, we have $dI/dt = -0.5$. We also have $dR/dt = 2$. We desire to find dP/dt when $I = 15$ and $R = 30$. For an electric circuit $P = I^2R$, then

$$\frac{dP}{dt} = I^2\frac{dR}{dt} + 2I\frac{dI}{dt}R$$

$$\therefore \quad \frac{dP}{dt} = (15)^2(2) + 2(15)(-0.5)(30)$$

and

$$\frac{dP}{dt} = 450 - 450 = 0$$

which says that the power is constant at this instant.

***Warning** After you take the derivative, substitute the values for the variable quantities.*

PROBLEM 2. Water is flowing into a conical container, vertex down, at the rate of 15 ft³/min. Find the rate at which the water is rising when $r = 2$ ft, $h = 5$ ft, and $dr/dt = 0.5$ ft/min.

EXAMPLE 3. The end of a 20-ft ladder slides down a vertical wall at a speed of 5 ft/s. The other end slides along the horizontal ground. How fast is the center of the ladder approaching the ground when it is 1 ft above the ground?

Solution. See Fig. 7.12.1, where $l = 20$ and $dy/dt = -5$. We want to find dh/dt when $h = 1$. We have

$$y^2 + x^2 = 400 \qquad \text{and} \qquad h = \tfrac{1}{2}y$$

Therefore,

$$4h^2 + x^2 = 400$$

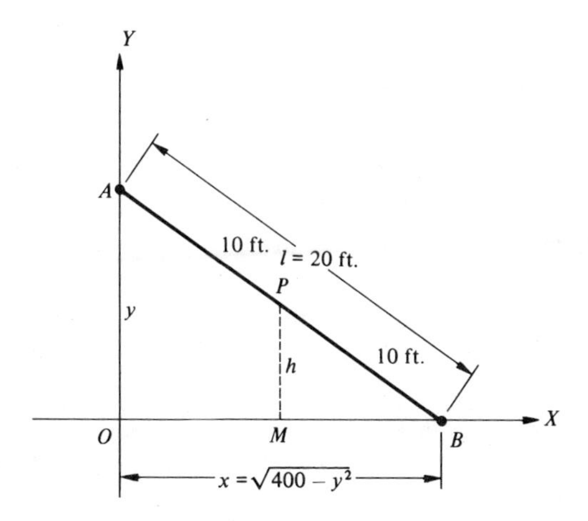

Figure 7.12.1 Example 3.

Differentiating this equation with respect to t, we obtain

$$8h\frac{dh}{dt} + 2x\frac{dx}{dt} = 0 \qquad \text{or} \qquad \frac{dh}{dt} = -\frac{1}{4}\frac{x}{h}\frac{dx}{dt}$$

But from $y^2 + x^2 = 400$ we find that

$$\frac{dx}{dt} = \frac{-y}{x}\frac{dy}{dt}$$

and so we have

$$\frac{dh}{dt} = -\frac{1}{4}\frac{x}{h}\left(-\frac{y}{x}\frac{dy}{dt}\right) = \frac{1}{4}\frac{y}{h}\frac{dy}{dt}$$

Finally, when $h = 1$, $dy/dt = -5$ and $y = 2$; thus

$$\frac{dh}{dt} = \frac{1}{4}\left(\frac{2}{1}\right)(-5) = -\frac{5}{2}\text{ ft/s}$$

PROBLEM 3. A 25-ft ladder leans against a vertical wall. How fast must the base of the ladder be pulled away from the wall in order to have the top of the ladder slide down the wall at a rate of 3 ft/s when the top is 15 ft above the ground?

EXERCISES 7.12

Group A

1. Gas is let into a spherical container at its center and expands uniformly in a sphere whose radius increases at the rate of 5 in./s. How fast is the volume of the space occupied by the gas increasing when the radius is 10 in.?

2. In an electric circuit the current decreases at the rate of 1 A/s, while the resistance increases at a rate of 4 Ω/s. Find how fast the power changes when the current, I, is 20 A and the resistance, R, is 40 Ω.

3. The end of a 10-ft ladder slides down a vertical wall at a speed of 2 ft/s. The other end slides along the horizontal ground. How fast is the center of the ladder approaching the ground when it is 1 ft above the ground?

4. The side of a square is increasing at the rate of 2 ft/min. How fast is the area changing when the side is 8 ft long?

5. Water is stored in a cylindrical reservoir of 10-ft radius. If the height of the water level decreases at the rate of 2 ft/min, find how fast the volume of the water in the reservoir is decreasing.

6. At noon on a given day ship A is 30 mi due north of ship B. If ship A sails west at 10 mi/h and ship B sails north at 6 mi/h, find how rapidly the distance between them is changing 1 h later.

7. A street lamp is 15 ft above the sidewalk. A man 6 ft tall walks away from the base of the light at a rate of 5 ft/s. How fast is his shadow lengthening when he is 20 ft away from the base of the lamp?

8. The area of a rectangle is increasing at the rate of 10 ft²/s. If the length is always twice the width, how fast are the sides changing (in ft/s) when the width is 5 ft?

9. A solution is being poured into a conical filter at the rate of 6 cm³/s and is running out at the rate of 1 cm³/s. How many centimeters per second is the level of liquid rising in the filter when the liquid is halfway to the top? The height of the filter is 20 cm and the radius is 10 cm.

10. The force of attraction between two certain bodies is given by $f = 25/r^2$. How fast is the force changing when $r = 5$ ft and is decreasing at $\frac{1}{2}$ ft/s?

Group B

1. The kinetic energy of an object in motion is given by the formula $\text{K.E.} = \frac{1}{2}mv^2$, where $m = w/g$, is the mass of the object and v is its speed. If a 10-lb object is accelerating at 16 ft/s², how fast is the kinetic energy changing when its speed is 5 ft/s?

2. A woman is running over a bridge at a rate of 10 ft/s, while a boat passes under the bridge directly below her at a rate of 20 ft/s. The boat's direction is at right angles to the woman's direction and 25 ft below the bridge. How fast are the boat and woman separating 2 s later? [*Hint*: $d^2 = x^2 + y^2 + (25)^2$.]

3. In elementary physics we learn that the period of a simple pendulum consisting of a particle of mass m supported by a wire of length l and of negligible mass is given by $T = 2\pi\sqrt{l/g}$, where g is the acceleration due to gravity. If for a particular pendulum the length of the wire decreases at the rate of $\frac{1}{2}$ in./s, what is the rate of change of the period of the pendulum when its length is 8 ft?

4. The current in a particular single-loop circuit is given by $I = 20/(R_1 + R_2)$, where R_1 is a resistance that changes constantly at a rate of 0.1 ohm/sec and R_2 is a resistance that changes constantly at a rate of 0.9 Ω/s. Find dI/dt when $R_1 = 10$ ohms and $R_2 = 15$ Ω.

5. Water is flowing at the rate of 4 ft³/min out of a tank shaped like a cone with vertex down, altitude 20 ft, and base radius 10 ft. How fast is the water level changing when the water is 8 ft deep?

6. The intensity of heat equals the strength of the source divided by the square of the distance from the source. If a man approaches a heated furnace of strength 100 units at a rate of 3 ft/s, how fast is the intensity of the heat changing when he is 6 ft from the furnace?

7. An ideal gas obeying Boyle's law (the pressure in a gas varies inversely as the volume if the temperature is constant) occupies 800 in.³ when the pressure is 3 atm and the volume is increasing at the rate of 40 in.³/min. Find how fast the pressure is changing when the volume is 1000 in.³.

8. One leg of a right triangle is 12 in. long and increasing at the rate of 2 in./s. The other leg, 5 in. long, is decreasing at the rate of 1 in./s. At what rate is the hypotenuse changing?

9. The speed, v ft/s, of a bullet passing through a certain material is given as $v = 750\sqrt{1 - 4x}$. How fast is the speed changing when the depth of penetration x is 2 in. and changing at the rate of 4 in./s?

10. Water is running into a hemispherical bowl with a 10-in. radius at the rate of 2 in.³/s. How fast is the height of the water increasing when it is 5 in. deep? [The volume of a spherical segment with one base is $V = (\pi h^2/3)(3r - h)$.]

7.13 INTEGRATION OF $u^n\,du$

In Sec. 4.3 we saw that

$$\int x^n\,dx = \frac{x^{n+1}}{n+1} + C \qquad n \neq -1 \tag{4.3.1}$$

Now if u is any differentiable function of x, we have from Eq. 7.6.6, $d(u^n) = nu^{n-1}\,du$, that

$$\boxed{\int u^n\,du = \frac{u^{n+1}}{n+1} + C \qquad n \neq -1} \tag{7.13.1}$$

Equation 7.13.1 enables us to integrate many more functions than did Eq. 4.3.1. It is often useful when integrating a sum or difference of terms raised to a power.

Example 1. Find $\int (\frac{1}{2}x^2 + 3)^3\, x\,dx$.

Solution. If we let $u = \frac{1}{2}x^2 + 3$, we have $du = x\,dx$, and the integral becomes $\int u^3\,du = \frac{1}{4}u^4 + C$ by Eq. 7.13.1.

$$\therefore \quad \int (\tfrac{1}{2}x^2 + 3)x\,dx = \tfrac{1}{4}(\tfrac{1}{2}x^2 + 3)^4 + C$$

PROBLEM 1. Find $\int (\frac{1}{3}x^3 - 5)^2 x^2\,dx$.

To make full use of Eq. 7.13.1 we note that the integrand consists of two factors, namely u^n and du. The expression for u can be any differentiable function as long as du is the other factor. We could write Eq. 7.13.1 as

$$\int (*)^n\,d(*) = \frac{(*)^{n+1}}{n+1} + C \qquad n \neq -1$$

EXAMPLE 2. Find $\int \sqrt{x^3 + 5}\,x^2\,dx$.

Solution. One factor of the integrand is a sum of terms raised to a power. We see this by rewriting the integral in the form $\int (x^3 + 5)^{1/2} x^2\,dx$.

Since this integral is not precisely in the form of Eq. 7.13.1, we shall try to change it to that form by a suitable substitution. Let $u = x^3 + 5$; then $du = 3x^2\,dx$ and $x^2\,dx = \frac{1}{3}du$. Upon substitution the given integral becomes

$$\int u^{1/2}\frac{1}{3}\,du = \frac{1}{3}\int u^{1/2}\,du = \frac{1}{3}\frac{u^{3/2}}{(3/2)} + C$$

$$\therefore \quad \int \sqrt{x^3 + 5}\,x^2\,dx = \frac{2}{9}(x^3 + 5)^{3/2} + C$$

One nice thing about integration is that the answer can always be checked by differentiation. You should check the answer to Example 2.

PROBLEM 2. Find $\int (x^2 - 6)^{-1/2} x\,dx$.

EXAMPLE 3. Evaluate $\int_1^2 (x + 1)^3\,dx$.

Solution. Let $u = x + 1$; then $du = dx$ and $\int (x + 1)^3\,dx = \int u^3\,du = \frac{1}{4}u^4 + C$. Hence

$$\begin{aligned}\int_1^2 (x + 1)^3\,dx &= [\tfrac{1}{4}(x + 1)^4]_1^2 \\ &= \tfrac{1}{4}(3)^4 - \tfrac{1}{4}(2)^4 \\ &= \tfrac{65}{4} = 16.25\end{aligned}$$

PROBLEM 3. Evaluate $\int_{-1}^{0.5} (2x^2 + 3)^4 x\,dx$.

EXERCISES 7.13

Group A

1. Find each of the following integrals.

(a) $\int (3x^2 + 2x + 1)^2(6x + 2)\,dx$. (b) $\int (x^2 + 1)^{1/2} x\,dx$.

(c) $\int (x^2 + 2)^{-1/2} x\, dx.$

(d) $\int \frac{x^3}{(x^4 + 2)^2}\, dx.$

(e) $\int \frac{x}{\sqrt[3]{x^2 + 3}}\, dx.$

(f) $\int x^3 (x^4 - 16)^{3/2}\, dx.$

(g) $\int x^3 \sqrt{16 - x^4}\, dx.$

(h) $\int (x + 2)\sqrt{16 - (x + 2)^2}\, dx.$

2. Evaluate each of the following definite integrals.

(a) $\int_0^1 x\sqrt{1 - x^2}\, dx.$

(b) $\int_0^a x\sqrt{a^2 - x^2}\, dx.$

(c) $\int_1^2 (x^4 + 2)^3 x^3\, dx.$

(d) $\int_{-2}^2 x\sqrt{36 - 9x^2}\, dx.$

(e) $\int_0^2 x^3 \sqrt{16 - x^4}\, dx.$

(f) $\int_0^2 \frac{x^2}{\sqrt[3]{x^3 + 1}}\, dx.$

3. Integrate $\int \sin x \cos x\, dx$ three different ways. (a) Let $u = \sin x$. (b) Let $u = \cos x$. (c) Let $\sin x \cos x = \frac{1}{2} \sin 2x$. (d) Are the answers all equivalent? Why? (Recall your study of trigonometry.)

SUMMARY OF IMPORTANT WORDS AND CONCEPTS

For each of the following, state in your own words your understanding of the statement or word.

1. Algebraic functions.

2. (a) Limit of a sum. (b) Limit of a product. (c) Limit of a constant times a function. (d) Limit of a quotient.

3. Derivatives of: (a) A constant. (b) A constant times a function. (c) A sum. (d) A product. (e) A quotient.

4. (a) Composite functions. (b) The chain rule.

5. Derivative of a function raised to a power.

6. (a) Rational functions. (b) Graphs of rational functions (seven steps). (c) Infinity. (d) Asymptote.

7. (a) Implicit function. (b) Implicit differentiation.

8. Differential of arc.

9. (a) Parametric equations. (b) Parameter.

10. (a) Vectors. (b) Position vector.

11. (a) Motion along a curve. (b) Velocity. (c) Acceleration: tangential and normal acceleration components.

12. Related rates (five steps to solve related rate problems).

13. Integration of $u^n\, du$, $n \neq -1$.

Chapter 8

Exponential and Logarithmic Functions

8.1 INTRODUCTION

In previous chapters we have discussed differentiation and some integration of algebraic functions. Recall that the use of the formula $\int u^n\,du$ was restricted, namely $n \neq -1$, because we cannot divide by zero. Furthermore, we found that there is no algebraic function which when differentiated would yield $1/x$. These difficulties are overcome by using the logarithmic function, which in turn comes from the exponential function. We discuss these functions next.

8.2 EXPONENTIAL FUNCTIONS

Exponential functions are of special importance in certain fields of science. For example, in physics they describe transient electrical currents, radioactive decay, and many other phenomena. In biology they describe the growth rate of bacteria. In pharmacokinetics they describe drug concentrations. In business, exponential functions describe the "growth" of money invested at compound interest. They belong to a class of functions called *transcendental functions.*[1]

> ***Definition 8.2.1*** *The exponential function, f, has the rule of correspondence $f(x) = b^x$ with its base b a positive number other than* 1. *Its*

[1]Transcendental functions are nonalgebraic functions.

domain is the set of all real numbers and its range is the set of all positive real numbers.

Note carefully that in an exponential function, the exponent is a variable.

In a previous course in algebra you studied exponents that were positive and negative rational numbers. The following definitions and properties were discussed at that time. We list them for a review, where n and m are any integers and a and b are nonzero real numbers.

$$a^n = \underbrace{a \cdot a \cdot a \cdots a}_{n \text{ factors}} \qquad (n \text{ factors of } a)$$

$$a^0 = 1 \qquad \text{for } a \neq 0$$

$$a^m \cdot a^n = a^{m+n}$$

$$(a^m)^n = a^{mn}$$

$$\frac{a^m}{a^n} = a^{m-n}$$

$$(ab)^n = a^n b^n$$

$$\left(\frac{a}{b}\right)^n = \frac{a^n}{b^n}$$

$$a^{-n} = \frac{1}{a^n}$$

$$a^{1/n} = \sqrt[n]{a}$$

$$a^{m/n} = (a^{1/n})^m = (a^m)^{1/n} = (\sqrt[n]{a})^m = \sqrt[n]{a^m}$$

It is possible to prove that the theorems on exponents hold for all real numbers. In particular we state:

Theorem 8.2.1 *If x_1 and x_2 are any real numbers and a and b are positive real numbers, then*

(a) $b^{x_1} \cdot b^{x_2} = b^{x_1+x_2}$.

(b) $(b^{x_1})^{x_2} = b^{x_1 x_2}$.

(c) $\dfrac{b^{x_1}}{b^{x_2}} = b^{x_1-x_2}$.

(d) $(ab)^{x_1} = a^{x_1} b^{x_1}$.

(e) $\left(\dfrac{a}{b}\right)^{x_1} = \dfrac{a^{x_1}}{b^{x_1}}$.

EXAMPLE 1. If f is an exponential function, prove that $f(x_1 + x_2) = f(x_1)f(x_2)$.

Solution. Since f is an exponential function, $f(x) = b^x$. Thus

$$f(x_1 + x_2) = b^{x_1+x_2} = b^{x_1}b^{x_2} = f(x_1)f(x_2)$$

Note that this says that an exponential function of a sum is equal to the product of exponential functions.

PROBLEM 1. Let $f(x) = 2^x$. Verify the result of Example 1 for $x_1 = 2$ and $x_2 = 3$.

The graphs of exponential functions fall into two categories. For $x > 0$ they either rise rapidly for an increase in x (this occurs for $b > 1$) or they decrease slowly for an increase in x (which occurs for $b < 1$). This is shown in the following example.

EXAMPLE 2. Draw the graph of (a) $y = 2^x$ and (b) $y = (\frac{1}{2})^x$.

Solution. Construct a table of pairs of points and plot them.

(a) $y = 2^x$

x	-3	-2	-1	0	1	2	3
y	$\frac{1}{8}$	$\frac{1}{4}$	$\frac{1}{2}$	1	2	4	8

(b) $y = (\frac{1}{2})^x$

x	-3	-2	-1	0	1	2	3
y	8	4	2	1	$\frac{1}{2}$	$\frac{1}{4}$	$\frac{1}{8}$

The graphs are shown in Fig. 8.2.1(a) and (b). The graphs of all exponential functions pass through the point (0, 1).

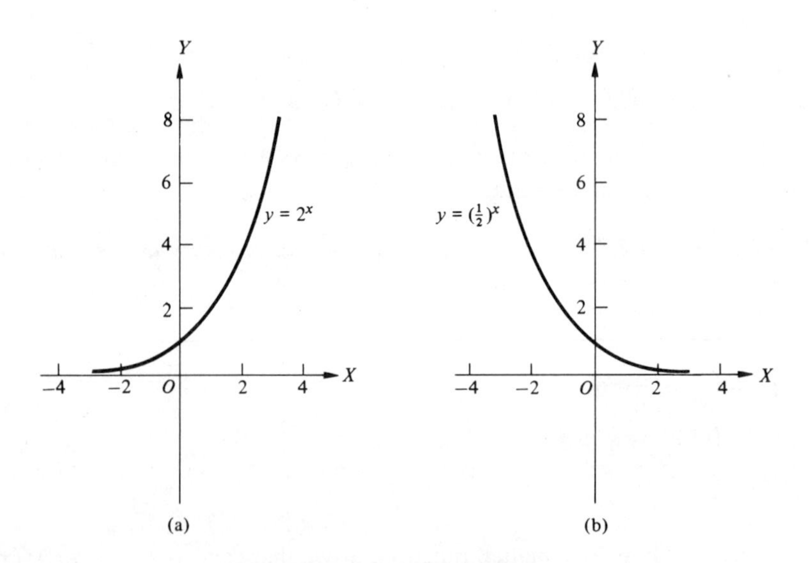

Figure 8.2.1 Example 2.

PROBLEM 2. Draw the graph of (a) $y = 10^x$; (b) $y = (\frac{1}{10})^x$.

EXERCISES 8.2

Group A

1. Compute each of the following.

(a) $2^2 \cdot 3^2$. (b) $\dfrac{6^{10}}{6^5}$. (c) $\dfrac{2^3}{6^3}$.

(d) $2^5 \cdot 3^2$. (e) $4^2 \cdot 4^3$. (f) $a^5 \cdot b^5$.

2. Graph the function $f(x) = 3^x$.

3. Graph the function $f(x) = (\frac{1}{3})^x$.

Group B

1. If f is an exponential function with base c, prove that $f(x_1 - x_2) = f(x_1)/f(x_2)$.

2. If f is an exponential function with base a and g is an exponential function with base b, find an exponential function h such that $h(x) = f(x)g(x)$.

3. (a) If $b = 1$, why can it not be used to define an exponential function? (b) If b is negative, why can it not be used to define a continuous exponential function?

4. Explain the difference between the algebraic function $f(x) = x^2$ and the exponential function $f(x) = 2^x$.

5. Why do the graphs of all exponential functions pass through the point (0, 1)?

8.3 THE IRRATIONAL NUMBER *e*

In Definition 8.2.1 we stated that any positive number other than one may be used as the base of an exponential function. Thus we could consider such numbers as $(\sqrt{3})^{1.7}$ whose value may be approximated by logarithms (Sec. 8.5). A particular irrational number, denoted by the symbol e, is used extensively in mathematics as a base. Its value is approximately $e = 2.7183\ldots$. The exponential function whose base is e is written as

$$f(x) = e^x \tag{8.3.1}$$

This function is so important it is called *the exponential function.* It is frequently written as $f(x) = \exp x$ when the exponent is complicated. We now define e.

Definition 8.3.1 *The irrational number*

$$e = \lim_{h \to 0} (1 + h)^{1/h}$$

The limit in Definition 8.3.1 will occur in the development of the derivative of the logarithmic function (Sec. 8.6).

The value of e may be determined in the following manner. Expand $(1 + h)^{1/h}$ by the binomial theorem to obtain

$$(1 + h)^{1/h} = 1 + \frac{1}{h}h + \frac{\frac{1}{h}\left(\frac{1}{h} - 1\right)}{2!}h^2 + \frac{\frac{1}{h}\left(\frac{1}{h} - 1\right)\left(\frac{1}{h} - 2\right)}{3!}h^3 + \dots$$

$$= 1 + 1 + \left(\frac{1 - h}{2!}\right) + \frac{(1 - h)(1 - 2h)}{3!} + \dots$$

As $h \longrightarrow 0$ we have

$$\lim_{h \to 0} (1 + h)^{1/h} = 1 + \frac{1}{1!} + \frac{1}{2!} + \frac{1}{3!} + \dots = e \tag{8.3.2}$$

Equation (8.3.2) is an infinite series for e (see Chap. 17). By it the value of e can be found to any number of decimal places. Recall that $1! = 1$, $2! = 2 \cdot 1 = 2$, $3! = 3 \cdot 2 \cdot 1 = 6$, etc. To find e correct to five decimal places we would need to use the first 10 terms of the series to find $e = 2.71828\dots$. Values of e^x may be obtained from a calculator. You should have such an electronic calculator. We may sketch the graph of $f(x) = e^x$ by constructing a table of pairs of points and then plotting them. Thus

x	-3	-2	-1	0	1	2	3
$y = f(x)$	0.05	0.14	0.37	1.00	2.72	7.39	20.1

See Fig. 8.3.1 for the graph of $y = e^x$. We can discuss this in detail after finding the derivative of e^x (Sec. 8.7).

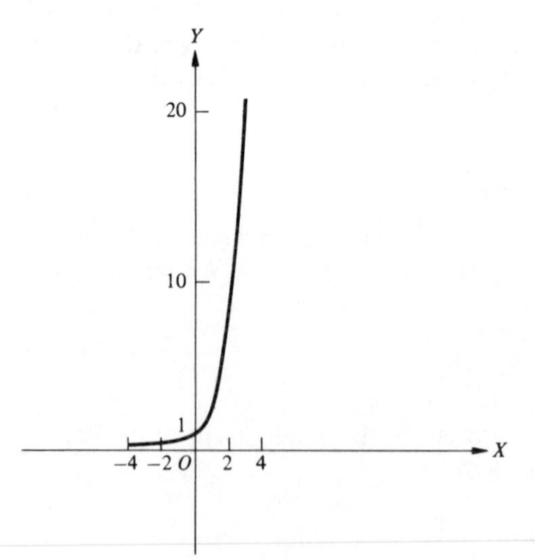

Figure 8.3.1 Graph of $y = e^x$.

The derivative of $f(x) = e^x$ is most easily obtained by considering the logarithmic function, which is the inverse of the exponential function, as discussed in Secs. 8.4 and 8.5.

EXERCISES 8.3

Group A

1. Simplify each of the following.

(a) e^2e^3. (b) e^xe^{-x}. (c) e^te^{-u}. (d) e^t/e^{t^2}.

2. If $e = 2.71828$ is correct to five decimal places, find the approximate error in using five terms of Eq. 8.3.2 to compute e.

3. Sketch the graph of each of the following.

(a) $y = e^{-x}$. (b) $y = e^{2x}$. (c) $y = e^{x/2}$.

4. Write each of the following in the form $f(x) = \exp x$.

(a) $f(x) = e^{3x}$. (b) $f(x) = e^{x^2+2x-3}$.

5. Simplify.

(a) $f(x) = \exp(4)\exp(x)$. (b) $f(x) = \exp(x^2 + 3)/\exp(x + 7)$.

Group B

1. If $f(x) = e^x$ and $f(y) = e^y$, show that $f(x + y) = f(x)f(y)$.

2. State the value of c if $e^{c_1x+c_2} = ce^{c_1x}$.

3. Use the delta process to compute $[f(x + \Delta x) - f(x)]/\Delta x$ for $f(x) = e^x$. Can you now find

$$\lim_{\Delta x \to 0} \frac{f(x + \Delta x) - f(x)}{\Delta x}?$$

4. Sketch the graph of $f(x) = (-e)^x$. Do you see why negative numbers are not used as a base for exponential functions?

5. The hyperbolic sine function of t is defined to be $\sinh t = \frac{1}{2}(e^t - e^{-t})$, and the hyperbolic cosine function of t is defined to be $\cosh t = \frac{1}{2}(e^t + e^{-t})$. A uniform flexible cable whose ends are in the same horizontal plane forms a curve called a *catenary* (from the Latin meaning "chain") when hanging under its own weight. Its equation is $y = (H/2w)(e^{wx/H} + e^{-wx/H})$. Write this equation using the definition of $\sinh t$ or $\cosh t$.

6. Sketch the graph of (a) $y = \sinh t$; (b) $y = \cosh t$.

7. Show that $\cosh^2 t - \sinh^2 t = 1$.

8.4 INVERSE FUNCTIONS

In Sec. 1.3 we stated that a function assigned one and only one element of the range to each element of the domain. If the function also has the property that each element of the range corresponds to only one element of the domain, such a function is said to set up a one-to-one correspondence between the elements of the domain and the elements of the range. When functions establish such a one-to-one correspondence, we can define a new function which is the inverse of the original function, and the domain of the new function is the range of the original function, and the range of the new function is the domain of the original function. Each function is said to be the inverse of the other.

> ***Definition 8.4.1*** *If f is a function such that $f(c) \neq f(d)$, where c and d are two different elements in the domain of f, and if $f(a) = b$, then the inverse function of f, sometimes denoted by f^{-1}, is the function which assigns a as the image of b. The domain of f^{-1} is the range of f. The range of f^{-1} is the domain of f.*

Definition 8.4.1 tells us that if $f(a) = b$, then $f^{-1}(b) = a$. The -1 above the symbol f is not to be treated as an exponent; it is merely a symbol to denote the inverse function.

A geometric interpretation of Definition 8.4.1 is that if each line parallel to the X-axis intersects the graph of $f(x)$ in no more than one point, then the function has an inverse. In Fig. 8.4.1(a) the function has an inverse. In Fig. 8.4.1(b) the function does not have an inverse. The working rule for finding the inverse of a function is to solve the equation $y = f(x)$ for the independent variable and then replace y by x.

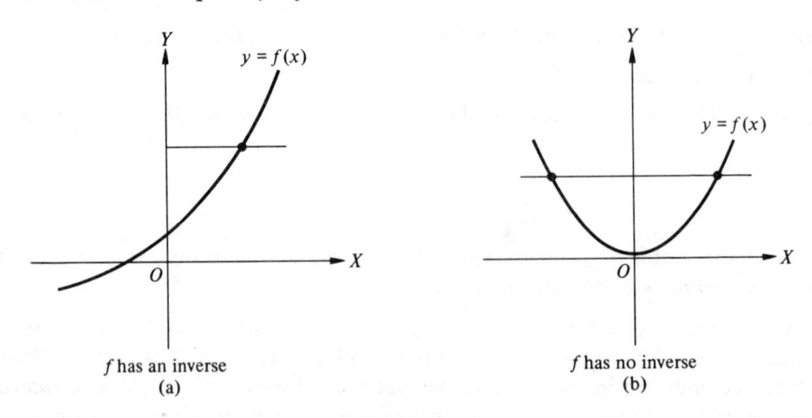

Figure 8.4.1

EXAMPLE 1. Find the inverse of $f(x) = x^3$.

Solution. Let $y = x^3$. Solve for x. Thus $x = y^{1/3}$. Then interchange the variable to obtain $f^{-1}(x) = x^{1/3}$ as the inverse to $f(x) = x^3$.

PROBLEM 1. Find the inverse function of $f(x) = 3x + 2$.

EXAMPLE 2. Find the inverse of $f(x) = x^2$.

Solution. Let $y = x^2$. Solving for x we find $x = \pm\sqrt{y}$. Since for each value of y there is more than one value of x, the function does not have an inverse function.

PROBLEM 2. Does the function $f(x) = 4x^2 + 9$ have an inverse function? If not, why?

EXERCISES 8.4

Group A

1. Sketch each of the following functions and determine which of them has an inverse. When one has an inverse, write a formula for it, if possible.

(a) $f(x) = 6x + 2$. (b) $f(x) = x^5$. (c) $f(x) = x^4$.
(d) $f(x) = 2^x$. (e) $f(x) = 2x^2 + 2$. (f) $f(x) = e^x$.

2. The graphs of $f(x)$ and its inverse function $f^{-1}(x)$ are reflections of each other about the line $y = x$. In each of the following, verify this by sketching the graph of $f(x)$ and $f^{-1}(x)$ on the same set of axes.

(a) $f(x) = 3x + 1$ and $f^{-1}(x) = (x - 1)/3$.
(b) $f(x) = \sqrt{x}$ and $f^{-1}(x) = x^2$, $x \geqq 0$.
(c) $f(x) = x^3$ and $f^{-1}(x) = x^{1/3}$.

8.5 THE LOGARITHMIC FUNCTION

From the graph of $f(x) = b^x$ (see Fig. 8.2.1 for $y = 2^x$) it is seen that b^x has an inverse. This inverse function is so important that it has a special name, the *logarithmic function of base b*.

Let us consider the exponential function $y = b^x$ and find its inverse. We use the "working rule" for finding inverse functions as stated in Sec. 8.4. Interchanging x and y, we obtain

$$b^y = x \tag{8.5.1}$$

Now we must solve for y. This we cannot do with our previous mathematical operations, so let us *define* y to be the logarithmic function to the base b and denote it by the symbol

$$y = \log_b x \tag{8.5.2}$$

(This is a favorite device of mathematicians to define something that cannot be solved for. Then they use this definition and see to where it leads.) Equation 8.5.1 is called the *exponential form* and Eq. 8.5.2 is called the *logarithmic form*.

These two equations are used interchangeably; that is,

$$y = \log_b x$$

implies that

$$b^y = x$$

If we substitute Eq. 8.5.2 into Eq. 8.5.1, we obtain the important fact that

$$b^{\log_b x} = x$$

EXAMPLE 1. (a) $2^3 = 8$ may be written in logarithmic form as $3 = \log_2 8$.
(b) $\log_{10} 100 = 2$ may be written in exponential form as $10^2 = 100$.

PROBLEM 1. (a) Write $2^5 = 32$ in logarithmic form.
(b) Write $\log_{10} 1000 = 3$ in exponential form.

We now give a formal definition of logarithms.

Definition 8.5.1 *The inverse of the exponential function* b^x, $b > 0$ *and* $b \neq 1$, *is called the logarithmic function to the base* **b**. *Its domain is the set of positive real numbers, and its range is the set of all real numbers. It is denoted by* $\log_b x$.

A common way of expressing logarithms is to say that the logarithm to the base b of a number N is the power to which b is raised to yield N; that is, if

$$L = \log_b N$$

then

$$b^L = N$$

Thus we see that logarithms are exponents.

The following fundamental properties of logarithms can be proved by using Definition 8.5.1 and properties of exponents.

$$\log_b(MN) = \log_b M + \log_b N \tag{8.5.3}$$

$$\log_b \left(\frac{M}{N}\right) = \log_b M - \log_b N \tag{8.5.4}$$

$$\log_b N^p = p \log_b N \tag{8.5.5}$$

$$\log_b 1 = 0 \tag{8.5.6}$$

$$\log_b \frac{1}{N} = -\log_b N \tag{8.5.7}$$

Note that in Definition 8.5.1 any positive number other than 1 can be used for the base b. There are two numbers widely used for the base of logarithms. One of them is 10 and the other is the number e. Logarithms using 10 as the base are called *common* or *Briggs'* logarithms and are written as $\log N$. (Note that the base $b = 10$ is not wirtten.) Logarithms using e as the base are called *natural* or *Napier's* logarithms and are written as ln N. Most of the time we shall use natural logarithms. (You will see why the word *natural* is used when we study Sec. 8.6.)

Logarithms of numbers may be obtained from an electronic calculator, which you should have. You may use your calculator to verify the graph of $y = \ln x$ as shown in Fig. 8.5.1.

Occasionally, we know the logarithm of a number to a given base and desire to find the logarithm of that number to another base. We may then use Theorem 8.5.1.

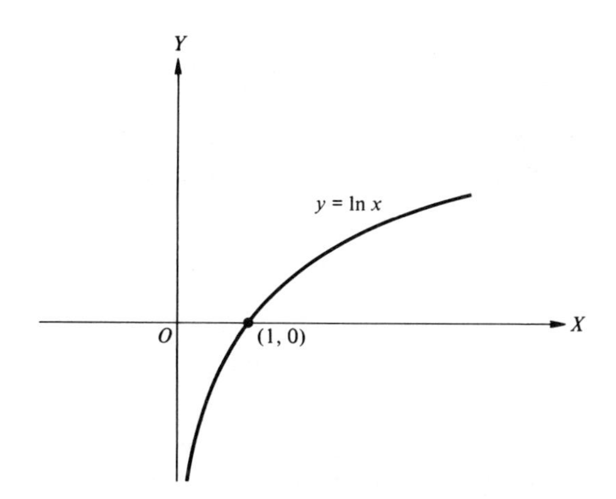

Figure 8.5.1 Graph of $y = \ln x$.

Theorem 8.5.1 $\log_a N = \log_b N \log_a b$.

Proof. Let $x = \log_b N$. Then $N = b^x$.
Taking logs to the base a, we have

$$\log_a N = \log_a b^x = x \log_a b$$

Therefore,

$$\log_a N = \log_b N \log_a b$$

EXAMPLE 2. Find a relation between common and natural logarithms.

Solution. Let $x = \ln N$. Then

$$e^x = N$$

Taking logs to the base 10, we have

$$x \log e = \log N$$

or

$$\log N = \ln N \log e$$

From a calculator we find that

$$\log e = 0.4343$$

Therefore,

$$\log N = 0.4343 \ln N \tag{8.5.8}$$

or

$$\ln N = 2.303 \log N \tag{8.5.9}$$

EXAMPLE 3. Find $\log_3 8$.

Solution. Using Theorem 8.5.1, we have

$$\log_b N = \frac{\log_a N}{\log_a b}$$

Here $b = 3$, $N = 8$, and let $a = 10$. Then

$$\log_3 8 = \frac{\log_{10} 8}{\log_{10} 3} = \frac{0.9031}{0.4771} = 1.893$$

(We could just as well have let $a = e$ and have written $\log_3 8 = \ln 8/\ln 3 = 2.0794/1.0986 = 1.893$.)

PROBLEM 2. Find $\log_2 9$.

EXERCISES 8.5

Group A

1. For each of the following equations, change those in exponential form to logarithmic form and those in logarithmic form to exponential form.

(a) $3^3 = 27$. (b) $5^{-2} = \frac{1}{25}$. (c) $(10)^0 = 1$.
(d) $\log_4 (64) = 3$. (e) $\log_{32} (2) = \frac{1}{5}$. (f) $\log_{1/3} (27) = -3$.
(g) $(\frac{1}{2})^3 = 8^{-1}$. (h) $\log_{1/8} (\frac{1}{2}) = \frac{1}{3}$. (i) $e^3 = 20.1$.
(j) $\ln 2 = 0.6931$.

2. In each of the following, determine the numerical value of x.

(a) $\log_x 100 = 2$. (b) $\log_4 (x) = \frac{1}{2}$. (c) $\log_3 (\frac{1}{81}) = x$.

3. Find the inverse function for each of the following functions.

(a) $f(x) = e^x$. (b) $f(x) = 3^x$.
(c) $f(x) = \log x$. (d) $f(x) = \ln x$.

4. Use Eqs. 8.5.3 through 8.5.7 to express each of the following as a logarithm with one argument.

(a) $\log 4 + \log 5$. (b) $\ln 8 - \ln 2$. (c) $4 \log_b 2 + \log_b 8$.
(d) $5 \log_b 4 - 10 \log_b 2$. (e) $\frac{1}{3} \log 8 + 4 \log 5$. (f) $6 \ln 5 + 6 \ln 6$.

5. Use a calculator to find each of the following, and verify Eqs. 8.5.8 and 8.5.9.

(a) $\ln 2$. (b) $\ln 0.2$. (c) $\log 2$.
(d) $\log 0.2$. (e) $\ln 15$. (f) $\ln 20$.

6. Utilize Theorem 8.5.1 to find each of the following in two different ways.

(a) $\log_{14} 48$. (b) $\log_2 20$.

7. Solve each of the following equations by using logarithms.

(a) $e^x = 10$. (b) $e^{3x} = 6$. (c) $e^{x+1} = 2$.
(d) $3^x = 10$. (e) $(2.1)^x = 3.4$. (f) $(1.6)^{x+1} = 0.6$.
(g) $x = (\sqrt{3})^{1.7}$.

8. On Fig. 8.5.1 draw the graphs of $y = e^x$ and $y = x$. What can you say about the graphs of $y = e^x$ and $y = \ln x$ with respect to the graph of $y = x$?

Group B

1. A certain differential equation has the solution $\ln(x^2 + 2) = \ln(y^3 + 3) + \ln c$. Show that $(x^2 + 2)/(y^3 + 3) = c$.

2. Show that $\ln y = x + \ln c$ is equivalent to $y = ce^x$.

3. A simple electrical circuit has the following equation for the current as a function of time: $I = 10(1 - e^{-5t})$. Solve the given equation for t.

4. (a) Show that $e^{\ln t} = t$. (b) What is $e^{\ln 2}$? $e^{2 \ln 3}$?

5. Why can we say that logarithms of zero and negative numbers do not exist?

6. Drug concentration in a first-order system is given by $C(t) = A_0(1 - e^{-kt})$. (a) Write this equation in logarithmic form. (b) Solve this equation for t.

Group C

1. Prove that $\log_a(NM) = \log_a N + \log_a M$.

2. Prove that $\log_a(N/M) = \log_a N - \log_a M$.

3. Prove that $\log_a N^p = p \log_a N$.

4. (a) Should $\ln N \ln M$ be equal to $\ln(NM)$? Why? (b) Should $\ln N \ln M$ be equal to $\ln N + \ln M$? (c) Should $(\ln N)/(\ln M)$ be equal to $\ln N - \ln M$? Why?

5. Why does one use "characteristics" and "mantissas" in working with common logarithms but not in working with natural logarithms?

8.6 DERIVATIVE OF THE LOGARITHMIC FUNCTION

Let us find the derivative of the logarithmic function by applying the delta process to the logarithmic function $y = \log_b u$, where u is any differentiable function of x. Consider

$$y = \log_b u$$

$$y + \Delta y = \log_b(u + \Delta u)$$

$$\Delta y = \log_b(u + \Delta u) - \log_b u$$

$$= \log_b\left(1 + \frac{\Delta u}{u}\right)$$

$$\frac{\Delta y}{\Delta u} = \frac{1}{\Delta u}\log_b\left(1 + \frac{\Delta u}{u}\right) \tag{8.6.1}$$

The next step is to take the limit as $\Delta u \longrightarrow 0$, but this does not do much for us on the right-hand side of Eq. 8.6.1, as we would get 0/0. Let us rewrite Eq. 8.6.1 as

$$\frac{\Delta y}{\Delta u} = \frac{1}{\Delta u} \log_b \left(1 + \frac{\Delta u}{u}\right)^{(u/\Delta u)(\Delta u/u)}$$

and then apply property 8.5.5 of logarithms to arrive at

$$\frac{\Delta y}{\Delta u} = \frac{1}{\Delta u}\frac{\Delta u}{u} \log_b \left(1 + \frac{\Delta u}{u}\right)^{u/\Delta u}$$

or

$$\frac{\Delta y}{\Delta u} = \frac{1}{u} \log_b \left(1 + \frac{\Delta u}{u}\right)^{u/\Delta u} \tag{8.6.2}$$

Now if we let $\Delta u/u = h$, then as $\Delta u \longrightarrow 0$, $h \longrightarrow 0$, and $\lim\limits_{\Delta u \to 0} [1 + (\Delta u/u)]^{u/\Delta u} = \lim\limits_{h \to 0} (1 + h)^{1/h} = e$ by Definition 8.3.1. Then taking the limit of Eq. 8.6.2, we have

$$\lim_{\Delta u \to 0} \frac{\Delta y}{\Delta u} = \lim_{\Delta u \to 0} \left[\frac{1}{u} \log_b \left(1 + \frac{\Delta u}{u}\right)^{u/\Delta u}\right]$$

which becomes

$$\frac{dy}{du} = \frac{1}{u} \log_b e \tag{8.6.3}$$

If for convenience we choose the base of the logarithm function to be e, then we have $y = \log_e u = \ln u$ and for

$$\boxed{\begin{aligned} y &= \ln u \\ \frac{dy}{du} &= \frac{1}{u} \end{aligned}} \tag{8.6.4}$$

(This simple form is why $\log_e u = \ln u$ is called the natural logarithm.)

In the function $y = \ln u$, if u is any differentiable function of x, we resort to Eq. 7.5.1, the chain rule, $dy/dx = (dy/du)(du/dx)$, to obtain for

$$\boxed{\begin{aligned} y &= \ln u \\ \frac{dy}{dx} &= \frac{1}{u}\frac{du}{dx} \end{aligned}} \tag{8.6.5}$$

It also follows that

$$d(\ln u) = \frac{1}{u}\, du \tag{8.6.6}$$

EXAMPLE 1. For $y = \ln(x^2 + 3)$, find dy/dx.

Solution. Let $u = x^2 + 3$. Then

$$\frac{dy}{dx} = \underbrace{\frac{1}{x^2+3}}_{1/u}\underbrace{\frac{d(x^2+3)}{dx}}_{du/dx} = \underbrace{\frac{1}{x^2+3}}_{1/u}\underbrace{2x}_{du/dx} = \frac{2x}{x^2+3}$$

PROBLEM 1. Find dy/dt for $y = \ln(2t^3 - 5)$.

EXAMPLE 2. For $y = \ln\sqrt{(1+2x)/(1-x^2)}$, find dy/dx.

Solution. If we use properties 8.5.5 and 8.5.4 of logarithms, we may write

$$y = \tfrac{1}{2}\ln(1+2x) - \tfrac{1}{2}\ln(1-x^2)$$

Then

$$\frac{dy}{dx} = \frac{1}{2}\underbrace{\frac{1}{1+2x}}_{1/u}\underbrace{2}_{du/dx} - \frac{1}{2}\underbrace{\frac{1}{1-x^2}}_{1/u}\underbrace{(-2x)}_{du/dx}$$

or

$$\frac{dy}{dx} = \frac{1}{1+2x} + \frac{x}{1-x^2}$$

which you can simplify.

PROBLEM 2. Find dI/dt for $I = 2\ln[(t^3 + t)^4(5t - 7)^3]$.

EXAMPLE 3. For $y = \log(3x^2 + x)$, find dy/dx.

Solution. Here the base is 10, so we make use of Eq. 8.6.3 with $b = 10$. Thus since

$$\frac{dy}{dx} = \frac{dy}{du}\frac{du}{dx}$$

we have

$$\frac{dy}{dx} = \underbrace{\frac{1}{3x^2+x}}_{1/u}\underbrace{(\log e)}_{\log_b e}\underbrace{(6x+1)}_{du/dx} = \frac{6x+1}{3x^2+x}(0.4343)$$

Notice that when the base of the logarithm is 10, we need to multiply by the factor (0.4343). This is why e is a natural choice for the base of logarithms.

PROBLEM 3. Find dy/dx for $y = 5\log(2x^3 - 5x)$.

EXAMPLE 4. Discuss and sketch the graph of $y = \ln x$.

Solution. For $y = \ln x$,

$$\frac{dy}{dx} = \frac{1}{x} \qquad \text{and} \qquad \frac{d^2y}{dx^2} = -\frac{1}{x^2}$$

There are no critical points and no points of inflection because dy/dx and d^2y/dx^2 cannot be zero. (Why?) As $x \longrightarrow \infty$, $dy/dx \longrightarrow 0$, and the slope is always positive. At $x = 1$, the slope is 1. Since d^2y/dx^2 is always negative, the curve is concave down (see Fig. 8.6.1).

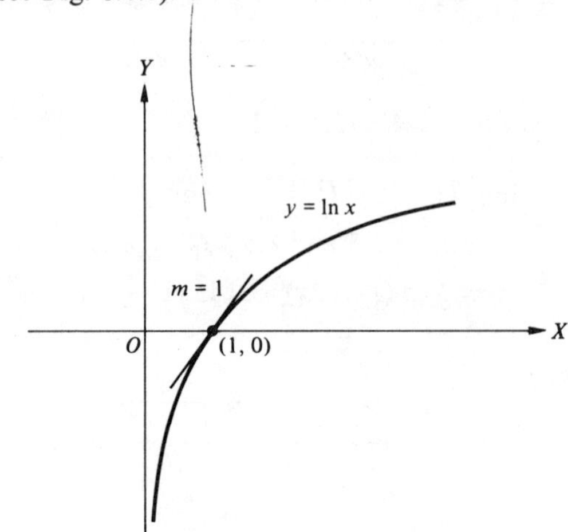

Figure 8.6.1 Graph of $y = \ln x$.

PROBLEM 4. Discuss and sketch the graph of $y = \ln x^3$.

PROBLEM 5. Find dy/dx for $y = x(\ln x - 1) + C$.

Note carefully that from the answer of Problem 5, we see that the antiderivative of $\ln u$ is $u(\ln u - 1) + C$. We make use of this in Sec. 8.8.

Applications of the derivative of the logarithmic function are given in Sec. 8.9.

EXERCISES 8.6

Group A

1. Differentiate with respect to their independent variable each of the following functions.

(a) $f(x) = \ln x^3$. (b) $f(x) = \ln (x^2 + 2x + 5)$.
(c) $f(t) = \ln t^4$. (d) $f(u) = \ln (2u + 8)$.
(e) $f(x) = \ln (x^3 - 3x + 2)$. (f) $f(x) = \ln (x^5 + x^3 + 2)$.

2. Differentiate with respect to x each of the following functions.

(a) $f(x) = \log x^2$. (b) $f(x) = \log_2 (x^3 + 1)$.
(c) $f(x) = \log (x^5 + 2x + 1)$. (d) $f(x) = \log_4 (x^4 + 1)$.

3. Find the differential of y in each of the following.

(a) $y = \ln x^3$. (b) $y = \ln (u + 1)$.
(c) $y = 3 \ln t$. (d) $y = -\frac{1}{3} \ln (x^2 + 2)$.

4. Find dy/dx for each of the following.

(a) $y = \ln 3x$.
(b) $y = \ln 3x^2$.
(c) $y = \ln (3x)^2$.
(d) $y = (\ln 3x)^2$.
(e) $y = x^2 \ln x^2 - x^2$.
(f) $y = (x + b)[\ln (x + b) - 1]$.

5. Find dy/dx for $y = x(\log x - \log e)$.

Group B

1. Differentiate with respect to the independent variable each of the following functions.

(a) $f(x) = \ln (x^2 + 2x)^{1/3}$.
(b) $f(x) = \ln \sqrt{x^2 + 2x}$.
(c) $f(x) = \ln \left(\dfrac{\sqrt{x^3 + 1}}{x^4 + 5x}\right)$.
(d) $f(t) = \ln t^3(t^2 + 2)^{1/2}$.
(e) $f(x) = \dfrac{(\ln x)^2}{1 + x^2}$.
(f) $f(x) = x^3 \ln (x^5 + 5)$.

2. Differentiate with respect to x each of the following functions.

(a) $f(x) = \log (x^2 + 2x)^{1/3}$.
(b) $f(x) = \log_3 (x^2 + 5)^{3/2}$.
(c) $f(x) = \log \left(\dfrac{\sqrt{x^7 + 3x + 2}}{(x^3 + 2)^{1/3}}\right)$.
(d) $f(x) = x^2 \log_5 (2x + 2)$.

3. Find the derivative with respect to x of the function $y = e^x$. (*Hint*: First take natural logarithms of each side of the equation.)

4. Find the derivative with respect to x of the function $y = 2^x$.

5. Find the derivative with respect to x of the function $f(x) = x^x$.

6. Use implicit differentiation to find dy/dx for the function given by (a) $\ln (x^2 + 2) = \ln (y^3 + 3) + 10$; (b) $x \ln y + y \ln x = 0$.

7. Find dy/dx for (a) $y = (\ln 3x)(\ln 2x)$; (b) $y = (\ln 3x)/\ln (2x)$.

8. Find the derivative with respect to the independent variable of each of the following functions.

(a) $y = x \ln 2x$.
(b) $y = \ln^2 3x = (\ln 3x)^2$.
(c) $s = (t^2 - t) \ln (t - 1)$.
(d) $s = t \ln^3 t$.

9. Find d^2y/dx^2 for each of the following.

(a) $y = \ln 2x$.
(b) $y = x \ln 2x$.

10. Sketch the graph of each of the following.

(a) $y = \ln 3x$.
(b) $y = \ln x^2$.
(c) $y = x \ln x$.
(d) $y = \dfrac{\ln x}{x}$.

11. Explain how dy/dx for $y = \ln x$ and $y = \log x$ are alike and how they are different.

12. Show that $\lim_{x \to 0} (1 + x)^{1/x} = e$. This limit is of the form 1^∞, which is an indeterminate form. Let $y = (1 + x)^{1/x}$ and show that $\ln y = [\ln (1 + x)]/x$. Apply

L'Hôpital's rule (see Exercise 2 of Group C in Exercises 7.4) to show that $\lim_{x\to 0} \ln y = 1$ and that the desired result then follows.

13. Find $\lim_{x\to 1}\left[\frac{1}{\ln x} - \frac{1}{x-1}\right]$. This is an indeterminate form of $\infty - \infty$. Write in the form 0/0 and apply L'Hôpital's rule.

8.7 DERIVATIVE OF EXPONENTIAL FUNCTIONS

Now we can develop the formula for the derivative of an exponential function.

Consider the exponential function $y = b^u$, where u is any differentiable function of x. If $y = b^u$, then by logarithms we have

$$\ln y = u \ln b$$

Differentiating with respect to x, we obtain

$$\frac{1}{y}\frac{dy}{dx} = \frac{du}{dx}\ln b \qquad \text{(from Eq. 8.6.5)}$$

Hence

$$\frac{dy}{dx} = y \ln b \frac{du}{dx}$$

or

$$\frac{dy}{dx} = b^u \ln b \frac{du}{dx} \tag{8.7.1}$$

If we let $b = e$, then for $y = e^u$ we have

$$\frac{dy}{dx} = e^u \ln e \frac{du}{dx}$$

or

$$\frac{dy}{dx} = e^u \frac{du}{dx} \tag{8.7.2}$$

It also follows that

$$d(e^u) = e^u \, du \tag{8.7.3}$$

In summary, for

$$y = b^u$$
$$\frac{dy}{dx} = b^u \ln b \frac{du}{dx} \tag{8.7.1}$$

and for

$$y = e^u$$
$$\frac{dy}{dx} = e^u \frac{du}{dx} \tag{8.7.2}$$

If $f(x) = e^x$, we have

$$f'(x) = e^x$$
$$f''(x) = e^x$$
$$f'''(x) = e^x$$

and so forth for higher-order derivatives.

The function e^x is the only function having the property that *the function and all of its derivatives are the same.*

EXAMPLE 1. For $y = e^{-x^3}$, find dy/dx.

Solution. For $y = e^{-x^3}$, $u = -x^3$; hence

$$\frac{dy}{dx} = \underbrace{e^{-x^3}}_{e^u}\underbrace{(-3x^2)}_{du/dx} \qquad \text{(by Eq. 8.7.2)}$$

or

$$\frac{dy}{dx} = -3x^2e^{-x^3}$$

PROBLEM 1. Find ds/dt for $s = 3e^{t^2}$.

EXAMPLE 2. Find dy/dx for $y = 10^{x^4}$.

Solution. Since the base is 10, we use formula 8.7.1 where $u = x^4$. We obtain

$$\frac{dy}{dx} = 10^{x^4}(\ln 10)(4x^3)$$

or

$$\frac{dy}{dx} = 9.21x^3(10^{x^4})$$

PROBLEM 2. Find dP/dt for $P = 5^{3t^2}$.

***Warning* When the base is not e, then the natural logarithm of the base is a factor in the derivative.**

EXAMPLE 3. If $Q = te^{-t}$, find I.

Solution. Since $I = dQ/dt$, we have

$$I = te^{-t}(-1) + e^{-t}(1)$$

or

$$I = e^{-t}(1 - t)$$

PROBLEM 3. Find $f'(x)$ for $f(x) = x^2e^{3x}$.

EXAMPLE 4. Discuss and sketch the graph of $y = e^x$.

Solution. For $y = e^x$,

$$\frac{dy}{dx} = e^x \qquad \text{and} \qquad \frac{d^2y}{dx^2} = e^x$$

Since dy/dx and d^2y/dx^2 cannot be zero, there are no critical points and no points of inflection. If $x > 0$, $e^x > 0$. If $x < 0$, $e^x > 0$, since $e^{-x} = 1/e^x$. Thus y is always positive, and the slope is always positive. At $x = 0$, $e^x = 1$, and the slope at (0, 1) is 1. Since $(d^2y/dx^2) > 0$ for all x, the curve is concave up (see Fig. 8.7.1 for the sketch).

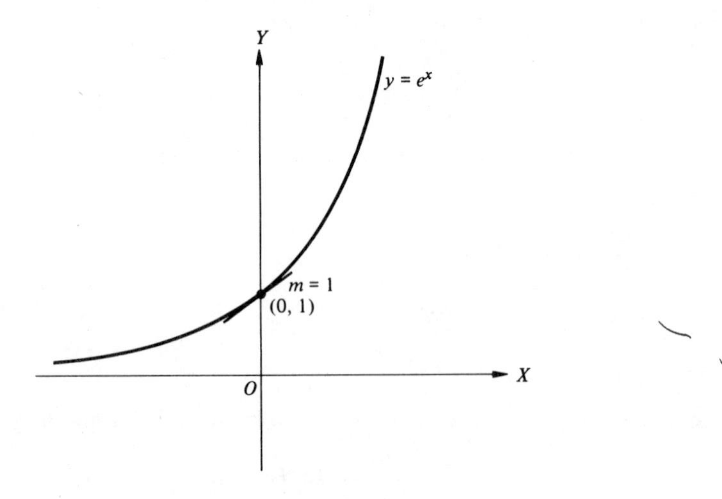

Figure 8.7.1 Graph of $y = e^x$.

PROBLEM 4. Sketch the graph of $y = xe^x$.

Applications of the derivative of exponential functions will be found in Sec. 8.9.

EXERCISES 8.7

Group A

1. Differentiate with respect to x each of the following functions.

(a) $f(x) = e^{2x}$.
(b) $f(x) = e^{x^2}$.
(c) $f(x) = e^{x^2+2x+1}$
(d) $f(x) = e^{x^{1/2}}$.
(e) $f(x) = e^{x^5+2x^4+3x^2+1}$.
(f) $f(x) = e^{(x^2+2x)^{1/2}}$.
(g) $f(x) = e^{-3x^2}$.
(h) $f(x) = 7e^{-(x^2+1)}$.
(i) $f(x) = e^3$.
(j) $f(x) = 3e^{-2}$.

2. Find $f'(x)$, $f''(x)$, $f'''(x)$, and $f^{IV}(x)$ for (a) $f(x) = e^{2x}$; (b) $f(x) = e^{x^2}$.

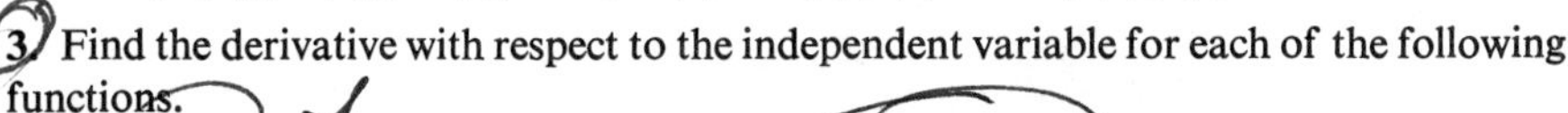

3. Find the derivative with respect to the independent variable for each of the following functions.

(a) $y = 3^x$. (b) $y = 10^{2x}$. (c) $s = 2^{t^2}$. (d) $u = 5^{3v-1}$.

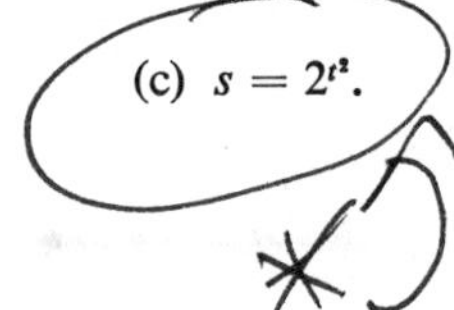

4. Sketch the graph of $y = e^{2x}$.

5. Sketch the graph of $y = e^{-x}$.

6. Sketch the graph of $y = xe^{-x}$.

Group B

1. Differentiate with respect to x each of the following functions.

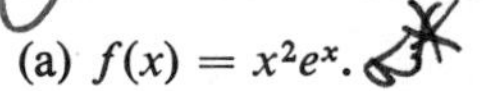

(a) $f(x) = x^2e^x$. (b) $f(x) = xe^{(x^2+2)^{1/2}}$. (c) $f(x) = \dfrac{e^{5x+2}}{x^2+1}$.

(d) $f(x) = x^4e^{x^3+2}$. (e) $f(x) = e^{(x^3+2x^2+1)^{2/3}}$. (f) $f(x) = x^3e^{x^2} + 2e^{5x}$.

2. Find the third derivative with respect to x of the function $f(x) = x^2e^{2x}$.

3. The current in an electrical circuit is described by $I = 10 - e^{-5t}$. Find dI/dt at $t = 2$.

4. Find the slope of the line tangent to the curve $f(x) = xe^{2x}$ at $x = 1$.

5. Use implicit differentiation to find dy/dx from the equation given by $x^2 + 2 = e^{(2y+3)}$.

6. Find dy/dx and d^2y/dx^2 given that $e^{xy} = xy$.

7. Find dy/dx for $\ln xy = e^{xy}$.

8. If $y = x^2e^{x^2}\ln(x^2 + 3)$, find dy/dx.

8.8 INTEGRATION YIELDING LOGARITHMIC AND EXPONENTIAL FUNCTIONS

We are now able to integrate u^n, where $n = -1$. From Eq. 8.6.6 we know that $d(\ln u) = (1/u)\,du = du/u$. Thus we have

$$\int \frac{du}{u} = \ln u + C$$

for $u > 0$, since logarithms are defined only for positive numbers. If $u < 0$, then $-u > 0$ and

$$\int \frac{d(-u)}{(-u)} = \ln(-u) + C$$

but $d(-u) = -du$ and therefore

$$\int \frac{d(-u)}{(-u)} = \int \frac{-du}{-u} = \int \frac{du}{u} = \ln u + C$$

Hence, in general,

$$\boxed{\int \frac{du}{u} = \ln|u| + C} \tag{8.8.1}$$

Formula 8.8.1 together with formula 5.4.1 now enables us to integrate x^n for n any real number.

EXAMPLE 1. Find $\int x^{-1}\,dx$.

Solution. We use formula 8.8.1 to obtain

$$\int x^{-1}\,dx = \int \frac{dx}{x} = \ln|x| + C$$

Warning ***To use formula 8.8.1 the exponent of u in the denominator must be +1.***

PROBLEM 1. Find $\int (t^{-1} + t^{-2})\,dt$.

If we desire to integrate a fraction where the numerator is the differential of the denominator, we may use Eq. 8.8.1.

EXAMPLE 2. Find

$$\int_1^3 \frac{x\,dx}{x^2+3}$$

Solution. We note that the numerator is similar to the differential of the denominator. Therefore let $u = x^2 + 3$; then $du = 2x\,dx$ and on substituting we obtain

$$\int_4^{12} \frac{\frac{1}{2}\,du}{u} = \frac{1}{2}\int_4^{12} \frac{du}{u}$$

Carefully note that when the variable of integration is changed from x to u the limits on the integral must be changed also. This is accomplished by our choice for u, namely $u = x^2 + 3$.

When $x = 1$, $u = (1)^2 + 3 = 4$; and when $x = 3$, $u = (3)^2 + 3 = 12$. Proceeding, we have

$$\frac{1}{2}\int_4^{12} \frac{du}{u} = \frac{1}{2}[\ln|u|]_4^{12} = \frac{1}{2}[\ln 12 - \ln 4] = \frac{1}{2}\ln 3$$

$$\therefore \int_1^3 \frac{x\,dx}{x^2+3} = \frac{1}{2}\ln 3 = 0.549$$

Note that it is not necessary to change back to the variable x when evaluating a definite integral.

PROBLEM 2. Find

$$\int_2^4 \frac{x^2\,dx}{x^3+2}$$

From Eq. 8.7.3 we know that $d(e^u) = e^u\,du$. Thus we have

$$\boxed{\int e^u\,du = e^u + C} \tag{8.8.2}$$

EXAMPLE 3. Find $\int e^{3x}\,dx$.

Solution. If we let $u = 3x$, then $du = 3\,dx$ and $\int e^{3x}\,dx = \int e^u(\frac{1}{3}\,du) = \frac{1}{3}\int e^u\,du$, and from Eq. 8.8.2 we have

$$\tfrac{1}{3}\int e^u\,du = \tfrac{1}{3}e^u + C$$

or

$$\int e^{3x}\,dx = \tfrac{1}{3}e^{3x} + C$$

***Warning** Before using the formulas of integration, the integrand must be exactly like the integrand in the formula. This is accomplished by substitution.*

PROBLEM 3. Find $\int_{0.5}^{1} e^{2x}\,dx$.

EXAMPLE 4. If in an electric circuit, the current I at any time t is given as $I = 10 - te^{-t^2}$, find the charge Q at time t if $Q = 0$ at $t = 0$.

Solution. Since $I = dQ/dt$, we have $Q = \int I\,dt$. Thus we have

$$Q = \int (10 - te^{-t^2})\,dt$$

or

$$Q = \int 10\,dt - \int te^{-t^2}\,dt$$

The first integral is $\int 10\,dt = 10t + C_1$. In the second integral let $u = -t^2$; then $du = -2t\,dt$. Thus

$$\int te^{-t^2}\,dt = \int e^{-t^2}t\,dt = \int e^u(-\tfrac{1}{2}\,du)$$

or

$$-\tfrac{1}{2}\int e^u\,du = -\tfrac{1}{2}e^u + C_2$$

and

$$\int te^{-t^2}\,dt = -\tfrac{1}{2}e^{-t^2} + C_2$$

Adding these results, we find that $Q = 10t + C_1 - \frac{1}{2}e^{-t^2} + C_2$.

Whenever two constants are added, the sum is also a constant. Thus we may write $Q = 10t - \frac{1}{2}e^{-t^2} + C$, where $C = C_1 + C_2$.

To evaluate the constant of integration C we use the given initial conditions, namely, when $t = 0$, $Q = 0$. Hence

$$0 = 10(0) - \tfrac{1}{2}e^{-(0)^2} + C$$

or

$$0 = 0 - \tfrac{1}{2} + C$$

Therefore,

$$C = \tfrac{1}{2}$$

and

$$Q = 10t - \tfrac{1}{2}e^{-t^2} + \tfrac{1}{2}$$

PROBLEM 4. Find the charge Q at $t = 1$ s in an electric circuit where $I = 5 - 2te^{-t^2}$ if at $t = 0$, $Q = 0$.

Using the result of Prob. 5 in Sec. 8.6 we may write the formula

$$\int \ln u \, du = u(\ln u - 1) + C \tag{8.8.3}$$

Formula 8.8.3 is obtained using the method of integration by parts discussed in Sec. 11.3.

PROBLEM 5. Find the value of $\int_1^3 \ln x \, dx$.

Applications of integration of the logarithmic and the exponential functions are given in Sec. 8.9.

EXERCISES 8.8

Group A

1. Evaluate each of the following indefinite integrals.

(a) $\int \frac{dx}{4x + 1}$.

(b) $\int 3xe^{x^2} \, dx$.

(c) $\int (6x^2 + 12)e^{x^3+6x} \, dx$.

(d) $\int \frac{6x \, dx}{x^2 + 2}$.

(e) $\int \frac{x \, dx}{ax^2 + b}$.

(f) $\int xe^{ax^2} \, dx$.

(g) $\int e^{2x}(e^{2x} + 3)^{-1/3} \, dx$.

(h) $\int \frac{4x + 5}{2x^2 + 5x} \, dx$.

(i) $\int \frac{e^{4x}}{e^{4x} + 1} \, dx$.

(j) $\int \frac{12x^2 + 16x \, dx}{x^3 + 2x^2 + 3}$

2. Evaluate each of the following definite integrals.

(a) $\int_0^1 \frac{x\,dx}{x^2+1}$.

(b) $\int_1^2 \frac{(x+1)\,dx}{x^2+2x+1}$.

(c) $\int_3^6 \frac{dx}{x-1}$.

(d) $\int_0^2 xe^{x^2}\,dx$.

(e) $\int_{-1}^1 (x^3+2)e^{x^4+8x}\,dx$.

(f) $\int_0^1 \frac{e^x\,dx}{2e^x+3}$.

(g) $\int_0^2 e^{x/2}\,dx$.

(h) $\int_0^1 \frac{5-e^{-x}}{e^x}\,dx$.

3. Evaluate each of the following.

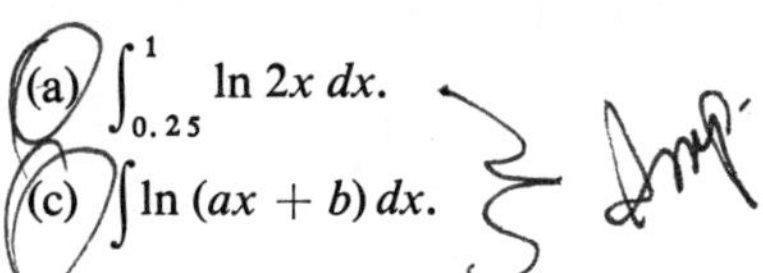

(a) $\int_{0.25}^1 \ln 2x\,dx$.

(b) $\int x \ln x^2\,dx$.

(c) $\int \ln(ax+b)\,dx$.

(d) $\int_1^2 \ln x^3\,dx$.

Group B

1. Evaluate each of the following indefinite integrals. (*Hint*: If numerator is of same or higher degree than denominator, try division before integrating.)

(a) $\int \frac{dx}{e^x}$.

(b) $\int \frac{x^2+3x+2}{x^2+2}\,dx$.

(c) $\int \frac{e^{x^2}}{e^{x^2}+2}\,x\,dx$.

(d) $\int \frac{x^3+3x}{x+1}\,dx$.

(e) $\int \frac{x^2+4}{x+2}\,dx$.

(f) $\int \frac{e^{\sqrt{x}}}{\sqrt{x}}\,dx$.

2. Evaluate each of the following definite integrals.

(a) $\int_0^1 \frac{(x^2+x+2)\,dx}{x+1}$.

(b) $\int_0^{1/3} \frac{dx}{e^{3x}}$.

(c) $\int_0^1 xe^{x^2}\,dx$.

(d) $\int_1^2 \frac{6x-1}{3x-2}\,dx$.

(e) $\int_0^3 \frac{2x+5}{2x+3}\,dx$.

(f) $\int_0^3 (e^{x/3}+e^{-(x/3)})\,dx$.

(g) $\int_0^1 \frac{x^2+3x+3}{x+2}\,dx$.

(h) $\int_0^1 (e^x+e^{-x})^2\,dx$.

(i) $\int_0^1 \frac{x\,dx}{x^2-4}$.

(j) $\int_0^2 \frac{x^2-2x-2}{x-3}\,dx$.

3. In an electric circuit a useful formula is $I = (1/L)\int E\,dt$. In each of the following, find I at time t for the given values of L, E, and the initial value of I.

(a) $L = 12$, $E = 6e^{-2t}$; $I = 0$ at $t = 0$.
(b) $L = 1$, $E = 12 - te^{-0.5t^2}$; $I = 0$ at $t = 0$.

4. Given that $E = -(1/RC) \int e^{-t/RC}\, dt$, find a formula for E in terms of t if at $t = 0$, $E = E_0$.

5. Show that (a) $\int [(e^u - e^{-u})/2]\, du = \cosh u + C$; (b) $\int [(e^u + e^{-u})/2]\, du = \sinh u + C$ (see Exercise 5 of Group B in Exercises 8.3). What is $\int \sinh x\, dx$ and $\int \cosh x\, dx$?

Group C

1. Show that $\int b^u\, du = (1/\ln b)b^u + C$.

2. Evaluate each of the following.

(a) $\int 5^x\, dx$. (b) $\int 10^x\, dx$. (c) $\int a^x e^x\, dx$.

(d) $\int 2^{3x}\, dx$. (e) $\int_0^1 3^x\, dx$. (f) $\int_1^2 \frac{dx}{x^4}$

3. State how $\int b^u\, du$ and $\int e^u\, du$ are alike and how they are different.

4. Obtain a formula for $\int \log u\, du$ by using dy/du for $y = u \log u - u \log e$. Evaluate $\int_1^2 \log x\, dx$.

8.9 APPLICATIONS

In the preceding sections of this chapter we developed the differentiation and integration of two new functions, the exponential and logarithmic functions. The exercises given at the end of each section were meant primarily for drill in the skills of differentiating and integrating these new types of function. We now give a set of exercises that reviews various applications of the calculus and uses the new functions.

EXERCISES 8.9

Group A

1. Find equations of the lines tangent and normal to the curve of the function given at the point corresponding to the given value of x.

(a) $f(x) = x \ln x$ at $x = e$. (b) $f(x) = x^2 e^x$ at $x = 1$.
(c) $f(x) = 5^x$ at $x = -1$. (d) $f(x) = x^2 \ln x$ at $x = 1$.
(e) $f(x) = e^{3x}$ at $x = 0$.

2. Graph each of the following functions.

(a) $f(x) = \ln(1/x)$. (b) $f(x) = e^{1/x}$. (c) $f(x) = xe^x$.
(d) $f(x) = e^{-x^2}$. (e) $f(x) = ex^{-x^2}$. (f) $f(x) = x^2 e^{-x^2}$.

3. Determine the critical points of the function $f(x) = x^2 e^x$. Draw its graph.

4. The amount of radium present in a certain quantity is given by $A = 60e^{-0.000418t}$ (grams), where the 60 g was the original amount of radium. Determine how fast the radium is distintegrating when $t = 1000$ years.

5. Determine the critical points of the function $f(x) = x^3 \ln x$.

6. Find the area above the X-axis and below the curve $y = xe^{x^2}$ between $x = 0$ and $x = 1$.

7. Find the center of mass of the region bounded by the curve $y = 1/x$ and the lines $x = 1$, $x = 3$, and $y = 0$.

8. The normal probability curve is given by $y = (1/\sqrt{2\pi}\sigma)e^{-(1/2)[(x-\mu)/\sigma]^2}$, where σ and μ are constants. Find its maximum point. Find its inflection points. Sketch the curve.

9. Find the radius of curvature at the point for the indicated value of x.

(a) $y = \ln x$, $x = 1$.
(b) $y = \ln 2x$, $x = 1$.
(c) $y = x \ln x$, $x = 1$.
(d) $y = \ln^2 x$, $x = 1$.
(e) $y = e^x$, $x = 0$.
(f) $y = xe^x$, $x = 0$.
(g) $y = e^{-x}$, $x = 1$.
(h) $y = x^2e^{-x}$, $x = 1$.

10. Find the average value of y with respect to x.

(a) $y = \dfrac{1}{x}$, $1 \leqq x \leqq 2$.
(b) $y = \dfrac{x}{x^2 + 3}$, $1 \leqq x \leqq 3$.
(c) $y = e^x$, $0 \leqq x \leqq 2$.
(d) $y = xe^{x^2}$, $0 \leqq x \leqq 2$.

Group B

1. A package is dropped from a plane with a parachute attached. The package and the parachute weigh 100 lb and fall according to the differential equation $[100/(100 - 2v)]\, dv = 32\, dt$, where v denotes velocity and t time. If $v = 0$ at $t = 0$, find an equation for v by integrating both sides of the given equation. As $t \longrightarrow \infty$, what does the velocity v approach?

2. The current in an electrical system is described by $I = 10 - 50te^{-5t^2}$. Find the equation for the charge Q if $Q = 0$ at $t = 0$.

3. If the current $I = te^{-t^2}$ find the average charge Q over the first 2 s. Also find the maximum values of I.

4. Prove that $(d/dx) \cosh x = \sinh x$ and that $(d/dx) \sinh x = \cosh x$. (See Exercise 5 of Group B in Exercises 8.3.)

5. Find the area of each of the following regions R.

(a) R bounded by $y = x/(x^2 + 1)$ and the lines $x = 1$, $x = 2$, and $y = 0$.
(b) R bounded by $y = x^2e^{x^3}$ and the lines $x = 0$, $x = 2$, and $y = 0$.

6. In a cylindrical cable in which the inner conductor has a radius r and the outer conductor has a radius R, the maximum electric intensity in the insulation is $E_m = V/[r \ln (R/r)]$ and occurs at the surface of the inner conductor. Determine the value of r/R that makes E_mR/V a maximum or minimum. (*Hint*: $E_mR/V = R/[r \ln (R/r)]$. Let $y = E_mR/V$ and $x = r/R$.) (Cell.)

7. The work that an ideal gas does when it expands is expressed by the differential equation $dW = (nGT/V)\,dV$, where n is the number of molecules, G a gas constant, and T the constant temperature. Find the work done as the gas expands from V_0 to V_T by integrating both sides of the given equation.

8. The velocity of a particle is $2/t$ ft/s. Find the displacement s when $t = 6$ if when $t = 2$, $s = 1$.

9. The acceleration of a particle is given as $a = e^{-t}$. Find $s = f(t)$ if when $t = 0$, $v = 1$, $s = 0$.

10. Find the approximate value of

(a) $e^{0.1}$. (b) $e^{-0.1}$. (c) $\ln 1.1$. (d) $\ln 0.9$.

11. Use Newton's method to find the smallest value of x for

(a) $x - \ln 2x = 0$. (b) $3x^2 - e^x = 0$.

12. (a) Show that $ds = \frac{1}{2}(e^x + e^{-x})\,dx$ for $y = \frac{1}{2}(e^x + e^{-x})$.
(b) Find the length of the curve $y = \frac{1}{2}(e^x + e^{-x})$ from $x = -1$ to $x = 1$.
(c) Find the area of the surface obtained by revolving the arc of part (b) about the X-axis. See Exercise 1 of Group C in Exercises 7.9.

13. Work done by a gas at p lb/ft² expanding from volume v_0 ft³ to v_1 ft³ is given by $W = \int_{v_0}^{v_1} p\,dv$. If $pv^n = c$, find W for the gas, where
(a) $v_0 = 16$ ft³, $p_0 = 60$ lb/in.², $n = 1$; $v_1 = 32$ ft³.
(b) $v_0 = 16$ ft³, $p_0 = 60$ lb/in.², $n = 1.4$; $p_1 = 30$ lb/in.².
(c) Show that in general for $n > 1$, $W = \dfrac{p_0v_0 - p_1v_1}{n - 1}$.

14. A rectangle has two sides along the XY-axes and one vertex, P, on the curve $y = e^{-x}$. If P moves such that y decreases at the rate of $\frac{1}{2}$ unit/sec, how fast is the area of the rectangle decreasing when $x = 2$?

15. In a certain electric cable the speed, v, of the signal is given by $v = x^2 \ln(1/x)$. Find x for maximum speed. (x is the ratio of the core to the thickness of the insulation.)

16. In an inductive circuit the current I at time t is given as $I = I_0e^{-Rt/L}$. It is said that I "falls to zero." Show that this is true. Sketch the graph.

17. In an RL circuit the voltage E at time t is given as $E = E_T(1 - e^{-Rt/L})$. It is said that E "rises to the steady state E_T." Show that this is true. Sketch the graph.

18. Given that a drug concentration is determined by the equation $C = A_0e^{-kt}$, explain how one can estimate k using logarithmic graph paper and known concentrations of the drug in the system.

SUMMARY OF IMPORTANT WORDS AND CONCEPTS

For each of the following, state in your own words your understanding of the statement or word.

1. Transcendental functions.

2. (a) Exponential function. (b) Rules for computing with exponents.

3. Irrational number e.

4. Inverse functions.

5. (a) Logarithmic function. (b) Common logarithms. (c) Natural logarithms. (d) Change of base.

6. Derivatives of logarithmic and exponential functions.

7. Integration of logarithmic and exponential functions.

Chapter 9

Trigonometric Functions

9.1 INTRODUCTION

In recent years the emphasis of a trigonometry course has changed from one of solving triangles to one of trigonometric analysis. This tendency is due in part to the increasing study of periodic functions. Many phenomena are periodic, that is, they repeat themselves; for example, the rotation of the earth about its axis, the movement of the earth about the sun, the tides of the ocean, a vibrating violin string, the motion of a swinging pendulum, and the electric current in an alternating-current system. To describe the motion of such periodic movements a class of functions that are periodic is needed. The trigonometric functions comprise such a class. We shall discuss periodic functions in general in Chap. 18, where Definition 18.1.1 defines periodic functions.

The trigonometric functions are subtle. The domain of the algebraic, exponential, and logarithmic functions consists of real numbers. But the domain (as we shall see) of the trigonometric functions may be either a real number or an angle. If it is an angle, the units of measurement may be either degrees or radians. It is this subtlety that sometimes creates difficulty, so be alert and be sure of the nature of the argument of the functions we are about to discuss. In Example 1 of Sec. 10.3 it will become apparent why we need trigonometric functions of real numbers.

9.2 TRIGONOMETRIC FUNCTIONS OF REAL NUMBERS

It is assumed that you have previously studied some trigonometry. Probably the trigonometric functions were defined for the angles of a right triangle and an angle in standard position. Now we would like to extend your knowledge of trigonometry to include the trigonometric functions of a real number.[1] Such functions are based on the terminal point of an arc on the circumference of a unit circle (a circle of radius 1 unit) whose equation is $x^2 + y^2 = 1$.

Let us consider the unit circle as shown in Fig. 9.2.1. Indicate by u the arc AP with initial point A at $(1, 0)$ and terminal point P at (x, y). The length of arc AP measured counterclockwise represents a positive real number just as the length from the origin on the X- and Y-coordinate axes represents a real number. If $u = \text{arc } AB$, then $u = \pi = 3.14$. The length of the entire arc $APBCA$

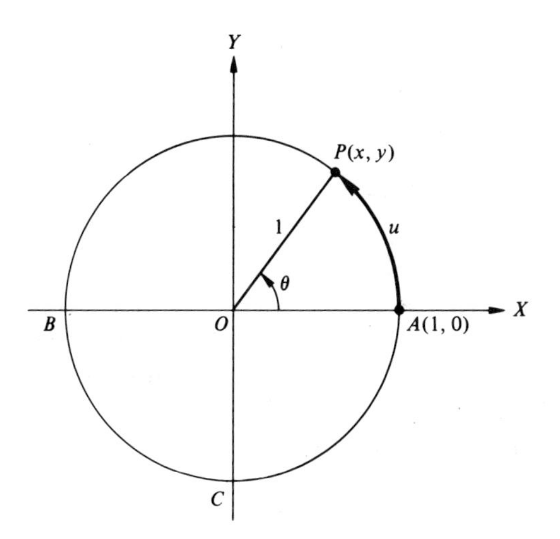

Figure 9.2.1 Unit circle.

represents $2\pi = 6.28$, which is the circumference of the circle. For numbers greater than 2π, simply continue one or more revolutions. Negative numbers are represented by arcs measured from A in a clockwise direction. The number 0 is represented by point A. For each real number, u, there is one and only one pair of coordinates (x, y) which are the coordinates of the terminal point of the arc representing u on the unit circle. Thus we can form functions of u in terms of x and y. The trigonometric functions of u are listed in Table 9.2.1.

[1]For a more thorough discussion of the trigonometric functions of real numbers, see *Plane Trigonometry*, 3rd ed., by N. O. Niles, John Wiley & Sons, Inc., New York, 1976.

TABLE 9.2.1

Function and Rule of Correspondence	Domain	Range
$\sin u = y$	all real numbers	$-1 \leqq$ all real numbers $\leqq 1$
$\cos u = x$	all real numbers	$-1 \leqq$ all real numbers $\leqq 1$
$\tan u = \frac{y}{x}$	all real numbers for which $x \neq 0$	all real numbers
$\cot u = \frac{x}{y}$	all real numbers for which $y \neq 0$	all real numbers
$\sec u = \frac{1}{x}$	all real numbers for which $x \neq 0$	all real numbers $\leqq -1$ or $\geqq 1$
$\csc u = \frac{1}{y}$	all real numbers for which $y \neq 0$	all real numbers $\leqq -1$ or $\geqq 1$

EXERCISES 9.2

Group A

1. Draw a unit circle and locate each of the following real numbers on it.

(a) 0. (b) $\pi/2$. (c) 1. (d) 3. (e) 6.
(f) -2. (g) -4. (h) 10. (i) -7. (j) -25.

2. Make use of the unit circle to find the value of each of the following.

(a) $\sin(\pi/2)$. (b) $\cos \pi$. (c) $\tan(-\pi)$. (d) $\sin(-3\pi/2)$.

Group B

1. Find the coordinates of the terminal point of the arc representing the real number $\pi/4$. Then find $\sin(\pi/4)$, $\cos(\pi/4)$, and $\tan(\pi/4)$. (*Hint*: Does $x = y$?)

2. The x-coordinate of the terminal point for $u = 2\pi/3$ is $-\frac{1}{2}$. Find $\sin(2\pi/3)$, $\cos(2\pi/3)$, and $\tan(2\pi/3)$.

3. The x-coordinate of the terminal point for $\pi/3$ is $1/2$. Find $\sin(\pi/3)$, $\cos(\pi/3)$, $\tan(\pi/3)$.

4. The x-coordinate of the terminal point for $\pi/6$ is $\sqrt{3}/2$. Find $\sin(\pi/6)$, $\cos(\pi/6)$, $\tan(\pi/6)$.

5. The current I in an electric circuit is given as $I = 5 \sin 60\pi t$, find I when $t = 2$ s.

Group C

1. Use the equation of the unit circle to show that for all real numbers u each of the following is true for values of u where they exist.

(a) $\sin^2 u + \cos^2 u = 1$. (b) $1 + \tan^2 u = \sec^2 u$. (c) $1 + \cot^2 u = \csc^2 u$.

2. Determine the algebraic sign of each trigonometric function of u for u in each quadrant.

3. The reference *arc*, u_r, of a given real number, u, is the nonnegative arc less than or equal to $\pi/2$ between the terminal point of the representative arc u and the X-axis. If $T(u)$ represents any of the trigonometric functions, show that $|T(u)| = |T(u_r)|$. (*Hint*: Draw u and u_r and make use of similar triangles.)

4. Make use of Exercise 3 to find each of the following. (*Hint*: The algebraic sign is determined by the quadrant of u.)

(a) $\sin(5\pi/6)$. (b) $\cos(5\pi/6)$. (c) $\sin(-\pi/4)$.

5. Express each of the following in terms of the same trigonometric function of a nonnegative real number less than $\pi/2$. Recall that $\pi/2 = 1.57$, $\pi = 3.14$, and $2\pi = 6.28$ approximately.

(a) sin 1.8. (b) cos 2.8. (c) tan 3.6. (d) cot 5.
(e) sec 4.2. (f) csc 4.9. (g) sin 30. (h) tan 10.

9.3 SUMMARY OF FORMULAS

In geometry it was shown that the length, s, of an arc on a circle is $s = r\theta$, where r is the radius and θ is the central angle measured in radians. Referring to Fig. 9.2.1 and using $s = r\theta$, we have that

$$u = (1)\theta$$

or

$$u = \theta$$

By using $u = \theta$ it can be shown that the trigonometric functions of a real number are equivalent to the corresponding trigonometric functions of an angle θ measured in radians. Thus all of the formulas in trigonometry that are valid for an angle θ measured in radians are valid for the real number u. Hence $\sin 180° = \sin \pi$ (radians) $= \sin \pi$, $\tan 3 = \tan 3$ (radians) $= \tan[3(180°/\pi)] = \tan 171.9°$. (See Table 9.3.2 on page 278 for the relation between degrees and radians.)

These relations point out the fact that we must be extremely careful in indicating the proper units of measurement of the argument of the trigonometric functions. If the argument is an angle measured in degrees, minutes, and seconds, the symbols °, ′, and ″ *must be used.* Otherwise, when no symbol is used, the argument is an angle measured in radians or is a real number. Thus sin 30° is "the sine of an angle of 30 degrees," whereas sin 30 is either "the sine of an angle of 30 radians" or its equivalent, "the sine of a real number 30."

 **Warning** ***If T represents any trigonometric function, then $T(u) = T(\theta)$ where u is a real number and θ is an angle measured in radians. u and θ have the same measure.***

We now list some of the formulas of trigonometry.

Reciprocal

$$\sin u \csc u = 1 \tag{9.3.1}$$

$$\cos u \sec u = 1 \tag{9.3.2}$$

$$\tan u \cot u = 1 \tag{9.3.3}$$

Ratio

$$\tan u = \frac{\sin u}{\cos u} \tag{9.3.4}$$

$$\cot u = \frac{\cos u}{\sin u} \tag{9.3.5}$$

Squared

$$\sin^2 u + \cos^2 u = 1 \tag{9.3.6}$$

$$1 + \tan^2 u = \sec^2 u \tag{9.3.7}$$

$$1 + \cot^2 u = \csc^2 u \tag{9.3.8}$$

Sum and Difference

(In formulas with $\pm$ or $\mp$ symbols, use all top signs or all bottom signs.)

$$\sin (u \pm v) = \sin u \cos v \pm \cos u \sin v \tag{9.3.9}$$

$$\cos (u \pm v) = \cos u \cos v \mp \sin u \sin v \tag{9.3.10}$$

$$\tan (u \pm v) = \frac{\tan u \pm \tan v}{1 \mp \tan u \tan v} \tag{9.3.11}$$

Double-Argument

$$\sin 2u = 2 \sin u \cos u \tag{9.3.12}$$

$$\cos 2u = \cos^2 u - \sin^2 u \tag{9.3.13}$$

$$\cos 2u = 1 - 2 \sin^2 u \tag{9.3.14}$$

$$\cos 2u = 2 \cos^2 u - 1 \tag{9.3.15}$$

$$\tan 2u = \frac{2 \tan u}{1 - \tan^2 u} \tag{9.3.16}$$

Half-Argument

$$\sin \frac{u}{2} = \pm\sqrt{\frac{1 - \cos u}{2}} \quad \text{(the correct sign is determined by the quadrant of } u/2\text{)} \tag{9.3.17}$$

$$\cos \frac{u}{2} = \pm\sqrt{\frac{1 + \cos u}{2}} \tag{9.3.18}$$

$$\tan \frac{u}{2} = \frac{1 - \cos u}{\sin u} \tag{9.3.19}$$

$$\tan \frac{u}{2} = \frac{\sin u}{1 + \cos u} \tag{9.3.20}$$

Product and Sum

$$\sin u \cos v = \tfrac{1}{2}[\sin (u + v) + \sin (u - v] \tag{9.3.21}$$

$$\cos u \sin v = \tfrac{1}{2}[\sin (u + v) - \sin (u - v)] \tag{9.3.22}$$

$$\cos u \cos v = \tfrac{1}{2}[\cos (u + v) + \cos (u - v)] \tag{9.3.23}$$

$$\sin u \sin v = -\tfrac{1}{2}[\cos (u + v) - \cos (u - v)] \tag{9.3.24}$$

$$\sin u + \sin v = 2 \sin \tfrac{1}{2}(u + v) \cos \tfrac{1}{2}(u - v) \tag{9.3.25}$$

$$\sin u - \sin v = 2 \cos \tfrac{1}{2}(u + v) \sin \tfrac{1}{2}(u - v) \tag{9.3.26}$$

$$\cos u + \cos v = 2 \cos \tfrac{1}{2}(u + v) \cos \tfrac{1}{2}(u - v) \tag{9.3.27}$$

$$\cos u - \cos v = -2 \sin \tfrac{1}{2}(u + v) \sin \tfrac{1}{2}(u - v) \tag{9.3.28}$$

Negative Arguments

$$\sin (-u) = -\sin u \tag{9.3.29}$$

$$\cos (-u) = \cos u \tag{9.3.30}$$

$$\tan (-u) = -\tan u \tag{9.3.31}$$

Reduction

$$\sin \left(\frac{\pi}{2} - u\right) = \cos u \tag{9.3.32}$$

$$\cos \left(\frac{\pi}{2} - u\right) = \sin u \tag{9.3.33}$$

$$\tan \left(\frac{\pi}{2} - u\right) = \cot u \tag{9.3.34}$$

$$\sin (\pi - u) = \sin u \tag{9.3.35}$$

$$\cos (\pi - u) = -\cos u \tag{9.3.36}$$

$$\tan (\pi - u) = -\tan u \tag{9.3.37}$$

$$\sin (\pi + u) = -\sin u \tag{9.3.38}$$

$$\cos(\pi + u) = -\cos u \tag{9.3.39}$$

$$\tan(\pi + u) = \tan u \tag{9.3.40}$$

$$\sin(2\pi - u) = -\sin u \tag{9.3.41}$$

$$\cos(2\pi - u) = \cos u \tag{9.3.42}$$

$$\tan(2\pi - u) = -\tan u \tag{9.3.43}$$

$$\sin(2\pi + u) = \sin u \tag{9.3.44}$$

$$\cos(2\pi + u) = \cos u \tag{9.3.45}$$

Table 9.3.1 lists the algebraic signs of $\sin u$, $\cos u$, and $\tan u$ in quadrants I, II, III, and IV.

TABLE 9.3.1

Quadrant	I	II	III	IV
$\sin u$	+	+	−	−
$\cos u$	+	−	−	+
$\tan u$	+	−	+	−

It is convenient to express the radian measure of an angle in terms of π. Table 9.3.2 lists some relations between radians (rad) and degrees.

$$1 \text{ rad} = \frac{180°}{\pi} = 57.3°$$

$$1° = \frac{\pi}{180} = 0.0175 \text{ rad}$$

TABLE 9.3.2

Radians	Degrees
0	0°
$\pi/6$	30°
$\pi/4$	45°
$\pi/3$	60°
$\pi/2$	90°
π	180°
$3\pi/2$	270°
2π	360°

In Table 9.3.3 are listed some special values of $\sin u$, $\cos u$, and $\tan u$.

TABLE 9.3.3

u (real number):	0	$\pi/6$	$\pi/4$	$\pi/3$	$\pi/2$	π	$3\pi/2$	2π
θ (radians):	0	$\pi/6$	$\pi/4$	$\pi/3$	$\pi/2$	π	$3\pi/2$	2π
θ (degrees):	0°	30°	45°	60°	90°	180°	270°	360°
sin	0	1/2	$\sqrt{2}/2$	$\sqrt{3}/2$	1	0	−1	0
cos	1	$\sqrt{3}/2$	$\sqrt{2}/2$	1/2	0	−1	0	1
tan	0	$\sqrt{3}/3$	1	$\sqrt{3}$	does not exist	0	does not exist	0

EXAMPLE 1. Find sin 3.

Solution. Your electronic calculator should have the sine, cosine, and tangent functions. Be sure that it is in radian mode. Then find $\sin 3 = 0.141$.

If the number is too large for your calculator, you must make use of the reduction formulas.

PROBLEM 1. Find cos 3.5 by using a calculator.

EXERCISES 9.3

Group A

1. State which reduction formulas should be used when the terminal point of u is in the (a) second quadrant; (b) third quadrant; (c) fourth quadrant. Given that $|u| < 2\pi = 6.28$.

2. Find the value of each of the following.

(a) sin 2.54. (b) cos 4.19. (c) tan 3.
(d) sin (−4.59). (e) cos (−4.59). (f) sin 30.

3. Verify each of the following identities.

(a) $\sin 3u \cos 2u - \cos 3u \sin 2u = \sin u$.
(b) $\cos 3u \cos 2u - \sin 3u \sin 2u = \cos 5u$.
(c) $\sin^2 u(\cot^2 u + 1) = 1$.
(d) $\dfrac{\csc u + \cot u}{\sin u + \tan u} = \csc u \cot u$.
(e) $\sin (u + \Delta u) - \sin u = 2 \cos \left(u + \dfrac{\Delta u}{2}\right) \sin \dfrac{\Delta u}{2}$.
(f) $2 \sin^2 \dfrac{h}{2} = 1 - \cos h$.

Group B

1. Use Eq. 9.3.14 to prove Eq. 9.3.17.

2. Prove Eq. 9.3.19.

3. (a) Verify that $a \cos \omega t + b \sin \omega t = \sqrt{a^2 + b^2} \sin (\omega t + \phi)$, where $\sin \phi = a/\sqrt{a^2 + b^2}$ and $\cos \phi = b/\sqrt{a^2 + b^2}$.
(b) Express $3 \cos 2t + 4 \sin 2t$ in terms of a sine function.
(c) Express $5 \cos t - 12 \sin t$ in terms of a sine function.

4. How must the formula in Exercise 3(a) be modified to express $a \cos \omega t + b \sin \omega t$ in terms of $\cos (\omega t + \phi)$? Express $3 \cos 2t - 4 \sin 2t$ in terms of a cosine function.

5. Show that (a) $\sin (\pi/2 + u) = \cos u$; (b) $\cos (3\pi/2 - u) = -\sin u$; (c) $\cos (3\pi/2 + u) = \sin u$.

9.4 GRAPHS OF TRIGONOMETRIC FUNCTIONS

The graphs of the trigonometric functions may be drawn by plotting points obtained from a table of values. However, it is usually simpler to consider the zeros of the function and the maximum and minimum points. Usually, real numbers are used for the argument of the functions.

EXAMPLE 1. Sketch one period of the graph of $y = \sin x$.

Solution. From Eq. 9.3.44, $\sin (2\pi + u) = \sin u$. We recall from trigonometry that $\sin x$ has period 2π. (If necessary, see Definition 18.1.1.) From Table 9.3.3 we see that $y = 0$ when $x = 0, \pi, 2\pi$.

The maximum value of $y = 1$ occurs at $x = \pi/2$. The minimum value of $y = -1$ occurs at $x = 3\pi/2$. Plot these points on a coordinate system where 1 unit on the Y-axis is approximately $\pi/3$ units on the X-axis and connect them with a smooth curve (see Fig. 9.4.1). (We can discuss this curve further after we learn how to obtain the derivative of $y = \sin x$.)

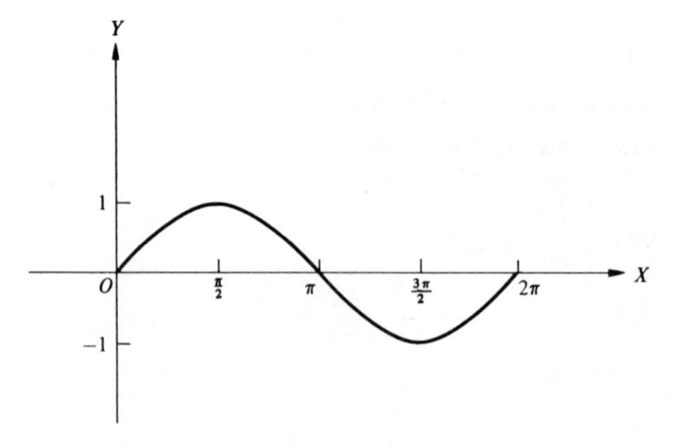

Figure 9.4.1 Graph of $y = \sin x$ (one period).

PROBLEM 1. Sketch the graph of $y = \sin x$ for $-2\pi \leqq x \leqq 3\pi$.

The graph of the more general function $y = a \sin (bx + c)$ is similar to the graph of the sine curve. It has an *amplitude* of $|a|$, a *period* of $2\pi/|b|$ and a *displacement* of $|c/b|$ from the sine curve—to the right for $c/b < 0$ and to the left for $c/b > 0$.

EXAMPLE 2. Sketch one cycle (or period) of $y = 3 \sin [2x + (\pi/4)]$.

Solution. The amplitude is 3 and the period is $2\pi/2 = \pi$. The graph (Fig. 9.4.2) crosses the X-axis when the argument $2x + (\pi/4)$ has the values

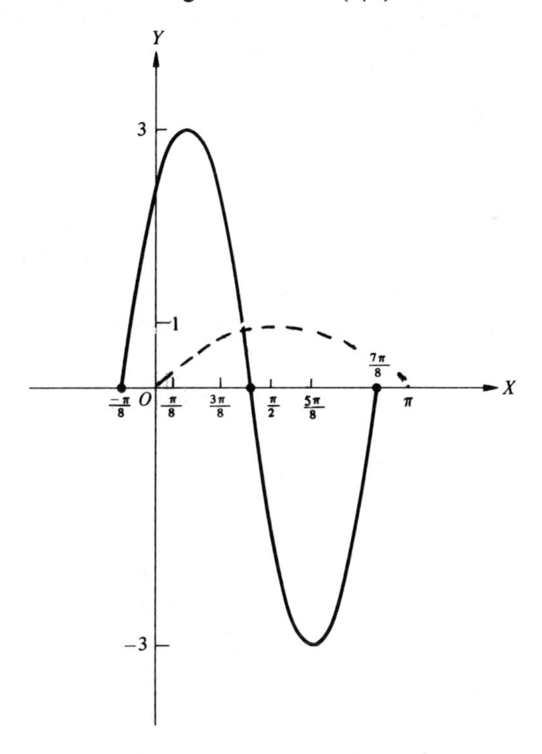

Figure 9.4.2 Graph of $y = 3 \sin [2x + (\pi/4)]$.

$$2x + \frac{\pi}{4} = 0, \quad \pi, \quad \text{and} \quad 2\pi$$

or when

$$x = \frac{-\pi}{8}, \quad \frac{3\pi}{8}, \quad \text{and} \quad \frac{7\pi}{8}$$

The maximum value is reached when

$$2x + \frac{\pi}{4} = \frac{\pi}{2}$$

or when

$$x = \frac{\pi}{8}$$

The minimum value is reached when

$$2x + \frac{\pi}{4} = \frac{3\pi}{2}$$

or when

$$x = \frac{5\pi}{8}$$

When $x = 0$, $y = 3 \sin (\pi/4) = 2.12$. The graph is shown in Fig. 9.4.2. Note that the graph is displaced $\pi/8$ units to the left of the sine curve. The dashed curve is part of the graph of $y = \sin x$ for comparison.

PROBLEM 2. Sketch one period of $y = 2 \sin [3x - \pi/2]$. Also sketch $y = \sin x$ for comparison.

It can be shown that $\cos x = \sin (x + \pi/2)$ by using Eq. 9.3.9. Then the graph of $y = \cos x$ may be obtained by considering the graph of $y = \sin (x + \pi/2)$. This graph is displaced $\pi/2$ units to the left of the sine curve. It has an amplitude of 1 and a period of 2π. Its graph is shown in Fig. 9.4.3. The "darker" portion of the curve is usually referred to as *one cycle of the cosine curve.*

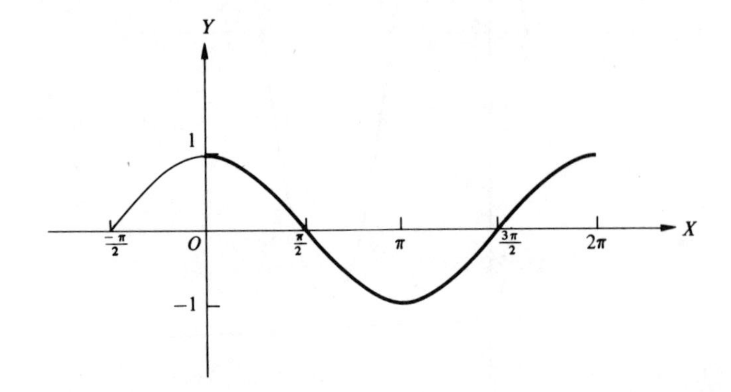

Figure 9.4.3 Graph of $y = \cos x$.

For reference one cycle of the graphs of $y = \tan x$, $y = \cot x$, $y = \sec x$, and $y = \csc x$ are shown in Fig. 9.4.4.

EXERCISES 9.4

Group A

1. In each of the following, sketch the graph for the indicated region.

(a) $y = \sin x, \; -2\pi \leqq x \leqq 2\pi$.

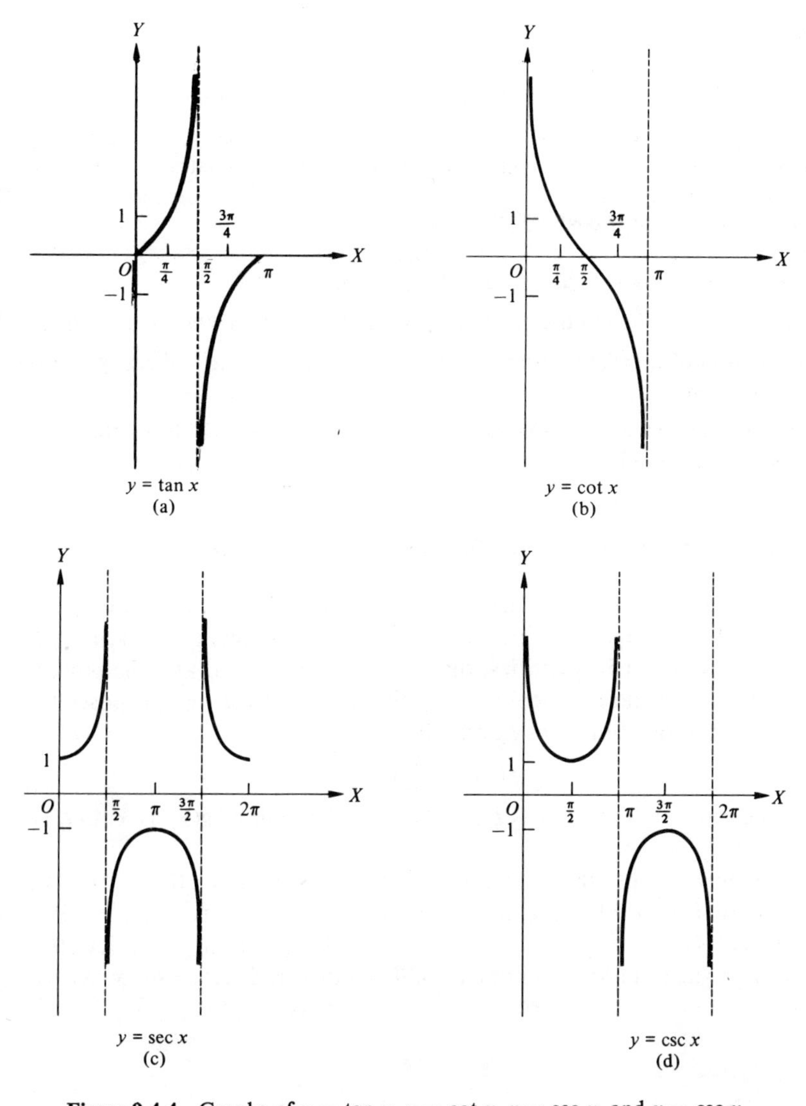

Figure 9.4.4 Graphs of $y = \tan x$, $y = \cot x$, $y = \sec x$, and $y = \csc x$.

(b) $y = 3 \sin 2x$, $0 \leqq x \leqq 2\pi$. State the amplitude and period.

(c) $y = \sin\left(x - \frac{\pi}{2}\right)$, $-\pi \leqq x \leqq 2\pi$.

(d) $y = 2 \sin\left(x + \frac{\pi}{4}\right)$, $\left(\frac{-\pi}{4}\right) \leqq x \leqq 2\pi$.

(e) $y = 2 \cos x$, $0 \leqq x \leqq 3\pi$.

(f) $y = 3 \cos 2x$, for two periods.

(g) $y = \tan 2x$, $0 \leqq x \leqq \pi$.

2. Discuss and sketch the graph of $y = \sin(x + \pi/2)$ as in Example 2.

Group B

1. In each of the following, sketch the graph for the indicated interval.

(a) $f(x) = \sin x + \cos x$, $-2\pi \leqq x \leqq 2\pi$. [*Hint*: Let $y_1 = \sin x$, $y_2 = \cos x$. Sketch the graphs of y_1 and y_2 and then take the sum of the ordinates at each abscissa; that is, $f(x) = y_1 + y_2$.]
(b) $f(x) = 3 \sin 2x + 2 \cos x$, $0 \leqq x \leqq \pi$.
(c) $f(x) = 2 \sin x \cos x$, $(-\pi/2) \leqq x \leqq (\pi/2)$.
(d) $f(x) = \sin 2x \cos x + \cos 2x \sin x$, $0 \leqq x \leqq \pi$.

2. The voltage E in an electric circuit is given as $E = 120 \sin 60\pi t$. Sketch the graph.

3. The motion of a weight on a vibrating spring is given as $x = 4 \cos 8t + 3 \sin 8t$. Sketch its graph.

4. A certain electric current is described by $I = e^{-t} \sin t$. Sketch its graph. Note that as t increases the amplitude e^{-t} decreases.

9.5 DERIVATIVES OF THE TRIGONOMETRIC FUNCTIONS

To obtain formulas for the derivatives of the trigonometric functions we will make use of some right-triangle trigonometry, the "squeeze" theorem, some of the basic trigonometric identities, the chain rule, Eq. 7.5.1, and the power rule, Eq. 7.5.2. The "squeeze" theorem very simply stated for our purposes tells us that for continuous functions f and g if

$$f(x) > g(x) > f(b)$$

and if the limit of $f(x)$ as $x \longrightarrow b$ is $f(b)$, then the limit of $g(x)$ as $x \longrightarrow b$ is also $f(b)$.

We now will use the delta process to obtain the derivative of the tangent function. In Fig. 9.5.1, triangles ACB and ACF are right triangles. Regions $ABDA$ and $AEFA$ are sectors of circles. In triangle ACB, if we let $AC = 1$, then from right triangle trigonometry $CB = \tan x$ and $AB = \sec x$, where x is

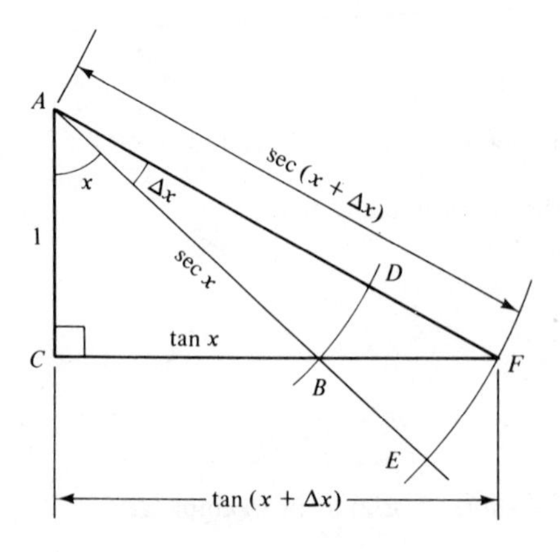

Figure 9.5.1 Derivative of tan x.

an angle measured in radians or the equivalent real number. In triangle ACF, $CF = \tan(x + \Delta x)$ and $AF = \sec(x + \Delta x)$. Since the area of a triangle is $(\frac{1}{2})$ (base) (height) and the area of a sector of a circle is $(\frac{1}{2})$ (radius)^2 (central angle in radians), we have

$$\text{area of } \Delta ACB = \tfrac{1}{2}(\tan x)(1) = \tfrac{1}{2}\tan x$$
$$\text{area of } \Delta ACF = \tfrac{1}{2}[\tan(x + \Delta x)](1) = \tfrac{1}{2}\tan(x + \Delta x)$$
$$\text{area of } \Delta ABF = \tfrac{1}{2}\tan(x + \Delta x) - \tfrac{1}{2}\tan x$$
$$\text{area of sector } ABDA = \tfrac{1}{2}(\sec^2 x)(\Delta x)$$
$$\text{area of sector } AEFA = \tfrac{1}{2}[\sec^2(x + \Delta x)](\Delta x)$$

Comparing these areas, we find that

$$\text{area } AEFA > \text{area } \Delta ABF > \text{area } ABDA$$

or

$$\tfrac{1}{2}(\sec^2(x + \Delta x)](\Delta x) > [\tfrac{1}{2}\tan(x + \Delta x) - \tfrac{1}{2}\tan x] > \tfrac{1}{2}(\sec^2 x)(\Delta x)$$

Upon dividing by $\frac{1}{2}(\Delta x)$, we obtain

$$\sec^2(x + \Delta x) > \frac{\tan(x + \Delta x) - \tan x}{\Delta x} > \sec^2 x$$

Taking the limit of this expression as $\Delta x \to 0$ and applying the "squeeze" theorem, we have

$$\lim_{\Delta x \to 0} \frac{\tan(x + \Delta x) - \tan x}{\Delta x} = \sec^2 x$$

Now using the definition of the derivative of a function, Definition 2.2.2(b), we find that

$$\boxed{\frac{d(\tan x)}{dx} = \sec^2 x} \qquad (9.5.1)$$

Formula 9.5.1 was derived[2] for $0 < x < \pi/2$. But, by using the formulas $\tan(-x) = -\tan x$ and $\tan(x + \pi) = \tan x$, it can be generalized for all permissible values of x.

If $y = \tan u$, where u is any differentiable function of x, we use the chain rule, $dy/dx = (dy/du)(du/dx)$, to write for

$$\boxed{\begin{aligned} y &= \tan u \\ \frac{dy}{dx} &= \sec^2 u \frac{du}{dx} \end{aligned}} \qquad (9.5.2)$$

[2]This derivation of the derivative of tan x is based on an article by Norman Schaumberger, Bronx Community College, New York, published in the September 1979 issue of the *Two-Year College Mathematics Journal* (Mathematical Association of America) and is used with their permission.

EXAMPLE 1. Find dy/dx for (a) $y = 2\tan 3x$; (b) $y = \tan^5 7x$.

Solution. For part (a) we use Eq. 9.5.2, where $u = 3x$. Then for $y = 2\tan 3x$, we have

$$\frac{dy}{dx} = 2\underbrace{(\sec^2 3x)}_{\sec^2 u}\underbrace{(3)}_{du/dx}$$

or

$$\frac{dy}{dx} = 6\sec^2 3x$$

For part (b) we may rewrite it as $y = (\tan 7x)^5$; then we have a function raised to a power of the form $y = u^n$, where $u = \tan 7x$. Applying the power rule, we have

$$\frac{dy}{dx} = \underbrace{5}_{n}\underbrace{(\tan 7x)^4}_{u^{n-1}}\underbrace{(\sec^2 7x)(7)}_{du/dx}$$

or

$$\frac{dy}{dx} = 35\tan^4 7x\sec^2 7x$$

PROBLEM 1. Find dy/dx for (a) $y = 3\tan 2x$; (b) $y = 9\tan^2 4x$.

The derivatives of the other five trigonometric functions are obtained by using the fact that cofunctions of complementary angles (or numbers) are equal, and by using the power rule on some of the reciprocal and squared identities. For the derivative of the cotangent we use the identity $\cot u = \tan(\pi/2 - u)$. Thus

$$\frac{d(\cot u)}{dx} = \frac{d}{dx}\left[\tan\left(\frac{\pi}{2} - u\right)\right]$$

$$= \left[\sec^2\left(\frac{\pi}{2} - u\right)\right]\left(0 - \frac{du}{dx}\right) \qquad \text{(from Eq. 9.5.2)}$$

or

$$\boxed{\frac{d(\cot u)}{dx} = -\csc^2 u\frac{du}{dx}} \qquad (9.5.3)$$

since $\sec(\pi/2 - u) = \csc u$.

EXAMPLE 2. (a) For $y = 3\cot 5x$,

$$\frac{dy}{dx} = 3\underbrace{(-\csc^2 5x)}_{-\csc^2 u}\underbrace{(5)}_{du/dx} = -15\csc^2 5x$$

(b) For $y = \cot^3 2x$, we use the power rule to obtain

$$\frac{dy}{dx} = \underbrace{3(\cot^2 2x)}_{nu^{n-1}}\underbrace{(-\csc^2 2x)(2)}_{[d(\cot v)]/dx = -\csc^2 v\,(dv/dx)}$$

or

$$\frac{dy}{dx} = -6\cot^2 2x \csc^2 2x$$

PROBLEM 2. (a) Find dy/dx for $y = 4\cot 3x$. (b) Find ds/dt for $s = 2\cot^2 7t$.

Using the squared identity $\sec^2 u = 1 + \tan^2 u$, we can obtain the derivative of the secant function. Thus

$$\frac{d(\sec^2 u)}{dx} = \frac{d(1 + \tan^2 u)}{dx}$$

or

$$2\sec u \frac{d(\sec u)}{dx} = 0 + 2\tan u \frac{d(\tan u)}{dx}$$

$$= 2\tan u \sec^2 u \frac{du}{dx} \qquad \text{(from Eq. 9.5.2)}$$

Dividing by $2\sec u$ (which is permissible since $\sec u$ is never equal to zero), we obtain

$$\boxed{\frac{d(\sec u)}{dx} = \sec u \tan u \frac{du}{dx}} \qquad (9.5.4)$$

EXAMPLE 3. For $y = \sec 5x$,

$$\frac{dy}{dx} = \underbrace{(\sec 5x}_{\sec u}\ \underbrace{\tan 5x)}_{\tan u}\ \underbrace{(5)}_{du/dx}$$

or

$$\frac{dy}{dx} = 5\sec 5x \tan 5x$$

PROBLEM 3. Find dy/dt for (a) $y = 7\sec 2t$; (b) $y = 3\sec^2 4t$.

The derivative of the cosine function can be obtained from the identity $\cos u = 1/\sec u = (\sec u)^{-1}$. Thus

$$\frac{d(\cos u)}{dx} = \frac{d(\sec u)^{-1}}{dx}$$

$$= (-1)(\sec u)^{-2}\frac{d(\sec u)}{dx}$$

$$= -\frac{1}{\sec^2 u}\cdot \sec u \tan u \frac{du}{dx} \qquad \text{(from Eq. 9.5.4)}$$

$$= -\frac{1}{\sec u}\cdot \frac{\sin u}{\cos u}\frac{du}{dx}$$

Using the identity $\sec u \cos u = 1$, we have

$$\boxed{\frac{d(\cos u)}{dx} = -\sin u \frac{du}{dx}} \tag{9.5.5}$$

EXAMPLE 4. For $y = 2 \cos 13x$,

$$\frac{dy}{dx} = 2\underbrace{(-\sin 13x)}_{-\sin u}\underbrace{(13)}_{du/dx}$$

or

$$\frac{dy}{dx} = -26 \sin 13x$$

PROBLEM 4. Find dy/dx for (a) $y = \cos 7x$; (b) $y = 3 \cos^2 5x$.

Now we can obtain the derivative formula for the sine function by using the identity $\sin u = \cos(\pi/2 - u)$. Thus

$$\frac{d(\sin u)}{dx} = \frac{d}{dx}\left[\cos\left(\frac{\pi}{2} - u\right)\right]$$

$$= \left[-\sin\left(\frac{\pi}{2} - u\right)\right]\left(0 - \frac{du}{dx}\right) \qquad \text{(from Eq. 9.5.5)}$$

or

$$\boxed{\frac{d(\sin u)}{dx} = \cos u \frac{du}{dx}} \tag{9.5.6}$$

since $\sin(\pi/2 - u) = \cos u$.

EXAMPLE 5. For $y = 3 \sin 7t$,

$$\frac{dy}{dt} = 3\underbrace{(\cos 7t)}_{\cos u}\underbrace{(7)}_{du/dt}$$

or

$$\frac{dy}{dt} = 21 \cos 7t$$

PROBLEM 5. Find dx/dt for (a) $x = -2 \sin 5t$; (b) $x = \sin^3 2t$.

The derivative of the cosecant function can be obtained from the identity $\csc u = \sec(\pi/2 - u)$ and formula 9.5.4. Thus

$$\frac{d(\csc u)}{dx} = \frac{d}{dx}\left[\sec\left(\frac{\pi}{2} - u\right)\right]$$

$$= \left[\sec\left(\frac{\pi}{2} - u\right)\tan\left(\frac{\pi}{2} - u\right)\right]\left[0 - \frac{du}{dx}\right]$$

or

$$\frac{d(\csc u)}{dx} = -\csc u \cot u \frac{du}{dx} \tag{9.5.7}$$

since $\sec(\pi/2 - u) = \csc u$ and $\tan(\pi/2 - u) = \cot u$.

EXAMPLE 6. For $y = 5 \csc(x/2)$,

$$\frac{dy}{dx} = 5\underbrace{\left(-\csc\frac{x}{2}\cot\frac{x}{2}\right)}_{-\csc u \cot u}\underbrace{\left(\frac{1}{2}\right)}_{du/dx}$$

or

$$\frac{dy}{dx} = -\frac{5}{2}\csc\frac{x}{2}\cot\frac{x}{2}$$

PROBLEM 6. Find dy/dx for (a) $y = -3 \csc 2x$; (b) $y = \csc^2 3x$.

***Warning** When you use the derivative formulas for the trigonometric functions the argument of the function must be a real number or the equivalent angle measured in radians.*

The differentials of the trigonometric functions are readily obtained from their derivatives. The following summary of formulas may be used for a quick reference.

Derivatives	*Differentials*
$\frac{d(\sin u)}{dx} = \cos u \frac{du}{dx}$	$d(\sin u) = \cos u \, du$
$\frac{d(\cos u)}{dx} = -\sin u \frac{du}{dx}$	$d(\cos u) = -\sin u \, du$
$\frac{d(\tan u)}{dx} = \sec^2 u \frac{du}{dx}$	$d(\tan u) = \sec^2 u \, du$
$\frac{d(\cot u)}{dx} = -\csc^2 u \frac{du}{dx}$	$d(\cot u) = -\csc^2 u \, du$
$\frac{d(\sec u)}{dx} = \sec u \tan u \frac{du}{dx}$	$d(\sec u) = \sec u \tan u \, du$
$\frac{d(\csc u)}{dx} = -\csc u \cot u \frac{du}{dx}$	$d(\csc u) = -\csc u \cot u \, du$

***Warning** Note carefully that the "cofunctions" are the only ones whose derivatives have a minus sign.*

The trigonometric functions may be combined with other functions in many ways. The following examples illustrate a few of them.

EXAMPLE 7. Find dy/dx for $y = 3x \sin x^3$.

Solution. Here we have a product. Also the argument of the sine function is x^3. If we let $u = 3x$ and $v = \sin x^3$, we may apply the product rule for differentiation to obtain

$$\frac{dy}{dx} = \underbrace{(3x)}_{u}\underbrace{\frac{d(\sin x^3)}{dx}}_{dv/dx} + \underbrace{(\sin x^3)}_{v}\ \underbrace{(3)}_{du/dx}$$

Now if we let $w = x^3$, we can write $(d/dx) \sin x^3$ as $(d/dx) \sin w$ and obtain

$$\frac{dy}{dx} = (3x)\underbrace{(\cos x^3)}_{\cos w}\underbrace{(3x^2)}_{dw/dx} + (\sin x^3)(3)$$

Rearranging, we have

$$\frac{dy}{dx} = 9x^3 \cos x^3 + 3 \sin x^3$$

Usually, we make the substitutions for u, v, and w mentally and write down the derivative as

$$\frac{dy}{dx} = (3x)(\cos x^3)(3x^2) + (\sin x^3)(3)$$

and then simplify.

PROBLEM 7. Find dx/dt for $x = 2t^3 \cos t^2$.

EXAMPLE 8. Find dy/dx for $y = \sin^2 7x \cos^3 5x$.

Solution. Here we have a product of functions raised to a power. Thus $dy/dx = (\sin^2 7x)[3(\cos 5x)^2(-\sin 5x)(5)] + (\cos^3 5x)[2(\sin 7x)(\cos 7x)(7)]$ and $dy/dx = -15 \sin^2 7x \sin 5x \cos^2 5x + 14 \cos^3 5x \sin 7x \cos 7x$.

PROBLEM 8. Find dy/dx for $y = 3 \tan^2 5x \sec^4 2x$.

EXAMPLE 9. In an electric circuit it was found that the charge Q was given by

$$Q = \tfrac{1}{10}e^{-6t}(4 \cos 8t + 3 \sin 8t) - \tfrac{2}{5} \cos 10t$$

Find I.

Solution. Since $I = dQ/dt$, we have

$$I = \tfrac{1}{10}e^{-6t}(-32 \sin 8t + 24 \cos 8t) + \tfrac{1}{10}(4 \cos 8t + 3 \sin 8t)(-6e^{-6t}) + 4 \sin 10t$$

Thus

$$I = \tfrac{1}{10}e^{-6t}(-50 \sin 8t) + 4 \sin 10t$$

Note that the term with e^{-6t} will soon become negligible as t increases. For this reason this term is called the *transient current*, while the term $4 \sin 10t$, which remains, is called the *steady-state current.*

PROBLEM 9. Find I in an electric circuit if the charge Q is given by $Q = \frac{1}{2}e^{-3t}(2 \cos 5t - \sin 5t) - \frac{2}{3} \sin 2t$.

EXAMPLE 10. Use implicit differentiation to find dy/dx for $y = \sin (x + y^2)$.

Solution

$$\frac{dy}{dx} = [\cos (x + y^2)]\left(1 + 2y\frac{dy}{dx}\right)$$

or

$$\frac{dy}{dx}[1 - 2y \cos (x + y^2)] = \cos (x + y^2)$$

and

$$\frac{dy}{dx} = \frac{\cos (x + y^2)}{[1 - 2y \cos (x + y^2)]}$$

PROBLEM 10. Use implicit differentiation to find dy/dx for $y = 3 \cos (x^2 - y)$.

EXERCISES 9.5

Group A

1. For each of the following functions, find the derivative with respect to the independent variable.

(a) $y = \sin 3x$.
(b) $y = 7 \sin 2x$.
(c) $y = \cos 5x$.
(d) $y = 9 \cos x$.
(e) $y = \sin 3x + 2 \cos x$.
(f) $y = 2 \sin 7x - 3 \cos 4x$.
(g) $y = x^2 + \sin 3x$.
(h) $Q = e^{-3t} + \cos 2t$.
(i) $Q = 7e^t - 5 \sin 110t$.
(j) $s = 4 \cos 7t + \ln 3t$.

2. In each of the following, find the derivative of the function with respect to the independent variable.

(a) $y = 3 \tan 2x$.
(b) $y = 2 \cot 3x$.
(c) $y = -5 \sec 7x$.
(d) $y = -3 \csc 2x$.
(e) $s = \tan t^2$.
(f) $y = 2 \tan (1/x)$.
(g) $s = 2 \cot t$.
(h) $s = \tan^2 3t$.
(i) $y = \sec 2x + \csc 2x$.
(j) $y = \sec^2 3x$.

3. For each of the following functions, find the derivative with respect to the independent variable.

(a) $y = 3x \sin x^2$.
(b) $y = x^2 \sin 3x$.
(c) $y = (x^2 + 3) \sin 5x$.
(d) $y = (x^2 - x) \sin 2x^3$.
(e) $y = \sin^2 3x$.
(f) $y = 3 \sin^3 (2x - 1)$.

(g) $y = 2x \cos x^3$.
(h) $y = x^2 \cos x^2$.
(i) $y = (x^2 - 3x) \cos x^2$.
(j) $y = (3x - 2) \cos^3 2x$.
(k) $Q = e^{-3t} \sin 110t$.
(l) $Q = 3e^{-10t} \cos 12t$.

4. For each of the following functions, find the derivative with respect to the independent variable.

(a) $y = \sin 2x \cos 2x$.
(b) $y = \sin^3 7x \cos^2 5x$.
(c) $y = \sin^2 (3x + 1) \cos^2 x$.
(d) $y = \sin^3 2x \cos 3x$.

5. Find the derivative with respect to the independent variable of each of the following functions.

(a) $y = \sec x \tan x$.
(b) $y = x \tan 3x$.
(c) $y = \ln \sec x$.
(d) $y = \ln \csc 2x$.
(e) $s = e^{-\tan t}$.
(f) $s = e^{\ln \sec t}$.
(g) $y = x \sec^2 3x$.
(h) $y = x \tan^3 2x$.
(i) $y = \ln (\sec x + \tan x)$.
(j) $u = \ln (\csc 2v - \cot 2v)$.

6. Find the second derivative with respect to the independent variable for each of the following.

(a) $y = 3 \sin 2x$.
(b) $x = 5 \cos 7t$.
(c) $y = 4 \sin 2x + 3 \cos 2x$.
(d) $s = t \sin t$.
(e) $y = \sin^2 x \cos 3x$.
(f) $y = \sin^2 x$.

7. In each of the following, find the second derivative of the function with respect to the independent variable.

(a) $y = \tan x$.
(b) $y = \sec 3x$.
(c) $u = \tan^2 3v$.
(d) $u = \tan 3v^2$.
(e) $s = t \tan 3t$.
(f) $s = t^2 \cot t^2$.

8. Find dy for each of the following.

(a) $y = 3 \sin 2x$.
(b) $y = 5 \cos 7x$.
(c) $y = x \sin 3x$.
(d) $y = (x^2 + 1) \cos 10x$.

9. Find dy for each of the following.

(a) $y = 3 \tan 5x$.
(b) $y = x \cot 2x$.
(c) $y = x^2 \sec x$.
(d) $y = x^2 \csc^2 x$.

10. Find the transient current and the steady-state current for $Q = e^{-3t} \sin 2t + 5 \cos 10t$.

Group B

1. For $Q = te^{-3t} + \sin 5t$, find I at $t = 0.2$ sec.

2. Find the derivative with respect to the independent variable of each of the following functions.

(a) $y = \ln \sin 3x$.
(b) $y = \ln \sqrt{\cos 2x}$.
(c) $y = \dfrac{\sin x}{x}$.
(d) $y = \ln \sqrt{\dfrac{1 - \sin x}{1 + \cos x}}$.

(e) $s = \sqrt{t} \cos \sqrt{t}$. (f) $s = e^{\sin 2t}$.

(g) $u = \cos e^{v}$. (h) $u = \sin \left(\frac{2}{v}\right)$.

3. In each of the following, find the derivative with respect to the independent variable of the function given.

(a) $y = x^{\tan x}$. (b) $y = \ln \sqrt{\frac{1 + \cot x}{1 - \cot x}}$.

(c) $y = (\cot x)^{x}$. (d) $y = e^{2x} \tan 4x$.

(e) $r = \tan^2 \frac{\theta}{2}$. (f) $s = e^{-8t} \cot 2t$.

4. For each of the following, find the value of the derivative with respect to x at the given value of the independent variable.

(a) $y = x + \sin x$, $x = 1$. (b) $y = \frac{\sin x}{x}$, $x = 0$.

(c) $y = \ln \sin x$, $x = 0.5$. (d) $y = 5e^{-x} \sin \pi x$, $x = \frac{1}{2}$.

5. In each of the following, find the value of the derivative with respect to x at the given value of x.

(a) $y = e^{x} \tan x$, $x = 2$. (b) $y = e^{2x} \tan \pi x$, $x = 1$.

(c) $y = \ln \cot \frac{\pi}{2} x$, $x = \frac{1}{2}$ (d) $y = \ln \sqrt{\tan x}$, $x = \frac{\pi}{4}$.

(e) $y = \ln^2 \sec x$, $x = 0$. (f) $y = 10e^{-x/10} \csc 3x$, $x = 1$.

6. For each of the following functions, find the second derivative with respect to the independent variable.

(a) $y = \frac{\sin x}{x}$. (b) $y = \ln \cos 2x$.

(c) $s = e^{\sin 2t}$. (d) $s = t \sin t^2$.

(e) $y = e^{x} \sin x$. (f) $y = e^{2x} \cos 3x$.

7. In each of the following, find the second derivative with respect to x of the function given.

(a) $y = e^{x} \tan x$. (b) $y = \frac{\cot x}{x}$.

(c) $y = \cot 2x$. (d) $y = e^{-x} \tan 3x$.

(e) $y = e^{ax} \cot bx$. (f) $y = \ln (\tan^2 2x)$.

8. Use implicit differentiation to find dy/dx in each of the following.

(a) $y \sin x + x \sin y = 0$. (b) $y \sin x + x \cos y = 0$.

(c) $\sin xy + \cos xy = 0$. (d) $e^{xy} \sin xy = 0$.

9. Find d^2y/dx^2 for (a) $\sin xy + \cos xy = 0$; (b) $e^{xy} \sin xy = 0$.

10. In each of the following, find dy/dx using implicit differentiation.

(a) $y = \tan (x - y)$. (b) $y = \cot (x + y)$.

(c) $\ln y = \tan (x + y)$. (d) $\tan y = \ln (x + y)$.

(e) $e^{y} = \cot (x + y)$. (f) $e^{y} = \tan (2x + y)$.

Group C

1. Use Fig. 9.5.2, where region $OABO$ is a sector of a circle, and show that $\lim_{h\to 0} [(\sin h)/h] = 1$.

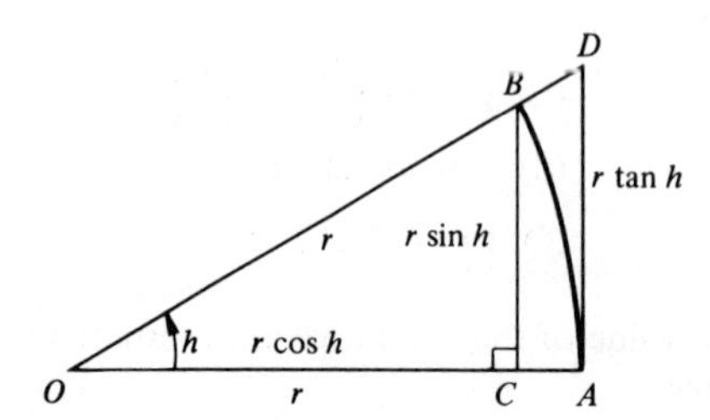

Figure 9.5.2 $\lim_{h\to 0} [(\sin h)/h] = 1$.

2. Use the result of Exercise 1, Group C, and Eq. 9.3.17 to show that $\lim_{h\to 0} [(1 - \cos h)/h] = 0$. (*Hint*: Let $u = h/2$.)

3. Use the delta process, and the results of Exercises 1 and 2 of Group C to obtain the derivative of $y = \sin x$.

4. Obtain the formula for $d(\cot u)/dx$ from the identity $\cot u = 1/\tan u$ and formula 9.5.1.

5. Obtain the formula for $d(\sin u)/dx$ from the identity $\sin^2 u + \cos^2 u = 1$ and formula 9.5.5.

6. Obtain the formula for $d(\csc u)/dx$ from the identity $\csc u = 1/\sin u$ and formula 9.5.6.

9.6 APPLICATIONS

The applications we studied with reference to algebraic functions may be extended to the trigonometric functions. Some of these will be illustrated by examples.

EXAMPLE 1. Sketch the curve of $y = e^x \sin x$ for $x \geqq 0$.

Solution. For $y = e^x \sin x$, we have

$$\frac{dy}{dx} = e^x \cos x + (\sin x)e^x = e^x (\cos x + \sin x)$$

and

$$\frac{d^2y}{dx^2} = e^x (-\sin x + \cos x) + (\cos x + \sin x)e^x$$
$$= 2e^x \cos x$$

To find the relative maxima and minima, set $dy/dx = 0$ and find the critical values.

$$\frac{dy}{dx} = e^x(\cos x + \sin x) = 0$$

Since e^x can never be zero, we have

$$\cos x + \sin x = 0$$

or

$$\tan x = -1 \qquad x = \frac{3\pi}{4}, \frac{7\pi}{4}, \frac{11\pi}{4}, \ldots, \text{etc.}$$

Testing the signs of the derivatives, we find

x	$3\pi/4$	$7\pi/4$	$11\pi/4$
dy/dx	0	0	0
d^2y/dx^2	$-$	$+$	$-$

Thus

$$\left(\frac{3\pi}{4}, \frac{1}{\sqrt{2}}e^{3\pi/4}\right) = (2.4, 7.46) \text{ is a relative maximum}$$

$$\left(\frac{7\pi}{4}, -\frac{1}{\sqrt{2}}e^{7\pi/4}\right) = (5.5, -173) \text{ is a relative minimum}$$

$$\left(\frac{11\pi}{4}, \frac{1}{\sqrt{2}}e^{11\pi/4}\right) = (8.6, 3995) \text{ is a relative maximum}$$

The graph is sketched in Fig. 9.6.1. Note that as x increases so does the amplitude.

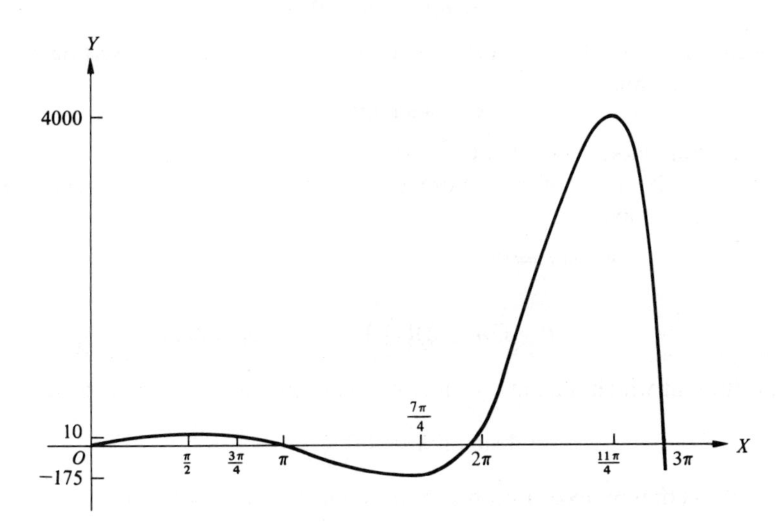

Figure 9.6.1 Graph of $y = e^x \sin x$.

PROBLEM 1. Sketch the graph of $y = e^x \cos x$ for $x \geqq 0$.

EXAMPLE 2. Simple harmonic motion is described by an equation of the form $x = a \sin bt$. Given $x = 4 \sin t$, find (a) the time t for maximum displacement;

(b) the time and position for maximum velocity; (c) the time and position for maximum acceleration.

Solution. For $x = 4 \sin t$, we find that

$$v = 4 \cos t$$

and

$$a = -4 \sin t$$

(a) For the time of maximum x, set $v = 0$. Thus

$$4 \cos t = 0$$

$$\cos t = 0$$

and

$$t = \frac{\pi}{2}, \frac{3\pi}{2}, \frac{5\pi}{2}, \ldots$$

or

$$t = (2n + 1)(\pi/2) \qquad n = 0, 1, 2, \ldots$$

(b) For the time of maximum v, set $a = 0$. Thus

$$-4 \sin t = 0$$

$$\sin t = 0$$

and

$$t = 0, \pi, 2\pi, \ldots$$

or

$$t = n\pi \qquad n = 0, 1, 2, \ldots$$

The position at which maximum velocity occurs is found by setting $t = n\pi$ in $x = 4 \sin t$. Thus

$$x = 4 \sin n\pi = 0$$

and maximum velocity occurs at $x = 0$.

(c) For the time of maximum acceleration, set $da/dt = 0$. Since $da/dt = -4 \cos t$, we have

$$-4 \cos t = 0$$

and

$$t = (2n + 1)\left(\frac{\pi}{2}\right) \qquad n = 0, 1, 2, \ldots$$

The position at which maximum acceleration occurs is then found to be

$$x = 4 \sin\left[(2n + 1)\left(\frac{\pi}{2}\right)\right] = \pm 4$$

Note that maximum acceleration occurs when the velocity is zero.

PROBLEM 2. A certain weight vibrating on a spring has its motion characterized by $x = 4 \cos t$.

(a) Find the time for maximum displacement.
(b) Find the time and position for maximum velocity.
(c) Find the time and position for maximum acceleration.

EXAMPLE 3. A person uses the formula $h = b \tan \theta$ to compute the height of an object whose angle of elevation is θ and is distance b from the person. Find the approximate error made in computing h if an error of 0.2° is made in the angle of elevation when θ is 32° and $b = 120$ m.

Solution. The approximate error in h is given by dh. For

$$h = b \tan \theta$$

$$dh = b \sec^2 \theta \, d\theta$$

We must change the angle measure to radians. Thus $32° = 0.56$ and $d\theta = 0.2° = 0.0035$. Then we find that

$$dh = 120(\sec^2 0.56)(0.0035) = 0.59 \text{ m}$$

PROBLEM 3. Find the approximate error in computing v from the formula $v = 25 \sec \theta$ when $\theta = 23°$ is in error by 0.1°.

EXERCISES 9.6

Group A

1. Find the slope of each curve at $t = 0, t = \pi/2, t = \pi, t = 3\pi/2$, and $t = 2\pi$. Use these values to sketch the graph.

(a) $x = \sin t$. (b) $x = \cos t$.

2. Find the slope of $y = \tan x$ at $x = 0$, $x = \pi/4$, and $x = \pi/2$. Is the slope of $y = \tan x$ ever negative? Sketch the graph.

3. Make use of the first two derivatives to show that a relative minimum of $y = \sec x$ is 1.

4. Sketch the graph of $I = e^{-t} \sin t$. Why is the graph bounded by the graph of $I = |e^{-t}|$?

5. Find an approximate value for each of the following.

(a) sin 31°. (b) cos 29°. (c) tan 47°.
(d) sin 2.1 ($2\pi/3 = 2.09$). (e) cos 3.6 ($7\pi/6 = 3.66$).

6. Find the angle of intersection of the curves $y = 2 \tan (x/2)$ and $y = 4 \sin (x/3)$ at the point $(\pi/2, 2)$.

7. Find the maximum value of $y = a \sin x + b \cos x$.

Group B

1. Sketch the graph of

$$y = \begin{cases} x \sin \dfrac{1}{x}, & x > 0 \\ 0 & x = 0 \end{cases}$$

2. A piston moving in a cylinder has a stroke of 4 in. The equation of motion is given by $x = 2 \sin 8t$, where x is the displacement in inches at time t seconds. The origin of axes is at the middle of the cylinder.

(a) Find its velocity when $t = \pi/4$.
(b) Find its acceleration when $t = \pi/4$.
(c) Find its velocity when it is 1 in. from the end and moving in a negative direction.
(d) Why do we say that this piston has periodic motion?
(e) When and where will the piston have maximum velocity?
(f) Find its velocity and acceleration when $t = 0.5$.

3. Show that $x = t \sin t$ is a solution of the differential equation $d^2x/dt^2 + x = 2 \cos t$. (*Hint*: Does the substitution for x and d^2x/dt^2 make the left side equal to the right side?) This expression occurs in a vibrating spring problem when there is resonance. The spring will break when $x = 6$. Find the value of t when the spring breaks. (*Hint*: Let $x_1 = 6/t$ and $x_2 = \sin t$. Sketch their graphs and determine an approximate value of t at their point of intersection. Then use a calculator to find t such that $x_1 = x_2$.)

4. In a certain electric circuit $Q = e^{-t} \sin 2t$. Find the first time I has a relative maximum. As t increases, what happens to I?

5. The displacement x at time t of a certain weight vibrating on an idealized spring is given by $x = 8 \sin 2t$. This represents simple harmonic motion.

(a) Find its displacement, velocity, and acceleration at (1) $t = 0.96$ s $(= \pi/4)$; (2) $t = 3\pi/4$ s; (3) $t = 6.28$ s; (4) $t = 1$ s.
(b) Find t the first time $x = 4$ ft.
(c) Explain why this type of motion is called periodic motion.

6. A certain electric circuit in which resonance occurs has $Q = e^{-6t}(4 \cos 8t + 3 \sin 8t) - t \cos 10t$. Find I when $t = 0.1$ s.

7. A projectile is fired from a gun at an angle of α above the horizontal with velocity of v_0. Neglecting air resistance the horizontal range R is given by $R = (v_0^2/g) \sin 2\alpha$. Find α for maximum R.

8. A tangent galvanometer is used to measure the electric current. The percentage of error due to a small error in reading is proportional to $\tan x + \cot x$. For what value of x will this percentage error be least?

9. A weight W is dragged along the ground by a force F that makes an angle θ with the ground. If μ is the coefficient of friction,

$$F = \frac{\mu W}{\cos \theta + \mu \sin \theta}$$

Find θ for a minimum F.

10. A 27-ft-long ladder is to be moved from a corridor, 8 ft wide, into a hall at right angles to the corridor. If the ladder is always horizontal, how wide must the hall be so that the ladder may just go around the corner?

11. Find the differential of arc for the curve given by these parametric equations (a) $x = \cos t$, $y = \sin t$; (b) $x = 3 \cos t$, $y = 4 \sin t$.

12. Let the position of a particle moving on a circle be given by $\mathbf{R}(t) = 5 \cos t\,\mathbf{i} + 5 \sin t\,\mathbf{j}$. (a) Find $\mathbf{V}(t)$. (b) Find $\mathbf{A}(t)$. (c) What is the relation of $\mathbf{A}(t)$ to $\mathbf{V}(t)$? (d) Find the tangential and normal components of $\mathbf{A}(t)$.

Group C

1. An inductor L, capacitor C, and resistor R are connected in series. At $t = 0$, the charge on the capacitor is Q_0, and the current is zero. The charge Q at any subsequent time t is given by $Q = (Q_0/2\omega L)e^{-Rt/2L}\sqrt{R^2 + \omega^2 L^2} \sin(\omega t + \phi)$, where $\omega = \sqrt{(1/LC) - (R^2/4L^2)}$, $\tan \phi = 2\omega L/R$. Show that $I = \{[-Q_0(R^2 + 4\omega^2 L^2)]/4\omega L^2\}e^{-Rt/2L} \sin \omega t$.

9.7 INTEGRALS OF TRIGONOMETRIC FUNCTIONS

We may extend our list of integral formulas by considering the antiderivatives of the trigonometric functions. These formulas may be verified by differentiating their right-hand side.

$$\int \sin u\, du = -\cos u + C \tag{9.7.1}$$

$$\int \cos u\, du = \sin u + C \tag{9.7.2}$$

$$\int \sec^2 u\, du = \tan u + C \tag{9.7.3}$$

$$\int \csc^2 u\, du = -\cot u + C \tag{9.7.4}$$

$$\int \sec u \tan u\, du = \sec u + C \tag{9.7.5}$$

$$\int \csc u \cot u\, du = -\csc u + C \tag{9.7.6}$$

In using formulas of integration the integrand must be *exactly like* the integrand of the formula. This is usually accomplished by a suitable substitution.

EXAMPLE 1. Find $\int \sin 3x\, dx$.

Solution. This looks like Eq. 9.7.1. To make it exactly like Eq. 9.7.1 let $u = 3x$ (the argument of the function); then $du = 3dx$ and $dx = \frac{1}{3}du$. Upon substitution we have

$$\int \sin 3x\, dx = \int \sin u \left(\frac{du}{3}\right) = \frac{1}{3}\int \sin u\, du = \frac{-1}{3}\cos u + C$$

$$\therefore \quad \int \sin 3x\, dx = -\frac{1}{3}\cos 3x + C$$

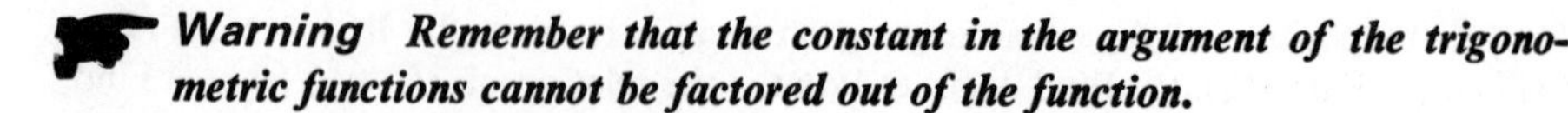

Warning ***Remember that the constant in the argument of the trigonometric functions cannot be factored out of the function.***

PROBLEM 1. Find $\int \cos(8x + 5)\, dx$.

EXAMPLE 2. Find $\int (\cot 5x/\sin 5x)\, dx$.

Solution. This is not like any of the formulas. However, recalling that $1/\sin 5x = \csc 5x$, we have

$$\int \frac{\cot 5x}{\sin 5x}\, dx = \int \csc 5x \cot 5x\, dx$$

Then this looks like Eq. 9.7.6. Let $u = 5x$ (the argument of the functions); then $du = 5\, dx$ and $dx = \frac{1}{5}\, du$. Hence

$$\int \csc 5x \cot 5x\, dx = \tfrac{1}{5}\int \csc u \cot u\, du$$

$$= -\tfrac{1}{5}\csc u + C \qquad \text{(by Eq. 9.7.6)}$$

and finally

$$\int \frac{\cot 5x}{\sin 5x}\, dx = -\frac{1}{5}\csc 5x + C$$

PROBLEM 2. Find $\int (\tan 7x/\cos 7x)\, dx$.

EXAMPLE 3. Find $\int e^x \cos(e^x)\, dx$.

Solution. Let $u = e^x$ (the argument of the cosine function); then $du = e^x\, dx$ and substitution yields

$$\int e^x \cos(e^x)\, dx = \int \cos u\, du$$

$$= \sin u + C$$

Therefore,

$$\int e^x \cos(e^x)\, dx = \sin(e^x) + C$$

PROBLEM 3. Find $\displaystyle\int \frac{1}{x} \sin(\ln x)\, dx$.

The integrals of the tangent, cotangent, secant, and cosecant functions are obtained by using some trigonometric identities. Thus

$$\int \tan u\, du = \int \frac{\sin u}{\cos u}\, du$$

Now we note that the numerator is the differential of the denominator (except for the sign), so let $z = \cos u$, then $dz = -\sin u\, du$, and we have

$$\int \tan u\, du = \int \frac{\sin u}{\cos u}\, du = -\int \frac{dz}{z}$$

Then

$$\int \tan u\, du = -\int \frac{dz}{z} = -\ln|z| + C \qquad \text{(by Eq. 8.8.1)}$$

and

$$\int \tan u\, du = -\ln|\cos u| + C \tag{9.7.7}$$

Similarly,

$$\int \cot u\, du = \int \frac{\cos u}{\sin u}\, du = \ln|\sin u| + C \tag{9.7.8}$$

The formula for $\int \sec u\, du$ is found by a tricky device[3], namely multiplying by $(\sec u + \tan u)/(\sec u + \tan u)$. Thus

$$\int \sec u\, du = \int \frac{\sec u(\sec u + \tan u)\, du}{\sec u + \tan u}$$

Let $z = \sec u + \tan u$; then

$$dz = (\sec u \tan u + \sec^2 u)\, du = \sec u(\sec u + \tan u)\, du$$

and

$$\int \sec u\, du = \int \frac{\sec u(\sec u + \tan u)}{\sec u + \tan u}\, du = \int \frac{dz}{z} = \ln|z| + C$$

Therefore,

$$\int \sec u\, du = \ln|\sec u + \tan u| + C \tag{9.7.9}$$

Similarly,

$$\int \csc u\, du = \int \frac{\csc u(\csc u - \cot u)}{\csc u - \cot u}\, du$$
$$= \ln|\csc u - \cot u| + C \tag{9.7.10}$$

In summary, we have

$$\int \tan u\, du = -\ln|\cos u| + C \tag{9.7.7}$$

$$\int \cot u\, du = \ln|\sin u| + C \tag{9.7.8}$$

$$\int \sec u\, du = \ln|\sec u + \tan u| + C \tag{9.7.9}$$

$$\int \csc u\, du = \ln|\csc u - \cot u| + C \tag{9.7.10}$$

[3]There are two fundamental "tricks" in mathematics used to change the form of an expression but not its value. One is to multiply by 1 expressed in a form that makes the problem at hand manageable. The other trick is to add zero in a suitable form. Experience teaches one when and in what form these tricks should be played.

EXAMPLE 4. Evaluate $\int_0^{\pi/6} [(1 + \cos 2x)/\cot 2x]\, dx$.

Solution

$$\int_0^{\pi/6} \frac{1 + \cos 2x}{\cot 2x}\, dx = \int_0^{\pi/6} \tan 2x\, dx + \int_0^{\pi/6} \frac{(\cos 2x)}{\cos 2x}(\sin 2x)\, dx$$

$$= \int_0^{\pi/6} \tan 2x\, dx + \int_0^{\pi/6} \sin 2x\, dx$$

Let $u = 2x$; then $du = 2dx$ and $dx = \frac{1}{2}\, du$. Thus

$$\int_0^{\pi/6} \frac{1 + \cos 2x}{\cot 2x}\, dx = \frac{1}{2}\int_0^{\pi/3} \tan u\, du + \frac{1}{2}\int_0^{\pi/3} \sin u\, du$$

(Note the change in the limits since we changed the variable of integration from x to u.) Hence

$$\int_0^{\pi/6} \frac{1 + \cos 2x}{\cot 2x}\, dx = \frac{1}{2}[-\ln|\cos u|]_0^{\pi/3} + \frac{1}{2}[-\cos u]_0^{\pi/3}$$

$$= -\frac{1}{2}\left[\ln\left|\cos\left(\frac{\pi}{3}\right)\right| - \ln|\cos 0|\right] - \frac{1}{2}\left[\cos\left(\frac{\pi}{3}\right) - \cos 0\right]$$

$$= -\frac{1}{2}\ln\cos\left(\frac{\pi}{3}\right) - \frac{1}{2}\cos\left(\frac{\pi}{3}\right) + \frac{1}{2}$$

$$= -\tfrac{1}{2}\ln\left(\tfrac{1}{2}\right) - \tfrac{1}{2}\left(\tfrac{1}{2}\right) + \tfrac{1}{2}$$

$$= \tfrac{1}{2}\ln 2 + \tfrac{1}{4} = 0.597$$

PROBLEM 4. Evaluate $\displaystyle\int_{\pi/6}^{\pi/3} \frac{1 + \sin 2x}{\tan 2x}\, dx$.

EXAMPLE 5. Find $\int \tan^5 x \sec^2 x\, dx$.

Solution. Since $d(\tan x) = \sec^2 x\, dx$, we here let $u = \tan x$ and obtain

$$\int u^5\, du = \tfrac{1}{6}u^6 + C$$

Hence

$$\int \tan^5 x \sec^2 x\, dx = \tfrac{1}{6}\tan^6 x + C$$

PROBLEM 5. Find $\int \cot^3 2x \csc^2 2x\, dx$.

Some applications are discussed in Sec. 9.9.

Powers and products of some of the trigonometric functions are discussed in Exercises 9.7. The $\int \sec^3 x\,dx$ needs a very special treatment and is discussed in Sec. 11.3.

EXERCISES 9.7

Group A

1. Integrate each of the following.

(a) $\int \sin 2x\,dx.$ (b) $\int \cos 5x\,dx.$

(c) $\int \sec^2 7x\,dx.$ (d) $\int \csc^2 2x\,dx.$

(e) $\int \sec 2x \tan 2x\,dx.$ (f) $\int \csc 3x \cot 3x\,dx.$

(g) $\int \tan 3x\,dx.$ (h) $\int \cot 5x\,dx.$

(i) $\int \sec 5x\,dx.$ (j) $\int \csc 10x\,dx.$

2. Integrate each of the following.

(a) $\int \frac{\tan 2x}{\cos 2x}\,dx.$ (b) $\int \frac{\csc 3x}{\sin 3x}\,dx.$

(c) $\int e^{2x} \sin e^{2x}\,dx.$ (d) $\int \frac{\cos 2x}{\sin 2x}\,dx.$

(e) $\int \frac{1 + \sin 3x}{\tan 3x}\,dx.$ (f) $\int \sec (2x - 1) \tan (2x - 1)\,dx.$

(g) $\int \sin u \cos u\,du$ (use Eq. 9.3.12).

3. Evaluate each of the following.

(a) $\int_0^{\pi/2} \sin 2x\,dx.$ (b) $\int_{-\pi/2}^{\pi/2} \cos x\,dx.$

(c) $\int_0^{\pi} e^t \sin e^t\,dt.$ (d) $\int_0^{\sqrt{\pi}} x \cos x^2\,dx.$

(e) $\int_{\pi/6}^{\pi/3} \frac{\cot 2x}{\sin 2x}\,dx.$ (f) $\int_0^{1/3} \cos 3x\,dx.$

4. Integrate each of the following.

(a) $\int \frac{\cos^2 x}{\sin x}\,dx.$ $\left[\textit{Hint}: \int \frac{1 - \sin^2 x}{\sin x}\,dx = \int (\csc x - \sin x)\,dx.\right]$

(b) $\int \frac{\tan^2 x}{\sec x}\,dx.$

(c) $\int \tan^2 x\,dx.$ (*Hint*: $\tan^2 x = \sec^2 x - 1$.)

(d) $\int \cot^2 x \, dx.$

5. (a) Show that for a a constant $\int \sin au \, du = -(1/a) \cos au + C$. (b) Write similar formulas for the remaining trigonometric functions when the argument is au. (c) Tell how these formulas may be of use. Do Exercise 1 using these formulas.

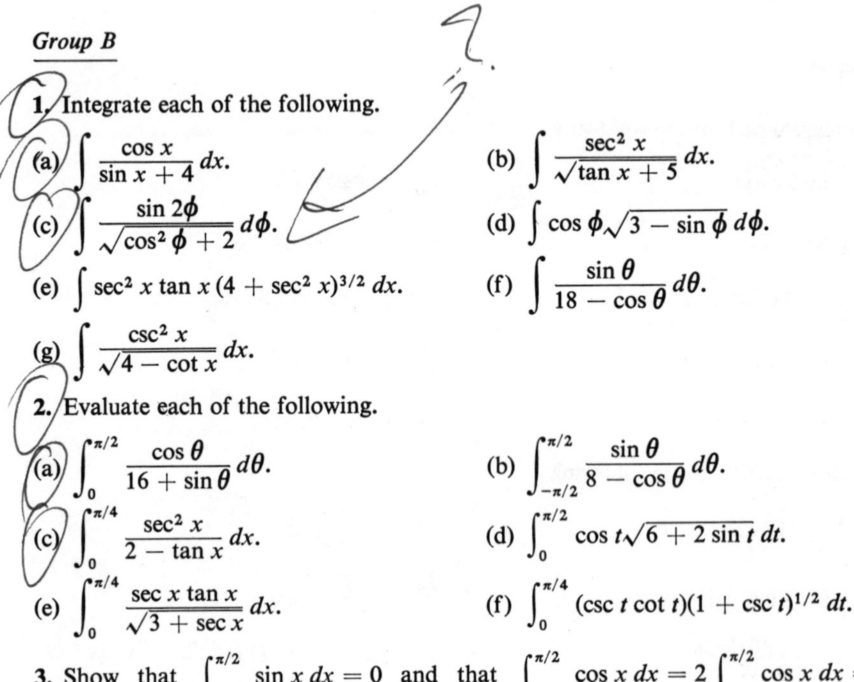

Group B

1. Integrate each of the following.

(a) $\displaystyle\int \frac{\cos x}{\sin x + 4} \, dx.$ (b) $\displaystyle\int \frac{\sec^2 x}{\sqrt{\tan x + 5}} \, dx.$

(c) $\displaystyle\int \frac{\sin 2\phi}{\sqrt{\cos^2 \phi + 2}} \, d\phi.$ (d) $\displaystyle\int \cos \phi \sqrt{3 - \sin \phi} \, d\phi.$

(e) $\displaystyle\int \sec^2 x \tan x \, (4 + \sec^2 x)^{3/2} \, dx.$ (f) $\displaystyle\int \frac{\sin \theta}{18 - \cos \theta} \, d\theta.$

(g) $\displaystyle\int \frac{\csc^2 x}{\sqrt{4 - \cot x}} \, dx.$

2. Evaluate each of the following.

(a) $\displaystyle\int_0^{\pi/2} \frac{\cos \theta}{16 + \sin \theta} \, d\theta.$ (b) $\displaystyle\int_{-\pi/2}^{\pi/2} \frac{\sin \theta}{8 - \cos \theta} \, d\theta.$

(c) $\displaystyle\int_0^{\pi/4} \frac{\sec^2 x}{2 - \tan x} \, dx.$ (d) $\displaystyle\int_0^{\pi/2} \cos t \sqrt{6 + 2 \sin t} \, dt.$

(e) $\displaystyle\int_0^{\pi/4} \frac{\sec x \tan x}{\sqrt{3 + \sec x}} \, dx.$ (f) $\displaystyle\int_0^{\pi/4} (\csc t \cot t)(1 + \csc t)^{1/2} \, dt.$

3. Show that $\displaystyle\int_{-\pi/2}^{\pi/2} \sin x \, dx = 0$ and that $\displaystyle\int_{-\pi/2}^{\pi/2} \cos x \, dx = 2 \int_0^{\pi/2} \cos x \, dx = 2.$ Note that $\sin x$ is an odd function and that $\cos x$ is an even function (see Exercises 2 and 3 of Group B in Exercises 1.3).

(a) If $f(x)$ is an odd function, show geometrically that $\displaystyle\int_{-a}^{a} f(x) \, dx = 0.$

(b) If $f(x)$ is an even function, show geometrically that $\displaystyle\int_{-a}^{a} f(x) \, dx = 2 \int_0^{a} f(x) \, dx.$

4. Integrate each of the following by making use of $\int u^n \, du$.

(a) $\displaystyle\int \sin^2 x \cos x \, dx.$ (b) $\displaystyle\int \sin^6 x \cos x \, dx.$

(c) $\displaystyle\int \sin^3 x \cos x \, dx.$ (d) $\displaystyle\int \cos^2 x \sin x \, dx.$

(e) $\displaystyle\int \cos^6 x \sin x \, dx.$ (f) $\displaystyle\int \cos^3 2x \sin 2x \, dx.$

5. Evaluate $\displaystyle\int_0^{\pi/2} \sin x \cos x \, dx$ by (a) letting $u = \sin x$; (b) letting $u = \cos x$; (c) using the identity $\sin x \cos x = \frac{1}{2} \sin 2x$. Are the answers the same for all three parts?

6. Integrate each of the following.

(a) $\int \tan x \sec^2 x \, dx.$ (b) $\int \tan^2 x \sec^2 x \, dx.$ (c) $\int \tan^3 2x \sec^2 2x \, dx.$

(d) $\int \cot x \csc^2 x \, dx.$ (e) $\int \cot^2 3x \csc^2 3x \, dx.$ (f) $\int \cot^4 5x \csc^2 5x \, dx.$

7. Integrate each of the following.

(a) $\int \sec^3 x(\sec x \tan x) \, dx.$ (b) $\int \sec^2 5x(\sec 5x \tan 5x) \, dx.$

(c) $\int \csc^4 x(\csc x \cot x) \, dx.$ (d) $\int \csc^2 7x(\csc 7x \cot 7x) \, dx.$

Group C

1. Verify Eq. 9.7.3.

2. Verify Eq. 9.7.4.

3. Verify Eq. 9.7.6.

4. Find $\int \sin^6 x \cos^3 x \, dx$. [*Hint*: $\cos^2 x = 1 - \sin^2 x$; therefore, the given integrand may be written $\int \sin^6 x(1 - \sin^2 x) \cos x \, dx = \int \sin^6 x \cos x \, dx - \int \sin^8 x \cos x \, dx$.]

5. Using Exercise 4 as a model, formulate a method for integrating $\int \sin^m x \cos^n x \, dx$ when n is odd.

6. Find $\int \sin^7 x \cos^2 x \, dx$. [*Hint*: $\sin^2 x = 1 - \cos^2 x$; therefore, the given integrand may be written $\int (1 - \cos^2 x)^3 \sin x \cos^2 x \, dx$.]

7. Using Exercise 6 as a model, formulate a method for integrating $\int \sin^m x \cos^n x \, dx$ when m is odd.

8. In Exercises 4 through 7, the scheme was essentially to isolate, as a factor, the differential of one of the functions and express the other factor as a sum of this function. For $\int \sin^m x \cos^n x \, dx$ when m and n are both even, we use one or more of the identities $\sin^2 x = \frac{1}{2}(1 - \cos 2x)$, $\cos^2 x = \frac{1}{2}(1 + \cos 2x)$, and $\sin x \cos x = \frac{1}{2} \sin 2x$ (see Example 4 of Sec. 6.6). Find (a) $\int \sin^2 x \cos^2 x \, dx$. (b) $\int \sin^4 x \cos^2 x \, dx$. (c) $\int \sin^4 x \, dx$. (d) $\int \cos^4 3x \, dx$.

9. The formulas $d(\tan x) = \sec^2 x \, dx$, $d(\cot x) = -\csc^2 x \, dx$, $\tan^2 x = \sec^2 x - 1$, and $\cot^2 x = \csc^2 x - 1$, are employed in integrating some tangent and cotangent terms. Evaluate each of the following (see Exercise 6 of Group B).

(a) $\int \tan^2 x \, dx = \int (\sec^2 x - 1) \, dx.$ (b) $\int \tan^3 x \, dx = \int \tan x(\tan^2 x) \, dx.$

(c) $\int \tan^4 x \, dx = \int \tan^2 x \sec^2 x \, dx - \int \tan^2 x \, dx.$

(d) $\int \cot^2 x \, dx.$ (e) $\int \cot^4 x \, dx.$

(f) $\int \tan^2 x \sec^4 x \, dx = \int \tan^2 x(1 + \tan^2 x) \sec^2 x \, dx.$

10. For products of $\tan x$ and $\sec x$, and of $\cot x$ and $\csc x$, the formulas $d(\sec x) = \sec x \tan x \, dx$ and $d(\csc x) = -\csc x \cot x \, dx$ are sometimes useful. Evaluate each of the following (see Exercise 7 of Group B).

(a) $$\int \sec^3 x \tan^3 x \, dx = \int \sec^2 x \tan^2 x(\sec x \tan x \, dx)$$
$$= \int (\sec^4 x - \sec^2 x) \sec x \tan x \, dx.$$

(b) $\int \csc^3 x \cot x \, dx.$

11. Evaluate each of the following integrals. Make use of at least one of the Eqs. 9.3.22, 9.3.23, or 9.3.24.

(a) $\int \sin 4x \cos 2x \, dx.$ (b) $\int \cos 3x \cos 5x \, dx.$ (c) $\int \sin 3x \sin x \, dx.$

12. In developing formulas for the coefficients in the Fourier series of a periodic function (Sec. 18.2) the following integrals are used. If m and n are positive integers and $2p$ is the period, show that each of the following are true.

(a) $\int_{-p}^{p} \cos \frac{n\pi}{p} t \, dt = 0, n \neq 0.$

(b) $\int_{-p}^{p} \sin \frac{n\pi}{p} t \, dt = 0.$

(c) $\int_{-p}^{p} \cos \frac{m\pi}{p} t \cos \frac{n\pi}{p} t \, dt = 0, m \neq n.$ (Use Eq. 9.3.23.)

(d) $\int_{-p}^{p} \cos^2 \frac{n\pi}{p} t \, dt = p, n \neq 0.$

(e) $\int_{-p}^{p} \cos \frac{m\pi}{p} t \sin \frac{n\pi}{p} t \, dt = 0.$ (Use Eq. 9.3.22.)

(f) $\int_{-p}^{p} \sin \frac{m\pi}{p} t \sin \frac{n\pi}{p} t \, dt = 0, m \neq n.$ (Use Eq. 9.3.24.)

(g) $\int_{-p}^{p} \sin^2 \frac{n\pi}{p} t \, dt = p, n \neq 0.$

9.8 INTEGRALS BY TRIGONOMETRIC SUBSTITUTION

Algebraic functions of the type $\sqrt{a^2 + x^2}$, $\sqrt{a^2 - x^2}$, and $\sqrt{x^2 - a^2}$ may sometimes be integrated by making a trigonometric substitution.

(1) Integrand Involving $\sqrt{a^2 + x^2}$

EXAMPLE 1. Find $\int dx/\sqrt{4 + x^2}$.

Solution. The expression $\sqrt{4 + x^2}$ suggests a right triangle with legs 2 and x and hypotenuse $\sqrt{4 + x^2}$. See Fig. 9.8.1. From the triangle (Fig. 9.8.1) we have

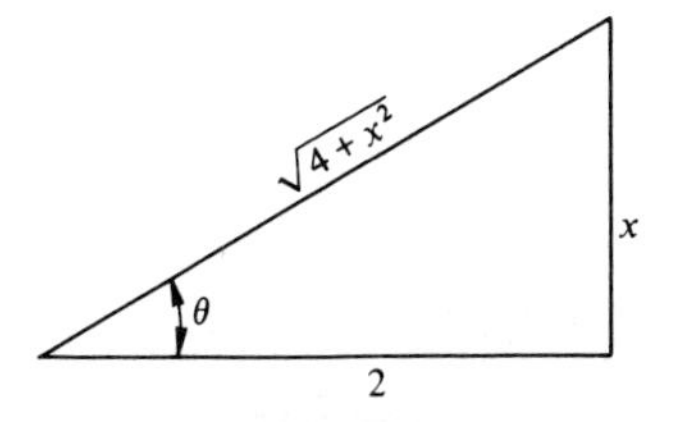

Figure 9.8.1 Example 1.

$$x = 2\tan\theta$$
$$dx = 2\sec^2\theta\, d\theta$$
$$\sqrt{4+x^2} = 2\sec\theta$$

Upon substitution the integral becomes

$$\int \frac{dx}{\sqrt{4+x^2}} = \int \frac{2\sec^2\theta\, d\theta}{2\sec\theta} = \int \sec\theta\, d\theta$$
$$= \ln|\sec\theta + \tan\theta| + C \qquad \text{(by Eq. 9.7.9)}$$

From the triangle (Fig. 9.8.1) we see that

$$\sec\theta = \tfrac{1}{2}\sqrt{4+x^2} \qquad \text{and} \qquad \tan\theta = \tfrac{1}{2}x$$

Hence

$$\int \frac{dx}{\sqrt{4+x^2}} = \ln\left|\frac{\sqrt{4+x^2}}{2} + \frac{x}{2}\right| + C_1$$
$$= \ln|\sqrt{4+x^2} + x| - \ln 2 + C_1$$

and

$$\int \frac{dx}{\sqrt{4+x^2}} = \ln|\sqrt{4+x^2} + x| + C$$

where $C = C_1 - \ln 2$. (Recall that ln 2 is a constant and that a constant minus a constant is another constant.) For $\int \sqrt{a^2+x^2}\, dx$, see Exercise 11 of Group B in Exercises 11.3.

PROBLEM 1. Find $\int 7dx/\sqrt{25+x^2}$.

Warning ***Always use the Pythagorean theorem to check the labeling of the sides of the right triangle.***

(2) Integrand Involving $\sqrt{a^2 - x^2}$

EXAMPLE 2. Evaluate $\int_0^{3/2} \sqrt{9-x^2}\, dx$.

Solution. The expression $\sqrt{9-x^2}$ suggests a right triangle with hypotenuse of 3 and a leg of x (see Fig. 9.8.2). From the triangle (Fig. 9.8.2), we have

$$x = 3\sin\theta$$
$$dx = 3\cos\theta\, d\theta$$
$$\sqrt{9-x^2} = 3\cos\theta$$

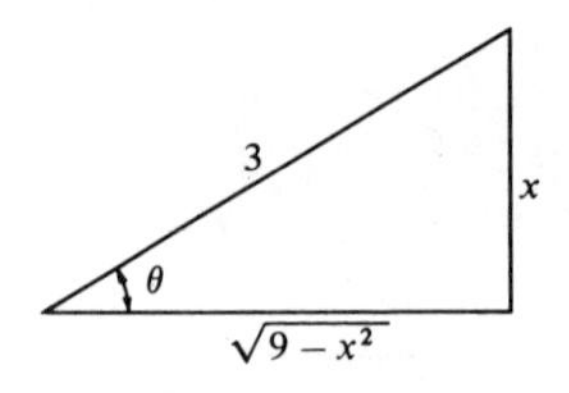

Figure 9.8.2 Example 2.

Upon substitution the integral becomes

$$\int_0^{3/2} \sqrt{9 - x^2}\, dx = \int_0^{\pi/6} (3 \cos \theta) 3 \cos \theta \, d\theta$$

(The new limits come from the relation $x = 3 \sin \theta$. When $x = 0$, we have $0 = 3 \sin \theta$; thus $\theta = 0$. When $x = \frac{3}{2}$, we have $\frac{3}{2} = 3 \sin \theta$ or $\sin \theta = \frac{1}{2}$; hence $\theta = \pi/6$.) Thus

$$\begin{aligned}
\int_0^{3/2} \sqrt{9 - x^2}\, dx &= 9 \int_0^{\pi/6} \cos^2 \theta \, d\theta \\
&= \frac{9}{2} \int_0^{\pi/6} (1 + \cos 2\theta)\, d\theta \qquad \text{(by Eq. 9.3.15)} \\
&= \frac{9}{2} \left[\theta + \frac{1}{2} \sin 2\theta \right]_0^{\pi/6} \\
&= \frac{9}{2} \left[\frac{\pi}{6} + \frac{1}{2} \sin \left(\frac{\pi}{3} \right) - 0 - 0 \right] \\
&= \frac{9}{2} \left[\frac{\pi}{6} + \frac{1}{2} \left(\frac{\sqrt{3}}{2} \right) \right] \\
&= \frac{9}{24} (2\pi + 3\sqrt{3}) = 4.3
\end{aligned}$$

PROBLEM 2. In Fig. 9.8.2 label the leg adjacent to angle θ by x, then solve the integral of Example 2. Is there an advantage to the labeling used in Example 2? If so, what is it?

(3) Integrand Involving $\sqrt{x^2 - a^2}$

EXAMPLE 3. Find $\int x^3 \sqrt{x^2 - 25}\, dx$.

Solution. The expression $\sqrt{x^2 - 25}$ suggests a right triangle with hypotenuse x and one leg 5 (see Fig. 9.8.3). From the triangle (Fig. 9.8.3), we have

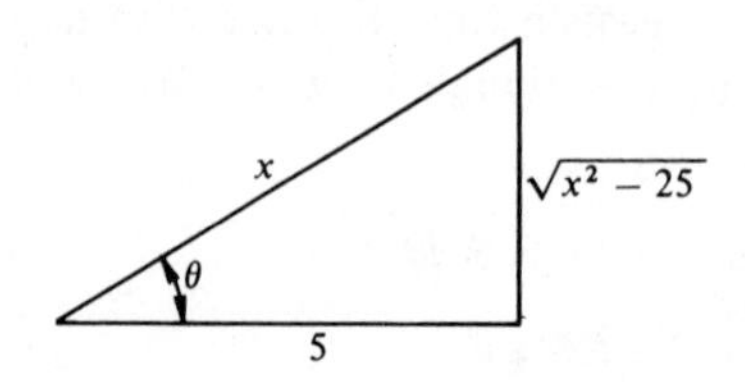

Figure 9.8.3 Example 3.

$$x = 5 \sec \theta$$
$$dx = 5 \sec \theta \tan \theta \, d\theta$$
$$\sqrt{x^2 - 25} = 5 \tan \theta$$

Upon substitution the integral becomes

$$\int x^3 \sqrt{x^2 - 25}\, dx = \int (5)^3 \sec^3 \theta \,(5 \tan \theta)(5 \sec \theta \tan \theta \, d\theta)$$
$$= (5)^5 \int \tan^2 \theta \sec^4 \theta \, d\theta$$
$$= (5)^5 \int \tan^2 \theta \,(1 + \tan^2 \theta) \sec^2 \theta \, d\theta \quad \text{(by Eq. 9.3.7)}$$
$$= (5)^5 \int (\tan^2 \theta + \tan^4 \theta) \sec^2 \theta \, d\theta$$

Since $\sec^2 \theta \, d\theta$ is the differential of $\tan \theta$, we let $u = \tan \theta$; then $du = \sec^2 \theta \, d\theta$, and upon substitution we obtain

$$(5)^5 \int (u^2 + u^4)\, du = (5)^5 \left[\frac{u^3}{3} + \frac{u^5}{5}\right] + C$$

Therefore,

$$\int x^3 \sqrt{x^2 - 25}\, dx = (5)^5 \left[\frac{1}{3} \tan^3 \theta + \frac{1}{5} \tan^5 \theta\right] + C$$

From the triangle (Fig. 9.8.3), we note that $\tan \theta = \sqrt{x^2 - 25}/5$, and finally we have

$$\int x^3 \sqrt{x^2 - 25}\, dx = (5)^5 \left[\frac{(x^2 - 25)^{3/2}}{(3)(5)^3} + \frac{(x^2 - 25)^{5/2}}{(5)(5)^5}\right] + C$$
$$= \frac{\sqrt{(x^2 - 25)^3}}{15}(50 + 3x^2) + C$$

PROBLEM 3. Find $\int 5t^3 \sqrt{4t^2 - 9}\, dt$.

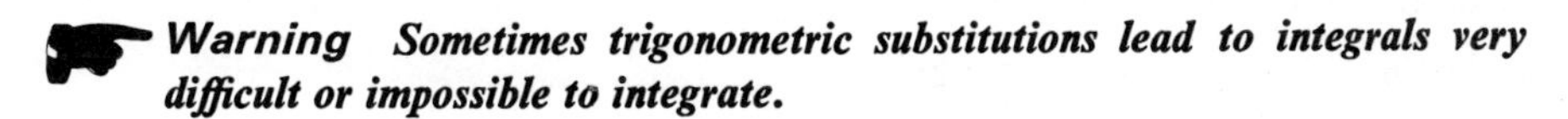

Warning ***Sometimes trigonometric substitutions lead to integrals very difficult or impossible to integrate.***

EXERCISES 9.8

Group A

1. Use trigonometric substitution to integrate each of the following.

(a) $\displaystyle\int \frac{dx}{\sqrt{9 + x^2}}$.

(b) $\displaystyle\int \frac{dx}{\sqrt{4 + 9x^2}}$.

(c) $\displaystyle\int_0^{2\sqrt{3}} \sqrt{16 - x^2}\, dx$.

(d) $\displaystyle\int_0^{3/2} \sqrt{9 - 4x^2}\, dx$.

(e) $\int \frac{dx}{\sqrt{x^2 - 25}}$.

(f) $\int \frac{x\,dx}{\sqrt{25 - x^2}}$.

(g) $\int \frac{dx}{(x^2 + 9)^{3/2}}$.

(h) $\int \frac{x\,dx}{(x^2 + 16)^{3/2}}$.

(i) $\int \frac{dx}{x^2\sqrt{x^2 + 1}}$.

(j) $\int_3^5 \frac{dx}{\sqrt{x^2 - 9}}$.

Group B

1. Use trigonometric substitution to integrate each of the following.

(a) $\int_0^1 \sqrt{2 - x^2}\,dx$.

(b) $\int \frac{dx}{\sqrt{1 + 3x^2}}$.

(c) $\int_{-\sqrt{154}/22}^{\sqrt{154}/22} \sqrt{7 - 11x^2}\,dx$.

(d) $\int \frac{dx}{\sqrt{\frac{1}{4}x^2 - 1}}$.

(e) $\int \frac{x + 1}{\sqrt{3 - 2x - x^2}}\,dx$. [*Hint*: complete the square (Sec. A.1) under the radical to obtain $3 - (x^2 + 2x) = 4 - (x^2 + 2x + 1) = 4 - (x + 1)^2$.]

(f) $\int_1^3 \frac{\sqrt{x^2 + 2x - 3}\,dx}{x + 1}$.

(g) $\int_a^{1.5a} \frac{x^2\,dx}{\sqrt{2ax - x^2}}$.

(h) $\int_{-2}^0 \frac{x\sqrt{4 - x^2}}{2 - x}\,dx$.

2. Show that the integral $\int \sqrt{25 + x^2}\,dx$ leads to $\int \sec^3 \theta\,d\theta$. This last integral can be evaluated by the method of integration by parts (see Sec. 11.3). How would you find $\int x\sqrt{25 + x^2}\,dx$?

9.9 APPLICATIONS

EXAMPLE 1. If the current I in an electric circuit is $I = 5 \sin 60t$, find the charge Q at any time t if $Q = 0$ when $t = 0$.

Solution. Since $Q = \int I\,dt$, we have

$$Q = \int 5 \sin 60t\,dt$$

or

$$Q = \frac{-5}{60} \cos 60t + C$$

Using the initial conditions of $Q = 0$ at $t = 0$, we have

$$0 = \frac{-5}{60}(1) + C$$

$$\therefore \quad C = \frac{5}{60} = \frac{1}{12}$$

and

$$Q = \frac{1}{12}(1 - \cos 60t)$$

PROBLEM 1. Find the charge Q at $t = 0.1$ s in an electric circuit if $I = 15 \cos 60t$ and $Q = 0$ at $t = 0$.

EXAMPLE 2. Find the area bounded by $y = \sin x$, $y = 0$, $x = 0$, and $x = \pi$.

Solution. Since the total area is the sum of all the little areas, we need to find dA and then integrate. The required area is shown in Fig. 9.9.1.

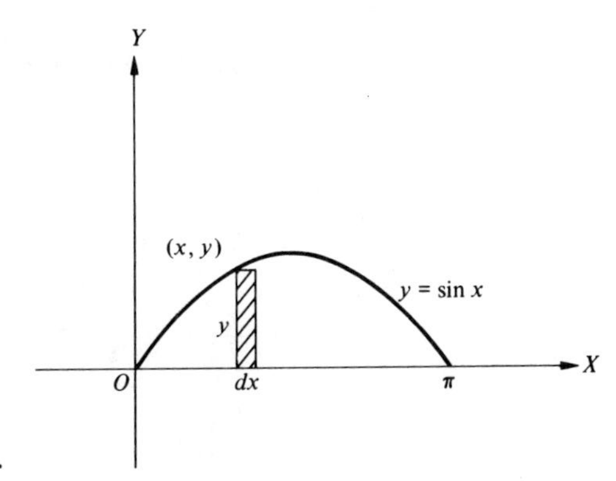

Figure 9.9.1 Area under sine curve.

$$dA = y\,dx = \sin x\,dx$$

Thus

$$\begin{aligned} A &= \int_0^\pi \sin x\,dx \\ &= [-\cos x]_0^\pi = [-\cos \pi + \cos 0] \\ &= -(-1) + 1 \end{aligned}$$

Therefore,

$$A = 2$$

PROBLEM 2. Find the area bounded by one arch of the cosine curve $y = \cos x$ and the X-axis. How does this compare with the answer in Example 2?

EXAMPLE 3. Find the volume of the solid of revolution formed by revolving the area bounded by $y = \cos x$, $x = 0$, $y = 0$, and $x = \pi/2$ about the X-axis.

Solution. The volume is shown in Fig. 9.9.2.

$$dV = \pi y^2\,dx = \pi \cos^2 x\,dx$$

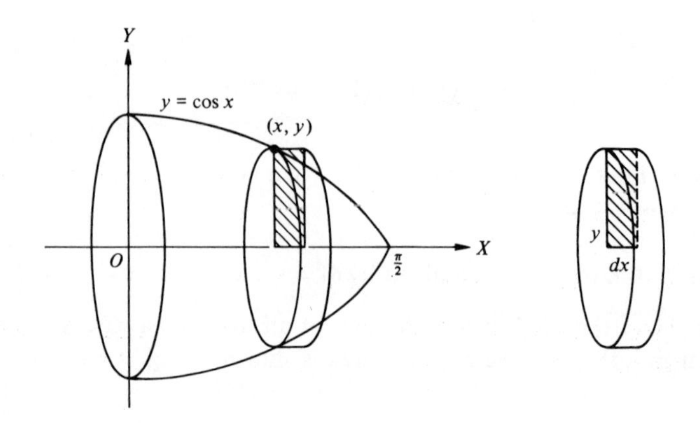

Figure 9.9.2 Example 3.

$$V = \pi \int_0^{\pi/2} \cos^2 x \, dx$$

$$= \frac{\pi}{2} \int_0^{\pi/2} (1 + \cos 2x) \, dx$$

$$= \frac{\pi}{2} \left[x + \frac{1}{2} \sin 2x \right]_0^{\pi/2}$$

$$= \frac{\pi}{2} \left[\frac{\pi}{2} + \frac{1}{2} \sin \pi - 0 - 0 \right]$$

$$\therefore \quad V = \frac{\pi^2}{4} = 2.47$$

PROBLEM 3. Find the volume of the solid of revolution formed by revolving the area bounded by $y = 2 \sin (x/2)$, $x = 0$, $y = 0$, and $x = \pi$, about the X-axis.

For problems involving average value and root-mean-square values, see Sec. 6.6.

EXERCISES 9.9

Group A

1. In each of the following, find the area indicated.

(a) Under $y = \cos x$ between $x = -\pi/2$, $x = \pi/2$, and $y = 0$.
(b) Under $y = \tan x$ between $x = 0$, $x = \pi/4$, and $y = 0$.
(c) Under $y = \sec (x/2)$ between $x = 0$, $x = \pi/2$, and $y = 0$.
(d) Under $y = \csc x$ between $x = \pi/6$, $x = \pi/2$, and $y = 0$.

2. If the current in an electric circuit is given by $I = 6 (\sin 2t + \cos 2t)$, find the charge Q at any time t if $Q = 0$ when $t = 0$.

3. Find the volume of the solids of revolution formed by revolving each of the regions about the designated axis.

(a) $y = 3 \cos x$, $x = 0$, $x = \pi/2$, and $y = 0$ about the X-axis.
(b) $y = \sin [x + (\pi/2)]$, $x = 0$, $x = \pi/2$, and $y = 0$ about the X-axis.
(c) $y = \tan x$, $x = 0$, $x = \pi/4$, and $y = 0$ about the X-axis.

4. The acceleration of a moving particle is given by $dx^2/dt^2 = 4 \cos t$. If $dx/dt = 0$ when $t = 0$ and $x = 5$ when $t = 0$, find expressions for the velocity v at any time t and the position x at any time t.

5. Find the area bounded by $y = 1/\sqrt{x^2 - 4}$, $x = \sqrt{5}$, $x = \sqrt{8}$, and $y = 0$.

Group B

1. Find the area bounded by the circle $x^2 + y^2 = 4$ using integration.

2. In each of the following, find the area indicated.

(a) Bounded by $y = x^3/\sqrt{x^2 - 6}$, $x = \sqrt{10}$, $x = \sqrt{15}$, and $y = 0$.
(b) Bounded by $y = \sqrt{16 - x^2}/x^2$, $x = 2$, $x = 2\sqrt{3}$, and $y = 0$.
(c) Bounded by $y = 1/(x^2\sqrt{x^2 - 9})$, $x = 4$, $x = 6$, and $y = 0$.

3. The velocity of a particle is given by $dx/dt = \sqrt{t^2 - 25}/t^2$. If $x = 0$ when $t = 5$, find an equation for the position x of the particle at any time t.

4. Find the average ordinate of $y = \sin x$ from $x = 0$ to $x = \pi$ (see Sec. 6.6).

5. The voltage in a certain electric circuit is given as $E = 120 \sin 60t$. Find the average voltage with respect to time as t changes from 0 to $\pi/60$.

6. If $I = \sin^2 t$, find the average value of I between $t = 0$ and $t = \pi$. (Use Eq. 9.3.14.)

7. If T is the time for one period, find I_{rms} for each of the following currents (see Sec. 6.6).

(a) $I = \sin t$.
(b) $I = 2 \sin \pi t$.
(c) $I = 120 \sin 60t$.
(d) $I = a \sin \omega t$.

8. Find the root-mean-square voltage for one period if

(a) $E = 12 \sin 2\pi t$.
(b) $E = E_0 \cos \omega t$.

9. Find the length of arc of the curve given by the parametric equations $x = 2 \cos t$, $y = 2 \sin t$ between the points where $t = 0$ and $t = \pi/2$.

10. Find the area of the surface obtained by revolving the arc of Exercise 9 about the X-axis (see Exercise 1 of Group C in Exercises 7.9).

SUMMARY OF IMPORTANT WORDS AND CONCEPTS

For each of the following, state in your own words your understanding of the statement or word.

1. (a) Trigonometric functions of real numbers. (b) Unit circle.

2. (a) Geometric representation of real numbers as arc lengths. (b) Relation between a real number u and an angle θ.

3. (a) Graphs of trigonometric functions. (b) Amplitude. (c) Period.

4. Derivatives of the trigonometric functions.

5. (a) Transient current. (b) Steady-state current. (c) Simple harmonic motion.

6. Integrals of the trigonometric functions.

7. Integration by trigonometric substitution.

Chapter 10

Inverse Trigonometric Functions

10.1 INVERSE TRIGONOMETRIC FUNCTIONS

In Sec. 8.4 we discussed inverse functions. We saw that for a function to have an inverse it must provide a one-to-one correspondence between the elements of its domain and the elements of its range.

The graph of $f(x) = \sin x$ (Fig. 10.1.1) shows that for each element x in the domain there corresponds only one image $f(x)$. The line l, parallel to the X-axis through any point in the range, intersects the curve in many places. Thus $\sin x$ does not establish a one-to-one correspondence between the elements of its range and the elements of its domain. Similar statements may be made about the other trigonometric functions. Thus the trigonometric functions, strictly speaking, do not have inverses. However, we get around this difficulty by restricting the domain so that there will be a one-to-one correspondence between the elements of the restricted domain and the range. Let us denote the trigonometric functions with this restricted domain by writing the first letters of the trigonometric functions as capital letters, Sin, Cos, Tan, Cot, Sec, and Csc (read "Cap-Sine," etc.).

We then write their inverse functions with the symbols Sin^{-1}, Cos^{-1}, Tan^{-1}, Cot^{-1}, Sec^{-1}, and Csc^{-1} (read "inverse Cap-Sine," etc.). These functions are called the *inverse trigonometric functions*. Another common symbol used to denote the inverse sine is *arc Sin* (read "arc Cap-Sine"). This notation emphasizes the relationship between a number in the range of the sine function and the arc length which is assigned to it. Similar expressions are used for the other inverse trigonometric functions.

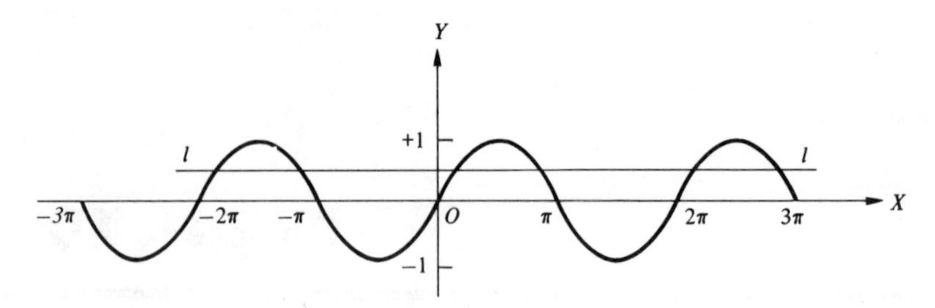

Figure 10.1.1 Graph of $f(x) = \sin x$.

In Table 10.1.1 we have listed the restricted domains of the trigonometric functions of real numbers or of angles measured in radians.

TABLE 10.1.1

Function	Function Value	Domain	Range
Sin	$y = \text{Sin } x$	$-\frac{\pi}{2} \leqq x \leqq \frac{\pi}{2}$	$-1 \leqq y \leqq 1$
Cos	$y = \text{Cos } x$	$0 \leqq x \leqq \pi$	$-1 \leqq y \leqq 1$
Tan	$y = \text{Tan } x$	$-\frac{\pi}{2} < x < \frac{\pi}{2}$	all real numbers
Cot	$y = \text{Cot } x$	$0 < x < \pi$	all real numbers
Sec	$y = \text{Sec } x$	$\begin{cases} 0 \leqq x < \frac{\pi}{2} \\ -\pi \leqq x < -\frac{\pi}{2} \end{cases}$	$y \geqq 1$ $y \leqq -1$
Csc	$y = \text{Csc } x$	$\begin{cases} 0 < x \leqq \frac{\pi}{2} \\ -\pi < x \leqq -\frac{\pi}{2} \end{cases}$	$y \geqq 1$ $y \leqq -1$

Note carefully that these "capital" functions are not the same as the trigonometric functions because their *domains are not the same* as the domains for the trigonometric functions, which are all real numbers (or angles).

In Table 10.1.2, we have listed the domain and range of the inverse trigonometric functions. There is no universal agreement among mathematicians as to the range of the inverse trigonometric functions. The listed ranges are quite common.[1]

If the domain of the trigonometric functions consists of angles, then the range of the inverse trigonometric functions consists of angles. That is, if $y = \text{Sin } x$, x an angle, then its inverse, $\text{Sin}^{-1} x$, is said to be "an angle whose Sine is x." If $y = \text{arc Sin } u$, u a real number, then "u is the number (arc) whose sine is y."

It follows from the definition of inverse functions that if $y = \text{Sin } \theta$, then $\theta = \text{Sin}^{-1} y$. Thus for $\theta = \text{Sin}^{-1} (\frac{1}{2})$, we may write $\text{Sin } \theta = \frac{1}{2}$.

[1]Some people call the *range* of Sin^{-1} the *principal values* of $\sin^{-1}$ (which is not a function in the modern sense) and similarly for the other inverse functions.

TABLE 10.1.2

Function	Function Value	Domain	Range
Sin^{-1}	$y = \text{Sin}^{-1}\, x$	$-1 \leqq x \leqq 1$	$-\frac{\pi}{2} \leqq y \leqq \frac{\pi}{2}$
Cos^{-1}	$y = \text{Cos}^{-1}\, x$	$-1 \leqq x \leqq 1$	$0 \leqq y \leqq \pi$
Tan^{-1}	$y = \text{Tan}^{-1}\, x$	all real numbers	$-\frac{\pi}{2} < y < \frac{\pi}{2}$
Cot^{-1}	$y = \text{Cot}^{-1}\, x$	all real numbers	$0 < y < \pi$
Sec^{-1}	$y = \text{Sec}^{-1}\, x$	$\begin{cases} x \geqq 1 \\ x \leqq -1 \end{cases}$	$0 \leqq y < \frac{\pi}{2}$; $-\pi \leqq y < -\frac{\pi}{2}$
Csc^{-1}	$y = \text{Csc}^{-1}\, x$	$\begin{cases} x \geqq 1 \\ x \leqq -1 \end{cases}$	$0 < y \leqq \frac{\pi}{2}$; $-\pi < y \leqq -\frac{\pi}{2}$

EXAMPLE 1. Find the exact value of $\text{Cos}^{-1}(-\frac{1}{2})$.

Solution. Let $\theta = \text{Cos}^{-1}(-\frac{1}{2})$. Then $\text{Cos}\,\theta = -\frac{1}{2}$. We must find a number (or an angle in radians) whose cosine is $-\frac{1}{2}$. There are many such numbers, for example, $(2\pi/3) \pm 2n\pi$ and $(4\pi/3) \pm 2n\pi$, n an integer.

However, the *range* of $\text{Cos}^{-1}\, x$ tells us that $0 \leqq \theta \leqq \pi$; hence

$$\theta = \frac{2\pi}{3} \qquad \text{or} \qquad \text{Cos}^{-1}\left(-\frac{1}{2}\right) = \frac{2\pi}{3}$$

PROBLEM 1. Find the exact value of arc Sin $(-\frac{1}{2})$.

EXAMPLE 2. Find cos [arc Tan $(-\frac{3}{4})$].

Solution. If we let $\theta = \text{arc Tan}\,(-\frac{3}{4})$, then we are to find the value of $\cos\theta$. $\theta = \text{arc Tan}\,(-\frac{3}{4})$ implies that $\text{Tan}\,\theta = -\frac{3}{4}$. The range of the inverse tangent tells us that θ must be a negative fourth-quadrant number (or angle). Hence from the sketch (Fig. 10.1.2) we have

$$\cos\theta = \tfrac{4}{5}$$

Thus

$$\cos\,[\text{arc Tan}\,(-\tfrac{3}{4})] = \tfrac{4}{5}$$

PROBLEM 2. Find sin [$\text{Cos}^{-1}(-\frac{1}{2})$].

The following examples show a few methods of obtaining some relations among inverse trigonometric functions.

EXAMPLE 3. Express $\text{Cos}^{-1}\, y$ in terms of an inverse sine.

Solution. Let $\theta = \text{Cos}^{-1}\, y$; then $y = \text{Cos}\,\theta$. From the fundamental identity $\text{Sin}^2\,\theta + \text{Cos}^2\,\theta = 1$, we have

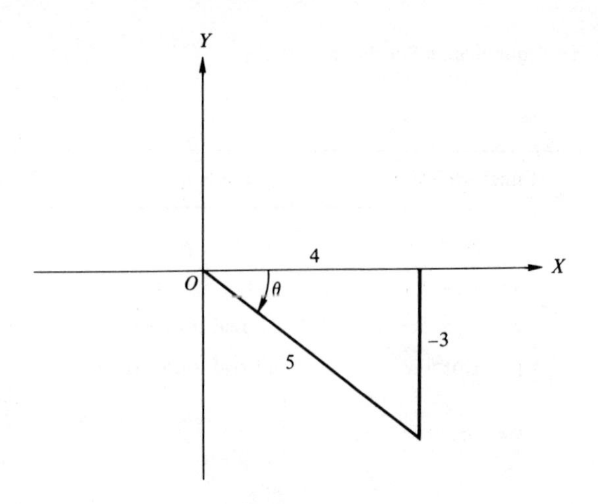

Figure 10.1.2 Example 2.

$$\text{Sin } \theta = \pm\sqrt{1 - \text{Cos}^2 \theta}$$

For Cos θ, $0 \leqq \theta \leqq \pi$; hence Sin $\theta > 0$. Thus

$$\text{Sin } \theta = \sqrt{1 - y^2}$$

or

$$\theta = \text{Sin}^{-1} \sqrt{1 - y^2}$$

Hence

$$\text{Cos}^{-1} y = \text{Sin}^{-1} \sqrt{1 - y^2}$$

PROBLEM 3. Express $\text{Cos}^{-1} y$ in terms of an inverse tangent.

EXAMPLE 4. Find the value of sin (2 arc Tan x) for $x > 0$.

Solution. Let $\theta = \text{arc Tan } x$; then $\text{Tan } \theta = x$. We wish to find $\sin 2\theta = 2 \sin \theta \cos \theta$. For $x > 0$, $0 < \theta < \pi/2$. (Why?) From Fig. 10.1.3, we find

$$\sin 2\theta = 2\left(\frac{x}{\sqrt{1 + x^2}}\right)\left(\frac{1}{\sqrt{1 + x^2}}\right)$$

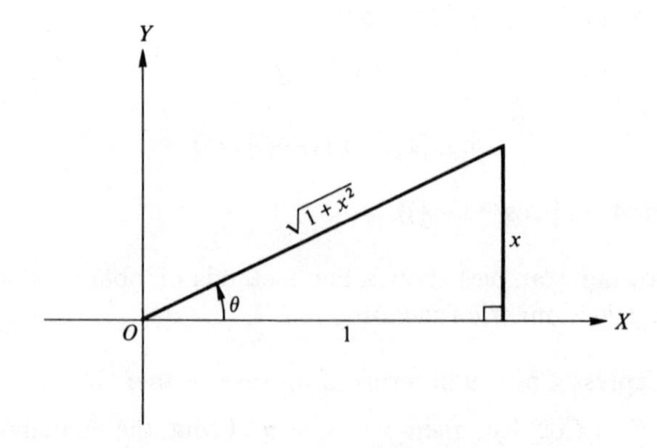

Figure 10.1.3 Example 4.

Thus

$$\sin(2 \text{ arc Tan } x) = \frac{2x}{1 + x^2}$$

PROBLEM 4. Find the value of $\cos(2 \text{ Tan}^{-1} x)$ for $x > 0$.

EXAMPLE 5. Find the value of $\tan [\text{Sin}^{-1} \frac{3}{5} - \text{Cos}^{-1} (-\frac{4}{5})]$.

Solution. Let $A = \text{Sin}^{-1} \frac{3}{5}$ and $B = \text{Cos}^{-1} (-\frac{4}{5})$. Then

$$\text{Sin } A = \tfrac{3}{5} \qquad \text{and} \qquad \text{Cos } B = -\tfrac{4}{5}$$

From the sketches (Fig. 10.1.4), we have

$$\tan A = \tfrac{3}{4} \qquad \text{and} \qquad \tan B = -\tfrac{3}{4}$$

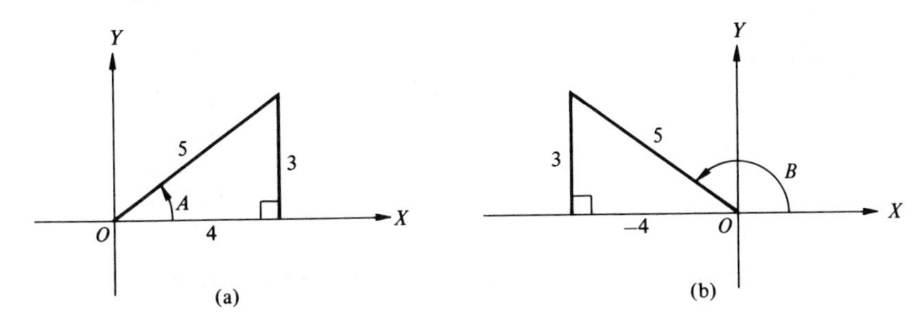

Figure 10.1.4 Example 5.

(Why is B not a third-quadrant number?) Since $\tan(A - B) \equiv (\tan A - \tan B)/(1 + \tan A \tan B)$, Eq. 9.3.11 we have

$$\tan\left[\text{Sin}^{-1} \frac{3}{5} - \text{Cos}^{-1}\left(-\frac{4}{5}\right)\right] = \frac{\frac{3}{4} - (-\frac{3}{4})}{1 + (\frac{3}{4})(-\frac{3}{4})} = \frac{24}{7} = 3.43$$

PROBLEM 5. Find the value of $\cos [\text{Cos}^{-1} (-\frac{3}{5}) + \text{Tan}^{-1} (-\frac{4}{3})]$.

Trigonometric equations must not be confused with inverse trigonometric functions. The inverse trigonometric functions are a class of functions with restricted domains and ranges. For example, $\tan u = -1$ has as its solution $u = (3\pi/4) \pm n\pi$, $n = 0, 1, 2, \ldots$; whereas $u = \text{Tan}^{-1}(-1)$ has as its only solution $u = -\pi/4$.

Warning ***You must remember that an inverse trigonometric equation has only one solution. A regular trigonometric equation has many solutions.***

PROBLEM 6. (a) Solve the equations $\cos u = -0.5$ and $u = \text{Cos}^{-1}(-0.5)$.

(b) Is it correct to solve the equation $\cos u = -0.5$ by writing $u = \text{Cos}^{-1}(-0.5)$? Explain your answer.

The graphs of the inverse trigonometric functions can be drawn quite easily by comparing them with the graphs of their inverses. Since $y = \text{Sin } x$ is a restricted branch of the sine curve which oscillates about the X-axis, and $y = \text{Sin}^{-1} x$ can be written as $x = \text{Sin } y$, it is clear that the graph of $y = \text{Sin}^{-1} x$ is a branch of the sine curve which oscillates about the Y-axis. The graphs of $y = \text{Sin}^{-1} x$, $y = \text{Cos}^{-1} x$, and $y = \text{Tan}^{-1} x$ are shown in Fig. 10.1.5. The dashed lines represent the graphs of $y = \text{Sin } x$, $y = \text{Cos } x$, and $y = \text{Tan } x$. (Note that the inverse trigonometric functions are not periodic.)

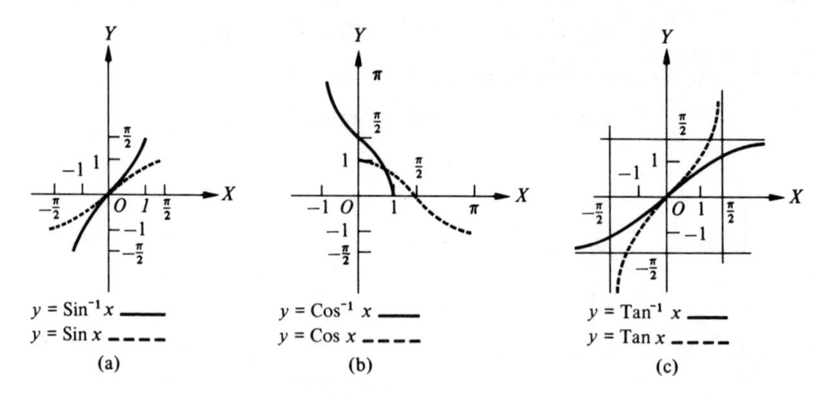

Figure 10.1.5 Graphs of the inverse trigonometric functions.

PROBLEM 7. Explain why the inverse trigonometric functions are not periodic.

EXERCISES 10.1

Group A

1. Make a table listing the restricted domains and the ranges of the trigonometric functions if their argument is an angle measured in degrees. Also state the inverses of these functions.

2. Find the exact value of the number (or angle measured in radians) expressed as an inverse function.

(a) $\text{Sin}^{-1} (\frac{1}{2})$. (b) $\text{Sin}^{-1} \left(-\frac{\sqrt{3}}{2}\right)$. (c) $\text{Cos}^{-1} 1$.

(d) $\text{arc Cos} (-\frac{1}{2})$. (e) $\text{arc Sin} (-1)$. (f) $\text{arc Sin} \frac{\sqrt{3}}{2}$.

(g) $\text{Tan}^{-1} (-1)$. (h) $\text{Tan}^{-1} \sqrt{3}$. (i) $\text{arc Tan} \left(-\frac{\sqrt{3}}{3}\right)$.

(j) $\text{arc Cot} (-\sqrt{3})$. (k) $\text{Cot}^{-1} (-1)$. (l) $\text{Cot}^{-1} 0$.

3. Use a calculator to find the value of the real number.

(a) $\text{Sin}^{-1} (0.7513)$. (b) $\text{arc Sin} (-0.1494)$. (c) $\text{arc Cos} (-0.3624)$.
(d) $\text{Cos}^{-1} (0.4265)$. (e) $\text{Tan}^{-1} (-1.398)$. (f) $\text{arc Tan} (0.7602)$.

4. Find the exact value of each of the following.

(a) $\sin[\text{Tan}^{-1}(\frac{5}{12})]$.
(b) $\sin[\text{Sin}^{-1}(\frac{1}{2})]$.
(c) $\text{Sin}^{-1}[\sin(\frac{1}{2})]$.
(d) $\cos[\text{arc Sin}(-\frac{8}{17})]$.
(e) $\cos[\text{arc Tan}(\frac{7}{24})]$.
(f) $\cos[2\,\text{Sin}^{-1}(\frac{15}{17})]$.
(g) $\tan[\text{arc Cos}(-\frac{3}{5})]$.
(h) $\tan[2\,\text{arc Tan}(\frac{4}{3})]$.
(i) $\tan[\text{Cot}^{-1}(-\frac{24}{7})]$.
(j) $\cot[\text{Sin}^{-1}(-\frac{5}{13})]$.
(k) $\cot[\text{arc Tan}(\frac{8}{15})]$.
(l) $\cot[\text{arc Cos}(-\frac{24}{25})]$.

5. Find the value of the following trigonometric functions.

(a) $\sin(\text{Sin}^{-1} 2x)$, $x > 0$.
(b) $\cos(\text{Cos}^{-1} u)$, $u > 0$.
(c) $\tan(\text{arc Sin}\, y)$, $y < 0$.
(d) $\sin(\text{arc Sin}\, v)$, $v < 0$.
(e) $\cos(2\,\text{Sin}^{-1} u)$, $u < 0$.
(f) $\cos(2\,\text{arc Cos}\, x)$, $x > 0$.
(g) $\sin(2\,\text{arc Tan}\, v)$, $v < 0$.
(h) $\sin(2\,\text{Cos}^{-1} y)$, $y < 0$.
(i) $\cos[\text{Sin}^{-1}\frac{4}{5} + \text{Cos}^{-1}\frac{12}{13}]$.
(j) $\tan[\text{Cos}^{-1}\frac{4}{5} + \text{Tan}^{-1}(-\frac{1}{2})]$.
(k) $\sin[\text{Sin}^{-1}\frac{12}{13} + \text{Cos}^{-1}(-\frac{3}{5})]$.
(l) $\tan[\text{arc Tan}(-\frac{8}{15}) - \text{arc Cos}(-\frac{3}{5})]$.
(m) $\sin[\text{arc Cot}(-\frac{24}{7}) - \text{arc Sin}(-\frac{5}{13})]$.
(n) $\cos[\text{arc Sin}\frac{15}{17} + \text{arc Cos}(-\frac{12}{13})]$.

6. (a) Express $y = \text{Cot}^{-1} x$ in terms of Tan^{-1}.
(b) Express $y = \text{Cos}^{-1} x$ terms of Sec^{-1}.
(c) Express $y = \text{Sec}^{-1} x$ in terms of Cos^{-1}.
(d) Express $y = \text{Cot}^{-1} 2x$ in terms of Tan^{-1}.
(e) Express $y = \text{Csc}^{-1} 3x$ in terms of Sin^{-1}.
(f) Express $y = \text{Sec}^{-1}(1/\sqrt{x})$ in terms of Cos^{-1}.

Group B

1. Prove each of the following relations.

(a) $2\,\text{Sin}^{-1} x = \text{Cos}^{-1}(1 - 2x^2)$.

(b) $\text{Sin}^{-1} y = \text{Cos}^{-1}\sqrt{1 - y^2}$.

(c) $\text{Sin}^{-1} u = \text{Tan}^{-1}\dfrac{u}{\sqrt{1 - u^2}}$, $1 - u^2 \neq 0$.

(d) $2\,\text{Cot}^{-1} x = \text{Csc}^{-1}\dfrac{1 + x^2}{2x}$, $x \neq 0$.

(e) $\text{Tan}^{-1} u + \text{Tan}^{-1} v = \text{Tan}^{-1}\dfrac{u + v}{1 - uv}$.

(f) $\sin(2\,\text{Sin}^{-1} u) = 2u\sqrt{1 - u^2}$.

(g) $2\,\text{Tan}^{-1} u = \text{Tan}^{-1}\dfrac{2u}{1 - u^2}$.

(h) $\text{Tan}^{-1}\dfrac{5}{7} + \text{Tan}^{-1}\dfrac{1}{6} = \dfrac{\pi}{4}$.

(i) $\text{Cos}^{-1}\dfrac{1}{2} + 2\,\text{Sin}^{-1}\dfrac{1}{2} = \dfrac{2\pi}{3}$.

(j) $\text{Cot}^{-1} 3 + \text{Csc}^{-1}\sqrt{5} = \dfrac{\pi}{4}$.

(k) $3\,\text{Sin}^{-1} u = \text{Sin}^{-1}(3u - 4u^3)$.

2. Sketch the graph of each of the following.

(a) $y = \text{Cot}^{-1} x$.
(b) $y = \text{Sec}^{-1} x$.
(c) $y = \text{Csc}^{-1} x$.
(d) $y = \text{Cot}\, x$.
(e) $y = \text{Sec}\, x$.
(f) $y = \text{Csc}\, x$.

3. In Fig. 10.1.5 draw a straight line through the origin making an angle of 45° with the positive X-axis. What conclusion, with respect to this line, can you draw concerning the graphs of the restricted trigonometric functions and their inverses?

4. Sketch the graph of each of the following.

(a) $y = \text{Sin}^{-1} 2x$. (b) $y = \text{arc Cos } 2x$. (c) $y = \text{Tan}^{-1} 2x$.
(d) $y = 2 \text{ Sin}^{-1} x$. (e) $y = \frac{1}{2} \text{Cos}^{-1} 2x$. (f) $y = 3 \text{ arc Tan } \frac{1}{2}x$.
(g) $y - \frac{\pi}{2} = \text{Sin}^{-1} x$. (h) $y = \frac{1}{3} \text{Cos}^{-1} 3x$. (i) $y = 2 \text{ Sin}^{-1} (\frac{1}{2}x)$.

5. State whether each of the following is sometimes, always, or never true.

(a) $\sin (\text{Sin}^{-1} x) = x, \ -1 \leqq x \leqq 1$. (b) $\text{Sin}^{-1} (\sin x) = x$.
(c) $\cos \left[\text{Cos}^{-1} \left(-\frac{\pi}{2}\right)\right] = -\frac{\pi}{2}$. (d) $\text{Cos}^{-1} \left(\cos \frac{3\pi}{2}\right) = \frac{\pi}{2}$.
(e) $\text{Sin}^{-1} (\sin 2\pi) = 0$. (f) $\cos (\text{Cos}^{-1} x) = x, \ -1 \leqq x \leqq 1$.

6. Explain clearly the difference (if any) (a) between $\sin \theta = \frac{1}{2}$ and $\theta = \text{Sin}^{-1} (\frac{1}{2})$ and (b) between $\sin \theta = -\frac{1}{2}$ and $\theta = \text{Sin}^{-1} (-\frac{1}{2})$.

7. Explain why $\text{Cos}^{-1} (\cos 2\pi/3) = 2\pi/3$, but $\text{Sin}^{-1} (\sin 2\pi/3) \neq 2\pi/3$.

10.2 DERIVATIVES OF INVERSE TRIGONOMETRIC FUNCTIONS

To find the derivative of $y = \text{Sin}^{-1} u$, where u is a differentiable function of x, we first write $u = \text{Sin } y$ and then take the derivative with respect to x. Thus for $y = \text{Sin}^{-1} u$,

$$u = \text{Sin } y$$

Then

$$\frac{du}{dx} = \text{Cos } y \frac{dy}{dx}$$

and

$$\frac{dy}{dx} = \frac{1}{\text{Cos } y} \frac{du}{dx}$$

Recalling that $\text{Cos}^2 y + \text{Sin}^2 y = 1$, we have by algebra

$$\text{Cos } y = \pm\sqrt{1 - \text{Sin}^2 y} = \pm\sqrt{1 - u^2}$$

But due to the range of $\text{Sin}^{-1} u$, we have $(-\pi/2) \leqq y \leqq \pi/2$ and thus $\text{Cos } y$ must be positive. Therefore,

$$\text{Cos } y = \sqrt{1 - u^2}$$

and

$$\boxed{\frac{d(\text{Sin}^{-1} u)}{dx} = \frac{1}{\sqrt{1 - u^2}} \frac{du}{dx}} \tag{10.2.1}$$

EXAMPLE 1. For $y = \text{Sin}^{-1} (3x)$, we have

$$\frac{dy}{dx} = \frac{1}{\sqrt{1 - 9x^2}}(3) = \frac{3}{\sqrt{1 - 9x^2}}$$

PROBLEM 1. Find dx/dt for $x = \mathrm{Sin}^{-1} 5t$.

The development of the derivatives of $\mathrm{Cos}^{-1} u$ and $\mathrm{Tan}^{-1} u$ follow a similar procedure. We shall list their formulas and leave their proofs as Exercises.

$$\frac{d(\mathrm{Cos}^{-1} u)}{dx} = -\frac{1}{\sqrt{1-u^2}}\frac{du}{dx} \qquad (10.2.2)$$

$$\frac{d(\mathrm{Tan}^{-1} u)}{dx} = \frac{1}{1+u^2}\frac{du}{dx} \qquad (10.2.3)$$

The derivatives of the other inverse trigonometric functions are not utilized much in applied problems, so we shall not discuss them. They can be avoided by using reciprocal functions. It is interesting to note that the derivatives of the inverse functions are algebraic functions. This implies that the integrals of certain algebraic functions are inverse trigonometric functions (see Sec. 10.3).

EXAMPLE 2. For $y = \mathrm{Tan}^{-1} (e^{\sin 3x})$, find dy/dx.

Solution. Here $u = e^{\sin 3x}$.

$$\frac{dy}{dx} = \frac{1}{1 + (e^{\sin 3x})^2}[e^{\sin 3x}(\cos 3x)(3)]$$

or

$$\frac{dy}{dx} = \frac{3(\cos 3x)e^{\sin 3x}}{1 + e^{2 \sin 3x}}$$

PROBLEM 2. Find dy/dx for $y = \mathrm{Cos}^{-1} (\ln \cos 2x)$.

Some applications of the derivatives of inverse trigonometric functions are included in Sec. 10.4.

The differentials of $\mathrm{Sin}^{-1} u$, $\mathrm{Cos}^{-1} u$, and $\mathrm{Tan}^{-1} u$ are

$$d(\mathrm{Sin}^{-1} u) = \frac{1}{\sqrt{1-u^2}} du \qquad (10.2.4)$$

$$d(\mathrm{Cos}^{-1} u) = -\frac{1}{\sqrt{1-u^2}} du \qquad (10.2.5)$$

$$d(\mathrm{Tan}^{-1} u) = \frac{1}{1+u^2} du \qquad (10.2.6)$$

For integration of inverse trigonometric functions, see Sec. 11.3.

EXERCISES 10.2

Group A

1. Find the derivative with respect to x of each of the following.

(a) $y = \text{Sin}^{-1}(2x)$. (b) $y = \text{Sin}^{-1}(3x + 1)$. (c) $y = \text{Cos}^{-1}(3x)$.
(d) $y = \text{Cos}^{-1}(5x - 7)$. (e) $y = \text{Tan}^{-1}(2x)$. (f) $y = \text{Tan}^{-1}(3x + 2)$.

2. Find the derivative with respect to x of each of the following.

(a) $y = \text{Sin}^{-1}\dfrac{3x - 1}{2}$. (b) $y = \text{Cos}^{-1}\dfrac{x^2}{x + 1}$.

(c) $y = \text{Tan}^{-1}\dfrac{3x - 2}{x + 6}$. (d) $y = \text{Sin}^{-1} x + \text{Cos}^{-1} x$.

Group B

1. Find the derivative with respect to x of each of the following.

(a) $y = \text{Sin}^{-1} x^2$. (b) $y = x^2 + \text{Cos}^{-1} x^2$.
(c) $y = \text{Tan}^{-1}(\ln x)$. (d) $y = \text{Sin}^{-1}(\cos x)$.
(e) $y = x\,\text{Sin}^{-1} x$. (f) $y = 3x\,\text{Cos}^{-1} 3x$.
(g) $y = \ln \text{Tan}^{-1} 2x$. (h) $y = e^{\text{Sin}^{-1} 3x}$.

2. Find the value of dy/dx at the given value of x.

(a) $y = x \text{ arc Sin } x,\ x = \dfrac{\sqrt{2}}{2}$. (b) $y = x\,\text{Cos}^{-1} x,\ x = \dfrac{-\sqrt{2}}{2}$.

(c) $y = \dfrac{\text{arc Tan } x}{x},\ x = 1$.

3. Find the radius of curvature of

(a) $y = \text{Sin}^{-1} x$ at $x = \frac{1}{2}$. (b) $y = x\,\text{Sin}^{-1} x$ at $x = 1$.

Group C

1. Prove each of the following formulas.

(a) $$\frac{d}{dx}(\text{Cos}^{-1} u) = -\frac{1}{\sqrt{1 - u^2}}\frac{du}{dx}. \qquad (10.2.2)$$

(b) $$\frac{d}{dx}(\text{Tan}^{-1} u) = \frac{1}{1 + u^2}\frac{du}{dx} \qquad (10.2.3)$$

(c) $$\frac{d}{dx}(\text{Cot}^{-1} u) = -\frac{1}{1 + u^2}\frac{du}{dx}. \qquad (10.2.7)$$

10.3 INTEGRALS THAT LEAD TO INVERSE TRIGONOMETRIC FUNCTIONS

If we apply Eq. 10.2.4 to $\text{Sin}^{-1}(u/a)$, where a is constant, we obtain

$$d\left(\text{Sin}^{-1}\frac{u}{a}\right) = \frac{1}{\sqrt{1 - (u^2/a^2)}}\frac{du}{a} = \frac{a}{\sqrt{a^2 - u^2}}\frac{du}{a} = \frac{du}{\sqrt{a^2 - u^2}}$$

Integrating this differential, we obtain

$$\boxed{\int \frac{du}{\sqrt{a^2 - u^2}} = \mathrm{Sin}^{-1}\frac{u}{a} + C} \qquad (10.3.1)$$

Applying Eq. 10.2.6 to $\mathrm{Tan}^{-1}(u/a)$, where a is constant, we obtain

$$d\left(\mathrm{Tan}^{-1}\frac{u}{a}\right) = \frac{1}{1 + (u^2/a^2)}\frac{du}{a} = \frac{a\,du}{a^2 + u^2}$$

Integrating this differential, we obtain

$$\boxed{\int \frac{du}{a^2 + u^2} = \frac{1}{a}\mathrm{Tan}^{-1}\frac{u}{a} + C} \qquad (10.3.2)$$

Note that we are thus able to express the integral of some algebraic functions in terms of some inverse trigonometric functions. This is an important use of inverse trigonometric functions.

EXAMPLE 1. Find $\int_{1/2}^{\sqrt{3}} dx/\sqrt{4 - x^2}$.

Solution. Apply Eq. 10.3.1. Let $a = 2$ and $u = x$; then $du = dx$. Thus

$$\begin{aligned}\int_{1/2}^{\sqrt{3}} \frac{dx}{\sqrt{4 - x^2}} &= \left[\mathrm{Sin}^{-1}\frac{x}{2}\right]_{1/2}^{\sqrt{3}} = \mathrm{Sin}^{-1}\frac{\sqrt{3}}{2} - \mathrm{Sin}^{-1}\frac{1}{4}\\ &= \mathrm{Sin}^{-1} 0.866 - \mathrm{Sin}^{-1} 0.25\\ &= 1.047 - 0.253\\ &= 0.794\end{aligned}$$

Note very carefully that Example 1 illustrates the importance of the trigonometric functions of real numbers. In the given integral the limits are real numbers and the answer must be a real number. We can interpret the integral as the area bounded by $y = 1/\sqrt{4 - x^2}$, $x = \frac{1}{2}$, $x = \sqrt{3}$, and $y = 0$, and areas are measured by real numbers.

PROBLEM 1. Find $\int_0^4 dx/\sqrt{25 - x^2}$.

EXAMPLE 2. Find $\int dx/(16x^2 + 9)$.

Solution. Here let $a = 3$ and $u = 4x$; then $du = 4\,dx$. Thus

$$\begin{aligned}\int \frac{dx}{16x^2 + 9} &= \int \frac{du/4}{u^2 + a^2} = \frac{1}{4}\int \frac{du}{u^2 + a^2}\\ &= \frac{1}{4}\left[\frac{1}{a}\mathrm{Tan}^{-1}\frac{u}{a}\right] + C \qquad \text{(by Eq. 10.3.2)}\\ &= \frac{1}{12}\mathrm{Tan}^{-1}\frac{4x}{3} + C\end{aligned}$$

PROBLEM 2. Find $\int dx/(36 + 16x^2)$.

EXAMPLE 3. Evaluate $\int_{-5}^{1} dx/(x^2 + 4x + 13)$.

Solution. This does not look anything like the integral formulas which we have studied so far. But suppose we "complete the square" (Sec. A.1) on the denominator and rewrite it as

$$x^2 + 4x + 13 = (x^2 + 4x + 4) + 9 = (x + 2)^2 + (3)^2$$

Then we have

$$\int_{-5}^{1} \frac{dx}{(x + 2)^2 + (3)^2}$$

Now let $a = 3$ and $u = x + 2$; then $du = dx$. Being careful to change the limits for the variable u, we obtain

$$\begin{aligned}\int_{-5}^{1} \frac{dx}{(x + 2)^2 + (3)^2} = \int_{-3}^{3} \frac{du}{u^2 + a^2} &= \left[\frac{1}{a}\,\mathrm{Tan}^{-1}\,\frac{u}{a}\right]_{-3}^{3} \\ &= \frac{1}{3}\left[\mathrm{Tan}^{-1}\left(\frac{3}{3}\right) - \mathrm{Tan}^{-1}\left(\frac{-3}{3}\right)\right] \\ &= \frac{1}{3}[\mathrm{Tan}^{-1}(1) - \mathrm{Tan}^{-1}(-1)] \\ &= \frac{1}{3}\left[\frac{\pi}{4} - \left(-\frac{\pi}{4}\right)\right] \\ &= \frac{\pi}{6} = 0.52\end{aligned}$$

PROBLEM 3. Evaluate $\int_0^2 dx/(x^2 - 2x + 5)$.

EXERCISES 10.3

Group A

1. Integrate each of the following.

(a) $\displaystyle\int \frac{dx}{\sqrt{9 - x^2}}$.

(b) $\displaystyle\int \frac{dx}{\sqrt{4 - 9x^2}}$.

(c) $\displaystyle\int \frac{dx}{x^2 + 4}$.

(d) $\displaystyle\int \frac{2\,dx}{4x^2 + 9}$.

(e) $\displaystyle\int \frac{dx}{\sqrt{3 - 2x^2}}$.

(f) $\displaystyle\int \frac{2\,dx}{3x^2 + 5}$.

(g) $\displaystyle\int \frac{dx}{(x + 2)^2 + 9}$.

(h) $\displaystyle\int \frac{dx}{\sqrt{4 - (x - 1)^2}}$.

(i) $\displaystyle\int \frac{dx}{x^2 + 4x + 20}$.

(j) $\displaystyle\int \frac{dx}{\sqrt{5 - x^2 + 4x}}$.

2. Evaluate each of the following.

(a) $\int_0^1 \frac{dx}{\sqrt{4-x^2}}$.

(b) $\int_{-1/4}^{1/4} \frac{3\,dx}{\sqrt{1-4x^2}}$.

(c) $\int_0^2 \frac{dx}{x^2+4}$.

(d) $\int_{-5/3}^{0} \frac{dx}{9x^2+25}$.

(e) $\int_{-1/2}^{\sqrt{3}/2} \frac{dx}{\sqrt{1-x^2}}$.

(f) $\int_{-2}^{2} \frac{-dx}{\sqrt{4-x^2}}$.

Group B

1. Find each of the following integrals.

(a) $\int \frac{dx}{(9-8x-x^2)^{1/2}}$.

(b) $\int \frac{-dx}{(16-6x-x^2)^{1/2}}$.

(c) $\int \frac{dx}{x^2+4x+6}$.

(d) $\int \frac{dx}{x^2+10x+50}$.

2. Evaluate each of the following.

(a) $\int_{-1}^{2} \frac{dx}{\sqrt{8-x^2-2x}}$.

(b) $\int_0^2 \frac{dx}{\sqrt{1-x^2+2x}}$.

(c) $\int_{-2}^{2} \frac{dx}{x^2+4x+8}$.

(d) $\int_0^2 \frac{dx}{x^2-2x+2}$.

Group C

1. Use the method of trigonometric substitution of Sec. 9.8 to derive Eqs. 10.3.1 and 10.3.2.

2. Use the method of trigonometric substitution to show that $\int dx/x\sqrt{x^2-a^2} = (1/a)\,\text{Sec}^{-1}\,(x/a) + C$.

3. Integrate each of the following.

(a) $\int_{2\sqrt{3}/3}^{2} \frac{dx}{x\sqrt{x^2-1}}$.

(b) $\int \frac{dx}{x\sqrt{4x^2-9}}$.

10.4 APPLICATIONS

The various applications of the calculus studied in previous chapters are used in this set of problems on the inverse trigonometric functions.

EXERCISES 10.4

Group A

1. Use differentials to approximate each of the following.

(a) $\text{Sin}^{-1}\left(\frac{1}{10}\right)$. (b) $\text{Cos}^{-1}\left(-\frac{1}{16}\right)$. (c) $\text{Tan}^{-1}(1.1)$. (d) $\text{Sin}^{-1}(0.9)$.

2. The base of an isosceles triangle is 6 ft. If the altitude is 4 ft and is increasing at the rate of $\frac{1}{6}$ ft/min, at what rate is the vertex angle changing?

3. Find equations of the lines tangent to and normal to the following curves at the given value of x.

(a) $y = \text{Sin}^{-1} x$, $x = 0$. (b) $y = \text{Tan}^{-1}$, $x = \sqrt{3}$.
(c) $y = \text{Cos}^{-1} x$, $x = -\frac{1}{2}$. (d) $y = \text{Cot}^{-1} x$, $x = \sqrt{3}$.
[*Hint*: Consider $y = \text{Tan}^{-1}(1/x)$].

4. Find $\bar{x}$ of the region bounded by $y = (16 - x^2)^{-1/2}$ and the lines $x = 0$, $x = 2$, and $y = 0$.

Group B

1. Find the centroid of the solid generated by rotating the region bounded by $y = (x^2 + 1)^{-1/2}$, $x = 0$, $x = 2$, and $y = 0$ about the X-axis.

2. Sketch each of the following curves.

(a) $y = x\,\text{Sin}^{-1}(x)$. (b) $y = x\,\text{Tan}^{-1}(x)$.

3. In the right triangle of Fig. 10.4.1, P is the midpoint of side BC whose length is $2x$ and θ is the angle BAP. Express θ as a function of x. Find the value of x that makes θ a maximum.

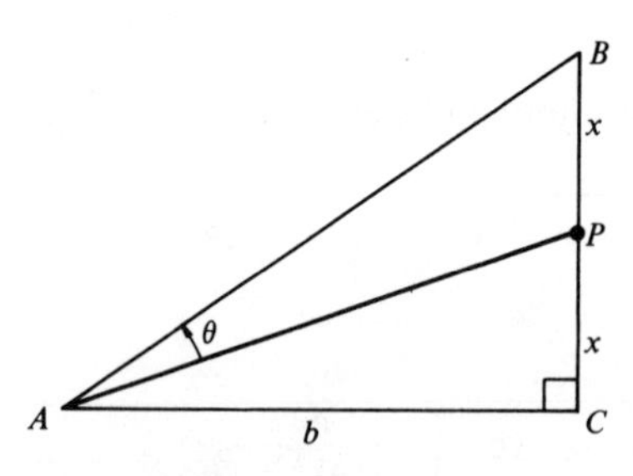

Figure 10.4.1 Exercise 3.

4. A woman approaches a bridge in her car while driving along the horizontal road at 60 mi/h. The bridge rises 500 ft above the level of the woman's eyes. How fast is the angle subtended by the bridge at the woman's eyes increasing when the woman is 1200 ft from the bridge?

5. The bottom of a 7-ft-high window is 9 ft above the level of an observer's eye. How far from the wall should he stand so that his angle of vision of the window will be a maximum?

6. Find the area bounded by the coordinate axes $y = 1/\sqrt{4 - x^2}$ and $x = 1$.

7. Find the area bounded by the X-axis, $y = 1/(x^2 + 4)$, $x = -2$, and $x = 2$.

8. A weather balloon that is initially 400 ft from an observer rises (straight up) at a speed of 100 ft/min. How fast is the angle of elevation changing when the balloon is 400 ft high?

9. The legs of a right triangle are 3 ft and 4 ft. Find the approximate change in the smaller angle when the shorter leg is increased by 0.1 ft.

10. An airplane is flying at a constant altitude of 400 ft at a speed of 300 ft/s. How fast must the angle of a direction indicator on the ground change when the plane is $400\sqrt{2}$ ft away and will pass over the indicator?

SUMMARY OF IMPORTANT WORDS AND CONCEPTS

For each of the following, state in your own words your understanding of the statement or word.

1. Restricted domain of trigonometric functions.

2. (a) Inverse trigonometric functions. (b) Domain. (c) Range.

3. Trigonometric equations.

4. Graphs of inverse trigonometric functions.

5. Derivatives of inverse trigonometric functions.

6. Integral of some algebraic functions in terms of inverse trigonometric functions.

Chapter 11

More Integration Methods

11.1 INTRODUCTION

We have seen that integration formulas may be obtained by reversing the process of differentiation. In this manner we have learned several formulas for integration. The derivatives of functions are relatively easy to obtain; however, we cannot integrate every function. For example, $\int \sin x^2\,dx$ cannot be integrated exactly. That is, there is no function which when differentiated will yield $\sin x^2$. Every time you perform a differentiation you have at hand a formula for an integration. As we have seen, many functions may be changed to an integrable function by a suitable substitution. In this chapter we shall study some more techniques to help us in finding integrals of some functions.

11.2 SUMMARY OF INTEGRATION FORMULAS

We now bring together the integral formulas we have developed in the previous chapters.

$$\int u^n\,du = \frac{u^{n+1}}{n+1} + C, \qquad n \neq -1 \tag{7.13.1}$$

$$\int \frac{du}{u} = \ln|u| + C \tag{8.8.1}$$

$$\int e^u\,du = e^u + C \tag{8.8.2}$$

$$\int \sin u\,du = -\cos u + C \tag{9.7.1}$$

$$\int \cos u \, du = \sin u + C \tag{9.7.2}$$

$$\int \sec^2 u \, du = \tan u + C \tag{9.7.3}$$

$$\int \csc^2 u \, du = -\cot u + C \tag{9.7.4}$$

$$\int \sec u \tan u \, du = \sec u + C \tag{9.7.5}$$

$$\int \csc u \cot u \, du = -\csc u + C \tag{9.7.6}$$

$$\int \tan u \, du = -\ln |\cos u| + C \tag{9.7.7}$$

$$\int \cot u \, du = \ln |\sin u| + C \tag{9.7.8}$$

$$\int \sec u \, du = \ln |\sec u + \tan u| + C \tag{9.7.9}$$

$$\int \csc u \, du = \ln |\csc u - \cot u| + C \tag{9.7.10}$$

$$\int \frac{du}{\sqrt{a^2 - u^2}} = \operatorname{Sin}^{-1} \frac{u}{a} + C \tag{10.3.1}$$

$$\int \frac{du}{a^2 + u^2} = \frac{1}{a} \operatorname{Tan}^{-1} \frac{u}{a} + C \tag{10.3.2}$$

The main trick in integrating is to spot the correct formula to use and to apply a suitable substitution. This, of course, becomes easier as more and more problems are done. There is no quick way always to find the proper formula. We have shown in examples certain methods of substitution. We now extend this a bit. Substitution is merely a method of changing the variable to make the integral exactly like one of the formulas.

In Sec. 7.13 we discussed $\int u^n \, du$, where we let u be the factor raised to a power and du was the other factor. Sometimes Eq. 7.13.1 can be used for other integrals.

EXAMPLE 1. Find $\int x\sqrt{x + 4} \, dx$.

Solution. Here we have a factor raised to a power, so we let $u = x + 4$. Now $du = dx$ is not the other factor ($x \, dx$), but we still will make a substitution and see what happens. Follow the steps carefully.

For $u = x + 4$, $du = dx$, and $x = u - 4$. Substitution yields

$$\begin{aligned} \int x\sqrt{x + 4} \, dx &= \int (u - 4)u^{1/2} \, du \\ &= \int (u^{3/2} - 4u^{1/2}) \, du \\ &= \tfrac{2}{5}u^{5/2} - 4(\tfrac{2}{3})u^{3/2} + C \qquad \text{(by Eq. 7.13.1)} \end{aligned}$$

For an indefinite integral (no limits on the integral) we must return to the original variable. Thus

$$\int x\sqrt{x+4}\,dx = \tfrac{2}{5}(x+4)^{5/2} - \tfrac{8}{3}(x+4)^{3/2} + C$$

which simplifies to

$$\tfrac{2}{15}(x+4)^{3/2}(3x-8) + C$$

PROBLEM 1. Find $\int 3x(x-2)^{1/3}\,dx$.

For integrals involving fractions, we always check to see if the numerator is the differential of the denominator. If it is, we use Eq. 8.8.1 $\int du/u$. If it is not, we try other methods.

EXAMPLE 2. Find $\int [(x+3)/(4x^2+25)]\,dx$.

Solution. The numerator is not the differential of the denominator. However, if we rewrite the integral as

$$\int \frac{x+3}{4x^2+25}\,dx = \int \frac{x}{4x^2+25}\,dx + \int \frac{3\,dx}{4x^2+25}$$

we then might spot the correct formulas. In the first integral let $u = 4x^2 + 25$; then $du = 8x\,dx$ and $x\,dx = \tfrac{1}{8}du$. Therefore

$$\int \frac{x\,dx}{4x^2+25} = \frac{1}{8}\int \frac{du}{u} = \frac{1}{8}\ln|u| + C_1 = \frac{1}{8}\ln|4x^2+25| + C_1$$

For the second integral let $v = 2x$; then $dv = 2dx$. Also, let $a = 5$. Then

$$\int \frac{3\,dx}{4x^2+25} = \frac{3}{2}\int \frac{dv}{v^2+a^2} = \frac{3}{2}\left(\frac{1}{a}\operatorname{Tan}^{-1}\frac{v}{a}\right) + C_2$$

and

$$\int \frac{3dx}{4x^2+25} = \frac{3}{10}\operatorname{Tan}^{-1}\frac{2x}{5} + C_2 \qquad \text{(by Eq. 10.3.2)}$$

Putting these together, we have

$$\int \frac{x+3}{4x^2+25}\,dx = \frac{1}{8}\ln|4x^2+25| + \frac{3}{10}\operatorname{Tan}^{-1}\frac{2x}{5} + C$$

where $C = C_1 + C_2$.

PROBLEM 2. Find $\int [(2x-5)/(9x^2+16)]\,dx$.

EXAMPLE 3. Find $\int dx/(x^2 - 6x + 13)$.

Solution. The denominator may be rewritten as the sum of two squares by the process of "completing the square" (Sec. A.1). Group the second- and first-degree terms of x and write

$$x^2 - 6x + 13 = (x^2 - 6x \quad\quad) + 13$$

Now make the terms in the parentheses a perfect square by adding the square of one half of the coefficient of x. We also must subtract this number from 13 to keep the value of the expression the same. Thus, since $\frac{1}{2}(-6) = -3$ and $(-3)^2 = 9$, we write

$$x^2 - 6x + 13 = (x^2 - 6x + 9) + 13 - 9$$

or

$$x^2 - 6x + 13 = (x - 3)^2 + 4$$

We now have

$$\int \frac{dx}{x^2 - 6x + 13} = \int \frac{dx}{(x-3)^2 + 2^2}$$

Let $a = 2$ and $u = x - 3$, then $du = dx$, and we have

$$\begin{aligned}\int \frac{dx}{x^2 - 6x + 13} &= \int \frac{dx}{(x-3)^2 + (2)^2} \\ &= \int \frac{du}{u^2 + a^2} \\ &= \frac{1}{a}\operatorname{Tan}^{-1}\frac{u}{a} + C \\ &= \frac{1}{2}\operatorname{Tan}^{-1}\frac{x-3}{2} + C\end{aligned}$$

PROBLEM 3. Find $\int dt/(t^2 + 4t + 20)$.

Sometimes the methods of both Examples 2 and 3 are necessary.

EXAMPLE 4. Find $\int [(2x + 1)/(x^2 - 6x + 13)]\, dx$.

Solution. First complete the square in the denominator:

$$\int \frac{2x+1}{x^2 - 6x + 13}\, dx = \int \frac{2x+1}{(x-3)^2 + (2)^2}\, dx$$

Now let $a = 2$ and $u = x - 3$; then $x = u + 3$ and $dx = du$, so

$$\int \frac{2x+1}{(x-3)^2 + (2)^2}\, dx = \int \frac{2(u+3)+1}{u^2 + a^2}\, dx = \int \frac{2u+7}{u^2 + a^2}\, du$$

Then follow Example 2 to obtain

$$\begin{aligned}\int \frac{2u+7}{u^2+a^2}\, du &= \int \frac{2u\, du}{u^2 + a^2} + 7\int \frac{du}{u^2 + a^2} \\ &= \ln|u^2 + a^2| + \frac{7}{a}\operatorname{Tan}^{-1}\frac{u}{a} + C\end{aligned}$$

Hence

$$\int \frac{2x+1}{x^2 - 6x + 3}\, dx = \ln|x^2 - 6x + 13| + \frac{7}{2}\operatorname{Tan}^{-1}\frac{x-3}{2} + C$$

PROBLEM 4. Find $\int [(3t - 1)/(t^2 + 4t + 20)]\, dt$.

Warning ***When using integral formulas you must make the given integral exactly like the formula.***

EXERCISES 11.2

Group A

1. Integrate each of the following.

(a) $\int x\sqrt{x^2+2}\,dx.$ (b) $\int (x^3-7)x^2\,dx.$

(c) $\int x^3\sqrt{x^4-16}\,dx.$ (d) $\int x\sqrt{25-x}\,dx.$

(e) $\int x\sqrt{25-x^2}\,dx.$ (f) $\int x^2\sqrt{9-4x}\,dx.$

(g) $\int x\sqrt{7-x^2}\,dx.$ (h) $\int x(25+x^2)^{3/2}\,dx.$

(i) $\int x\sqrt{25+x^2}\,dx.$ (j) $\int x(16+9x^2)^{3/2}\,dx.$

(k) $\int \frac{\sqrt{x^2-4}}{x}\,dx.$ (l) $\int \frac{\sqrt{9x^2-16}}{x}\,dx.$

(m) $\int (e^{3x}+4)^2 e^{3x}\,dx.$ (n) $\int (e^{-2x}-3)^3 e^{-2x}\,dx.$

(o) $\int \frac{x+4}{x^2+16}\,dx.$ (p) $\int \frac{3x-5}{\sqrt{25-x^2}}\,dx.$

(q) $\int \frac{2x-16}{9x^2+25}\,dx.$ (r) $\int \frac{dx}{x^2+4x+13}.$

(s) $\int \frac{3dx}{x^2-4x+8}.$ (t) $\int \frac{dx}{\sqrt{3-x^2+2x}}.$

(u) $\int \frac{dx}{\sqrt{24-x^2-2x}}.$ (v) $\int \frac{x+2}{x^2+4x+13}\,dx.$

(w) $\int \frac{x\,dx}{x^2+4x+13}.$ (x) $\int \frac{\sec^2 2x}{3+\tan 2x}\,dx.$

Group B

1. Integrate each of the following.

(a) $\int x\sqrt{25-16x^4}\,dx.$ (Would Sec. 9.8 help?)

(b) $\int \frac{\sqrt{25-16x^4}}{x}\,dx.$ (c) $\int x^3\sqrt{25+16x^4}\,dx.$

(d) $\int \frac{\sqrt{x^4-9}}{x}\,dx.$ (e) $\int \frac{dx}{x^2+3x+3}.$

(f) $\int \frac{3\,dx}{2x^2+x+1}.$

Group C

1. Verify each of the following by differentiation.

(a) $\int \frac{du}{u^2 - a^2} = \frac{1}{2a} \ln \left| \frac{u-a}{u+a} \right| + C.$

(b) $\int \frac{du}{\sqrt{u^2 + a^2}} = \ln |u + \sqrt{u^2 + a^2}| + C.$

(c) $\int \frac{du}{\sqrt{u^2 - a^2}} = \ln |u + \sqrt{u^2 - a^2}| + C.$

2. Make use of the formulas in Exercise 1 to find the value of each of the following.

(a) $\int \frac{dx}{x^2 - 25}$.

(b) $\int \frac{dx}{4x^2 - 9}$.

(c) $\int \frac{dx}{\sqrt{x^2 + 9}}$.

(d) $\int \frac{dx}{\sqrt{9x^2 + 25}}$.

(e) $\int \frac{dx}{\sqrt{x^2 - 9}}$.

(f) $\int \frac{dx}{\sqrt{4x^2 - 9}}$.

11.3 INTEGRATION BY PARTS

Another method of integration called *integration by parts* is a powerful method. It is developed from the formula for the differential of a product. Consider $d(uv) = u\,dv + v\,du$. If we integrate both sides, we have

$$uv = \int u\,dv + \int v\,du$$

or

$$\boxed{\int u\,dv = uv - \int v\,du} \qquad (11.3.1)$$

Note that in Eq. 11.3.1 the variables in the integrand and the differential became interchanged. The purpose in using integration by parts is to make the integral on the right simpler than the original integral.

EXAMPLE 1. Find $\int xe^x\,dx$.

Solution. This is not one of the forms previously studied. If we let

$$u = x \qquad \text{and} \qquad dv = e^x\,dx$$

we have

$$du = dx \qquad \text{and} \qquad v = e^x$$

(We shall use only one constant of integration in the answer, so we do not include a C here in the expression for v as it would only combine with the final constant.) Thus using the integration by parts Eq. 11.3.1, we obtain

$$\int xe^x\,dx = \underbrace{xe^x}_{uv} - \underbrace{\int e^x\,dx}_{\int v\,du}$$
$$= xe^x - e^x + C$$

and

$$\int xe^x\,dx = e^x(x-1) + C$$

PROBLEM 1. Find $\int x \ln x\,dx$.

The trick in using integration by parts is in making the choice of u and dv. It is desirable to choose dv such that v is easily found.

EXAMPLE 2. Evaluate $\int_0^{\pi/2} x \sin x\,dx$.

Solution. Let $u = x$, $dv = \sin x\,dx$,

$$du = dx \qquad \text{and} \qquad v = -\cos x$$

Then

$$\int x \sin x\,dx = \underbrace{-x\cos x}_{uv} - \underbrace{\int (-\cos x)\,dx}_{\int v\,du}$$
$$= -x\cos x + \int \cos x\,dx$$
$$= -x\cos x + \sin x + C$$

Hence

$$\int_0^{\pi/2} x\sin x\,dx = [-x\cos x + \sin x]_0^{\pi/2}$$
$$= \left[\left(-\frac{\pi}{2}\right)\cos\left(\frac{\pi}{2}\right) + \sin\left(\frac{\pi}{2}\right)\right] - [0+0]$$
$$= 1$$

If in Example 2 we had made the choice

$$u = \sin x \qquad \text{and} \qquad dv = x\,dx$$
$$du = \cos x\,dx \qquad \text{and} \qquad v = \frac{x^2}{2}$$

we would have arrived at

$$\int x\sin x\,dx = \underbrace{\frac{x^2}{2}\sin x}_{uv} - \underbrace{\int \frac{x^2}{2}\cos x\,dx}_{\int v\,du}$$

which makes the new integral worse than the original because of the x^2 in the new integral.

PROBLEM 2. Evaluate $\int_0^{\pi} x \cos x \, dx$.

Some integrals require one or more applications of integration by parts.

EXAMPLE 3. Find $\int e^x \sin x \, dx$.

Solution. Let $u = e^x$, $dv = \sin x \, dx$,

$$du = e^x \, dx \qquad \text{and} \qquad v = -\cos x$$

Then

$$\int e^x \sin x \, dx = \underbrace{-e^x \cos x}_{uv} - \underbrace{\int (-\cos x)e^x \, dx}_{\int v \, du}$$

$$= -e^x \cos x + \int e^x \cos x \, dx$$

The new integral does not look any simpler than the original, nor does it look any worse. So let us now apply integration by parts on $\int e^x \cos x \, dx$ and see what happens. Let

$$u = e^x \qquad \text{and} \qquad dv = \cos x \, dx$$

$$du = e^x \, dx \qquad \text{and} \qquad v = \sin x$$

Then

$$\int e^x \cos x \, dx = \underbrace{e^x \sin x}_{uv} - \underbrace{\int e^x \sin x \, dx}_{\int v \, du}$$

This gives us the original integral but of opposite sign. Thus we may write

$$\int e^x \sin x \, dx = -e^x \cos x + \left(e^x \sin x - \int e^x \sin x \, dx\right) + C_1$$

or

$$2\int e^x \sin x \, dx = -e^x \cos x + e^x \sin x + C_1$$

and

$$\int e^x \sin x \, dx = \tfrac{1}{2}e^x(\sin x - \cos x) + C$$

PROBLEM 3. Evaluate $\int_0^{\pi/2} e^t \cos t \, dt$.

EXAMPLE 4. Find $\int \operatorname{Sin}^{-1} x \, dx$.

Solution. Let $u = \operatorname{Sin}^{-1} x$, $dv = dx$; $du = (1/\sqrt{1 - x^2}) \, dx$, and $v = x$. Then

$$\int \operatorname{Sin}^{-1} x \, dx = x \operatorname{Sin}^{-1} x - \int \frac{x}{\sqrt{1 - x^2}} \, dx$$

In the new integral let $z = 1 - x^2$; then $dz = -2x\,dx$. We then find

$$\int \frac{x\,dx}{\sqrt{1-x^2}} = -\frac{1}{2}\int z^{-1/2}\,dz = -z^{1/2} + C$$
$$= -\sqrt{1-x^2} + C$$

Thus

$$\int \mathrm{Sin}^{-1}\,x\,dx = x\,\mathrm{Sin}^{-1}\,x + \sqrt{1-x^2} + C$$

PROBLEM 4. Find $\int \mathrm{Cos}^{-1}\,x\,dx$.

Some integrals may be integrated by more than one method. It is hoped that through familiarity with the various methods you will be able to choose the easiest one.

EXERCISES 11.3

Group A

1. Integrate each of the following.

(a) $\int xe^{2x}\,dx$.

(b) $\int 2xe^{-x}\,dx$.

(c) $\int x\cos 2x\,dx$.

(d) $\int t\sin 3t\,dt$.

(e) $\int e^{-t}\cos t\,dt$.

(f) $\int e^{2x}\sin x\,dx$.

(g) $\int \mathrm{Sin}^{-1}\,2x\,dx$.

(h) $\int \mathrm{Tan}^{-1}\,x\,dx$.

(i) $\int x^2\ln x\,dx$.

(j) $\int \ln x\,dx$.

2. Evaluate $\int_0^{\pi/6} x\sin 2x\,dx$.

3. Evaluate $\int_0^1 te^{-3t}\,dt$.

Group B

1. Integrate each of the following.

(a) $\int t^2e^t\,dt$. (*Hint*: Apply Eq. 11.3.1 twice.)

(b) $\int e^{ax}\sin bx\,dx$.

(c) $\int e^{ax}\cos bx\,dx$.

(d) $\int x\sin^2 3x\,dx$.

(e) $\int \sec^3 x\, dx$. $\left[\textit{Hint}: \int \sec^3 x\, dx = \int \sec x(\sec^2 x\, dx).\right.$ Let $dv = \sec^2 x\, dx$. Also, $\left.\int \sec x \tan^2 x\, dx = \int (\sec^3 x - \sec x)\, dx.\right]$

2. Evaluate each of the following.

(a) $\int_0^{\pi} x^2 \sin x\, dx$.

(b) $\int_0^{\pi/2} t^2 \sin 3t\, dt$.

(c) $\int_0^1 (\ln \cos x) \sin x\, dx$.

(d) $\int_0^1 x \operatorname{Tan}^{-1} x\, dx$.

(e) $\int_0^2 x\sqrt{1 + x}\, dx$.

(f) $\int \sqrt{1 + x^2}\, dx$. [*Hint*: Use trigonometric substitution and then see Exercise 1(e).]

3. Find the volume formed by rotating about the Y-axis the area bounded by $y = \sin x$, $y = 0$, $x = 0$, and $x = \pi$. Use the shell method.

4. The current in an electric circuit is $I = e^{-3t} \sin 2t$. Find the charge Q at $t = 1$ if $Q = 0$ at $t = 0$.

5. Find the average value of the current $I = e^{-t} \sin t$ for $t = 0$ to $t = 3$.

6. Find the x-coordinate of the centroid of the area bounded by $y = xe^{-x}$, $y = 0$, and $x = 2$.

Group C

1. Show that $\int x^n e^{ax}\, dx = (1/a)x^n e^{ax} - (n/a) \int x^{n-1} e^{ax}\, dx$. Use this *reduction formula* twice to find $\int x^2 e^{3x}\, dx$.

2. Consider the integral $\int x^4 \sin x\, dx$. Write x^4 and its derivatives in one column and $\sin x$ and its integrals in a second column, continuing until you reach a row in which the derivative equals zero. Thus

Differentiate		*Integrate*
x^4	$+$	$\sin x$
$4x^3$	$-$	$-\cos x$
$12x^2$	$+$	$-\sin x$
$24x$	$-$	$\cos x$
24	$+$	$\sin x$
0		$-\cos x$

Multiply the factors at the ends of a given arrow and the sign of the arrow, giving the term the indicated algebraic sign, and add the resulting terms. Thus $\int x^4 \sin x\, dx = -x^4 \cos x + 4x^3 \sin x + 12x^2 \cos x - 24x \sin x - 24 \cos x + C$. (a) Find $\int x^2 \sin x\, dx$ using the process just described. (b) Check the answer to part (a) by using integration by parts. This method may be used to integrate any product of a power of x by a function which can be integrated successively. (c) Find each of the following integrals: (1) $\int x^4 e^{2x}\, dx$; (2) $\int x^5 \cos x\, dx$.

11.4 PARTIAL FRACTIONS

A fraction whose numerator and denominator are polynomials is called a *rational fraction.* Rational fractions can sometimes be integrated after expressing them as a sum of partial fractions. It may be shown in algebra that any polynomial with real coefficients can be expressed as a product of linear and/or quadratic factors with real coefficients. We shall illustrate the method of partial fractions wherein the numerator is of lower degree than the denominator. (If it is not, division can be performed which will yield a polynomial plus a fraction whose numerator is of lower degree than the denominator.)

Type I. Linear Factors Nonrepeated

EXAMPLE 1. Integrate $\int [(x + 5)/(x^2 + x - 2)]\, dx$.

Solution. Factor the denominator to obtain

$$x^2 + x - 2 = (x + 2)(x - 1)$$

Then we assume that

$$\frac{x + 5}{(x + 2)(x - 1)} = \frac{A}{x + 2} + \frac{B}{x - 1} \tag{11.4.1}$$

where A and B are constants to be determined. Multiply both sides of Eq. 11.4.1 by $(x + 2)(x - 1)$ to obtain

$$(x + 5) = A(x - 1) + B(x + 2) \tag{11.4.2}$$

Equation 11.4.2 is true for all values of x. If we choose $x = -2$ (to make the factor after B zero), then we have

$$-2 + 5 = A(-2 - 1) + B(-2 + 2)$$
$$3 = -3A + B(0)$$
$$\therefore A = -1$$

Now let $x = 1$ (to make the factor after A zero):

$$1 + 5 = (-1)(1 - 1) + B(1 + 2)$$
$$6 = 0 + 3B$$
$$\therefore B = 2$$

Hence we may write

$$\frac{x + 5}{x^2 + x - 2} = \frac{-1}{x + 2} + \frac{2}{x - 1}$$

and

$$\int \frac{x + 5}{x^2 + x - 2}\, dx = \int \frac{-dx}{x + 2} + \int \frac{2dx}{x - 1}$$
$$= -\ln|x + 2| + 2\ln|x - 1| + C$$
$$= \ln\left|\frac{(x - 1)^2}{x + 2}\right| + C$$

PROBLEM 1. Find $\int [(-x + 8)/(x^2 - x - 2)]\, dx$.

Type II. Quadratic Factors Nonrepeated

EXAMPLE 2. Integrate $\int [(x^2 + x - 1)/(x^3 - 2x^2 + x - 2)]\, dx$.

Solution. Factor the denominator. Thus

$$x^3 - 2x^2 + x - 2 = (x^2 + 1)(x - 2)$$

Then

$$\frac{x^2 + x - 1}{(x - 2)(x^2 + 1)} = \frac{A}{x - 2} + \frac{Bx + C}{x^2 + 1} \tag{11.4.3}$$

where A, B, and C are constants to be determined. Note carefully that for the linear denominator the numerator is a constant. For the quadratic denominator the numerator is a general linear expression. Now clear Eq. 11.4.3 of fractions to obtain

$$x^2 + x - 1 = A(x^2 + 1) + (Bx + C)(x - 2) \tag{11.4.4}$$

For $x = 2$,

$$(2)^2 + 2 - 1 = A(5) + (2B + C)(0)$$

$$\therefore A = 1$$

There is no real value of x which makes the factor following A equal to zero, so just choose some values of x easy to work with. For $x = 1$,

$$(1)^2 + 1 - 1 = (1)(1 + 1) + (B + C)(-1)$$

$$\therefore B + C = 1 \tag{11.4.5}$$

For $x = -1$,

$$(-1)^2 + (-1) - 1 = (1)(1 + 1) + (-B + C)(-3)$$

$$\therefore -B + C = 1 \tag{11.4.6}$$

We now solve Eqs. 11.4.5 and 11.4.6 simultaneously:

$$\begin{array}{r} B + C = 1 \\ -B + C = 1 \\ \hline 2C = 2 \\ \therefore C = 1 \end{array}$$

and

$$B = 0$$

Hence

$$\frac{x^2 + x - 1}{(x - 2)(x^2 + 1)} = \frac{1}{x - 2} + \frac{1}{x^2 + 1}$$

and

$$\int \frac{x^2 + x - 1}{x^3 - 2x^2 + x - 2}\, dx = \int \frac{dx}{x - 2} + \int \frac{dx}{x^2 + 1}$$

$$= \ln|x - 2| + \text{Tan}^{-1} x + C$$

Warning ***The numerator is a constant for a linear denominator. The numerator is of the form $Ax + B$ for a nonfactorable quadratic denominator.***

PROBLEM 2. Find $\int [(-3x^2 + 2x - 14)/(x^2 + 4)(x - 1)]\, dx$.

There are two other types: *linear factors repeated* and *quadratic factors repeated.* They will not bc discussed here (see Exercises 1 and 2 of Group C).

EXERCISES 11.4

Group A

1. Integrate each of the following.

(a) $\int \frac{3x + 3}{x^2 + x - 2}\, dx.$

(b) $\int \frac{-x - 4}{x^2 - x - 2}\, dx.$

(c) $\int \frac{x + 3}{x^2 + 3x + 2}\, dx.$

(d) $\int \frac{2x^2 - 10x - 18}{(x + 1)(x - 2)(x + 3)}\, dx.$

(e) $\int \frac{-2x^2 - 2x + 6}{(x - 1)(x + 1)(x - 2)}\, dx.$

(f) $\int \frac{3x^2 - 2x + 1}{(x - 1)(x^2 + 1)}\, dx.$

(g) $\int \frac{-x^2 + x}{(x + 1)(x^2 + 1)}\, dx.$

(h) $\int \frac{x^2 - x + 3}{x^3 - 2x^2 + x - 2}\, dx.$

(i) $\int \frac{5x^2 + 2x + 22}{(x - 1)(4x^2 + 25)}\, dx.$

(j) $\int \frac{x^2 - 2x + 1}{(x^2 + 1)(x + 1)}\, dx.$

Group B

1. Integrate each of the following.

(a) $\int \frac{x^2 + 11x + 14}{(x + 3)(x^2 - 4)}\, dx.$ Factors

(b) $\int \frac{x^2 - 2x - 2}{(x - 1)(x^2 + x + 1)}\, dx.$

(c) $\int \frac{x^2 + 2x}{(x + 1)(x^2 + x + 1)}\, dx.$?

(d) $\int \frac{5x^2 - 6x + 4}{(x - 1)(x^2 + 2)}\, dx.$

(e) $\int \frac{3x + 7}{(x + 1)(x + 2)(x + 3)}\, dx.$

2. Evaluate each of the following.

(a) $\int_2^3 \frac{3 - x}{x^3 + 4x^2 + 3x}\, dx.$ take x out Factors

(b) $\int_0^1 \frac{3x^2 + 7x}{(x + 1)(x + 2)(x + 3)}\, dx.$

(c) $\int_1^3 \frac{2 - x^2}{x^3 + 3x^2 + 2x}\, dx.$

(d) $\int_0^2 \frac{4x^2 + 5x}{(x + 1)(x^2 - x - 6)}\, dx.$

(e) $\int_0^2 \frac{x^2 + x + 5}{(x + 1)(x^2 + 4)}\, dx.$

3. The time t to create an amount x of a substance by a chemical reaction is given by

$$\int_0^x \frac{du}{(a-u)(b-u)} = k\int_0^t dv. \text{ Find } t = f(x).$$

Group C

1. Linear factors of the type $(x-a)^r$ are called *repeated linear factors*. For each repeated linear factor we write r fractions such as

$$\frac{A_1}{x-a} + \frac{A_2}{(x-a)^2} + \ldots + \frac{A_r}{(x-a)^r}$$

Thus

$$\frac{x+1}{(x-1)(x-2)^2} = \frac{A}{x-1} + \frac{B}{x-2} + \frac{C}{(x-2)^2}$$

Find (a) $\int \{(-x^2+3x-1)/[(x-1)(x-1)^2]\}\,dx$ and (b) $\int \{(x^2+5x+2)/[(x^2+2x+1)(x-1)]\}\,dx$.

2. Quadratic factors of the type $(x^2+ax+b)^r$ are called *quadratic repeated factors*. For each repeated quadratic factor we write r fractions such as

$$\frac{A_1x+B_1}{x^2+ax+b} + \frac{A_2x+B_2}{(x^2+ax+b)^2} + \ldots + \frac{A_rx+B_r}{(x^2+ax+b)^r}$$

Find

$$\int \frac{x^4-x^3+x^2+x-1}{(x+2)(x^2+1)^2}\,dx$$

11.5 USE OF TABLE OF INTEGRALS AND APPROXIMATE INTEGRATION

As we have seen, integration depends on recognition of forms, certain reductions, and substitutions. Tables of integrals have been formed (some are very extensive) that help in integrating. A table of integrals is given as Table II in the Appendix. To use such a table it is frequently necessary to make a substitution that will change the given integral to the exact form listed.

You should look at this table of integrals very carefully to note how the integrals are grouped and the type of integrals it contains.

EXAMPLE 1. Find $\int x\sqrt{3+5x}\,dx$.

Solution. In Table II, we note that formula 25 seems to fit. Let $u = x$, $du = dx$, $a = 3$, and $b = 5$, Then

$$\int x\sqrt{3+5x}\,dx = \int u\sqrt{a+bu}\,du$$

$$= \frac{2(3bu-2a)}{15b^2}(a+bu)^{3/2} + C$$

$$= \frac{2(15x-6)}{375}(3+5x)^{3/2} + C$$

PROBLEM 1. Find $\int 3x^2\sqrt{5-2x}\,dx$.

***Warning** Remember that a constant factor can be placed before the integral and after a substitution the integral must be exactly like the one in the table of integrals.*

EXAMPLE 2. Find $\int [x\,dx/(16+25x^4)]$.

Solution. Let $u = 5x^2$, $du = 10x\,dx$, and $a = 4$. Then by formula 16 we have

$$\int \frac{x\,dx}{16+25x^4} = \frac{1}{10}\int \frac{du}{a^2+u^2} = \frac{1}{10}\frac{1}{a}\,\mathrm{Tan}^{-1}\frac{u}{a} + C$$

$$= \frac{1}{40}\,\mathrm{Tan}^{-1}\frac{5x^2}{4} + C$$

PROBLEM 2. Find $\int 5x\sqrt{16x^4-9}\,dx$.

EXAMPLE 3. Evaluate $\int_{\pi/6}^{\pi/3} \cos^3 2x\,dx$.

Solution. Let $u = 2x$ then $du = 2dx$, and

$$\int_{\pi/6}^{\pi/3} \cos^3 2x\,dx = \frac{1}{2}\int_{\pi/3}^{2\pi/3} \cos^3 u\,du \qquad \text{(note the change in the limits)}$$

$$= \frac{1}{2}\left[\sin u - \frac{1}{3}\sin^3 u\right]_{\pi/3}^{2\pi/3} \qquad \text{(by formula 86)}$$

$$= \frac{1}{2}\left[\frac{\sqrt{3}}{2} - \frac{1}{3}\left(\frac{\sqrt{3}}{2}\right)^3\right]$$

$$- \frac{1}{2}\left[\frac{\sqrt{3}}{2} - \frac{1}{3}\left(\frac{\sqrt{3}}{2}\right)^3\right] = 0$$

PROBLEM 3. Evaluate $\int_0^{\pi/2} \sin^3 2x\,dx$.

EXAMPLE 4. Find Q if $I = e^{-3t}\sin 8t$.

Solution. Since $Q = \int I\,dt$, we have

$$Q = \int e^{-3t}\sin 8t\,dt$$

Using formula 113, we have

$$Q = \frac{e^{-3t}(-3\sin 8t - 8\cos 8t)}{73} + C$$

PROBLEM 4. Find Q at $t = 0.1$ if $I = e^{-2t}\cos 3t$ and $Q = 0$ at $t = 0$.

Sometimes there are integrals that cannot be integrated exactly by any means. However, the value of all definite integrals may be approximated by several methods, one of which is the trapezoidal rule—Eq. 5.2.1.

EXAMPLE 5. Find an approximate value of $\int_0^{\pi/2} \sqrt{\cos x}\, dx$ by using the trapezoidal rule.

Solution. Refer to Eq. 5.2.1. For convenience let $n = 4$. Then

$$\Delta x = \frac{(\pi/2) - 0}{4} = \frac{\pi}{8}$$

Let $f(x) = \sqrt{\cos x}$.

$$x_0 = 0 \qquad f(x_0) = \sqrt{\cos 0} = 1 \qquad f(x_0) = 1.00$$

$$x_1 = \frac{\pi}{8} \qquad f(x_1) = \sqrt{\cos\left(\frac{\pi}{8}\right)} = 0.96 \qquad 2f(x) = 1.92$$

$$x_2 = \frac{\pi}{4} \qquad f(x_2) = \sqrt{\cos\left(\frac{\pi}{4}\right)} = 0.84 \qquad 2f(x_2) = 1.68$$

$$x_3 = \frac{3\pi}{8} \qquad f(x_3) = \sqrt{\cos\left(\frac{3\pi}{8}\right)} = 0.62 \qquad 2f(x_3) = 1.24$$

$$x_4 = \frac{\pi}{2} \qquad f(x_4) = \sqrt{\cos\left(\frac{\pi}{2}\right)} = 0 \qquad f(x_4) = 0.00$$

$$\text{Sum} = 5.84$$

Thus

$$\int_0^{\pi/2} \sqrt{\cos x}\, dx = \frac{\pi/8}{2}[5.84] = 1.15 \quad \text{(approximately)}$$

Other methods include Simpson's rule (see Exercise 9 of Group C in Exercises 12.4) and infinite series (see Chap. 17).

PROBLEM 5. Use a calculator and the trapezoidal rule to find an approximate value of $\int_0^2 \sin (x^2)\, dx$. Let $n = 4$.

EXERCISES 11.5

Group A

1. Use appropriate formulas in the table of integrals to find each of the following integrals.

(a) $\int \frac{dx}{4 - x^2}$.

(b) $\int \frac{dx}{8 - 4x^2}$.

(c) $\int \frac{dx}{\sqrt{x^2 - 9}}$.

(d) $\int \frac{dx}{x^2(3 + 2x)}$.

(e) $\int \sin^2 4x\, dx$.

(f) $\int \frac{dx}{x\sqrt{3 + x}}$.

(g) $\int \tan^2 3x\, dx$.

(h) $\int \sec^4 5x\, dx$.

(i) $\displaystyle\int \frac{4x^2}{(16 - 4x^2)^{3/2}}\,dx.$

(j) $\displaystyle\int x^2\sqrt{x^2 - 16}\,dx.$

2. Use the trapezoidal rule to find an approximate value for each of the following.

(a) $\displaystyle\int_0^{\pi/2} \sqrt{\sin x}\,dx,\ n = 4.$

(b) $\displaystyle\int_0^1 e^{x^2}\,dx,\ n = 5.$

(c) $\displaystyle\int_1^4 \ln x^2\,dx,\ n = 6.$

(d) $\displaystyle\int_0^{-2} \sqrt{\frac{64 - 9x^2}{16 - x^2}}\,dx,\ n = 2.$

Group B

1. Use appropriate formulas in the table of integrals to find each of the following integrals.

(a) $\displaystyle\int \frac{dx}{(3 - 4x^2)^{3/2}}.$

(b) $\displaystyle\int \frac{\sqrt{3x^2 + 10}}{3x^2}\,dx.$

(c) $\displaystyle\int 10x^4 e^{2x}\,dx.$

(d) $\displaystyle\int e^{3x}\sin 2x\,dx.$

(e) $\displaystyle\int \frac{5x^2}{(x^2 + 3)^{3/2}}\,dx.$

(f) $\displaystyle\int \frac{dx}{4x^2 + 5x + 1}.$

(g) $\displaystyle\int \frac{x\,dx}{4x^2 + 5x + 1}.$

(h) $\displaystyle\int x^4 \ln 2x\,dx.$

(i) $\displaystyle\int \sin 3x \cos 2x\,dx.$

(j) $\displaystyle\int \frac{dx}{3 + 2\cos x}.$

(k) $\displaystyle\int x^2 \sin 3x\,dx.$

(l) $\displaystyle\int x^3 e^{2x}\,dx.$

(m) $\displaystyle\int_0^{\pi/2} \sin^4 x\,dx.$

(n) $\displaystyle\int_0^{\pi/2} \sin^3 x \cos^4 x\,dx.$

(o) $\displaystyle\int_0^{\pi/6} \cos^4 \theta\,d\theta.$

(p) $\displaystyle\int_0^1 \sqrt{1 + 4x^2}\,dx$ (see Example 1 of Sec. 7.9).

2. The velocity of an object after falling s ft is given by $v = \sqrt{v_0^2 + 2gs}$, where v_0 is the initial velocity. If the body falls from rest, find the average velocity with respect to s during (a) the first 50 ft; (b) the first 100 ft.

3. The integral $2L\displaystyle\int_0^h \sqrt{a^2 - (y - a)^2}\,dy$ occurs in finding the volume of liquid in a horizontal right circular cylindrical tank of radius a and length L, where h is the depth of the liquid. Evaluate the integral. (*Hint*: Let $u = y - a$.) Show that its value is (a) 0 for $h = 0$; (b) $\pi a^2 L/2$ for $h = a$; (c) $\pi a^2 L$ for $h = 2a$. (d) Explain how the evaluation of this integral requires the trigonometric functions of numbers.

4. Evaluate $\displaystyle\int_{-2}^{3} x^2\sqrt{25 - x^2}\,dx$. Explain why trigonometric functions of numbers are necessary for this problem.

11.6 IMPROPER INTEGRALS

Sometimes it is necessary to use infinity for one or both of the limits of a definite integral, or sometimes the integrand does not exist within the bounds of the limits. These integrals are then called *improper integrals.* For an integral with upper limit infinity and lower limit a real number we define

$$\int_a^\infty f(x)\,dx = \lim_{b\to\infty} \int_a^b f(x)\,dx$$

whenever the latter integral exists.

EXAMPLE 1. Find the charge Q if $I = e^{-t} \sin t$ as t increases without bound and $Q = 0$ when $t = 0$.

Solution

$$\begin{aligned}
Q &= \int I\,dt \\
&= \int_0^\infty e^{-t} \sin t\,dt = \lim_{b\to\infty} \int_0^b e^{-t} \sin t\,dt \\
&= \lim_{b\to\infty} \left[\frac{e^{-t}(\sin t - \cos t)}{2}\right]_0^b \\
&= \lim_{b\to\infty} \left[\frac{e^{-b}}{2}(\sin b - \cos b)\right] - \frac{e^{-0}}{2}(\sin 0 - \cos 0) \\
&= \lim_{b\to\infty} \left[\frac{\sin b - \cos b}{2e^b}\right] - \frac{1}{2}(0 - 1)
\end{aligned}$$

As $b \longrightarrow \infty$, $\sin b$ and $\cos b$ are bounded by $+1$ and -1, e^b will increase without bound, and in the limit the expression in brackets will become zero. Hence

$$Q = \tfrac{1}{2} = 0.5$$

PROBLEM 1. Find the charge Q if $I = e^{-2t} \cos 10t$ as t increases without bound and $Q = 0$ at $t = 0$.

EXAMPLE 2. Find the area bounded by $y = 1/x^3$, $x = 1$, and the X-axis.

Solution. See Fig. 11.6.1 for the required area.

$$\begin{aligned}
A &= \int dA \\
dA &= y\,dx = \frac{1}{x^3}\,dx \\
A &= \int_1^\infty \frac{dx}{x^3} = \lim_{b\to\infty} \int_1^b x^{-3}\,dx \\
&= \lim_{b\to\infty} \left[\frac{x^{-2}}{-2}\right]_1^b = \lim_{b\to\infty} \left[\frac{1}{-2b^2} + \frac{1}{2}\right] \\
\therefore A &= \frac{1}{2} = 0.5
\end{aligned}$$

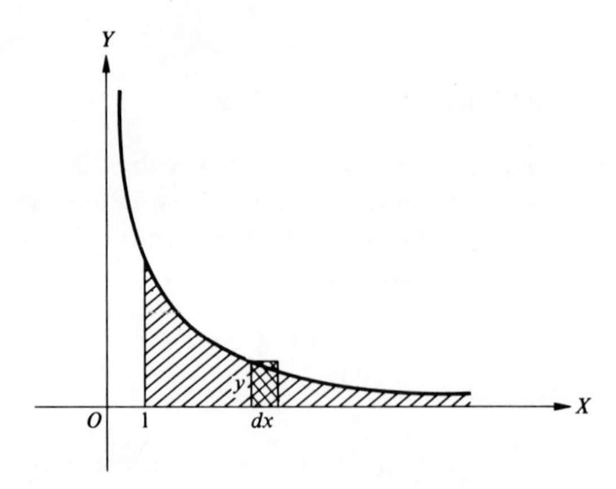

Figure 11.6.1 Area under $y = 1/x^3$.

Note that even though the area is not bounded on the right, the area does exist.

PROBLEM 2. Find the volume of revolution formed by revolving about the X-axis the area bounded by $y = 1/x^{3/2}$, $x = 1$, and $y = 0$.

EXAMPLE 3. Find the area bounded by $y = 1/x$, $x = 1$, and the X-axis.

Solution. See Fig. 11.6.2 for the required area.

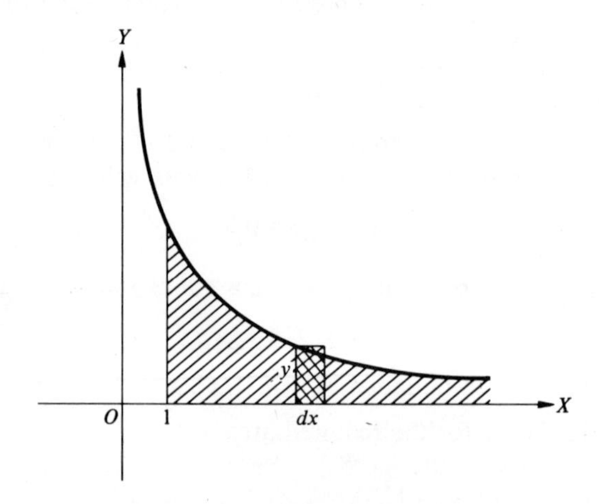

Figure 11.6.2 Area under $y = 1/x$.

$$A = \int dA$$

$$dA = y\,dx$$

$$A = \int_1^\infty \frac{1}{x}\,dx = \lim_{b\to\infty} \int_1^b \frac{dx}{x}$$

$$= \lim_{b\to\infty} [\ln|x|]_1^b = \lim_{b\to\infty} [\ln b - \ln 1]$$

As $b \longrightarrow \infty$, $\ln b \longrightarrow \infty$ and A increases without bound. Therefore, A does not exist.

Warning ***Even though the areas in Figs. 11.6.1 and 11.6.2 look about the same, you cannot conclude as to their existence without evaluating the improper integrals.***

PROBLEM 3. Find the area bounded by $y = 1/\sqrt{x}$, $x = 1$, and $y = 0$.

For the case where the integrand does not exist within the bounds of the limits, we again use a limit process to examine the improper integral. This case is illustrated in Example 4.

EXAMPLE 4. Evaluate $\int_1^3 dx/(x-3)^2$.

Solution. Since $f(x) = 1/(x-3)^2$ does not exist at $x = 3$, we use the following method:

$$\int_1^3 \frac{dx}{(x-3)^2} = \lim_{h \to 0} \int_1^{3-h} \frac{dx}{(x-3)^2}$$

$$= \lim_{h \to 0} \left[-\frac{1}{(x-3)} \right]_1^{3-h}$$

$$= \lim_{h \to 0} \left[\frac{-1}{-h} - \frac{1}{-2} \right]$$

As $h \longrightarrow 0$, the expression $1/h$ does not exist; therefore, $\int_1^3 dx/(x-3)^2$ does not exist.

PROBLEM 4. Evaluate $\int_2^4 dx/\sqrt{16 - x^2}$.

EXERCISES 11.6

Group A

1. In each of the following, find the value of the improper integral if it exists.

(a) $\int_1^\infty \frac{1}{x^2}\,dx$.

(b) $\int_1^\infty \frac{1}{x^4}\,dx$.

(c) $\int_1^\infty \frac{1}{x^p}\,dx,\ p > 1$.

(d) $\int_{-1}^2 \frac{dx}{(2-x)^2}$.

(e) $\int_0^3 \frac{dx}{\sqrt{9-x^2}}$.

(f) $\int_{-2}^0 \frac{dx}{(x+1)^{3/2}}$.

Group B

1. In each of the following, find the value of the improper integral if it exists.

(a) $\int_0^1 \frac{1}{x}\,dx.$

(b) $\int_0^1 \frac{1}{\sqrt{x}}\,dx.$

(c) $\int_0^1 \frac{1}{x^p}\,dx,\ p < 1.$

(d) $\int_1^\infty \frac{1}{x^p}\,dx,\ p < 1.$

(e) $\int_0^\infty \frac{dx}{x^2+1}.$

(f) $\int_{-\infty}^0 xe^x\,dx.$ (*Hint*: $\lim_{t\to\infty} te^{-t} = 0.$)

(g) $\int_{-4}^0 \frac{dx}{\sqrt{16-x^2}}.$

(h) $\int_0^3 \frac{dx}{\sqrt{3x-x^2}}.$

(i) $\int_{-1}^\infty \frac{dx}{(2+x)^2}.$

(j) $\int_{-\infty}^\infty x^2 e^{-x^3}\,dx.$

2. Find the area, if it exists, to the right of the Y-axis, above the X-axis and below $y = xe^{-x^2}$.

3. Find the volume formed by revolving about the X-axis the area bounded by $y = 1/x$, $x = 1$, and $y = 0$. Explain why this volume exists but the area involved, as in Example 3, does not exist. (*Note*: The Mathematical Association of America has an interesting series of films on integrals. This exercise is described in the film "Infinite Acres.")

Group C

1. (a) Let $F(t) = t$. Define a new function $f(s) = \int_0^\infty e^{-st}t\,dt$. For what values of s is $f(s)$ defined and what is the formula for $f(s)$ for these values of s? (*Hint*: Consider s as a constant and integrate with respect to t, also $\lim_{b\to\infty} b/e^b = 0$.) (b) Let $F(t)$ be a continuous function that satisfies $|F(t)| < Ke^{at}$, as $t \longrightarrow \infty$, for some real constant K and for some real constant a. Show that $f(s) = \int_0^\infty e^{-st}F(t)\,dt$ exists for suitable values of s. The function $f(s)$ is called the *Laplace transform* of the function $F(t)$. It has far-reaching applications in solving differential equations (see Sec. 14.8).

2. Evaluate $\int_1^5 dx/(x-3)^2$. Since the integrand is not defined at $x = 3$, this is an improper integral. Consider

$$\lim_{h\to 0}\left[\int_1^{3-h} \frac{dx}{(x-3)^2} + \int_{3+h}^5 \frac{dx}{(x-3)^2}\right].$$

SUMMARY OF IMPORTANT WORDS AND CONCEPTS

For each of the following, state in your own words your understanding of the statement or word.

1. Integral of a function raised to a power.

2. Integration by substitution.

3. Integration by parts.

4. (a) Partial fractions. (b) Linear factors nonrepeated. (c) Quadratic factors nonrepeated.

5. Use of table of integrals.

6. Approximate integration.

7. Improper integrals.

Chapter 12

Conic Sections

12.1 INTRODUCTION

Up to this point we have studied derivatives and integrals of algebraic, exponential, logarithmic, trigonometric, and inverse trigonometric functions. These five functions are usually called the *elementary functions.* At this point we would like to shift our emphasis to some very special equations whose graphs are called *conic sections* or just *conics.* They are the *circle,* the *parabola,* the *ellipse,* and the *hyperbola.* They are called conic sections because each of them can be obtained by the intersection of a plane and a cone (see Fig. 12.1.1). We shall develop the equations for each of these conics and it is important for you to be able to recognize the conics from their equations.

The equation of each conic is a special form of the *general quadratic equation*

$$Ax^2 + Bxy + Cy^2 + Dx + Ey + F = 0 \tag{12.1.1}$$

where A, B, C, D, E, and F are constants and at least one of A, B, or C is not zero.

12.2 SYMMETRY

The property of symmetry is sometimes helpful in drawing graphs because when one part is drawn its corresponding symmetric part may be drawn quickly. There are three steps to follow to test for symmetry. If in an equation the same equation is obtained when

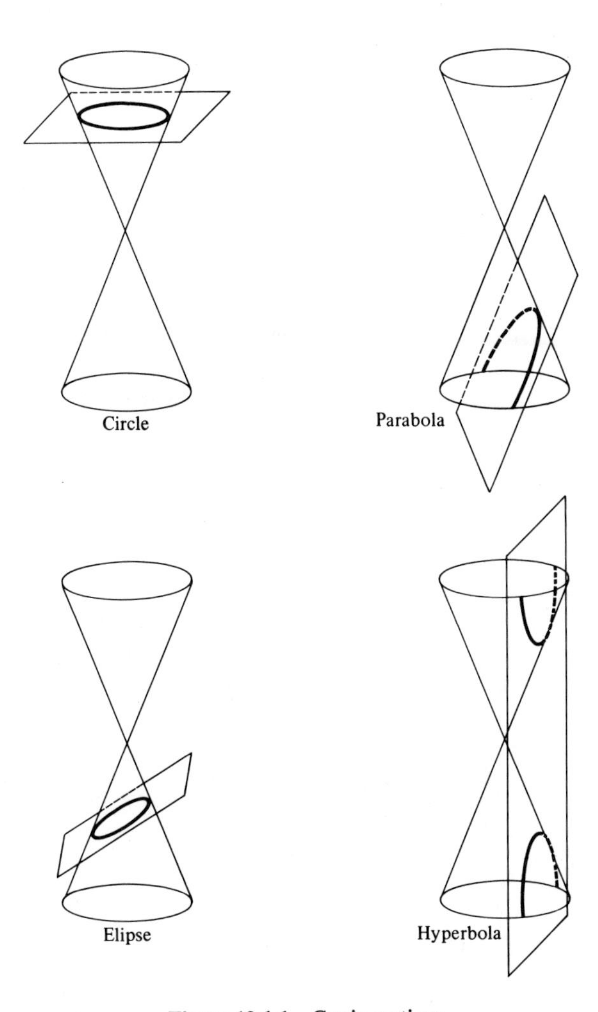

Figure 12.1.1 Conic sections.

1. x is replaced by $-x$, the graph is symmetric with respect to the Y-axis. [This means that the right-hand side of the curve is a reflection of the left-hand side, and vice versa (see Fig. 12.2.1).]
2. y is replaced by $-y$, the graph is symmetric with respect to the X-axis. [The upper half of the curve is a reflection of the lower half and vice versa (see Fig. 12.2.2).]
3. x is replaced by $-x$ and y is replaced by $-y$, the graph is symmetric with respect to the origin (see Fig. 12.2.3).

EXAMPLE 1. Discuss the symmetry of (a) $x^2 + y = 7$, (b) $x + y^2 = 7$, and (c) $xy = 2$.

Solution. For (a) $x^2 + y = 7$, we have:

1. For $x = -x$, $(-x)^2 + y = 7$ or $x^2 + y = 7$, therefore symmetric with respect to the Y-axis.

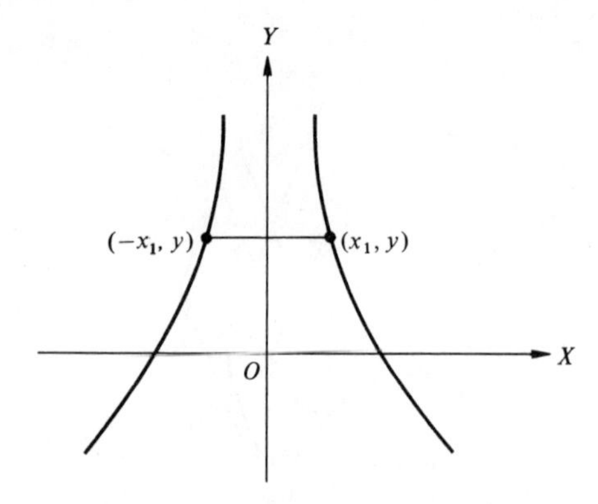

Figure 12.2.1 Symmetry with respect to Y-axis.

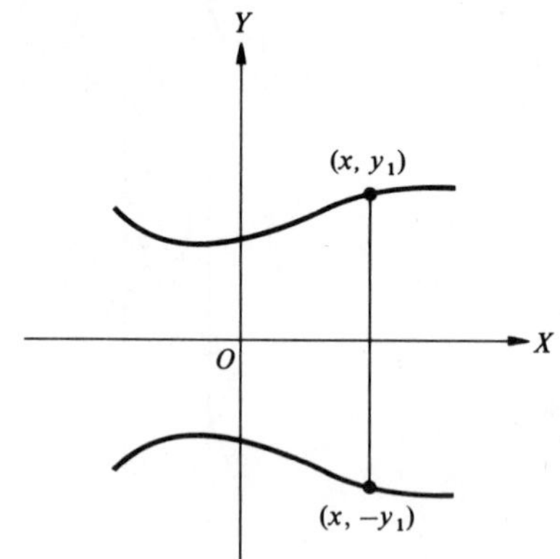

Figure 12.2.2 Symmetry with respect to X-axis.

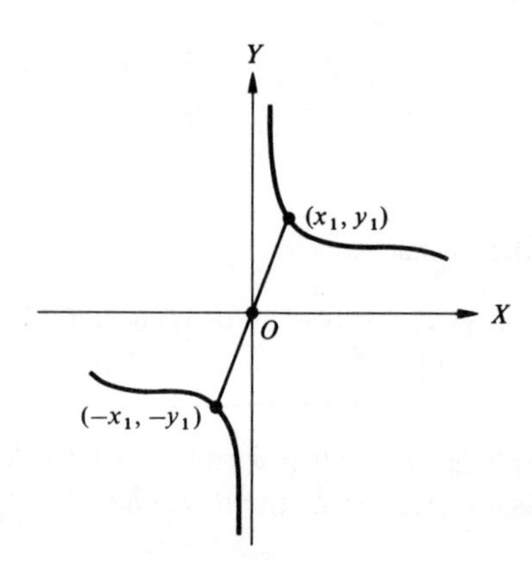

Figure 12.2.3 Symmetry with respect to the origin.

2. For $y = -y$, $x^2 + (-y) = 7$ or $x^2 - y = 7$, therefore not symmetric with respect to the X-axis.
3. For $x = -x$ and $y = -y$, $(-x)^2 + (-y) = 7$ or $x^2 - y = 7$ therefore not symmetric with respect to the origin (see Fig. 12.2.4).

For (b), $x + y^2 = 7$, we have:

1. For $x = -x$, $(-x) + y^2 = 7$ or $-x + y^2 = 7$, therefore not symmetric with respect to the Y-axis.

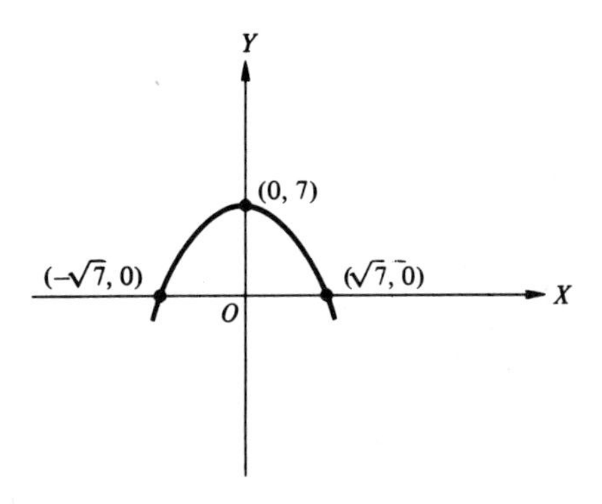

Figure 12.2.4 $x^2 + y = 7$.

2. For $y = -y$, $x + (-y)^2 = 7$ or $x + y^2 = 7$, therefore symmetric with respect to the X-axis.
3. For $x = -x$ and $y = -y$, $(-x) + (-y)^2 = 7$ or $-x + y^2 = 7$, therefore not symmetric with respect to the origin (see Fig. 12.2.5).

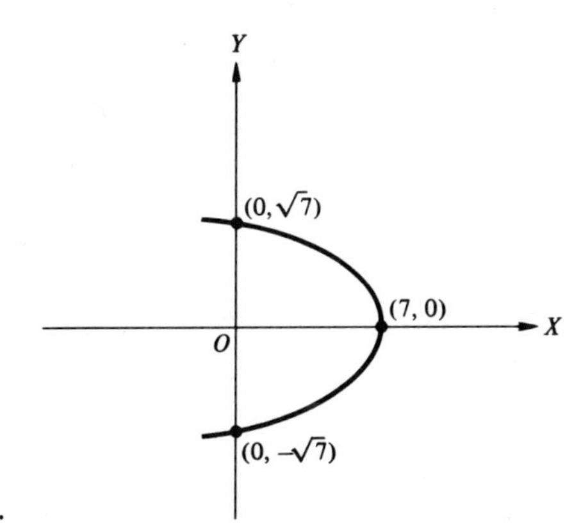

Figure 12.2.5 $x + y^2 = 7$.

For (c), $xy = 2$, we have:

1. For $x = -x$, $(-x)y = 2$ or $-xy = 2$, therefore not symmetric with respect to the Y-axis.
2. For $y = -y$, $x(-y) = 2$ or $-xy = 2$, therefore not symmetric with respect to the X-axis.
3. For $x = -x$ and $y = -y$, $(-x)(-y) = 2$ or $xy = 2$, therefore symmetric with respect to the origin (see Fig. 12.2.6).

PROBLEM 1. Discuss the symmetry of (a) $3x^2 - y = 5$; (b) $2x + 3y^2 = 4$; (c) $xy = -4$.

EXERCISES 12.2

Group A

1. Discuss the symmetry of each of the following.

(a) $x^2 + y^2 = 4$. (b) $x^2 - y^2 = 4$.

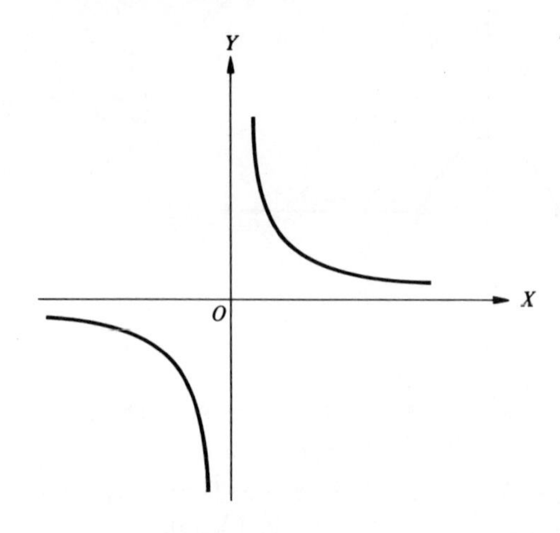

Figure 12.2.6 $xy = 2$.

(c) $xy^2 = 7$.

(d) $x^3y = 7$.

(e) $x - y = 0$.

(f) $x - y = 3$.

(g) $x^2 + 4y^2 = 4$.

(h) $3x - 2y^2 = 4$.

(i) $3x - 2y^3 = 4$.

(j) $3x - 2y^3 = 0$.

Group B

1. Discuss the symmetry of each of the following.

(a) $y = \cos x$.

(b) $y = \sin x$.

(c) $y = \text{Cos}^{-1} x$.

(d) $y = \text{Sin}^{-1} x$.

Group C

1. (a) If an equation contains only the even powers of x, what can you say about its symmetry? (b) If an equation contains only the even powers of y, what can you say about its symmetry? (c) If an equation contains only the even powers of both x and y, what can you say about its symmetry?

2. Use the results of Exercise 1 to discuss the symmetry of each of the following.

(a) $x^2 = y$.

(b) $y^2 = x$.

(c) $x^2 + y^2 = 25$.

(d) $x^2 - y^2 = 25$.

(e) $xy^2 - x^3 = 7$.

(f) $x^2y - x^4 = 11$.

12.3 THE CIRCLE

There are two fundamental types of problems associated with analytic geometry. They are:

1. For a given geometric description, find a corresponding equation.
2. For a given equation, describe the corresponding geometric figure and its basic properties.

Definition 12.3.1 *A circle is defined to be the set of points in a plane such that each point is a constant distance, called the radius, from a fixed point, called the center.*

Let the center be the fixed point C at (h, k) in Fig. 12.3.1. Let $P(x, y)$ be any point on the circle. Then the radius, a, is the constant distance $|PC|$. Thus

$$|PC| = a$$

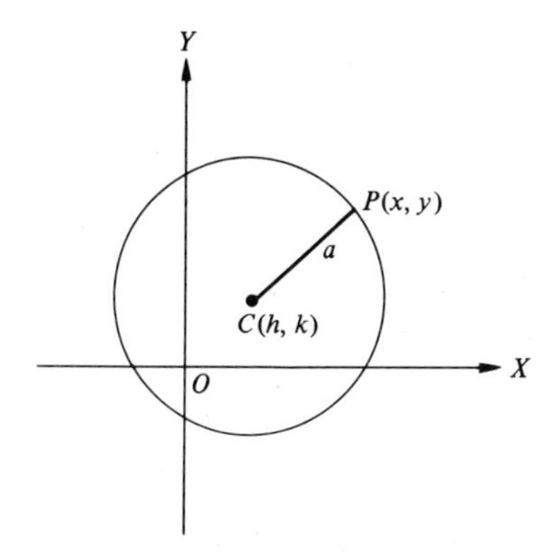

Figure 12.3.1 Circle: $(x-h)^2 + (y-k)^2 = a^2$.

Using the distance formula, Eq. 2.6.1, between two points we have

$$\sqrt{(x-h)^2 + (y-k)^2} = a$$

or

$$\boxed{(x-h)^2 + (y-k)^2 = a^2} \tag{12.3.1}$$

Equation 12.3.1 is called the *standard form* of the equation of a circle of radius a with center at the point (h, k). If the center is at the origin, then $h = 0$, $k = 0$, and Eq. 12.3.1 becomes

$$\boxed{x^2 + y^2 = a^2} \tag{12.3.2}$$

Using set notation, the circle with its center at the origin and radius a may be described as

$$\{(x, y) \,|\, x^2 + y^2 = a^2\}$$

You may show that Eq. 12.3.2 is symmetric with respect to the X-axis, the Y-axis, and the origin.

There are two basic properties of a circle: the *location of the center* and the *length of the radius.*

EXAMPLE 1. Discuss the equation $x^2 + y^2 = 25$.

Solution. From Eq. 12.3.2 we know that $x^2 + y^2 = 25$ is a circle with its center at the origin and a radius of 5.

PROBLEM 1. Discuss the equation $2x^2 + 2y^2 = 18$.

EXAMPLE 2. Discuss the equation $(x - 3)^2 + (y + 4)^2 = 49$.

Solution. From Eq. 12.3.1 we know that $(x - 3)^2 + (y + 4)^2 = 49$ is a circle. Since $h = 3$, $k = -4$, and $a = 7$, we know that the center is at $(3, -4)$ and the radius is 7.

PROBLEM 2. Discuss the equation $(x + 2)^2 + (y - 1)^2 = 15$.

If we expand Eq. 12.3.1, we obtain $x^2 - 2hx + h^2 + y^2 - 2ky + k^2 = a^2$. Since h, k, and a are constants, we may rewrite this equation as

$$\boxed{x^2 + y^2 + Dx + Ey + F = 0} \qquad (12.3.3)$$

where $D = -2h$, $E = -2k$, and $F = a^2 - h^2 - k^2$.

Equation 12.3.3 is called the *general equation of a circle.* Note that the coefficients of x^2 and y^2 are both $+1$. Equation 12.3.3 may be obtained from Eq. 12.1.1 if $A = C$ and $B = 0$. Whenever we have an equation of the form of Eq. 12.3.3 we may find the center and radius by the method of *completing the square* on the x terms and on the y terms.

EXAMPLE 3. Find the center and radius of $x^2 + y^2 + 4x - 6y + 5 = 0$.

Solution. We recognize this equation as the general equation of a circle, Eq. 12.3.3. To find its center and radius let us complete the square in order to rewrite the equation in the form of Eq. 12.3.1. We proceed as follows. Group the x^2 and x terms and the y^2 and y terms. Thus

$$(x^2 + 4x + \quad) + (y^2 - 6y + \quad) = -5$$

Next, take one-half of the coefficient of x, square it, and add to both sides of the equation. Do a similar process for the coefficient of y. Thus

$$(x^2 + 4x + 4) + (y^2 - 6y + 9) = -5 + 4 + 9$$

Next, factor the perfect squares. Thus

$$(x + 2)^2 + (y - 3)^2 = 8$$

By comparison with Eq. 12.3.1 we see that $h = -2$, $k = 3$, and $a = \sqrt{8}$. The center is at $(-2, 3)$ and the radius is $\sqrt{8} = 2.83$ (see Fig. 12.3.2).

If the equation is of the form $Ax^2 + Ay^2 + Cx + Dy + F = 0$, we must divide through by A before completing the square.

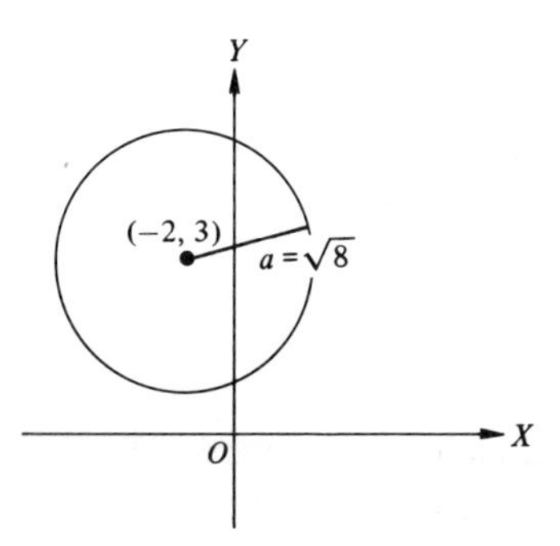

Figure 12.3.2 Circle: $x^2 + y^2 + 4x - 6y + 5 = 0$

Warning ***For a circle the coefficients of the squared terms must both be $+1$.***

PROBLEM 3. Find the center and radius of $3x^2 + 3y^2 - 6x + 12y - 12 = 0$.

Sometimes we know some conditions on the circle and would like to write its equation.

EXAMPLE 4. Write an equation of the circle tangent to the X-axis at $(3, 0)$ whose center is on the line $x - 2y = 1$.

Solution. See Fig. 12.3.3. We need to find h, k, and a. Since the circle is tangent to the X-axis at $(3, 0)$, $h = 3$. Since the center is on $x - 2y = 1$, we have

$$h - 2k = 1$$

or

$$k = \tfrac{1}{2}(h - 1)$$

and

$$k = \tfrac{1}{2}(3 - 1) = 1$$

Here $a = k = 1$. So from Eq. 12.3.1 the equation of the circle is

$$(x - 3)^2 + (y - 1)^2 = 1$$

or

$$x^2 + y^2 - 6x - 2y + 9 = 0$$

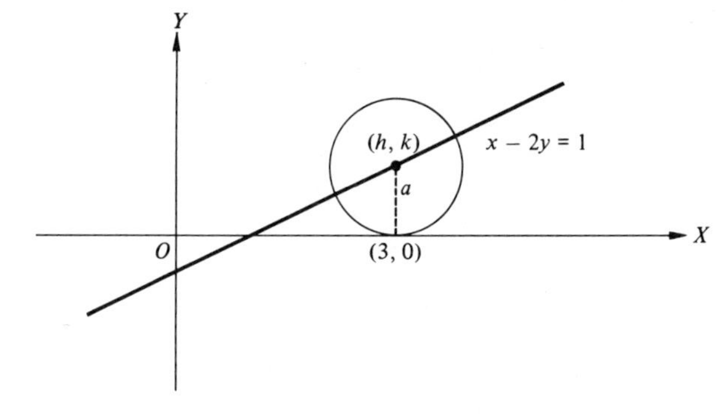

Figure 12.3.3 Example 4.

PROBLEM 4. Write an equation of the circle tangent to the X-axis at $(-2, 0)$ whose center is on the line $x + 2y - 2 = 0$.

If we solve the equation $x^2 + y^2 = a^2$ for y, we have $y = \pm\sqrt{a^2 - x^2}$. We then note that a circle is made up of two different functions $y = \sqrt{a^2 - x^2}$ and $y = -\sqrt{a^2 - x^2}$. Recall that a function has only one value of y for each x. That is, a circle has one function for its upper half and another function for its lower half. Note that y has a meaning only for $|x| \leqq a$.

EXAMPLE 5. Use integration to find the area enclosed by the circle $x^2 + y^2 = 16$.

Solution. The circle has its center at (0, 0) and a radius of $a = 4$ (see Fig. 12.3.4). Knowing that $A = \int dA$, we must find dA.

$$dA = (y_u - y_l)\,dx$$

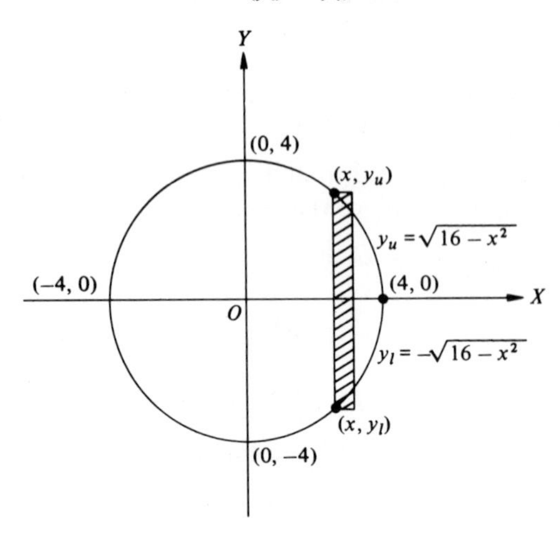

Figure 12.3.4 Area bounded by $x^2 + y^2 = 16$.

Since $y_u = \sqrt{16 - x^2}$ and $y_l = -\sqrt{16 - x^2}$, we have

$$dA = \sqrt{16 - x^2} - (-\sqrt{16 - x^2})\,dx$$
$$= 2\sqrt{16 - x^2}\,dx$$

and

$$A = 2\int_{-4}^{4} \sqrt{16 - x^2}\,dx \qquad (12.3.4)$$

Using integral formula 57, we have

$$A = 2\left[\frac{x}{2}\sqrt{16 - x^2} + \frac{16}{2}\,\text{Sin}^{-1}\,\frac{x}{4}\right]_{-4}^{4}$$
$$= [4(0) + 16\,\text{Sin}^{-1}\,1] - [(-4)(0) + 16\,\text{Sin}^{-1}\,(-1)]$$
$$= 16\left(\frac{\pi}{2}\right) - 16\left(\frac{-\pi}{2}\right) = 16\pi = 50.3$$

By noting the symmetry we could have expressed Eq. 12.3.4 as $A = 4\int_0^4 \sqrt{16 - x^2}\,dx$, that is, four times the area in the first quadrant.

PROBLEM 5. Use a y-integral to find the area enclosed by $x^2 + y^2 = 49$.

EXAMPLE 6. Use integration to find the circumference of a circle of radius a.

Solution. For convenience let the equation of the circle be $x^2 + y^2 = a^2$. The circumference, s, is the sum of a lot of little arc lengths. That is, $s = \int ds$. In Fig. 12.3.5, ds is a typical differential of arc. From Eq. 7.9.3,

$$ds = \sqrt{1 + \left(\frac{dy}{dx}\right)^2}\,dx$$

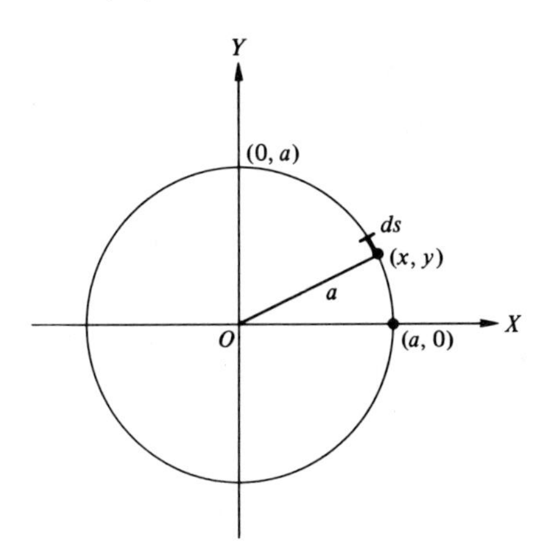

Figure 12.3.5 Circumference of a circle.

To find dy/dx we use implicit differentiation on $x^2 + y^2 = a^2$. Thus

$$2x + 2y\frac{dy}{dx} = 0$$

or

$$\frac{dy}{dx} = \frac{-x}{y}$$

Thus

$$ds = \sqrt{1 + \frac{x^2}{y^2}}\,dx = \sqrt{\frac{y^2 + x^2}{y^2}}\,dx = \frac{a}{y}\,dx$$

and

$$ds = \frac{a}{\sqrt{a^2 - x^2}}\,dx$$

Noting the symmetry we need only to integrate over the first quadrant and multiply by 4. Thus

$$s = 4a\int_0^a \frac{dx}{\sqrt{a^2 - x^2}}$$

Using Eq. 10.3.1, we then find that

$$s = 4a\left[\operatorname{Sin}^{-1}\frac{x}{a}\right]_0^a$$
$$= 4a[\operatorname{Sin}^{-1} 1 - \operatorname{Sin}^{-1} 0]$$

and

$$s = 4a\left(\frac{\pi}{2}\right) = 2\pi a$$

PROBLEM 6. Use a y-integral to find the circumference of a circle of radius 6.

EXERCISES 12.3

Group A

1. Discuss each of the following equations; that is, give the name of the equation and give its important properties.

(a) $x^2 + y^2 = 4.$
(b) $x^2 + y^2 = 7.$
(c) $3x^2 + 3y^2 = 16.$
(d) $(x - 2)^2 + (y - 4)^2 = 9.$
(e) $(x + 3)^2 + y^2 = 16.$
(f) $x^2 + (y + 7)^2 = 11.$
(g) $x^2 + y^2 = 0.$
(h) $y^2 = 13 - x^2.$

2. Write each equation of Exercise 1 in the form of Eq. 12.1.1, stating the values of A, B, C, D, E, and F.

3. Find the center and radius of each of the following.

(a) $x^2 + y^2 - 4x - 6y + 5 = 0.$
(b) $x^2 + y^2 + 2x - 4y - 4 = 0.$
(c) $2x^2 + 2y^2 - 4x + 8y + 2 = 0.$
(d) $36x^2 + 36y^2 - 36x + 24y - 23 = 0.$
(e) $x^2 + y^2 + 4y - 16 = 0.$
(f) $x^2 + y^2 - 4x - 16 = 0.$

4. Write an equation of a circle for each of the following.

(a) Center $(0, 0)$, $a = 2.$
(b) Center $(0, 3)$, $a = 3.$
(c) Center $(\frac{1}{3}, 0)$, $a = 1.$
(d) Center $(-2, -\frac{1}{3})$, $a = 2.$
(e) Center $(-\frac{1}{4}, -3)$, $a = \frac{1}{2}.$
(f) Center $(2, -3)$, $a = \sqrt{5}.$

5. Write an equation of the circle whose end points of a diameter are $(4, 5)$ and $(-2, -3)$.

6. Write an equation of the circle tangent to the Y-axis at $(0, 3)$ whose center is on the line $2x - y = 1$.

7. Write an equation of the circle tangent to the X-axis whose center is 2 units below it and 2 units to the right of the Y-axis.

Group B

1. Find the slope of the circle $x^2 + y^2 = 25$ at $(-3, 4)$.

2. Find an equation of the tangent to the circle $x^2 + y^2 = 169$ at $(-12, 5)$.

3. Use set notation to describe the circle of radius a whose center is at (h, k).

4. Find an equation of the circle passing through the three points $(1, 1)$, $(1, -1)$, and $(-1, 1)$. (*Hint*: Use Eq. 12.1.1 with $A = C = 1$, three times, once for each point. Then solve the three simultaneous equations for D, E, and F.)

5. Does the equation $x^2 + y^2 + 2x + 4y + 5 = 0$ represent a circle? What about $x^2 + y^2 + 2x + 4y + 6 = 0$? (Be careful.)

6. Solve the equation $x^2 + y^2 = a^2$ for y. Is y a function of x? Why? What values of x are not allowed to be used?

7. Let the motion of a particle be given by the parametric equations

$$\begin{cases} x = a \cos t \\ y = a \sin t \end{cases}$$

where a is a fixed positive number and $t \geqq 0$. Using trigonometric identities, show that the particle moves around a circle of radius a.

8. Use integration to find the area enclosed by $x^2 + y^2 = 25$.

9. Use integration to find the area enclosed by $x^2 + y^2 = a^2$. (Does the answer agree with the formula you know for the area enclosed by a circle?)

10. Find the area between the upper part of $x^2 + y^2 = 25$ and $x - 3y + 5 = 0$.

11. Find the length of arc of $x^2 + y^2 = 25$ from the point $(0, 5)$ to the point $(4, 3)$.

12. Find the centroid of the first-quadrant region bounded by $x^2 + y^2 = 16$ and the coordinate axes.

13. Find the moment of inertia of a circular disc with respect to its diameter. (*Hint*: Find i_Y for the region bounded by $x^2 + y^2 = a^2$.)

14. Find the slope of the circle $x^2 + y^2 = a^2$ at each point where it crosses the coordinate axes.

15. A particle moves on the circle $x^2 + y^2 = 25$. Find its speed at the point $(3, 4)$ if it is moving in such a manner that its speed in the X-direction is $\frac{4}{3}$ ft/s.

16. Find the radius of curvature of $x^2 + y^2 = a^2$.

17. Find the circumference of a circle of radius a using the parametric equations $x = a \cos t$, $y = a \sin t$. (*Hint*: Make use of Eq. 7.9.2.)

18. Find the surface area of a sphere obtained by revolving a circle about its diameter (see Exercise 1 of Group C in Exercises 7.9).

(a) Consider $x^2 + y^2 = a^2$ revolved about the X-axis.
(b) Consider the parametric equations $x = a \cos t$, $y = a \sin t$.

19. A fuel tank is a horizontal right circular cylinder of radius a and length L. Obtain a formula for the volume of fuel at a height h on a dipstick. (*Hint*: Let the circular cross section be a circle tangent to the X-axis and center on the Y-axis. Let $dV = L\, dA$, where dA is an element of area of the circular cross section.) (See Exercise 3 of Group B in Exercises 11.5.) Find the length L of the fuel tank of radius 6 in. that will have a capacity of 20 gal. (1 gal = 231 in.^{3}.) Find the amount of fuel in a tank of radius 6 in., length 3.5 ft when the dip stick indicates 3 in. of fuel.

12.4 THE PARABOLA

Definition 12.4.1 *A parabola is defined to be the set of points in a plane such that each point is equidistant from a fixed point called the focus and a fixed line called the directrix.*

For convenience in obtaining a simple equation of the parabola let the focus, F, be at $(p, 0)$, and let the directrix be the line $x = -p$ (see Fig. 12.4.1).

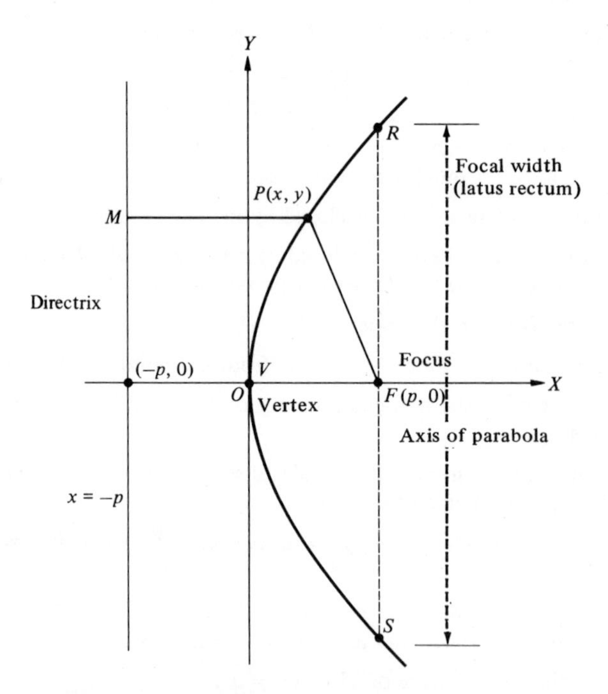

Figure 12.4.1 Parabola.

From Definition 12.4.1, we have $|PF| = |PM|$.

$$|PF| = \sqrt{(x-p)^2 + (y-0)^2}$$

$$|PM| = x - (-p) = x + p$$

Thus

$$\sqrt{(x-p)^2 + (y-0)^2} = x + p$$

or

$$x^2 - 2px + p^2 + y^2 = x^2 + 2px + p^2$$

which becomes

$$\boxed{y^2 = 4px} \tag{12.4.1}$$

The line through the focus and perpendicular to the directrix is called the *axis* of the parabola. The point of intersection of the parabola and its axis is called the *vertex* of the parabola. Equation 12.4.1 is called the *standard form* of the parabola whose vertex is at the origin and whose axis is the X-axis. It is symmetric with respect to the X-axis. (If y is replaced by $-y$, the same equation is obtained.) Note that Eq. 12.1.1 reduces to Eq. 12.4.1 if $A = 0$, $B = 0$, $C = 1$, $D = 4p$, $E = 0$, and $F = 0$.

EXAMPLE 1. Discuss the equation $y^2 = 8x$.

Solution. $y^2 = 8x$ is similar to Eq. 12.4.1 with $4p = 8$ or $p = 2$; thus the equation is a parabola with its focus at (2, 0) and its vertex at (0, 0) whose directrix is $x = -2$ (see Fig. 12.4.2).

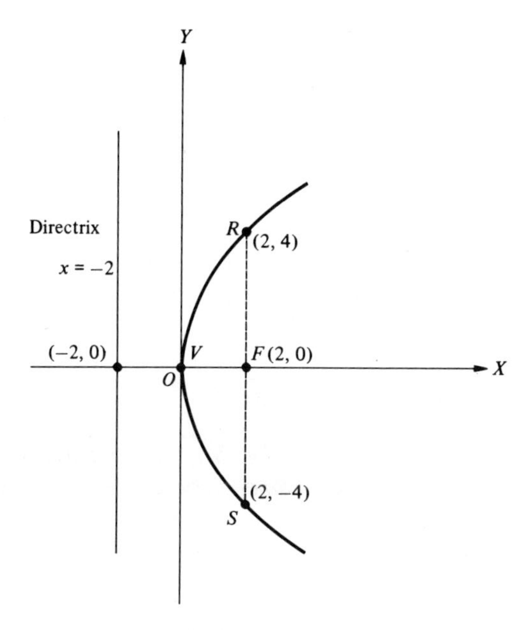

Figure 12.4.2 Graph of $y^2 = 8x$.

Note carefully that y is squared and x is not. This means that for one value of x, there are two values of y. Also note that the curve opens around its focus.

PROBLEM 1. Discuss the equation $y^2 = -12x$.

In sketching the graph of a parabola, it is useful to know the width of the parabola at the focal point. This distance is called the *focal width* or the *latus rectum*. In Fig. 12.4.1 it is the distance $|RS|$. When $x = p$, $y^2 = 4p^2$ or $y = \pm 2p$. Hence R is at $(p, 2p)$ and S is at $(p, -2p)$; therefore, $|RS| = |2p - (-2p)| = |4p|$. Note that this tells us that the focal width of any parabola is *four times the distance from the focus to the vertex.* If the parabola opens to the left, p is negative.

EXAMPLE 2. Sketch the parabola $y^2 + 4x = 0$.

Solution. Here $y^2 = -4x$, so $4p = -4$ and $p = -1$. This means that the parabola opens to the left. The focus is at $(-1, 0)$. The focal width is $|4p| = |-4| = 4$. The vertex is at the origin (see Fig. 12.4.3).

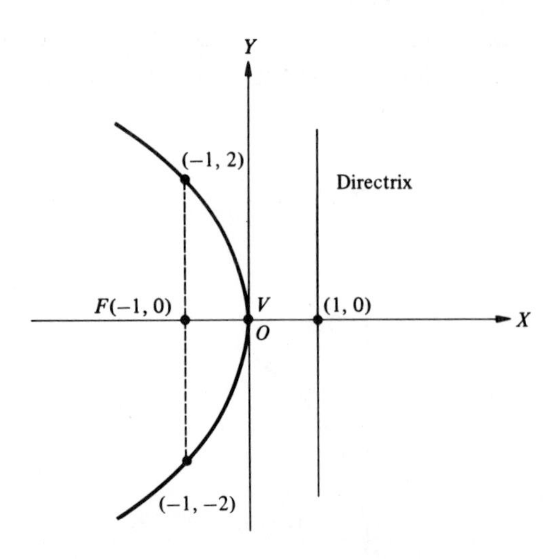

Figure 12.4.3 Graph of $y^2 + 4x = 0$.

Warning ***To give the correct sketch of a parabola, remember that the focal width is four times the distance between the vertex and the focus.***

PROBLEM 2. Sketch the graph of $y^2 - 3x = 0$. State the focus and the focal width.

If the parabola has as its axis the Y-axis, the equation becomes

$$\boxed{x^2 = 4py} \tag{12.4.2}$$

(*Note:* There are two values of x for one value of y.) If p is positive, it opens up. If p is negative, it opens down [see Fig. 12.4.4 (a) and (b)].

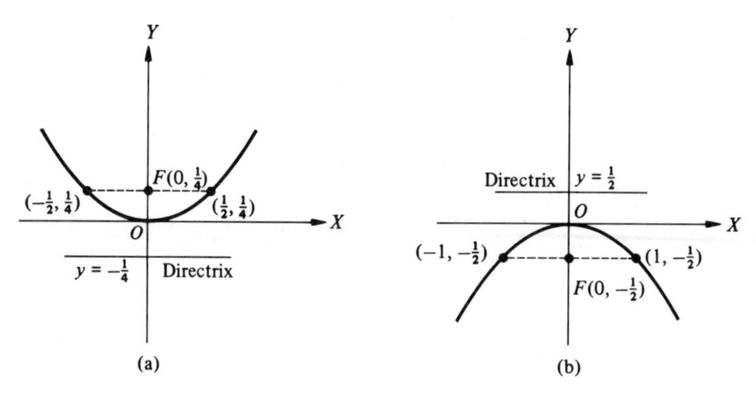

Figure 12.4.4 Graph of (a) $x^2 = y$. (b) $x^2 = -2y$.

If the *vertex* is at the point (h, k), it may be shown that the equation of the parabola is either Eq. 12.4.3 or 12.4.4.

$$(x - h)^2 = 4p(y - k) \tag{12.4.3}$$

For Eq. 12.4.3 the axis is parallel to the Y-axis. If $p > 0$, it opens up; for $p < 0$, it opens down.

$$(y - k)^2 = 4p(x - h) \tag{12.4.4}$$

For Eq. 12.4.4 the axis is parallel to the X-axis. If $p > 0$, it opens to the right; for $p < 0$, it opens to the left.

Upon expanding Eqs. 12.4.3 and 12.4.4 it may be shown that the general equation of a parabola is

$$Ax^2 + Dx + Ey + F = 0 \tag{12.4.5}$$

or

$$Cy^2 + Dx + Ey + F = 0 \tag{12.4.6}$$

Note that the equation of a parabola contains the square of either x or y, but not both, and the first power of the other variable. Equations 12.4.5 and 12.4.6 are special forms of Eq. 12.1.1.

EXAMPLE 3. Find the vertex, focus, and focal width of $4x^2 + 8x - 16y - 12 = 0$.

Solution. Noting that x is squared and y is not, we see we have a parabola (compare with Eq. 12.4.5). We now complete the square.

$$
\begin{aligned}
4(x^2 + 2x \qquad) &= 16y + 12 \\
4(x^2 + 2x + 1) &= 16y + 12 + 4 \\
4(x + 1)^2 &= 16(y + 1) \\
(x + 1)^2 &= 4(y + 1)
\end{aligned}
$$

Comparing with Eq. 12.4.3, we have $h = -1, k = -1$, $4p = 4$, and $p = 1$. The axis of the parabola is parallel to the Y-axis (since x is squared).

Vertex: $(-1, -1)$

Focus: $(-1, 0)$ (since F is p units from V on the axis)

Focal width: 4.

See Fig. 12.4.5.

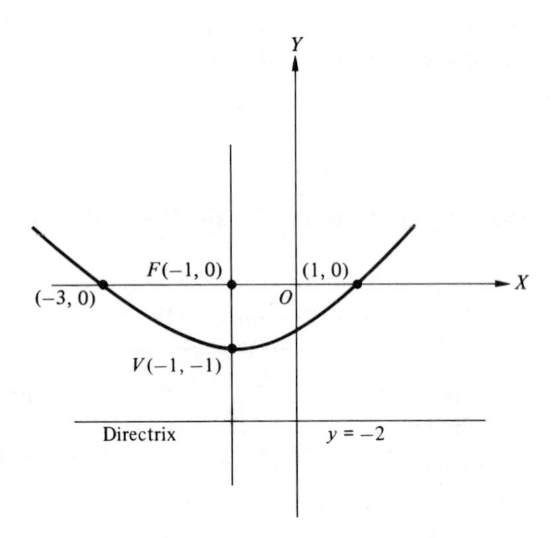

Figure 12.4.5 Graph of $4x^2 + 8x - 16y - 12 = 0$.

Problem 3. Find the vertex, focus, and focal width of $2x^2 - 4x - 16y + 34 = 0$. Sketch the graph.

We can find an equation of a parabola if we know or can find the positions of the vertex and the focus.

Example 4. Find an equation of the parabola with its axis parallel to the X-axis whose vertex is at (2, 3) and whose focus is on the line $x + y - 1 = 0$.

Solution. See Fig. 12.4.6. Since the focus F is on the axis of the parabola, its coordinates are $(2 + p, 3)$, because F is p units from V and p is negative. Since F is on the line $x + y - 1 = 0$, we have

$$2 + p + 3 - 1 = 0$$

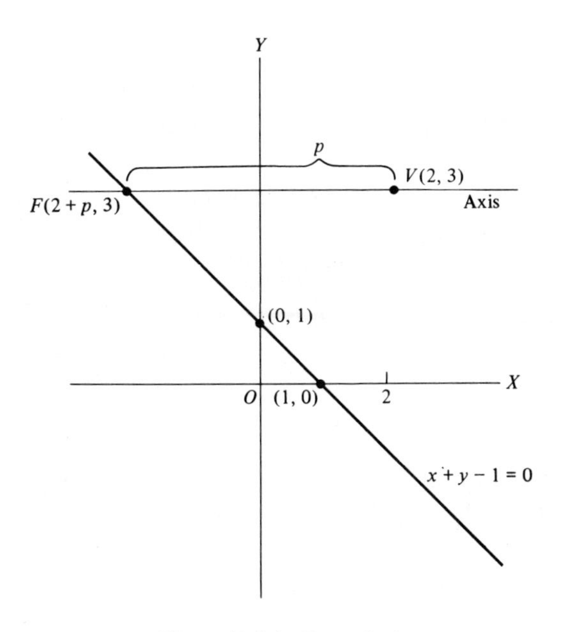

Figure 12.4.6 Example 4.

or

$$p = -4$$

Thus the required equation is

$$(y - 3)^2 = -16(x - 2) \qquad \text{(from Eq. 12.4.4)}$$

or

$$y^2 + 16x - 6y - 23 = 0$$

PROBLEM 4. Find an equation of the parabola with its axis parallel to the Y-axis whose vertex is at (3, 3) and whose focus is on the line $x - y - 1 = 0$.

Many problems involve a parabola of the form

$$y = ax^2 + bx + c \tag{12.4.7}$$

which you should be able to sketch quickly. The relative maximum or minimum is found by applying the derivative. Thus

$$\frac{dy}{dx} = 2ax + b$$

$$2ax + b = 0$$

and

$$x = -\frac{b}{2a}$$

Thus the high (or low) point occurs at $x = -b/2a$. The curve crosses the X-axis when $y = 0$. The second derivative $d^2y/dx^2 = 2a$ tells us that the curve opens up if $a > 0$ and down if $a < 0$. Using this information, the graph may be drawn.

EXAMPLE 5. Write the integral to find the length of the first-quadrant portion of $y = 4x - x^2$.

Solution. The equation $y = 4x - x^2$ is a parabola opening downward (the negative x^2 term). When $y = 0$, $x(4 - x) = 0$ or $x = 0$ and $x = 4$. Since $dy/dx = 4 - 2x$, it has a maximum at (2, 4) (see Fig. 12.4.7). Since the total length $s = \int ds$, we must find ds.

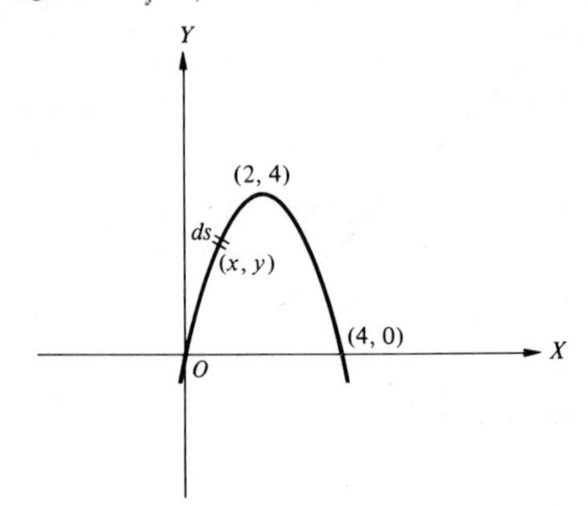

Figure 12.4.7 Example 5.

$$\begin{aligned} ds &= \sqrt{1 + \left(\frac{dy}{dx}\right)^2}\, dx \\ &= \sqrt{1 + (4 - 2x)^2}\, dx \end{aligned}$$

and

$$s = \int_0^4 \sqrt{1 + (4 - 2x)^2}\, dx$$

(You may evaluate this integral as Exercise 10 of Group B.)

PROBLEM 5. Write a y-integral to find the length of the first quadrant portion of $x = 3y - y^2$.

An interesting property of parabolas is that sound waves and light waves gathered in by a parabolic receiver are reflected to the focus of the parabola. See Exercise 8 of Group C in Exercises 12.4.

EXERCISES 12.4

Group A

1. For each of the following, give the name of the curve and find the vertex, focus, and focal width. Sketch the curve.

(a) $y^2 = 4x$. (b) $y^2 = 9x$.
(c) $y^2 = -8x$. (d) $y^2 = -x$.

(e) $x^2 = 4y$.
(f) $x^2 = 9y$.
(g) $x^2 = -y$.
(h) $x^2 = -2y$.

2. For each of the following, give the name of the curve and find the vertex, focus, and focal width. Sketch the curve.

(a) $x^2 - 2x - 4y + 5 = 0$.
(b) $x^2 + 2x - 4y + 5 = 0$.
(c) $x^2 - 2x + 8y - 7 = 0$.
(d) $x^2 + 6x - y + 10 = 0$.
(e) $x - y^2 + 2y - 2 = 0$.
(f) $y^2 - 2y - 4x - 3 = 0$.
(g) $x^2 - 4x + y = 0$.
(h) $y - x^2 + 4 = 0$.

3. Use Eq. 12.4.3 or 12.4.4 to write an equation of the parabola determined by the data in each of the following.

(a) Vertex $(1, -2)$; focus $(2, -2)$.
(b) Vertex $(-1, 3)$; focus $(0, 3)$.
(c) Vertex $(2, -1)$; focus $(2, 3)$.
(d) Vertex $(5, -3)$; focus $(5, -7)$.

4. Find the maximum or minimum points of each of the following and sketch the curve.

(a) $y = 9x - x^2$.
(b) $y = -4x + x^2$.
(c) $y = x^2 - 4$.
(d) $y = x^2 + x - 2$.
(e) $y = x^2 + 4$.
(f) $y = 4 - x^2$.
(g) $y = x^2 + 3x + 2$.
(h) $y = 2 + x - x^2$.

5. Find the area of the region R described by each of the following boundaries.

(a) $y = x^2 - 4$, the X-axis.
(b) $y = x^2 + 3x + 2$, the X-axis.
(c) $y = 4x - x^2$, the X-axis.
(d) $y = 2 + x - x^2$, the X-axis.

6. Find the centroid of the area bounded by $y = 4x - x^2$ and the X-axis.

7. Find the centroid of the area bounded by $y = x^2 - 4$ and the X-axis.

8. Find the moment of inertia with respect to the Y-axis of the area bounded by $y = x^2 - 4$ and the X-axis.

Group B

1. Find the equation of each of the following parabolas with vertex at the origin, axis the X-axis, and the curve passing through (a) $(4, 4)$ and (b) $(-4, 4)$.

2. Find the equation of each of the following parabolas with vertex at the origin, axis the Y-axis, and the curve passing through (a) $(4, 4)$ and (b) $(-4, -4)$.

3. State the important parts of a parabola.

4. Can a parabola be sketched if we know only the vertex and focus? Do we need to know the principal axis? Why?

5. Use set notation to define a parabola.

6. Solve the equation $y^2 = 4px$ for y. Is y a function of x? Why? Are all values of x allowed? If not, state the values of x that are allowed.

7. Use trigonometric identities to show that the parametric equations $x = \sin t$, $y = 2 \cos^2 t$ are the equations of a parabola.

8. Find the area between the parabola $y = 4x - x^2$ and the line $y = x$.

9. Find the first-quadrant area between $x^2 + y^2 = 25$ and $4x^2 = 9y$.

10. Complete the integration of Example 5.

11. Prove that the graph of the position of a body moving vertically under the attraction of gravity only is represented by a parabola. [*Hint*: Use $F = ma = mg$ (that is, $dv/dt = 32 \text{ ft/s}^2$) and integrate twice.]

12. Find the radius of curvature of $y = x^2$ at $(1, 1)$.

Group C

1. Show that $y^2 = 8x$ is a special form of Eq. 12.1.1.

2. Show that $(x - h)^2 = 4p(y - k)$ is a special form of Eq. 12.1.1.

3. Derive the equation of the parabola with vertex at $(0, 0)$ and focus at $(-2, 0)$. Compare the result with Eqs. 12.4.5 and 12.4.6.

4. Derive the equation of the parabola with vertex at $(1, 2)$ and focus at $(1, 3)$. Compare the result with Eqs. 12.4.5 and 12.4.6.

5. If the vertex of a parabola is at (h, k), with its focus at $(h, k + p)$, show that its equation is Eq. 12.4.3.

6. Referring to Eq. 12.4.5 or 12.4.6, we note that there are only three constants needed to completely determine the equation of a parabola. (We can divide by A or by C.) Find the equation of the parabola that passes through the points $(1, -4)$, $(0, -1)$, and $(-2, -1)$ whose axis is parallel to the Y-axis. (*Hint*: Write Eq. 12.4.5 as $x^2 + ax + by + c = 0$. Then use each of the given points to find three equations in terms of a, b, and c.)

7. A uniform flexible wire hanging between two poles takes the form of a curve called a *catenary* (see Exercise 5 of Group B in Exercises 8.3). If the wire is loaded uniformly along the horizontal, it takes the form of a parabola. The cable on a suspension bridge is hung between two towers 200 ft apart. The top of each tower is 50 ft above the river. It is uniformly loaded (by the bridge deck) such that the middle of the cable is 20 ft above the river. Find the equation of the cable and the length of the cable.

8. One of the properties of a parabola is that a light at the focus will send out parallel rays. Consider the parabola $y^2 = 4x$, as shown in Fig. 12.4.8. Show that $\theta_1 = \theta_2$ at $P(1, 2)$. If $P_1(x, y)$ is any general point on the parabola, show that $\alpha_1 = \alpha_2$.

9. The trapezoidal rule, Eq. 5.2.1, may be used to approximate the value of a definite integral (see Sec. 11.5). If arcs of parabolas are used as upper boundaries of small elements of area, then $\int_a^b y\,dx$ may be approximated by *Simpson's rule*, which is

$$\int_a^b y\,dx = \frac{h}{3}(y_0 + 4y_1 + 2y_2 + 4y_3 + 2y_4 + \ldots + 4y_{n-1} + y_n)$$

where $h = (b - a)/n$ and n is even.

Use Simpson's rule to evaluate to three decimal places.

(a) $\int_0^4 (x^3 + 1)\,dx$, $n = 4$.

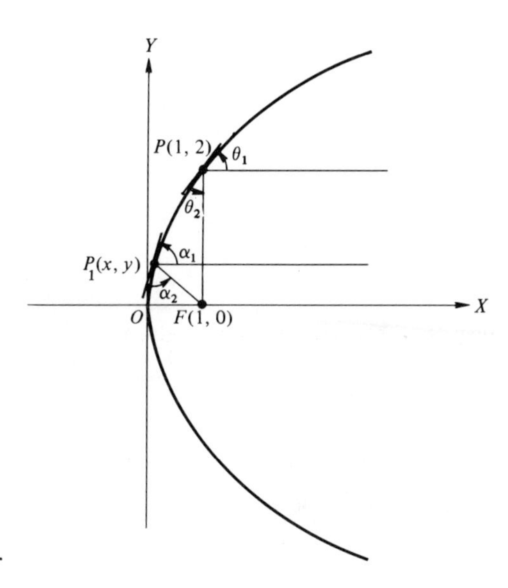

Figure 12.4.8 Exercise 8.

(b) $\int_1^5 \frac{dx}{x}$, $n = 4$. (How does this compare to ln 5?)

(c) $\int_0^{\pi/2} \sqrt{\sin x}\, dx$, $n = 4$.

(d) $\int_1^5 \ln x^2\, dx$, $n = 4$.

(e) $\int_0^1 \sin x^2\, dx$, $n = 4$.

(f) $\int_0^2 \sqrt{1 - \sin x}\, dx$, $n = 8$.

10. Use Simpson's rule to approximate the average value of Q from $t = 0$ to $t = 2$ for the data

t	0	0.5	1.0	1.5	2.0
I	10	9	11	10	8

12.5 THE ELLIPSE

Definition 12.5.1 *An ellipse is defined to be the set of points in a plane such that the sum of the distances from each point to two fixed points, called the foci, is a constant.*

For convenience in obtaining a simple equation of an ellipse, let the foci F and F' be at $(c, 0)$ and $(-c, 0)$, respectively (see Fig. 12.5.1). Let the constant sum be denoted by $2a$ (for convenience). If P is any point on the ellipse, then from Definition 12.5.1 we have $|PF| + |PF'| = 2a$. Upon applying the distance formula, we have

$$\sqrt{(x - c)^2 + (y - 0)^2} + \sqrt{(x + c)^2 + (y - 0)^2} = 2a$$

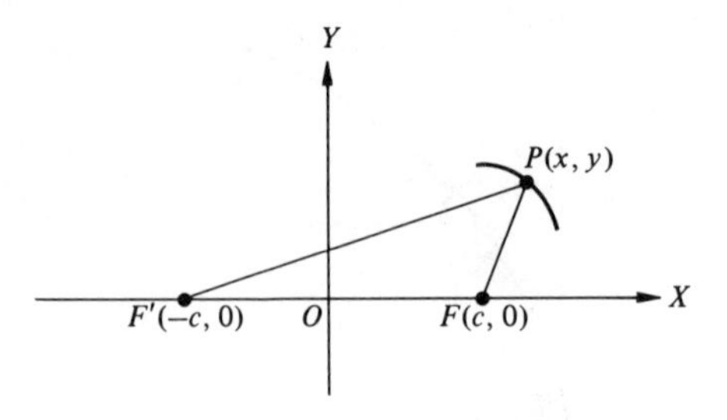

Figure 12.5.1 Construction of ellipse.

Transposing the second radical and squaring, we obtain

$$x^2 - 2cx + c^2 + y^2 = 4a^2 - 4a\sqrt{(x + c)^2 + y^2} + x^2 + 2cx + c^2 + y^2$$

or

$$4a\sqrt{(x + c)^2 + y^2} = 4a^2 + 4cx$$

Now divide by 4 (the reason for choosing $2a$ instead of just a) to obtain

$$a\sqrt{(x + c)^2 + y^2} = a^2 + cx$$

Squaring both sides, we have

$$a^2(x^2 + 2cx + c^2 + y^2) = a^4 + 2a^2cx + c^2x^2$$

or

$$a^2x^2 + a^2c^2 + a^2y^2 = a^4 + c^2x^2$$

This may be regrouped as

$$(a^2 - c^2)x^2 + a^2y^2 = a^2(a^2 - c^2)$$

From Fig. 12.5.1 we see that $2a > 2c$; hence $a > c$. If $a > c$, then $a^2 - c^2$ is positive. Let us denote this by b^2 (a positive number). Thus

$$b^2 = a^2 - c^2 \tag{12.5.1}$$

Our equation then becomes $b^2x^2 + a^2y^2 = a^2b^2$. Dividing by a^2b^2, we have

$$\boxed{\frac{x^2}{a^2} + \frac{y^2}{b^2} = 1 \qquad a^2 > b^2} \tag{12.5.2}$$

as the equation of the ellipse.

The line on which the foci are located is called the *major axis*. The point on the major axis halfway between the foci is called the *center*. The points where the ellipse crosses the major axis are called the *vertices*. The distance between the vertices is called the *length* of the major axis. Equation 12.5.2 is called the *standard* form of an ellipse with its center at the origin and its major axis along the X-axis.

The graph of $x^2/a^2 + y^2/b^2 = 1$ is shown in Fig. 12.5.2. When $y = 0$, $x = \pm a$. Thus the vertices are at $(a, 0)$ and $(-a, 0)$, and the length of the major axis is $2a$. When $x = 0$, $y = \pm b$. The points $(0, b)$ and $(0, -b)$ are called the *end points of the minor axis*. The length of the minor axis is $2b$. Note that the graph lies entirely within the rectangle formed by the lines $x = \pm a$

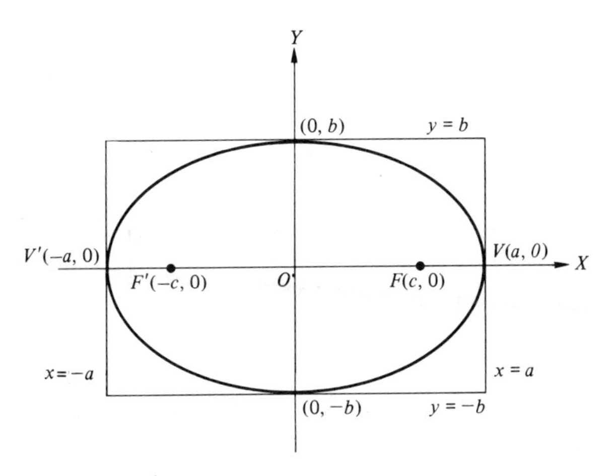

Figure 12.5.2 Graph of $(x^2/a^2) + (y^2/b^2) = 1$.

and $y = \pm b$. You may show that it is symmetric with respect to the X-axis, the Y-axis, and the origin.

EXAMPLE 1. Find the center, foci, and vertices of $9x^2 + 16y^2 - 144 = 0$. Sketch the graph.

Solution. If we transpose and divide by 144, we have

$$\frac{x^2}{16} + \frac{y^2}{9} = 1$$

This is the standard form of an ellipse. Comparing it with Eq. 12.5.2 we have $a = 4$ and $b = 3$. From Eq. 12.5.1 we find that

$$c^2 = a^2 - b^2 = 16 - 9 = 7$$

Thus we have

Center:	$(0, 0)$
Foci:	$(\sqrt{7}, 0), (-\sqrt{7}, 0)$
Vertices:	$(4, 0), (-4, 0)$

The graph is shown in Fig. 12.5.3.

PROBLEM 1. Find the center, foci, and vertices of $x^2 + 4y^2 - 4 = 0$. Sketch the graph.

If the major axis is on the Y-axis, the equation is

$$\boxed{\frac{x^2}{b^2} + \frac{y^2}{a^2} = 1 \qquad a^2 > b^2} \qquad (12.5.3)$$

Warning ***The major axis is always the axis associated with the quotient having the larger denominator when the equation is in standard form. The larger denominator is always a^2.***

EXAMPLE 2. Locate the vertices and foci and sketch the graph of the equation $(x^2/9) + (y^2/25) = 1$.

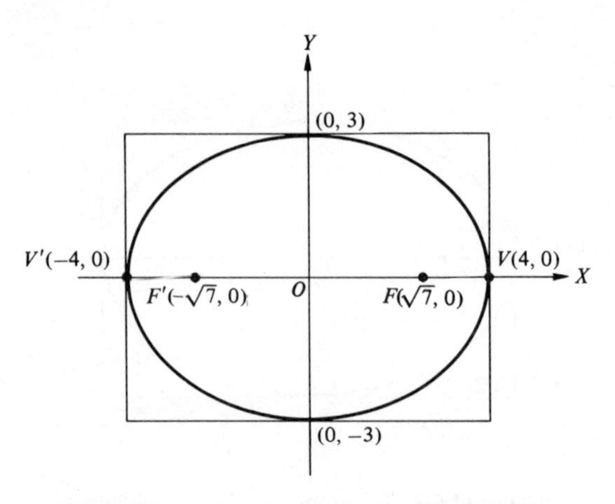

Figure 12.5.3 Graph of $9x^2 + 16y^2 - 144 = 0$.

Solution. This equation is in the standard form of an ellipse. Since $25 > 9$, we let $a^2 = 25$ and $b^2 = 9$. Thus $a = 5$ and $b = 3$. The major axis is on the Y-axis (the larger number is under y^2). Since $c^2 = a^2 - b^2$, $c = 4$.

Vertices: $(0, \pm 5)$
Foci: $(0, \pm 4)$

The graph is shown in Fig. 12.5.4.

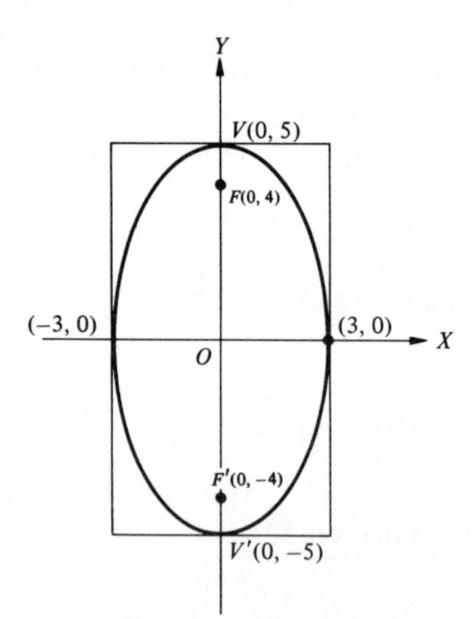

Figure 12.5.4 Graph of $(x^2/9) + (y^2/25) = 1$.

PROBLEM 2. Locate the vertices and foci and sketch the graph of the equation $25x^2 + 4y^2 - 100 = 0$.

It is interesting to note that if $b \to a$, the ellipse reduces to a circle of radius a. Also note that Eq. 12.5.2 may be obtained from Eq. 12.1.1, where $A = 1/a^2$, $B = 0$, $C = 1/b^2$, $D = 0$, $E = 0$, and $F = -1$.

It may be shown, after heroic feats of algebraic manipulation, that if the center is at (h, k), the following equations are ellipses. For

$$\boxed{\frac{(x-h)^2}{a^2} + \frac{(y-k)^2}{b^2} = 1 \qquad a^2 > b^2} \tag{12.5.4}$$

we have

Major axis:	parallel to X-axis
Center:	(h, k)
Vertices:	$(h \pm a, k)$
Foci:	$(h \pm c, k)$

For

$$\boxed{\frac{(x-h)^2}{b^2} + \frac{(y-k)^2}{a^2} = 1 \qquad a^2 > b^2} \tag{12.5.5}$$

we have

Major axis:	parallel to Y-axis
Center:	(h, k)
Vertices:	$(h, k \pm a)$
Foci:	$(h, k \pm c)$

Upon expanding and rearranging terms, Eqs. 12.5.4 and 12.5.5 may be reduced to the form of Eq. 12.1.1 with $B = 0$. The important thing to note is that for an ellipse the coefficients of *both* the x^2 and y^2 terms must be of the *same sign.*

EXAMPLE 3. Find the center, vertices, and foci of $16x^2 + 25y^2 - 32x + 100y = 284$ and sketch the curve.

Solution. Since the coefficients of both the x^2 and the y^2 terms are positive, the equation is an ellipse. We must complete the square on the x terms and also on the y terms. Regrouping, we have

$$16(x^2 - 2x \qquad) + 25(y^2 + 4y \qquad) = 284$$

$$16(x^2 - 2x + 1) + 25(y^2 + 4y + 4) = 284 + 16 + 100$$

$$16(x-1)^2 + 25(y+2)^2 = 400$$

$$\frac{(x-1)^2}{25} + \frac{(y+2)^2}{16} = 1 \tag{12.5.6}$$

Since $25 > 16$ and 25 is under the $(x-1)^2$ term, we compare Eq. 12.5.6 with Eq. 12.5.4, where $a^2 = 25$ and $b^2 = 16$. Thus $h = 1, k = -2, a = 5,$ and $b = 4$. Since $c^2 = a^2 - b^2$, we have $c = 3$. The major axis is parallel to the X-axis. The vertices are five units from the center on the major axis.

Center: $(1, -2)$
Vertices: $(6, -2), (-4, -2)$
Foci: $(4, -2), (-2, -2)$

The graph is shown in Fig. 12.5.5.

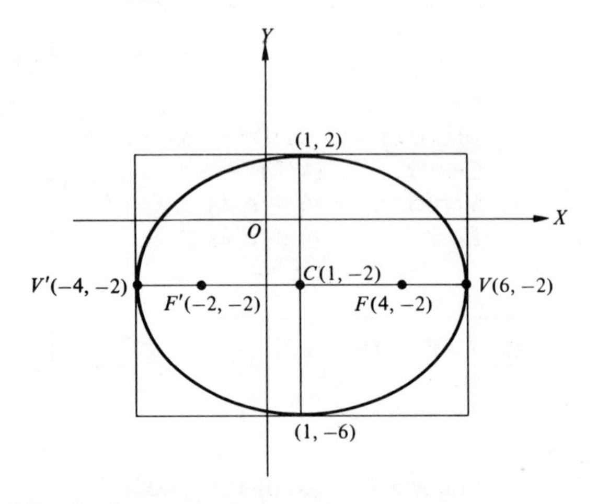

Figure 12.5.5 Graph of $16x^2 + 25y^2 - 32x + 100y = 284$.

PROBLEM 3. Find the center, vertices, and foci of $16x^2 + 7y^2 + 32x - 28y - 68 = 0$ and sketch the curve.

EXAMPLE 4. Find an equation of the ellipse whose major axis is parallel to the Y-axis and is 8 units long, whose foci are 3 units from the center, and whose center is at $(-1, 2)$.

Solution. Since the major axis is parallel to the Y-axis, we make use of Eq. 12.5.5. We have $h = -1$, $k = 2$, and $2a = 8$; therefore, $a = 4$. Since $c = 3$ and $c^2 = a^2 - b^2$, we find $b^2 = (4)^2 - (3)^2 = 7$. Thus the required equation is $[(x + 1)^2/7] + [(y - 2)^2/16] = 1$. You may expand this to obtain the general form of Eq. 12.1.1.

PROBLEM 4. Find an equation of the ellipse whose major axis is parallel to the X-axis and is 6 units long, whose foci are 2 units from the center, and whose center is at $(-1, -2)$.

In doing problems involving an ellipse we should note that if we solve the equation $x^2/a^2 + y^2/b^2 = 1$ for y, we obtain $y = \pm(b/a)\sqrt{a^2 - x^2}$. That is, an ellipse is made up of two different functions, $y = (b/a)\sqrt{a^2 - x^2}$ and $y = -(b/a)\sqrt{a^2 - x^2}$. Caution should be used in determining the proper function for calculations. Note that y has meaning only if $|x| \leqq a$.

An elliptical reflector has the property that a sound originating at one focal point is reflected to the other focal point (see Exercise 1 of Group C in Exercises 12.5). This property is sometimes used in designing large auditoriums.

EXERCISES 12.5

Group A

1. In each of the following, state the name of the curve and find the center, vertices, and foci. Sketch the graph.

(a) $\dfrac{x^2}{25} + \dfrac{y^2}{16} = 1.$ (b) $\dfrac{x^2}{25} + \dfrac{y^2}{9} = 1.$

(c) $3x^2 + 4y^2 - 12 = 0.$ (d) $x^2 + 4y^2 - 4 = 0.$

(e) $\dfrac{x^2}{9} + \dfrac{y^2}{16} = 1.$ (f) $\dfrac{x^2}{16} + \dfrac{y^2}{25} = 1.$

(g) $4x^2 + y^2 - 4 = 0.$ (h) $9x^2 + 4y^2 - 36 = 0.$

2. In each of the following, state the name of the curve and find the center, vertices, and foci. Sketch the graph.

(a) $16x^2 + 25y^2 - 32x - 100y = 284.$ (b) $9x^2 + 25y^2 + 36x - 50y = -36.$

(c) $16x^2 + 7y^2 + 32x - 28y = 68.$ (d) $9x^2 + 4y^2 - 36x + 8y + 4 = 0.$

3. In each of the following, find the equation of the ellipse defined by the given data.

(a) Vertices at $(\pm 4, 0)$; ends of minor axis at $(0, \pm 3)$.
(b) Vertices at $(\pm 4, 0)$; foci at $(\pm 3, 0)$.
(c) Vertices at $(0, \pm 5)$; foci at $(0, \pm 3)$.
(d) Foci at $(0, \pm 4)$; ends of minor axis at $(\pm 3, 0)$.
(e) Center at $(1, 2)$; vertices at $(6, 2)$ and $(-4, 2)$; foci at $(4, 2)$ and $(-2, 2)$.
(f) Center at $(2, -1)$; major axis is 8 units long; foci at $(2, 2)$ and $(2, -4)$.

Group B

1. Derive Eq. 12.5.4.

2. Derive Eq. 12.5.5.

3. Show that Eqs. 12.5.4 and 12.5.5 may be reduced to Eq. 12.1.1 and state the values of A, B, C, D, E, and F.

4. State the important parts of an ellipse.

5. Does the equation $ax^2 + by^2 + cx + dy + e = 0$ always represent an ellipse if $a > 0$ and $b > 0$? Be careful. Consider the equation $16x^2 + 25y^2 - 32x + 100y + 284 = 0$.

6. Use set notation to define an ellipse.

7. Solve the equation $x^2/a^2 + y^2/b^2 = 1$ for y. Is y a function of x? Why? Are there any restrictions on the allowable values of x? What are they?

8. Show that the parametric equations $x = \frac{1}{3} \sin t$, $y = \frac{1}{4} \cos t$ define an ellipse. Find the foci, vertices, major axis, minor axis, and center of this ellipse.

9. Find the area enclosed by the ellipse $9x^2 + 16y^2 - 144 = 0$. (*Hint*: Make use of symmetry.)

10. Find the centroid of the first-quadrant region bounded by $9x^2 + 16y^2 - 144 = 0$ and the coordinate axes.

11. Find the moment of inertia of an elliptic disc with respect to its minor axis. (*Hint*: Find i_Y for the region bounded by $x^2/a^2 + y^2/b^2 = 1$.)

12. Set up an integral to find the circumference of an ellipse. Use $x^2/a^2 + y^2/b^2 = 1$. [Note that this integral cannot be evaluated by direct integration. (Approximate methods such as the trapezoidal rule or infinite series are needed.) Integrals of this type lead to the study of "elliptic integrals."]

13. (a) Find a formula for the area enclosed by the ellipse $x^2/a^2 + y^2/b^2 = 1$. If $a = b$, how does this formula compare with the formula for the area enclosed by a circle of radius a? (b) Solve part (a) using the parametric equations $x = a \cos t$, $y = b \sin t$. (*Hint*: Show that $A = 4ab \int_0^{\pi/2} \sin^2 t \, dt$.)

14. A satellite is put into an elliptical orbit around the earth with the center of the earth at one focus. Assuming the radius of the earth to be 4000 miles (the equatorial semidiameter is 3963.296 miles), find the equation of the satellite's orbit if apogee is 100 miles and perigee is 50 miles.

15. In mechanical design one sometimes uses elliptical-shaped cams rather than circles when a slow, powerful stroke followed by a short quick return is desired. Suppose that an elliptical-shaped cam whose equation is $(x^2/16) + (y^2/9) = 1$ always touches a circle as in Fig. 12.5.6 and that the circle's center is allowed to slide to keep the cams in contact. What is the maximum distance the circle's center will slide?

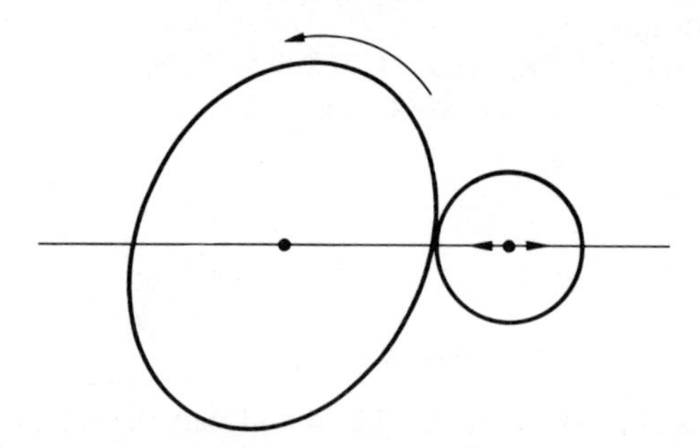

Figure 12.5.6 Exercise 15.

16. Find a formula for the focal width of an ellipse.

17. Find the radius of curvature of the ellipse $(x^2/4) + (y^2/9) = 1$ at the point where $x = 1$.

Group C

1. Consider the ellipse $x^2/25 + y^2/16 = 1$ as shown in Fig. 12.5.7. Find the equation of the tangent line TT at the point P where $x = 4$. Find α_1 and α_2. If line NN is normal to the ellipse at P, show that NN bisects the angle between the lines connecting point P and the foci.

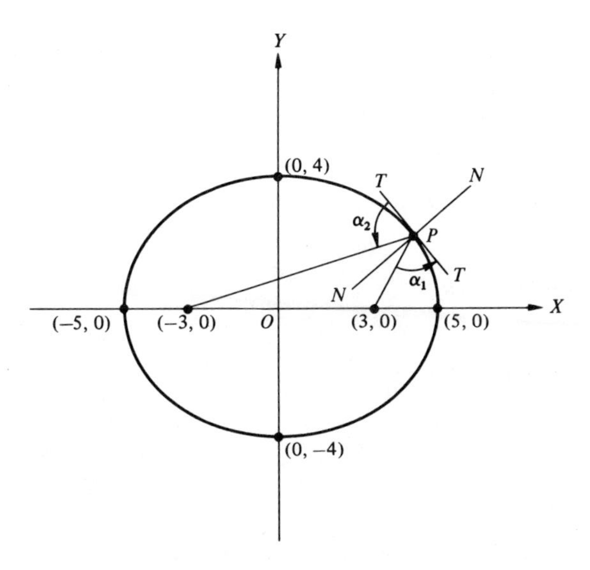

Figure 12.5.7 Exercise 1.

12.6 THE HYPERBOLA

Definition 12.6.1 *A hyperbola is defined to be the set of points in a plane such that the difference of the distances from each point to two fixed points, called the foci, is constant.*

For convenience in obtaining a simple equation of a hyperbola, let the foci F and F' be at $(c, 0)$ and $(-c, 0)$, respectively (see Fig. 12.6.1). Let the

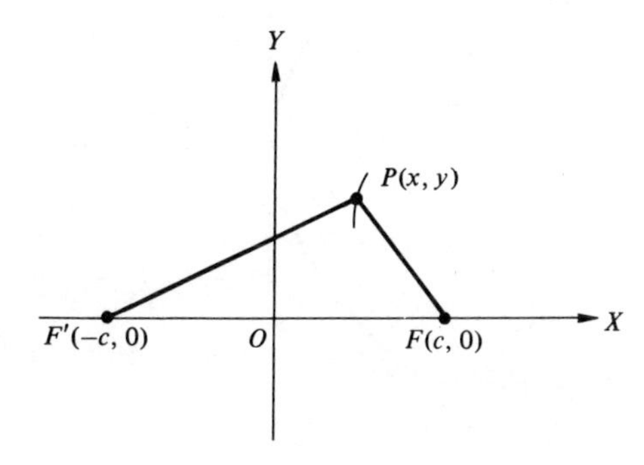

Figure 12.6.1 Definition of hyperbola.

constant difference be denoted by $2a$ (for convenience). If P is any point on the hyperbola, then from Definition 12.6.1 we have $|PF'| - |PF| = 2a$. Upon applying the distance formula Eq. 2.6.1, we obtain

$$\sqrt{(x + c)^2 + (y - 0)^2} - \sqrt{(x - c)^2 + (y - 0)^2} = 2a$$

If we transpose the second radical, square, simplify, transpose, square, and simplify again, as we did for the ellipse, we shall eventually arrive at

$$(c^2 - a^2)x^2 - a^2y^2 = a^2(c^2 - a^2)$$

From Fig. 12.6.1 it may be seen that

$$2a < 2c \qquad \text{or} \qquad a < c$$

This means that $c^2 - a^2$ is positive. We then let

$$b^2 = c^2 - a^2 \tag{12.6.1}$$

and our equation becomes

$$b^2x^2 - a^2y^2 = a^2b^2$$

Dividing by a^2b^2, we obtain

$$\boxed{\frac{x^2}{a^2} - \frac{y^2}{b^2} = 1} \tag{12.6.2}$$

The line on which the foci are located is called the *transverse axis*. The point on the transverse axis halfway between the foci is called the *center*. The points where the hyperbola crosses the transverse axis are called the *vertices*. The distance between the vertices is called the *length of the transverse axis*. Equation 12.6.2 is called the *standard form* of a hyperbola with its center at the origin and its transverse axis along the X-axis. The graph of $x^2/a^2 - y^2/b^2 = 1$ is shown in Fig. 12.6.2. When $y = 0$, $x = \pm a$. Thus the vertices are at

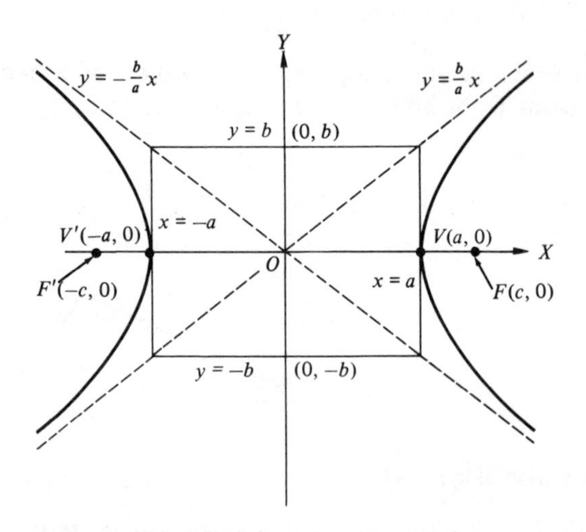

Figure 12.6.2 Graph of $(x^2/a^2) - (y^2/b^2) = 1$.

$(a, 0)$ and $(-a, 0)$ and the length of the transverse axis is $2a$. When $x = 0$, $y^2 = -b^2$, and thus there is no Y-intercept. If we solve for

$$y = \pm\frac{b}{a}\sqrt{x^2 - a^2} \tag{12.6.3}$$

and note that as $x \rightarrow \infty$, $\sqrt{x^2 - a^2} \rightarrow x$, we see that the hyperbola will approach but never reach the lines

$$y = \pm \frac{b}{a} x \tag{12.6.4}$$

These lines are called the *asymptotes* of the hyperbola. They are shown in Fig. 12.6.2 as dashed lines. You may show that the hyperbola is symmetric with respect to both the X- and Y-axis and to the origin.

The points $(0, b)$ and $(0, -b)$ are called *the end points of the conjugate axis*. Its length is $2b$. In sketching a hyperbola, it is best to draw the rectangle formed by the lines $y = \pm b$ and $x = \pm a$ and then to draw asymptotes which are the diagonals of the rectangle. The hyperbola lies outside the rectangle and inside the asymptotes.

If the foci are on the Y-axis, the equation becomes

$$\boxed{\frac{y^2}{a^2} - \frac{x^2}{b^2} = 1} \tag{12.6.5}$$

Note that a^2 is always associated with the positive term. Equation 12.6.5 may be obtained from Eq. 12.1.1 if $A = -1/b^2$, $B = 0$, $C = 1/a^2$, and $D = E = F = 0$. A hyperbola is recognized by the algebraic *signs* of the squared terms being *different*.

EXAMPLE 1. Find the center, vertices, foci, and asymptotes and sketch the graph of $16x^2 - 9y^2 - 144 = 0$.

Solution. This equation is a hyperbola and may be written as $(x^2/9) - (y^2/16) = 1$. It is then in the standard form of Eq. 12.6.2, where $a^2 = 9$ and $b^2 = 16$. Since $c^2 = a^2 + b^2$, from Eq. 12.6.1, we have $a = 3$, $b = 4$, and $c = 5$.

Center:	$(0, 0)$
Vertices:	$(\pm 3, 0)$
Foci:	$(\pm 5, 0)$
Asymptotes:	$y = \pm \frac{4}{3}x$

The graph is shown in Fig. 12.6.3.

Warning ***A hyperbola is recognized by the signs of the squared terms being different. The transverse axis always is associated with the positive sign as is a^2.***

PROBLEM 1. Find the center, vertices, foci, and asymptotes and sketch the graph of $9x^2 - 4y^2 - 36 = 0$.

If the center of the hyperbola is at (h, k), its equation may take one of the following forms. For

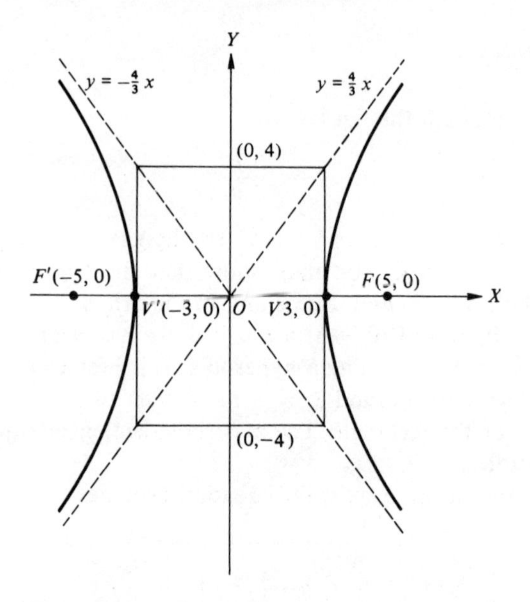

Figure 12.6.3 Graph of $16x^2 - 9y^2 - 144 = 0$.

$$\boxed{\frac{(x-h)^2}{a^2} - \frac{(y-k)^2}{b^2} = 1} \tag{12.6.6}$$

we have the transverse axis parallel to the X-axis:

Center: (h, k).
Vertices: $(h \pm a, k)$.
Foci: $(h \pm c, k)$.

For

$$\boxed{\frac{(y-k)^2}{a^2} - \frac{(x-h)^2}{b^2} = 1} \tag{12.6.7}$$

we have the transverse axis parallel to the Y-axis:

Center: (h, k)
Vertices: $(h, k \pm a)$
Foci: $(h, k \pm c)$

The general quadratic equation 12.1.1, with $B = 0$, may be written as one of the forms of Eq. 12.6.6 or 12.6.7 by completing the square on the x and on the y terms.

Example 2. Find the center, vertices, and foci and sketch the graph of $6x^2 - 5y^2 + 24x - 10y + 49 = 0$.

Solution. Rewrite the equation as

$$6(x^2 + 4x + \quad) - 5(y^2 + 2y + \quad) = -49$$

and complete the square. Thus

$$6(x^2 + 4x + 4) - 5(y^2 + 2y + 1) = -49 + 24 - 5$$
$$6(x + 2)^2 - 5(y + 1)^2 = -30$$
$$-\frac{(x + 2)^2}{5} + \frac{(y + 1)^2}{6} = 1$$

Then $a^2 = 6$, $b^2 = 5$, and $c^2 = 11$:

Center: $C(-2, -1)$
Vertices: $V(-2, 1.4)$, $V'(-2, -3.4)$
Foci: $F(-2, 2.3)$, $F'(-2, -4.3)$

The graph is shown in Fig. 12.6.4.

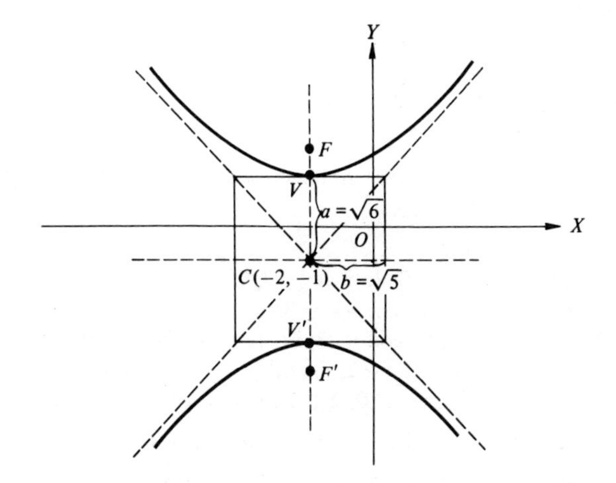

Figure 12.6.4 Graph of $6x^2 - 5y^2 + 24x - 10y + 49 = 0$.

PROBLEM 2. Find the center, vertices, and foci and sketch the graph of $2x^2 - 3y^2 - 4x - 6y + 5 = 0$.

Just as we found that the circle and the ellipse consisted of two functions, so also does the hyperbola. For if we solve $x^2/a^2 - y^2/b^2 = 1$ for y, we obtain

$$y = \pm\frac{b}{a}\sqrt{x^2 - a^2}$$

Thus we have two functions $y = (b/a)\sqrt{x^2 - a^2}$ and $y = -(b/a)\sqrt{x^2 - a^2}$. We thus must use caution in working problems involving hyperbolas. Note also that y has meaning only if $|x| \geqq a$.

It is of interest to note that the graph of the equation $xy = 1$ is a hyperbola whose transverse axis is the line making an angle of 45° with the X-axis. It may be changed to the standard form of a hyperbola by a rotation of axes (see Exercise 1 of Group C in Exercises 12.7).

EXERCISES 12.6

Group A

1. In each of the following, state the name of the curve and find the center, vertices, foci, and asymptotes. Sketch the graph.

(a) $9x^2 - 16y^2 - 144 = 0.$ (b) $\dfrac{x^2}{144} - \dfrac{y^2}{25} = 1.$
(c) $x^2 - y^2 - 1 = 0.$ (d) $3x^2 - y^2 - 3 = 0.$
(e) $16x^2 - 9y^2 + 144 = 0.$ (f) $y^2 - x^2 = 1.$
(g) $25x^2 - 144y^2 + (144)(25) = 0.$ (h) $3x^2 - y^2 + 3 = 0.$

2. In each of the following, state the name of the curve and find the center, vertices, and foci. Sketch the graph.

(a) $5x^2 - 6y^2 - 10x + 24y - 49 = 0.$
(b) $9x^2 - 16y^2 + 18x + 32y - 151 = 0.$
(c) $16x^2 - 9y^2 - 32x - 18y + 151 = 0.$
(d) $x^2 - y^2 - 4x - 2y + 4 = 0.$

3. In each of the following, find the equation of the hyperbola defined by the given data.

(a) Vertices at $(\pm 4, 0)$; foci at $(\pm 5, 0)$.
(b) Length of transverse axis is 10; foci at $(\pm 7, 0)$.
(c) Foci at $(\pm 5, 0)$; one asymptote is $4y = 3x$.
(d) Center at $(1, 6)$; foci at $(5, 6)$ and $(-3, 6)$; length of transverse axis is 6.
(e) Center at $(-2, -1)$; vertices at $(-2, 1)$ and $(-2, -3)$; slope of one asymptote is $\frac{3}{2}$.
(f) Center at $(-4, 2)$; one focus at $(0, 2)$; the length of the conjugate axis is 6.

4. Find the area bounded by $x^2/16 - y^2/9 = 1$ and $x = 6$.

5. Find $\bar{x}$ for the area bounded by $x^2/16 - y^2/9 = 1$ and $x = 6$ and above the X-axis.

6. Find the slope of $x^2/16 - y^2/9 = 1$ in the first quadrant when $x = 10$. How does this compare with the slope of the asymptote at $x = 10$?

7. Find the volume of revolution obtained by revolving the area bounded by $x^2 - y^2 = 1$, $y = 0$, $x = 1$, and $x = 2$, about the X-axis.

Group B

1. Does the equation $ax^2 + by^2 + cx + dy + e = 0$ always represent a hyperbola if a and b are of different algebraic sign?

2. Use set notation to define a hyperbola.

3. Derive Eq. 12.6.6.

4. Derive Eq. 12.6.7.

5. Show that Eqs. 12.6.6 and 12.6.7 may be reduced to Eq. 12.1.1 and state the values of A, B, C, D, E, and F.

6. State the important parts of a hyperbola.

7. Solve the equation $x^2/a^2 - y^2/b^2 = 1$ for y. Is y a function of x? Why? State any restrictions on allowable values of x.

8. (a) Show that the parametric equations $x = \cosh t$, $t = \sinh t$ represent a hyperbola (see Exercises 5 and 7 of Group B in Exercises 8.3). (b) Show that the parametric equations $x = 3 \sec t$, $y = 2 \tan t$ represent a hyperbola. (c) Find the area bounded by the hyperbola of part (b) and $x = 6$.

9. In the study of strength of materials one has the formula $S_t = (s_t/2) + \frac{1}{2}(s_t^2 + 4s_s^2)^{1/2}$, which gives the maximum tensile unit stress S_t when a bar is subjected to a combined tensile and twisting load. In this formula s_t is the tensile unit stress due to the axial load P lb and s_s is the shearing unit stress due to the twisting load. (a) Assuming that $s_s = 400$ lb/in.^{2}, sketch the graph of S_t as a function of s_t. Use the positive value of the square root. (b) Determine the relation approached between S_t and s_t as s_t increases without limit through positive values and give the geometric significance of this result. (c) Determine the relation approached between S_t and s_t as s_t decreases without limit through negative values. Give the geometric significance. (d) In this problem, the minimum tensile unit stress (or maximum compressive unit stress) is given by $S_C = (s_t/2) - \frac{1}{2}(s_t^2 + 4s_s^2)^{1/2}$ and the maximum shearing unit stress is given by $S_s = \frac{1}{2}(s_t^2 + 4s_s^2)^{1/2}$. Sketch these two curves on the same graph ($s_s = 400$ lb/in.^{2}). (Cell.)

10. A listening post at B is 4 miles east and 3 mi south of a listening post at A. The explosion of a gun is heard at B 10 s before it is heard at A, and the explosion sound comes from an easterly direction. Since sound travels at about 1086 ft/s, the gun is about 10,860 ft closer to B than to A. Choose the X-axis through the two listening posts and the Y-axis as the perpendicular bisector of the line segment AB and determine the equation of the hyperbola. (*Note*: The principle involved here is used in the Loran system of navigation.) (Cell.)

11. Find the radius of curvature of $x^2 - y^2 = 1$ at a point where $x = 2$.

12. Set up an integral that will give the length of the hyperbola $x^2 - y^2 = 1$ from $(1, 0)$ to $(2, \sqrt{3})$.

Group C

1. Find the vertical distance between a hyperbola and its asymptote. Then show that this distance approaches zero as $x \longrightarrow \infty$. (*Hint*: Use that portion of $x^2/a^2 - y^2/b^2 = 1$ that lies in QI.)

2. Find the horizontal distance between a point in the first quadrant on the hyperbola $(x^2/a^2) - (y^2/b^2) = 1$ and the corresponding asymptote. Then show that this distance approaches zero as $x \longrightarrow \infty$.

3. The tangent to a hyperbola at P meets the tangent at one vertex at Q. Prove that the line joining the other vertex to P is parallel to the line joining the center to Q.

4. Prove that the product of the perpendicular distances from any point on a hyperbola to the asymptotes is constant.

5. (a) Sketch the conic given by the parametric equations $x = \cos t$, $y = \sin t$. Sketch the branch of the conic given by the parametric equations $x = \cosh t$, $y = \sinh t$ for $x \geq 0$ (see Exercise 8 of Group B). (b) In Fig. 12.6.5(a) find the area of the sector of the circle AOP where P has coordinates $(\cos t, \sin t)$. (c) In Fig. 12.6.5(b) find the area of the region AOP where P has coordinates $(\cosh t, \sinh t)$. (*Hint*: The area desired is the area of triangle OBP minus the area bounded by the hyperbola $x^2 - y^2 = 1$, the X-axis, and the vertical line at $x = \cosh t$.) (d) Compare areas found in part (b) and (c). Using parts (a), (b), and (c), draw analogies between the trigonometric functions and the hyperbolic functions.

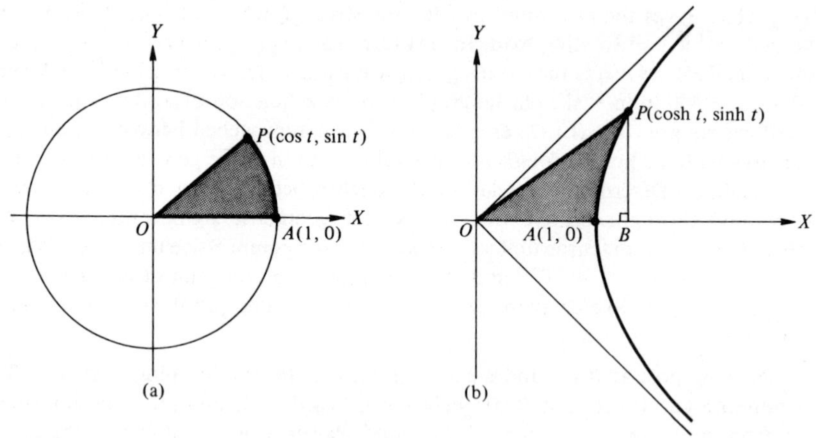

Figure 12.6.5 Exercise 5.

12.7 TRANSLATION OF AXES

We have seen that the general quadratic equation

$$Ax^2 + Bxy + Cy^2 + Dx + Ey + F = 0 \qquad (12.1.1)$$

represents many different curves. We now summarize some of our results.

Coefficients	Equation 12.1.1 Represents:
$A = B = C = 0$	straight line
$A = B = 0$	parabola
$B = C = 0$	parabola
$B = 0$, $A = C$	circle
$B = 0$, A and C same sign	ellipse
$B = 0$, A and C opposite signs	hyperbola

We found that when both the squared and the first-power terms were present we could complete the square to obtain a recognizable form. A method called *translation of axes* also helps us to simplify some equations. Translation of axes involves two coordinate systems whose axes are parallel. Let the point (h, k) in the XY-coordinate system be chosen as the origin of the $X'Y'$-coordinate system. Then any point, P, in the plane has two sets of coordinates, an (x, y) set and an (x', y') set. In Fig. 12.7.1, let (h, k) be the origin of the $X'Y'$-coordinate system. Then

$$x = x' + h \qquad y = y' + k \tag{12.7.1}$$

Also,

$$x' = x - h \qquad y' = y - k \tag{12.7.2}$$

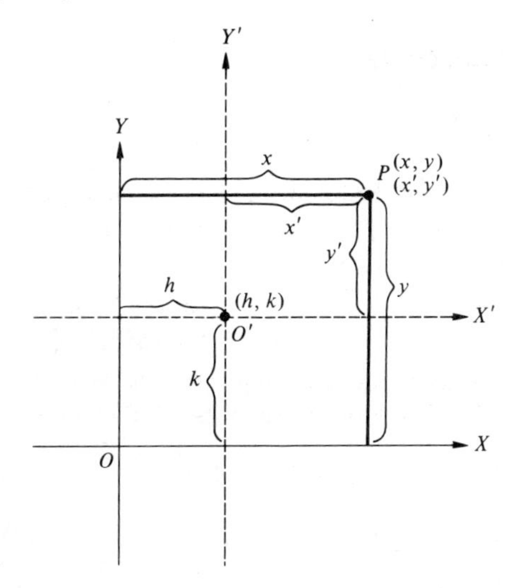

Figure 12.7.1 Translation of axes.

Example 1. What are the coordinates of $P(3, -4)$ in the $X'Y'$-coordinate system if the origin O' is at $(1, -5)$?

Solution. We have that $h = 1$ and $k = -5$. From Eq. 12.7.2 we find that

$$x' = 3 - 1 = 2$$
$$y' = -4 - (-5) = 1$$

Hence in the $X'Y'$-coordinate system the coordinates are $(2, 1)$.

Problem 1. What are the coordinates of $P(-2, 1.5)$ in the $X'Y'$-coordinate system if the origin O' is at $(-1, 3)$?

Example 2. The equation of the ellipse $[(x - 2)^2/4] + [(y + 1)^2/9] = 1$ may be simplified by translation of axes. Let $x' = x - 2$ and $y' = y + 1$; then the equation becomes

$$\frac{(x')^2}{4} + \frac{(y')^2}{9} = 1$$

which is an ellipse with center at the origin of the $X'Y'$-coordinate system. The origin of the new system is at $(2, -1)$ of the old system.

PROBLEM 2. The equation of a hyperbola is $[(x + 3)^2/4] - [(y - 2)^2/9] = 1$. Write the equation in terms of the $X'Y'$-coordinate system that would simplify the equation by a translation of axes. What are the coordinates of O' in the XY-coordinate system?

The xy term in the general quadratic equation may be eliminated by a rotation of axes. For a discussion of this, see textbooks on analytic geometry.

EXERCISES 12.7

Group A

1. In each of the following, find the coordinates of the given point in terms of (x', y') for the origin O' at $(1, -5)$. Plot each point in the XY-coordinate system and the $X'Y'$-coordinate system.

(a) $P_1(-3, 4)$. (b) $P_2(2, 3)$. (c) $P_3(-1, -1)$.
(d) $P_4(2, -1)$. (e) $P_5(-1, 5)$. (f) $P_6(0, 0)$.

2. In each of the following, write the given equation in terms of x' and y' if O' is at $(2, -1)$. Sketch the curve in the $X'Y'$-coordinate system.

(a) $(x - 2)^2 + (y + 1)^2 = 4$. (b) $\dfrac{(x - 2)^2}{9} + \dfrac{(y + 1)^2}{4} = 1$.

(c) $(x - 2)^2 = 4(y + 1)$. (d) $\dfrac{(x - 2)^2}{9} - \dfrac{(y + 1)^2}{4} = 1$.

3. In each of the following, find (h, k) such that the equation is simplified by a translation of axes.

(a) $x^2 + y^2 + 4x - 6y + 5 = 0$.
(b) $4x^2 + 8x - 16y - 12 = 0$.
(c) $16x^2 + 25y^2 - 32x + 100y - 284 = 0$.
(d) $6x^2 - 5y^2 + 24x - 10y + 49 = 0$.

Group B

1. Does translation of axes change the shape of the curve?

2. In the equation $x^2 + xy + y^2 - 2x + 2y + 1 = 0$, the linear terms may be removed by a translation of axes. Find h and k for such a translation. (*Hint*: Let $x = x' + h$ and $y = y' + k$. Substitute and after simplifying, equate the coefficients of x' and y' equal to zero and solve for h and k.)

Group C

1. A rotation of axes is used to eliminate the xy term in the general quadratic equation 12.1.1. When the XY-axes are rotated through an angle θ to an $X'Y'$ set of axes, the equations of rotation are

$$x = x' \cos \theta - y' \sin \theta$$
$$y = x' \sin \theta + y' \cos \theta$$

Show that $xy = 1$ becomes the standard form of a hyperbola in the $X'Y'$-coordinate system when $\theta = 45°$. Sketch the curve in the $X'Y'$-coordinate system.

2. Simplify the equation $x^2 - xy + y^2 = 1$ by rotating the XY-axes through an angle of $\theta = 45°$. What familiar curve does this equation represent?

SUMMARY OF IMPORTANT WORDS AND CONCEPTS

For each of the following, state in your own words your understanding of the statement or word.

1. (a) Conic sections. (b) General quadratic equation.

2. Symmetry with respect to (a) X-axis. (b) Y-axis. (c) Origin.

3. Circle: center, radius, standard form.

4. Parabola: vertex, focus, axis, directrix, focal width, standard form.

5. Ellipse: center, foci, vertices, major axis, minor axis, standard form.

6. Hyperbola: center, foci, vertices, transverse axis, asymptotes, conjugate axis, standard form.

7. Translation of axes.

Chapter 13

Polar Coordinates

13.1 THE POLAR COORDINATE SYSTEM

The position of a point in a plane may be determined in many ways. All that is necessary is that the system determines only one point for a set of data. One of the most useful coordinate systems is the Cartesian or XY-coordinate system which we have used so far. This system determines a point by means of two distances x and y. Another very useful system uses a direction and a distance. This system is called the *polar coordinate system.*

In Fig. 13.1.1 let the fixed line OM be the *polar axis* and O the *origin.* Then point P may be determined by angle θ measured from the polar axis to the line OP and the length of the line segment OP denoted by r called the *polar distance*. We then say that point P has polar coordinates (r, θ).

If θ is positive, it is measured in a counterclockwise direction. If θ is negative, it is measured in a clockwise direction. If r is positive, it is measured from O out along the terminal side of θ. If r is negative, it is measured from O along the backward extension of the terminal side of θ. It thus follows that many

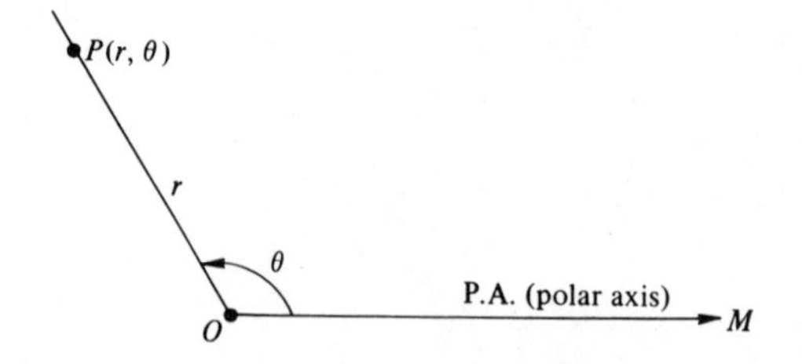

Figure 13.1.1 Polar coordinates.

pairs of coordinates may be used to determine a point in the polar coordinate systems, but only one point is determined by any one pair of polar coordinates. For convenience, we usually use positive θ.

EXAMPLE 1. Plot each of the given points in the polar coordinate system (a) $(3, \pi/6)$; (b) $(2, -\pi/2)$; (c) $(-1, \pi/3)$; (d) $(2, 11\pi/4)$.

Solution. The points are shown in Fig. 13.1.2.

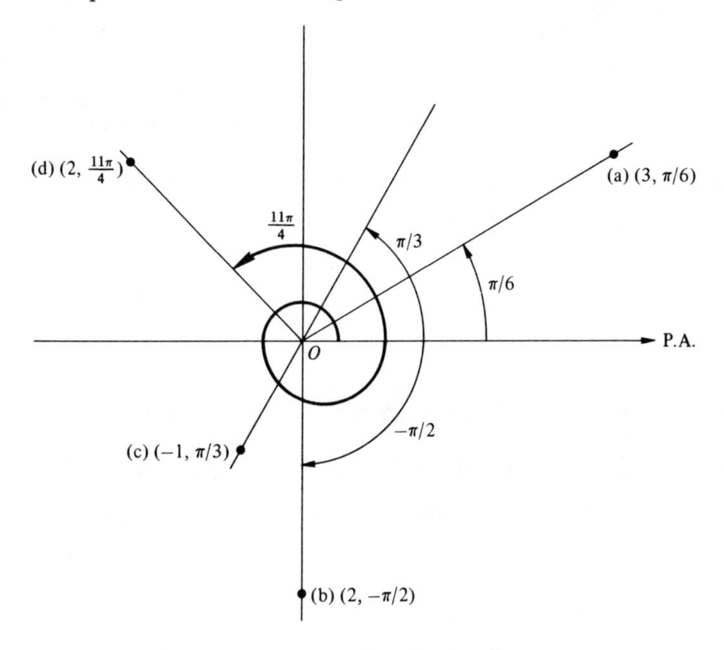

Figure 13.1.2 Points in polar coordinates.

PROBLEM 1. Plot each of the given points in the polar coordinate system. (a) $(1, \pi/4)$; (b) $(3, -\pi/2)$; (c) $(2, 2.5)$; (d) $(-1, 7\pi/6)$.

If the polar axis coincides with the positive X-axis and the origins of each axis are the same, then the relation between (r, θ) and (x, y) are given by

$$\boxed{\begin{aligned} x &= r\cos\theta \\ y &= r\sin\theta \end{aligned}} \tag{13.1.1}$$

and

$$\boxed{\begin{aligned} r &= \sqrt{x^2 + y^2} \\ \tan\theta &= \frac{y}{x} \end{aligned}} \tag{13.1.2}$$

These are obtained by considering the definitions of the trigonometric functions and Fig. 13.1.3.

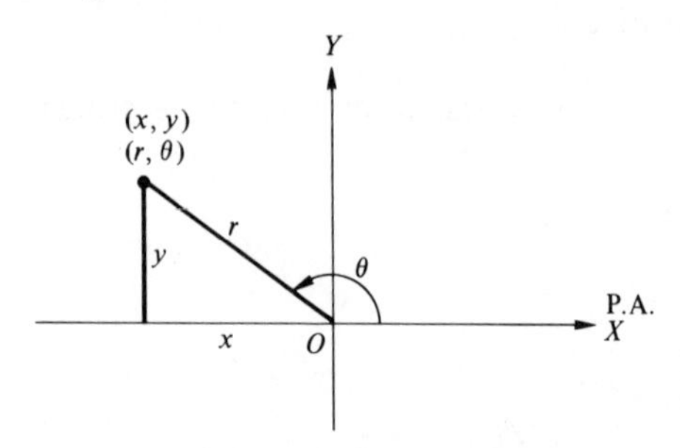

Figure 13.1.3 Relation between Cartesian coordinates and polar coordinates.

EXAMPLE 2. Find the Cartesian coordinates of $P(r, \theta)$ when P is at $(5, 150°)$.

Solution. Using Eqs. 13.1.1, we find that

$$x = 5 \cos 150° = -5 \cos 30° = \frac{-5\sqrt{3}}{2} = -4.33$$

$$y = 5 \sin 150° = 5 \sin 30° = \frac{5}{2} = 2.5$$

PROBLEM 2. Find the Cartesian coordinates of $P(r, \theta)$ when P is at $(3, 210°)$.

EXAMPLE 3. Find the polar coordinates of $P(x, y)$ when P is at $(5, -12)$.

Solution. A sketch helps (see Fig. 13.1.4). Using Eqs. 13.1.2, we find that

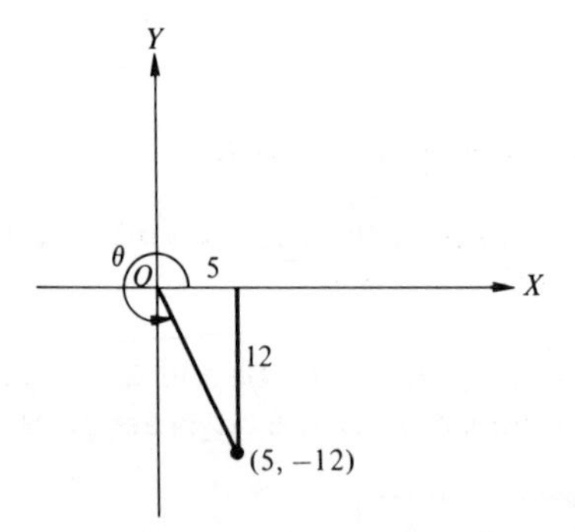

Figure 13.1.4 Example 3.

$$r = \sqrt{(5)^2 + (-12)^2} = 13$$

and

$$\tan \theta = \frac{-12}{5}$$

This implies that θ is either a second- or fourth-quadrant angle. But since x is positive and y is negative, θ must be in the fourth quadrant. Hence

$$\theta = 292.6°$$

PROBLEM 3. Find the polar coordinates of $P(x, y)$ when P is at $(-3, -4)$.

EXAMPLE 4. Transform the Cartesian equation $x^2 + 3y^2 = 4$ into a polar equation.

Solution. Since $x = r\cos\theta$ and $y = r\sin\theta$, we have

$$r^2\cos^2\theta + 3r^2\sin^2\theta = 4$$

or

$$r^2 = \frac{4}{\cos^2\theta + 3\sin^2\theta}$$

and

$$r^2 = \frac{4}{1 + 2\sin^2\theta}$$

PROBLEM 4. Transform the Cartesian equation $x^2 - 8y = 0$ into a polar equation.

Warning ***For polar coordinates r is always listed as the first member of the ordered pair and θ the second member. θ may be measured in radians or degrees.***

EXERCISES 13.1

Group A

1. Plot each of the given points in the polar coordinate system.

(a) $(2, \pi/6)$. (b) $(1, \pi/2)$. (c) $(2, 3\pi/2)$.
(d) $(-2, \pi)$. (e) $(2, 0)$. (f) $(1, 7\pi/2)$.
(g) $(-2.8, 3\pi/4)$. (h) $(-4, -\pi)$. (i) $(-1, \pi/2)$.

2. What do the pairs of coordinates $(0, 0)$, $(0, \pi/2)$, $(0, -\pi)$, and $(0, 3\pi/11)$ have in common?

3. Find the Cartesian coordinates for each of the following points given in polar coordinates.

(a) $(5, 120°)$. (b) $(3, 45°)$. (c) $(\frac{1}{2}, 300°)$.
(d) $(7, \pi)$. (e) $(6, 3\pi/2)$. (f) $(-4, 2\pi)$.
(g) $(6, -\pi)$. (h) $(-2, -\pi/2)$. (i) $(3, -400°)$.

4. Find a set of polar coordinates for each of the following points given in Cartesian coordinates.

(a) $(5, 12)$. (b) $(\sqrt{3}, -1)$. (c) $(-2, 1)$.
(d) $(-3, 0)$. (e) $(0, 6)$. (f) $(0, -6)$.

5. Find the polar coordinates for each of the following points given in Cartesian coordinates such that $r > 0$ and $-\pi < \theta \leqq \pi$.

(a) $(2, 2)$. (b) $(0, 1)$. (c) $(0, -1)$.

(d) $(-2, 2)$. (e) $(-2, 0)$. (f) $(-2, -2)$.

6. Transform each of the following Cartesian equations into a polar equation.

(a) $x^2 + y^2 = 16$. (b) $3x^2 + y^2 = 4$. (c) $y^2 = 4x$.
(d) $3x + y = 4$. (e) $x^2 - y^2 = 4$. (f) $xy - 7 = 0$.

7. Transform each of the following polar equations into a Cartesian equation.

(a) $r = 9$. (b) $r = 4 \cos \theta$. (c) $\theta = \pi/4$.

(d) $r = 3 \sin 2\theta$. (e) $r^2 = \dfrac{4}{1 + 2\cos^2\theta}$ (f) $r = 1 - 2\cos\theta$.

(g) $r^2 = 4 \cos 2\theta$. (h) $r = 5 \sec \theta$.

8. Explain how polar coordinates of a point and the position vector of a point are alike and how they are different.

9. Sometimes r is called the *radius vector* and θ is called the *vectorial angle* of point P. Explain why this terminology might be appropriate.

13.2 GRAPHS IN POLAR COORDINATES

When an equation is given in polar coordinates its curve may be drawn by plotting several points and drawing a smooth curve through them. The plotting of points in polar coordinates is helped by using polar coordinate graph paper, as shown in Fig. 13.2.1. The angle θ is determined by the number at the end of the lines and r is determined by use of the concentric circles.

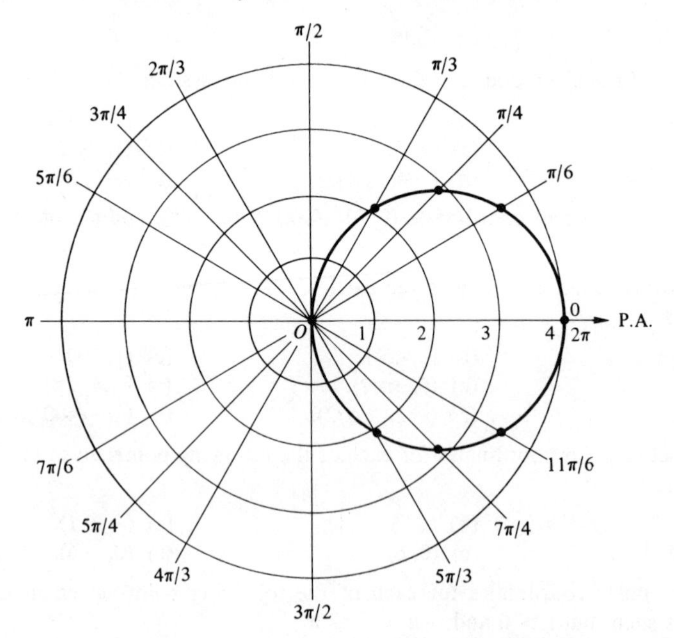

Figure 13.2.1 Graph of $r = 4 \cos \theta$ (circle).

EXAMPLE 1. Draw the graph of $r = 4 \cos \theta$.

Solution. Construct a table of values where θ is the independent variable:

θ	$0°$ 0	$30°$ $\pi/6$	$45°$ $\pi/4$	$60°$ $\pi/3$	$90°$ $\pi/2$	$120°$ $2\pi/3$	$135°$ $3\pi/4$	$150°$ $5\pi/6$	$180°$ π
r	4	3.5	2.8	2.0	0	−2.0	−2.8	−3.5	−4

The graph is drawn in Fig. 13.2.1. We stopped the table at $\theta = \pi$ because as θ increases between π and 2π, the points (r, θ) will be the same points as obtained for θ between 0 and π. That is, for $\theta = 7\pi/6$, $r = -3.5$. This point $(-3.5, 7\pi/6)$ coincides with the point $(3.5, \pi/6)$. Incidentally, the Cartesian equation of this curve is $x^2 + y^2 - 4x = 0$, which is a circle with its center at $(2, 0)$ and radius 2.

PROBLEM 1. Draw the graph of $r = 2 \sin \theta$. Obtain the corresponding Cartesian equation.

EXAMPLE 2. Draw the graph of $r = 4$.

Solution. Since $r = 4$ for all θ, the graph is a circle of radius 4 (see Fig. 13.2.2).

PROBLEM 2. Draw the graph of $r = 5$. Obtain the corresponding Cartesian equation. Which equation is simpler?

EXAMPLE 3. Draw the graph of $\theta = \pi/3$.

Solution. Since $\theta = \pi/3$ for all r, the graph is a straight line passing through the origin (see Fig. 13.2.3).

PROBLEM 3. Draw the graph of $\theta = -\pi/6$. What is the corresponding Cartesian equation?

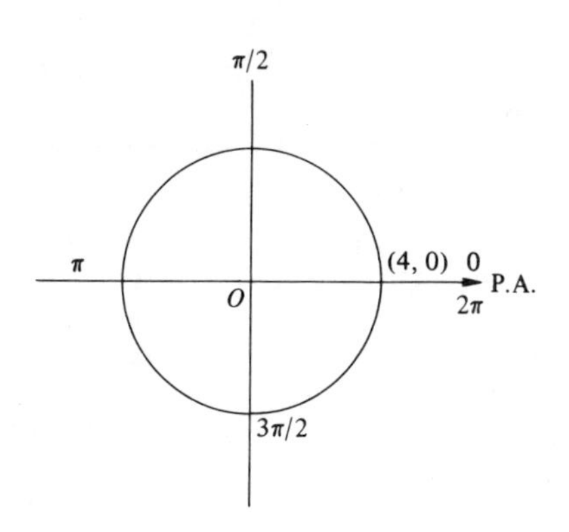

Figure 13.2.2 Graph of $r = 4$ (circle).

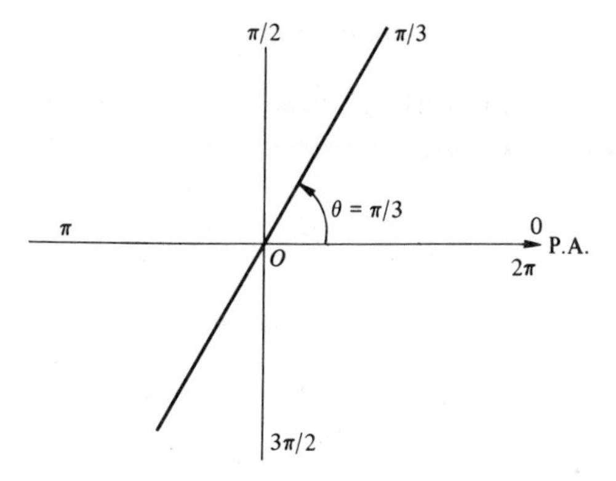

Figure 13.2.3 Graph of $\theta = \pi/3$ (straight line).

EXAMPLE 4. Draw the graph of $r = 3 + 2 \sin \theta$.

Solution

θ	0° 0	45° $\pi/4$	90° $\pi/2$	135° $3\pi/4$	180° π	225° $5\pi/4$	270° $3\pi/2$	315° $7\pi/4$	360° 2π
r	3	3.4	5	3.4	3	1.6	1	1.6	3

The graph is shown in Fig. 13.2.4. Graphs of the equations $r = a + b \sin \theta$ or $r = a + b \cos \theta$ for $|a| > |b|$ are called *cardioids*.

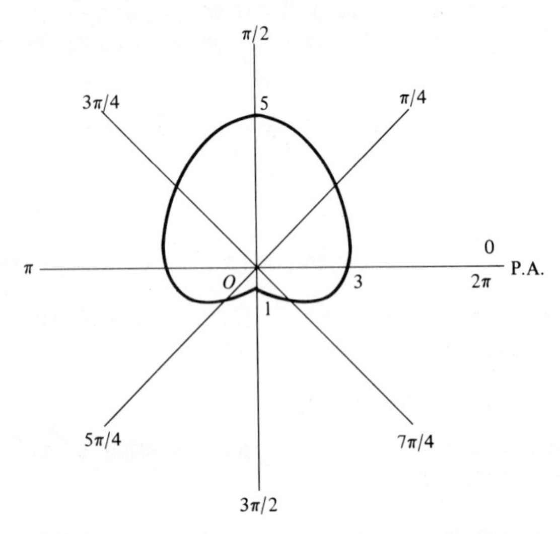

Figure 13.2.4 Graph of $r = 3 + 2 \sin \theta$ (cardioid).

PROBLEM 4. Draw the graph of $r = 3 + 2 \cos \theta$. Can you find the corresponding Cartesian equation?

EXAMPLE 5. Draw the graph of $r = 1 - 2 \cos \theta$.

Solution

θ	0° 0	45° $\pi/4$	60° $\pi/3$	90° $\pi/2$	135° $3\pi/4$	180° π	225° $5\pi/4$	270° $3\pi/2$	300° $5\pi/3$	315° $7\pi/4$	360° 2π
r	−1	−0.4	0	1	2.4	3	2.4	1	0	−0.4	−1

The graph is shown in Fig. 13.2.5. Graphs of the equations $r = a + b \sin \theta$ or $r = a + b \cos \theta$, where $|a| < |b|$ are called *limaćons*.

PROBLEM 5. Draw the graph of $r = 1 - 2 \sin \theta$. Try to obtain the corresponding Cartesian equation.

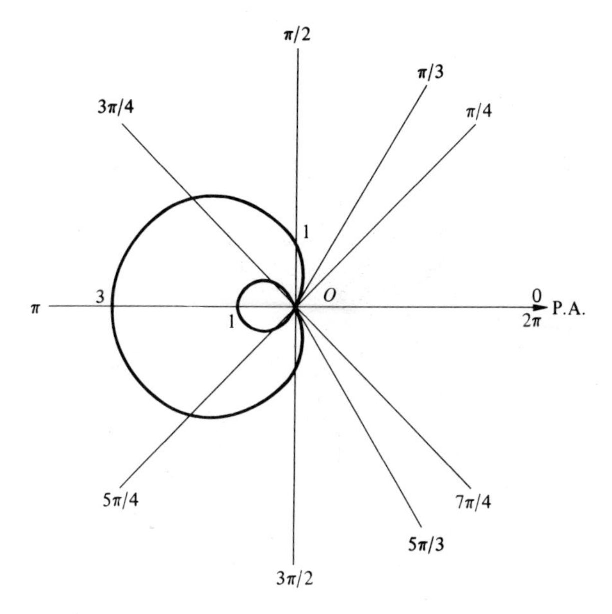

Figure 13.2.5 Graph of $r = 1 - 2\cos\theta$ (limaćon).

EXAMPLE 6. Draw the graph of $r = 2 \sin 3\theta$.

Solution

θ	0° 0	15° $\pi/12$	30° $\pi/6$	45° $\pi/4$	60° $\pi/3$	75° $5\pi/12$	90° $\pi/2$	105° $7\pi/12$	120° $2\pi/3$	135° $3\pi/4$	150° $5\pi/6$	165° $11\pi/12$	180° π
r	0	1.4	2	1.4	0	−1.4	−2	−1.4	0	1.4	2	1.4	0

The graph is shown in Fig. 13.2.6. As θ increases from π to 2π the values of r will be those already found because $\sin 3(\pi + \theta) = -\sin 3\theta$.

Graphs of the equations $r = a \sin b\theta$ or $r = a \cos b\theta$ are called *roses*. They will have b leaves for b odd and $2b$ leaves for b even.

PROBLEM 6. Draw the graph of $r = 4 \cos 3\theta$. Try to obtain the corresponding Cartesian equation.

EXAMPLE 7. Draw the graph of $r^2 = 4 \cos 2\theta$.

Solution. $r = \pm 2\sqrt{\cos 2\theta}$:

θ	0° 0	30° $\pi/6$	45° $\pi/4$	⟵⟶	135° $3\pi/4$	150° $5\pi/6$	180° π
r	±2	±1.4	0	no real value for r	0	±1.4	±2

For $\pi/4 < \theta < 3\pi/4$, $\cos 2\theta < 0$; hence there will be no real value of r for these values of θ (why?). As θ increases from π to 2π, no new values of r will be found. The graph is shown in Fig. 13.2.7. Graphs of the equations $r^2 = a \cos 2\theta$ and $r^2 = a \sin 2\theta$ are called *lemniscates*.

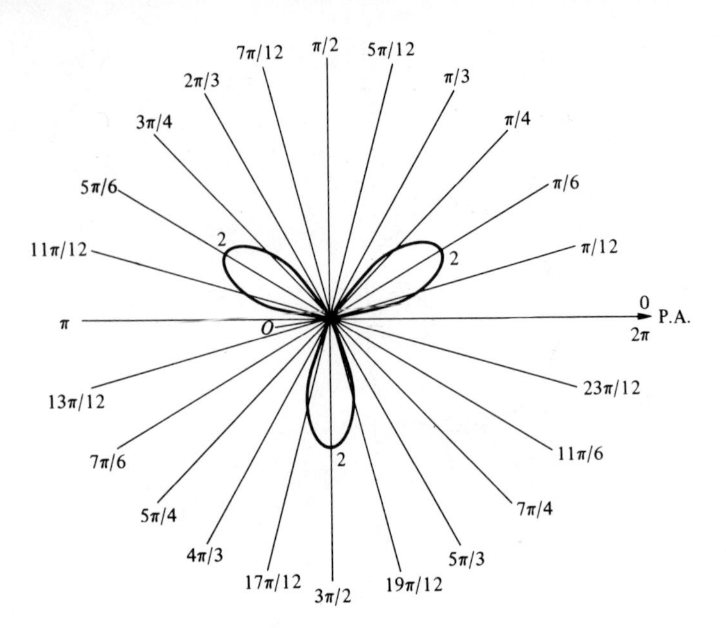

Figure 13.2.6 Graph of $r = 2 \sin 3\theta$ (three-leaf rose).

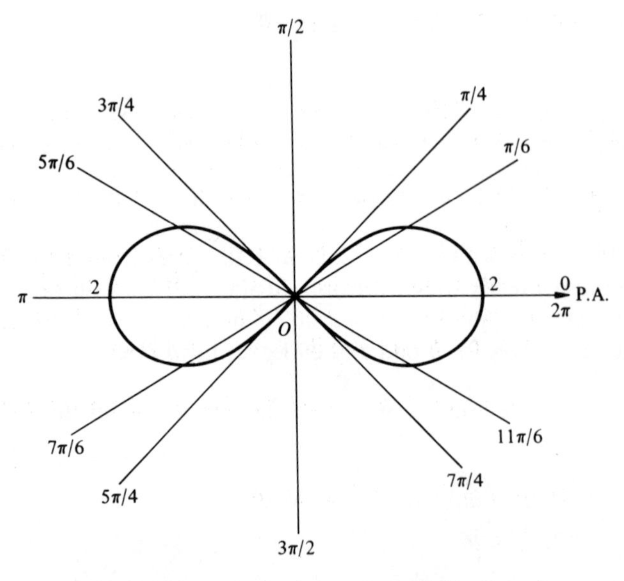

Figure 13.2.7 Graph of $r^2 = 4 \cos 2\theta$ (lemniscate).

PROBLEM 7. Draw the graph of $r^2 = 9 \sin 2\theta$. Try to obtain the corresponding Cartesian equation.

EXAMPLE 8. The graph of $r = a\theta$ is shown in Fig. 13.2.8. It is called the *spiral of Archimedes*. The solid line is for $\theta > 0$; the dashed line is for $\theta < 0$.

PROBLEM 8. Draw the graph of $r = \theta/2$. Can you obtain the corresponding Cartesian equation?

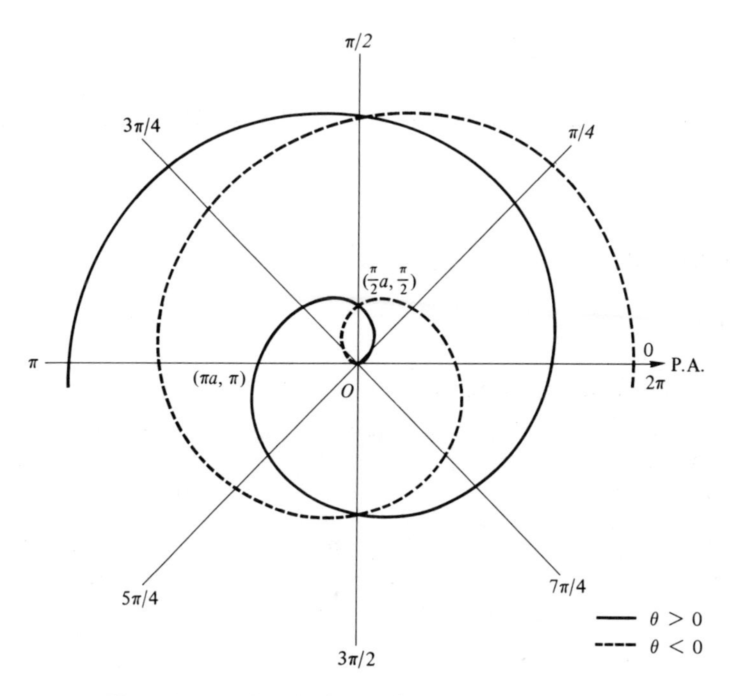

Figure 13.2.8 Graph of $r = a\theta$ (spiral of Archimedes).

EXAMPLE 9. The graph of $r = e^{a\theta}$ is shown in Fig. 13.2.9. It is called the *logarithmic spiral*. The solid portion is for $\theta > 0$; the dashed portion is for $\theta < 0$.

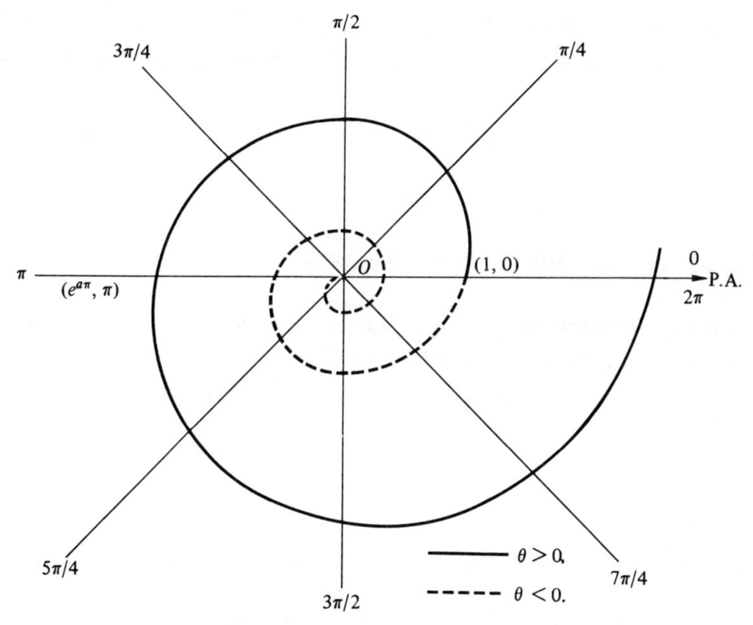

Figure 13.2.9 Graph of $r = e^{a\theta}$ (logarithmic spiral).

PROBLEM 9. Draw the graph of $r = e^{0.2\theta}$. Can you obtain the corresponding Cartesian equation?

EXERCISES 13.2

Group A

1. Draw the graph of each of the following.

(a) $r = 2 \cos \theta$.
(b) $r = 4 \sin \theta$.
(c) $r = 3$.
(d) $\theta = 3\pi/4$.
(e) $r = 2 + \sin \theta$.
(f) $r = 2 - \sin \theta$.
(g) $r = 2 + \cos \theta$.
(h) $r = 3 - 2 \cos \theta$.
(i) $r = 1 + 2 \cos \theta$.
(j) $r = 2 - 3 \sin \theta$.
(k) $r = 2 + 5 \sin \theta$.
(l) $r = 4 \sin 3\theta$.
(m) $r = 4 \sin 2\theta$.
(n) $r = 2 \cos 3\theta$.
(o) $r = 2 \cos 2\theta$.
(p) $r^2 = 9 \cos 2\theta$.
(q) $r^2 = 4 \sin 2\theta$.
(r) $r = \theta$.
(s) $r = e^{\theta}$.
(t) $r = e^{-\theta}$.

Group B

1. How do the graphs of $y = \sin x$ and $r = \sin \theta$ differ? Does the graph of $y = \sin x$ help in drawing the graph of $r = \sin \theta$?

2. Draw the graph of $y = 2 \sin 3x$, $0 \leqq x \leqq 2\pi$. Compare this graph with Fig. 13.2.6. Do you see how this graph could help you in drawing the graph of $r = 2 \sin 3\theta$?

3. State the coordinate system in which each of the following curves have the simplest equation.

(a) Circles.
(b) Straight lines.
(c) Cardioids.
(d) Parabolas.
(e) Spirals.

13.3 POLAR EQUATIONS OF THE CONICS

The Cartesian equation of a straight line is $ax + by + c = 0$. By using Eqs. 13.1.1 and changing to polar coordinates it becomes

$$a(r \cos \theta) + b(r \sin \theta) + c = 0$$

or

$$r = \frac{-c}{a \cos \theta + b \sin \theta} \tag{13.3.1}$$

We also can find the equation of a line by using trigonometry.

EXAMPLE 1. Write the equation of the line perpendicular to the polar axis and 3 units to the right of the origin.

Solution. In Fig. 13.3.1, let $P(r, \theta)$ be any point on line L. Then we find

$$r = 3 \sec \theta$$

You may show that Eq. 13.3.1 reduces to this equation.

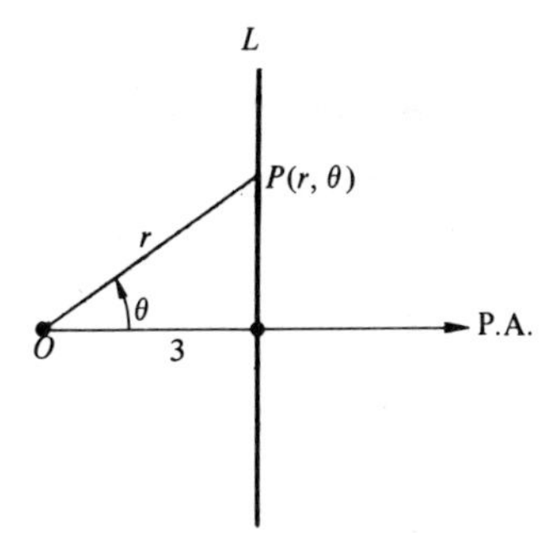

Figure 13.3.1 Example 1.

PROBLEM 1. Write the equation of the line parallel to the polar axis and 2 units below it.

The polar equations for some circles were given in Sec. 13.2 as examples. In summary,

$r = a$ is a circle of radius a	center at origin
$r = a \cos \theta$ is a circle of radius $a/2$	center at $(a/2, 0)$
$r = a \sin \theta$ is a circle of radius $a/2$	center at $(a/2, \pi/2)$

In general, a circle of radius a and center at (c, α) is $r^2 - 2cr \cos(\theta - \alpha) + c^2 = a^2$.

In a more comprehensive study of the conics than we gave in Chap. 12 the concept of eccentricity is used in defining parabolas, ellipses, and hyperbolas.

In Fig. 13.3.2, let O be the focus and line L be the directrix intersecting the the polar axis extended at (p, π). The eccentricity, denoted as e, is given by

$$e = \frac{|OP|}{|PM|}$$

where this ratio is a constant for any given curve. Then it can be shown that

$$r = \frac{ep}{1 - e \cos \theta} \tag{13.3.2}$$

is *a parabola* if $e = 1$, an *ellipse* if $0 < e < 1$, and a *hyperbola* if $e > 1$.

Equation 13.3.2 is called the *polar equation for the conics*, where e is the eccentricity and p is the distance from the focus to the directrix. For other forms of the conics see Exercise 1 Group B.

EXAMPLE 2. Identify the conic $r = 3/(2 - \cos \theta)$.

Solution. To compare this equation with Eq. 13.3.2 we must divide by 2 and

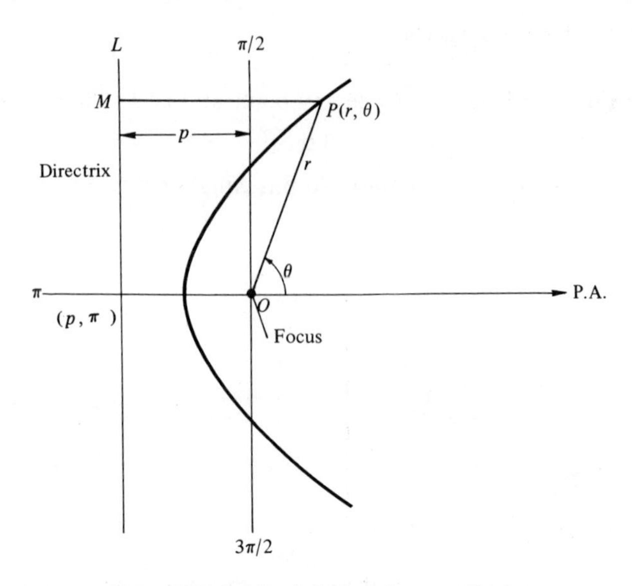

Figure 13.3.2 Parabola in polar coordinates.

rewrite it as

$$r = \frac{\frac{3}{2}}{1 - \frac{1}{2}\cos\theta}$$

Then $e = \frac{1}{2}$, $p = 3$, and the curve is an ellipse.

PROBLEM 2. Identify the conic $r = 5/(2 - 4\cos\theta)$.

EXERCISES 13.3

Group A

1. In each of the following, write the polar equation of the specified curve.

(a) A line perpendicular to the polar axis and 4 units to the left of the origin.
(b) A line parallel to the polar axis and 3 units above it.
(c) A line passing through the origin with a slope of $1/\sqrt{3}$.
(d) A circle of radius 5 and center at the origin.
(e) A circle of radius 4 and center at $(8, \pi/2)$.
(f) A circle of radius 2 and center at $(2, \pi/4)$.

2. Identify each of the following conics and state the values of e and p.

(a) $r = \dfrac{2}{1 - \cos\theta}$. (b) $r = \dfrac{2}{3 - \cos\theta}$.

(c) $r = \dfrac{3}{2 - 3\cos\theta}$. (d) $r = \dfrac{2}{1 - 3\cos\theta}$.

3. Identify and sketch each of the following conic sections. (*Hint*: Let $\theta = 0, \pi/2, \pi,$ etc. and compute the value of r.)

(a) $r = \dfrac{2}{1 - \cos\theta}$.

(b) $r = \dfrac{12}{3 - \cos\theta}$.

(c) $r = \dfrac{1}{1 - \sqrt{2}\cos\theta}$.

(d) $r = \dfrac{4}{1 - \cos\theta}$.

Group B

1. If the directrix is perpendicular to the polar axis at $(p, 0)$, then

$$r = \frac{ep}{1 + e\cos\theta}$$

If the directrix is parallel to the polar axis and above it,

$$r = \frac{ep}{1 + e\sin\theta}$$

If the directrix is parallel to the polar axis and below it,

$$r = \frac{ep}{1 - e\sin\theta}$$

Sketch each of the following conics.

(a) $r = \dfrac{2}{1 + \cos\theta}$.

(b) $r = \dfrac{2}{1 + \sin\theta}$.

(c) $r = \dfrac{2}{1 - \sin\theta}$.

(d) $r = \dfrac{12}{3 + \cos\theta}$.

(e) $r = \dfrac{1}{1 + \sqrt{2}\sin\theta}$.

(f) $r = \dfrac{4}{1 - 2\sin\theta}$.

13.4 VELOCITY IN POLAR COORDINATES

In Sec. 7.11 we discussed the velocity of a particle as it moved along a curve in the XY-plane. We saw that $v^2 = v_x^2 + v_y^2$, where v_x and v_y are components of the velocity in the X- and Y-directions, respectively. Now we shall discuss the velocity of a particle as it moves along a curve in polar coordinates.

Using Eqs. 13.1.1,

$$x = r\cos\theta \qquad \text{and} \qquad y = r\sin\theta$$

we differentiate with respect to time t to obtain

$$\frac{dx}{dt} = v_x = r(-\sin\theta)\frac{d\theta}{dt} + \frac{dr}{dt}\cos\theta$$

and

$$\frac{dy}{dt} = v_y = r\cos\theta\frac{d\theta}{dt} + \frac{dr}{dt}\sin\theta$$

Squaring these equations, we have

$$v_x^2 = r^2\sin^2\theta\left(\frac{d\theta}{dt}\right)^2 - 2r\sin\theta\cos\theta\frac{d\theta}{dt}\frac{dr}{dt} + \left(\frac{dr}{dt}\right)^2\cos^2\theta$$

and

$$v_y^2 = r^2 \cos^2 \theta \left(\frac{d\theta}{dt}\right)^2 + 2r \sin \theta \cos \theta \frac{d\theta}{dt}\frac{dr}{dt} + \left(\frac{dr}{dt}\right)^2 \sin^2 \theta$$

Now adding and recalling that $\sin^2 \theta + \cos^2 \theta = 1$, we obtain

$$v_x^2 + v_y^2 = \left(r\frac{d\theta}{dt}\right)^2 + \left(\frac{dr}{dt}\right)^2 \tag{13.4.1}$$

If we define

$$r\frac{d\theta}{dt} = v_\theta$$

and

$$\frac{dr}{dt} = v_r$$

we have

$$v^2 = v_x^2 + v_y^2 = v_\theta^2 + v_r^2 \tag{13.4.2}$$

where v_θ is called the *transverse velocity* and v_r is called the *radial velocity*. Note that v_r is the component of the velocity in the direction of r and that v_θ is the component of the velocity in the direction perpendicular to r (see Fig. 13.4.1). The expression $d\theta/dt$ is the rate of change of θ with respect to t and is called the *angular* velocity of the line from the origin to the particle. It is denoted by the lowercase Greek letter omega (ω). Thus

$$\omega = \frac{d\theta}{dt} \tag{13.4.3}$$

(ω is sometimes called the angular velocity of the particle.) Equation 13.4.1 may then be written as

$$v^2 = r^2\omega^2 + \left(\frac{dr}{dt}\right)^2$$

Note that if r is constant, then $v = r\omega$.

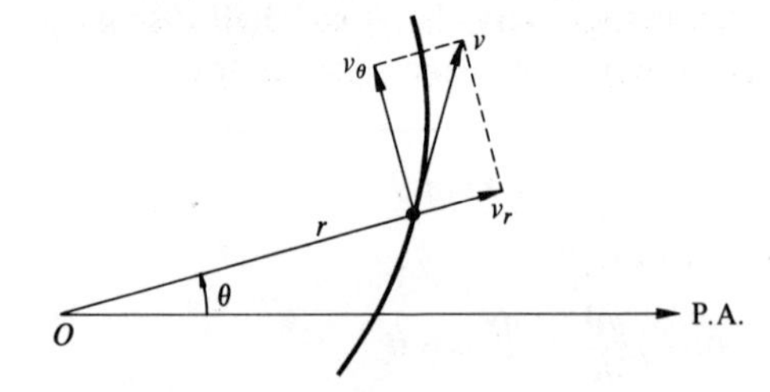

Figure 13.4.1 Radial and transverse components of velocity.

EXAMPLE 1. A particle is moving on the cardioid $r = 5 + 3\cos\theta$ with a constant angular velocity of 2 rad/s. Find its speed when $\theta = \pi/3$.

Solution. We desire to find $|v|$. To do this we need v_r and v_θ. For $r = 5 + 3\cos\theta$,

$$v_r = \frac{dr}{dt} = -3\sin\theta\frac{d\theta}{dt}$$

and

$$v_\theta = r\frac{d\theta}{dt} = (5 + 3\cos\theta)\frac{d\theta}{dt}$$

Since $\omega = d\theta/dt = 2$, then at $\theta = \pi/3$ we have

$$v_r = -3\left[\sin\left(\frac{\pi}{3}\right)\right](2) = -3\sqrt{3}$$

and

$$v_\theta = \left[5 + 3\cos\left(\frac{\pi}{3}\right)\right](2) = 13$$

Then $v^2 = v_r^2 + v_\theta^2 = (-3\sqrt{3})^2 + (13)^2 = 196$ and $|v| = \sqrt{196} = 14$ units/s.

PROBLEM 1. A particle is moving on the limaçon $r = 1 - 2\cos\theta$ with a constant angular velocity of 3 rad/s. Find its speed when $\theta = 3\pi/4$.

EXERCISES 13.4

Group A

1. A particle is moving on the circle $r = 2\cos\theta$ with a constant angular velocity of 4 rad/s. Find its speed when $\theta = \pi/6$.

2. A particle is moving on the circle $r = 4$ with a constant speed of 2 units/s in a counterclockwise direction. Find the angular velocity of the particle. (*Hint*: Since the particle is constrained to move in a circle, there is no radial velocity.)

3. If a particle moves counterclockwise on $r = 10\sin\theta$ with speed 30 units/min, find $d\theta/dt$ when $\theta = \pi/4$.

4. A particle moves on $r = e^\theta$ with constant angular velocity $\omega = 2\pi$ rad/min. Find its speed at $\theta = 0$.

5. A particle moves counterclockwise on $r = 4(1 + \cos\theta)$ with constant speed 4 units/s.

(a) Find ω for arbitrary θ. (b) Find ω when $\theta = 0$.
(c) Find ω when $\theta = \pi/2$.

6. A particle's motion is described by the parametric equations $r = t^2$, $\theta = 2t$. Find v_r, v_θ, and the speed (a) at any time t; (b) at $t = 1$.

7. A particle's motion is described by the parametric equations $r = 3t^2$, $\theta = \sin(\pi t/4)$. Find v_r, v_θ, and the speed at $t = 1$.

13.5 AREA

The area of a region that is bounded by curves in polar coordinates is given by $A = \int_\alpha^\beta dA$, where

$$dA = \tfrac{1}{2}r^2\,d\theta \tag{13.5.1}$$

is the differential area of a sector of a circle.[1] Consider Fig. 13.5.1. In geometry it is shown that the area of sectors in a circle are proportional to their central angle. Thus

$$\frac{\pi r^2}{2\pi} = \frac{dA}{d\theta}$$

and

$$dA = \tfrac{1}{2}r^2\, d\theta$$

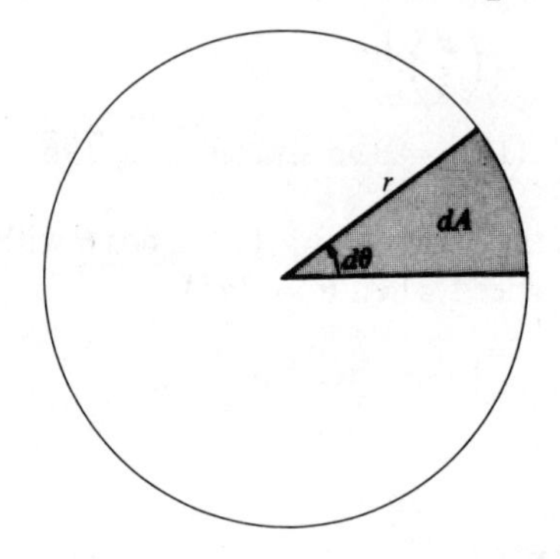

Figure 13.5.1 Area of a sector of a circle.

EXAMPLE 1. Set up an integral to find the area of the cardioid $r = 3 + 2\sin\theta$.

Solution. See Fig. 13.5.2.

$$dA = \tfrac{1}{2}r^2\, d\theta = \tfrac{1}{2}(3 + 2\sin\theta)^2\, d\theta$$

$$A = \tfrac{1}{2}\int_0^{2\pi} (3 + 2\sin\theta)^2\, d\theta$$

We could also write

$$A = 2(\tfrac{1}{2})\int_{-\pi/2}^{\pi/2} (3 + 2\sin\theta)^2\, d\theta \qquad (13.5.2)$$

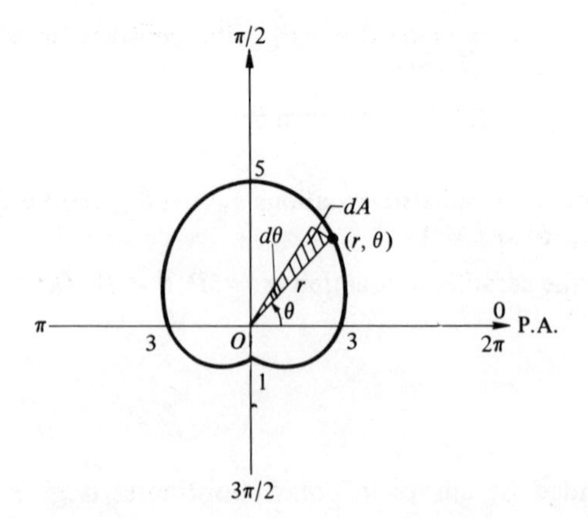

Figure 13.5.2 Area of $r = 3 + 2\sin\theta$.

[1]The element of area in polar coordinates may be remembered by recalling that a pie is usually cut into wedges.

Note that in Eq. 13.5.2 we make use of symmetry and that the lower limit is $-\pi/2$. We do not use $3\pi/2$ because $d\theta$ is positive and we cannot get to $\pi/2$ from $3\pi/2$ in a positive direction. [You may evaluate the integral by expanding then integrating (see Exercise 7 in Group B).]

PROBLEM 1. Find the area bounded by the cardioid $r = 3 + 2\cos\theta$.

In the XY-coordinate system we found

$$(ds)^2 = (dx)^2 + (dy)^2 \qquad \text{(see Eq. 7.9.2)}$$

In polar coordinates we have for the differential of arc

$$(ds)^2 = (r\,d\theta)^2 + (dr)^2 \tag{13.5.3}$$

This may be obtained in a manner similar to obtaining Eq. 13.4.1.

EXAMPLE 2. Set up an integral to find the length of arc of $r = 3 + 2\cos\theta$ between $\theta = 0$ and $\theta = \pi/2$.

Solution

$$s = \int ds$$

$$ds = \sqrt{(r\,d\theta)^2 + (dr)^2} = \sqrt{r^2 + \left(\frac{dr}{d\theta}\right)^2}\,d\theta \qquad \text{(from Eq. 13.5.3)}$$

$$\frac{dr}{d\theta} = -2\sin\theta$$

Therefore,

$$ds = \sqrt{(3 + 2\cos\theta)^2 + (-2\sin\theta)^2}\,d\theta$$

and

$$s = \int_0^{\pi/2} \sqrt{13 + 12\cos\theta}\,d\theta$$

Although we cannot evaluate this integral here, it may be evaluated by numerical methods (trapezoidal rule, Simpson's rule, etc.). What we desired to show is that even though a problem may be easy to formulate, its solution may be quite involved.

PROBLEM 2. Set up an integral to find the length of arc of $r = 3 + 2\sin\theta$ between $\theta = -\pi/2$ and $\theta = \pi/2$.

EXERCISES 13.5

Group A

1. In each of the following, use integration to find the area of the required region.

(a) Area enclosed by the circle $r = 2\sin\theta$.
(b) Area enclosed by the circle $r = 2\cos\theta$.

(c) Area enclosed by $r = 2(1 - \sin\theta)$.
(d) Area enclosed by $r = 4\sin^2\theta$.
(e) Area enclosed by $r = 3\cos 2\theta$.
(f) Area enclosed by one loop of $r^2 = 9\sin\theta$.
(g) Area enclosed by the small loop of the limaćon $r = 1 + 2\sin\theta$.

2. Set up integrals to find what is required in each of the following.

(a) The length of arc of $r = 4 + 3\cos\theta$ between $\theta = 0$ and $\theta = \pi/2$.
(b) The length of arc of $r = 2\sin\theta$ between $\theta = \pi/2$ and $\theta = \pi$.
(c) The length of arc of $r = 2/(1 + \cos\theta)$ between $\theta = 0$ and $\theta = 3\pi/5$.
(d) The length of arc of $r = 4e^{-\theta/2}$ between $\theta = 0$ and $\theta = 4\pi$.
(e) The total length of the cardioid $r = 2(1 + \cos\theta)$.

Group B

1. Find the area enclosed by $r = 2\tan\theta$ between $\theta = 0$ and $\theta = \pi/3$.

2. Find the area enclosed by $r^2 = \sin\theta$.

3. Find the area enclosed by $r = \cos n\theta$, where n is an odd positive integer.

4. Find the arc length of $r = 3\cos\theta$ between $\theta = 0$ and $\theta = \pi/4$.

5. Find the arc length of $r = 4(1 + \cos\theta)$ between $\theta = 0$ and $\theta = \pi/2$. (*Hint*: See Eq. 9.3.18.)

6. Find the arc length of $r = 4e^{-\theta/2}$ between $\theta = 0$ and $\theta = 4\pi$.

7. Evaluate the integral in Example 1.

Group C

1. Given two curves in polar form one finds that an element of area between the curves is given by $dA = \frac{1}{2}(r^2_{\text{outer}} - r^2_{\text{inner}})\,d\theta$. Using this, find the following areas.

(a) Area between $r = 4\cos\theta$ and $r = 2$.
(b) Area between $r = 3 + 3\cos\theta$ and $r = 3$.
(c) Area between $r = 2a\cos\theta$ and $r = a$.
(d) Area between $r = 2/(1 - \cos\theta)$ and the line on which $\cos\theta = 0$.
(e) Area inside both the circle $r = 2$ and the cardioid $r = 2(1 + \sin\theta)$.

SUMMARY OF IMPORTANT WORDS AND CONCEPTS

For each of the following, state in your own words your understanding of the statement or word.

1. (a) Polar coordinates. (b) Polar axis. (c) Origin.

2. Relation between polar coordinates and Cartesian coordinates.

3. Graphs in polar coordinates.

4. (a) Polar equations of the conics. (b) Eccentricity. (c) Parabola. (d) Ellipse. (e) Hyperbola.

5. (a) Transverse velocity. (b) Radial velocity. (c) Angular velocity.

6. (a) Area in polar coordinates. (b) Differential of area. (c) Length of arc.

Chapter 14

Differential Equations

14.1 SOME GENERAL COMMENTS

Differential equations comprise one of the most important classes of equations in applied problems because so many things involve the rate of change of one variable with another. Any equation that contains a derivative or a differential is called a *differential equation.*

There are many types of differential equations. We classify them as to *order* and *degree.*

The *order* of a differential equation is the order of the highest-order derivative which is present. For example,

$$\frac{dy}{dx} + y = x \text{ is of } \textit{first order}$$

$$\frac{d^2y}{dx^2} + \frac{dy}{dx} = 0 \text{ is of } \textit{second order}$$

and

$$\frac{d^4y}{dx^4} + x = y \text{ is of } \textit{fourth order}$$

If a differential equation can be rationalized and cleared of fractions with regard to all derivatives present, then the *degree* of a differential equation is the degree of the highest-order derivative present. For example,

$$\frac{dy}{dx} + x = y \text{ is of } \textit{first order first degree}$$

$\sqrt{dy/dx + 3} = y$ becomes $dy/dx + 3 = y^2$ and hence is of *first order first degree.*

$$\frac{d^2y}{dx^2} + \left(\frac{dy}{dx}\right)^3 = x \text{ is of } \textit{second order first degree}$$

$$(y'')^2 + (y') = \cos y'' \text{ is of } \textit{second order} \text{ but has } \textit{no degree}$$

Note that every differential equation has an order but not every differential equation has a degree. In this book we study some differential equations of first order first degree and of second order first degree only.

The *solution* of a differential equation is any function, free of derivatives and differentials, which satisfies the differential equation exactly.

EXAMPLE 1. Show that $y = c_1 \sin x$ is a solution of $(d^2y/dx^2) + y = 0$, where c_1 is a constant.

Solution. For $y = c_1 \sin x$,

$$\frac{dy}{dx} = c_1 \cos x$$

and

$$\frac{d^2y}{dx^2} = -c_1 \sin x$$

Substituting in the given differential equation, we obtain

$$(-c_1 \sin x) + (c_1 \sin x) = 0$$

or

$$0 = 0$$

Thus $y = c_1 \sin x$ is a solution of $d^2y/dx^2 + y = 0$.

PROBLEM 1. Show that $y = c_1 \cos x + c_2 \sin x$ is a solution of $(d^2y/dx^2) + y = 0$, where c_1 and c_2 are constants.

Not every differential equation has a solution. Many differential equations have more than one solution.

The constant c_1 in the solution of Example 1 is called an *arbitrary constant.* The expression $y = c_1 \cos 3x + c_2 \sin 3x$ has two arbitrary constants, c_1 and c_2. If an expression involves a certain set of constants that cannot be replaced by a smaller set, then the constants are called *essential arbitrary constants.* When we talk about arbitrary constants we shall mean essential arbitrary constants.

EXAMPLE 2. How many arbitrary constants are there in the expression $y = c_1 e^{c_2 + c_3 x}$?

Solution. Upon first glance it looks like c_1, c_2, and c_3 are arbitrary constants. However, we may rewrite the expression as $y = c_1 e^{c_2} e^{c_3 x}$ (from Theorem

8.2.1). Since e is a constant, e^{c_2} is a constant, as is $c_1 e^{c_2}$. Thus we can let

$$c = c_1 e^{c_2}$$

and write

$$y = ce^{c_3 x}$$

and the expression has only two arbitrary constants, c and c_3.

PROBLEM 2. How many arbitrary constants are there in the expression $y = c_1 \ln e^{c_2 x}$?

The *general solution* of a differential equation contains the same number of arbitrary constants as the order of the equation. Thus the general solution of a first-order differential equation has one arbitrary constant. The general solution of a second-order differential equation has two arbitrary constants.

Warning ***The order of a differential equation tells us how many arbitrary constants must be in the general solution.***

If we know initial or boundary conditions, values for the arbitrary constants may be found, and then we have a *particular solution* of the differential equation.

EXAMPLE 3. The general solution of a differential equation is $y = c_1 \cos t + c_2 e^t$ subject to the intitial conditions at $t = 0$, $y = 1$, and $dy/dt = 3$. Find the particular solution.

Solution. For $y = c_1 \cos t + c_2 e^t$,

$$\frac{dy}{dt} = -c_1 \sin t + c_2 e^t$$

For $t = 0$ and $y = 1$ we have

$$1 = c_1 \cos 0 + c_2 e^0$$

or

$$1 = c_1 + c_2$$

For $t = 0$ and $dy/dt = 3$ we have

$$3 = -c_1 \sin 0 + c_2 e^0$$

or

$$3 = c_2$$

Thus

$$c_1 = 1 - c_2 = 1 - 3 = -2$$

and the particular solution is

$$y = -2 \cos t + 3e^t$$

PROBLEM 3. The general solution of a differential equation is $y = (c_1 + c_2 t)e^{-t}$.

Find the particular solution subject to the initial conditions at $t = 0$, $y = 3$, and $dy/dt = -1$.

EXERCISES 14.1

Group A

1. Classify each of the following as to order and degree.

(a) $\dfrac{dy}{dx} + y^2 = x$.

(b) $\dfrac{d^2y}{dx^2} + y = x$.

(c) $\left(\dfrac{dy}{dx}\right)^2 + \left(\dfrac{dy}{dx}\right) = x$.

(d) $\dfrac{d^2y}{dx^2} + \dfrac{dy}{dx} = x$.

2. Show that $y = c_2 \cos x$ is a solution of $(d^2y/dx^2) + y = 0$.

3. Show that $y = c_1 e^x$ is a solution of $(d^2y/dx^2) - y = 0$.

4. Show that $y = c_2 e^{-x}$ is a solution of $(d^2y/dx^2) - y = 0$.

5. Show that $y = c_1 e^x + c_2 e^{-x}$ is a solution of $(d^2y/dx^2) - y = 0$.

6. State the number of arbitrary constants in each of the following.

(a) $y = c_1 e^{c_2 - c_3 x}$.

(b) $y = c_1 e^x + c_2 e^{-x}$.

(c) $\ln y = \ln x + \ln c_1 + \ln c_2$.

(d) $y = c_1 + c_2 + c_3 x$.

7. The general solution of a differential equation is given subject to the stated initial conditions. In each of the following, find the particular solution.

(a) $y = ce^t$; at $t = 0$, $y = 5$.
(b) $y = c_1 \cos t + c_2 \sin t$; at $t = 0$, $y = 4$, $dy/dt = 0$.
(c) $y = c_1 \cos t + c_2 \sin t$; at $t = 0$, $y = 0$, $dy/dt = 3$.
(d) $y = c_1 \cos t + c_2 \sin t$; at $t = 0$, $y = 0$; at $t = \pi/2$, $y = 1$.
(e) $y = c_1 \cos t + c_2 \sin t$; at $t = 0$, $y = 1$, $dy/dt = 2$.
(f) $y = c_1 e^x + c_2 e^{-x}$; at $x = 0$, $y = 0$, $dy/dx = 1$.
(g) $Q = 10 + e^{-t}(c_1 \cos 2t + c_2 \sin 2t)$; at $t = 0$, $Q = 0$, $I = 0$.

8. How many arbitrary constants are in the general solution of each of the following?

(a) $d^2y/dx^2 + (dy/dx)^3 + y = 0$.

(b) $(2dy/dx) - y^3 = x$.

9. Find the values of m for which $y = e^{mx}$ is a solution of $(d^2y/dx^2) - (dy/dx) - 2y = 0$.

14.2 SEPARATION OF VARIABLES

Whenever we integrate we are actually solving a differential equation. For example, the expression $A = \int y\, dx$ is a solution to $dA/dx = y$. We found that in order to perform the integration, y had to be a function of x. Some first-order first-degree differential equations may be solved by a method called *separation of variables*. This involves getting the coefficient of dx as a function of x and the coefficient of dy as a function of y.

EXAMPLE 1. Solve $y\,dx + 3x\,dy = 0$.

Solution. This may be rewritten as

$$y\,dx = -3x\,dy$$

or

$$\frac{dx}{x} = \frac{-3dy}{y}$$

Now the variables have been "separated" and we can integrate[1] to obtain

$$\ln x + c_1 = -3 \ln y + c_2$$

or

$$\ln x = -3 \ln y + c_2 - c_1 = -3 \ln y + c_3$$

Since c_1 and c_2 combine to form c_3, it is customary to place the one arbitrary constant of integration on the right-hand side. To make this solution neater in form we may let $c_3 = \ln c$ and obtain

$$\begin{aligned} \ln x &= -3 \ln y + \ln c \\ &= \ln y^{-3} + \ln c = \ln cy^{-3} \end{aligned}$$

or

$$x = cy^{-3} \qquad \text{(because if } \ln A = \ln B\text{, then } A = B\text{)}$$

and, finally, $xy^3 = c$. Note that the solution has only one arbitrary constant since the equation is of first order.

PROBLEM 1. Solve $x\,dy + y\,dx = 0$. (Does this equation remind you of a differential formula? Could you have written the answer without separating the variables?)

In general, any first-order first-degree differential equation may be written as

$$M(x, y)\,dx + N(x, y)\,dy = 0 \tag{14.2.1}$$

where $M(x, y)$ indicates that M is a function of both x and y and similarly for $N(x, y)$. If by valid algebraic manipulations we can make the coefficient of dx a function of x only, and the coefficient of dy a function of y only, then we have separated the variables and we may integrate each expression. Remember that a constant may be considered as a function of x or as a function of y; that is $3 = 3x^0 = 3y^0$.

EXAMPLE 2. Find the solution of $dy/dx = (y^2 + 9)/(x^2y + 4y)$ subject to the conditions $x = 2$ and $y = 1$.

Solution. Noting that the equation is of first order first degree, we try to separate

[1]In this chapter we use $\int (du/u) = \ln u + c$, which implies that $u > 0$. The use of Eq. 8.8.1, $\int (du/u) = \ln |u| + c$, should be used in cases where $u > 0$ or $u < 0$.

the variables. Thus

$$y(x^2 + 4)\,dy = (y^2 + 9)\,dx$$

and

$$\frac{y\,dy}{y^2 + 9} = \frac{dx}{x^2 + 4}$$

We now integrate to obtain

$$\tfrac{1}{2}\ln(y^2 + 9) = \tfrac{1}{2}\,\mathrm{Tan}^{-1}\frac{x}{2} + c_1$$

or

$$\ln(y^2 + 9) = \mathrm{Tan}^{-1}\frac{x}{2} + c$$

where $c = 2c_1$. For $x = 2$ and $y = 1$ we have

$$\ln 10 = \mathrm{Tan}^{-1} 1 + c$$

Thus

$$c = \ln 10 - \frac{\pi}{4} = 2.3 - 0.78 = 1.52$$

and

$$\ln(y^2 + 9) = \mathrm{Tan}^{-1}\frac{x}{2} + 1.52$$

is the required solution.

PROBLEM 2. Solve $(x^2 + 1)\,dy - (x^2y^2 + 25x)\,dx = 0$ subject to the conditions $y = -5$ when $x = 0$.

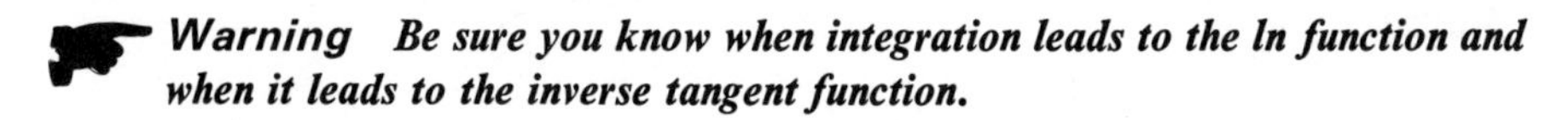

Warning ***Be sure you know when integration leads to the ln function and when it leads to the inverse tangent function.***

We study some applications in Sec. 14.4.

EXERCISES 14.2

Group A

1. Solve each of the following differential equations.

(a) $y\,dx - 3x\,dy = 0$.

(b) $2y\,dx + 5x\,dy = 0$.

(c) $x\,dy - (3y - 6)\,dx = 0$.

(d) $(x^2y + 4y)\,dy - (y^2 - 2)\,dx = 0$.

(e) $\dfrac{dy}{dx} = \dfrac{y^2 + 9}{\sqrt{9 - x^2}}$.

(f) $\dfrac{dy}{dx} = x$.

(g) $\dfrac{dI}{dt} = 2\cos t + 3\sin t$.

(h) $\dfrac{dI}{dt} = e^{-t}$.

(i) $\dfrac{dI}{dt} = e^{-t}\cos t$.

2. Solve each of the following differential equations subject to the given conditions.

(a) $\dfrac{dy}{dx} = \dfrac{y^2 - 8}{yx^2 + 16y}$; $x = 4$, $y = 3$.

(b) $v\dfrac{dv}{ds} = 4$; $v = 0$, $s = 2$.

(c) $\dfrac{d(xy)}{dx} = e^{3x}$; $x = 0$, $y = 1$. (*Hint*: Let $u = xy$.)

(d) $\dfrac{dy}{dx} = \dfrac{\cos x}{\sin y}$; $x = \dfrac{\pi}{2}$, $y = 0$.

(e) $\dfrac{d(x^2y)}{dx} = \dfrac{1}{x}$; $x = 1$, $y = 2$.

(f) $d\left(\dfrac{x}{y^2}\right) = x^2\, dx$; $x = 0$, $y = 1$.

Group B

1. Solve each of the following differential equations.

(a) $(x^2y^2 + y^2)\, dy + (y^3x + x)\, dx = 0$.
(b) $3x^2y\, dx + x^3\, dy = e^x\, dx$. (*Hint*: The left-hand side of the equation is the differential of some function u.)
(c) $2xy^2\, dx + 2x^2y\, dy = y\, dy$. [*Hint*: See part (b).]
(d) $xy\, dx + \sqrt{1 + x^2}\, dy = 0$.
(e) $(y + 3)\, dx + \cot x\, dy = 0$.
(f) $\sin x \cos y\, dx + \cos x \sin y\, dy = 0$.
(g) $\sin x \cos^2 y\, dx + \cos^2 x\, dy = 0$.

Group C

1. Differential equations of the form $d^2y/dx^2 = f(x)$ may be solved by separation of variables. Since $d^2y/dx^2 = (d/dx)(dy/dx)$, we have $d(dy/dx) = f(x)\, dx$, which upon integration becomes $dy/dx = \int f(x)\, dx + c_1$, from which y can be found by another integration. Solve each of the following.

(a) $\dfrac{d^2y}{dx^2} = x$.

(b) $\dfrac{d^2s}{dt^2} = g$ subject to $t = 0$, $v = v_0$, $s = s_0$. (g is constant.)

(c) $\dfrac{d^2Q}{dt^2} = e^{-t}$.

(d) $\dfrac{d^2Q}{dt^2} = \sin 3t$.

(e) $d^2y/dx^2 = x^2 - 2x + 3$ subject to $x = 0$, $y = 1$, $dy/dx = 0$.

(f) $\dfrac{d^4y}{dx^4} = -x^2$.

2. Let u and v be functions of the independent variable x. Using formulas for the differentials of such functions, we have $d(u + v) = du + dv$, $d(uv) = u\, dv + v\, du$, and $d(u/v) = (v\, du - u\, dv)/v^2$. Reviewing Exercises 2(c), (e), and (f) of Group A and Exercises 1(b) and (c) of Group B, we see that it was exactly these formulas which enabled us to solve them. Use these formulas to solve each of the following differential equations.

(a) $\dfrac{y\,dx - x\,dy}{y^2} = 6x\,dx.$

(b) $\dfrac{x\,dy - y\,dx}{x^2} = 0.$

(c) $2x^3y\,dy + 3x^2y^2\,dx = 0.$

(d) $\dfrac{y^3\,dx - 3y^2x\,dy}{y^6} = 0.$

(e) $\dfrac{2x^4y\,dy - 4x^3y^2\,dx}{x^8} = 0.$

14.3 FIRST-ORDER LINEAR EQUATIONS

A very special type of first-order first-degree differential equation is

$$\frac{dy}{dx} + P(x)y = Q(x) \qquad (14.3.1)$$

where P and Q are functions of x, denoted as $P(x)$ and $Q(x)$. It is called the *first-order linear* differential equation because it is of first order and both the derivative, dy/dx, and the dependent variable, y, are of first degree. In more comprehensive books on differential equations it is shown that the solution of the first-order linear differential equation is obtained by multiplying both sides of Eq. 14.3.1 by $e^{\int P\,dx}$. Thus

$$e^{\int P\,dx}\frac{dy}{dx} + e^{\int P\,dx}Py = e^{\int P\,dx}Q \qquad (14.3.2)$$

The left-hand side of Eq. 14.3.2 may be written as $(d/dx)\,ye^{\int P\,dx}$. This may be checked by performing the indicated differentiation. Thus

$$\frac{d}{dx}(ye^{\int P\,dx}) = y\frac{d}{dy}(e^{\int P\,dx}) + \frac{dy}{dx}e^{\int P\,dx}$$

$$= ye^{\int P\,dx}(P) + \frac{dy}{dx}e^{\int P\,dx}$$

which is the left-hand side of Eq. 14.3.2. We may thus separate the variables and write Eq. 14.3.2 as

$$d(ye^{\int P\,dx}) = Qe^{\int P\,dx}\,dx \qquad (14.3.3)$$

Since the integral of a differential of a quantity is just that quantity we obtain by integrating Eq. 14.3.3

$$ye^{\int P\,dx} = \int Qe^{\int P\,dx}\,dx \qquad (14.3.4)$$

which may be solved for y. The expression $e^{\int P\,dx}$ is called an *integrating factor*. In solving first-order linear differential equations we proceed as shown in the following examples.

EXAMPLE 1. Solve $(dy/dx) + (2/x)y = x$.

Solution. Comparing this equation with Eq. 14.3.1 we recognize this equation as first-order linear with $P = 2/x$ and $Q = x$. The integrating factor is

$$e^{\int P\,dx} = e^{\int (2/x)\,dx} = e^{2 \ln x} = e^{\ln x^2}$$

From the definition of logarithms we know that $e^{\ln u} = u$ (see Sec. 8.5). Hence our integrating factor $e^{\ln x^2} = x^2$. We next multiply through the given equation by this factor to obtain

$$x^2 \frac{dy}{dx} + 2xy = x^3$$

The left-hand side is always the derivative of y times the integrating factor, so we obtain

$$\frac{d}{dx}(yx^2) = x^3$$

$$d(yx^2) = x^3\,dx$$

and

$$yx^2 = \frac{x^4}{4} + c$$

and

$$y = \frac{1}{4}x^2 + \frac{c}{x^2}$$

is the required solution.

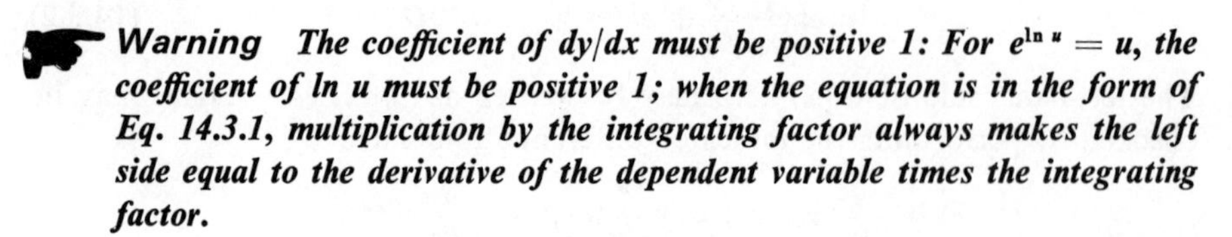

Warning ***The coefficient of dy/dx must be positive 1: For $e^{\ln u} = u$, the coefficient of $\ln u$ must be positive 1; when the equation is in the form of Eq. 14.3.1, multiplication by the integrating factor always makes the left side equal to the derivative of the dependent variable times the integrating factor.***

PROBLEM 1. Solve $x(dy/dx) + 5y = x^3$.

EXAMPLE 2. The differential equation for a certain electric circuit is $dI/dt + 20I = 10$. If the current I is zero when $t = 0$, find I at any time $t \geqq 0$.

Solution. For $(dI/dt) + 20I = 10$, the integrating factor is $e^{\int 20\,dt} = e^{20t}$. Thus

$$e^{20t} \frac{dI}{dt} + 20e^{20t}I = 10e^{20t}$$

$$\frac{d}{dt}(Ie^{20t}) = 10e^{20t}$$

$$d(Ie^{20t}) = 10e^{20t}\,dt$$

$$Ie^{20t} = \tfrac{1}{2}e^{20t} + c$$

and

$$I = \tfrac{1}{2} + ce^{-20t}$$

For $t = 0$, $I = 0$ then $0 = \frac{1}{2} + c(1)$; therefore, $c = -\frac{1}{2}$ and $I = \frac{1}{2}(1 - e^{-20t})$.

PROBLEM 2. Find the current I at $t = 3$ for an electric circuit whose differential equation is $2(dI/dt) + I = 12$. Assume at $t = 0$, $I = 0$.

We study some more applications in Sec. 14.4.

There are many first-order first-degree equations that cannot be solved by separation of variables or the method of this section. Texts on differential equations discuss various methods of solution.

EXERCISES 14.3

Group A

1. Solve each of the following.

(a) $\dfrac{dy}{dx} + \dfrac{3}{x}y = x.$

(b) $\dfrac{dy}{dx} + \dfrac{2}{x}y = x^2 + 1.$

(c) $x\dfrac{dy}{dx} - y = 3.$

(d) $\dfrac{dy}{dx} - \dfrac{2}{x}y = x.$

(e) $\dfrac{dy}{dx} + y = e^x.$

(f) $\dfrac{dI}{dx} + 10I = 20.$

(g) $\cos x\dfrac{dy}{dx} + y\sin x = \cos^2 x.$

2. Solve each of the following subject to the given conditions.

(a) $\dfrac{dy}{dx} - \dfrac{2}{x}y = x^2 - 1$; when $x = 1$, $y = 1$.

(b) $\dfrac{dI}{dt} + 10I = 10$; when $t = 0$, $I = 0$.

(c) $\dfrac{dI}{dt} + 20I = \sin 60t$; when $t = 0$, $I = 0$. Find I at $t = 0.01$.

(d) $\dfrac{dI}{dt} - 20I = 10$; when $t = 0$, $I = 0$.

(e) $\dfrac{dQ}{dt} + \dfrac{75}{8}Q = \dfrac{5}{2}$; when $t = 0$, $Q = 0$.

(f) $\dfrac{dQ}{dt} + Q = \dfrac{5}{2}\sin 2t$; when $t = 0$, $Q = 0$. Find Q at $t = 1$.

Group B

1. Solve each of the following.

(a) $\dfrac{dy}{dx} + y\cot x = \sec x.$

(b) $(x^2 + 1)\dfrac{dy}{dx} - xy = 1.$

(c) $\dfrac{dy}{dx} = e^{-x^2} - 2xy.$

(d) $\dfrac{dy}{dx} = \cos^3 x - y\tan x.$

(e) $\dfrac{dy}{dx} = x^7 - 4x^3y.$

Group C

1. If one finds a differential equation of the form $(dx/dy) + P(y)x = Q(y)$, [$P(y)$ denotes that P is a function of y], it can be thought of as a first-order linear differential equation in x and the method of Sec. 14.3 is applicable with the roles of x and y interchanged. Solve each of the following.

(a) $\dfrac{dx}{dy} + x = 1.$ (b) $\dfrac{dx}{dy} + yx = 2y.$

(c) $\dfrac{dx}{dy} + \dfrac{1}{y}x = \sin y.$

2. Sometimes a change in variable will reduce an equation to the form of a linear equation. A case in point is Bernoulli's equation, $(dy/dx) + P(x)y = Q(x)y^k$, where k is a constant. (a) Divide Bernoulli's equation by y^k and let $v = y^{-k+1}$. Show that the resulting equation becomes

$$\frac{dv}{dx} + (1 - k)P(x)v = (1 - k)Q(x)$$

(b) Solve the equation $3(1 + x^2)(dy/dx) + 2xy = 2xy^4$.

14.4 APPLICATIONS

(1) Families of Curves

Equations with arbitrary constants represent families of curves. For example, $y = mx$ represents the family of straight lines all passing through the origin. In Fig. 14.4.1 a few of them are shown. A particular member of the family is obtained by assigning a value to m. Thus if we want the member of this family passing through the point (3, 4), we would have $4 = m(3)$ or $m = \frac{4}{3}$ and $y = \frac{4}{3}x$ or $4x - 3y = 0$.

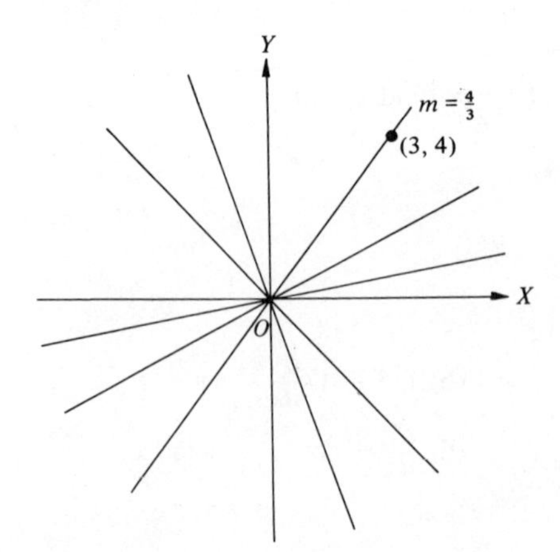

Figure 14.4.1 Some members of the family $y = mx$.

EXAMPLE 1. Find the family of curves whose slope is y/x.

Solution. Since the slope is dy/dx, we have

$$\frac{dy}{dx} = \frac{y}{x}$$

$$\frac{dy}{y} = \frac{dx}{x}$$

$$\ln y = \ln x + \ln c$$

$$\therefore \quad y = cx$$

PROBLEM 1. Find the family of curves whose slope at any point is twice the abscissa. In particular, find the member that passes through the point (1, 4).

(2) Orthogonal Trajectories

If each member of one family of curves intersects at right angles each member of another family of curves, we say that the families are *mutually orthogonal* or that either family forms a set of *orthogonal trajectories* of the other family.

EXAMPLE 2. Find the orthogonal trajectories of the circle $x^2 + y^2 = a^2$.

Solution. The key to this problem is that the slope of the orthogonal trajectories is the negative reciprocal of the slope of the given family of curves. Thus we proceed to find the slope of the given family of curves without the arbitrary constant. Taking the derivative with respect to x, we have

$$2x + 2y\frac{dy}{dx} = 0$$

and

$$\frac{dy}{dx} = -\frac{x}{y}$$

Then the slope of the orthogonal trajectories is

$$\frac{dy}{dx} = \frac{y}{x}$$

and we solve this differential equation:

$$\frac{dy}{y} = \frac{dx}{x}$$

$$\ln y = \ln x + \ln c$$

$$y = cx$$

Hence the family of straight lines passing through the origin and the concentric circles with centers at the origin are orthogonal trajectories of each other (see Fig. 14.4.2). (What can you now say about the polar coordinate system?)

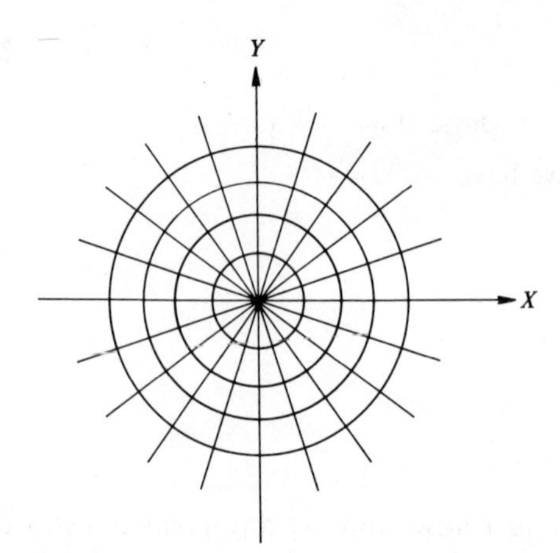

Figure 14.4.2 Example 2: Orthogonal trajectories.

PROBLEM 2. Find the orthogonal trajectories of the parabolas $y = x^2 + c$.

EXAMPLE 3. Find the orthogonal trajectories of $x^2 + cy^2 = 1$. Then find the particular one passing through the point (1, 2).

Solution

$$2x + 2cy\frac{dy}{dx} = 0$$

$$\frac{dy}{dx} = -\frac{x}{cy}$$

We must eliminate the arbitrary constant c. This is accomplished by noting, from the given equation, that $c = (1 - x^2)/y^2$. Hence

$$\frac{dy}{dx} = \frac{-xy^2}{y(1 - x^2)} = \frac{-xy}{1 - x^2}$$

The slope of the orthogonal trajectories is

$$\frac{dy}{dx} = \frac{1 - x^2}{xy}$$

Solving this equation, we have

$$y\,dy = \left(\frac{1 - x^2}{x}\right) dx$$

or

$$y\,dy = \frac{dx}{x} - x\,dx$$

Integrating, we obtain

$$\frac{y^2}{2} = \ln x - \frac{x^2}{2} + \ln c_1$$

and

$$x^2 + y^2 = \ln cx^2$$

or

$$e^{(x^2+y^2)} = cx^2$$

is the required family of orthogonal trajectories. When $x = 1$ and $y = 2$, we have $e^5 = c(1)$. Thus the member passing through (1, 2) is

$$e^{(x^2+y^2)} = e^5 x^2$$

or

$$x^2 = e^{(x^2+y^2-5)}$$

PROBLEM 3. Find the orthogonal trajectories of $y = cx^2$. Then find the particular one passing through the point (3, 4).

(3) Electric Circuits

Consider a simple electric circuit consisting of a voltage E, a resistor R, and an inductor L, connected in series. The differential equation for this circuit relating the current I and these components is

$$L\frac{dI}{dt} + RI = E \tag{14.4.1}$$

In a course in electric circuits you will find that when Kirchhoff's law (the algebraic sum of all voltage drops around an electric loop or circuit is zero) is applied to a circuit of a voltage E, a resistor R, and a capacitor C, connected in series, the equation relating the current and these components is

$$RI + \frac{Q}{C} = E$$

This is not a differential equation. However, recalling that $I = dQ/dt$, we have

$$R\frac{dQ}{dt} + \frac{Q}{C} = E \tag{14.4.2}$$

The circuit that contains a resistor, an inductance, and a capacitor is discussed in Sec. 14.5.

EXAMPLE 4. A generator having an electromotive force (emf) of 20 sin $10t$ volts is connected in series with a 15-Ω resistor and an inductor of 2 H. If at $t = 0$, $I = 0$, find the current as a function of t.

Solution. Using Eq. 14.4.1, we have

$$2\frac{dI}{dt} + 15I = 20 \sin 10t$$

or

$$\frac{dI}{dt} + \frac{15}{2}I = 10 \sin 10t$$

This is recognized as a first-order linear equation with integrating factor $e^{(15/2)t}$. Thus we have

$$d[Ie^{(15/2)t}] = 10e^{(15/2)t} \sin 10t\, dt$$

or

$$Ie^{(15/2)t} = 10\left\{\frac{e^{(15/2)t}(\frac{15}{2}\sin 10t - 10\cos 10t)}{(\frac{15}{2})^2 + (10)^2}\right\} + C \qquad \text{(Integral 113)}$$

and

$$I = \tfrac{40}{625}(\tfrac{15}{2}\sin 10t - 10\cos 10t) + Ce^{-(15/2)t}$$

When $t = 0$, $I = 0$ and $C = \frac{400}{625}$. Therefore,

$$I = \tfrac{40}{625}(\tfrac{15}{2}\sin 10t - 10\cos 10t + 10e^{-(15/2)t})$$

PROBLEM 4. A generator having an emf of 10 sin $2t$ volts is connected in series with a 5-Ω resistor and an inductor of 2 H. If at $t = 0$, $I = 0$, find the current I as a function of t. Then find I when $t = 2$.

(4) Growth and Decay

In many problems the time rate of change of a quantity is proportional to the quantity present. This may be expressed by the differential equation

$$\frac{dy}{dt} = ky \qquad (14.4.3)$$

where k is the proportionality constant. If k is positive, y increases; if k is negative, y decreases.

EXAMPLE 5. On a spherical surface the electric charge, in coulombs (C), leaks off at a rate proportional to the instantaneous charge. Initially, the charge is 9 C. If 3 C leaks off in 15 min, when will there be 1 C remaining?

Solution. The differential equation is $dQ/dt = kQ$. Separating variables, we have

$$\frac{dQ}{Q} = k\,dt$$

and

$$\ln Q = kt + c$$

or

$$Q = c^{kt+c_1} = ce^{kt}$$

When $t = 0$, $Q = 9$ or $9 = ce^0 = c$. Thus

$$Q = 9e^{kt}$$

When $t = 15$, $Q = 6$ or $6 = 9e^{15k}$, and

$$e^{15k} = \tfrac{2}{3}$$

Taking logarithms, we obtain

$$15k = \ln 2 - \ln 3 = 0.693 - 1.096 = -0.403$$

and

$$k = -0.0279$$

Thus

$$Q = 9e^{-0.0279t}$$

is the charge at any time t. When $Q = 1$, we have $1 = 9e^{-0.0279t}$ or

$$e^{-0.0279t} = \tfrac{1}{9}$$

$$-0.0279t = -\ln 9 = -2.197$$

and

$$t = 78.8 \text{ min}$$

PROBLEM 5. Bacteria reproduce at a rate proportional to the amount present. If the amount doubles in 2 h, how many are present at the end of 4 h?

(5) Mechanics

The basic formula in the study of mechanics is Newton's law, $f = ma$, where f is the resultant force (pounds) acting on a body, m is the mass (slugs) of the body, and a (feet per square second) is the acceleration of the body. For a body of weight W lb, the equation of motion is

$$f = \frac{W}{g} a \tag{14.4.4}$$

where g is the gravitational acceleration of approximately 32 ft/sec².

EXAMPLE 6. A 16-lb body falls from rest through a resisting medium that exerts a resistance force proportional to the velocity. If the resisting force is $2v$, find the velocity at $t = 1$ sec.

Solution. Since $a = dv/dt$, we have, from Eq. 14.4.4, the differential equation of motion:

$$16 - 2v = \frac{16}{32}\frac{dv}{dt}$$

Separating variables, we have

$$\frac{dv}{16 - 2v} = 2\,dt$$

Integrating, we obtain

$$-\tfrac{1}{2}\ln(16 - 2v) = 2t + C_1$$

or

$$16 - 2v = ce^{-4t}$$

and

$$v = \tfrac{1}{2}(16 - ce^{-4t})$$

Since the body falls from rest, $v = 0$ when $t = 0$. Thus

$$0 = \tfrac{1}{2}(16 - c)$$

and

$$c = 16$$

Therefore,

$$v = 8 - 8e^{-4t} = 8(1 - e^{-4t})$$

is the velocity of the body at time t. When $t = 1$, we have

$$v = 8(1 - e^{-4})$$

or

$$v = 8(1 - 0.0183)$$

and

$$v = 7.85 \text{ ft/s}$$

Note that since $v = 8 - 8e^{-4t}$, the limiting velocity is 8 ft/s because as $t \to \infty$, $e^{-4t} \to 0$.

PROBLEM 6. A 64-lb body falls from rest in a resisting medium. If the resisting force is $4v$ pounds, where v is the velocity in ft/s, find v when $t = 2$ s.

EXERCISES 14.4

Group A

1. Find the family of curves determined by the differential equation $dy/dx = \frac{1}{2}(y/x)$.

2. Find the family of curves determined by the differential equation $(dy/dx) + y = (1 + x)$.

3. Find the orthogonal trajectories of the family $y = ce^{-x}$ and determine the particular member of this family that passes through (0, 4).

4. Find the orthogonal trajectories of the family $cx^2 + y^2 = 1$ and determine the particular member of this family that passes through the point (1, 2).

5. An emf of 30 V is applied to a circuit consisting of an inductor of 3 H in series with a 60-ohm resistor. If the current is zero at $t = 0$, find a function that gives the current at any time $t \geqq 0$.

6. Do Exercise 5 if (a) the emf is 90 sin $9t$; (b) the emf is $30e^{-3t}$.

7. A capacitor of 4×10^{-3} F is in series with a 25-Ω resistor and an emf of 50 V. Assuming that the charge on the capacitor is zero at $t = 0$, find the charge at any time $t \geqq 0$.

8. Do Exercise 7 if (a) the emf is 50 sin $2t$; (b) the emf is $50e^{-t}$.

9. The electric charge, in coulombs, on a spherical surface leaks off at a rate proportional to the instantaneous charge. Initially, 7 C is present and one-half leaks off in 28 min. When will there be 1 C remaining?

10. Newton's law of cooling is given by the differential equation $dU/dt = -k(U - U_0)$, where $k > 0$ and U_0 is the temperature of the surrounding medium. Using this law, solve the following problem. Water is heated to 90°C. The water is removed from the heat and allowed to cool in a room whose temperature is kept constant at 60°C. After 3 min the water temperature is 80°C. Find the water temperature after 5 min.

11. Water at temperature 100°C cools in 5 min to 75°C in a room of temperature 20°C. (a) Find the temperature of the water after 10 min. (b) When is the temperature of the water 45°C?

12. Bacteria in a certain culture increase at a rate proportional to the number present. If the original number doubles in $\frac{1}{2}$ h, how many hours will it take to have 20 times the original number?

13. A ball is thrown vertically upward with an initial velocity of 96 ft/s. How long will it take to return to the ground?

14. A mass of $30g$ slugs falls from rest under the influence of gravity. Find the distance traveled and the velocity attained 3 s after the motion has begun.

15. The force of water resistance acting on a boat is proportional to its instantaneous velocity and is such that at 30 ft/s the water resistance is 60 lb. The boat weighs 180 lb and the only passenger weighs 140 lb. The motor can exert a steady force of 40 lb in the direction of motion. (a) Find the maximum velocity at which the boat will travel. (b) Find the distance traveled and the velocity at any time t assuming the boat starts from rest.

16. Bacteria reproduce at a rate proportional to the amount present. If 10,000 bacteria are present at the end of 3 h and 40,000 at the end of 5 h, how many were there initially?

Group B

1. The rate of decomposition of a certain chemical substance is proportional to the amount of the substance still unchanged. If this substance changes from 1000 to 500 g in 2 h, find an equation for the amount of substance left after t hours.

2. The rate of decay of a radioactive substance is proportional to the amount present at any time. If amount A_0 is the initial amount present, find a general expression for the amount present at any time t. If the *half-life* of the substance is defined to be the time required for the decay of half of the amount initially present, find an expression for the half-life.

3. A paratrooper and parachute weigh 200 lb. When the parachute opens he is traveling vertically downward at 40 ft/s. If air resistance varies directly as the instantaneous velocity and the air resistance is 80 lb when the velocity is 20 ft/s, find equations for the velocity and position at any time t. The *limiting velocity* of a moving object is defined to be that velocity attained as $t \longrightarrow \infty$. Find the limiting velocity for the paratrooper.

4. A 100-lb object is dropped vertically downward from a building high above New York City. The law of resistance is given by $0.01v^2$, where v is the instantaneous velocity. Find (a) the velocity as a function of time; (b) the velocity as a function of distance (*Hint*: $a = dv/dt = v\, dv/ds$); (c) the limiting velocity.

5. A resistance of R ohms varies with time t according to $R = 1 + 0.1t$. It is connected in series with a 0.1-F capacitor and a 100-V emf. If the initial charge on the capacitor is 1 C, find the charge and current as functions of time.

6. An inductance of L henries varies with time t according to $L = 1 + 0.005t$, $0 \leqq t \leqq 10{,}000$. It is connected in series with a 5-Ω resistor and an emf of 50 V. Find the current I at any time $t \geqq 0$ if $I(0) = 0$.

7. A rocket moves forward due to the backward expulsion of a mass of gas formed by burning fuel. As the fuel burns, the total mass of the rocket changes. The motion of a rocket can be studied by applying Newton's second law of motion, which states that the time rate of change of momentum of a body is equal to the net force acting on the body. Hence if mv is the momentum, then $[d(mv)]/dt = F$. By considering the momentums of the gas and of the rocket it can be shown that the basic equation for rocket motion is $M(dV/dt) - v(dM/dt) = F$, where M is the total mass of the rocket at time t, V is the velocity of the rocket relative to the earth at time t, $-v$ is the magnitude of the velocity of the gas relative to the rocket, and F is the external force acting on the rocket in the direction of the rocket's motion. (a) A rocket fires its gas backward at a constant rate of k kg/s and at a constant speed of b m/s relative to the rocket. Let V_0 be the initial velocity of the rocket and let M_0 be its initial mass. Use $M = M_0 - kt$ and the method of separation of variables to show that $V = V_0 - [b + (F/k)] \ln [1 - (k/M_0)t]$ when F is a constant force. (b) A rocket has a mass of 30,000 kg, which includes 25,000 kg of fuel. The rocket expels combustion products at the rate of 1500 kg/s at a speed of 500 m/s relative to the rocket. If the rocket starts from rest and flies vertically, find its speed 10 s after lift off. Assume that the only external force is the constant force of gravity ($g = 9.8 \text{ m/s}^2$).

8. The total constant force F that a muscle must expand against elasticity and viscosity is

$$F = al + b\frac{dl}{dt}$$

where l is the length of extension at time t. Find l as a function of t.

9. The rate at which a certain substance is dissolved in water is

$$\frac{dx}{dt} = kx\left(\frac{x + 20}{100}\right)$$

where x is the amount of undissolved substance at t hours. If $x = 30$ g when $t = 0$, find the amount dissolved after 5 h for $k = 0.05$. (*Hint*: Use partial fractions.)

10. The rate constant k of a certain chemical reaction that depends on the temperature T is given by $d(\ln k)/dT = Q/RT^2$, where Q and R are constants. Find k.

14.5 SECOND-ORDER LINEAR DIFFERENTIAL EQUATIONS

The differential equation

$$a_0 \frac{d^2y}{dx^2} + a_1 \frac{dy}{dx} + a^2 y = 0 \tag{14.5.1}$$

where a_0, a_1, and a_2 are constants, is called the *second-order linear differential equation with constant coefficients and right-hand side equal to zero*. Note that it is linear in its derivatives and dependent variable.

Upon examining Eq. 14.5.1, we note that the solution of Eq. 14.5.1 must be a function with two arbitrary constants (because of the second-order derivative). Furthermore, it must be such a function that itself and its first two derivatives are the same (except for coefficients) so that their sum is zero. The only function meeting these requirements is $y = e^{mx}$, m a constant real or complex number. To find m, we proceed as follows. For $y = e^{mx}$, $dy/dx = me^{mx}$ and $dy^2/dx^2 = m^2e^{mx}$. Substitution in Eq. 14.5.1 yields

$$a_0m^2e^{mx} + a_1me^{mx} + a_2e^{mx} = 0$$

and

$$e^{mx}(a_0m^2 + a_1m + a_2) = 0$$

Since e^{mx} can never be zero, we have

$$a_0m^2 + a_1m + a_2 = 0 \qquad (14.5.2)$$

Equation 14.5.2 is called *the auxiliary equation.* Its roots may be found by factoring or by the quadratic formula

$$m = \frac{-a_1 \pm \sqrt{a_1^2 - 4a_0a_2}}{2a_0} \qquad (14.5.3)$$

Let its two roots be m_1 and m_2. Then $y = e^{m_1x}$ and $y = e^{m_2x}$ are both solutions of Eq. 14.5.1. In the theory of differential equations it is shown that the *general solution* of Eq. 14.5.1 is

$$y = c_1e^{m_1x} + c_2e^{m_2x} \qquad (14.5.4)$$

EXAMPLE 1. Find the general solution of $(d^2y/dx^2) - (dy/dx) - 2y = 0$.

Solution. This is a second-order linear differential equation of the form of Eq. 14.5.1. The auxiliary equation is

$$m^2 - m - 2 = 0$$

Factoring, we have

$$(m + 1)(m - 2) = 0$$

Therefore,

$$m = -1 \qquad \text{and} \qquad m = 2$$

Thus

$$y = c_1e^{-x} + c_2e^{2x}$$

is the required general solution. (You should verify this by substitution.)

PROBLEM 1. Find the general solution of $2(d^2y/dx^2) + (dy/dx) - 3 = 0$.

The nature of the roots of the auxiliary equation fall into three categories:

1. Real and distinct.
2. Real and equal.
3. Complex.

Table 14.5.1 summarizes the type of solution for the different type of roots.

TABLE 14.5.1

Roots of Auxiliary Equation	Type of Solution of Eq. 14.5.1
real and unequal: m_1, m_2	$y = c_1e^{m_1x} + c_2e^{m_2x}$
real and equal: $m = m_1 = m_2$	$y = (c_1 + c_2x)e^{mx}$
complex: $m = a \pm bi$ (where $i = \sqrt{-1}$)	$y = e^{ax}(c_1 \cos bx + c_2 \sin bx)$

The case of real and unequal roots was discussed in Example 1. The case of real and equal roots is discussed in Example 2.

EXAMPLE 2. The differential equation for a certain electric circuit is $(d^2Q/dt^2) + (4dQ/dt) + 4Q = 0$. Solve for $Q = f(t)$.

Solution. The auxiliary equation is

$$m^2 + 4m + 4 = 0$$

Thus

$$(m + 2)^2 = 0$$

and the roots are $m_1 = -2$ and $m_2 = -2$, which are real and equal. If we let $Q = c_1e^{-2t} + c_2e^{-2t}$, this would lead to $Q = c_3e^{-2t}$, and there would be only one arbitrary constant. (We need two for a second-order equation.) Referring to Table 14.5.1, we see that the solution is

$$Q = (c_1 + c_2t)e^{-2t}$$

This may be verified by direct substitution in the given equation.

PROBLEM 2. Find the general solution of $4(d^2Q/dt^2) - 4(dQ/dt) + 1 = 0$.

Before discussing the case of imaginary roots we will show that

$$e^{iu} = \cos u + i \sin u \qquad (14.5.5)$$

where $i = \sqrt{-1}$ and $i^2 = -1$. Equation 14.5.5 is known as *Euler's formula.* It ties together trigonometric and exponential functions. Let $z = \cos u + i \sin u$. Then

$$\frac{dz}{du} = -\sin u + i \cos u$$

Multiplying by i, we obtain

$$i\frac{dz}{du} = -i \sin u - \cos u = -z$$

Multiplying by i again, we have

$$\frac{dz}{du} = iz$$

Separating variables, we obtain

$$\frac{dz}{z} = i\,du$$

and

$$\ln z = iu + C$$

When $u = 0$, $z = \cos 0 + i \sin 0 = 1$. Thus we may find C. For $u = 0$ and $z = 1$, $\ln 1 = i(0) + C$ and $C = 0$. Thus

$$\ln z = iu$$

and

$$z = e^{iu}$$

or

$$e^{iu} = \cos u + i \sin u \tag{14.5.5}$$

Now we may proceed with the case of imaginary roots of the auxiliary equation. If $m = a \pm bi$ are the roots, we would have $y = Ae^{(a+bi)x} + Be^{(a-bi)x}$, where A and B are arbitrary constants. This can be written as

$$y = e^{ax}(Ae^{bxi} + Be^{-bxi})$$

Making use of Eq. 14.5.5 we have

$$\begin{aligned} y &= e^{ax}\{A(\cos bx + i \sin bx) + B[\cos(-bx) + i \sin(-bx)]\} \\ &= e^{ax}[A(\cos bx + i \sin bx) + B(\cos bx - i \sin bx)] \end{aligned}$$

Upon regrouping, we have

$$y = e^{ax}[(A + B) \cos bx + i(A - B) \sin bx]$$

Since A, B, and i are all constants we may let $c_1 = A + B$ and $c_2 = i(A - B)$. Then we have

$$y = e^{ax}(c_1 \cos bx + c_2 \sin bx)$$

as the solution to Eq. 14.5.1 for the case of imaginary roots of the auxiliary equation.

EXAMPLE 3. Solve $(d^2y/dx^2) - 4(dy/dx) + 13y = 0$.

Solution. The auxiliary equation is

$$m^2 - 4m + 13 = 0$$

and

$$\begin{aligned} m &= \frac{4 \pm \sqrt{16 - 52}}{2} = \frac{4 \pm \sqrt{-36}}{2} \\ &= 2 \pm 3i \end{aligned}$$

Referring to Table 14.5.1, we have

$$y = e^{2x}(c_1 \cos 3x + c_2 \sin 3x)$$

PROBLEM 3. Solve $(d^2y/dx^2) + 6(dy/dx) + 13y = 0$.

EXAMPLE 4. Find the solution of $(d^2x/dt^2) + 16x = 0$ subject to the initial conditions when $t = 0$, $x = 4$, and $dx/dt = 0$.

Solution. The auxiliary equation is

$$m^2 + 4 = 0$$

and

$$m = \pm 2i$$

Thus

$$x = c_1 \cos 2t + c_2 \sin 2t$$

since

$$m = \pm 2i = 0 \pm 2i \qquad \text{and} \qquad e^0 = 1$$

We need dx/dt before finding the values of c_1 and c_2. Thus

$$\frac{dx}{dt} = -2c_1 \sin 2t + 2c_2 \cos 2t$$

For $t = 0$ and $x = 4$, $4 = c_1(1) + c_2(0)$ or $c_1 = 4$. For $t = 0$ and $dx/dt = 0$, $0 = -2(4)(0) + 2c_2(1)$ or $c_2 = 0$. Thus the required solution is

$$x = 4 \sin 2t$$

PROBLEM 4. Find the solution of $2(d^2x/dt^2) = -50x$ subject to the initial conditions when $t = 0$, $x = 0$, and $dx/dt = 3$.

We discuss some applications in Sec. 14.7.

EXERCISES 14.5

Group A

1. In each of the following, use Table 14.5.1 to write the solution of the differential equation whose auxiliary equation has the given roots and no others.

(a) 2, 3. (b) -2, 5. (c) 1, 0. (d) 2, 2.
(e) $-3, -3$. (f) $\frac{1}{2}, \frac{1}{2}$. (g) $-2 \pm 3i$. (h) $3 \pm 2i$.
(i) $-\frac{1}{2} \pm \frac{1}{3}i$. (j) $\pm 10i$. (k) $\pm 5i$. (l) $\pm 3.25i$.

2. Find the general solution of each of the following.

(a) $\dfrac{d^2y}{dx^2} + \dfrac{dy}{dx} - 2y = 0$. (b) $2\dfrac{d^2x}{dt^2} - 3\dfrac{dx}{dt} + x = 0$.

(c) $\dfrac{d^2y}{dx^2} - 4\dfrac{dy}{dx} = -4y$. (d) $\dfrac{d^2y}{dx^2} = -6\dfrac{dy}{dx} - 9y$.

(e) $\dfrac{d^2x}{dt^2} + 16x = 0$. (f) $\dfrac{d^2s}{dt^2} + 7s = 0$.

(g) $\dfrac{d^2y}{dx^2} + 4\dfrac{dy}{dx} + 13y = 0$. (h) $\dfrac{d^2y}{dx^2} + 6\dfrac{dy}{dx} + 13y = 0$.

(i) $\dfrac{d^2Q}{dt^2} + 12\dfrac{dQ}{dt} + 100Q = 0$. (j) $\dfrac{d^2Q}{dt^2} + 10\dfrac{dQ}{dt} + 60Q = 0$.

3. In each of the following, find the solution subject to the stated conditions. Recall that $y(0) = 1$ means that $y = 1$ when $x = 0$.

(a) $\dfrac{d^2y}{dx^2} + 3\dfrac{dy}{dx} + 2y = 0$; $y(0) = 1$, $y'(0) = 0$.

(b) $4\dfrac{d^2y}{dx^2} - 4y = 0$; $y(0) = 0$, $y'(0) = 1$.

(c) $\dfrac{d^2x}{dt^2} + \dfrac{dx}{dt} - 12x = 0$; $x(0) = 1$, $x'(0) = 1$.

(d) $\dfrac{d^2y}{dt^2} + 4\dfrac{dy}{dt} + 4y = 0$; $y(0) = 2$, $y(1) = 1$.

(e) $\dfrac{d^2y}{dx^2} + 2\sqrt{2}\,\dfrac{dy}{dx} + y = 0$; $y(0) = 3$, $y'(0) = 2$.

(f) $\dfrac{d^2x}{dt^2} + x = 0$; $x(0) = 1$, $x(\pi/2) = 1$.

(g) $\dfrac{d^2y}{dt^2} + 2\dfrac{dy}{dt} + 2y = 0$; $y(0) = 2$, $y'(0) = -2$.

Group B

1. Given an electric circuit with a resistance of R ohms, an inductance of L henries, and a capacitance of C farads, connected in series, one has, using Kirchhoff's laws, $L(d^2Q/dt^2) + R(dQ/dt) + (1/C)Q = 0$, where Q is the charge at any time t. Given that $Q = Q_0$ and $I = 0$ at time $t = 0$, find Q and I as functions of t.

2. If a spring is allowed to vibrate vertically from a support with a weight W on the free end in a medium that dampens the motion, one has the equation $(W/g)(d^2x/dt^2) + \beta(dx/dt) + kx = 0$, which describes its motion, where g is the gravity constant, β the damping coefficient, and k the spring constant. If the spring is set in motion with $v = v_0$ from the equilibrium position ($x = 0$), find v and x as functions of t.

14.6 RIGHT-HAND SIDE NOT ZERO

The equation

$$a_0\frac{d^2y}{dx^2} + a_1\frac{dy}{dx} + a_2y = F(x) \tag{14.6.1}$$

differs from Eq. 14.5.1 in that the right-hand side is not zero. The method of Sec. 14.5 yields a y that reduces the left-hand side of Eq. 14.6.1 to zero. This y is called the *complementary solution* and is denoted as y_c. What we need is a *particular* y denoted as y_p which will reduce the left side to the right side. Then, since y_c contains the necessary arbitrary constants, our general solution would be $y = y_c + y_p$. There are several methods that may be used to solve equations like Eq. 14.6.1. The method we use is called *the method of undetermined coefficients.* It is applicable when $F(x)$ of Eq. 14.6.1 contains a polynomial, terms of the form $\sin ax$, $\cos ax$, e^{ax}, or combinations of sums and products of these, where a is a constant. The method is shown in the following examples. The scheme is that we assume y_p to be a function such that a combination of its derivatives and itself will form $F(x)$.

EXAMPLE 1. Solve $(d^2y/dx^2) + 16y = 5x^2$.

Solution. We first find the complementary solution y_c for $(d^2y/dx^2) + 16y = 0$. Here

$$m^2 + 16 = 0$$

and

$$m = \pm 4i$$

so

$$y_c = c_1 \cos 4x + c_2 \sin 4x$$

To find the particular solution, y_p, we note that $F(x) = 5x^2$ is a polynomial of second degree. We then assume a general polynomial of the same degree. Thus assume that

$$y_p = ax^2 + bx + c$$

Then

$$\frac{dy_p}{dx} = 2ax + b$$

and

$$\frac{d^2y_p}{dx^2} = 2a$$

Substituting in the given differential equation, we have

$$2a + 16(ax^2 + bx + c) = 5x^2$$

or

$$16ax^2 + 16bx + (2a + 16c) = 5x^2$$

For this to be an identity we have

$$16a = 5$$

$$16b = 0$$

and

$$2a + 16c = 0$$

Thus

$$a = \tfrac{5}{16}$$

$$b = 0$$

and

$$c = -\tfrac{5}{128}$$

Therefore,

$$y_p = \tfrac{5}{16}x^2 - \tfrac{5}{128}$$

Since the general solution is

$$y = y_c + y_p$$

we have

$$y = c_1 \cos 4x + c_2 \sin 4x + \tfrac{5}{16}x^2 - \tfrac{5}{128}$$

You may verify this solution by substituting in the given differential equation.

PROBLEM 1. Solve $(d^2y/dx^2) - (dy/dx) - 2y = 4x$.

EXAMPLE 2. Solve $(d^2y/dx^2) + (dy/dx) - 2y = \sin 2x$.

Solution. For the complementary equation

$$\frac{d^2y}{dx^2} + \frac{dy}{dx} - 2y = 0$$

the auxiliary equation is $m^2 + m - 2 = 0$, whose roots are $m = 1$ and $m = -2$. Thus

$$y_c = c_1 e^x + c_2 e^{-2x}$$

Since $F(x) = \sin 2x$, we assume that

$$y_p = a \sin 2x + b \cos 2x$$

(Note that we shall obtain a $\sin 2x$ upon differentiating $\cos 2x$.) Then

$$\frac{dy_p}{dx} = 2a \cos 2x - 2b \sin 2x$$

and

$$\frac{d^2y_p}{dx^2} = -4a \sin 2x - 4b \cos 2x$$

Substituting in the given differential equation, we have

$$(-4a \sin 2x - 4b \cos 2x) + (2a \cos 2x - 2b \sin 2x) - 2(a \sin 2x + b \cos 2x) = \sin 2x$$

Regrouping, this becomes

$$(-6a - 2b) \sin 2x + (2a - 6b) \cos 2x = \sin 2x$$

Thus

$$-6a - 2b = 1$$

and

$$2a - 6b = 0$$

From these we find that $a = -\frac{3}{20}$, $b = -\frac{1}{20}$, and $y_p = -\frac{3}{20} \sin 2x - \frac{1}{20} \cos 2x$. Thus

$$y = c_1 e^x + c_2 e^{-2x} - \tfrac{3}{20} \sin 2x - \tfrac{1}{20} \cos 2x$$

You may verify this solution.

PROBLEM 2. Solve $(d^2y/dx^2) + 9y = 2 \sin 5x$.

EXAMPLE 3. Find the solution of $(dQ^2/dt^2) - 4Q = 2e^{-3t}$ subject to the initial conditions $Q = 0$ and $I = 0$ when $t = 0$.

Solution. The complementary equation is

$$\frac{d^2Q}{dt^2} - 4t = 0$$

The auxiliary equation is

$$m^2 - 4 = 0$$
$$m = \pm 2$$

Thus

$$Q_c = c_1 e^{2t} + c_2 e^{-2t}$$

Since $F(t) = 2e^{-3t}$, we assume that

$$Q_p = ae^{-3t}$$

Then

$$\frac{dQ_p}{dt} = -3ae^{-3t}$$

and

$$\frac{d^2Q_p}{dt^2} = 9ae^{-3t}$$

Substituting in the given differential equation, we have

$$9ae^{-3t} - 4ae^{-3t} = 2e^{-3t}$$

or

$$5ae^{-3t} = 2e^{-3t}$$

Thus

$$5a = 2 \qquad \text{and} \qquad a = \tfrac{2}{5}$$

and

$$Q_p = \tfrac{2}{5}e^{-3t}$$

Therefore,

$$Q = c_1 e^{2t} + c_2 e^{-2t} + \tfrac{2}{5}e^{-3t}$$

To satisfy the initial conditions we have, for $t = 0$ and $Q = 0$, $0 = c_1 + c_2 + \frac{2}{5}$. Since $I = dQ/dt$, we have

$$I = 2c_1 e^{2t} - 2c_2 e^{-2t} - \tfrac{6}{5}e^{-3t}$$

For $t = 0$ and $I = 0$, $0 = 2c_1 - 2c_2 - \frac{6}{5}$. We then solve the simultaneous equations

$$c_1 + c_2 = -\tfrac{2}{5}$$

and

$$2c_1 - 2c_2 = \tfrac{6}{5}$$

to find $c_1 = \frac{1}{10}$ and $c_2 = -\frac{1}{2}$. We then find the required solution to be

$$Q = \tfrac{1}{10}e^{2t} - \tfrac{1}{2}e^{-2t} + \tfrac{2}{5}e^{-3t}$$

PROBLEM 3. Find the solution of $(d^2Q/dt^2) + (dQ/dt) + Q = 2 \sin t$ subject to the initial conditions at $t = 0$, $Q = I = 0$.

In summary, to solve a second-order linear differential equation with constant coefficients with right-hand side not zero:

1. Find the complementary solution y_c.
2. For the particular solution y_p, if the right-hand side is
   (a) a polynomial of degree n, assume a general polynomial of degree n.
   (b) an expression containing $\sin mx$ and/or $\cos mx$, assume that $y_p = a \sin mx + b \cos mx$.
   (c) an expression containing e^{bx}, assume that $y_p = ae^{bx}$.
   (d) A sum of parts (a), (b), and (c), assume y_p to be a sum of the corresponding terms.
3. Determine the value of the coefficients for y_p.
4. For the general solution add y_p to y_c.

(We might note that if any terms in the assumed particular solution already occur in the complementary solution, this method needs further refinement. You are referred to texts on differential equations. Also see the exercises in Group B.)

Warning ***The initial conditions are used only in the general solution not in the complementary.***

EXERCISES 14.6

Group A

1. Solve each of the following differential equations.

(a) $\dfrac{d^2y}{dx^2} + \dfrac{dy}{dx} - 6y = 24.$

(b) $\dfrac{d^2y}{dx^2} - \dfrac{dy}{dx} - 12y = e^{2x}.$

(c) $\dfrac{d^2y}{dx^2} + \dfrac{2dy}{dx} + 5y = \sin x.$

(d) $\dfrac{d^2y}{dx^2} + \dfrac{2dy}{dx} + 5y = x + \sin x.$

(e) $\dfrac{d^2y}{dt^2} + \dfrac{2dy}{dt} - 15y = t^2 + 2t + e^{-t}.$

(f) $\dfrac{d^2x}{dt^2} + 4x = 8 + e^{-t} + \sin t.$

(g) $\dfrac{d^2x}{dt^2} + 9x = 16t^2 + 8 \cos 2t.$

(h) $\dfrac{d^2x}{dt^2} - 9x = e^{-2t} + 2e^t.$

(i) $\dfrac{d^2y}{dt^2} + 25\dfrac{dy}{dt} = e^{-5t}.$

(j) $\dfrac{d^2y}{dx^2} - 2\dfrac{dy}{dx} = 3 \sin 2x + \cos 2x.$

2. Solve each of the following differential equations subject to the given conditions.

(a) $\dfrac{d^2x}{dt^2} + \dfrac{dx}{dt} - 6x = 12;\ x(0) = 1,\ x'(0) = 0.$

(b) $\dfrac{d^2y}{dx^2} + 6\dfrac{dy}{dx} + 8y = \sin x;\ y(0) = 0,\ y'(0) = 1.$

(c) $\dfrac{d^2x}{dt^2} - x = \cos 2t;\ x(0) = 0,\ x'(0) = 1.$

(d) $\frac{d^2x}{dt^2} - 4\frac{dx}{dt} = 12e^{2t}$; $x(0) = 1$, $x'(0) = 2$.

(e) $\frac{d^2x}{dt^2} + x = t^2 - e^{-t}$; $x(0) = 0$, $x'(0) = 0$.

Group B

1. (a) Find the complementary solution of $(d^2x/dt^2) + 2(dx/dt) - 15x = e^{3t}$. (b) Show that $x_p = ae^{3t}$ is *not* a particular solution for the given equation. (c) Show that $x_p = \frac{1}{8}te^{3t}$ is a particular solution for the given equation.

2. (a) Find the complementary solution of $(d^2x/dt^2) + x = \sin t$. (b) Show that no A and B exist such that $x_p = A \sin t + B \cos t$ is a particular solution for the given equation. (c) Show that $x_p = -\frac{1}{2}t \cos t$ is a particular solution for the given equation. [*Observation*: If we add to rule 2(a), (b), and (c) a rule 2′ as follows, then we may solve problems such as Exercises 1 and 2 of Group B.]

> ***Rule 2′*** *Multiply the assumed particular solution y_p by the independent variable to the least power that removes any repetition of the complementary solution.*

3. Use rule 2′ of Exercise 2 to solve each of the following.

(a) $\frac{d^2x}{dt^2} + \frac{dx}{dt} - 6x = e^{2t}$.

(b) $\frac{d^2x}{dt^2} + 4x = \cos 2t$.

(c) $\frac{d^2x}{dt^2} + 2\frac{dx}{dt} + x = e^{-t}$.

14.7 APPLICATIONS

The second-order linear differential equation with constant coefficients occurs frequently in applied problems. Two such problems are vibratory motion and electric circuits. We shall discuss both of these problems.

If a weight of W lb suspended by a spring with a spring constant of k lb/ft is set in vibrating motion in a resisting medium that opposes the motion with a force proportional to the velocity of the weight, it may be shown, by using Newton's law of $f = ma$, that the differential equation of motion is

$$\frac{W}{g}\frac{d^2x}{dt^2} + \beta\frac{dx}{dt} + kx = 0 \tag{14.7.1}$$

In Eq. 14.7.1 g is the gravitational force, β the constant of proportionality called the *damping constant*, and x the displacement of the weight at time t sec. If $\beta = 0$ (no resistance to the motion), we have *simple harmonic motion.*

EXAMPLE 1. A weight of 64 lb is hung on a spring with $k = 50$ lb/ft. It is released from rest when $x = 4$ ft. Find its displacement as a function of t.

Solution. Use Eq. 14.7.1, with $W = 64$, $\beta = 0$, $k = 50$, and $g = 32$ (this value of g is a convenient approximation). Thus

$$\frac{64}{32}\frac{d^2x}{dt^2} + 50x = 0 \qquad \text{or} \qquad \frac{d^2x}{dt^2} + 25x = 0$$

The auxiliary equation is $m^2 + 25 = 0$, and $m = \pm 5i$. The general solution is $x = c_1 \cos 5t + c_2 \sin 5t$. Also,

$$\frac{dx}{dt} = -5c_1 \sin 5t + 5c_2 \cos 5t$$

For $t = 0$ and $x = 4$, $4 = c_1(1) + c_2(0)$, or $c_1 = 4$. For $t = 0$ and $dx/dt = 0$ (released from rest), $0 = 0 + 5c_2(1)$ and $c_2 = 0$. Therefore,

$$x = 4 \cos 5t$$

This solution tells us that the weight vibrates with an amplitude of 4 ft. Its motion is oscillatory with period $2\pi/5$ s (see Fig. 14.7.1). This is an example of simple harmonic motion.

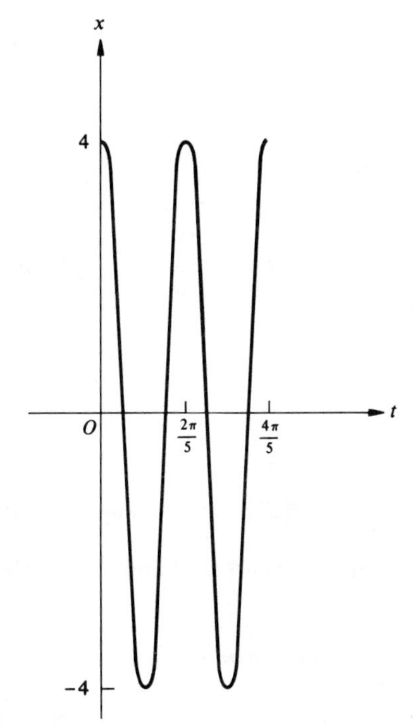

Figure 14.7.1 $x = 4 \cos 5t$.

PROBLEM 1. A weight of 16 lb is hung on a spring with $k = 8$ lb/ft. The damping coefficient is zero and the motion starts with $v = 5$ ft/s when $x = 0$. Find its displacement x as a function of t. What is the amplitude of motion? How fast is it moving when $t = 1$ s?

Let us consider the general solution of Eq. 14.7.1. Its auxiliary equation is

$$\frac{W}{g}m^2 + \beta m + k = 0$$

with roots

$$m = \frac{-\beta \pm \sqrt{\beta^2 - 4(kW/g)}}{2(W/g)}$$

The nature of the general solution depends on the roots m, which in turn depends on the terms under the radical sign.

1. If $\beta^2 < 4(kW/g)$, the roots are complex. The solution is of the form $x = e^{-at}(c_1 \cos bt + c_2 \sin bt)$. This represents a *damped oscillatory motion* (see Fig. 14.7.2).
2. If $\beta^2 > 4(kW/g)$, the roots are real and distinct. The solution is of the form

$$x = c_1e^{-at} + c_2e^{-bt}$$

This represents *overdamped* motion. Note that there is no sine or cosine term which would give vibratory motion (see Fig. 14.7.2).

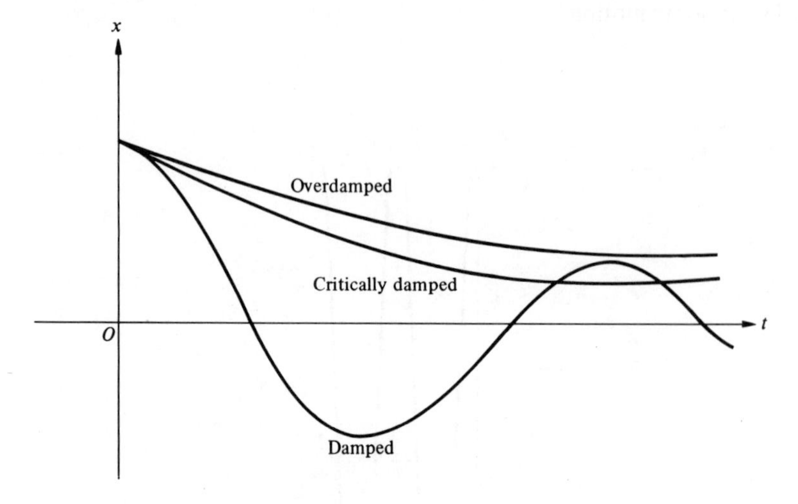

Figure 14.7.2 Motion with damping.

3. If $\beta^2 = 4(kW/g)$, the roots are real and equal. The solution is of the form

$$x = (c_1 + c_2t)e^{-at}$$

This represents *critically damped* motion (see Fig. 14.7.2). If in Eq. 14.7.1 the right-hand side is not zero, we then have *forced* motion which might lead to resonance (see Exercises 2 and 3 of Group B).

The differential equation 14.7.1 of mechanical systems has an analogy in electric circuits. If an electric circuit has a voltage source E, an inductance L, a resistance R, and a capacitance C, connected in series, its differential equation may be shown to be

$$L\frac{d^2Q}{dt^2} + R\frac{dQ}{dt} + \frac{1}{C}Q = E \tag{14.7.2}$$

where E may be zero or a function of time t. (Recall that $I = dQ/dt$.) The auxiliary equation for Eq. 14.7.2 is

$$Lm^2 + Rm + \frac{1}{C} = 0$$

whose roots are

$$m = \frac{-R \pm \sqrt{R^2 - 4L/C}}{2L} = \frac{-R}{2L} \pm \sqrt{\frac{R^2}{4L^2} - \frac{1}{LC}}$$

Usually, the quantity $(R^2/4L^2 - 1/LC)$ is negative and m is complex. For convenience let $\omega = \sqrt{1/LC - R^2/4L^2}$. Then

$$Q_c = e^{-(R/2L)t}(c_1 \cos \omega t + c_2 \sin \omega t)$$

is the complementary solution of Eq. 14.7.2. (Note that the presence of the sine and cosine terms indicate an oscillatory charge.)

EXAMPLE 2. A series circuit consists of a 2-H inductance, a 20-Ω resistor, a capacitor of 0.002 F, and a voltage source of 30 cos 10t V. Find the charge and the current at time t if the charge on the capacitor is zero when the switch is closed at $t = 0$. Also find the steady-state current.

Solution. Using Eq. 14.7.2, we have

$$2\frac{d^2Q}{dt^2} + 20\frac{dQ}{dt} + \frac{1}{0.002}Q = 30 \cos 10t$$

or

$$\frac{d^2Q}{dt^2} + 10\frac{dQ}{dt} + 250Q = 15 \cos 10t \tag{14.7.3}$$

The complementary solution of Eq. 14.7.3 is

$$Q_c = e^{-5t}(c_1 \cos 15t + c_2 \sin 15t)$$

For the particular solution we assume that

$$Q_p = a \cos 10t + b \sin 10t$$

and find $a = \frac{9}{130}$ and $b = \frac{3}{65}$. Thus the general solution is

$$Q = e^{-5t}(c_1 \cos 15t + c_2 \sin 15t) + \tfrac{9}{130} \cos 10t + \tfrac{3}{65} \sin 10t$$

Using the initial conditions, $t = 0$, $Q = 0$, and $I = dQ/dt = 0$, we find $c_1 = -\frac{9}{130}$ and $c_2 = -\frac{7}{130}$. Hence

$$Q = e^{-5t}(-\tfrac{9}{130} \cos 15t - \tfrac{7}{130} \sin 15t) + \tfrac{9}{130} \cos 10t + \tfrac{3}{65} \sin 10t$$

and

$$I = e^{-5t}(\tfrac{17}{13} \sin 15t - \tfrac{6}{13} \cos 15t) - \tfrac{9}{13} \sin 10t + \tfrac{6}{13} \cos 10t \tag{14.7.4}$$

Note that as t increases, $e^{-5t} \longrightarrow 0$. The term with e^{-5t} is called the *transient solution*. The term $(-\frac{9}{13}\sin 10t + \frac{6}{13}\cos 10t)$ that remains as the transient term becomes negligible is called the *steady-state solution* or the *steady-state current*. The steady-state current may also be obtained by differentiating the particular solution of Q.

PROBLEM 2. A series circuit consists of a 2-H inductor, a 12-Ω resistor, a capacitor of 0.02 F, and a voltage source of 20 sin $5t$. Find the charge and the current at time t if the charge on the capacitor is zero when the switch is closed at $t = 0$. Find the current at $t = 1$. Also find the steady-state current.

EXERCISES 14.7

Group A

1. A 2-lb weight is suspended from the end of a spring with a spring constant of 16 lb/ft. If the damping coefficient β is zero, find the position and velocity as functions of time if the motion is started from rest at $x = \frac{1}{4}$ ft.

2. A 3-lb weight is suspended from the end of a spring with a spring constant of 6 lb/ft and damping coefficient $\beta = 0$. If the motion is started at $x = 0$ with a velocity of 2 ft/s, find the position and velocity as functions of time.

3. A 4-lb weight is suspended from a spring with a spring constant of 16 lb/ft in a medium whose damping force is equal to $2v$ pounds ($\beta = 2$) at any time. If at time $t = 0$, $v = 0$ and $x = \frac{1}{2}$ ft, find the position as a function of time. Is the motion damped oscillatory, critically damped, or overdamped?

4. A 2-lb weight is suspended from a spring with a spring constant of 16 lb/ft in a medium whose damping force is equal to $2v$ pounds at any time. If at time $t = 0$, $v = 0$ and $x = -\frac{1}{2}$ ft, find the position as a function of time. Is the motion damped oscillatory, critically damped, or overdamped?

5. An emf of 300 V is in series with an inductance of 1 H, a resistance of 25 Ω, and a capacitance of 0.01 F. If at time $t = 0$, the current I and charge Q are zero, find the charge and current for $t \geqq 0$. Indicate the steady-state and transient parts of the solution.

6. An emf of 80 sin $10t$ volts is in series with a 2-H inductor, an 8-Ω resistor, and a 0.0125-F capacitor. (a) Find Q and I if at time $t = 0$, $Q = 0$ and $I = 0$. (b) Indicate the steady-state and transient parts of the solutions.

7. An emf of 50 V is in series with a 2-H inductor, an 8-Ω resitor, and a 0.1-F capacitor. If at time $t = 0$, $I = 0$ and $Q = 6$ C, find Q and I at any time $t \geqq 0$.

8. A 10^{-1}-H inductor is in series with a 10^{-3}-F capacitor and an emf of 50 V. Find Q and I at any time $t \geqq 0$ if Q and I are zero at time $t = 0$.

Group B

1. A vertical spring having a spring constant of 3.5 lb/ft has a 16-lb weight suspended from it. An external force given by $F(t) = 50 \sin t$, $t \geqq 0$, is applied. A damping force of $4v$ lb is assumed to act. Initially, the velocity is zero and $x = 0$. (a) Find the displacement at any time t. (b) Find the velocity at $t = 1$ s. (c) Indicate the transient and steady-state parts of this solution. (d) Find the velocity at $t = 12.56$ s.

2. (a) A vertical spring having a spring constant of 2 lb/ft has a 16-lb weight suspended from it. An external force given by $F(t) = \sin 2t$, $t \geqq 0$, is applied and it is assumed that there is no damping. At time $t = 0$, $x = 0$, and $v = 0$. Show that after a short period of time the spring will break. (*Hint*: See Exercise 2 of Group B in Exercises 14.6.) This phenomenon is known as ***resonance***. Resonance occurs when the frequency of the external force is the same as the frequency of the natural system. (b) Show that the solution to the undamped equation $(d^2x/dt^2) + \omega^2 x = A_0 \sin \omega t$ always involves the phenomenon of resonance.

3. (a) A 1-H inductor and a 0.25-F capacitor are in series with an emf of $16 \cos 2t$ V, $t \geqq 0$. If $Q = 0$ and $I = 0$ at $t = 0$, find the charge Q at any time $t \geqq 0$. Show that resonance is involved. (b) Show that the solution to the equation $(d^2x/dt^2) + \omega^2 x = A_0 \cos \omega t$ always involves the phenomenon of resonance.

14.8 THE LAPLACE TRANSFORM

When a differential equation is to be solved subject to certain initial conditions, the method of using the Laplace transform is sometimes useful. This method transforms a differential equation into an algebraic equation. This algebraic equation is then solved. An inverse transform is applied to this solution to obtain the solution of the differential equation. This method is very powerful for certain types of equations. Before using it we must know something about the Laplace transform. We shall just barely touch the subject. Those wishing a more comprehensive treatment may consult one of the many texts on the subject. The symbol used in discussing Laplace transforms is a script L, thus $\mathcal{L}$.

> ***Definition 14.8.1*** *The Laplace transform of a function* $F(t)$, *denoted by* $\mathcal{L}\{F(t)\}$ *or* $f(s)$, *is defined by the integral*
>
> $$\mathcal{L}\{F(t)\} = f(s) = \int_0^\infty e^{-st} F(t)\, dt \qquad (14.8.1)$$
>
> *if this integral exists.*
>
> In Eq. 14.8.1, s is a parameter, which is to be treated like a constant.

EXAMPLE 1. Find the Laplace transform of $F(t) = 1$.

Solution. Using Eq. 14.8.1, we have

$$\mathcal{L}\{(1)\} = \int_0^\infty e^{-st}(1)\,dt = \lim_{b\to\infty}\int_0^b e^{-st}\,dt \qquad \text{(see Section 11.6)}$$

$$= -\frac{1}{s}\lim_{b\to\infty}[e^{-st}]_0^b = -\frac{1}{s}\lim_{b\to\infty}(e^{-sb} - 1)$$

As $b \longrightarrow \infty$, $e^{-sb} \longrightarrow 0$; thus for $s > 0$ we have

$$\mathcal{L}\{(1)\} = -\frac{1}{s}(0 - 1) = \frac{1}{s}$$

You may discover what happens for $s \leq 0$ (see Exercise 3 of Group B).

PROBLEM 1. Use Eq. 14.8.1 to find $\mathcal{L}\{5\}$.

EXAMPLE 2. Find $\mathcal{L}\{(t)\}$.

Solution. Here $F(t) = t$; hence we have

$$\mathcal{L}\{(t)\} = \int_0^\infty e^{-st}t\,dt = \lim_{b\to\infty}\left[\frac{e^{-st}}{s^2}(-st - 1)\right]_0^b$$

As $b \longrightarrow \infty$, the value at the upper limit is zero. Hence for $s > 0$,

$$\mathcal{L}\{(t)\} = \frac{1}{s^2}$$

PROBLEM 2. Use Eq. 14.8.1 to find $\mathcal{L}\{5t\}$.

EXAMPLE 3. Find $\mathcal{L}\{(\sin at)\}$.

Solution

$$\mathcal{L}\{(\sin at)\} = \int_0^\infty e^{-st}\sin at\,dt$$

$$= \lim_{b\to 0}\left[\frac{e^{-st}}{a^2 + s^2}(-s\sin at - a\cos at)\right]_0^b$$

$$= 0 - \frac{1}{a^2 + s^2}(0 - a)$$

$$= \frac{a}{s^2 + a^2} \qquad \text{for} \qquad s > 0$$

PROBLEM 3. Use Eq. 14.8.1 to find $\mathcal{L}\{\cos at\}$.

It appears then that if the improper integral in the definition of the Laplace transform exists, we can find the transform and it is an algebraic function of s. The restrictions on s are useful in theoretical discussions. We shall assume in the following discussions that s is always such that the Laplace transform is meaningful.

It can be shown that the Laplace transform is a linear transformation; that is,

$$\mathfrak{L}\{c_1F_1(t) + c_2F_2(t)\} = c_1\mathfrak{L}\{F_1(t)\} + c_2\mathfrak{L}\{F_2(t)\} \tag{14.8.2}$$

where c_1 and c_2 are constants and $F_1(t)$ and $F_2(t)$ are functions whose Laplace transforms exist. By means of Eq. 14.8.2 the Laplace transforms of many functions may be found.

EXAMPLE 4. Find $\mathfrak{L}\{2 - 5 \sin 3t\}$.

Solution. By use of Eq. 14.8.2 we have

$$\mathfrak{L}\{2 - 5 \sin 3t\} = 2\mathfrak{L}\{1\} - 5\mathfrak{L}\{\sin 3t\}$$

Using the results of Examples 1 and 3, we have

$$\mathfrak{L}\{2 - 5 \sin 3t\} = \frac{2}{s} - \frac{15}{s^2 + 9}$$

PROBLEM 4. Use Eq. 14.8.2 and Probs. 1 and 3 to find $\mathfrak{L}\{5 - 3 \cos 2t\}$.

***Warning** Equation 14.8.2 is not valid for the product of two variable quantities.*

The Laplace transforms of many functions have been tabulated. We list a few of them in Table 14.8.1.

We need two more Laplace transforms before applying them to the solution of differential equations. They are the Laplace transforms for dy/dt and d^2y/dt^2.

To find the Laplace transform for $dy/dt = F'(t)$ we proceed as follows:

$$\mathfrak{L}\{F'(t)\} = \int_0^\infty e^{-st}F'(t)\,dt$$

Applying integration by parts (Eq. 11.3.1) we have $u = e^{-st}$, $dv = F'(t)\,dt$, $du = -se^{-st}\,dt$, $v = F(t)$, and

$$\int_0^\infty e^{-st}F'(t)\,dt = \lim_{b\to\infty} [e^{-st}F(t)]_0^b + s\int_0^\infty e^{-st}F(t)\,dt$$
$$= 0 - F(0) + s\mathfrak{L}\{F(t)\}$$

(The last term comes from Definition 14.8.1.) Thus

$$\mathfrak{L}\{F'(t)\} = s\mathfrak{L}\{F(t)\} - F(0) \tag{14.8.3}$$

In a similar manner $\mathfrak{L}\{F''(t)\}$ may be found. It is

$$\mathfrak{L}\{F''(t)\} = s^2\mathfrak{L}\{F(t)\} - sF(0) - F'(0) \tag{14.8.4}$$

Note that in Eqs. 14.8.3 and 14.8.4 that we must know the initial values of F and F'.

EXAMPLE 5. If $F(0) = 3$ and $F'(0) = 2$, find an expression for $\mathfrak{L}\{F''(t)\}$.

TABLE 14.8.1 Short Table of Laplace Transforms

	$F(t) = \mathcal{L}^{-1}\{f(s)\}$	$f(s) = \mathcal{L}\{F(t)\}$
1.	$F(t)$	$f(s) = \int_0^\infty e^{-st}F(t)\,dt$
2.	$c_1F_1(t) + c_2F_2(t)$	$c_1\mathcal{L}\{F_1(t)\} + c_2\mathcal{L}\{F_2(t)$
3.	1	$\frac{1}{s}$
4.	$t^n, n = 1, 2, 3, \ldots$	$\frac{n!}{s^{n+1}}$
5.	e^{at}	$\frac{1}{s-a}$
6.	$\sin \omega t$	$\frac{\omega}{s^2+\omega^2}$
7.	$\cos \omega t$	$\frac{s}{s^2+\omega^2}$
8.	$F'(t)$	$s\mathcal{L}\{F(t)\} - F(0)$
9.	$F''(t)$	$s^2\mathcal{L}\{F(t)\} - sF(0) - F'(0)$
10.	$\frac{1}{a}(e^{at} - 1)$	$\frac{1}{s(s-\mathrm{a})}$
11.	te^{at}	$\frac{1}{(s-\mathrm{a})^2}$
12.	$t^ne^{at}, n = 1, 2, 3, \ldots$	$\frac{n!}{(s-a)^{n+1}}$
13.	$e^{at}(1 + at)$	$\frac{s}{(s-a)^2}$
14.	$\sin \omega t - \omega t \cos \omega t$	$\frac{2\omega^3}{(s^2+\omega^2)^2}$
15.	$\sin \omega t + \omega t \cos \omega t$	$\frac{2\omega s^2}{(s^2+\omega^2)^2}$
16.	$t \sin \omega t$	$\frac{2\omega s}{(s^2+\omega^2)^2}$
17.	$t \cos \omega t$	$\frac{s^2-\omega^2}{(s^2+\omega^2)^2}$
18.	$\frac{b \sin at - a \sin bt}{ab(b^2-a^2)}$	$\frac{1}{(s^2+a^2)(s^2+b^2)}$
19.	$\frac{\cos at - \cos bt}{b^2-a^2}$	$\frac{s}{(s^2+a^2)(s^2+b^2)}$
20.	$\frac{a \sin at - b \sin bt}{a^2-b^2}$	$\frac{s^2}{(s^2+a^2)(s^2+b^2)}$
21.	$e^{-bt} \sin \omega t$	$\frac{\omega}{(s+b)^2+\omega^2}$
22.	$e^{-bt} \cos \omega t$	$\frac{s+b}{(s+b)^2+\omega^2}$
23.	$1 - \cos \omega t$	$\frac{\omega^2}{s(s^2+\omega^2)}$

Solution. Using Eq. 14.8.4, we have

$$\mathcal{L}\{F''(t)\} = s^2\mathcal{L}\{F(t)\} - 3s - 2$$

(Note that if $\mathcal{L}\{F(t)\}$ is an algebraic function of s, as it is, then $\mathcal{L}\{F''(t)\}$ is an algebraic function of s.)

PROBLEM 5. Find an expression for $\mathcal{L}\{F'(t)\}$ if $F(0) = -2$.

☛ ***Warning*** ***We can find the Laplace transforms of $F'(t)$ and of $F''(t)$ only if we have values for $F(0)$, and for $F(0)$ and $F'(0)$, respectively.***

The next question is: If we know the Laplace transform of a function, can we find the function? This is accomplished by finding the inverse transform, which is denoted by

$$\mathcal{L}^{-1}\{f(s)\} = F(t) \tag{14.8.5}$$

The process is best shown by some examples.

EXAMPLE 6. Evaluate $\mathcal{L}^{-1}\{4/(s^2 + 16)\}$.

Solution. From number 6 in the table of Laplace transforms (Table 14.8.1), we have

$$\mathcal{L}^{-1}\left\{\frac{4}{s^2 + 16}\right\} = \sin 4t$$

PROBLEM 6. Evaluate $\mathcal{L}^{-1}\{s/(s^2 + 9)\}$.

EXAMPLE 7. Evaluate $\mathcal{L}^{-1}\{5/(s - 7)^4\}$.

Solution. This looks like number 12 in Table 14.8.1. Here $n + 1 = 4$, and so $n = 3$. Hence the numerator should be $3! = 6$. We take care of this by writing

$$\mathcal{L}^{-1}\left\{\frac{5}{(s - 7)^4}\right\} = \tfrac{5}{6}\mathcal{L}^{-1}\left\{\frac{6}{(s - 7)^4}\right\}$$

Thus

$$\mathcal{L}^{-1}\left\{\frac{5}{(s - 7)^4}\right\} = \tfrac{5}{6}t^3e^{7t}$$

PROBLEM 7. Evaluate $\mathcal{L}^{-1}\{3/(s^2 + 7)\}$.

EXAMPLE 8. Evaluate $\mathcal{L}^{-1}\{(s + 1)/(s^2 + 6s + 25)\}$.

Solution. This does not look like anything in Table 14.8.1. But the method of completing the square comes to our rescue. We may write

$$\begin{aligned} s^2 + 6s + 25 &= (s^2 + 6s + \quad) + 25 \\ &= (s^2 + 6s + 9) + 16 \\ &= (s + 3)^2 + (4)^2 \end{aligned}$$

Then

$$\mathcal{L}^{-1}\left\{\frac{s + 1}{s^2 + 6s + 25}\right\} = \mathcal{L}^{-1}\left\{\frac{s + 1}{(s + 3)^2 + (4)^2}\right\}$$

By adding and subtracting 2 in the numerator, we have

$$\begin{aligned} &\mathcal{L}^{-1}\left\{\frac{s + 3}{(s + 3)^2 + (4)^2} + \frac{-2}{(s + 3)^2 + (4)^2}\right\} \\ &\qquad = \mathcal{L}^{-1}\left\{\frac{s + 3}{(s + 3)^2 + (4)^2}\right\} - \tfrac{1}{2}\mathcal{L}^{-1}\left\{\frac{4}{(s + 3)^2 + (4)^2}\right\} \end{aligned}$$

Then from numbers 22 and 21 in Table 14.8.1 we have

$$\mathcal{L}^{-1}\left\{\frac{s+1}{s^2+6s+25}\right\} = e^{-3t}\cos 4t - \tfrac{1}{2}e^{-3t}\sin 4t$$

Thus

$$F(t) = e^{-3t}(\cos 4t - \tfrac{1}{2}\sin 4t)$$

Warning ***To use Table 14.8.1, we must make the function exactly like the entry in the table.***

PROBLEM 8. Evaluate $\mathcal{L}^{-1}\{(s-1)/(s^2-4s+13)\}$.

EXAMPLE 9. Evaluate $\mathcal{L}^{-1}\{2/(s^2-7s+12)\}$.

Solution. Once again this does not look like anything in Table 14.8.1. Using a different algebraic technique, we may arrive at the solution. We notice that the denominator is factorable into $(s-4)(s-3)$. Using partial fractions, (see Sec. 11.4)

$$\frac{2}{s^2-7s+12} = \frac{A}{s-4} + \frac{B}{s+3}$$

and multiplying both sides by $s^2-7s+12$, we obtain

$$2 = A(s-3) + B(s-4)$$

Let $s = 4$: $2 = A(4-3) + B(4-4)$ or $A = 2$. Let $s = 3$: $2 = A(3-3) + B(3-4)$ or $B = -2$. Thus

$$\mathcal{L}^{-1}\left\{\frac{2}{s^2-7s+12}\right\} = \mathcal{L}^{-1}\left\{\frac{2}{s-4} - \frac{2}{s-3}\right\}$$

$$= 2\mathcal{L}^{-1}\left\{\frac{1}{s-4}\right\} - 2\mathcal{L}^{-1}\left\{\frac{1}{s-3}\right\}$$

and using number 5 in Table 14.8.1, we have

$$\mathcal{L}^{-1}\left\{\frac{2}{s^2-7s+12}\right\} = 2e^{4t} - 2e^{3t}$$

PROBLEM 9. Evaluate $\mathcal{L}^{-1}\{(-2s-13)/(s^2+s-2)\}$.

EXERCISES 14.8

Group A

1. Use Table 14.8.1 to find each of the following Laplace transforms.

(a) $\mathcal{L}\{t^2+3t+1\}$
(b) $\mathcal{L}\{(\sin 2t + \cos 4t\}$.
(c) $\mathcal{L}\{t^5 + te^{2t}\}$.
(d) $\mathcal{L}\{e^{-2t}\sin 4t\}$.
(e) $\mathcal{L}\{2e^{3t} - 4e^{-3t}\}$.
(f) $\mathcal{L}\{2e^{4t} - 2te^{3t}\}$.

(g) $\mathcal{L}\{\frac{1}{2}(e^{2t} - 1)\}$.

(h) $\mathcal{L}\{t \sin t\}$.

(i) $\mathcal{L}\{\sin 3t - 3t \cos 3t\}$.

(j) $\mathcal{L}\{t^4 + 6t^2 + \cos t\}$.

2. Use Table 14.8.1 to find each of the following inverse Laplace transforms.

(a) $\mathcal{L}^{-1}\left\{\frac{1}{s}\right\}$.

(b) $\mathcal{L}^{-1}\left\{\frac{2}{s^2 + 4}\right\}$.

(c) $\mathcal{L}^{-1}\left\{\frac{2s}{s^2 + 16}\right\}$.

(d) $\mathcal{L}^{-1}\left\{\frac{1}{s - 5}\right\}$.

(e) $\mathcal{L}^{-1}\left\{\frac{1}{s + 4} + \frac{6}{(s - 3)^4}\right\}$.

(f) $\mathcal{L}^{-1}\left\{\frac{16s}{(s^2 + 4)^2}\right\}$.

(g) $\mathcal{L}^{-1}\left\{\frac{\sqrt{6}}{(s - 2)^2 + 6}\right\}$.

(h) $\mathcal{L}^{-1}\left\{\frac{s + 2}{(s + 2)^2 + 5}\right\}$.

(i) $\mathcal{L}^{-1}\left\{\frac{s}{(s^2 + 4)(s^2 + 9)}\right\}$.

(j) $\mathcal{L}^{-1}\left\{\frac{1}{(s + 16)^2}\right\}$.

3. Use formulas 8 and 9 in Table 14.8.1 to find the Laplace transforms of each of the following equations.

(a) $y'' + 2y' + 3y = 0;\ y(0) = 1,\ y'(0) = 2$.
(b) $y'' + 16y = \sin t;\ y(0) = 0,\ y'(0) = 4$.
(c) $2y'' + 3y' + y = t^2 + 3;\ y(0) = 1,\ y'(0) = 0$.
(d) $\frac{3}{2}y'' + \frac{1}{2}y' + \frac{1}{4}y = e^{-5t};\ y(0) = 0,\ y'(0) = 0$.
(e) $y'' + 8y' + 2y = \cos 2t;\ y(0) = 2,\ y'(0) = 1$.

Group B

1. Use Table 14.8.1 and appropriate algebraic techniques to find each of the following inverse Laplace transforms.

(a) $\mathcal{L}^{-1}\left\{\frac{1}{s^2 - 2s}\right\}$.

(b) $\mathcal{L}^{-1}\left\{\frac{3}{(s - 2)(s + 1)}\right\}$.

(c) $\mathcal{L}^{-1}\left\{\frac{2s + 4}{(s - 2)(s^2 + 4)}\right\}$.

(d) $\mathcal{L}^{-1}\left\{\frac{s + 2}{s^2 + 6s + 25}\right\}$.

(e) $\mathcal{L}^{-1}\left\{\frac{3s - 1}{s^2 + 2s + 5}\right\}$.

(f) $\mathcal{L}^{-1}\left\{\frac{5 + 7s - 2s^2}{s^3}\right\}$.

(g) $\mathcal{L}^{-1}\left\{\frac{2s^2 + 15s + 7}{(s + 1)(s - 1)(s - 2)}\right\}$.

(h) $\mathcal{L}^{-1}\left\{\frac{8}{(s - 2)(s + 4)(s - 3)}\right\}$.

(i) $\mathcal{L}^{-1}\left\{\frac{2s - 10}{s^2 - 4s + 20}\right\}$.

(j) $\mathcal{L}^{-1}\left\{\frac{s + 5}{s^2 + 2s - 3}\right\}$.

2. (a) Use the definition of the Laplace transform to compute $\mathcal{L}\{e^{2t}\}$. (b) Use formula 8, Table 14.8.1, and the fact that $y = e^{2t}$ is a solution to the differential equation $y = \frac{1}{2}y'$, and $y(0) = 1$, to compute $\mathcal{L}\{e^{2t}\}$.

3. Show that in Example 1 if $s \leqq 0$, $\mathcal{L}\{1\}$ does not exist.

Group C

1. It can be shown that if $\mathcal{L}^{-1}\{f(s)\} = F(t)$ and $\mathcal{L}^{-1}\{g(s)\} = G(t)$, then under suitable restrictions on $F(t)$ and $G(t)$, $\mathcal{L}^{-1}(f(s)g(s)\} = \int_0^t F(u)G(t - u)\,du$. The integral on the right-hand side is called a *convolution of the functions* $F(t)$ *and* $G(t)$. (a) Let $F * G = \int_0^t F(u)\, G(t-u)\, du$. Prove that $F * G = G * F$. (*Hint*: Make the change of variable $r = t - u$.) Note that this identity says one may pick F and G in the most convenient way. (b) Find $\mathcal{L}^{-1}\{(1/s)[2/(s^2+4)]\}$ using the convolution of $F(t) = \sin 2t$ and $G(t) = 1$.

2. Given that $\mathcal{L}^{-1}\left\{\dfrac{25}{s(s^2+25)}\right\} = 1 - \cos 5t$, use the convolution method to find $\mathcal{L}^{-1}\left\{\dfrac{25}{s^2(s^2+25)}\right\}$.

3. Given that $\mathcal{L}^{-1}\left\{\dfrac{1}{s(s-1)}\right\} = e^t - 1$, use the convolution method to find $\mathcal{L}^{-1}\left\{\dfrac{4}{s^2(s-1)}\right\}$.

14.9 LAPLACE TRANSFORMS APPLIED TO DIFFERENTIAL EQUATIONS

Differential equations having initial conditions for the function and its derivatives may be solved by the use of Laplace transforms.

EXAMPLE 1. Solve $(d^2x/dt^2) + 9x = 2$ subject to the conditions when $t = 0$, $x = 0$, and $dx/dt = 0$.

Solution. For convenience we shall let x denote the function of t. Then taking the Laplace transform of each term, we obtain from Table 14.8.1

$$\underbrace{\mathcal{L}\{d^2x/dt^2\}}_{[s^2\mathcal{L}\{x\} - s(0) - 0]} + \underbrace{\mathcal{L}\{9x\}}_{9\mathcal{L}\{x\}} = \underbrace{\mathcal{L}\{2\}}_{\frac{2}{s}}$$

Solving for $\mathcal{L}\{x\}$, we obtain

$$(s^2+9)\mathcal{L}\{x\} = \frac{2}{s}$$

and

$$\mathcal{L}\{x\} = \frac{2}{s(s^2+9)}$$

To use number 23 of Table 14.8.1 we write

$$\mathcal{L}\{x\} = \left(\frac{2}{9}\right)\frac{(3)^2}{s[s^2+(3)^2]}$$

and

$$x = \tfrac{2}{9}(1 - \cos 3t)$$

PROBLEM 1. Solve $(d^2x/dt^2) + 16x = 5$ subject to the conditions when $t = 0$, $x = 0$, and $dx/dt = 0$.

EXAMPLE 2. Solve $(d^2x/dt^2) - 4(dx/dt) + 4x = 2e^{2t}$ if when $t = 0$, $x = 0$ and $dx/dt = 5$.

Solution

$$\overbrace{[s^2\mathcal{L}\{x\} - s(0) - 5]}^{\mathcal{L}\{d^2x/dt^2\}} - \overbrace{4[s\mathcal{L}\{x\} - 0]}^{4\mathcal{L}\{dx/dt\}} + \overbrace{4\mathcal{L}\{x\}}^{4\mathcal{L}\{x\}} = \overbrace{\frac{2}{s-2}}^{2\mathcal{L}\{e^{2t}\}}$$

$$(s^2 - 4s + 4)\mathcal{L}\{x\} = \frac{2}{s-2} + 5$$

$$\mathcal{L}\{x\} = \frac{2}{(s-2)^3} + \frac{5}{(s-2)^2}$$

To use number 12 of Table 14.8.1 we write $\mathcal{L}(x)$ as

$$\mathcal{L}\{x\} = \frac{2}{(s-2)^3} + (5)\frac{1}{(s-2)^2}$$

and obtain

$$x = t^2e^{2t} + 5te^{2t} = te^{2t}(t + 5)$$

PROBLEM 2. Solve $(d^2x/dt^2) - 6(dx/dt) + 9x = 5e^{3t}$ subject to the conditions when $t = 0$, $x = 0$, and $dx/dt = 4$.

EXAMPLE 3. Solve $(d^2Q/dt^2) + 10(dQ/dt) + 250Q = 10$ if when $t = 0$, $Q = 0$ and $dQ/dt = 0$.

Solution. Using Laplace transforms, we have

$$(s^2\mathcal{L}\{Q\} - 0 - 0) + 10(s\mathcal{L}\{Q\} - 0) + 250\mathcal{L}\{Q\} = \frac{10}{s}$$

$$(s^2 + 10s + 250)\mathcal{L}\{Q\} = \frac{10}{s}$$

We may write $s^2 + 10s + 150 = (s + 5)^2 + 225$. Then

$$\mathcal{L}\{Q\} = \frac{10}{s[(s+5)^2 + 225]}$$

This is not one of the forms listed in Table 14.8.1, but we can apply the method of partial fractions. Thus

$$\frac{10}{s[(s+5)^2 + 225]} = \frac{A}{s} + \frac{Bs + C}{(s+5)^2 + 225}$$

$$10 = A[(s+5)^2 + 225] + s(Bs + C)$$

For $s = 0$,

$$10 = A(250)$$

For $s = 1$,

$$10 = A(261) + B + C$$

For $s = -1$,

$$10 = A(241) + B - C$$

Therefore,

$$A = \tfrac{1}{25} \qquad B = -\tfrac{1}{25} \qquad C = -\tfrac{10}{25}$$

Thus

$$\mathcal{L}\{Q\} = \frac{1}{25s} + \frac{[-(1/25)]s - (10/25)}{[(s+5)^2 + (15)^2]}$$

$$= \frac{1}{25s} - \frac{1}{25}\frac{s+10}{[(s+5)^2 + (15)^2]}$$

$$= \frac{1}{25s} - \frac{1}{25}\frac{s+5}{[(s+5)^2 + (15)^2]} - \frac{5}{(25)(15)}\left[\frac{15}{[(s+5)^2 + (15)^2]}\right]$$

Then

$$Q = \tfrac{1}{25} - \tfrac{1}{25}(e^{-5t}\cos 15t) - \tfrac{1}{75}(e^{-5t}\sin 15t)$$

or

$$Q = \frac{e^{-5t}}{75}(-3\cos 15t - \sin 15t) + \tfrac{1}{25}$$

PROBLEM 3. Find Q at $t = 0.5$ for $(d^2Q/dt^2) + 8(dQ/dt) + 41Q = 5$ if at $t = 0$, $Q = 0$, and $dQ/dt = 0$.

EXERCISES 14.9

Group A

1. Solve each of the following using the Laplace transform method.

(a) $\dfrac{d^2x}{dt^2} + 64x = 0$, if at $t = 0$, $x = \frac{5}{4}$ and $x' = 0$.

(b) $\dfrac{d^2x}{dt^2} + 64x = 16\cos 4t$, if at $t = 0$, $x = 0$ and $x' = 0$.

(c) $\dfrac{d^2x}{dt^2} + 4\dfrac{dx}{dt} + 4x = 4e^{-2t}$; $x(0) = -1$, $x'(0) = 4$.

(d) $\dfrac{d^2x}{dt^2} + 6\dfrac{dx}{dt} + 9x = 0$; $x(0) = 3$, $x'(0) = 0$.

(e) $\dfrac{d^2x}{dt^2} + 6\dfrac{dx}{dt} + 9x = 6t^2e^{-3t}$; $x(0) = 0$, $x'(0) = 0$.

(f) $\dfrac{d^2x}{dt^2} + 2\dfrac{dx}{dt} - 15x = 0$, if at $t = 0$, $x = 1$ and $x' = 2$.

(g) $\dfrac{d^2x}{dt^2} + 2\dfrac{dx}{dt} - 15x = 16\cos t - 128\sin t$ if at $t = 0$, $x = 0$ and $x' = 0$.

(h) $\dfrac{d^2x}{dt^2} + 2\dfrac{dx}{dt} + 2x = 0$; $x(0) = 1$, $x'(0) = 1$.

(i) $\dfrac{d^2x}{dt^2} + 2\dfrac{dx}{dt} + 5x = 0$; $x(0) = 0$, $x'(0) = 1$.

(j) $\dfrac{d^2x}{dt^2} + \dfrac{dx}{dt} - 12x = 1$; $x(0) = 0$, $x'(0) = 0$.

Group B

1. In each of the following, use Laplace transforms to find the solution.

(a) A 20-Ω resistor and 5-H inductor are in series with an emf of e^{-4t} V. If at $t = 0$, $I = 3$, find I for any time $t \geqq 0$.

(b) A 40-Ω resistor and 0.005-F capacitor are in series with an emf of $4 \sin t$ V. If at $t = 0$, $Q = 4$ C, find Q for any time $t \geqq 0$.

(c) A 2-H inductor, a 0.125-F capacitor, and a generator having an emf given by $32 \sin 10t$ V, $t \geqq 0$, are connected in series. If at $t = 0$, $I = Q = 0$, find the charge Q and current I as functions of time.

(d) A 2-H inductor, a 0.125-F capacitor, and an 8-Ω resistor are in series with an emf of 40 V. If at $t = 0$, $I = Q = 0$, find the charge Q and current I as functions of time.

(e) A spring with spring constant 16 lb/ft is set in motion with a 2-lb weight on its free end in a medium whose damping force is $2v$ lb. If at time $t = 0$, $x = \frac{1}{2}$ ft and $v = 10$ ft/s, find the position and velocity at any time $t \geqq 0$.

(f) A horizontal, simply supported, uniform beam 6 ft long bends under its own weight, which is 2 lb/ft. If the beam's modulus of elasticity is 1, then the beam's deflection is described by the differential equation $y'' = x^2 - 6x$. If $y(0) = 0$ and $y'(0) = -0.216$, find $y(x)$.

Group C

1. When an electric network consists of more than one loop, a system of simultaneous equations occurs. The use of Laplace transforms may simplify their solution. Consider the system

$$\begin{cases} \dfrac{dI}{dt} - I + 10Q = 0 & (1) \\[2ex] -I + \dfrac{dQ}{dt} + Q = 0 & (2) \end{cases}$$

with the initial conditions $I(0) = 3$ and $Q(0) = 0$.

(a) Write the Laplace transform of Eqs. 1 and 2 and show that they are

$$\begin{cases} (s - 1)\mathcal{L}\{I\} + 10\mathcal{L}\{Q\} = 3 & (3) \\ -\mathcal{L}\{I\} + (s + 1)\mathcal{L}\{Q\} = 0 & (4) \end{cases}$$

(This system is similar to simultaneous equations in algebra with $\mathcal{L}\{I\} = x$ and $\mathcal{L}\{Q\} = y$.)

(b) Eliminate $\mathcal{L}\{I\}$ from the system in part (a) by multiplying each term in Eq. 4 by $(s - 1)$ and adding. Show that $\mathcal{L}\{Q\} = 3/(s^2 + 9)$.

(c) Find Q from part (b).

(d) Use the result of part (c) in Eq. 2 and solve for I.

2. Consider the system $I - dQ/dt = 2t - 1$, $dI/dt + Q = t - t^2$ with initial conditions $I(0) = 0$ and $Q(0) = -1$. Solve the system for I and Q using Laplace transforms. (*Hint*: After writing the Laplace transform of each equation eliminate $\mathcal{L}\{Q\}$, solve for $\mathcal{L}\{I\}$ and then I.)

3. Laplace transforms are useful in the field of pharmacokinetics. Consider a system of two compartments with a first-order exchange of a drug betwen the compartments. (A first-order exchange means that the *rate in change is proportional to the amount present*.) Also consider that there is a constant intravenous infusion rate K_{01} for time t in the interval $0 \leqq t \leqq a$ and there is clearance of the drug from compartment 1. (*Note*: $K_{01} = 0$ for $t > a$.) Schematically the situation is described by Figure 14.9.1

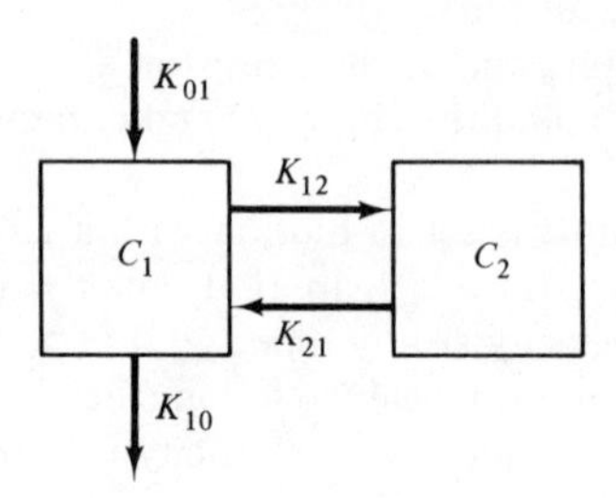

Figure 14.9.1 Exercise 3.

where K_{ij} is the first-order rate constant describing the transfer from compartment i to compartment j; $i = 0, 1, 2$; $j = 0, 1, 2$. The system of differential equations describing this system is

$$\begin{cases} \dfrac{dX_1(t)}{dt} = K_{01} + K_{21}X_2(t) - (K_{10} + K_{12})X_1(t) & (1) \\ \dfrac{dX_2(t)}{dt} = K_{12}X_1(t) - K_{21}X_2(t) & (2) \end{cases}$$

where $X_i(t)$ is the amount of drug in compartment i at time t. For the initial conditions $X_i(0) = 0$ we can use Laplace transforms to solve for the drug concentration, $X_i(t)$, in compartment 1, C_1. The system is transformed to

$$\begin{cases} s\mathfrak{L}\{X_1\} - 0 = \dfrac{K_{01}}{s}(1 - e^{-as}) + K_{21}\mathfrak{L}\{X_2\} - (K_{10} + K_{12})\mathfrak{L}\{X_1\} & (3) \\ s\mathfrak{L}\{X_2\} - 0 = K_{12}\mathfrak{L}\{X_1\} - K_{21}\mathfrak{L}\{X_2\} & (4) \end{cases}$$

(a) Solve for $\mathfrak{L}\{X_2\}$ in Eq. 4. Substitute the result in Eq. 3 and show that

$$\mathfrak{L}\{X_1\} = \frac{K_{01}}{s}(1 - e^{-as})\left[\frac{s + K_{21}}{(s + K_{21})(s + K_{10} + K_{12}) - K_{21}K_{12}}\right]$$

The result of part (a) is the Laplace transform of the rate of introduction of the drug into compartment 1 times the Laplace transform of the exchange within the system. The first Laplace transform is called the *input function* and the second is called the *disposition function*. We then may write $\mathfrak{L}\{X_1\} = (\text{in})_s(\text{disp})_s$.

Pharmacokineticists can study different reactions of the system to different types of input by changing the input function and solving accordingly. If we assume that the input was a simple injection of amount D, our initial condition would be $X_1(0) = D$ and we would obtain

$$\mathfrak{L}\{X_1\} = \frac{D(s + K_{21})}{(s + K_{21})(s + K_{10} + K_{12}) - K_{21}K_{12}}$$

$$= \frac{D(s + K_{21})}{(s + \alpha)(s + \beta)} \qquad (5)$$

(b) Find the expressions for α and β for Eq. 5. The inverse Laplace transform of Eq. 5 is

$$X_1(t) = D\left[\frac{(K_{21} - \alpha)e^{-\alpha t}}{\beta - \alpha} + \frac{(K_{21} - \beta)e^{-\beta t}}{\alpha - \beta}\right]$$

4. Find the Laplace transform for the central compartment C_1 in the first-order system of Fig. 14.9.2.

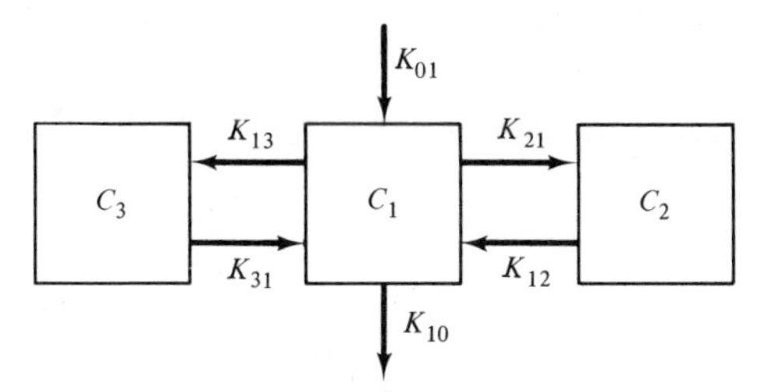

Figure 14.9.2 Exercise 4.

14.10 A NUMERICAL METHOD OF SOLVING DIFFERENTIAL EQUATIONS

In some applied problems in which x and y are related by a differential equation and we know the value of y for a value of x, it may be desired to construct a table of values for x and y or to find y for a certain x. If the exact solution to the equation is readily found, the table is easy to compute. But if the solution is not easy to come by, some numerical approximations would suffice. There are many numerical techniques available for approximating a solution to a differential equation. One of the simplest is called the *constant-slope method.* In this method we start at the known point for $x = a$ and work our way to the required point for $x = b$ by series of intervals over which we assume the slope is constant. Refer to Fig. 14.10.1 for the following discussion. Consider the differential equation $y' = 3x + y$ such that at $x = 1, y = 2$. Let it be required to find y at $x = 3$. For a start let us divide the interval $1 \leqq x \leqq 3$ into four subdivisions (for convenience) each of width $\Delta x = (3 - 1)/4 = 0.5$. The line segment $M_1N_1 = (\Delta x) \tan \theta = (\Delta x)y'$, where y' is computed at point P. The ordinate L_1N_1, called y_{new}, is equal to the ordinate LP, called y_{old}, plus M_1N_1. That is, $y_{\text{new}} = y_{\text{old}} + (\Delta x)y'$. We now repeat the process over the next interval. We then consider L_1N_1 as y_{old} and obtain

$$L_2N_2 = L_1N_1 + M_2N_2 = L_1N_1 + (\Delta x)y'$$

where y' is now computed at the value of x for point N_1. This process is repeated until we reach point Q. The work is best demonstrated in a table.

x	y	$y' = 3x + y$	$y_{\text{new}} = y_{\text{old}} + (\Delta x)y'$
1.0	2	$3(1.0) + 2 = 5$	$2 + (0.5)(5) = 4.5$
1.5	4.5	$3(1.5) + 4.5 = 9$	$4.5 + (0.5)(9) = 9$
2.0	9	$3(2.0) + 9 = 15$	$9 + (0.5)(15) = 16.5$
2.5	16.5	$3(2.5) + 16.5 = 24$	$16.5 + (0.5)(24) = 28.5$
3.0	28.5		

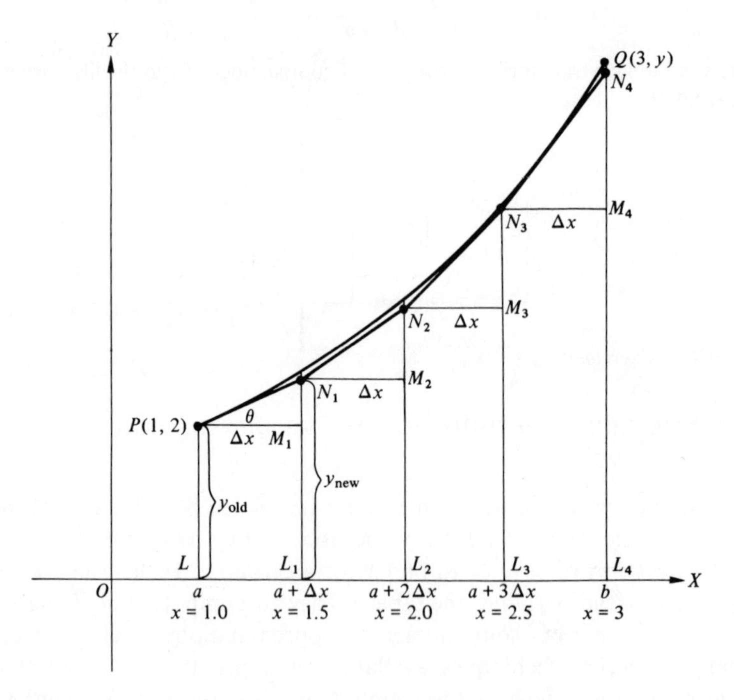

Figure 14.10.1 Constant-slope method.

The answer $y = 28.5$ may not be a good approximation as we took such a few number of subdivisions for demonstration purposes. Note carefully that the value of y in the second and succeeding lines is the value of y_{new} in the preceding line.

EXAMPLE 1. Find y at $x = 1.6$ for $y' = x^2 - y$ given that when $x = 1$, $y = 0$. Use six subdivisions.

Solution. $\Delta x = (1.6 - 1)/6 = 0.1$.

x	y	$y' = x^2 - y$	$y_{new} = y_{old} + (\Delta x)y'$
1.0	0.00	1.0	0.10
1.1	0.10	1.11	0.21
1.2	0.21	1.23	0.33
1.3	0.33	1.36	0.47
1.4	0.47	1.49	0.62
1.5	0.62	1.63	0.78
1.6	0.78		

You may show by solving the equation that the exact answer is $y = 0.81$.

PROBLEM 1. Find y at $x = 1.8$ for $y' = 2x - y^2$ given that when $x = 1$, $y = 0.5$. Use four subdivisions. (Can you solve this equation exactly?)

Numerical methods are suited for the high-speed computers. A course in numerical analysis discusses various methods and their reliabilities. The constant-slope method may also be utilized in solving a second-order differential equation.

EXERCISES 14.10

Group A

1. Use the constant slope method to solve each of the following differential equations with their associated conditions.

(a) $y' = x + y$; $y(0) = 0$. Find $y(1)$; use $n = 5$.
(b) $y' = 5x + 2y$; $y(0) = 0$. Find $y(0.5)$; use $n = 5$.
(c) $y' = e^x - y$; $y(0) = 1$. Find $y(0.3)$; use $n = 3$.
(d) $y' = x^3 - y$; $y(1) = 0$. Find $y(1.5)$; use $n = 5$.
(e) $y' = \dfrac{y-1}{x}$; $y(4) = 3$. Find $y(4.2)$; use $n = 4$.
(f) $y' = 4x^2 + y^2$; $y(0) = 1$. Find $y(0.6)$; use $n = 6$.
(g) $y' = \dfrac{x+y}{x-y}$; $y(5) = 3$. Find $y(4.5)$; use $n = 10$.
(h) $y' = \sqrt{3x - y}$; $y(5) = 1$. Find $y(15)$; use $n = 10$.
(i) $y' = y^3 + y^{1/3}$; $y(0) = 0.9$. Find $y(-0.3)$; use $n = 5$.
(j) $y' = x^2 - xy$; $y(1) = 0$. Find $y(1.4)$; use $n = 8$.

Group B

1. A 2-H inductor is in series with a 4-Ω resistor and an emf of 8 V. If $I(0) = 0$, find an approximate value for $I(0.5)$ using $n = 5$.

2. An 8-Ω resistor and a 0.125-F capacitor are in series with an emf of $16t^2$ V. If $Q(0) = 1$ C, find an approximate value for $Q(0.7)$ using $n = 7$.

SUMMARY OF IMPORTANT WORDS AND CONCEPTS

For each of the following, state in your own words your understanding of the statement or word.

1. (a) Differential equation. (b) Order. (c) Degree.

2. (a) Solution to differential equations. (b) Essential arbitrary constants. (c) General solution. (d) Particular solution.

3. (a) Methods of solving linear differential equations. (b) Separation of variables. (c) First-order linear (integrating factor).

4. Applications. (a) Families of curves. (b) Orthogonal trajectories. (c) Electric circuits. (d) Growth and decay. (e) Mechanics ($f = ma$).

5. (a) Methods of solving second-order linear differential equations. (b) Auxiliary equation. (c) Roots of auxiliary equation. (d) Types of solutions. (e) Euler's formula. (f) Complementary solution. (g) Particular solution (method of undetermined coefficients.) (h) General solution. (i) Initial conditions.

6. Applications. (a) Simple harmonic motion. (b) Vibratory motion. (c) Spring constant. (d) Damping constant. (e) Damped oscillatory motion. (f) Overdamped. (g) Critically damped. (h) Forced motion. (i) Resonance.

7. (a) Laplace transform. (b) Inverse Laplace transform. (c) Use of Lapace transform and inverse Laplace transform to solve differential equations.

8. Constant-slope method for a numerical solution of a differential equation.

Chapter 15

Functions of Two Variables and Partial Differentiation

15.1 DEFINITION OF A FUNCTION OF TWO VARIABLES

So far in this book we have discussed the elementary functions of one variable. That is, the argument has been only one real number. There are functions whose argument consists of pairs of numbers. The definition of a function of two variables is similar to the definition of a function of one variable. The only difference is in the domain. For a function of two variables the domain is a set of ordered number pairs.

Definition 15.1.1 *A function* $\boldsymbol{f}$ *of two variables with domain* $\boldsymbol{D}$ *and range* $\boldsymbol{R}$ *assigns to each ordered pair* $(\boldsymbol{x}, \boldsymbol{y})$ *of* $\boldsymbol{D}$ *exactly one element* $\boldsymbol{f}(\boldsymbol{x}, \boldsymbol{y})$ *of* $\boldsymbol{R}$*. The element* $\boldsymbol{f}(\boldsymbol{x}, \boldsymbol{y})$ *is said to be the value of the function at the argument* $(\boldsymbol{x}, \boldsymbol{y})$*.*

Remarks similar to those made about functions of a single variable can be made about functions of two variables. Unless otherwise stated, the range will consist of the set of all real numbers and the domain will consist of ordered pairs of real numbers for which the rule of correspondence makes sense for real numbers.

Functions of two variables may be denoted symbolically as $f(x, y)$. We may say that $z = f(x, y)$, where x and y are the independent variables and z is the dependent variable.

EXAMPLE 1. Given $f(x, y) = x^2 + y^2$, find (a) $f(2, -1)$ and (b) $z = f(x, y)$ at $(0, -7)$.

Solution. (a) $f(2, -1) = (2)^2 + (-1)^2 = 5$. (b) $z = (0)^2 + (-7)^2 = 49$.

PROBLEM 1. Given $f(x, y) = 3x - 4y$, find (a) $f(-2, 1)$ and (b) $z = f(x, y)$ at $(9, 0)$.

Fortunately, there is a rather simple way to visualize functions of two variables: with the aid of a three-dimensional coordinate system. At the origin of the rectangular XY-coordinate system in a plane, construct a line perpendicular to the XY-plane. On this line construct a number scale similar to the one on the X- and Y-axes. Call this the Z-axis (see Fig. 15.1.1). We must choose a positive

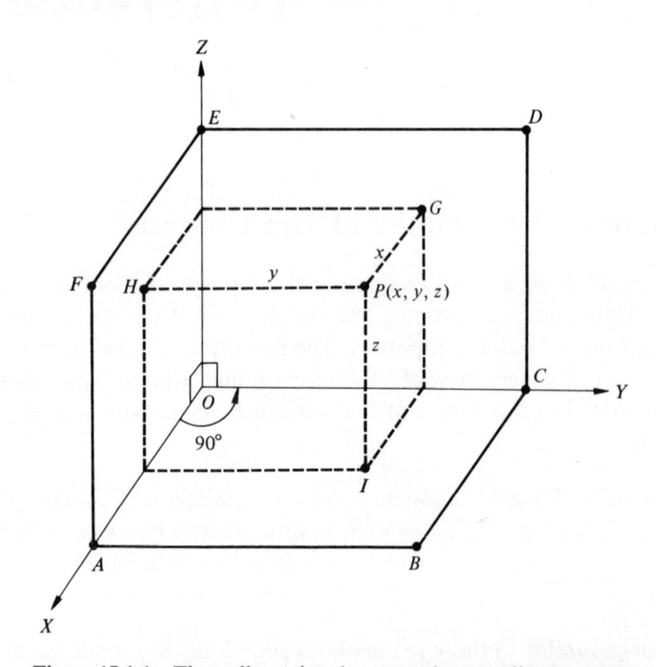

Figure 15.1.1 Three-dimensional rectangular coordinate system.

direction on the Z-axis. If we let the angle from the positive X-axis to the positive Y-axis be 90° in the direction shown in Fig. 15.1.1 and take positive Z as up, we have a right-handed coordinate system. (If we let the fingers of the right hand be closed in the direction of the smallest angle from the positive X-axis to the positive Y-axis, the direction of the positive Z-axis will be in the direction of the upraised thumb.) If positive Z is taken downward, then we have a left-handed coordinate system. We shall use the right-handed coordinate system.

The coordinate axes X, Y, and Z form three mutually perpendicular planes intersecting at O. These planes are called the *coordinate planes*. In Fig. 15.1.1, the XY-plane contains the points O, A, B, and C; the YZ-plane contains the points O, C, D, and E; and the XZ-plane contains the points O, A, F, and E.

A point, P, in three-space has as coordinates the ordered triple of numbers

(x, y, z), where x is the directed distance from the YZ-plane (the line segment, GP, Fig. 15.1.1), y is the directed distance from the XZ-plane (the line segment HP), and z is the directed distance from the XY-plane (the line segment IP). This system of coordinates establishes a one-to-one correspondence between the set of ordered triples of real numbers and the set of points in three-space.

The coordinate planes divide the three-space into eight octants. The first octant consists of the set of points whose x-, y-, and z-coordinates are positive. The other seven octants are not usually numbered.

EXAMPLE 2. Plot the following points: (a) (2, 5, 4), (b) (−4, 0, 2), (c) (−2, −3, 5), and (d) (3, 3, −4).

Solution. See Fig. 15.1.2.

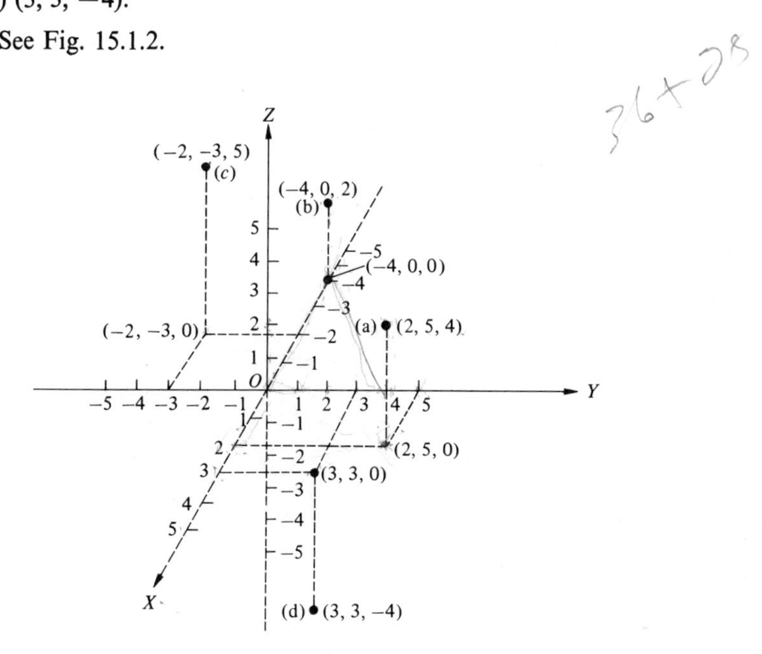

Figure 15.1.2 Points in three-space.

(a) Point (2, 5, 4) is 4 units above the point (2, 5, 0), which is in the XY-plane.

(b) Point (−4, 0, 2) is 2 units above the point (−4, 0, 0), which is on the negative X-axis.

(c) Point (−2, −3, 5) is 5 units above the point (−2, −3, 0), which is in the XY-plane.

(d) Point (3, 3, −4) is 4 units below the point (3, 3, 0), which is in the XY-plane.

PROBLEM 2. Plot the following points: (a) (3, 2, 1), (b) (4, −2, 1), (c) (4, −4, 2), (d) (2, −2, 0).

A formula for the distance between two points in space may be obtained by considering Fig. 15.1.3. A rectangular box consisting of six planes is shown therein. Faces of the box are parallel to coordinate planes. Points $P_1(x_1, y_1, z_1)$ and $P_2(x_2, y_2, z_2)$ are at opposite corners of the box. In right triangle P_1AB, $P_1A = x_2 - x_1$, $AB = y_2 - y_1$, and $|P_1B| = \sqrt{(P_1A)^2 + (AB)^2} = \sqrt{(x_2 - x_1)^2 + (y_2 - y_1)^2}$. In right triangle P_1BP_2, $BP_2 = z_2 - z_1$, and $|P_1P_2| = \sqrt{(P_1B)^2 + (BP_2)^2}$. Thus

$$|P_1P_2| = \sqrt{(x_2 - x_1)^2 + (y_2 - y_1)^2 + (z_2 - z_1)^2} \tag{15.1.1}$$

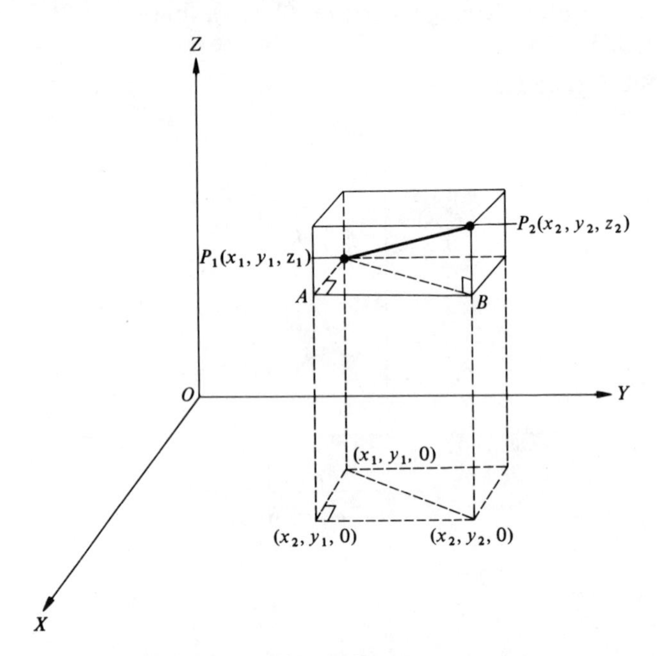

Figure 15.1.3 Distance between two points.

EXAMPLE 3. Find the distance between $P_1(2, 0, -3)$ and $P_2(-1, 1, 4)$.

Solution. Use Eq. 15.1.1 to obtain

$$|P_1P_2| = \sqrt{(-1 - 2)^2 + (1 - 0)^2 + [4 - (-3)]^2}$$

and

$$|P_1P_2| = \sqrt{9 + 1 + 49} = \sqrt{59} = 7.68$$

Warning ***In using formula 15.1.1 be sure that you subtract the x values, the y values, and the z values. Do not mix the different letters.***

PROBLEM 3. Find the distance between $P_1(-3, 2, 0)$ and $P_2(0, 5, -7)$.

EXERCISES 15.1

Group A

1. Let the function f have domain D as indicated and range R be the set of all real numbers. Let the rule of correspondence be given by $f(x, y)$. Find the corresponding elements in the range.

(a) $f(x, y) = x + y$; $D = \{(-2, 1), (1, 0), (0, 0), (3, -2)\}$.
(b) $f(x, y) = 3x - y^2$; $D = \{(0, 0), (1, -1), (-4, -2)\}$.
(c) $f(x, y) = x^2 - y$; $D = \{(7, -11), (\sqrt{2}, -1), (3, 9)\}$.
(d) $f(x, y) = \sqrt{x^2 + y^2 - 16}$; $D = \{(-4, 4), (4, -4), (4, 4), (0, 0)\}$.

2. State whether each of the following represents a function. Tell why.

(a) $z = x^2 + y^2$.
(b) $z = x^2 - y^2$.
(c) $z^2 = x^2 + y^2$.
(d) $3x + 2y + 5z = 0$.
(e) $z^3 + x + y^2 = 0$.
(f) $2x^2 + 3y^2 + 4z^2 = 0$.
(g) $z = \sqrt{x^2 - y^2}$.
(h) $x^2 - y^2 - z^2 = 0$.

3. Plot the following points on a rectangular Cartesian coordinate system.

(a) $(2, 1, 7)$.
(b) $(-3, 0, 2)$.
(c) $(-1, -2, 1)$.
(d) $(2, -3, 0)$.
(e) $(1, 2, -1)$.
(f) $(-2, 1, -2)$.
(g) $(-1, -3, -4)$.
(h) $(2, -1, -3)$.

4. Find the distance between P_1 and P_2 for each of the following.

(a) $P_1(5, 2, 4)$, $P_2(7, 4, 5)$.
(b) $P_1(10, -1, -3)$, $P_2(6, 7, 5)$.
(c) $P_1(7, -2, -1)$, $P_2(6, -4, 1)$.
(d) $P_1(6, -5, 5)$, $P_2(18, 7, -6)$.

15.2 GRAPH OF $z = f(x, y)$

The graphs of functions of two variables are generally three-dimensional surfaces. Sometimes it is difficult to indicate clearly a three-dimensional surface on a two-dimensional piece of paper.

Consider for a moment the graph of $x = 3$ in a one-dimensional, two-dimensional, and three-dimensional coordinate system. In a one-dimensional coordinate system (a number line) the graph of $x = 3$ is a point 3 units from the origin [see Fig. 15.2.1(a)]. In a two-dimensional coordinate system the graph of $x = 3$ is a straight line 3 units from the Y-axis [Fig. 15.2.1(b)]. In a three-dimensional coordinate system the graph of $x = 3$ is a plane 3 units from the YZ-coordinate plane [Fig. 15.2.1(c)].

It can be shown that the graph of a linear function in two variables such as $f(x, y) = 3x - y + 5$ is a plane. Thus if $z = f(x, y)$, the equation $ax + by + cz + d = 0$ represents a plane in three dimensions. If any three-dimensional surface intersects a plane, the curve of intersection is called a *trace* in that plane. (Curves of intersection might be straight lines.) The traces in the coordinate planes are helpful in sketching surfaces. To find the trace of $z = f(x, y)$ in the

1. XY-plane, let $z = 0$.

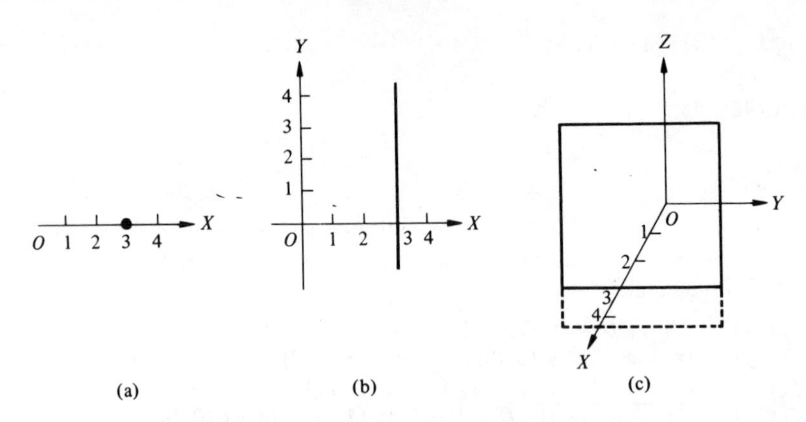

Figure 15.2.1 $x = 3$ represents a point, a line, or a plane.

2. YZ-plane, let $x = 0$.
3. XZ-plane, let $y = 0$.

The graph of $z = f(x, y)$ can be obtained by plotting points. However, the use of traces is simpler.

EXAMPLE 1. Find the traces of the plane $4x + 2y + 3z = 12$ in the coordinate planes and sketch them. Then sketch the plane in the first octant.

Solution. For the trace in the XY-plane let $z = 0$, and we obtain $4x + 2y = 12$. This is a straight line (Sec. 2.5) in the XY-plane. It intersects the X-axis (set $y = 0$) at $x = 3$ and the Y-axis (set $x = 0$) at $y = 6$. Its sketch is in Fig. 15.2.2(a). (For clarity we shall use only the first octant.)

For the trace in the YZ-plane let $x = 0$, and we obtain $2y + 3z = 12$. This is a straight line in the YZ-plane. It intersects the Y-axis (set $z = 0$) at $y = 6$. It intersects the Z-axis (set $y = 0$) at $z = 4$. Its sketch is in Fig. 15.2.2(b).

For the trace in the XZ-plane let $y = 0$, and we obtain $4x + 3z = 12$. This is a straight line in the XZ-plane. It intersects the X-axis (set $z = 0$) at $x = 3$. It intersects the Z-axis (set $x = 0$) at $z = 4$. Its sketch is in Fig. 15.2.2(c).

The sketch of the plane $4x + 2y + 3z = 12$ is in Fig. 15.2.2(d), which is a combination of parts (a), (b), and (c). [In practice, we usually sketch all traces and the surface in one figure, such as in part (d).]

PROBLEM 1. Find the traces of the plane $2x + y + 3z - 6 = 0$ in the coordinate planes. Then sketch the plane in the first octant.

EXAMPLE 2. Sketch the surface $f(x, y) = x^2 + y^2$.

Solution. Let $z = f(x, y)$; then we have $z = x^2 + y^2$. We next find the traces in the coordinate planes.

When $x = 0$, $z = y^2$, which is a parabola in the YZ-plane. We sketch this first. In Fig. 15.2.3 it is labeled (1).

When $y = 0$, $z = x^2$, which is a parabola in the XZ-plane. Its sketch is labeled (2).

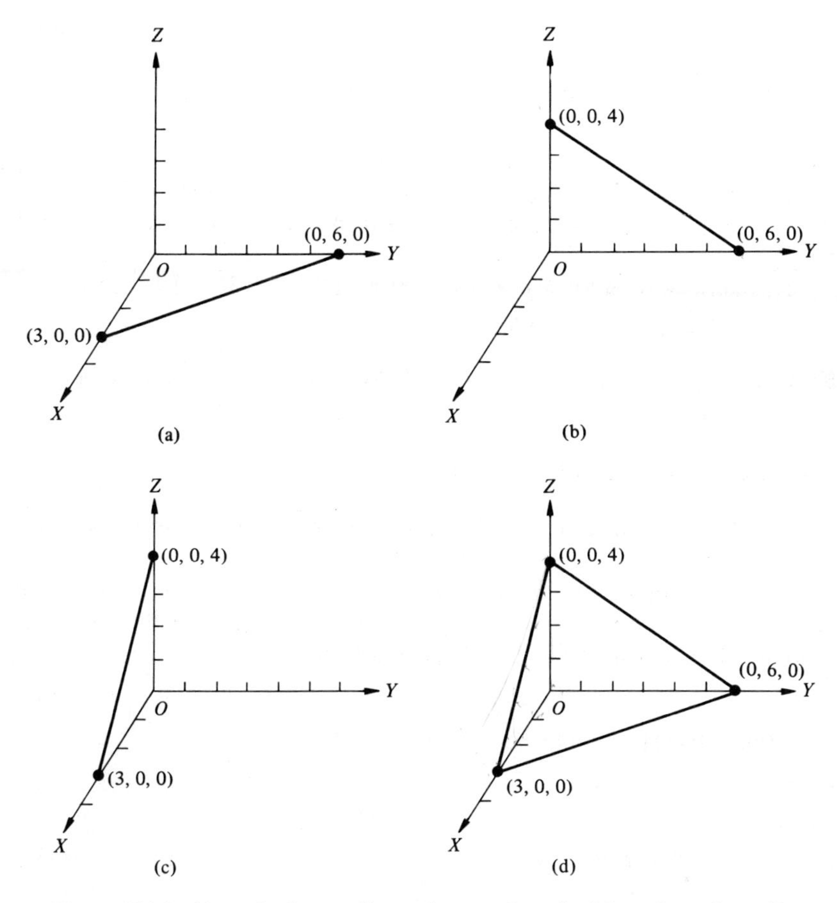

Figure 15.2.2 Traces in the coordinate planes and graph of $4x + 2y + 3z = 12$.

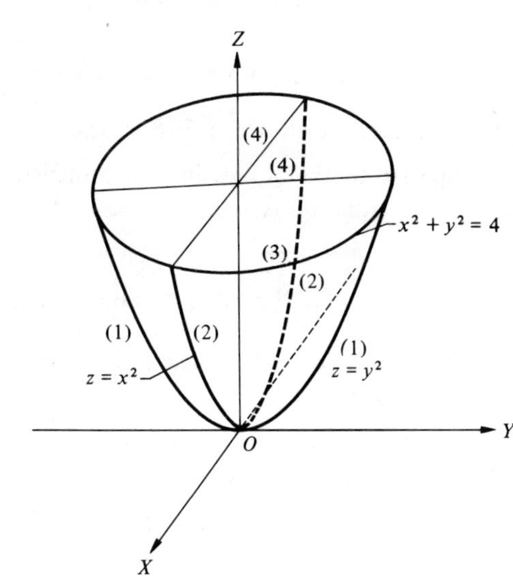

Figure 15.2.3 The paraboloid $z = x^2 + y^2$.

When $z = 0$, $x^2 + y^2 = 0$, which is the point (0, 0, 0), the origin.

If we let z equal some constant, we will obtain a trace in a plane parallel to the XY-plane. For convenience let $z = 4$.

Then the trace in the plane $z = 4$ is $x^2 + y^2 = 4$, which is a circle of radius 2. Its sketch is labeled (3). [The lines labeled (4) are drawn to help clarify the sketch.]

This surface is called a *paraboloid of revolution* as it can be obtained by rotating a parabola about its axis.

Warning ***A quick recognition of two-dimensional curves is helpful in sketching the traces.***

PROBLEM 2. Sketch the surface $z = 4x^2 + 4y^2$.

Equations similar to the conics represent some common surfaces in three spaces.

A sphere is defined as *the set of points each point of which is a constant distance, called the radius, from a fixed point called the center.* If the center is at the origin, then by use of Eq. 15.1.1 the equation of the sphere is

$$x^2 + y^2 + z^2 = a^2 \tag{15.2.1}$$

Other common surfaces are

ellipsoid: $$\frac{x^2}{a^2} + \frac{y^2}{b^2} + \frac{z^2}{c^2} = 1 \tag{15.2.2}$$

paraboloid: $$\frac{x^2}{a^2} + \frac{y^2}{b^2} = z \tag{15.2.3}$$

hyperboloid of one sheet: $$\frac{x^2}{a^2} + \frac{y^2}{b^2} - \frac{z^2}{c^2} = 1 \tag{15.2.4}$$

hyperboloid of two sheets: $$\frac{x^2}{a^2} - \frac{y^2}{b^2} - \frac{z^2}{c^2} = 1 \tag{15.2.5}$$

These surfaces are shown in Fig. 15.2.4.

Similar to the conics in two dimensions, if the center of a sphere is at (h, k, l), its equation is

$$(x - h)^2 + (y - k)^2 + (z - l)^2 = a^2 \tag{15.2.6}$$

Equations for the other surfaces with center not at the origin are similar.

EXAMPLE 3. Identify the surface $x^2 + y^2 + z^2 - 2x + 4y - 4 = 0$. Find its center and radius.

Solution. Since the coefficients of the squared terms are equal, we have a sphere. Completing the square, we have

$$(x^2 - 2x \quad) + (y^2 + 4y \quad) + z^2 = 4$$

$$(x^2 - 2x + 1) + (y^2 + 4y + 4) + z^2 = 4 + 5$$

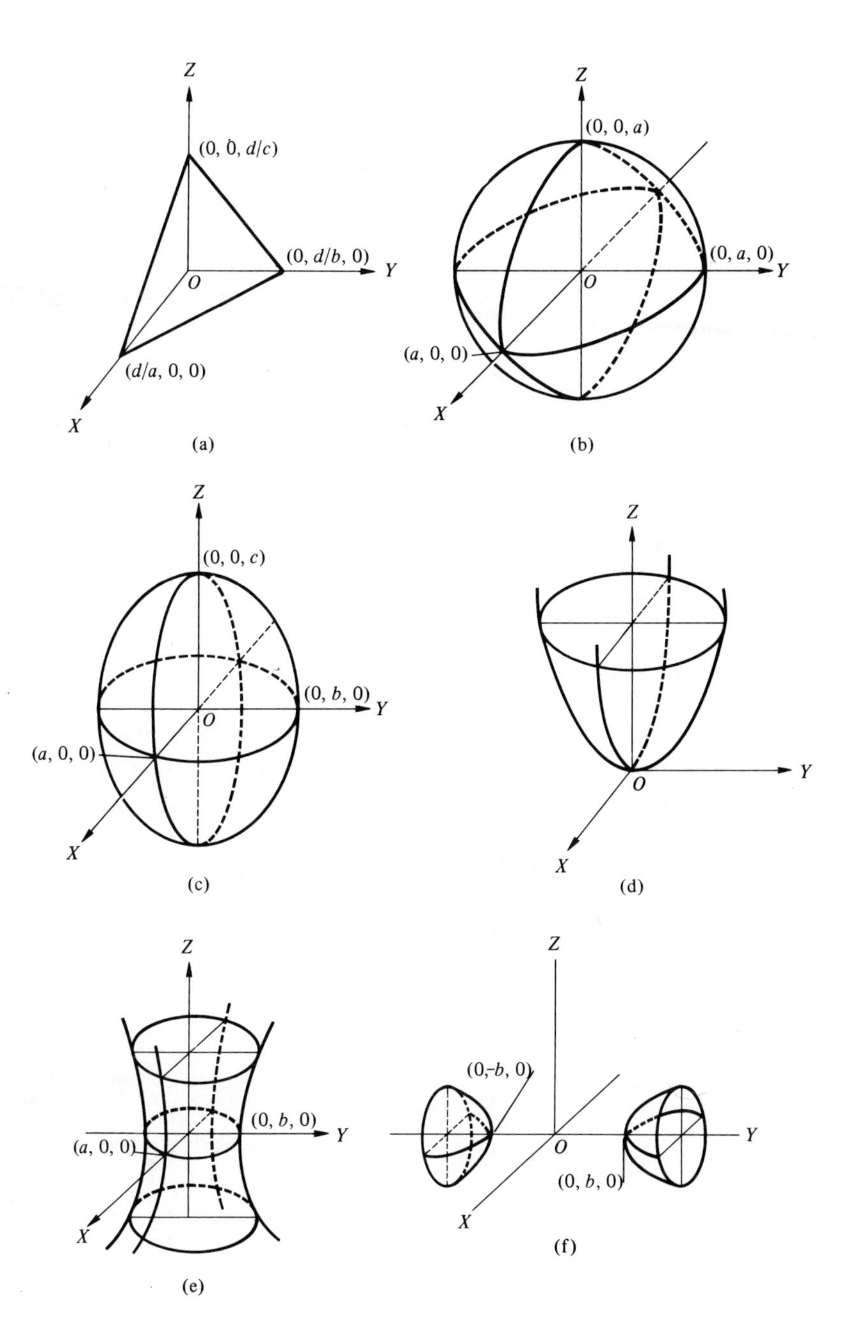

Figure 15.2.4

(a) Plane; $ax + by + cz = d$.

(b) Sphere; $x^2 + y^2 + z^2 = a^2$.

(c) Ellipsoid; $\frac{x^2}{a^2} + \frac{y^2}{b^2} + \frac{z^2}{c^2} = 1$.

(d) Paraboloid; $\frac{x^2}{a^2} + \frac{y^2}{b^2} = z$.

(e) Hyperboloid of one sheet; $\frac{x^2}{a^2} + \frac{y^2}{b^2} - \frac{z^2}{c^2} = 1$.

(f) Hyperboloid of two sheets; $\frac{y^2}{b^2} - \frac{x^2}{a^2} - \frac{z^2}{c^2} = 1$.

and

$$(x - 1)^2 + (y + 2)^2 + (z - 0)^2 = (3)^2$$

Comparing this with Eq. 15.2.6, we see that the center of the sphere is at (1, −2, 0) and its radius is 3.

PROBLEM 3. Identify the surface $x^2 + y^2 + z^2 - 4x + 6z + 4 = 0$. Find its center and radius.

The graph in three-space of an equation containing only two variables is called a *cylindrical surface*. For example, the three-dimensional graph of the equation $f(x, y) = 0$ is a cylindrical surface whose trace in the XY-plane is the curve $f(x, y) = 0$.

EXAMPLE 4. Sketch the graph of the surface $y = x^2$.

Solution. In the XY-plane the trace is the parabola $y = x^2$. This is drawn first in Fig. 15.2.5. It is labeled (1).

Note that if line l, intersecting the trace, is parallel to the Z-axis, that any point on l has coordinates (x, y, z). Hence the coordinates of any point $P(x, y, l)$ on l satisfy the equation $y = x^2$, and thus P is on the required surface. This is not true for any other line. The line l is called the *generator* of the cylindrical surface. (It generates the cylindrical surface by moving around the trace while remaining parallel to the Z-axis.)

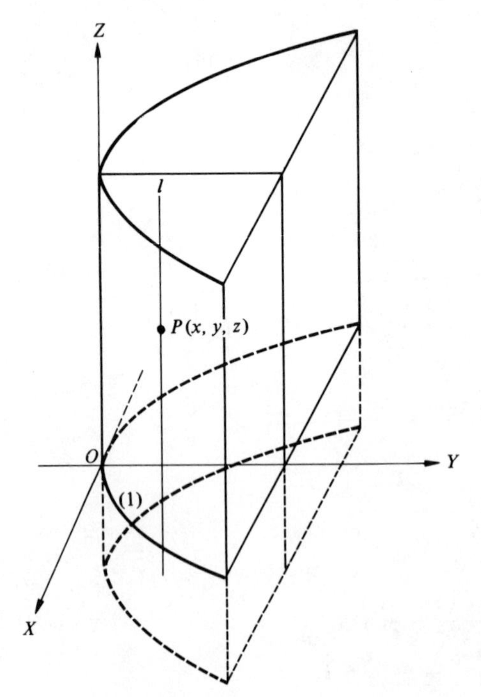

Figure 15.2.5 Parabolic cylinder, $y = x^2$.

We then complete the sketch by noting that the trace in any plane parallel to the XY-plane is the parabola $y = x^2$. This surface is called a *parabolic cylinder.*

PROBLEM 4. Sketch the graph of the surface $y - 4x^2 = 0$.

EXERCISES 15.2

Group A

1. Sketch the graph of each of the following planes.

(a) $x = 2$. (b) $x = -2$. (c) $y = 3$. (d) $y = -3$.
(e) $z = 4$. (f) $z = -4$.

2. Find the coordinate traces of each of the following planes and sketch the planes.

(a) $x + y + z = 3$. (b) $x + y - z = 3$.
(c) $2x - 2y + 3z = 6$. (d) $-3x + 2y + z = 6$.

3. For a particular function $z = f(x, y)$ the coordinate traces were found to be as given. Sketch each trace in the appropriate plane.

(a) $x = 0$, $z = 4y^2$. (b) $z = 0$, $y = -4x^2$.
(c) $z = 0$, $x^2 = 9y$. (d) $z = 0$, $x^2 = -9y$.
(e) $y = 0$, $x^2 + z^2 = 4$. (f) $y = 0$, $9x^2 + 4z^2 = 36$.
(g) $z = 0$, $9x^2 - 4y^2 = 36$. (h) $z = 0$, $-x^2 + y^2 = 4$.

4. Sketch each of the following surfaces. (If possible, identify the surface.)

(a) $x^2 + y^2 + z^2 = 9$. (b) $4z = x^2 + y^2$.
(c) $z = 4x^2 + 9y^2$. (d) $x^2 + y^2 = -z$.
(e) $9x^2 + 4y^2 + z^2 = 36$. (f) $z^2 = x^2 + y^2$
(g) $36x^2 + 64y^2 - 144z^2 = 1$. (h) $36x^2 - 144y^2 + 64z^2 = 1$.
(i) $-36x^2 + 64y^2 - 144z^2 = 1$.

5. Sketch each of the following cylindrical surfaces.

(a) $x^2 + y^2 = 1$. (b) $9x^2 + 4y^2 = 36$. (c) $y^2 = x$.
(d) $-4x^2 + 9y^2 = 36$. (e) $3x + 4y = 12$.

6. Sketch the surface represented by each of the following functions.

(a) $f(x, y) = 3x - 2y$. (b) $f(x, y) = -2x + 3y - 7$.
(c) $f(x, y) = x^2 + y^2$. (d) $f(x, y) = \sqrt{-x^2 - y^2 + 16}$.
(e) $f(x, y) = -\sqrt{-x^2 - y^2 + 16}$. (f) $f(x, y) = \sqrt{4x^2 + 9y^2 - 36}$.

15.3 VOLUMES AND PLANE SLICES

The volume of a solid is frequently described by bounding surfaces, for example, the volume bounded by the coordinate planes, the planes $x = 2$, $y = 3$, and $z = 2$. This volume is shown in Fig. 15.3.1.

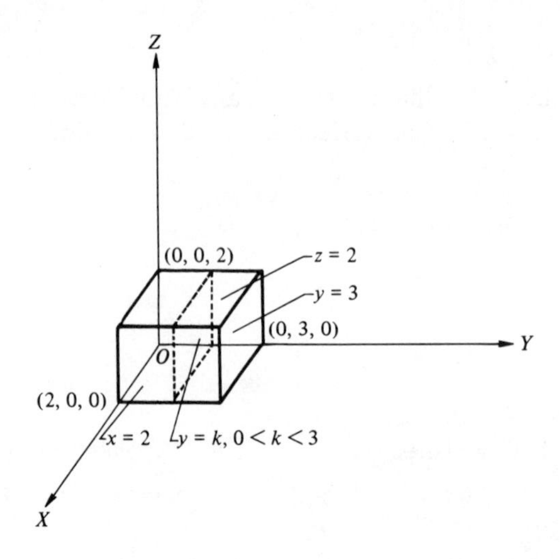

Figure 15.3.1 Volume bounded by the coordinate planes and $x = 2$, $y = 3$, $z = 1$.

Any plane parallel to one of the coordinate planes that intersects this volume "slices" the volume such that the area of this plane slice is bounded by a rectangle. The plane $y = k$, $0 < k < 3$, is such a plane. It is shown by dashed lines in Fig. 15.3.1.

EXAMPLE 1. Sketch the first octant volume bounded by $z = 4 - x^2 - y^2$. Describe the plane slices parallel to the coordinate planes.

Solution. If $z = 0$, we have the circle $x^2 + y^2 = 4$ [labeled (1) in Fig. 15.3.2].

If $x = 0$, we have the parabola $y^2 = 4 - z$. For $z = 0$ this gives $y = \pm 2$, and for the first octant we use only $y = 2$. For $y = 0$, we find $z = 4$. The YZ-trace is labeled (2).

If $y = 0$, we have the parabola $x^2 = 4 - z$. For $z = 0$, we have $x = 2$. For $x = 0$, we have $z = 4$. The XZ-trace is labeled (3). The required volume is shown in Fig. 15.3.2.

The plane slices may be obtained by considering any plane parallel to one of the coordinate planes that intersects the volume.

Consider $x = k$, $0 < k < 2$. The equation is $z = (4 - k^2) - y^2$ or $y^2 = (4 - k^2) - z$, which is a parabola parallel to the YZ-plane and k units from it. The cross section of the volume here is similar to the section in the YZ-plane. It is labeled (4).

If $y = k$, $0 < k < 2$, the cross section is similar to the section in the XZ-plane.

If $z = k$, $0 < k < 4$, the equation becomes $k = 4 - x^2 - y^2$ or $x^2 + y^2 = 4 - k$, which is a circle k units above the XY-plane. It is labeled (5).

PROBLEM 1. Sketch the first-octant volume bounded by $z = 9 - x^2 - y^2$. Describe the plane slices parallel to the coordinate planes.

Warning ***Plane slices help to picture the volume and give us information as to the shape of a cross section.***

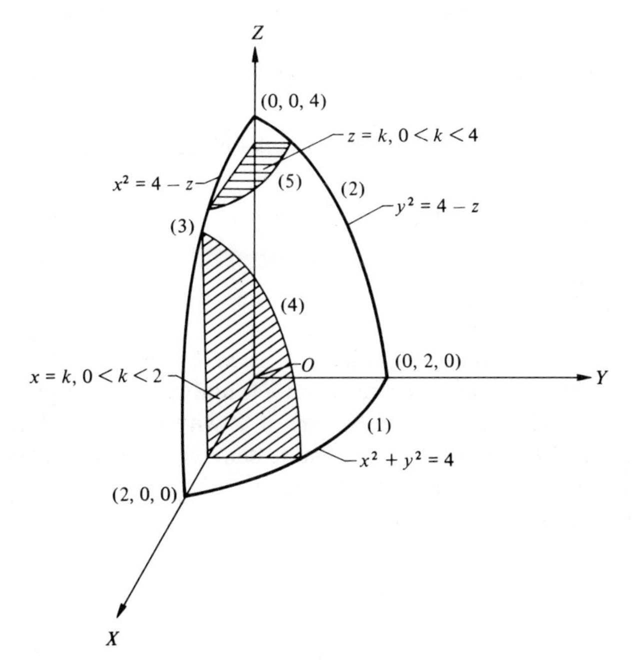

Figure 15.3.2 First octant volume bounded by $z = 4 - x^2 - y^2$.

In Chap. 16 we see how integration may be used to find volumes.

EXERCISES 15.3

Group A

1. Sketch the first-octant volume of each of the following. Describe cross sections parallel to the coordinate planes.

(a) $2x + 3y + 4z = 12$.
(b) $x^2 + y^2 + z^2 = 4$.
(c) $z = 36 - 4x^2 - 9y^2$.
(d) $\sqrt{x^2 + y^2} + z = 2$.

2. Sketch the volume bounded by the surfaces listed. For each volume describe cross sections parallel to the coordinate planes.

(a) $x = 0, x = 2, y = 3, z = 0, z = y$.
(b) $x = 0, y = 0, x^2 + y^2 = z, z = 4$.
(c) $x = 0, y = 0, z = \sqrt{x^2 + y^2}, z = 4$.
(d) $x = 0, y = 0, z = 0, z = 3, x^2 + y^2 + z^2 = 16$.
(e) $y = 0, z = 0, z = 4, x^2 + y - 4 = 0$.
(f) $x = 0, y = 0, z = 0, y = 4, x^2 + z = 4$.

3. Sketch the volume indicated for each of the following.

(a) Bounded by $x^2 + y^2 = 1$, $z = 2y + 3$, and $z = 0$.
(b) Under $z = x$ above $z = 0$, and the surface bounded by $x = y^2$ and $x = y + 2$.
(c) In the first octant bounded above by $x^2 + z - 4 = 0$ and below by $y - z = 0$.

(d) In the first octant bounded by the intersecting cylinders $x^2 + z^2 = 4$ and $y^2 + z^2 = 4$.

4. Sketch the volume in the first octant below the plane $z = 3$ and bounded by the surfaces indicated.

(a) $y = \sin x$, $x = \pi$. (b) $y = e^{-x^2}$, $x = 3$. (c) $y = \ln x$, $y = 4$.

15.4 PARTIAL DERIVATIVES

The volume of a right circular cylinder is $V = \pi r^2 h$, where r is the base radius and h is the height. If r is held constant and h is allowed to change, there would be a change in V. The ratio of the change of V with respect to h is called the *partial derivative of V with respect to h* and is denoted by $\partial V/\partial h$. (The symbol ∂ is called a "*round dee.*") Similarly, $\partial V/\partial r$ is the partial derivative of V with respect to r, where h is held constant. Thus for $V = \pi^2 rh$,

$$\frac{\partial V}{\partial h} = \pi r^2$$

and

$$\frac{\partial V}{\partial r} = 2\pi rh$$

EXAMPLE 1. If $z = x^2y + 3y^7x^5$,

$$\frac{\partial z}{\partial x} = 2xy + 15y^7x^4$$

and

$$\frac{\partial z}{\partial y} = x^2 + 21y^6x^5$$

PROBLEM 1. Find $\partial z/\partial x$ and $\partial z/\partial y$ for $z = 3x^4y^2 - y^3x$.

EXAMPLE 2. If $z = 3x^2y^5u^6v^7$,

$$\frac{\partial z}{\partial x} = 6xy^5u^6v^7$$

$$\frac{\partial z}{\partial y} = 15x^2y^4u^6v^7$$

$$\frac{\partial z}{\partial u} = 18x^2y^5u^5v^7$$

and

$$\frac{\partial z}{\partial v} = 21x^2y^5u^6v^6$$

Warning ***Note that for partial derivatives each time all variables are held constant except the one with respect to which the function is being differentiated.***

Problem 2. Find the partial derivatives of z with respect to each of the different independent variables of the function $z = 2x^3y^2u^{1/2}v^{-2/3}$.

Example 3. If $P = I^2R$, find $\partial P/\partial I$ when $I = 15$ and $R = 20$.

Solution

$$\frac{\partial P}{\partial I} = 21R$$

$$\left.\frac{\partial P}{\partial I}\right|_{(15,20)} = 2(15)(20) = 600$$

Problem 3. The formula $1/R = 1/R_1 + 1/R_2$ is used in electricity for two resistors in parallel. Find $\partial R/\partial R_1$ when $R_1 = 10$ and $R_2 = 15$.

The symbol $\partial z/\partial x$ is not to be considered as the quotient of two differentials. We use $\partial z/\partial x$ wholly and solely to indicate the operation of taking the partial derivative of the function z with respect to the variable x.

We now give a geometric interpretation of a partial derivative. Consider the surface $z = f(x, y)$, as shown in Fig. 15.4.1(a) and (b). When we take the partial derivative $\partial z/\partial x$, we hold y constant. This means that we are considering how z changes with respect to x in a plane parallel to the XZ-plane [see Fig. 15.4.1(a)], and $\partial z/\partial x$ represents the slope of the tangent line APB, that is,

$$\frac{\partial z}{\partial x} = \tan \alpha$$

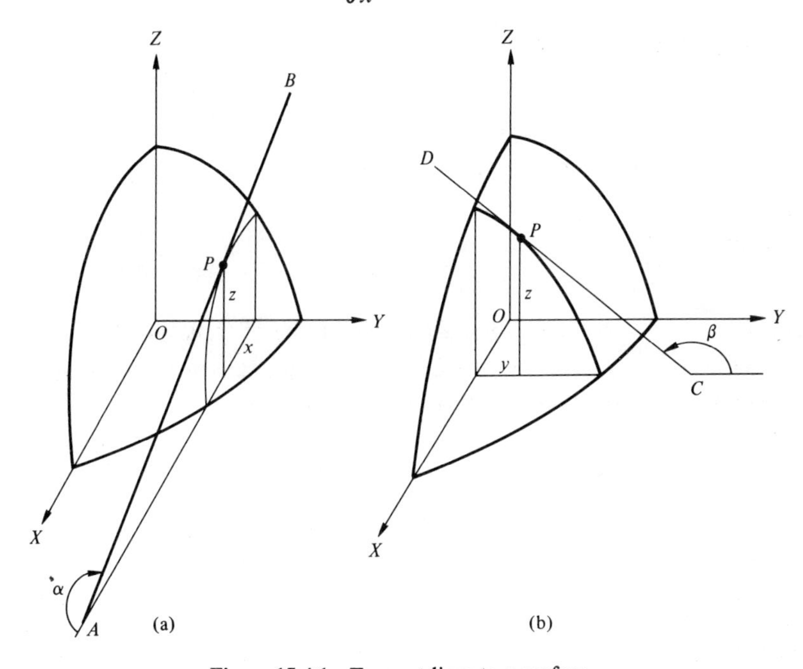

Figure 15.4.1 Tangent lines to a surface.

Similarly, $\partial z/\partial y$ represents the slope of the tangent line *CPD* [Fig. 15.4.1(b)], that is,

$$\frac{\partial z}{\partial y} = \tan \beta$$

EXAMPLE 4. Find the slope of the line tangent to the surface $x^2 + y^2 + z^2 = 9$ at the point (2, 1, 2) and parallel to the *YZ*-plane.

Solution. Parallel to the *YZ*-plane means that we desire to find $\partial z/\partial y$. This may be done by implicit differentiation. We treat x as a constant. Thus we obtain

$$0 + 2y + 2z\frac{\partial z}{\partial y} = 0$$

or

$$\frac{\partial z}{\partial y} = \frac{-y}{z}$$

At (2, 1, 2) we have

$$\frac{\partial z}{\partial y} = -\frac{1}{2}$$

PROBLEM 4. Find the slope of the line tangent to the surface $z = 4 - x^2 - y^2$ at the point (1, 1, 2) and parallel to the *XZ*-plane.

EXERCISES 15.4

Group A

1. For each of the following, find $\partial z/\partial x$ and $\partial z/\partial y$.

(a) $z = x^4 + x^3y + x^2y^2 - y^3$.
(b) $z = e^{x^2y}$.
(c) $z = \ln(x^2 + y^2)$.
(d) $z = x \sin xy$.
(e) $z = \text{Tan}^{-1}\frac{x}{y}$.

2. If $f(x, y) = x^2 - xy^2 + 3y^5$, find $\partial f/\partial x$ and $\partial f/\partial y$ at (1, 0, 1) and at (−1, 1, 5).

3. In each of the following, find the slope of the lines tangent to the given surface at the given point and parallel to the *XY*- and *XZ*-planes.

(a) $x^2 + y^2 + z^2 = 16$; $(1, 2, \sqrt{11})$.
(b) $x^2 + 2y^2 + z^2 = 18$; (0, 1, 4).
(c) $x^2 - y^2 - z^2 = 20$; (5, 2, 1).
(d) $2x^2 - y + z = 3$; (0, 1, 4).

Group B

1. The volume of a cone is $V = \frac{1}{3}\pi r^2 h$. If $r = 5$ in. and $h = 8$ in., find (a) the rate of change of the volume with respect to the radius; (b) the rate of change of the volume with respect to the height.

2. If $P = I^2R$, find the rate of change of P with respect to R when $I = 15$ and $R = 10$.

3. The volume of the frustrum of a right circular cone with base radius r_1, radius of the top r_2, and altitude h is given by $V = \pi(h/3)(r_1^2 + r_1r_2 + r_2^2)$. when $r_1 = 5$, $r_2 = 3$, and $h = 6$, find the rate in change in volume (a) with respect to h; (b) with respect to r_1; (c) with respect to r_2.

4. The "flow around a cylinder" may be characterized by the equation $\phi = ax/(x^2 + y^2)$. Determine the velocity components $v_x = \partial\phi/\partial x$ and $v_y = \partial\phi/\partial y$ and give the magnitude of the velocity for these components. (Cell.)

5. The equation for a perfect gas is $PV = kT$, where P is the pressure, V the volume, T the temperature on an absolute basis, and k a positive constant. Determine $\partial T/\partial P$, $\partial T/\partial V$, and $\partial P/\partial V$. (Cell.)

Group C

1. Higher-order partial derivatives may be found. If $z = f(x, y)$, the second-order partial derivatives may be denoted as $\partial^2z/\partial x^2$, $\partial^2z/\partial y^2$, and $(\partial^2z/\partial x\,\partial y)$, where

$$\frac{\partial^2 z}{\partial x^2} = \frac{\partial}{\partial x}\frac{\partial z}{\partial x}, \frac{\partial^2 z}{\partial y^2} = \frac{\partial}{\partial y}\frac{\partial z}{\partial y} \quad \text{and} \quad \frac{\partial^2 z}{\partial x\,\partial y} = \frac{\partial}{\partial x}\frac{\partial z}{\partial y}$$

Under certain conditions it may be shown that $(\partial^2z/\partial x\,\partial y) = (\partial^2z/\partial y\,\partial x)$. In each of the following, find $\partial^2z/\partial x^2$, $\partial^2z/\partial y^2$, and $(\partial^2z/\partial x\,\partial y)$.

(a) $z = x^2 + y^2$.
(b) $z = x^2y$.
(c) $z = e^y \sin x$.
(d) $z = x^3 \ln y$.

2. Verify that $(\partial^2z/\partial x\,\partial y) = (\partial^2z/\partial y\,\partial x)$ for the functions in Exercise 1.

15.5 SOME APPLICATIONS OF PARTIAL DERIVATIVES

(1) Total Differential

In Sec. 4.1 for a function of one variable we defined $dy = (dy/dx)\,dx = f'(x)\,dx$. For a function of two variables $z = f(x, y)$ the total differential dz is defined to be

$$dz = \frac{\partial z}{\partial x}dx + \frac{\partial z}{\partial y}dy \tag{15.5.1}$$

If we consider dz as an approximate change in z (see Sec. 4.2), we note from Eq. 15.5.1 that it is the sum of the change in the function due to the change in x and the change in the function due to the change in y.

EXAMPLE 1. For a particular gas the formula $PV = T$ relates the pressure P, volume V, and temperature T. Find the approximate change in the volume when $P = 200$ psi, $T = 300$ K, and the pressure increases by 20 psi while the temperature decreases by 15 K.

Solution. $V = T/P$.

$$dV = \frac{\partial V}{\partial P}\,dP + \frac{\partial V}{\partial T}\,dT = \frac{-T}{P^2}\,dP + \frac{1}{P}\,dT$$

$$= \frac{-300}{(200)^2}(20) + \frac{1}{200}(-15)$$

$$= \frac{-300}{2(10)^3} - \frac{15}{200} = \frac{-30}{200} - \frac{15}{200}$$

$$= \frac{-45}{200} = -\frac{9}{40} = -0.225 \text{ in.}^3$$

PROBLEM 1. For the formula $PV = T$ of Example 1, find the approximate change in the pressure when $V = 30$ in.^{3}, $T = 360$ K, and the volume decreases by 5 in.^{3} while the temperature increases by 10 K.

(2) Relative Maxima and Minima

By considering the geometric interpretation of the partial derivative, it may be shown that when a relative maximum (or minimum) occurs for $z = f(x, y)$, then

$$\frac{\partial z}{\partial x} = 0 \quad \text{and} \quad \frac{\partial z}{\partial y} = 0$$

at the same point.

The general theory of maximum and minimum points for functions of two or more variables is much more complicated than the general theory for functions of a single variable. If interested you may study this topic in a more comprehensive text. Here we shall limit ourselves to problems such that the criteria

$$\frac{\partial f}{\partial x} = 0 \quad \text{and} \quad \frac{\partial f}{\partial y} = 0$$

are sufficient conditions for a maximum or minimum value, and such that one may use the first derivative test on either $\partial f/\partial x$ or $\partial f/\partial y$ to find whether or not the function achieves a maximum or minimum value at the point in question.

EXAMPLE 2. The pressure exerted on each point of a flat plate is given by

$$P = x^2 + 2y^2 - 4x + 4y + 10$$

Find the minimum pressure.

Solution. $\partial P/\partial x = 2x - 4$ and $\partial P/\partial y = 4y + 4$. Setting each of these equal to zero, we have

$$2x - 4 = 0 \qquad 4y + 4 = 0$$
$$x = 2 \qquad y = -1$$

For $x < 2$, $\partial P/\partial x < 0$, and for $x > 2$, $\partial P/\partial x > 0$. Thus the minimum pressure occurs at (2, −1). It has a value of

$$P = (2)^2 + 2(-1)^2 - 4(2) + 4(-1) + 10$$

or

$$P = 4$$

PROBLEM 2. The pressure exerted on each point of a flat plate is given by $P = 2x^2 - y^2 + 4x + 4y + 20$. Find the minimum pressure.

If $\partial f/\partial x$ and $\partial f/\partial y$ contain both x and y, solve the equations $\partial f/\partial x = 0$ and $\partial f/\partial y = 0$ simultaneously for the critical values x_0 and y_0. Then perform the first derivative test on either $\partial f/\partial x$ or $\partial f/\partial y$. If you use $\partial f/\partial x$ let $y = y_0$ and if you use $\partial f/\partial y$ let $x = x_0$.

EXERCISES 15.5

Group A

1. If $P = I^2R$, find the total differential of P.

2. Find the approximate change in P of Exercise 1 when $I = 10$, $R = 5$, $dI = 0.5$, and $dR = 0.005$.

3. Let $w = x^2 + y^2 + z^2$. (a) Find dw. (b) Let $x = r\cos\theta$ and $y = r\sin\theta$. Find dx and dy. (c) Find dw in terms of r and θ.

4. Find and identify the maximum or minimum points of each of the following surfaces.

(a) $z = x^2 + 2y^2 - 4x + 4y - 3$.
(b) $z = x^2 + xy + y^2 + 3x - 3y + 4$.
(c) $z = x^2 + 2y^2 - 4x + 4y + 2xy - 3$.
(d) $z = -3x^2 - 4y^2 + 3x + 2y + 12xy$.
(e) $z = -5x^2 - 2y^2 + 10x + 4y + 16xy$.

5. Find the approximate change of each of the following at the points indicated and for the changes indicated.

(a) $z = x^2 - xy + 2y^2$ at $x = 2$ and $y = -1$ for $dx = -0.01$ and $dy = 0.02$.
(b) $z = x^3 + y^3 + 2xy$ at $x = 1$ and $y = 1$ for $dx = -0.01$ and $dy = 0.01$.
(c) $z = 2x^2 + 4y^2 + 2xy + 2x + 4y$ at $x = 2$ and $y = 1$ for $dx = 0.001$ and $dy = 0.002$.
(d) $z = \sin x + \sin y + \sin(x + y)$ at $x = \pi$ and $y = \pi$ for $dx = 0.0001$ and $dy = 0.002$.

Group B

1. Let $x = r\sin\phi\cos\theta$, $y = r\sin\phi\sin\theta$, and $z = r\cos\phi$. (a) Find dx, dy, and dz. (*Hint*: Extend Eq. 15.5.1.) (b) If $w = x^2 + y^2 + z^2$, find dw.

2. Use differentials to approximate $\sqrt{(2.97)^2 + (4.01)^2}$.

3. A crate has square ends of 12.02 in. on each side and has a length of 29.97 in. Find its approximate volume.

4. The charge, Q, on a capacitor which has voltage, V, across it, is given by $Q = CV$, where C is the capacitance. If the voltage changes from 40 V to 60 V and the capacitance changes from 0.005 F to 0.004 F, find the approximate change in the charge.

5. A variable resistance and a 4-H inductance are connected in series with an emf of 80 V. The current I at any time t and for any resistance R is given by $I = 80(1 - e^{(R/4)t})$. If R changes from 20 to 22 Ω and t changes from 0 to 0.1 s, find the approximate change in I.

6. Find the dimensions of an open rectangular box of largest volume with a surface area of 125 in.^{2}.

7. Find positive numbers x, y, and z such that xyz^2 is as large as possible if $x + y + z = 24$.

8. A manufacturer produces a product with three factors x, y, and z and has a budget constraint of $250 = 4x + 3y + 6z$. The output of the product is given by $Q = 3xyz$. For what inputs can the manufacturer maximize the output without violating budget constraints?

Group C

1. Suppose that one has a finite number of experimental data points $(x_1, y_1), \ldots, (x_n, y_n)$ and wishes to approximate these by some straight line which is in some way the "best" approximation. If we assume that the straight line is given by $y = mx + b$, then the deviation from the straight line at data point (x_k, y_k) is given by $d_k = y_k - (mx_k + b)$. Let $f(m, b) = d_1^2 + d_2^2 + \ldots + d_n^2$ and interpret best approximation to mean the values of m and b which minimize $f(m, b)$. Find the best straight-line approximation for each of the following sets of data points (see Sec. A.2).

(a) $(-1, 2), (0, 1), (3, -1)$.
(b) $(1, 1), (2, 4), (3, 2)$.
(c) $(0, 0), (1, 2), (2, 3)$.

SUMMARY OF IMPORTANT WORDS AND CONCEPTS

For each of the following, state in your own words your understanding of the statement or word.

1. (a) Function of two variables. (b) Domain. (c) Range. (d) Dependent variable. (e) Independent variables.

2. (a) Three-dimensional coordinate system. (b) Distance between two points.

3. (a) Graphs in three dimensions. (b) Trace. (c) Plane. (d) Paraboloid of revolution. (e) Sphere. (f) Cylindrical surface. (g) Generator. (h) Parabolic cylinder.

4. (a) Volumes. (b) Plane slices.

5. (a) Partial derivative. (b) Total differential. (c) Relative maxima and minima.

Chapter 16

Repeated Integration

16.1 DOUBLE INTEGRALS

Double integration is a process analogous to partial differentiation wherein we hold x constant and integrate with respect to y, or vice versa.

The symbol

$$\int_a^b \int_{y_1}^{y_2} f(x, y)\, dy\, dx \tag{16.1.1}$$

in which a and b are constants and y_1 and y_2 are either constants or functions of x, is called a *double integral*. It indicates that two integrations are to be performed—the first one with respect to y holding x constant and the second one with respect to x. The inside integral is integrated first.

EXAMPLE 1. Evaluate $\int_0^2 \int_1^3 x^2 y\, dy\, dx$.

Solution. We must perform two integrations. First we integrate with respect to y, because dy is the inside differential, holding x constant. Thus

$$\begin{aligned}
\int_0^2 \left(\int_1^3 x^2 y\, dy\right) dx &= \int_0^2 \left[x^2 \frac{y^2}{2}\right]_1^3 dx \\
&= \int_0^2 \left[x^2 \frac{(3)^2}{2} - x^2 \frac{(1)^2}{2}\right] dx \\
&= \int_0^2 \left(\frac{9}{2} x^2 - \frac{1}{2} x^2\right) dx \\
&= \int_0^2 4x^2\, dx
\end{aligned}$$

Next we integrate with respect to x. Thus

$$\int_0^2 4x^2\,dx = \left[\frac{4}{3}x^3\right]_0^2 = \left[\frac{4}{3}(2)^3 - \frac{4}{3}(0)^3\right] = \frac{32}{3}$$

Hence

$$\int_0^2 \int_1^3 x^2 y\,dy\,dx = \frac{32}{3} = 10.7$$

PROBLEM 1. Evaluate $\int_1^3 \int_{-1}^2 2xy^3\,dy\,dx$.

The limits on the inside integral may be constants, or variables of the outside differential. The limits on the first integral must always be constants.

EXAMPLE 2. Evaluate $\int_{-1}^2 \int_1^{1+x} (x+y)\,dy\,dx$.

Solution

$$\begin{aligned}\int_{-1}^2 \int_1^{1+x} (x+y)\,dy\,dx &= \int_{-1}^2 \left[xy + \frac{y^2}{2}\right]_1^{1+x} dx \\ &= \int_{-1}^2 \left\{\left[x(1+x) + \frac{1}{2}(1+x)^2\right] \right. \\ &\qquad \left. - \left[x + \frac{1}{2}\right]\right\} dx \\ &= \int_{-1}^2 \left(x + \frac{3}{2}x^2\right) dx \\ &= \left[\frac{x^2}{2} + \frac{1}{2}x^3\right]_{-1}^2 = \left[\frac{(2)^2}{2} + \frac{1}{2}(2)^3\right] \\ &\qquad - \left[\frac{(-1)^2}{2} + \frac{1}{2}(-1)^3\right] \\ &= 6\end{aligned}$$

PROBLEM 2. Evaluate $\int_0^2 \int_x^{x^2} (x^2 - y)\,dy\,dx$.

Sometimes it is necessary to perform the integration with respect to x first. Then we use the symbol

$$\int_c^d \int_{x_1}^{x_2} f(x, y)\,dx\,dy \tag{16.1.2}$$

where c and d are constants and x_1 and x_2 are either functions of y or constants.

EXAMPLE 3. Evaluate $\int_0^1 \int_{2y}^3 x^2 y\,dx\,dy$.

Solution. We integrate with respect to x first, holding y constant. Thus

$$\int_0^1 \int_{2y}^3 x^2 y \, dx \, dy = \int_0^1 \left[\frac{x^3 y}{3}\right]_{2y}^3 dy$$
$$= \int_0^1 \left(9y - \frac{8}{3}y^4\right) dy$$
$$= \left[\frac{9}{2}y^2 - \frac{8}{15}y^5\right]_0^1$$
$$= \left[\frac{9}{2}(1)^2 - \frac{8}{15}(1)^5\right] - \left[\frac{9}{2}(0)^2 - \frac{8}{15}(0)^5\right]$$
$$= \frac{9}{2} - \frac{8}{15} = \frac{135 - 16}{30} = \frac{119}{30}$$

Therefore

$$\int_0^1 \int_{2y}^3 x^2 y \, dx \, dy = \frac{119}{30} = 3.97$$

PROBLEM 3. Evaluate $\int_{-1}^2 \int_y^{y^2-1} xy^2 \, dx \, dy$.

Warning ***For repeated integration we always work with the inside integral first.***

We may extend this concept to polar coordinates.

EXAMPLE 4. Evaluate $\int_0^{\pi/2} \int_0^{1+\cos\theta} r \, dr \, d\theta$.

Solution

$$\int_0^{\pi/2} \int_0^{1+\cos\theta} r \, dr \, d\theta = \int_0^{\pi/2} \left[\frac{r^2}{2}\right]_0^{1+\cos\theta} d\theta$$
$$= \frac{1}{2}\int_0^{\pi/2} (1 + 2\cos\theta + \cos^2\theta) \, d\theta$$
$$= \frac{1}{2}\left[\theta - 2\sin\theta + \frac{\theta}{2} + \frac{\sin 2\theta}{4}\right]_0^{\pi/2}$$
$$= \frac{1}{2}\left(\pi - 2 + \frac{\pi}{4} + 0\right) - 0$$
$$= \frac{1}{2}\left(\frac{5\pi}{4} - 2\right)$$
$$= \frac{5\pi - 8}{8} = 0.96$$

PROBLEM 4. Evaluate $\int_0^{\pi/2} \int_1^{\sin\theta} \cos\theta \, r \, dr \, d\theta$.

See Exercise 1 of Group C for triple integrals.

EXERCISES 16.1

Group A

1. Evaluate each of the following.

(a) $\int_0^1 \int_1^2 xy\, dy\, dx.$

(b) $\int_0^1 \int_1^x xy\, dy\, dx.$

(c) $\int_{-1}^2 \int_x^{x^2} xy\, dy\, dx.$

(d) $\int_{-1}^1 \int_0^3 (x^2 + 2)\, dy\, dx.$

2. Evaluate each of the following.

(a) $\int_0^1 \int_0^2 xy\, dx\, dy.$

(b) $\int_0^1 \int_0^y xy\, dx\, dy.$

(c) $\int_{-1}^1 \int_{-y}^{y^2} xy\, dx\, dy.$

(d) $\int_0^2 \int_{-1}^y (y^2 + 2)\, dx\, dy.$

3. Evaluate each of the following.

(a) $\int_0^{\pi/2} \int_0^1 r\, dr\, d\theta.$

(b) $\int_0^{\pi/2} \int_0^{\cos\theta} r\, dr\, d\theta.$

(c) $\int_0^{\pi/6} \int_{\cos\theta}^{\sin\theta} r\, dr\, d\theta.$

(d) $\int_0^{\pi/2} \int_1^{\sin\theta} (r + 2)\, dr\, d\theta.$

Group B

1. Evaluate each of the following.

(a) $\int_0^1 \int_0^x (x^2 + y^2)\, dy\, dx.$

(b) $\int_0^2 \int_0^x e^{x^2}\, dy\, dx.$

(c) $\int_0^{\pi/6} \int_0^1 y \sin x\, dy\, dx.$

(d) $\int_0^a \int_0^{\sqrt{a^2-x^2}} (x + y)\, dy\, dx.$

2. Evaluate each of the following.

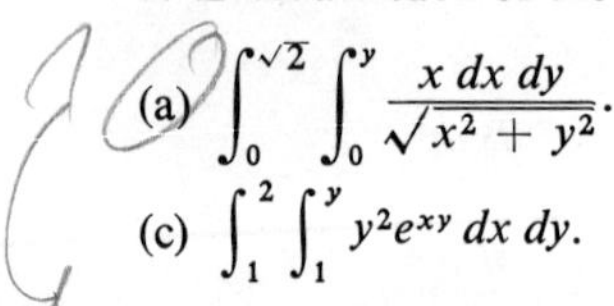

(a) $\int_0^{\sqrt{2}} \int_0^y \frac{x\, dx\, dy}{\sqrt{x^2 + y^2}}.$

(b) $\int_0^{\pi/2} \int_0^{y^2} \cos \frac{x}{y}\, dx\, dy.$

(c) $\int_1^2 \int_1^y y^2 e^{xy}\, dx\, dy.$

(d) $\int_0^1 \int_y^{y^2-1} (x - y)\, dx\, dy.$

3. Evaluate each of the following.

(a) $\int_0^{\pi/3} \int_0^{\sin 3\theta} r\, dr\, d\theta.$

(b) $\int_0^{\pi/2} \int_0^4 \sin\theta \frac{r\, dr\, d\theta}{\sqrt{16 - r^2}}.$

(c) $\int_0^{\pi/2} \int_0^4 r\sqrt{16 - r^2}\, dr\, d\theta.$

(d) $\int_0^{\pi} \int_0^{(1+\cos\theta)} r^2 \sin\theta\, dr\, d\theta.$

Group C

1. The symbol $\int_a^b \int_{y_1}^{y_2} \int_{z_1}^{z_2} f(x, y, z)\, dz\, dy\, dx$, where z_1 and z_2 are either constants or functions of x and/or y, and y_1 and y_2 are either constants or functions of x, is called a *triple integral.* It is evaluated starting with the inside integral first. (Here integrate with respect to z keeping x and y constant.) Then you continue as for double integrals. Evaluate each of the following.

(a) $\int_1^2 \int_{-2}^1 \int_0^1 xyz\, dz\, dy\, dx.$ (b) $\int_0^2 \int_1^x \int_{x+y}^{2y} dz\, dy\, dx.$

(c) $\int_0^1 \int_{-1}^1 \int_1^{xy} 2y^2 z\, dz\, dx\, dy.$ (d) $\int_0^{\pi/2} \int_0^1 \int_0^{1-r^2} r\, dz\, dr\, d\theta.$

16.2 AREA

A geometric interpretation of the double integral is illustrated by finding the area bounded by two or more curves. This, of course, could be done by single integration (see Sec. 6.1). Consider the area shown in Fig. 16.2.1. Let the area

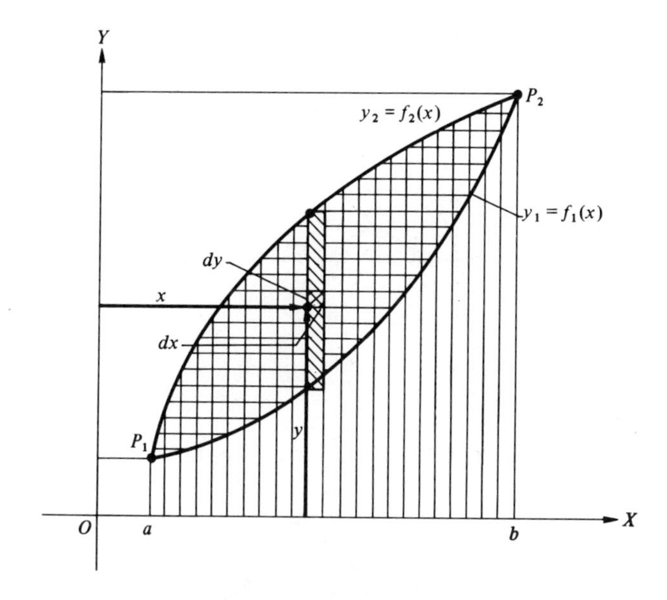

Figure 16.2.1 Area by double integration.

be bounded below by the curve $y_1 = f_1(x)$ and above by the curve $y_2 = f_2(x)$. Let the points of intersection be the points P_1 and P_2. Construct many rectangles in the area by drawing many lines parallel to the X-axis and many lines parallel to the Y-axis. At the point (x, y) an interior point of the region, consider the "double-hatched" rectangle of dimension dx by dy. This small area is

$$dA = dy\, dx$$

If we now hold x constant and sum all of the small rectangles from y_1 to y_2, we would obtain the area of the shaded vertical strip. This could have been done for any of the vertical strips from $x = a$ to $x = b$. Now if we sum all such vertical strips from $x = a$ to $x = b$, we would obtain an approximation to the required area. If we let the number of small rectangles increase without bound, we would obtain the exact area bounded by the curves. This is denoted by

$$A = \int_a^b \int_{y_1=f_1(x)}^{y_2=f_2(x)} dy\,dx \tag{16.2.1}$$

EXAMPLE 1. Find the area bounded by the parabolas $y = x^2$ and $y^2 = x$.

Solution. See Fig. 16.2.2. The points of intersection are (0, 0) and (1, 1). At any interior point (x, y), construct a small rectangle of dimension dx by dy. Then we have

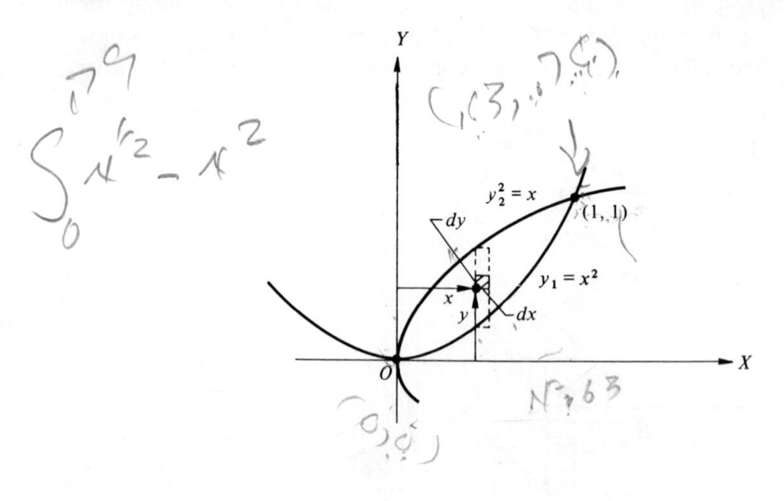

Figure 16.2.2 Area between two curves.

$$dA = dy\,dx$$

and

$$A = \int_0^1 \int_{x^2}^{\sqrt{x}} dy\,dx$$

(Note that in obtaining the limits on the integrals the sums are performed in the positive Y- and the positive X-directions.)

$$A = \int_0^1 [y]_{x^2}^{\sqrt{x}}\,dx = \int_0^1 (x^{1/2} - x^2)\,dx$$

$$= \left[\frac{2}{3}x^{3/2} - \frac{x^3}{3}\right]_0^1 = \left(\frac{2}{3} - \frac{1}{3}\right) - (0 - 0)$$

and

$$A = \tfrac{1}{3} = 0.33$$

In Example 1, we could have summed in the X-direction first. Then we would have

$$A = \int_0^1 \int_{y^2}^{\sqrt{y}} dx\, dy$$

which you may evaluate and obtain the same value for A (see Exercise 1 in Group B).

PROBLEM 1. Find the area bounded by the parabola $y^2 = x$ and the line $y = x$.

EXAMPLE 2. Find the area bounded by the parabola $y^2 = x$ and the line $y = x - 2$.

Solution. See Fig. 16.2.3. At any interior point (x, y) construct a small rectangle of dimension dx by dy. We now must make a choice. Do we use $dA = dx\, dy$ or $dA = dy\, dx$? Note carefully that if we sum in the Y-direction first (holding x constant) that the lower limit for y is not the same curve throughout the entire region. That is, to the left of $(1, -1)$ the lower limit would be $y = -\sqrt{x}$, and to the right of $(1, -1)$ the lower limit would by $y = x - 2$. This then is not a good choice. If we sum in the X-direction first, holding y constant, the limits would be the same functions throughout the region. Thus we choose

$$dA = dx\, dy$$

$$A = \int_{-1}^{2} \int_{y^2}^{y+2} dx\, dy$$

$$= \int_{-1}^{2} [x]_{y^2}^{y+2}\, dy = \int_{-1}^{2} (y + 2 - y^2)\, dy = \left[\frac{y^2}{2} + 2y - \frac{y^3}{3}\right]_{-1}^{2}$$

$$= \tfrac{9}{2} = 4.5$$

If in Example 2 we insist that $dA = dy\, dx$, then we would have

$$A = \int_0^1 \int_{-\sqrt{x}}^{\sqrt{x}} dy\, dx + \int_1^4 \int_{x-2}^{\sqrt{x}} dy\, dx$$

you may show that this gives the same result (see Exercise 2 in Group B).

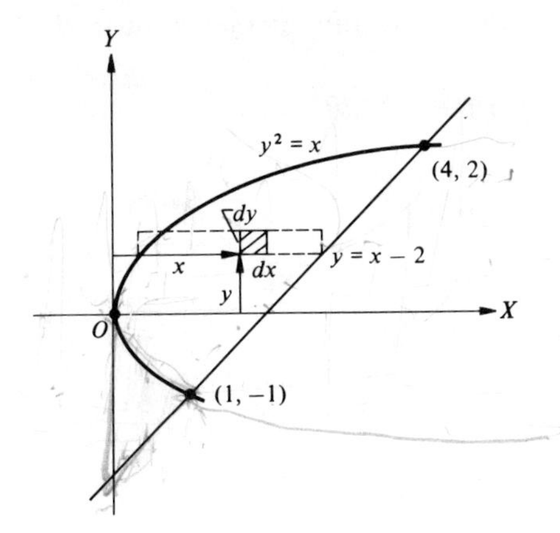

Figure 16.2.3 Example 2.

PROBLEM 2. Find the area bounded by $y^2 = -x$ and the line $y = x + 2$.

In polar coordinates the differential of area is

$$dA = r\,dr\,d\theta$$

and

$$A = \int_{\theta_1}^{\theta_2} \int_{r_1=f_1(\theta)}^{r_2=f_2(\theta)} r\,dr\,d\theta \qquad (16.2.2)$$

Consider Fig. 16.2.4. The region is subdivided into many small elements of area by many concentric circles and many rays from the origin.

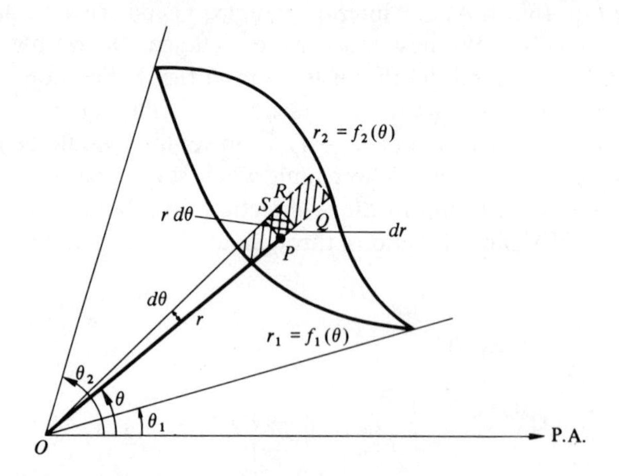

Figure 16.2.4 Area in polar coordinates by double integration.

Arc $PS = r\,d\theta$, and $PQ = dr$. Then by the use of differentials it may be shown that $dA = r\,dr\,d\theta$.

EXAMPLE 3. Find the area bounded by $r = 2\cos\theta$ from $\theta = 0$ to $\theta = \pi/2$.

Solution. See Fig. 16.2.5. At any interior point (r, θ), construct an element of area. Then

$$dA = r\,dr\,d\theta$$

and

$$A = \int_0^{\pi/2} \int_0^{2\cos\theta} r\,dr\,d\theta$$

$$= \int_0^{\pi/2} \left[\frac{r^2}{2}\right]_0^{2\cos\theta} d\theta = 2\int_0^{\pi/2} \cos^2\theta\,d\theta$$

$$= 2\left[\frac{\theta}{2} + \frac{\sin 2\theta}{4}\right]_0^{\pi/2}$$

$$= \frac{\pi}{2} = 1.57$$

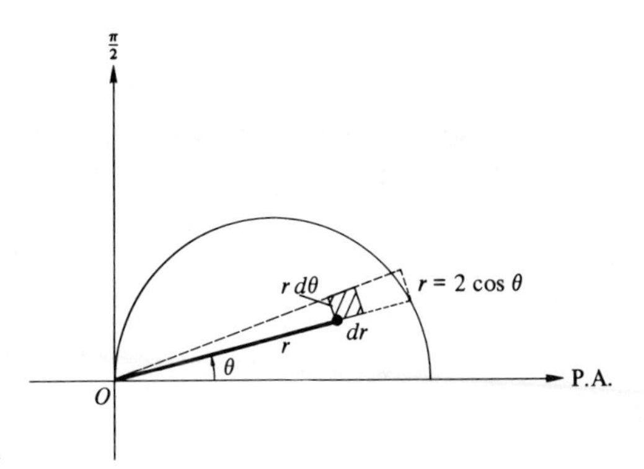

Figure 16.2.5 Area of $r = 2\cos\theta$.

PROBLEM 3. Find the area bounded by $r = \cos\theta$ from $\theta = -\pi/2$ to $\theta = \pi/2$.

EXERCISES 16.2

Group A

1. In each of the following, use double integration.

(a) Find the area bounded by the parabola $y = x^2$ and the line $y = x$.
(b) Find the area bounded by $y^2 = x$ and $x + y - 2 = 0$.
(c) Find the first quadrant area bounded by $y^2 = x^3$ and $y = x$.
(d) Find the area bounded by $y = 4x - x^2$ and $y = x$.
(e) Find the area bounded by $y^2 = 4x$ and $y = 4 - 2x$.
(f) Find the area bounded by $y^2 = 12x$ and $x^2 = 12y$.
(g) Find the first-quadrant area bounded by $x^2 + y^2 = 10$, $y^2 = 9x$, and $y = 0$.
(h) Find the first-quadrant area bounded by $y = \sin x$, $x = \pi/2$, and $y = 0$.
(i) Find the area bounded by $r = \sin\theta$, $\theta = 0$, and $\theta = \pi/2$.
(j) Find the area bounded by one loop of the lemniscate $r^2 = 4\cos 2\theta$.
(k) Find the area that is inside the circle $r = 4\cos\theta$ and to the right of the line $r\cos\theta = 3$.
(l) Find the area that is inside the circle $r = 4\cos\theta$ and outside the cardioid $r = 4(1 - \cos\theta)$.

Group B

1. Evaluate $\int_0^1 \int_{y^2}^{\sqrt{y}} dx\, dy$ and compare the result with the answer to Example 1.

2. Evaluate $\int_0^1 \int_{-\sqrt{x}}^{\sqrt{x}} dy\, dx + \int_1^4 \int_{x-2}^{\sqrt{x}} dy\, dx$ and compare the result with the answer to Example 2.

3. The product of inertia P of a plane area with respect to the rectangular coordinate axes is given by

$$P = \iint_S xy\, dA$$

where dA is $dx\,dy$ or $dy\,dx$ and S defines the region. For each of the following, find the product of inertia with respect to the rectangular coordinate axes of the area bounded by the given curves.

(a) $x = 0$, $x = 4$, $y = 0$, $y = 3$. (b) $x = 0$, $y = 0$, $x + y = 1$.
(c) $y = x^2$, $x = y^2$. (d) $x^2 + y^2 = 1$.

Group C

1. Show that the use of the double integral for finding the area bounded above by $y = f(x)$, below by the X-axis, on the left by $x = a$, and on the right by $x = b$ leads to Eq. 5.3.4.

2. Show that the use of Eq. 16.2.2 leads to Eq. 13.5.1.

3. State the advantages and disadvantages of using double integrals to find a bounded area.

4. The improper integral $\int_0^\infty e^{-x^2}\,dx$ may be evaluated by means of double integrals and polar coordinates. (a) Let $I = \int_0^\infty e^{-x^2}\,dx$. Assuming that $\int_0^\infty f(x)\,dx \int_0^\infty g(y)\,dy$ equals $\int_0^\infty \int_0^\infty f(x)g(y)\,dx\,dy$, show that $I^2 = \int_0^\infty \int_0^\infty e^{-(x^2+y^2)}\,dx\,dy$. (b) Let $dA = dx\,dy$, and noting that the limits of integration indicate a summation over the entire first quadrant, change the integral to polar coordinates and show that $I^2 = \int_0^{\pi/2} \int_0^\infty re^{-r^2}\,dr\,d\theta$. (c) Evaluate I^2 in part (b) and then show that $\int_0^\infty e^{-x^2}\,dx = \sqrt{\pi}/2$. (This integral arises in probability theory when one studies the normal distribution.)

16.3 CENTROID AND MOMENT OF INERTIA

(1) Centroid

In Sec. 6.2 we defined the centroid of a plane region to be the point $(\bar{x}, \bar{y})$ where

$$\bar{x} = \frac{\int x\,dA}{\int dA} \quad \text{and} \quad \bar{y} = \frac{\int y\,dA}{\int dA} \tag{16.3.1}$$

Sometimes the centroid is easier to obtain if double integration is used.

EXAMPLE 1. Find the centroid of the region bounded by the curve $y = \sin x$, $y = 0$, $x = 0$, and $x = \pi$.

Solution. See Fig. 16.3.1.

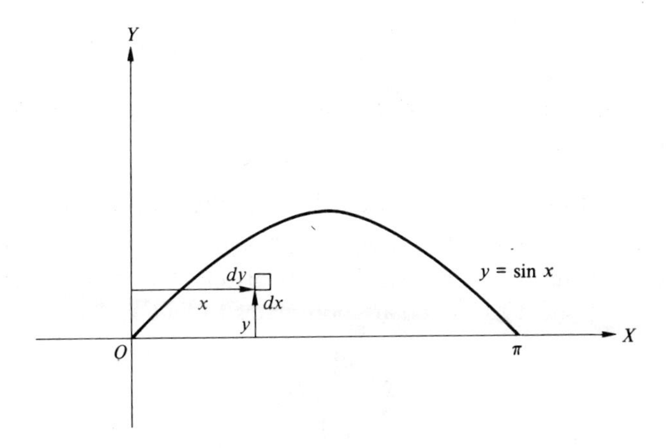

Figure 16.3.1 Example 1.

$$\begin{aligned} A = \int dA &= \int_0^{\pi}\int_0^{\sin x} dy\, dx \\ &= \int_0^{\pi} \sin x\, dx = [-\cos x]_0^{\pi} = 2 \end{aligned}$$

$$\begin{aligned} \int x\, dA &= \int_0^{\pi}\int_0^{\sin x} x\, dy\, dx \\ &= \int_0^{\pi} x \sin x\, dx \\ &= [\sin x - x\cos x]_0^{\pi} = [0 - \pi(-1)] - [0 - 0] \\ &= \pi \end{aligned}$$

$$\therefore\ \bar{x} = \frac{\int x\, dA}{\int dA} = \frac{\pi}{2} = 1.57$$

(You should expect this result from the symmetry of the area about the line $x = \pi/2$.)

$$\begin{aligned} \int y\, dA &= \int_0^{\pi}\int_0^{\sin x} y\, dy\, dx \\ &= \int_0^{\pi} \frac{1}{2}[y^2]_0^{\sin x}\, dx = \frac{1}{2}\int_0^{\pi} \sin^2 x\, dx \\ &= \frac{1}{2}\left[\frac{x}{2} - \frac{\sin 2x}{4}\right]_0^{\pi} \\ &= \frac{1}{2}\left(\frac{\pi}{2} - 0\right) - \frac{1}{2}(0 - 0) \\ &= \frac{\pi}{4} \end{aligned}$$

$$\therefore\ \bar{y} = \frac{\int y\, dA}{\int dA} = \frac{\pi/4}{2} = \frac{\pi}{8} = 0.39$$

Thus the centroid is at $(\pi/2, \pi/8)$ or $(1.57, 0.39)$.

PROBLEM 1. Find the centroid of the area bounded by one arch of the curve $y = \sin 2x$, $y = 0$, and $x = 0$.

EXAMPLE 2. Find the centroid of a quadrant of the area bounded by a circle.

Solution. For convenience let the circle be of radius a with center at the origin, and place the quadrant so that the X-axis bisects it (see Fig. 16.3.2). Then $\bar{y} = 0$. It is convenient to use polar coordinates to find $\bar{x}$. Thus

$$\bar{x} = \frac{\int x\, dA}{\int dA}$$

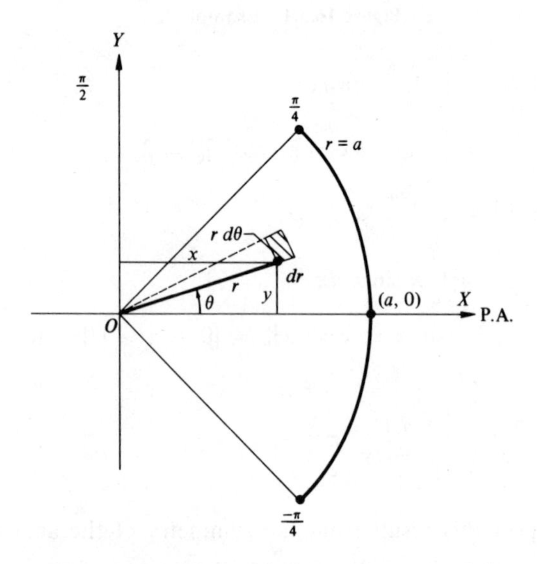

Figure 16.3.2 Centroid of a quadrant of a circle.

Since $\int dA = A$, we have that $A = \frac{1}{4}\pi(a^2)$.

$$\begin{aligned}
\int x\, dA &= \int_{-\pi/4}^{\pi/4} \int_0^a (r \cos \theta) r\, dr\, d\theta \\
&= \int_{-\pi/4}^{\pi/4} \left[\frac{r^3}{3}\right]_0^a \cos \theta\, d\theta \\
&= \frac{2}{3} a^3 \int_0^{\pi/4} \cos \theta\, d\theta \qquad \text{(new limits by use of symmetry)} \\
&= \frac{2}{3} a^3 [\sin \theta]_0^{\pi/4} \\
&= \frac{2}{3} a^3 \left(\frac{\sqrt{2}}{2} - 0\right) \\
&= \frac{\sqrt{2}}{3} a^3
\end{aligned}$$

and

$$\bar{x} = \frac{(\sqrt{2}/3)a^3}{\pi a^2/4} = \frac{4}{3}\frac{\sqrt{2}}{\pi}a = 0.6a$$

Thus the centroid is at $(4a\sqrt{2}/3\pi, 0) = (0.6a, 0)$.

PROBLEM 2. Use double integrals to find the centroid of the region bounded by $r = \sin\theta$, $\theta = 0$, and $\theta = \pi/2$.

(2) Moment of Inertia

In Sec. 6.3 we defined the differential moment of inertia of an area with respect to an axis as the product of an element of area and the square of the distance from the axis to that element of area. The use of double integration may simplify the process. For the area A of a plane region the moment of inertia with respect to the Y-axis is

$$i_Y = \iint_A x^2\,dy\,dx \qquad (16.3.2)$$

and with respect to the X-axis is

$$i_X = \iint_A y^2\,dy\,dx \qquad (16.3.3)$$

(The symbol $\iint_A$ indicates the integration is over the area A.) Since each element of area is a distance $r = \sqrt{x^2 + y^2}$ from the origin (or pole), its *polar moment* of inertia is

$$i_O = \iint_A (x^2 + y^2)\,dy\,dx \qquad (16.3.4)$$

Thus

$$\boxed{i_O = i_X + i_Y} \qquad (16.3.5)$$

EXAMPLE 3. For the area bounded above by the parabola $y = 4x - x^2$ and below by the line $y = x$, find i_X, i_Y, and i_O.

Solution. See Fig. 16.3.3.

$$i_X = \int_0^3 \int_x^{4x-x^2} y^2\,dy\,dx$$

$$= \int_0^3 \left[\frac{y^3}{3}\right]_x^{4x-x^2} dx = \frac{1}{3}\int_0^3 [(4x - x^2)^3 - x^3]\,dx$$

$$= \frac{1}{3}\int_0^3 (-x^6 + 12x^5 - 48x^4 + 63x^3)\,dx$$

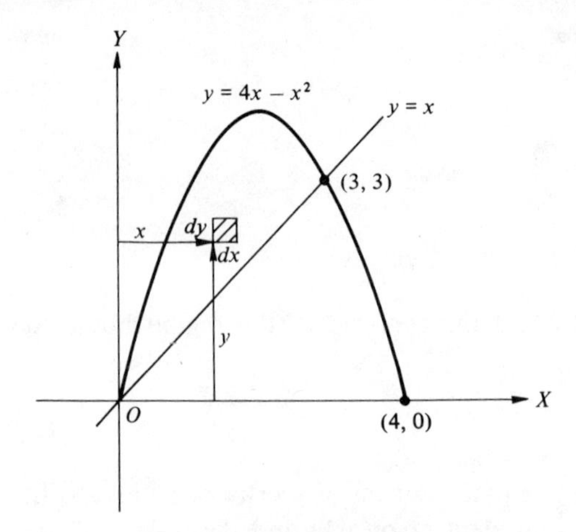

Figure 16.3.3 Moment of inertia.

$$= \frac{1}{3}\left[\frac{-x^7}{7} + 2x^6 - \frac{48}{5}x^5 + \frac{63}{4}x^4\right]_0^3$$

$$= \frac{1}{3}\left[(x^4)\left(\frac{-x^3}{7} + 2x^2 - \frac{48}{5}x + \frac{63}{4}\right)\right]_0^3$$

$$= \frac{1}{3}\left[(3^4)\left(\frac{-27}{7} + 18 - \frac{144}{5} + \frac{63}{4}\right)\right] - 0$$

$$= 45.9$$

$$i_Y = \int_0^3 \int_x^{4x-x^2} x^2\, dy\, dx$$

$$= \int_0^3 [x^2 y]_x^{4x-x^2}\, dx = \int_0^3 [x^2(4x - x^2) - x^2(x)]\, dx$$

$$= \int_0^3 (3x^3 - x^4)\, dx$$

$$= \left[\frac{3}{4}x^4 - \frac{x^5}{5}\right]_0^3 = (3)^4\left(\frac{3}{4} - \frac{3}{5}\right) - 0$$

$$= 81\left(\frac{3}{20}\right) = 12.1$$

Since $i_O = i_X + i_Y$,

$$i_O = 45.9 + 12.1 = 58.0$$

Problem 3. For the area bounded on the right by $x = 6y - y^2$ and on the left by the line $y = x$, find i_x, i_y, and i_o.

If we change Eq. 16.3.2 to polar coordinates, we have

$$i_Y = \iint r^3 \cos^2 \theta\, dr\, d\theta \qquad (16.3.6)$$

Similarly, Eq. 16.3.3 becomes

$$i_X = \iint r^3 \sin^2 \theta\, dr\, d\theta \qquad (16.3.7)$$

and Eq. 16.3.4 becomes

$$i_O = \iint r^3\, dr\, d\theta \tag{16.3.8}$$

An important theorem called the *parallel axis theorem* (or the *transfer theorem*) is given by

$$\boxed{i_L = i_G + md^2} \tag{16.3.9}$$

where L is a line parallel to a line through the centroid G and distance d from it and m is the mass of the body. Consider Fig. 16.3.4.

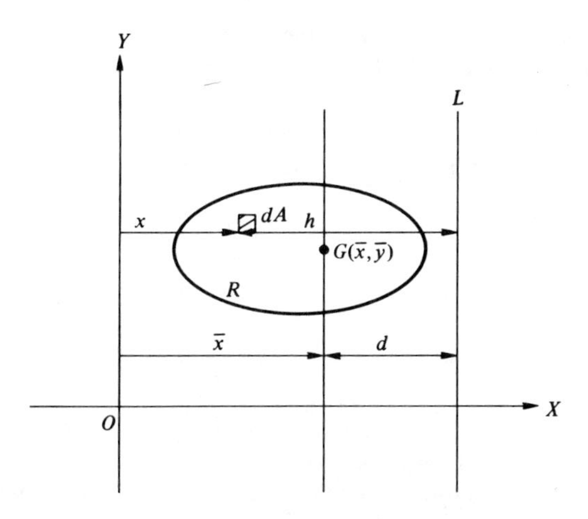

Figure 16.3.4 Parallel axis theorem.

If we let $\rho = 1$, then from Sec. 6.2 we have $m = \rho A = A$, and we shall prove the theorem for plane areas.

$$i_L = \iint h^2\, dA$$

$$x + h = \bar{x} + d$$

and so

$$\begin{aligned}
h &= \bar{x} + d - x \\
i_L &= \iint (\bar{x} + d - x)^2\, dA \\
&= \iint (\bar{x}^2 - 2x\bar{x} + x^2)\, dA + \iint d^2\, dA \\
&\quad + \iint 2\bar{x}d\, dA - 2\iint xd\, dA \\
&= \iint (\bar{x} - x)^2\, dA + d^2 \iint dA + 2\bar{x}d \iint dA - 2d \iint x\, dA
\end{aligned}$$

Since $(\bar{x} - x)$ is the distance of dA from G, $\iint (\bar{x} - x)^2\, dA = i_G$. Since $\bar{x}A = M_Y$ and $\iint x\, dA = M_Y$, we have

$$i_L = i_G + Ad^2$$

EXAMPLE 4. For a rectangle of base b and height h, we know that $i_{\text{base}} = \frac{1}{3}bh^3$ from Sec. 6.3. Since

$$i_G = i_{\text{base}} - Ad^2$$

we have

$$i_G = \frac{1}{3}bh^3 - bh\left(\frac{h}{2}\right)^2 = \frac{1}{12}bh^3$$

PROBLEM 4. A region of area 10 in.[2] has its centroid at (2, 5). If $i_G = 25$, find i_L where L is (a) the line $x = 4$; (b) the line $x = -3$.

Warning ***Remember that centroids are associated with the first moment, which is a distance times a mass, and that the moment of inertia is a distance squared times a mass.***

EXERCISES 16.3

Group A

1. Use double integrals in each of the following.
(a) Find the centroid of the area bounded by $y = 4x - x^2$ and $y = 0$.
(b) Find the centroid of the first-quadrant region of the area bounded by the circle $x^2 + y^2 = a^2$.
(c) Find the centroid of the area bounded by $y^2 = 4x$, and $x = 4$.
(d) Find the center of gravity of the triangular area bounded by $y = 0$, $x = 0$, and $y + x = 1$.
(e) Find the center of gravity of the area bounded by $y = 6x - x^2$ and $y = 2x$.
(f) Find i_Y and i_X for the rectangular area bounded by $x = 0$, $y = 0$, $x = b$, and $y = h$.
(g) Find i_Y and i_X for the triangular area bounded by $x = 0$, $y = 0$, and $x + y = 1$.
(h) Find i_X for the area above the X-axis and below $x^2 + y^2 = 4$.
(i) Find i_Y, i_X, and i_O for the area bounded by the ellipse $(x^2/a^2) + (y^2/b^2) = 1$.
(j) Find i_Y for the area bounded above by $x^2 + y^2 = 128$ and below by $x^2 = 8y$.
(k) Use polar coordinates to find i_{diameter} of a circular region. (*Hint*: Consider $r = a$ and find i_Y.)
(l) Use Eq. 16.3.9 and the result of Exercise 1(k) to find i with respect to a line tangent to a circle.
(m) Use Eq. 16.3.8 to find the polar moment of inertia of a circular region.
(n) Make use the results of Exercise 4 in Group A of Exercises 6.2 and Exercise 5 of Group A of Exercises 6.3 to find the moment of inertia of the area of a right triangle with respect to a line through its centroid and parallel to its base.
(o) Consider Fig. 16.3.5. If the area of region R is 4 and $i_{L_1} = 200$, find i_{L_2}.

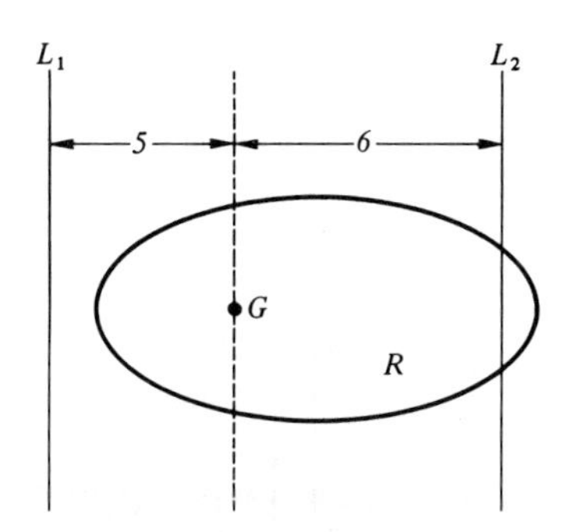

Figure 16.3.5 Exercise 1(o).

(p) Find i_O for the region bounded by $r = 1 + \cos 2\theta$.
(q) Find i_O for the region bounded by one loop of $r = 2 \cos 2\theta$.
(r) Find i_O for the region bounded by one loop of $r^2 = 9 \cos 2\theta$.
(s) Find $\bar{x}$ for the area bounded by $r = 4 \cos \theta$ between $\theta = 0$ and $\theta = \pi/6$.
(t) Find $\bar{x}$ for the area bounded by $r = 3 \sin \theta$ between $\theta = 0$ and $\theta = \pi/2$.
(u) Find the centroid of the area bounded by the cardioid $r = 1 + \cos \theta$.
(v) Find the centroid of a semicircular region. (Make use of polar coordinates.)

Group B

1. Use the appropriate substitution to derive Eqs. 16.3.6 through 16.3.8.

2. Determine the moment of inertia of the cross section shown in Fig. 16.3.6 with respect to its centroidal X-axis.

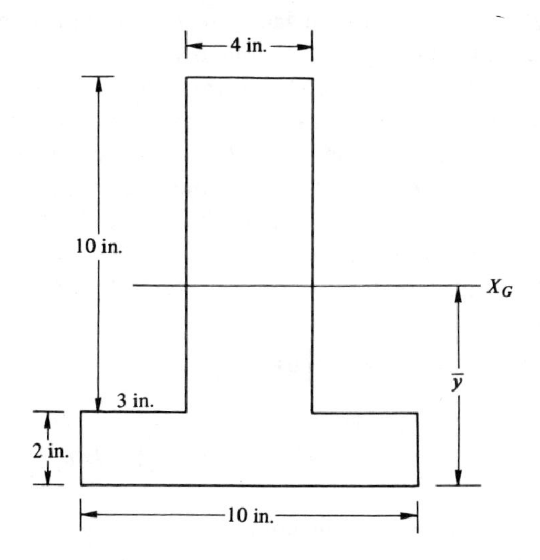

Figure 16.3.6 Exercise 2.

3. Determine the moment of inertia of the cross section shown in Fig. 16.3.7 with respect to its centroidal axis.

4. Find the centroid of the region outside the circle $r = 2$ and within the cardioid $r = 2(1 + \cos \theta)$.

5. Tell how the centroid and the radius of gyration of an area are alike and how they are different.

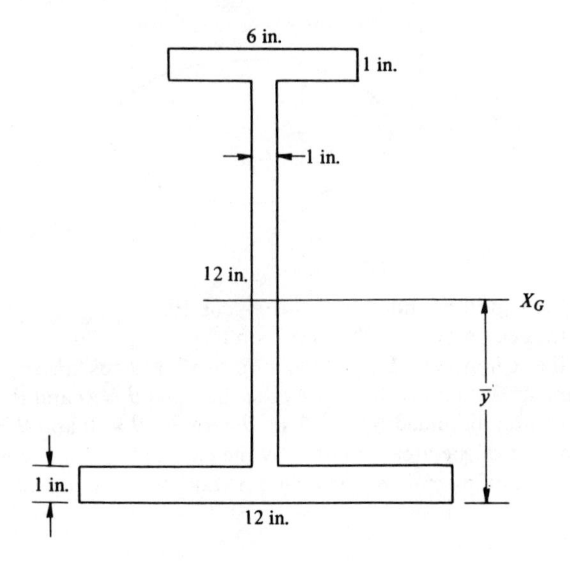

Figure 16.3.7 Exercise 3.

16.4 VOLUMES

Many volumes bounded by surfaces may be obtained by the use of double integrals. Consider the volume bounded above by $z = f(x, y)$ and below by the plane $z = 0$ (see Fig. 16.4.1). At any interior point $(x, y, 0)$, construct an element of area dA. On this base, dA, erect a column, parallel to the Z-axis, to the point (x, y, z) on the surface $z = f(x, y)$. This column represents a differential volume, $dV = z\,dA$. If we construct such columns at many points in the region of the XY-plane bounded by $f(x, y) = 0$ and sum their volumes, we will obtain an approximation to the required volume. We may obtain the exact sum by using integrals. Since the total volume is the sum of all the little volumes, we have

$$V = \int dV = \int z\,dA$$

or

$$V = \iint_A z\,dy\,dx = \iint_A z\,dx\,dy$$

In obtaining the limits on the integrals we always consider dx and dy to be positive. In Fig. 16.4.1, holding x constant and summing in the positive Y-direction, the volume of the columns, we obtain the volume of the slice parallel to the YZ-plane. Then we sum, in the positive X-direction, the volumes of all such slices to obtain the total volume.

EXAMPLE 1. Set up an integral to find the first-octant volume of the solid bounded above by the paraboloid $z = 16 - x^2 - y^2$.

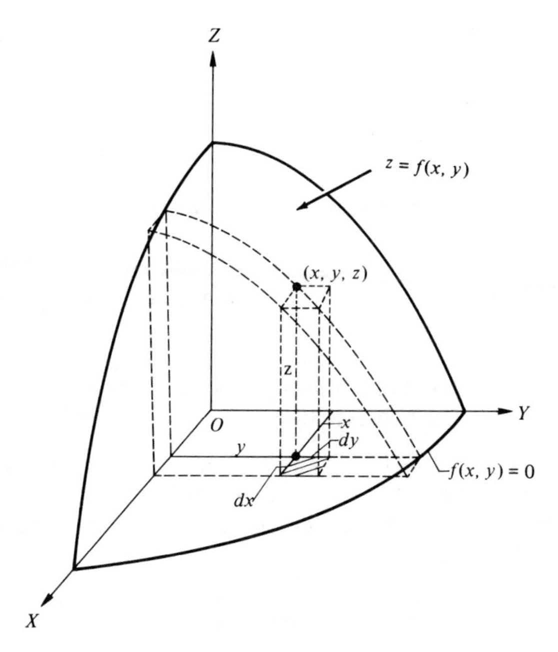

Figure 16.4.1 Volume by double integration.

Solution. See Fig. 16.4.2. At any interior point $(x, y, 0)$, construct a column with base $dA = dy\,dx$ and height z. Then

$$V = \int dV = \int_0^4 \int_0^{\sqrt{16-x^2}} z\,dy\,dx$$

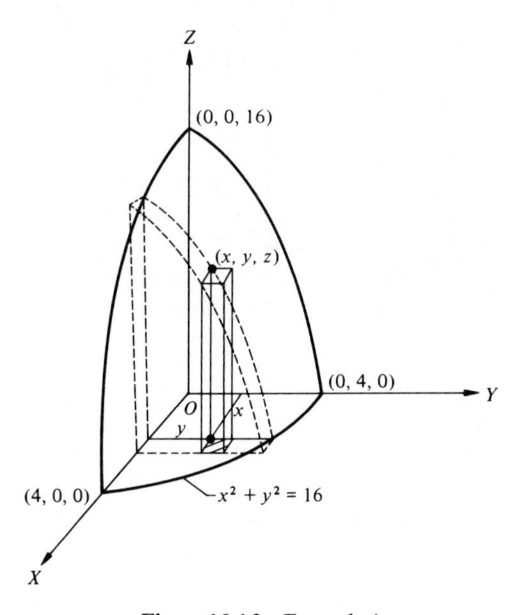

Figure 16.4.2 Example 1.

Since $z = 16 - x^2 - y^2$, we have

$$V = \int_0^4 \int_0^{\sqrt{16-x^2}} (16 - x^2 - y^2)\, dy\, dx$$

We shall leave the mechanics of evaluating this integral to you. See Exercise 11 in Group A.

PROBLEM 1. Find the first-octant volume bounded above by the surface $z = 36 - 4x^2 - 9y^2$.

EXAMPLE 2. Set up a double integral to find the first-octant volume bounded by the intersection of the two cylinders $x^2 + z^2 = a^2$ and $y^2 + z^2 = a^2$.

Solution. See Fig. 16.4.3. We note from Fig. 16.4.3 that if point $(x, y, 0)$ is chosen in front of the plane $y = x$, the height of the column z is a different function than if $(x, y, 0)$ is chosen behind the plane $y = x$. Here then we shall need two integrals. Thus

$$V = \int_0^a \int_0^x \sqrt{a^2 - x^2}\, dy\, dx + \int_0^a \int_x^a \sqrt{a^2 - y^2}\, dy\, dx$$

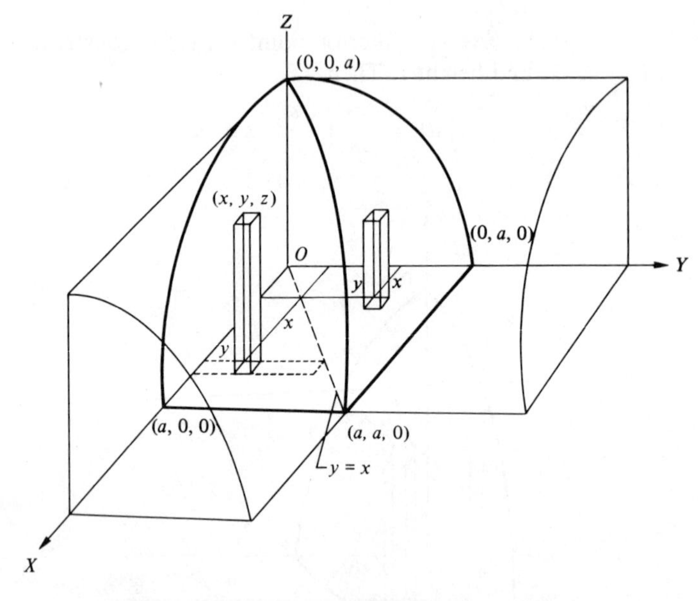

Figure 16.4.3 Volume bounded by two cylinders.

When $z = 0$ is not a bounding surface the height of the column may be expressed as $(z_{\text{upper}} - z_{\text{lower}})$.

PROBLEM 2. Find the first-octant volume of the solid bounded above by the surface $z = 16 - x^2 - y^2$ and below by $z = 9$. (Be careful of the limits for the base area.)

The polar coordinate system extended to three dimensions is called the *cylindrical coordinate system.* In this system a point has cylindrical coordinates (r, θ, z). The relation between the coordinates of a point in the Cartesian system and the cylindrical system is given by

$$x = r\cos\theta \qquad r = \sqrt{x^2 + y^2}$$

$$y = r\sin\theta \qquad \tan\theta = \frac{y}{x}$$

$$z = z \qquad z = z$$

The element of volume using cylindrical coordinates is

$$dV = zr\,dr\,d\theta$$

It is helpful to change Cartesian coordinate equations to cylindrical coordinate equations and then proceed. Thus for Example 1 we would have $z = 16 - r^2$ and in cylindrical coordinates the volume would be given by

$$V = \int_0^{\pi/2} \int_0^4 (16 - r^2)\, r\,dr\,d\theta$$

EXERCISES 16.4

Group A

1. Find the first-octant volume bounded by $x = 0$, $y = 0$, $z = 0$, and $x^2 + y^2 + z^2 = 1$.

2. Find the first-octant volume bounded by $x = 0$, $y = 0$, $z = 0$, and $x + y + z = 1$.

3. Find the volume bounded by the paraboloid $z = x^2 + y^2$ and the plane $z = 4$.

4. Find the volume under $z = 9 - x^2$, above $z = 0$, and within $y^2 = 9x$.

5. Find the volume bounded above by $y^2 = 16 - 4z$, below by $z = 0$, and within $x^2 + y^2 = 16$.

6. Using cylindrical coordinates find the volume above $z = 0$, below $z^2 = x^2 + y^2$, and within $x^2 + y^2 = 4x$.

7. Using cylindrical coordinates find the volume below $z = x$ and above $z = x^2 + y^2$. (*Hint*: The curve of intersection is $x^2 + y^2 = x$ or $r = \cos\theta$.)

8. Using cylindrical coordinates find the volume below $z = y$ and above $2z = x^2 + y^2$.

9. Using cylindrical coordinates find the volume bounded by the XY-plane, the cylinder $x^2 + y^2 = 3y$, and the paraboloid $x^2 + y^2 = 3z$.

10. Using cylindrical cordinates find the volume in the first octant bounded by $16z = xy^2$, $y = x$, and $x = 4$.

11. Evaluate the integral in Example 1.

Group B

1. Triple integrals of the form $\iiint\limits_V dz\,dy\,dx$, where $dV = dz\,dy\,dx$, are sometimes useful in finding volumes of solids. Use a triple integral to find the first-octant volume bounded above by the plane $x + y + z = 1$ (see Exercise 1 of Group C of Exercises 16.1).

2. Use triple integrals to evaluate Exercises 1 through 10 in Group A. For cylindrical coordinates $dV = r\,dz\,dr\,d\theta$.

SUMMARY OF IMPORTANT WORDS AND CONCEPTS

For each of the following, state in your own words your understanding of the statement or word.

1. (a) Double integral (rectangular and polar coordinate systems). (b) Area. (c) Centroid. (d) Moment of inertia. (e) Parallel axis theorem. (f) Volume.

2. Cylindrical coordinate system.

Chapter 17

Infinite Series

17.1 INTRODUCTION TO CONVERGENCE

Infinite series gives us a method of solving some problems that cannot be solved in other ways. For example, the integral $\int_0^x [(\sin u)/u]\,du$ occurs frequently in applied problems. It is called the *sine integral function of x*, and it cannot be expressed in terms of the elementary functions. However, the integrand can be expressed as an infinite series and then the integral can be evaluated.

In our study of limits in Sec. 1.6, we discussed the amount of chalk being placed in a box and saw that the amount could be expressed as

$$S = 1 + \tfrac{1}{2} + \tfrac{1}{4} + \tfrac{1}{8} + \tfrac{1}{16} + \cdots \qquad (17.1.1)$$

Equation 17.1.1 is an example of an *infinite series*. It is different from the expression

$$S = 1 + \tfrac{1}{2} + \tfrac{1}{4} + \tfrac{1}{8} + \tfrac{1}{16} \qquad (17.1.2)$$

Equation 17.1.2 is called a *finite series.* Note carefully that an infinite series is indicated by a plus sign and a row of three dots (ellipses) at the end. Every finite series has a sum. Some infinite series have a sum and some do not. Those that have a sum are called *convergent;* the others are called *divergent.*

Certain applications call for use of series with variable terms such as powers of the variable x or the variable $(x - a)$, where a is a fixed constant. Such series may converge for all values of x or only for certain values of x.

Special types of convergence determine particular properties of the function such as differentiability and integrability. Hence convergence is a very important question. For the present our examples of series and tests for convergence will be developed using series whose terms are constants. If we let $u_1, u_2, \ldots, u_n$ denote the first, second, ..., and nth (or general) terms of a sequence of numbers, respectively, then we may denote the sum of the first n terms of a series as

$$S_n = u_1 + u_2 + \ldots + u_n \tag{17.1.3}$$

If S_n approaches a limit, S, as $n \longrightarrow \infty$, we may write

$$S = u_1 + u_2 + \ldots + u_n + \ldots$$

and

$$S = \lim_{n\to\infty} S_n \tag{17.1.4}$$

S is called the *sum* of the infinite series.

Definition 17.1.1 *The infinite series* $\boldsymbol{S} = \boldsymbol{u}_1 + \boldsymbol{u}_2 + \ldots + \boldsymbol{u}_n + \ldots$ *is said to be convergent if* $\boldsymbol{S} = \lim_{n\to\infty} \boldsymbol{S}_n$, *where* $\boldsymbol{S}_n = \boldsymbol{u}_1 + \boldsymbol{u}_2 + \ldots + \boldsymbol{u}_n$ *and* $\lim_{n\to\infty} \boldsymbol{S}_n$ *exists.*

Sometimes S can be found exactly; most of the time it cannot. However, if we know a series is convergent, we can always approximate S to any desired accuracy by adding the first n terms.

In the study (in algebra) of the geometric series $a + ar + ar^2 + \ldots + ar^{n-1} + \ldots$, where a is a *constant*, r is the *geometric ratio*, and n is the *number of terms*, the sum of the first n terms was shown to be

$$S_n = \frac{a(1 - r^n)}{1 - r} \tag{17.1.5}$$

If $|r| < 1$, then $\lim_{n\to\infty} r^n = 0$, and the sum of the infinite geometric series is

$$S = \lim_{n\to\infty} S_n = \lim_{n\to\infty} a\left(\frac{1 - r^n}{1 - r}\right) = \frac{a}{1 - r} \tag{17.1.6}$$

Here then we can find the sum.

EXAMPLE 1. Find the sum of $1 + \frac{1}{2} + \frac{1}{4} + \frac{1}{8} + \ldots$.

Solution. This is an infinite geometric series with $a = 1$ and $r = \frac{1}{2}$. Hence from Eq. 17.1.6 we have

$$S = \frac{1}{1 - \frac{1}{2}} = 2$$

PROBLEM 1. Find the sum of $1 + \frac{2}{3} + \frac{4}{9} + \frac{8}{27} + \ldots$.

Usually, infinite series are indicated by an expression for their nth (or general) term. In the series

$$1 + \frac{1}{2} + \frac{1}{3} + \frac{1}{4} + \ldots + \frac{1}{n} + \ldots$$

$$u_n = \frac{1}{n}$$

and we could denote the series as $\sum_{n=1}^{\infty} 1/n$. This series is called the *harmonic series*. We shall show in Sec. 17.2 that it is divergent.

EXAMPLE 2. Write the first three terms and the general term of the infinite series $\sum_{n=1}^{\infty} 3/(n^2 + 1)$. Express the summation as an infinite series.

Solution. The first three terms are found to be

$$\frac{3}{(1)^2 + 1} = \frac{3}{2}, n = 1$$

$$\frac{3}{(2)^2 + 1} = \frac{3}{5}, n = 2$$

$$\frac{3}{(3)^2 + 1} = \frac{3}{10}, n = 3$$

The general term is

$$u_n = \frac{3}{n^2 + 1}$$

Also note that

$$u_{n+1} = \frac{3}{(n + 1)^2 + 1} \quad \text{and} \quad u_{n-1} = \frac{3}{(n - 1)^2 + 1}$$

As an infinite series we have

$$\sum_{n=1}^{\infty} \frac{3}{n^2 + 1} = \frac{3}{2} + \frac{3}{5} + \frac{3}{10} + \ldots + \frac{3}{n^2 + 1} + \ldots$$

PROBLEM 2. Write the first four terms and the general term of the infinite series $\sum_{n=1}^{\infty} n/(n^3 + 1)$. Express the summation as an infinite series. Also find u_{n-1} and u_{n+1}.

The general term of some infinite series involves factorial expressions. It is useful to be familiar with some algebra of factorials. Since

$$n! = n(n - 1)(n - 2) \ldots 1$$

we may also write $n! = n(n - 1)! = n(n - 1)(n - 2)!$, and so forth. In simplifying some factorial expressions it is sometimes helpful to express the largest factorials in terms of the smallest factorial. (By definition, $0! = 1$.)

EXAMPLE 3. Simplify the expression $n!/(n - 2)!$.

Solution. Since $n > n - 2$, we write

$$\frac{n!}{(n - 2)!} = \frac{n(n - 1)(n - 2)!}{(n - 2)!} = n(n - 1)$$

PROBLEM 3. Simplify $n!(n + 2)/(n + 3)!$.

EXERCISES 17.1

Group A

1. Which of the following are finite series and which are infinite series?

(a) $1 + 2 + 3 + 4$.

(b) $1 + 2 + 3 + 4 + \ldots.$

(c) $1 + \frac{1}{5} + \frac{1}{6} + \frac{1}{21} + \ldots.$

(d) $1 - \frac{1}{5} - \frac{1}{6} - \frac{1}{21}.$

2. Find the sum of each of the following geometric series if it exists.

(a) $1 + \frac{1}{3} + \frac{1}{9} + \frac{1}{27} + \ldots.$

(b) $1 - \frac{1}{2} + \frac{1}{4} - \frac{1}{8} + \ldots.$

(c) $1 - \frac{3}{2} + \frac{9}{4} - \frac{27}{8} + \ldots.$

(d) $1 - \frac{2}{3} + \frac{4}{9} - \frac{8}{27} + \ldots.$

3. Write the first four terms of each of the following infinite series.

(a) $\sum_{n=1}^{\infty} \frac{1}{n}.$

(b) $\sum_{n=1}^{\infty} \frac{(-1)^n}{n}.$

(c) $\sum_{n=1}^{\infty} \frac{(-1)^n}{n^2}.$

(d) $\sum_{n=1}^{\infty} \frac{(-1)^{n+1}}{n^2 + 1}.$

(e) $\sum_{n=1}^{\infty} \frac{1}{n!}.$

(f) $\sum_{n=1}^{\infty} \frac{n}{n + 1}.$

(g) $\sum_{n=1}^{\infty} \frac{(-1)^{n+1}x^{2n-1}}{(2n - 1)!}.$

(h) $\sum_{n=1}^{\infty} \frac{(-1)^{n+1}x^{2n-2}}{(2n - 2)!}.$

(i) $\sum_{n=1}^{\infty} \frac{x^{n-1}}{(n - 1)!}.$

4. Simplify each of the following.

(a) $\frac{(n + 2)!}{n!}.$

(b) $\frac{n!(n - 1)!}{3(n + 2)!}.$

(c) $\frac{(3n)!}{(3n + 1)!}.$

(d) $\frac{2^n n!}{2^{n+1}(n + 1)!}.$

Group B

1. If u_n is the general term of a series, find the indicated terms for each of the following.

(a) $u_n = \frac{1}{n}$; find u_{n-1}.

(b) $u_n = \frac{n!}{3n + 1}$; find u_{n+1}.

(c) $u_n = \frac{1}{n^3 - 1}$; find u_{n+1}.

(d) $u_n = \frac{3}{2^n - n}$; find u_{n-1}.

2. If $S_n = 1 + \frac{1}{4} + \frac{1}{9} + \ldots + 1/n^2$, find S_1, S_2, and S_3. Does $S_3 = S_1 + S_2$? Does $S_4 = S_3 + S_1$?

3. If $S_n = u_1 + u_2 + u_3 + \ldots + u_n$, show that $u_n = S_n - S_{n-1}$.

4. In each of the following, write the first five terms of the series $\sum_{n=1}^{\infty} u_n$ given the formula for S_n. [*Hint*: If $S_n = n/(n + 1)$, then $S_{n-1} = (n - 1)/n$.]

(a) $S_n = \frac{n}{n + 1}.$

(b) $S_n = \frac{n(n + 1)}{2}.$

(c) $S_n = \dfrac{n(n+1)(2n+1)}{6}$. (d) $S_n = \dfrac{1}{2n-1}$.

5. If the first term of a series is given and a formula is given by which each term may be constructed from the preceding term, such a formula is called a *recursion formula*. Write the first four terms of each series given the following information.

(a) $u_1 = 1,\ u_n = u_{n-1} + 2$. (b) $u_1 = 1,\ u_n = 2u_{n-1}$.

(c) $u_1 = 3,\ u_n = \left(\dfrac{3}{n}\right)u_{n-1}$. (d) $u_1 = -3,\ u_n = (-1)^n(\tfrac{1}{3})u_{n-1}$.

6. (a) Write out several terms of $\sum_{n=1}^{\infty}(3n)!$. (b) Write out several terms of $\sum_{n=1}^{\infty} 3n!$. (c) Are $\sum_{n=1}^{\infty}(3n)!$ and $\sum_{n=1}^{\infty} 3n!$ the same? (d) Can $\sum_{n=1}^{\infty}(3n)!$ be expressed in terms of $n!$?

Group C

1. How are finite and infinite series alike, and how are they different?

2. For a series to converge, $\lim_{n\to\infty} S_n$ must exist. Does this imply that $\lim_{n\to\infty} S_{n-1}$ must exist also? What about $\lim_{n\to\infty} S_{n+1}$?

3. For some infinite series it is possible to find a formula for S_n. Assume that for each of the following S_n has been determined. Tell whether the corresponding infinite series will converge or diverge.

(a) $S_n = \dfrac{n}{n^2+1}$. (b) $S_n = \dfrac{n^2}{n+1}$.

(c) $S_n = \dfrac{1}{n}$. (d) $S_n = \dfrac{n}{2}(2n+1)$.

17.2 SOME BASIC TESTS FOR CONVERGENCE

We now shall investigate a few ways to tell if an infinite series is convergent. There are many tests used for convergence. Some will work on some series; others will work on other series. There is no one convenient test that works on all series. Formulas for S_n are usually difficult or impossible to find. We thus work with u_n to determine convergence.

Recall that for convergence we must have $S = \lim_{n\to\infty} S_n$ and that this limit must exist. Now as $n \to \infty$ it makes no difference whether we sum n terms or $n-1$ terms. In either case in the limit their sum is S. This leads us to the

Necessary Condition for Convergence

A necessary condition that the infinite series $u_1 + u_2 + u_3 + \ldots + u_n + \ldots$ will converge is

$$\lim_{n\to\infty} u_n = 0 \tag{17.2.1}$$

This may be shown by considering the following:

$$S_n = u_1 + u_2 + \ldots + u_{n-1} + u_n$$
$$S_{n-1} = u_1 + u_2 + \ldots + u_{n-1}$$

For convergence

$$S = \lim_{n\to\infty} S_n = \lim_{n\to\infty} S_{n-1} \qquad \text{or} \qquad \lim_{n\to\infty} (S_n - S_{n-1}) = 0$$

Now

$$S_n - S_{n-1} = u_n$$

Thus

$$\lim_{n\to\infty} (S_n - S_{n-1}) = \lim_{n\to\infty} u_n = 0$$

EXAMPLE 1. (a) The series $1 + \frac{1}{2} + \frac{1}{3} + \ldots + 1/n + \ldots$ may converge because $u_n = 1/n$ and $\lim_{n\to\infty} (1/n) = 0$.

(b) The series $\sum_{n=1}^{\infty} n/(2n + 1)$ does not converge because $u_n = n/(2n + 1) = 1/2[1 + (1/2n)]$ and $\lim_{n\to\infty} u_n = \lim_{n\to\infty} 1/2[1 + (1/2n)] = \frac{1}{2}$, which is not zero.

PROBLEM 1. (a) Show that the series $\sum_{n=1}^{\infty} 3/5n$ may converge.

(b) Show that the series $\sum_{n=1}^{\infty} 3n/(5n + 1)$ diverges.

Note that in part (a) of Example 1, we said that the series *may* converge. To find out if it will converge we need further tests. One such test is the integral test.

Warning ***Just because $\lim_{n\to\infty} u_n = 0$ does not mean that the series must converge.***

Integral Test

Since a definite integral is a summation process, as is an infinite series, we are able to use integrals to test for convergence.

Form the function $f(n) = u_n$; then the series $\sum_{n=1}^{\infty} u_n$ will converge if the improper integral

$$\int_1^{\infty} f(x)\, dx$$

exists. Otherwise, it is divergent.

EXAMPLE 2. Test $\sum_{n\to 1}^{\infty} 1/n$ for convergence.

Solution. Here $u_n = 1/n$; let $f(x) = 1/x$. Then

$$\int_1^{\infty} f(x)\, dx = \int_1^{\infty} \frac{1}{x}\, dx$$

From Sec. 11.6, we have

$$\int_1^{\infty} \frac{dx}{x} = \lim_{b\to\infty} \int_1^b \frac{dx}{x} = \lim_{b\to\infty} [\ln x]_1^b$$
$$= \lim_{b\to\infty} (\ln b - 0)$$

which does not exist. Therefore, $\sum_{n=1}^{\infty} 1/n$ is *divergent*. The series $\sum_{n=1}^{\infty} 1/n$ is called the *harmonic series*.

PROBLEM 2. Use the integral test to determine if the series $\sum_{n=1}^{\infty} 5/(10n + 3)$ converges or diverges.

EXAMPLE 3. Test $\sum_{n=1}^{\infty} 1/n^2$ for convergence.

Solution

$$\int_1^{\infty} \frac{1}{x^2}\,dx = \lim_{b\to\infty} \int_1^b x^{-2}\,dx$$
$$= \lim_{b\to\infty} \left[\frac{-1}{x}\right]_1^b = \lim_{b\to\infty} \left(\frac{-1}{b} + \frac{1}{1}\right) = 1$$

which exists. Therefore, $\sum_{n=1}^{\infty} 1/n^2$ converges.

PROBLEM 3. Test $\sum_{n=1}^{\infty} 1/n^{3/2}$ for convergence.

Warning ***The integral test is useful only when we can perform the integration.***

We shall now give two more tests for convergence. Their proofs may be found in more comprehensive texts on calculus.

Alternating Series Test

The series $u_1 - u_2 + u_3 - u_4 + \dots + (-1)^{n+1}\, u_n + \dots$ is called an *alternating series*. It will converge if two conditions are satisfied.

1. $\lim_{n\to\infty} u_n = 0$.
2. $u_{n+1} < u_n$.

Note that u_n does not include the algebraic sign.

EXAMPLE 4. Does $1 - \frac{1}{2} + \frac{1}{3} - \frac{1}{4} + \dots + (-1)^{n+1}(1/n) + \dots$ converge?

Solution. The given series is an alternating series with $u_n = 1/n$.

1. $\lim_{n\to\infty} u_n = \lim_{n\to\infty} \left(\frac{1}{n}\right) = 0$.
2. $u_{n+1} = \frac{1}{n+1},\ u_n = \frac{1}{n}$

$$\frac{1}{n+1} \overset{?}{<} \frac{1}{n}$$

$$n \overset{?}{<} n+1$$

$$0 < 1$$

Thus

$$u_{n+1} < u_n$$

Since the given series is an alternating series and *both* requirements are met, the series does converge.

☛ ***Warning*** ***For an alternating series to converge both requirements must be satisfied.***

PROBLEM 4. Test $\sum_{n=1}^{\infty} (-1)^{n+1}/(3n+2)$ for convergence.

Ratio Test

The absolute value of the ratio of the $n+1$ term to the nth term is sometimes used to test the series $\sum_{n=1}^{\infty} u_n$ for convergence.

1. If $\lim_{n\to\infty} \left|\frac{u_{n+1}}{u_n}\right| < 1$, the series converges.
2. If $\lim_{n\to\infty} \left|\frac{u_{n+1}}{u_n}\right| > 1$, the series diverges.
3. If $\lim_{n\to\infty} \left|\frac{u_{n+1}}{u_n}\right| = 1$, the ratio test fails and other tests must be used.

EXAMPLE 5. Use the ratio test to test $\sum_{n=1}^{\infty} 1/n!$ for convergence.

Solution. Recall that $n! = n(n-1)(n-2)\ldots 1$ and $(n+1)! = (n+1)(n)! = (n+1)(n)(n-1)(n-2)\ldots 1$. Here

$$u_n = \frac{1}{n!} \qquad \text{and} \qquad u_{n+1} = \frac{1}{(n+1)!}$$

Then

$$\left|\frac{u_{n+1}}{u_n}\right| = \left|\frac{1/(n+1)!}{1/n!}\right| = \left|\frac{n!}{(n+1)!}\right|$$

$$= \left|\frac{n!}{(n+1)n!}\right| = \left|\frac{1}{n+1}\right|$$

Thus

$$\lim_{n\to\infty}\left|\frac{u_{n+1}}{u_n}\right| = \lim_{n\to\infty}\left|\frac{1}{n+1}\right| = 0$$

Since $0 < 1$, this series converges.

PROBLEM 5. Use the ratio test to test $\sum_{n=1}^{\infty} n!/2^n$ for convergence.

As mentioned in Sec. 17.1, some series have variable terms such as x^n and $(x - a)^n$, where a is a constant. The values of the variable x for which the series is convergent comprise the *interval of convergence*. For values of x within the interval of convergence a function represented by the series may be differentiated and integrated by differentiating and integrating the given series term by term. The ratio test is valuable for determining the interval of convergence for the series.

EXAMPLE 6. Find the interval of convergence for the series

$$\frac{x}{2} + \frac{x^2}{2^2 \cdot 2} + \frac{x^3}{2^3 \cdot 3} + \ldots + \frac{x^n}{2^n \cdot n} + \ldots$$

Solution. Here

$$u_n = \frac{x^n}{2^n(n)} \qquad \text{and} \qquad u_{n+1} = \frac{x^{n+1}}{2^{n+1}(n+1)}$$

$$\left|\frac{u_{n+1}}{u_n}\right| = \left|\frac{x^{n+1}/[2^{n+1}(n+1)]}{x^n/[2^n(n)]}\right| = \left|\frac{x^{n+1}}{2^{n+1}(n+1)}\frac{2^n(n)}{x^n}\right| = \left|\frac{n}{n+1}\frac{x}{2}\right|$$

$$\lim_{n\to\infty}\left|\frac{u_{n+1}}{u_n}\right| = \lim_{n\to\infty}\left|\frac{n}{n+1}\right|\left|\frac{x}{2}\right| = (1)\left|\frac{x}{2}\right| = \left|\frac{x}{2}\right|$$

For convergence

$$\left|\frac{x}{2}\right| < 1 \qquad \text{or} \qquad -2 < x < 2$$

For divergence

$$\left|\frac{x}{2}\right| > 1 \qquad \text{or} \qquad x > 2 \ \text{ and } \ x < -2.$$

For $|x/2| = 1$, or $x = 2$ and $x = -2$, the ratio test fails. Thus we must use other tests for these values of x. First let $x = 2$; then the series is

$$1 + \frac{1}{2} + \frac{1}{3} + \ldots + \frac{1}{n} + \ldots$$

which is the harmonic series and it diverges (see Example 2). For $x = -2$, the series is

$$-1 + \frac{1}{2} - \frac{1}{3} + \ldots + (-1)^n\frac{1}{n} + \ldots$$

which is an alternating series. Since $\lim_{n\to\infty} u_n = 0$ and $u_{n+1} < u_n$, this series converges.

Since we have tested all values of x, our conclusion is that the series

$$\sum_{n=1}^{\infty} \frac{x^n}{2^n \cdot n}$$

converges only for $-2 \leqq x < 2$. Hence the interval $-2 \leqq x < 2$ is the interval of convergence.

PROBLEM 6. Find the interval of convergence of $\sum_{n=1}^{\infty} (-1)^{n+1}x^n/n^{1/2}$.

EXERCISES 17.2

Group A

1. Apply the necessary condition for convergence on each of the following and tell whether or not a further test is needed to determine convergence.

(a) $\sum_{n=1}^{\infty} \frac{1}{n}$. (b) $\sum_{n=1}^{\infty} \frac{1}{n-2}$. (c) $\sum_{n=1}^{\infty} \frac{n}{n+1}$.

(d) $\sum_{n=1}^{\infty} \frac{n^2}{n^3+1}$. (e) $\sum_{n=1}^{\infty} \frac{1}{n^2}$. (f) $\sum_{n=1}^{\infty} n$.

2. Use the integral test on each of the following to determine convergence.

(a) $\sum_{n=1}^{\infty} \frac{1}{n^3}$. (b) $\sum_{n=1}^{\infty} \frac{1}{n^2+1}$. (c) $\sum_{n=1}^{\infty} \frac{2}{n}$.

(d) $\sum_{n=1}^{\infty} \frac{1}{5n}$. (e) $\sum_{n=1}^{\infty} \frac{1}{\sqrt{n}}$. (f) $\sum_{n=1}^{\infty} \frac{n}{e^{n^2}}$.

(g) $\sum_{n=1}^{\infty} \frac{1}{3n+1}$. (h) $\sum_{n=1}^{\infty} \frac{1}{(3n+1)^2}$. (i) $\sum_{n=1}^{\infty} \frac{n}{2+n^4}$.

(j) $\sum_{n=1}^{\infty} \frac{1}{n(n+1)}$.

3. Use the alternating series test on each of the following to determine convergence.

(a) $\sum_{n=1}^{\infty} (-1)^n \frac{1}{n^2}$. (b) $\sum_{n=1}^{\infty} (-1)^n \frac{1}{\sqrt{n}}$. (c) $\sum_{n=1}^{\infty} (-1)^n \frac{1}{\ln n}$.

(d) $\sum_{n=1}^{\infty} (-1)^n \frac{1}{2^n}$. (e) $\sum_{n=1}^{\infty} (-1)^n \frac{n+1}{n}$. (f) $\sum_{n=1}^{\infty} (-1)^n \frac{2^n}{n!}$.

4. Use the ratio test on each of the following to determine convergence.

(a) $\sum_{n=1}^{\infty} \frac{1}{(n+1)!}$. (b) $\sum_{n=1}^{\infty} \frac{n}{2^n}$. (c) $\sum_{n=1}^{\infty} \frac{3^n}{n \cdot 2^n}$.

(d) $\sum_{n=1}^{\infty} \frac{n!}{7^n}$. (e) $\sum_{n=1}^{\infty} \frac{1}{(2n-1)!}$. (f) $\sum_{n=1}^{\infty} \frac{1}{(2n-1)(2n)}$.

(g) $\sum_{n=1}^{\infty} \frac{1}{n}$. (h) $\sum_{n=1}^{\infty} \frac{1}{n^2}$.

5. Find the interval of convergence for each of the following.

(a) $\sum_{n=1}^{\infty} \frac{(1-)^{n+1}x^n}{n^2}$. (b) $\sum_{n=1}^{\infty} \frac{x^{2n}}{(2n)!}$. (c) $\sum_{n=1}^{\infty} x^{n-1}$.

(d) $\sum_{n=1}^{\infty} (-1)^{n+1} \frac{x^n}{n}$. (e) $\sum_{n=1}^{\infty} \frac{(-1)^{n+1}x^{2n-1}}{(2n-1)!}$. (f) $\sum_{n=1}^{\infty} \frac{(-1)^{n+1}x^{2n-2}}{(2n-2)!}$.

(g) $\sum_{n=1}^{\infty} \frac{x^{n-1}}{(n-1)!}$. (h) $\sum_{n=1}^{\infty} 2^n x^n$.

Group B

1. Show that $\sum_{n=1}^{\infty} 1/n^p$ is convergent only for $p > 1$. This series is called the p-series.

2. Find the interval of convergence for each of the following. [*Hint*: $|x - a| < 1$ implies $(-1 + a) < x < (1 + a)$.]

(a) $\sum_{n=1}^{\infty} (-1)^{n+1} \frac{(x+1)^n}{n}$.

(b) $\sum_{n=1}^{\infty} \frac{(x-1)^n}{\sqrt{n}}$.

(c) $\sum_{n=1}^{\infty} \frac{(n+1)(2x+1)}{n!}$.

(d) $\sum_{n=1}^{\infty} \frac{(-1)^{n+1}(x-1)^n}{n}$.

3. Does a constant times a series affect its convergence or divergence? Why?

Group C

1. For the following consider that both $\sum_{n=1}^{\infty} u_n$ and $\sum_{n=1}^{\infty} v_n$ are positive term series. (a) If $\sum_{n=1}^{\infty} u_n$ converges and $v_n < u_n$, can you conclude that $\sum_{n=1}^{\infty} v_n$ is convergent? Explain. (b) If $\sum_{n=1}^{\infty} u_n$ diverges and $v_n > u_n$, what can you conclude about $\sum_{n=1}^{\infty} v_n$? (c) If $\sum_{n=1}^{\infty} u_n$ converges and $v_n > u_n$, what can you conclude as to convergence or divergence of $\sum_{n=1}^{\infty} v_n$? (d) Do you think that parts (a) and (b) could be called the *comparison tests*?

2. Use Exercise 1 of Group B and Exercise 1 of Group C to test the following series for convergence.

(a) $\sum_{n=1}^{\infty} \frac{1}{n^3}$.

(b) $\sum_{n=1}^{\infty} \frac{1}{n^{1/2}}$.

(c) $\sum_{n=1}^{\infty} \frac{1}{n^3 + n^{1/2}}$.

(d) $\sum_{n=1}^{\infty} \frac{1}{(n-5)^{1/2}}$

17.3 MACLAURIN SERIES

As you studied trigonometry you found that the exact value of sin 30° could be obtained by geometric means but that sin 31° could not. At that time you probably were told that the tables of values of the trigonometric functions were computed by use of series. One such series that is used by high-speed computers is known as the *Maclaurin series.* When we compute values of functions by series it is hoped that the series will converge rapidly. That is, we need use only a few terms at the beginning of the series to obtain our desired accuracy. The Maclaurin series converges rapidly around $x = 0$. Let us assume that a function of x may be written as

$$f(x) = a_0 + a_1x + a_2x^2 + a_3x^3 + \ldots + a_nx^n + \ldots \tag{17.3.1}$$

where $a_0, a_1, \ldots, a_n$ are constants. Equation 17.3.1 is called a *power series in* x. If for a given $f(x)$ we can find the values of the constants $a_0, a_1, a_2, \ldots, a_n$, then we say that $f(x)$ is represented by that series. The values of these constants may be found in various ways. The method most useful to us is the method using derivatives. In using this method we assume that $f(x)$ and all of its derivatives exist at $x = 0$. Let $f(x) = a_0 + a_1x + a_2x^2 + a_3x^3 + a_4x^4 + a_5x^5 + \ldots + a_nx^n + \ldots$; then

$$f'(x) = a_1 + 2a_2x + 3a_3x^2 + 4a_4x^3 + 5a_5x^4 + \ldots + na_nx^{n-1} + \ldots$$

$$f''(x) = 2a_2 + 3 \cdot 2a_3x + 4 \cdot 3a_4x^2 + 5 \cdot 4a_5x^3 + \ldots + n(n-1)a_nx^{n-2} + \ldots$$

$$f'''(x) = 3 \cdot 2a_3 + 4 \cdot 3 \cdot 2a_4 x + 5 \cdot 4 \cdot 3a_5 x^2 + \ldots$$
$$+ n(n-1)(n-2)a_n x^{n-3} + \ldots$$
$$f^{iv}(x) = 4 \cdot 3 \cdot 2a_4 + 5 \cdot 4 \cdot 3 \cdot 2a_5 x +$$
$$+ n(n-1)(n-2)(n-3)a_n x^{n-4} + \ldots$$
$$f^{v}(x) = 5 \cdot 4 \cdot 3 \cdot 2a_5 + \ldots + n(n-1)(n-2)(n-3)(n-4)a_n x^{n-5}$$
$$+ \ldots$$

Now if we let $x = 0$, all of the terms except the first on the right of the equal signs will vanish; thus we find

$$f(0) = a_0, f'(0) = a_1, f''(0) = 2!\,a_2, f'''(0) = 3!\,a_3$$
$$f^{iv}(0) = 4!\,a_4, f^{v}(0) = 5!\,a_5, \ldots, f^{n}(0) = n!\,a_n$$

or

$$a_0 = f(0), a_1 = \frac{f'(0)}{1!}, a_2 = \frac{f''(0)}{2!}, a_3 = \frac{f'''(0)}{3!}$$
$$a_4 = \frac{f^{iv}(0)}{4!}, a_5 = \frac{f^{v}(0)}{5!}, \ldots, a_n = \frac{f^{n}(0)}{n!}$$

Substituting these values of the constants in Eq. 17.3.1 we have

$$f(x) = f(0) + \frac{f'(0)}{1!}x + \frac{f''(0)}{2!}x^2 + \frac{f'''(0)}{3!}x^5 + \ldots$$
$$+ \frac{f^{n}(0)}{n!}x^n + \ldots \quad (17.3.2)$$

Equation 17.3.2 is called the *Maclaurin series* for $f(x)$.

EXAMPLE 1. Develop the Maclaurin series for $\sin x$. Obtain at least three nonzero terms.

Solution. It is best to construct a systematic table. Obtain the derivatives first; then evaluate them at $x = 0$.

$$f(x) = \sin x \qquad f(0) = \sin 0 = 0$$
$$f'(x) = \cos x \qquad f'(0) = \cos 0 = 1$$
$$f''(x) = -\sin x \qquad f''(0) = -\sin 0 = 0$$
$$f'''(x) = -\cos x \qquad f'''(0) = -\cos 0 = -1$$
$$f^{iv}(x) = \sin x \qquad f^{iv}(0) = \sin 0 = 0$$
$$f^{v}(x) = \cos x \qquad f^{v}(0) = \cos 0 = 1$$

Substituting these values in Eq. 17.3.2 we have

$$f(x) = \sin x = 0 + \frac{1}{1!}x + \frac{0}{2!}x^2 + \frac{-1}{3!}x^3 + \frac{0}{4!}x^4 + \frac{1}{5!}x^5 + \ldots$$

or

$$\sin x = x - \frac{x^3}{3!} + \frac{x^5}{5!} + \ldots$$

The general term for the Maclaurin expansion of sin x is

$$\frac{(-1)^{n+1}x^{2n-1}}{(2n-1)!}$$

Thus we may write

$$\sin x = \sum_{n=1}^{\infty} \frac{(-1)^{n+1}x^{2n-1}}{(2n-1)!}$$

This series is convergent for all x, as may be shown by applying the ratio test of Sec. 17.2 (see Exercise 5(e) in Group A of Exercises 17.2).

PROBLEM 1. Write the first three nonzero terms of the Maclaurin series for cos x. Write the general term and express the series in summation notation.

Warning ***For $f(x)$ to have a Maclaurin series it must have an infinite number of derivatives and they must all exist at $x = 0$.***

EXAMPLE 2. Obtain the first three nonzero terms of the Maclaurin series for $\ln(1 + x)$.

Solution. Note that we cannot find the Maclaurin series for $\ln x$ as $\ln 0$ does not exist.

$$f(x) = \ln(1 + x) \qquad f(0) = \ln 1 = 0$$

$$f'(x) = (1 + x)^{-1} \qquad f'(0) = (1 + 0)^{-1} = 1$$

$$f''(x) = -1(1 + x)^{-2} \qquad f''(0) = -1(1 + 0)^{-2} = -1$$

$$f'''(x) = (-2)(-1)(1 + x)^{-3} \qquad f'''(0) = -2(-1)(1 + 0)^{-3} = 2$$

Thus

$$\ln(1 + x) = x - \frac{x^2}{2!} + \frac{2x^3}{3!} + \dots$$

or

$$\ln(1 + x) = x - \frac{x^2}{2} + \frac{x^3}{3} + \dots$$

In summation form

$$\ln(1 + x) = \sum_{n=1}^{\infty} \frac{(-1)^{n+1}x^n}{n}$$

[For the interval of convergence, see Exercise 5(d) in Group A of Exercises 17.2.]

PROBLEM 2. Obtain at least the first four nonzero terms for the Maclaurin series for $1/(1 - x)$. [*Hint*: Let $f(x) = (1 - x)^{-1}$.] Divide algebraically $1/(1 - x)$. Is the answer the same as the Maclaurin series?

Warning ***The coefficient of the term x^n is a constant whose value is the nth derivative evaluated at $x = 0$ and is divided by $n!$.***

EXERCISES 17.3

Group A

1. Develop the Maclaurin series for each of the following. Write at least the first three nonvanishing terms.

(a) $f(x) = e^x$.
(b) $f(x) = \cos 3x$.
(c) $f(x) = \sin 2x$.
(d) $f(x) = e^{2x}$.
(e) $f(x) = \ln (1 - x)$.
(f) $f(x) = e^{-x}$.

Group B

1. Develop the Maclaurin series for each of the following. Write at least three nonvanishing terms.

(a) $f(x) = \tan x$.
(b) $f(x) = \text{Sin}^{-1} x$.
(c) $f(x) = \text{Tan}^{-1} x$.
(d) $f(x) = \sqrt{1 + x}$.
(e) $f(x) = \sin (x + \pi/4)$.
(f) $f(x) = e^{x^2}$.
(g) $f(x) = \ln (\cos x)$.
(h) $f(x) = \ln (1 + \sin x)$.

2. Develop the Maclaurin series for $f(x) = (a + x)^n$, where a is a constant. Compare your results with the binomial expansion you studied in algebra.

3. Differentiate the Maclaurin series expansion for $\sin x$ term by term to find the Maclaurin series for $\cos x$. Compare with Problem 1.

4. Differentiate the Maclaurin series expansion for e^x term by term to show that $(d/dx)e^x = e^x$.

Group C

1. Some differential equations may be solved by the use of Maclaurin series. If $y = f(x)$ has a Maclaurin series, then the series

$$y = f(0) + \frac{f'(0)}{1!}x + \frac{f''(0)}{2!}x^2 + \frac{f'''(0)}{3!}x^3 + \dots$$

is a solution to a differential equation if we can find the values of $f(0), f'(0), f''(0), \dots$. Let us solve the equation $y' + xy = 0$ subject to the boundary conditions when $x = 0$, $y = 1$. From the differential equation and successive differentiations we have

$$\begin{aligned} y' + xy = 0 \quad &\text{or} \quad f'(x) = -xf(x) \\ y'' + xy' + y = 0 \quad &\text{or} \quad f''(x) = -xf'(x) - f(x) \\ y''' + xy'' + y' + y' = 0 \quad &\text{or} \quad f'''(x) = -xf''(x) - 2f'(x) \\ y^{\text{iv}} + xy''' + y'' + y'' + y'' = 0 \quad &\text{or} \quad f^{\text{iv}}(x) = -xf'''(x) - 3f''(x) \end{aligned}$$

We now evaluate $f(x)$ and its derivatives at $x = 0$ using the foregoing relations and the boundary condition. Thus, since $y = f(x)$,

$$f(0) = 1$$
$$f'(0) = -0(1) = 0$$
$$f''(0) = -0 - 1 = -1$$
$$f'''(0) = 0 - 2(0) = 0$$

and

$$f^{\text{iv}}(0) = 0 - 3(-1) = 3$$

We thus find the solution of the differential equation to be $y = 1 - \frac{1}{2}x^2 + \frac{1}{8}x^4 + \ldots$. Solve each of the following by the use of Maclaurin series.

(a) $y' - y = 0$, when $x = 0$, $y = 1$.
(b) $y' + 2x = y$, when $x = 0$, $y = 2$.
(c) $y'' - y = x$, when $x = 0$, $y = 1$, $y' = 0$.
(d) $y'' - y = 0$, when $x = 0$, $y = 1$, $y' = 1$.

2. Consider the differential equation $y'' + y = 0$, with initial conditions $y(0) = 1$ and $y'(0) = 0$. Assume a solution of the form $y = a_0 + a_1x + a_2x^2 + \ldots + a_nx^n + \ldots$ and find a_0, a_1, and a_2 to prove that the solution to this equation is $y = \cos x$. (*Hint*: Find y'' in series form from the series for y. Substitute the series for y and y'' in $y'' + y = 0$. Collect the coefficients of like powers of x, set these equal to zero, and solve for the coefficients.)

3. Use the first five terms of the Maclaurin series for $\sin x$ to obtain a polynomial approximation to the Fresnel sine integral $\int_0^x \sin(t^2)\,dt$.

17.4 TAYLOR SERIES

In Sec. 17.3 we said that Maclaurin series converges rapidly around $x = 0$. This is so because if $0 < x < 1$, x^n becomes small quickly as n gets large. For example, if $x = 0.1$, $x^2 = 0.01$, $x^3 = 0.001$, and so forth. Many times we desire to use series to evaluate a function at some x not near zero; then we resort to a series called *Taylor series*, which is a power series in $(x - a)$. This series converges rapidly for values of x near the constant a. The development of Taylor series follows closely that of the development of Maclaurin series. Thus

$$f(x) = a_0 + a_1(x - a) + a_2(x - a)^2 + a_3(x - a)^3 + \ldots + a_n(x - a)^n + \ldots \tag{17.4.1}$$

$$f'(x) = a_1 + 2a_2(x - a) + 3a_3(x - a)^2 + \ldots + na_n(x - a)^{n-1} + \ldots$$

$$f''(x) = 2a_2 + 3 \cdot 2a_3(x - a) + \ldots + n(n - 1)a_n(x - a)^{n-2} + \ldots$$

$$f'''(x) = 3 \cdot 2a_3 + \ldots + n(n - 1)(n - 2)a_n(x - a)^{n-3} + \ldots$$

Note that for $x = a$, every term except the first on the right-hand side of the equals sign will vanish and we will be able to compute $a_0, a_1, a_2, \ldots, a_n, \ldots$.

Thus

$$f(a) = a_0, f'(a) = a_1, f''(a) = 2a_2, f'''(a) = 3 \cdot 2a_3$$

or

$$a_0 = f(a), a_1 = f'(a), a_2 = \frac{f''(a)}{2!}, a_3 = \frac{f'''(a)}{3!}, \ldots$$

Substituting these values in Eq. 17.4.1, we have

$$f(x) = f(a) + \frac{f'(a)}{1!}(x - a) + \frac{f''(a)}{2!}(x - a)^2 + \frac{f'''(a)}{3!}(x - a)^3 +$$

$$\ldots + \frac{f^n(a)}{n!}(x - a)^n + \ldots \tag{17.4.2}$$

Equation 17.4.2 is called the *Taylor series* for $f(x)$. Note that it can be developed only if $f(x)$ and all of its derivatives exist at $x = a$. If $a = 0$, the Taylor series becomes a Maclaurin series.

EXAMPLE 1. Develop the Taylor series for $\cos x$ for $a = \pi/3$. Write at least the first three nonvanishing terms.

Solution

$$f(x) = \cos x \qquad f\left(\frac{\pi}{3}\right) = \cos\left(\frac{\pi}{3}\right) = \frac{1}{2}$$

$$f'(x) = -\sin x \qquad f'\left(\frac{\pi}{3}\right) = -\sin\left(\frac{\pi}{3}\right) = -\frac{\sqrt{3}}{2}$$

$$f''(x) = -\cos x \qquad f''(x) = -\cos\left(\frac{\pi}{3}\right) = -\frac{1}{2}$$

Thus using Eq. 17.4.2, we have

$$\cos x = \frac{1}{2} - \frac{\sqrt{3}}{2}\left(x - \frac{\pi}{3}\right) - \frac{1}{2 \cdot 2!}\left(x - \frac{\pi}{3}\right)^2 + \ldots$$

It may be shown that this series converges for all x. It converges rapidly for values of x near $\pi/3$.

Warning ***The coefficient of the term containing $(x - a)^n$ is a constant whose value is the nth derivative evaluated at a and is divided by n!.***

PROBLEM 1. Develop the Taylor series for $f(x) = \sin x$ for $a = \pi/3$. Write at least the first three nonvanishing terms. For what values of x does it converge rapidly?

EXAMPLE 2. Expand $\ln x$ in a Taylor's series with $a = 2$.

Solution

$$f(x) = \ln x \qquad f(2) = \ln 2$$

$$f'(x) = x^{-1} \qquad f'(2) = 2^{-1} = \tfrac{1}{2}$$

$$f''(x) = -x^{-2} \qquad f''(2) = -(2)^{-2} = -\tfrac{1}{4}$$

$$f'''(x) = 2x^{-3} \qquad f'''(2) = 2(2)^{-3} = \tfrac{1}{4}$$

Thus

$$\ln x = \ln 2 + \frac{1}{2}(x-2) - \frac{1}{4 \cdot 2!}(x-2)^2 + \frac{1}{4 \cdot 3!}(x-2)^3 + \dots$$

or

$$\ln x = \ln 2 + \tfrac{1}{2}(x-2) - \tfrac{1}{8}(x-2)^2 + \tfrac{1}{24}(x-2)^3 + \dots$$

It may be shown that this series converges for all $x > 0$. It converges rapidly for values of x near 2.

PROBLEM 2. Expand $\ln(x+1)$ in a Taylor's series with $a = 1$. For what values of x does it converge rapidly?

EXERCISES 17.4

Group A

1. Develop the Taylor series for each of the following. Write at least the first three nonvanishing terms.

(a) $f(x) = e^x$, $a = 2$.
(b) $f(x) = \sin x$, $a = \pi/6$.
(c) $f(x) = \ln x$, $a = 1$.
(d) $f(x) = \cos x$, $a = \pi/6$.
(e) $f(x) = \sqrt{x}$, $a = 1$.
(f) $f(x) = \sin 2x$, $a = \pi/6$.
(g) $f(x) = \sec x$, $a = \pi/3$.
(h) $f(x) = \ln(\sin x)$, $a = \pi/4$.
(i) $f(x) = \sin x$, $a = 2\pi/3$.
(j) $f(x) = \dfrac{1}{\sqrt{1-x^2}}$, $a = 0$.

Group B

1. Develop the Taylor series for each of the following. Write at least the first three nonvanishing terms.

(a) $f(x) = \text{Tan}^{-1} x$, $a = 1$.
(b) $f(x) = \tan x$, $a = \pi/3$.
(c) $f(x) = \sqrt{1+x^2}$, $a = 1$.
(d) $f(x) = \dfrac{1}{x}$, $a = 2$.
(e) $f(x) = \dfrac{1}{\sqrt{x^2-1}}$, $a = 2$.
(f) $f(x) = \ln(1 + e^x)$, $a = 0$.

Group C

1. The differential equation $x^2y' = 1$ has for its initial condition $y(1) = 1$. Use a Taylor series expansion about $a = 1$ to find the solution, y, of this equation (see Exercise 1 in Group C of Exercises 17.3).

2. The differential equation $x^2y'' = x + 1$ has for its initial conditions $y(1) = 0$ and $y'(1) = 0$. Use a Taylor series expansion about $a = 1$ to find the solution, y, of this equation.

17.5 COMPUTATION BY SERIES

Convergent infinite series may be used for computing the value of a function for a given value of x. We generally use Maclaurin series when x is near zero. If $x > 1$, we should use a Taylor series where a is chosen such that $(x - a)$ is less than 1. Then the series will converge rapidly. Some useful Maclaurin series are

$$\sin x = x - \frac{x^3}{3!} + \frac{x^3}{5!} - \frac{x^7}{7!} + \dots + \frac{(-1)^{n+1}x^{2n-1}}{(2n-1)!} + \dots \quad (17.5.1)$$

$$\cos x = 1 - \frac{x^2}{2!} + \frac{x^4}{4!} - \frac{x^6}{6!} + \dots + \frac{(-1)^{n+1}x^{2n-2}}{(2n-2)!} + \dots \quad (17.5.2)$$

$$e^x = 1 + x + \frac{x^2}{2!} + \frac{x^3}{3!} + \dots + \frac{x^{n-1}}{(n-1)!} + \dots \quad (17.5.3)$$

EXAMPLE 1. Compute sin 12° correct to four decimal places.

Solution. We must convert 12° to radians, since calculus is based on real numbers and the values of the same trigonometric functions of real numbers and radians are equal. $12° = 12\pi/180 = 0.20944$ rad. Since this number is less than unity, we shall use Maclaurin series. We shall carry five decimal places to ensure accuracy in the fourth place. Using the Maclaurin series for $\sin x$, Eq. 17.5.1, we have

$$\begin{aligned}\sin 0.20944 &= 0.20944 - \frac{(0.20944)^3}{6} + \frac{(0.20944)^5}{120} + \dots \\ &= 0.20944 - 0.00153 + 0.000003 + \dots \\ &= 0.20791\end{aligned}$$

Thus to four decimal places $\sin 12° = 0.2079$. (Note that only the first two terms were needed.)

PROBLEM 1. Compute $e^{0.2}$ correct to four decimal places.

EXAMPLE 2. Compute cos 59° to four-place decimal accuracy.

Solution. Let $x = 59° = 59\pi/180 = 1.02974$ rad. Since this is greater than unity, we shall use a Taylor series with a convenient value of a near $59° = 1.02974$ rad. Example 1 of Sec. 17.4 gives the Taylor series for $\cos x$ with $a = \pi/3 = 1.04720$. Thus

$$(x - a) = 1.02974 - 1.04720 = -0.01746$$

$$\begin{aligned}\cos 1.0297 &= 0.50000 - \frac{\sqrt{3}}{2}(-0.01746) - \frac{1}{4}(-0.01746)^2 + \dots \\ &= 0.50000 + 0.01512 - 0.00008 + \dots \\ &= 0.5150\end{aligned}$$

PROBLEM 2. Compute ln 2.2 correct to four decimal places.

Maclaurin series is a polynomial of infinite terms. Frequently, we can approximate solutions to problems by using Maclaurin series.

EXAMPLE 3. Evaluate $\int_0^1 (\sin x/x)\, dx$.

Solution. From Eq. 17.5.1 we have

$$\sin x = x - \frac{x^3}{3!} + \frac{x^5}{5!} + \dots$$

Hence

$$\frac{\sin x}{x} = 1 - \frac{x^2}{3!} + \frac{x^4}{5!} + \dots$$

Thus

$$\begin{aligned}\int_0^1 \frac{\sin x}{x}\, dx &= \int_0^1 \left(1 - \frac{x^2}{3!} + \frac{x^4}{5!} + \dots\right) dx \\ &= \left[x - \frac{x^3}{18} + \frac{x^5}{600} + \dots\right]_0^1 \\ &= 1 - \frac{1}{18} + \frac{1}{600} + \dots \\ &= 1.0000 - 0.0555 + 0.0017 \\ &= 0.9462\end{aligned}$$

PROBLEM 3. Evaluate $\int_{0.1}^1 (\cos x/x)\, dx$.

EXAMPLE 4. Integrate $\int e^{-x^2}\, dx$.

Solution. In Eq. 17.5.3 let x be replaced by $-x^2$; then

$$e^{-x^2} = 1 - x^2 + \frac{x^4}{2!} - \frac{x^6}{3!} + \dots$$

and

$$\begin{aligned}\int e^{-x^2}\, dx &= \int \left(1 - x^2 + \frac{x^4}{2!} - \frac{x^6}{3!} + \dots\right) dx \\ &= x - \frac{x^3}{3} + \frac{x^5}{10} - \frac{x^7}{42} + \dots + C\end{aligned}$$

PROBLEM 4. Integrate $\int e^{\sqrt{x}}\, dx$

In using infinite series for computation there is always some error involved as we do not use all the terms. The subject of errors in general is beyond the scope of this book. For *alternating series* the error in using only n terms is less than the absolute value of the $(n + 1)$ term.

EXAMPLE 5. If we use the first two terms of the alternating series

$$\sin x = x - \frac{x^3}{3!} + \frac{x^5}{5!} - \frac{x^7}{7!} + \dots$$

to compute sin (−0.1), the error is less than

$$\left|\frac{(-0.1)^5}{5!}\right| = \frac{0.00001}{120} = 0.00000008$$

PROBLEM 5. What is the error in computing $e^{-0.2}$ by a Maclaurin series when only the first three terms are used?

In using a Taylor series to compute a function at $x = b$, it can be shown that

$$f(b) = f(a) + \frac{f'(a)}{1!}(b - a) + \frac{f''(a)}{2!}(b - a)^2 + \dots + \frac{f^n}{n!}(b - a)^n + R_n$$

where

$$R_n = \frac{f^{n+1}(\xi)}{(n + 1)!}(b - a)^{n+1}, \qquad a < \xi < b$$

The maximum value of R_n gives an estimate of the error. For a further discussion, see more-advanced books on calculus.

EXERCISES 17.5

Group A

1. Compute sin 10° correct to four decimal places.

2. Compute cos 61° correct to four decimal places.

3. Given that $\text{Tan}^{-1} x = x - (x^3/3) + (x^5/5) - (x^7/7) + \dots$, compute the value of π correct to one decimal place. (*Hint*: Consider $\text{Tan}^{-1} 1$.) How many terms are needed to compute π correct to three decimal places?

4. Compute $e^{-0.1}$ correct to four decimal places.

5. Compute ln 1.2 correct to four decimal places (see Example 2 of Sec. 17.3).

6. Compute ln 1.9 using the first three terms of the series of Example 2 of Sec. 17.4.

7. What is the error if only the first three terms of the Maclaurin series for cos x is used in computing cos 0.5?

8. If $\ln(1 + x) = x - (x^2/2) + (x^3/3) - (x^4/4) + \dots$, how accurate is ln 1.5 computed by using the first five terms of this series?

9. Evaluate each of the following correct to three decimal places.

(a) $\displaystyle\int_0^1 \frac{\cos x}{\sqrt{x}}\,dx.$ (b) $\displaystyle\int_0^1 e^{-x^2}\,dx.$ (c) $\displaystyle\int_0^1 \sin\sqrt{x}\,dx.$

Group B

1. Obtain a series for $f(x) = \frac{1}{2}(e^x + e^{-x})$. (*Hint*: Make use of the Maclaurin series for e^x and the fact that convergent series may be added.)

2. Obtain a series for $f(x) = e^x \cos x$.

3. Obtain a series for $f(x) = \tan x$ by considering series for $\sin x$ and for $\cos x$.

4. Obtain a series for $f(x) = \cos x + \sin x$.

Group C

1. Show by use of series that $e^{ix} = \cos x + i \sin x$. Compare with Eq. 14.5.1.

SUMMARY OF IMPORTANT WORDS AND CONCEPTS

For each of the following, state in your own words your understanding of the statement or word.

1. (a) Series. (b) Infinite. (c) Finite. (d) Convergent. (e) Divergent. (f) General term. (g) Sum. (h) Harmonic. (i) Factorials.

2. (a) Tests for convergence. (b) A necessary condition for convergence. (c) Integral test. (d) Alternating series. (e) Ratio test.

3. (a) Power series. (b) Interval of convergence. (c) Maclaurin series. (d) Taylor series.

4. Computation by use of series.

Chapter 18

Fourier Series

18.1 PERIODIC FUNCTIONS

Many natural phenomena are periodic; that is, they repeat themselves after a certain time interval. Such phenomena may sometimes be represented by periodic functions. Some periodic functions are continuous, for example, the sine and cosine functions, and some are not continuous, for example, a pulsating electronic signal. Periodic functions, with a few exceptions, may be represented by an infinite series called the *Fourier series*. Its terms consist of sines and cosines. Before developing the Fourier series let us discuss periodic functions.

> ***Definition 18.1.1*** *If f is a function such that $f(x+p) = f(x)$, then f is a periodic function of period p.*

EXAMPLE 1. Since $\sin(x + 2\pi) = \sin x$, $\sin x$ is a periodic function of period 2π.

PROBLEM 1. What is the period of the function f if $f(x+3) = f(x)$?

For other than the trigonometric functions the period must be stated. (Most of you know the periods of the trigonometric functions.)

EXAMPLE 2. Plot two periods of the function

$$f(t) = \begin{cases} \sin t & 0 < t \leqq \pi \\ 0 & \pi < t \leqq 2\pi \end{cases} \qquad \text{period } 2\pi$$

Solution. See Fig. 18.1.1.

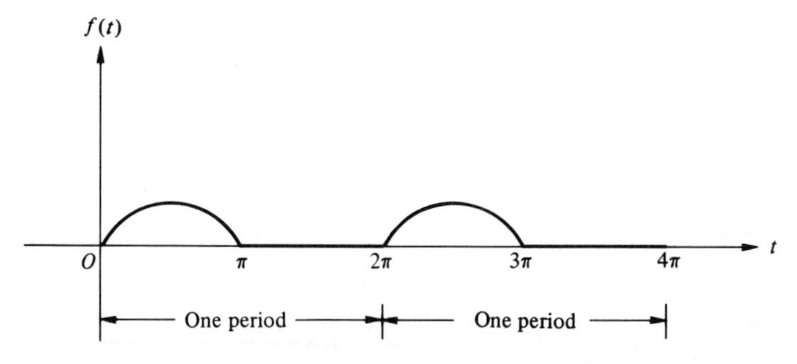

Figure 18.1.1 Graph of
$$f(t) = \begin{cases} \sin t, & 0 < t \leqq \pi, \\ 0, & \pi < t \leqq 2\pi, \end{cases} \quad \text{period } 2\pi.$$

PROBLEM 2. Plot three periods of the given function for the interval $-\pi < t \leqq 2\pi$.

$$f(t) = \begin{cases} \cos t & 0 < t \leqq \dfrac{\pi}{2} \\ 0 & \dfrac{\pi}{2} < t \leqq \pi \end{cases} \qquad \text{period } \pi$$

EXAMPLE 3. Plot several periods of the square-wave function

$$f(t) = \begin{cases} 1 & 0 \leqq t < \dfrac{\pi}{2} \\ 0 & \dfrac{\pi}{2} \leqq t < \pi \end{cases} \qquad \text{period } \pi$$

Find $f(1)$ and $f(8.5)$.

Solution. See Fig. 18.1.2.

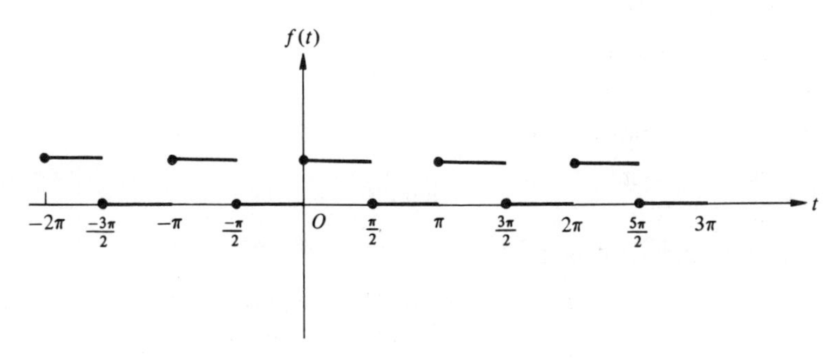

Figure 18.1.2 Square-wave function:
$$f(t) = \begin{cases} 1, & 0 \leqq t < \pi/2, \\ 0, & \pi/2 \leqq t < \pi, \end{cases} \quad \text{period } \pi.$$

To find $f(1)$, we note that $0 < 1 < \pi/2$; hence $f(1) = 1$.

To find $f(8.5)$ we need to find a value of t such that $0 \leqq t < \pi$ since $f(t)$ is defined for $0 \leqq t < \pi$. To do this we make use of Definition 18.1.1 and the fact that the period p is π. Thus

$$f(t) = f(t + \pi)$$

Since $8.5 = \pi + 5.36$ and $5.36 = \pi + 2.22$, we have

$$f(8.5) = f(5.36 + \pi) = f(5.36) = f(2.22 + \pi) = f(2.22)$$

and $f(2.22) = 0$ from the definition of the function. Thus $f(8.5) = 0$. This value may also be verified from the graph (Fig. 18.1.2) since $5\pi/2 < 8.5 < 3\pi$.

PROBLEM 3. Plot several periods of the function

$$f(t) = \begin{cases} t & 0 < t < 2 \\ 0 & 2 < t < 4 \end{cases} \qquad \text{period } 4$$

Find $f(1.5)$ and $f(7)$.

EXERCISES 18.1

Group A

1. Sketch three periods of each of the following periodic functions and find the required value of $f(t)$.

(a) $f(t) = \begin{cases} 1 & 0 < t \leqq 1 \\ 0 & 1 < t < 2 \end{cases}$ period 2.

Find $f(\frac{1}{2})$, $f(3.2)$, and $f(-1)$.

(b) $f(t) = \begin{cases} -1 & -1 < t < 0 \\ 1 & 0 < t < 1 \end{cases}$ period 2.

Find $f(\frac{1}{2})$, $f(\frac{5}{2})$, and $f(2)$.

(c) $f(t) = \begin{cases} \sin t & 0 < t < \pi \\ -1 & \pi < t < 2\pi \end{cases}$ period 2π.

Find $f(\pi/2)$, $f(3\pi/2)$, and $f(5\pi/2)$.

(d) $f(t) = \begin{cases} \sin t & -\pi < t < 0 \\ 0 & 0 < t < \pi \end{cases}$ period 2π.

Find $f(\pi/2)$, $f(3\pi/2)$, and $f(5\pi/2)$.

(e) $f(t) = t$, $0 < t < 2$, period 2. Find $f(1)$, $f(2)$, and $f(3)$.

(f) $f(t) = t^2$, $0 < t < 1$, period 1. Find $f(\frac{1}{2})$, $f(-\frac{3}{2})$, and $f(3)$.

Group B

1. Sketch two periods, one for negative t and one for positive t, of each of the following periodic functions.

(a) $f(t) = t^2, \quad -1 < t < 1$ period 2.

(b) $f(t) = \begin{cases} t & 0 < t < 1 \\ 0 & 1 < t < 2 \end{cases}$ period 2.

(c) $f(t) = e^t, \quad 0 < t < 1$ period 1.

(d) $f(t) = \begin{cases} 0 & 0 < t < \pi \\ \sin t & \pi < t < 2\pi \end{cases}$ period 2π.

(e) $f(t) = \begin{cases} -t^2 & -1 < t < 0 \\ t^2 & 0 < t < 1 \end{cases}$ period 2.

(f) $f(t) = t^3, \quad 0 < t < 1$ period 1.

(g) $f(t) = t^3, \quad -1 < t < 1$ period 2.

18.2 FOURIER SERIES

The infinite series

$$f(t) = a_0 + a_1 \cos\frac{\pi}{p}t + a_2 \cos\frac{2\pi}{p}t + \ldots + a_n \cos\frac{n\pi}{p}t + \ldots$$
$$+ b_1 \sin\frac{\pi}{p}t + b_2 \sin\frac{2\pi}{p}t + \ldots + b_n \sin\frac{n\pi}{p}t + \ldots \tag{18.2.1}$$

where $a_0, \ldots, a_n$ and $b_1, \ldots, b_n$ are constants and n is a positive integer, is called the *Fourier series* for the periodic function $f(t)$ of period $2p$. There are some[1] periodic functions which may not be represented by the series in Eq. 18.2.1, but for most practical applications the Fourier series represents the periodic function $f(t)$. All we need to do is to find the values of the constants a_0, $a_1, a_2, \ldots, a_n$ and $b_1, b_2, \ldots, b_n$. This is done by an ingenious method devised by Fourier. But first a few preliminaries on some integrals. For m and n positive integers it can be shown (see Exercise 12 in Group C of Exercises 9.7) by direct integration that

$$\int_{-p}^{p} \cos\frac{n\pi}{p}t\,dt = 0 \qquad n \neq 0 \tag{18.2.2}$$

$$\int_{-p}^{p} \sin\frac{n\pi}{p}t\,dt = 0 \tag{18.2.3}$$

$$\int_{-p}^{p} \cos\frac{m\pi}{p}t \cos\frac{n\pi}{p}t\,dt = 0 \qquad m \neq n \tag{18.2.4}$$

$$\int_{-p}^{p} \cos^2\frac{n\pi}{p}t\,dt = p \qquad n \neq 0 \tag{18.2.5}$$

$$\int_{-p}^{p} \cos\frac{m\pi}{p}t \sin\frac{n\pi}{p}t\,dt = 0 \tag{18.2.6}$$

$$\int_{-p}^{p} \sin\frac{m\pi}{p}t \sin\frac{n\pi}{p}t\,dt = 0 \qquad m \neq n \tag{18.2.7}$$

[1]If a function is not bounded in one period, it has no Fourier series. A function is said to be bounded if it does not exceed some finite value.

and

$$\int_{-p}^{p} \sin^2 \frac{n\pi}{p} t \, dt = p \qquad n \neq 0 \tag{18.2.8}$$

We now tackle the problem of obtaining the constants. To find a_0, integrate all terms in Eq. 18.2.1 from $-p$ to p, that is, over a complete period. Thus

$$\int_{-p}^{p} f(t)\, dt = a_0 \int_{-p}^{p} dt + a_1 \int_{-p}^{p} \cos \frac{\pi t}{p}\, dt + \ldots + a_n \int_{-p}^{p} \cos \frac{n\pi}{p} t \, dt$$
$$+ \ldots + b_1 \int_{-p}^{p} \sin \frac{\pi t}{p}\, dt + \ldots + b_n \int_{-p}^{p} \sin \frac{n\pi t}{p}\, dt + \ldots$$

Since $f(t)$ is a known function, the term on the left can be integrated. The first term on the right is

$$a_0 \int_{-p}^{p} dt = a_0 [t]_{-p}^{p} = 2a_0 p$$

Using Eq. 18.2.2, all terms on the right containing a cosine will vanish, and all terms containing a sine will vanish because of Eq. 18.2.3. Thus we have

$$\int_{-p}^{p} f(t)\, dt = 2a_0 p$$

or

$$\boxed{a_0 = \frac{1}{2p} \int_{-p}^{p} f(t)\, dt} \tag{18.2.9}$$

To find a_n ($n = 1, 2, 3, \ldots$), multiply all terms of Eq. 18.2.1 by $\cos(n\pi t/p)$ and integrate from $-p$ to p. Thus

$$\int_{-p}^{p} f(t) \cos \frac{n\pi}{p} t \, dt = a_0 \int_{-p}^{p} \cos \frac{n\pi}{p} t \, dt + a_1 \int_{-p}^{p} \cos \frac{\pi t}{p} \cos \frac{n\pi}{p} t \, dt$$
$$+ \ldots + a_n \int_{-p}^{p} \cos^2 \frac{n\pi}{p} t \, dt + \ldots$$
$$+ b_1 \int_{-p}^{p} \sin \frac{\pi}{p} t \cos \frac{n\pi}{p} t \, dt + \ldots$$
$$+ b_n \int_{-p}^{p} \sin \frac{n\pi}{p} t \cos \frac{n\pi}{p} t \, dt + \ldots$$

The integral on the left can be found when $f(t)$ is known. By using Eqs. 18.2.2 and 18.2.4, all terms on the right containing only cosine terms will vanish except the term $a_n \int_{-p}^{p} \cos^2 (n\pi/p)\, t \, dt$. By Eq. 18.2.5 this integral equals p. All terms containing both sine and cosine terms vanish because of Eq. 18.2.6.

Thus all we have left is

$$\int_{-p}^{p} f(t) \cos \frac{n\pi}{p} t \, dt = a_n \int_{-p}^{p} \cos^2 \frac{n\pi}{p} t \, dt = a_n p$$

Hence

$$\boxed{a_n = \frac{1}{p} \int_{-p}^{p} f(t) \cos \frac{n\pi}{p} t \, dt} \tag{18.2.10}$$

To find b_n ($n = 1, 2, 3, \ldots$), multiply all terms of Eq. 18.2.1 by $\sin (n\pi t/p)$ and integrate from $-p$ to p. Thus

$$\begin{aligned}\int_{-p}^{p} f(t) \sin \frac{n\pi}{p} t \, dt &= a_0 \int_{-p}^{p} \sin \frac{n\pi}{p} t \, dt + a_1 \int_{-p}^{p} \cos \frac{\pi}{p} t \sin \frac{n\pi}{p} t \, dt \\ &\quad + \ldots + a_n \int_{-p}^{p} \cos \frac{n\pi}{p} t \sin \frac{n\pi}{p} t \, dt + \ldots \\ &\quad + b_1 \int_{-p}^{p} \sin \frac{\pi}{p} t \sin \frac{n\pi}{p} t \, dt + \ldots \\ &\quad + b_n \int_{-p}^{p} \sin^2 \frac{n\pi}{p} t \, dt + \ldots\end{aligned}$$

The integral on the left can be determined when $f(t)$ is known. By using Eq. 18.2.3, and Eqs. 18.2.6 through 18.2.8, all terms on the right vanish except $\int_{-p}^{p} \sin^2 (n\pi/p) \, t \, dt$, which equals p. Thus

$$\int_{-p}^{p} f(t) \sin \frac{n\pi}{p} t \, dt = b_n \int_{-p}^{p} \sin^2 \frac{n\pi}{p} t \, dt = b_n p$$

Hence

$$\boxed{b_n = \frac{1}{p} \int_{-p}^{p} f(t) \sin \frac{n\pi t}{p} \, dt} \tag{18.2.11}$$

Thus we can find a_0, a_n, and b_n and can expand a periodic function into a Fourier series. We list these results in a form of a summary.

SUMMARY

If $f(t)$ is a periodic function with period $2p$, then $f(t)$ may be expanded into a Fourier series:

$$\begin{aligned}f(t) &= a_0 + a_1 \cos \frac{\pi}{p} t + a_2 \cos \frac{2\pi}{p} t + \ldots + a_n \cos \frac{n\pi}{p} t + \ldots \\ &\quad + b_1 \sin \frac{\pi}{p} t + b_2 \sin \frac{2\pi}{p} t + \ldots + b_n \sin \frac{n\pi}{p} t + \ldots\end{aligned} \tag{18.2.1}$$

where

$$a_0 = \frac{1}{2p}\int_{-p}^{p} f(t)\,dt \tag{18.2.9}$$

$$a_n = \frac{1}{p}\int_{-p}^{p} f(t)\cos\frac{n\pi}{p}t\,dt \tag{18.2.10}$$

and

$$b_n = \frac{1}{p}\int_{-p}^{p} f(t)\sin\frac{n\pi}{p}t\,dt \tag{18.2.11}$$

It may be shown that the limits on the integral must cover a complete period and not necessarily be $-p$ and p. We use this idea in Exercise 2(d) of Group A.

EXAMPLE 1. Write the Fourier series for

$$f(t) = \begin{cases} 2 & -1 < t < 0 \\ -2 & 0 < t < 1 \end{cases} \qquad \text{period } 2$$

Solution. Here $2p = 2$, and so $p = 1$. Sketch the curve at least from $-p$ to p (see Fig. 18.2.1). Since $f(t)$ has two different rules of correspondence between $t = -p$ and $t = p$, we shall need two integrals for each coefficient. From Eq. 18.2.9

$$a_0 = \frac{1}{2(1)}\int_{-1}^{0} 2dt + \frac{1}{2(1)}\int_{0}^{1} (-2)\,dt$$

$$= [t]_{-1}^{0} - [t]_{0}^{1}$$

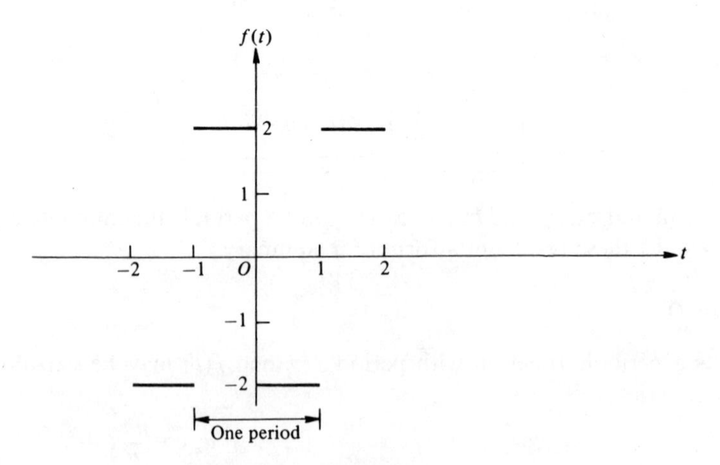

Figure 18.2.1 Graph of $f(t) = \begin{cases} 2, & -1 < t < 0, \\ -2, & 0 < t < 1, \end{cases}$ period 2.

and

$$a_0 = 0$$

From Eq. 18.2.10,

$$a_n = \frac{1}{1}\int_{-1}^{0} 2\cos n\pi t\, dt + \frac{1}{1}\int_{0}^{1} -2\cos n\pi t\, dt$$

$$= \frac{2}{n\pi}[\sin n\pi t]_{-1}^{0} - \frac{2}{n\pi}[\sin n\pi t]_{0}^{1}$$

and

$$a_n = \frac{2}{n\pi}[0 - \sin(-n\pi)] - \frac{2}{n\pi}(\sin n\pi - 0)$$

Since $\sin(-n\pi) = -\sin n\pi$ and $\sin n\pi = 0$, we have $a_0 = 0$.

From Eq. 18.2.11,

$$b_n = \frac{1}{1}\int_{-1}^{0} 2\sin n\pi t\, dt + \frac{1}{1}\int_{0}^{1} (-2)\sin n\pi t\, dt$$

$$= \frac{2}{n\pi}[-\cos n\pi t]_{-1}^{0} - \frac{2}{n\pi}[-\cos n\pi t]_{0}^{1}$$

$$= \frac{2}{n\pi}[-1 + \cos(-n\pi)] - \frac{2}{n\pi}(-\cos n\pi + 1)$$

$$= \frac{2}{n\pi}(-1 + \cos n\pi + \cos n\pi - 1)$$

$$= \frac{4}{n\pi}(\cos n\pi - 1)$$

For n odd, $\cos n\pi = -1$. For n even, $\cos n\pi = 1$. Thus

$$b_n = \frac{-8}{n\pi} \qquad n \text{ odd}$$

and

$$b_n = 0 \qquad n \text{ even}$$

from which we find $b_1 = -8/\pi$, $b_2 = 0$, $b_3 = -(8/3\pi)$, and so forth. Substituting in Eq. 18.2.1, we have

$$f(t) = \frac{-8}{\pi}\sin \pi t - \frac{8}{3\pi}\sin 3\pi t - \frac{8}{5\pi}\sin 5\pi t + \dots \tag{18.2.12}$$

PROBLEM 1. Write the first four nonzero terms of the Fourier series for

$$f(t) = \begin{cases} -1 & -2 < t < 0 \\ 1 & 0 < t < 2 \end{cases} \qquad \text{period } 4$$

EXAMPLE 2. In electricity a certain half-wave rectifier suppresses the negative half of the sinusoidal voltage and allows only the positive half to pass through.

Develop the Fourier series for the voltage

$$f(t) = \begin{cases} 0 & -\pi < t < 0 \\ \sin t & 0 < t < \pi \end{cases} \qquad \text{period } 2\pi$$

Solution. The graph of $f(t)$ is shown in Fig. 18.2.2. Since the period is 2π, we have $2p = 2\pi$ or $p = \pi$.

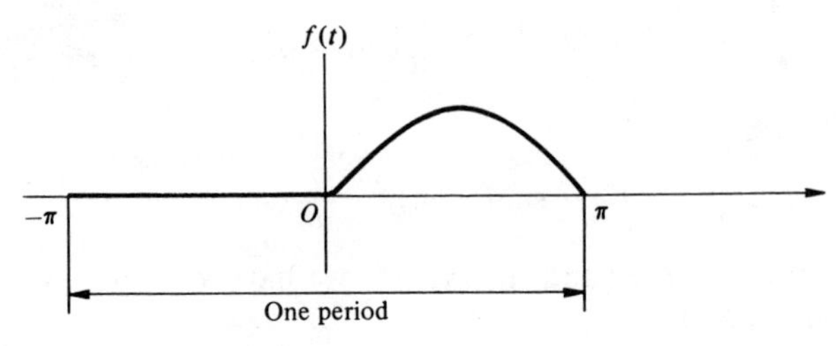

Figure 18.2.2 Graph of
$f(t) = \begin{cases} 0, & -\pi < t < 0, \\ \sin t, & 0 < t < \pi, \end{cases}$ period 2π.

From Eq. 18.2.9,

$$a_0 = \frac{1}{2\pi}\int_{-\pi}^{0} 0\,dt + \frac{1}{2\pi}\int_0^{\pi} \sin t\,dt$$

$$= \frac{1}{2\pi}[-\cos t]_0^{\pi} = \frac{1}{2\pi}(1+1) = \frac{1}{\pi}$$

From Eq. 18.2.10,

$$a_n = \frac{1}{\pi}\int_{-\pi}^{0} 0\,dt + \frac{1}{\pi}\int_0^{\pi} \sin t \cos nt\,dt$$

$$= \frac{1}{\pi}\left[-\frac{1}{2}\left(\frac{\cos(1-n)t}{1-n} + \frac{\cos(1+n)t}{1+n}\right)\right]_0^{\pi}$$

$$= -\frac{1}{2\pi}\left[\frac{\cos(1-n)\pi}{1-n} + \frac{\cos(1+n)\pi}{1+n} - \frac{1}{1-n} - \frac{1}{1+n}\right]$$

$$= -\frac{1}{2\pi}\left[\frac{\cos(\pi - n\pi)}{1-n} + \frac{\cos(\pi + n\pi)}{1+n} - \frac{2}{1-n^2}\right]$$

$$= -\frac{1}{2\pi}\left(\frac{-\cos n\pi}{1-n} + \frac{-\cos n\pi}{1+n} - \frac{2}{1-n^2}\right)$$

$$= \frac{1+\cos n\pi}{\pi(1-n^2)} \qquad n \neq 1$$

For n odd, we have $a_n = 0$; for n even, $a_n = 2/\pi(1-n^2)$. This formula is not valid for $n = 1$. (Why?) To find a_1, we resort to Eq. 18.2.10 and let $n = 1$. Thus

$$a_1 = \frac{1}{\pi}\int_{-\pi}^{0} 0\,dt + \frac{1}{\pi}\int_0^{\pi} \sin t \cos t\,dt$$

$$= \frac{1}{2\pi}[\sin^2 t]_0^{\pi} = 0$$

From Eq. 18.2.11,

$$b_n = \frac{1}{\pi}\int_{-\pi}^{0} 0\,dt + \frac{1}{\pi}\int_{0}^{\pi} \sin t \sin nt\,dt$$
$$= \frac{1}{\pi}\left[\frac{1}{2}\left(\frac{\sin(1-n)t}{1-n} - \frac{\sin(1+n)t}{1+n}\right)\right]_0^{\pi}$$

and $b_n = 0$ for all n except $n = 1$. (We cannot divide by zero.) To find b_1 use Eq. 18.2.11 with $n = 1$. Thus

$$b_1 = \frac{1}{\pi}\int_{-\pi}^{0} 0\,dt + \frac{1}{\pi}\int_{0}^{\pi} \sin^2 t\,dt$$
$$= \frac{1}{\pi}\left[\frac{t}{2} - \frac{\sin 2t}{4}\right]_0^{\pi}$$
$$= \frac{1}{2}$$

Substituting in Eq. 18.2.1, we have

$$f(t) = \frac{1}{\pi} + \frac{1}{2}\sin t - \frac{2}{\pi}\left(\frac{1}{3}\cos 2t + \frac{1}{15}\cos 4t + \frac{1}{35}\cos 6t + \dots\right) \qquad (18.2.13)$$

If we draw the graph of the function consisting of the first three terms of the Fourier series, Eq. 18.2.13, we may see that $f(t)$ of this example 2 is closely approximated by its Fourier series (see Fig. 18.2.3).

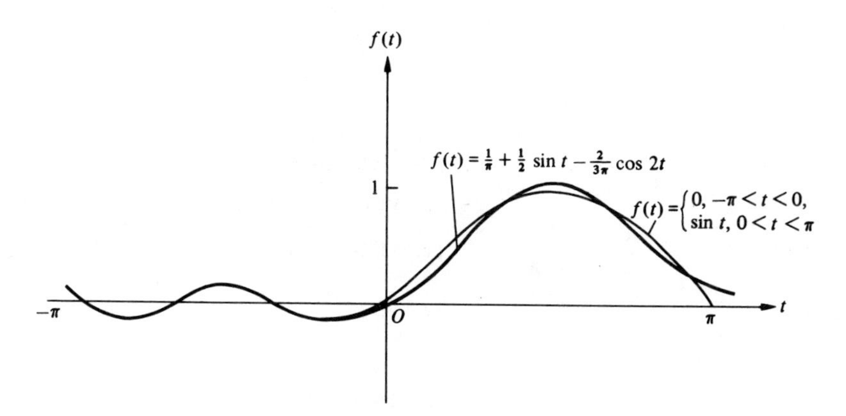

Figure 18.2.3 Comparison of Fourier series with $f(t)$.

PROBLEM 2. Develop the Fourier series for the voltage function

$$f(t) = \begin{cases} 0 & -\pi/2 < t < 0 \\ 12\sin 2t & 0 < t < \pi/2 \end{cases} \qquad \text{period } \pi$$

EXERCISES 18.2

Group A

1. Given the general expression for the coefficients and the period, write several terms of the Fourier series for each of the following.

(a) $a_0 = \frac{3}{2}$, $a_n = \frac{1}{n}$, $b_n = \frac{1}{n^2}$; $2p = 6$.

(b) $a_0 = 0$, $a_n = 0$, $b_n = \frac{(-1)^n}{n^2}$; period is 2.

(c) $a_0 = \frac{1}{2}$, $a_n = \frac{2}{n\pi}(\cos n\pi - 1)$, $b_n = 0$; period is 5.

(d) $a_0 = 0$, $a_n = 0$, $b_n = \sin \frac{n\pi}{2}$; $2p = \pi$.

2. Develop the Fourier series for each of the following.

(a) $f(t) = \begin{cases} 1 & 0 < t < 1 \\ 0 & 1 < t < 2 \end{cases}$ period 2.

(b) $f(t) = \begin{cases} 1 & 0 < t < 1 \\ -1 & 1 < t < 2 \end{cases}$ period 2.

(c) $f(t) = \begin{cases} 2 & 0 < t < \pi \\ 0 & \pi < t < 2\pi \end{cases}$ period 2π.

(d) $f(t) = t$, $0 < t < 1$ period 1.

[*Hint*: In part (d) consider 0 and 1 as limits.]

(e) $f(t) = \begin{cases} t & 0 < t < 1 \\ 0 & 1 < t < 2 \end{cases}$ period 2.

Group B

1. Develop the Fourier series for each of the following.

(a) $f(t) = e^t$, $0 < t < 1$ period 1.

(b) $f(t) = \begin{cases} 0 & -\pi < t < 0 \\ t^2 & 0 < t < \pi \end{cases}$ period 2π.

(c) $f(t) = \begin{cases} 0 & -\pi < t < 0 \\ t & 0 < t < \pi \end{cases}$ period 2π.

(d) $f(t) = \begin{cases} 0 & -2 < t < -1 \\ 1 & -1 < t < 1 \\ 0 & 1 < t < 2 \end{cases}$ period 4.

(e) $f(t) = \sin\left(\frac{t}{2}\right)$, $-\pi < t < \pi$ period 2π.

(f) $f(t) = t - t^2$, $0 < t < 1$ period 1.

(g) $f(t) = t - t^2$, $-1 < t < 1$ period 2.

18.3 FURTHER COMMENTS ON FOURIER SERIES

We might ask ourselves how a discontinuous function such as $f(t)$ in Example 1 of Sec. 18.2 can be represented by a series whose terms are continuous functions. That is, does the series converge, and if so, what does it converge to at a point of discontinuity? The German mathematician Peter Gustave Lejeune Dirichlet answered this question in his theorem called *Dirichlet conditions*. We now state the theorem for information only.

> ***Theorem 18.3.1*** *If $f(t)$ is a bounded periodic function which in any one period has at most a finite number of local maxima and minima and a finite number of points of discontinuity, then the Fourier series of $f(t)$ converges to $f(t)$ at all points, where $f(t)$ is continuous and converges to the average of the right- and left-hand limits of $f(t)$ at each point where $f(t)$ is discontinuous.*

Theorem 18.3.1 tells us that the Fourier series, Eq. 18.2.12 for Example 1, converges to -2 for $0 < t < 1$, to 2 for $1 < t < 2$, and to zero for $t = 1$.

We may obtain some interesting results by making use of Dirichlet conditions.

EXAMPLE 1. The Fourier series for

$$f(t) = \begin{cases} 0 & -\pi < t < 0 \\ t^2 & 0 < t < \pi \end{cases} \qquad \text{period } 2\pi$$

is

$$f(t) = \frac{\pi^2}{6} - 2\left(\cos t - \frac{1}{4}\cos 2t + \frac{1}{9}\cos 3t + \dots\right)$$
$$+ \left(\frac{\pi^2 - 4}{\pi}\sin t - \frac{\pi}{2}\sin 2t + \frac{9\pi^2 - 4}{27\pi}\sin 3t + \dots\right)$$

Show that

$$\frac{\pi^2}{6} = 1 + \frac{1}{2^2} + \frac{1}{3^2} + \frac{1}{4^2} + \dots$$

Solution. If $t = \pi$, all the sine terms will vanish and the cosine terms will be either $+1$ or -1. Using Fig. 18.3.1 and Dirichlet's conditions we note that $f(\pi) = \pi^2/2$. Hence we have

$$f(\pi) = \frac{\pi^2}{2} = \frac{\pi^2}{6} - 2\left(\cos \pi - \frac{\cos 2\pi}{4} + \frac{\cos 3\pi}{9} - \frac{\cos 4\pi}{16} + \dots\right) + 0$$

$$\frac{\pi^2}{2} = \frac{\pi^2}{6} - 2\left(-1 - \frac{1}{2^2} - \frac{1}{3^2} - \frac{1}{4^2} + \dots\right)$$

$$\frac{(\pi^2/2) - (\pi^2/6)}{2} = 1 + \frac{1}{2^2} + \frac{1}{3^2} + \frac{1}{4^2} + \dots$$

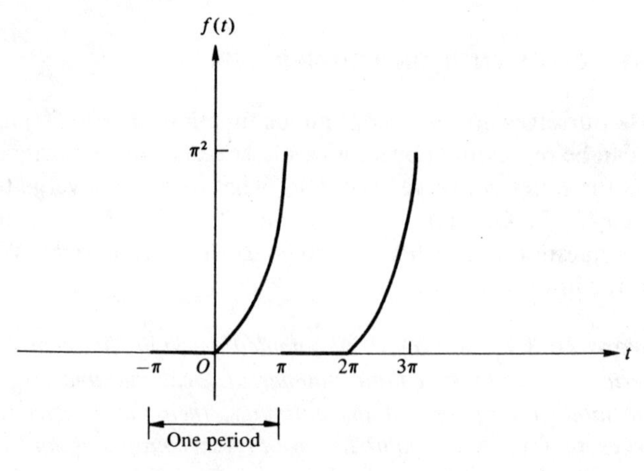

Figure 18.3.1 Graph of
$$f(t) = \begin{cases} 0, & -\pi < t < 0, \\ t^2, & 0 < t < \pi, \end{cases} \quad \text{period } 2\pi.$$

and

$$\frac{\pi^2}{6} = 1 + \frac{1}{2^2} + \frac{1}{3^2} + \frac{1}{4^2} + \ldots$$

PROBLEM 1. Use the Fourier series of Example 1 to show that

$$\frac{\pi^2}{12} = 1 - \frac{1}{2^2} + \frac{1}{3^2} - \frac{1}{4^2} + \ldots$$

(*Hint:* Consider $t = 0$.)

Using Euler's relation, $e^{iu} = \cos u + i \sin u$, Eq. 14.5.5, a complex form of Fourier series may be obtained. It is useful in some electrical problems. The complex form of Fourier series is given by

$$f(t) = \sum_{n=-\infty}^{n=\infty} c_n e^{ni\pi t/p} \tag{18.3.1}$$

where

$$c_n = \frac{1}{2p} \int_{-p}^{p} f(t) e^{-ni\pi t/p}\, dt$$

EXAMPLE 2. Find the complex form of the Fourier series for

$$f(t) = \begin{cases} 1 & 0 < t < 1 \\ 0 & 1 < t < 2 \end{cases} \quad \text{period } 2$$

Solution. The graph of $f(t)$ is shown in Fig. 18.3.2. Here $2p = 2$, and so $p = 1$.

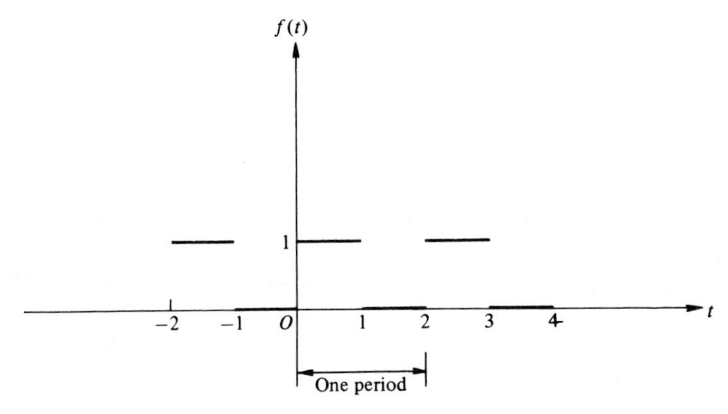

Figure 18.3.2 Graph of
$$f(t) = \begin{cases} 1, & 0 < t < 1 \\ 0, & 1 < t < 2 \end{cases} \quad \text{period 2.}$$

From Eq. 18.3.1,

$$\begin{aligned} c_n &= \frac{1}{2}\int_{-1}^{0} 0\,dt + \frac{1}{2}\int_{0}^{1} 1e^{-nint}\,dt \\ &= -\frac{1}{2ni\pi}[e^{-ni\pi t}]_0^1 \\ &= -\frac{1}{2ni\pi}(e^{-ni\pi} - 1) = \frac{i}{2n\pi}(e^{-ni\pi} - 1) \end{aligned}$$

Since $e^{-ni\pi} = \cos(-n\pi) + i\sin(-n\pi) = \cos n\pi = (-1)^n$, we have

$$c_n = \frac{i}{2n\pi}[(-1)^n - 1] \qquad n \neq 0$$

For n even, $c_n = 0$. For n odd, $c_n = -i/n\pi$, $n \neq 0$. For $n = 0$, we have from Eq. 18.3.1

$$c_0 = \frac{1}{2}\int_{-1}^{0} 0\,dt + \frac{1}{2}\int_{0}^{1} 1\,dt = \frac{1}{2}$$

Thus from Eq. 18.3.1 we may write the complex Fourier series for $f(t)$ as

$$f(t) = \ldots + \underbrace{\frac{i}{3\pi}e^{-3i\pi t}}_{n=-3} + \underbrace{\frac{i}{\pi}e^{-i\pi t}}_{n=-1} + \underbrace{\frac{1}{2}}_{n=0} - \underbrace{\frac{i}{\pi}e^{i\pi t}}_{n=1} - \underbrace{\frac{i}{3\pi}e^{3i\pi t}}_{n=3} + \ldots$$

PROBLEM 2. Find the complex form of the Fourier series for

$$f(t) = \begin{cases} 0 & 0 < t < 1 \\ 12 & 1 < t < 2 \end{cases} \qquad \text{period 2}$$

EXERCISES 18.3

Group A

1. Assume that a Fourier series for each of the following has been developed. Make use of Dirichlet conditions to determine the value to which the Fourier series would converge.

(a) $f(t) = \begin{cases} 0 & 0 < t < 2 \\ 1 & 2 < t < 3 \end{cases}$ period 3.

Find $f(1)$, $f(2)$, and $f(3)$.

(b) $f(t) = t^2$, $0 < t < 4$, period 4. Find $f(0)$, $f(1)$, $f(2)$, $f(4)$, and $f(6)$.

(c) $f(t) = t^2$, $-2 < t < 2$, period 2. Find $f(-2)$, $f(-1)$, $f(0)$, $f(2)$, and $f(3)$.

(d) $f(t) = \begin{cases} \sin t & 0 < t < \pi \\ 0 & \pi < t < 2\pi \end{cases}$ period 2π.

Find $f(0)$, $f(\pi/2)$, $f(\pi)$, and $f(3\pi/2)$.

2. For a certain $f(t)$ the Fourier series is $(8/\pi^2)[\sin(\pi t/20) - (1/3^2)\sin(3\pi t/20) + (1/5^2)\sin(5\pi t/20) + \dots]$. If $f(10) = 1$, show that

$$\frac{\pi^2}{8} = 1 + \frac{1}{3^2} + \frac{1}{5^2} + \frac{1}{7^2} + \dots$$

3. The Fourier series for $f(t) = t$, $0 < t < 2\pi$, period 2π is $f(t) = \pi - 2(\sin t + \frac{1}{2}\sin 2t + \frac{1}{3}\sin 3t + \dots)$. Show that $\pi/4 = 1 - \frac{1}{3} + \frac{1}{5} - \frac{1}{7} + \dots$.

4. Develop the complex form of the Fourier series of each of the following periodic functions whose definition over one period is given. Write out the terms corresponding to $n = -2, -1, 0, 1, 2$.

(a) $f(t) = \begin{cases} 0 & -2 < t < 0 \\ \pi & 0 < t < 2. \end{cases}$

(b) $f(t) = e^{-t}$, $-1 < t < 1$.

(c) $f(t) = \begin{cases} 1 & 0 < t < 1 \\ 0 & 1 < t < 3. \end{cases}$

(d) $f(t) = \begin{cases} -1 & -1 < t < 0 \\ 1 & 0 < t < 1. \end{cases}$

Group B

1. Develop the complex form of the Fourier series for each of the following periodic functions whose definition over one period is given. (*Hint*: The limits on the integral must cover one period and not necessarily be $-p$ and p.)

(a) $f(t) = t$, $0 < t < 1$.

(b) $f(t) = t$, $-1 < t < 1$.

(c) $f(t) = t^2$, $0 < t < 1$.

(d) $f(t) = \sin t$, $0 < t < \pi$. [*Hint*: The sine function may be expressed in terms of e by the relation $\sin u = (1/2i)(e^{iu} - e^{-iu})$.]

Group C

1. In electricity it has been shown that the steady-state current produced by a voltage of the form $E = Ae^{i\omega t}$, where A is a constant, may be found by dividing the voltage by the complex impedance Z, where $Z = R + i[\omega L - (1/\omega C)]$. Find the steady-state current for a circuit where $R = 150\ \Omega$, $L = 0.01$ H, $C = 1 \times 10^{-6}$ F, and

$$E(t) = \begin{cases} 12 & 0 < t < 0.005 \\ 0 & 0.005 < t < 0.010 \end{cases} \qquad \text{period } 0.010$$

[*Hint*: Expand $E(t)$ into a complex Fourier series. Note that ω may then be found to be $200n\pi$, n odd.]

SUMMARY OF IMPORTANT WORDS AND CONCEPTS

For each of the following, state in your own words your understanding of the statement or word.

1. (a) Periodic functions. (b) Period, cycle.

2. (a) Fourier series. (b) Computation of Fourier coefficients (Eqs. 18.2.9 through 18.2.11).

3. Complex form of Fourier series.

4. Dirichlet conditions.

Appendix

A.1 COMPLETING THE SQUARE

The solution of many problems is greatly facilitated by making an algebraic expression a perfect square. This is done by adding and subtracting a certain quantity. Consider the expression

$$ax^2 + bx + c \tag{A.1.1}$$

Group the x^2 and x terms and factor out a, the coefficient of x^2. Thus

$$a\left(x^2 + \frac{b}{a}x \qquad \right) + c \tag{A.1.2}$$

Now take one half of the coefficient of x, square it, and add to the quantity in the parentheses, that is, $b^2/4a^2$. To compensate for this we must subtract $a(b^2/4a^2)$ or $b^2/4a$ from c. Thus we write Eq. A.1.2 as

$$a\left(x^2 + \frac{b}{a}x + \frac{b^2}{4a^2}\right) + c - \frac{b^2}{4a} \tag{A.1.3}$$

(Equation A.1.3 simplifies to Eq. A.1.1.) The first term in Eq. A.1.3 may now be factored. Thus

$$a\left(x + \frac{b}{2a}\right)^2 + c - \frac{b^2}{4a} \tag{A.1.4}$$

We say that Eq. A.1.4 has been obtained from Eq. A.1.1 by *completing the square*.

EXAMPLE 1. Complete the square on $2x^2 - 6x - 3$.

Solution

$$2(x^2 - 3x \qquad) - 3$$

$$2\left(x^2 - 3x + \frac{9}{4}\right) - 3 - \frac{9}{2}$$

$$2\left(x - \frac{3}{2}\right)^2 - \frac{15}{2}$$

You may expand this last expression to show that it reduces to the original expression.

A.2 METHOD OF LEAST SQUARES

In many experiments the plot of the observed data seems to indicate that the data follow some simple mathematical law. The curve indicating their general trend is called an *empirical curve*. There are many methods used in finding the equation of an empirical curve. The method that gives the best results is the method of least squares. To use this method we make use of the *deviation* of a point which is the difference between the ordinate of a given point and the ordinate of the corresponding point on an empirical curve. The principle of least squares is as follows: *The equation of the empirical curve best fitting a set of data points is that equation in which the constants are so determined that they will make the sum of the squares of the deviations a minimum.* Before considering the problem in general let us consider a simple set of data. Suppose that in an experiment the relation between the resistance R in a wire and its temperature T was found to be as indicated in Table A.2.1.

TABLE A.2.1

T	R
1	2
3	3
5	6

Upon plotting the data, Fig. A.2.1, we note that the data seem to have a straight-line trend. Assume that the empirical curve is given by $R = mT + b$, where m and b are constants to be determined. The deviations are indicated by $d_1 = P_1Q_1$, $d_2 = P_2Q_2$, and $d_3 = P_3Q_3$, where the Q's are points on the line $R = mT + b$. Then we have in general $d_i = P_i - (mT_i + b)$ or

$$d_1 = 2 - (m + b)$$

$$d_2 = 3 - (3m + b)$$

and

$$d_3 = 6 - (5m + b)$$

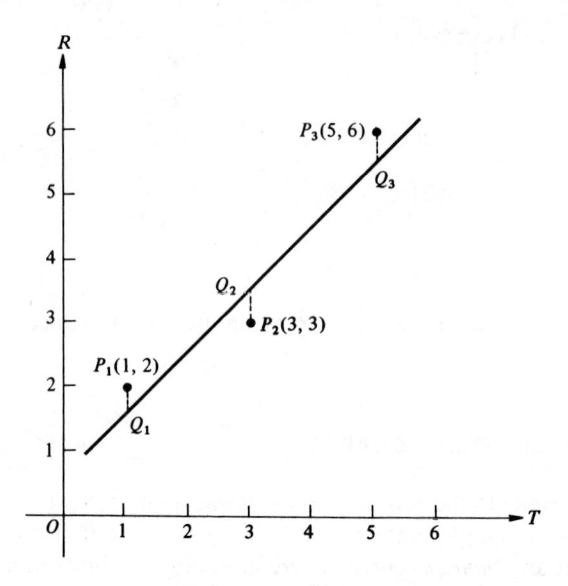

Figure A.2.1 Least squares.

We want the sum of the squares of the deviations to be a minimum. Let S denote this sum; then

$$S = d_1^2 + d_2^2 + d_3^2$$

or

$$S = [2 - (m + b)]^2 + [3 - (3m + b)]^2 + [6 - (5m + b)]^2$$

S is a minimum when $\partial S/\partial m = 0$ and $\partial S/\partial b = 0$. Now

$$\frac{\partial S}{\partial m} = 2[2 - (m + b)](-1) + 2[3 - (3m + b)](-3) + 2[6 - (5m + b)](-5)$$

and

$$\frac{\partial S}{\partial b} = 2[2 - (m + b)](-1) + 2[3 - (3m + b)](-1) + 2[6 - (5m + b)](-1)$$

Thus

$$\frac{\partial S}{\partial m} = 41 - 35m + 9b = 0$$

and

$$\frac{\partial S}{\partial b} = -11 + 9m + 3b = 0$$

These equations are solved simultaneously to find $m = 1$ and $b = \frac{2}{3}$. Thus the best empirical straight line is given by

$$R = T + \frac{2}{3}$$

In general, if

$$y = mx + b \tag{A.2.1}$$

is the required empirical equation for n data points, we may write the deviation of the ith point as

$$d_i = y_i - (mx_i + b)$$

where x_i and y_i are the values of the data. Then

$$S = \sum_{i=1}^{n} d_i^2 = \sum_{i=1}^{n} [y_i - (mx_i + b)]^2$$

$$\frac{\partial S}{\partial m} = \sum_{i=1}^{n} 2[y_i - (mx_i + b)](-x_i) = 0$$

and

$$\frac{\partial S}{\partial b} = \sum_{i=1}^{n} 2[y_i - (mx_i + b)](-1) = 0$$

These lead to the simultaneous equations

$$m \sum x^2 + b \sum x = \sum xy$$

and

$$m \sum x + nb = \sum y$$

where the subscript i has been dropped and the $\sum$ sign indicates the sum of all of the designated quantities. Solving for m and b, we find

$$m = \frac{n \sum xy - \sum x \sum y}{n \sum x^2 - (\sum x)^2}$$

and (A.2.2)

$$b = \frac{\sum x^2 \sum y - \sum x \sum xy}{n \sum x^2 - (\sum x)^2}$$

These values of m and b may be substituted in Eq. A.2.1 to obtain the best empirical equation.

When using Eqs. A.2.2 it is best to exhibit the work in tabular form.

EXAMPLE 1. Find the best straight line that fits the given data.

Solution

	x	y	x^2	xy
	1	2	1	2
	3	3	9	9
	5	6	25	30
$\sum$	9	11	35	41

$$n = 3, \quad \sum x = 9, \quad \sum y = 11, \quad \sum x^2 = 35, \quad \sum xy = 41$$

$$m = \frac{(3)(41) - (9)(11)}{(3)(35) - (9)^2} = 1$$

$$b = \frac{(35)(11) - (9)(41)}{(3)(35) - (9)^2} = \frac{2}{3}$$

Thus

$$y = x + \frac{2}{3}$$

When the plot of the data does not indicate a straight line, the method of least squares may be extended to include other functions, such as parabolas, hyperbolas, and exponentials. This may be accomplished by letting $f(x) = x^2$, $f(x) = 1/x$, or say $f(x) = 10^x$, and writing Eq. A.2.1 as

$$y = m[f(x)] + b \tag{A.2.3}$$

Thus if $f(x) = x^2$, then Eq. A.2.3 becomes $y = m(x^2) + b$, which is a parabola. The values of m and b can be obtained from Eqs. A.2.2 by replacing x by $f(x)$. The column headings for our table would then become

x	y	$f(x)$	$[f(x)]^2$	$[f(x)]y$

and

$$m = \frac{n\sum[f(x)]y - \sum f(x)\sum y}{n\sum[f(x)]^2 - [\sum f(x)]^2}$$

and

$$b = \frac{\sum[f(x)]^2\sum y - \sum f(x)\sum[f(x)]y}{n\sum[f(x)]^2 - [\sum f(x)]^2}$$

EXERCISES A.2

1. Find the equation of the straight line that best fits the following data:

x	1	3	5	7	9
y	1.5	6.2	12.8	18.5	22.0

2. The electrical resistance R of a copper rod at various temperatures T is given. Find the equation of the straight line that best fits the data. Find R at $T = 32.0$:

T	19.0	25.1	30.1	36.0	40.0
R	76.2	77.8	80.0	81.2	82.0

3. Find the equation $y = mx^2 + b$ that best fits the following data:

x	1	3	5	7	9
y	4	10	28	55	81

4. Find the equation $V = m(1/I) + b$ that best fits the following data:

I	2	4	6	8	10
V	13.1	7.9	6.0	5.6	5.0

5. Find the equation $I = me^{-0.5t} + b$ that best fits the following data:

t	0	2.0	4.0	6.0
I	7.0	3.8	2.7	2.2

TABLE I Table of Differentials*

1. $d(a) = 0.$
2. $d(au) = a\,du.$
3. $d(u + v) = du + dv$
4. $d(u \cdot v) = u\,dv + v\,du.$
5. $d\left(\frac{u}{v}\right) = \frac{v\,du - u\,dv}{v^2}.$
6. $d(u^n) = nu^{n-1}\,du.$
7. $d(e^u) = e^u\,du.$
8. $d(a^u) = a^u \ln a\,du.$
9. $d(\ln u) = \frac{du}{u}.$
10. $d(\log_a u) = \left(\frac{du}{u}\right)\left(\frac{1}{\ln a}\right).$
11. $d(\sin u) = \cos u\,du.$
12. $d(\cos u) = -\sin u\,du.$
13. $d(\tan u) = \sec^2 u\,du.$
14. $d(\cot u) = -\csc^2 u\,du.$
15. $d(\sec u) = \sec u \tan u\,du.$
16. $d(\csc u) = -\csc u \cot u\,du.$
17. $d(\mathrm{Sin}^{-1} u) = \frac{du}{\sqrt{1 - u^2}}.$
18. $d(\mathrm{Cos}^{-1} u) = -\frac{du}{\sqrt{1 - u^2}}.$
19. $d(\mathrm{Tan}^{-1} u) = \frac{du}{1 + u^2}.$
20. $d(\sinh u) = \cosh u\,du.$
21. $d(\cosh u) = \sinh u\,du.$
22. $d(\sinh^{-1} u) = \frac{du}{\sqrt{u^2 + 1}}.$
23. $d(\cosh^{-1} u) = \frac{du}{\sqrt{u^2 - 1}}$, $\cosh^{-1} u > 0.$
24. $d\left[\int_a^u f(u)\,du\right] = f(u)\,du.$
25. (Chain Rule) If $y = f(u)$ and $u = g(x)$, then $\frac{dy}{dx} = \frac{dy}{du}\frac{du}{dx}.$

*u and v are functions of an independent variable, such as x, or may themselves be independent variables; letters a and n denote constants.

To obtain a table of derivatives, divide both members of each formula by du or by dx.

TABLE II Table of Integrals

*Standard Integrals**

1. $\int u^n\,du = \frac{u^{n+1}}{n+1} + C, \qquad n \neq -1.$
2. $\int \frac{du}{u} = \ln|u| + C.$
3. $\int e^u\,du = e^u + C.$
4. $\int a^u\,du = \frac{a^u}{\ln a} + C.$

   (For $\int \ln u\,du$ see 124 with $n = 0$.)
5. $\int \sin u\,du = -\cos u + C.$
6. $\int \cos u\,du = \sin u + C.$
7. $\int \tan u\,du = -\ln|\cos|u + C.$
8. $\int \cot u\,du = \ln|\sin u| + C.$
9. $\int \sec^2 u\,du = \tan u + C.$
10. $\int \csc^2 u\,du = -\cot u + C.$
11. $\int \sec u \tan u\,du = \sec u + C.$
12. $\int \csc u \cot u\,du = -\csc u + C.$
13. $\int \sec u\,du = \ln|\sec u + \tan u| + C.$
14. $\int \csc u\,du = \ln|\csc u - \cot u| + C.$
15. $\int \frac{du}{\sqrt{a^2 - u^2}} = \text{Sin}^{-1}\frac{u}{a} + C, \qquad a > 0.$
16. $\int \frac{du}{a^2 + u^2} = \frac{1}{a}\text{Tan}^{-1}\frac{u}{a} + C, \qquad a > 0.$
17. $\int u\,dv = uv - \int v\,du.$

Forms Involving a + bu

18. $\int (a + bu)^n\,du.$ (Use 1 if $n \neq -1$.)
19. $\int \frac{du}{a + bu}.$ (Use 2.)
20. $\int \frac{du}{u(a + bu)} = \frac{1}{a}\ln\left|\frac{u}{a + bu}\right| + C.$
21. $\int \frac{du}{u^2(a + bu)} = -\frac{1}{au} + \frac{b}{a^2}\ln\left|\frac{a + bu}{u}\right| + C.$

*Letters a and b, denote constants, m and n denote positive integers. u and v denote functions of an independent variable such as x.

TABLE II Continued

22. $\int \frac{du}{u(a+bu)^2} = \frac{1}{a(a+bu)} - \frac{1}{a^2}\ln\left|\frac{a+bu}{u}\right| + C.$

23. $\int \frac{u\,du}{a+bu} = \frac{u}{b} - \frac{a}{b^2}\ln|a+bu| + C.$

24. $\int \frac{u\,du}{(a+bu)^2} = \frac{a}{b^2(a+bu)} + \frac{1}{b^2}\ln|a+bu| + C.$

25. $\int u\sqrt{a+bu}\,du = \frac{2(3bu-2a)(a+bu)^{3/2}}{15b^2} + C.$

26. $\int \frac{u\,du}{\sqrt{a+bu}} = \frac{2(bu-2a)\sqrt{a+bu}}{3b^2} + C.$

27. $\int u^2\sqrt{a+bu}\,du = \frac{2(15b^2u^2-12abu+8a^2)(a+bu)^{3/2}}{105b^3} + C.$

28. $\int \frac{u^2\,du}{\sqrt{a+bu}} = \frac{2(3b^2u^2-4abu+8a^2)\sqrt{a+bu}}{15b^3} + C.$

29. $\int \frac{du}{u\sqrt{a+bu}} = \frac{1}{\sqrt{a}}\ln\left|\frac{\sqrt{a+bu}-\sqrt{a}}{\sqrt{a+bu}+\sqrt{a}}\right| + C \quad \text{if } a>0.$

30. $\int \frac{du}{u\sqrt{a+bu}} = \frac{2}{\sqrt{-a}}\operatorname{Tan}^{-1}\sqrt{\frac{a+bu}{-a}} + C \quad \text{if } a<0.$

31. $\int \frac{\sqrt{a+bu}}{u}\,du = 2\sqrt{a+bu} + a\int\frac{du}{u\sqrt{a+bu}}.$

32. $\int \frac{du}{u^2\sqrt{a+bu}} = -\frac{\sqrt{a+bu}}{au} - \frac{b}{2a}\int\frac{du}{u\sqrt{a+bu}}.$

Forms Involving $u^2 \pm a^2$

$\int \frac{du}{u^2+a^2}.$ (See 16.)

33. $\int \frac{du}{u^2-a^2} = \frac{1}{2a}\ln\left|\frac{u-a}{u+a}\right| + C.$

33(a). $\int \frac{du}{a^2-u^2} = -\int\frac{du}{u^2-a^2}.$

34. $\int \frac{u\,du}{u^2\pm a^2}.$ (Use 2.)

35. $\int \frac{u^2\,du}{u^2-a^2} = u + \frac{a}{2}\ln\left|\frac{u-a}{u+a}\right| + C.$

36. $\int \frac{u^2\,du}{u^2+a^2} = u - a\operatorname{Tan}^{-1}\frac{u}{a} + C.$

37. $\int \frac{du}{u(u^2\pm a^2)} = \pm\frac{1}{2a^2}\ln\left|\frac{u^2}{u^2\pm a^2}\right| + C.$

Forms Involving $\sqrt{u^2 \pm a^2}$, $a > 0$

38. $\int \sqrt{u^2\pm a^2}\,du = \frac{u}{2}\sqrt{u^2\pm a^2} \pm \frac{a^2}{2}\ln|u+\sqrt{u^2\pm a^2}| + C.$

39. $\int u\sqrt{u^2\pm a^2}\,du = \frac{1}{3}(u^2\pm a^2)^{3/2} + C.$

40. $\int u^2\sqrt{u^2\pm a^2}\,du = \frac{u}{8}(2u^2\pm a^2)\sqrt{u^2\pm a^2} - \frac{a^4}{8}\ln|u+\sqrt{u^2\pm a^2}| + C.$

41. $\int \frac{\sqrt{u^2-a^2}}{u}\,du = \sqrt{u^2-a^2} - a\operatorname{Cos}^{-1}\frac{a}{|u|} + C.$

TABLE II Continued

42. $\int \frac{\sqrt{u^2 + a^2}}{u}\,du = \sqrt{u^2 + a^2} - a \ln \frac{a + \sqrt{u^2 + a^2}}{|u|} + C.$

43. $\int \frac{\sqrt{u^2 \pm a^2}}{u^2}\,du = -\frac{\sqrt{u^2 \pm a^2}}{u} + \ln |u + \sqrt{u^2 \pm a^2}| + C.$

44. $\int \frac{du}{\sqrt{a^2 + u^2}} = \ln (u + \sqrt{a^2 + u^2}) + C.$

45. $\int \frac{du}{\sqrt{u^2 - a^2}} = \ln |u + \sqrt{u^2 - a^2}| + C.$

46. $\int \frac{du}{\sqrt{u^2 \pm a^2}} = \sqrt{u^2 \pm a^2} + C.$

47. $\int \frac{u^2\,du}{\sqrt{u^2 \pm a^2}} = \frac{u}{2}\sqrt{u^2 \pm a^2} \mp \frac{a^2}{2} \ln |u + \sqrt{u^2 \pm a^2}| + C.$

48. $\int \frac{du}{u\sqrt{u^2 - a^2}} = \frac{1}{a} \operatorname{Cos}^{-1} \frac{a}{|u|} + C.$

49. $\int \frac{du}{u\sqrt{u^2 + a^2}} = \frac{1}{a} \ln \frac{|u|}{a + \sqrt{u^2 + a^2}} + C.$

50. $\int \frac{du}{u^2\sqrt{u^2 \pm a^2}} = \mp \frac{\sqrt{u^2 \pm a^2}}{a^2 u} + C.$

51. $\int \frac{du}{u^3\sqrt{u^2 - a^2}} = \frac{\sqrt{u^2 - a^2}}{2a^2u^2} + \frac{1}{2a^3} \operatorname{Cos}^{-1} \frac{a}{|u|} + C.$

52. $\int \frac{du}{u^3\sqrt{u^2 + a^2}} = -\frac{\sqrt{u^2 + a^2}}{2a^2u^2} + \frac{1}{2a^3} \ln \frac{a + \sqrt{u^2 + a^2}}{|u|} + C.$

53. $\int (u^2 \pm a^2)^{3/2}\,du = \frac{u}{8}(2u^2 \pm 5a^2)\sqrt{u^2 \pm a^2} + \frac{3a^4}{8} \ln |u + \sqrt{u^2 \pm a^2}| + C.$

54. $\int \frac{du}{(u^2 \pm a^2)^{3/2}} = \frac{\pm u}{a^2\sqrt{u^2 \pm a^2}} + C.$

55. $\int \frac{u\,du}{(u^2 \pm a^2)^{3/2}} = \frac{-1}{\sqrt{u^2 \pm a^2}} + C.$

56. $\int \frac{u^2\,du}{(u^2 \pm a^2)^{3/2}} = \frac{-u}{\sqrt{u^2 \pm a^2}} + \ln |u + \sqrt{u^2 \pm a^2}| + C.$

Forms Involving $\sqrt{a^2 - u^2}$, $a > 0$

57. $\int \sqrt{a^2 - u^2}\,du = \frac{u}{2}\sqrt{a^2 - u^2} + \frac{a^2}{2} \operatorname{Sin}^{-1} \frac{u}{a} + C.$

58. $\int u\sqrt{a^2 - u^2}\,du = -\frac{1}{3}(a^2 - u^2)^{3/2} + C.$

59. $\int u^2\sqrt{a^2 - u^2}\,du = \frac{u}{8}(2u^2 - a^2)\sqrt{a^2 - u^2} + \frac{a^4}{8} \operatorname{Sin}^{-1} \frac{u}{a} + C.$

60. $\int \frac{\sqrt{a^2 - u^2}}{u}\,du = \sqrt{a^2 - u^2} - a \ln \frac{a + \sqrt{a^2 - u^2}}{|u|} + C.$

61. $\int \frac{\sqrt{a^2 - u^2}}{u^2}\,du = -\frac{\sqrt{a^2 - u^2}}{u} - \operatorname{Sin}^{-1} \frac{u}{a} + C.$

$\int \frac{du}{\sqrt{a^2 - u^2}}$. (See 15.)

62. $\int \frac{u\,du}{\sqrt{a^2 - u^2}} = -\sqrt{a^2 - u^2} + C.$

63. $\int \frac{u^2\,du}{\sqrt{a^2 - u^2}} = -\frac{u}{2}\sqrt{a^2 - u^2} + \frac{a^2}{2} \operatorname{Sin}^{-1} \frac{u}{a} + C.$

TABLE II Continued

64. $\displaystyle\int \frac{du}{u\sqrt{a^2-u^2}} = \frac{1}{a}\ln\frac{a-\sqrt{a^2-u^2}}{|u|} + C.$

65. $\displaystyle\int \frac{du}{u^2\sqrt{a^2-u^2}} = -\frac{\sqrt{a^2-u^2}}{a^2u} + C.$

66. $\displaystyle\int \frac{du}{u^3\sqrt{a^2-u^2}} = -\frac{\sqrt{a^2-u^2}}{2a^2u^2} + \frac{1}{2a^3}\ln\frac{a-\sqrt{a^2-u^2}}{|u|} + C.$

67. $\displaystyle\int (a^2-u^2)^{3/2}\,du = \frac{u}{8}(5a^2-2u^2)\sqrt{a^2-u^2} + \frac{3a^4}{8}\operatorname{Sin}^{-1}\frac{u}{a} + C.$

68. $\displaystyle\int \frac{du}{(a^2-u^2)^{3/2}} = \frac{u}{a^2\sqrt{a^2-u^2}} + C.$

69. $\displaystyle\int \frac{u\,du}{(a^2-u^2)^{3/2}} = \frac{1}{\sqrt{a^2-u^2}} + C.$

70. $\displaystyle\int \frac{u^2\,du}{(a^2-u^2)^{3/2}} = \frac{u}{\sqrt{a^2-u^2}} - \operatorname{Sin}^{-1}\frac{u}{a} + C.$

Forms Involving $au^2 + bu + c$ ($a \neq 0$)

Let $R = au^2 + bu + c$, $D = b^2 - 4ac$.

71. $\displaystyle\int \frac{du}{R} = \frac{1}{\sqrt{D}}\ln\left|\frac{D-b-2au}{D+b+2au}\right| + C \qquad \text{if } D > 0.$

72. $\displaystyle\int \frac{du}{R} = \frac{2}{\sqrt{-D}}\operatorname{Tan}^{-1}\left(\frac{2au+b}{\sqrt{-D}}\right) + C \qquad \text{if } D < 0.$

73. $\displaystyle\int \frac{u\,du}{R} = \frac{1}{2a}\ln|R| - \frac{b}{2a}\int\frac{du}{R}.$

74. $\displaystyle\int \frac{du}{\sqrt{R}} = \frac{1}{\sqrt{a}}\ln|2au+b+2\sqrt{a}\sqrt{R}| + C \qquad \text{if } a > 0.$

75. $\displaystyle\int \frac{du}{\sqrt{R}} = -\frac{1}{\sqrt{-a}}\operatorname{Sin}^{-1}\left(\frac{2au+b}{\sqrt{D}}\right) + C \qquad \text{if } a < 0 \text{ and } D > 0.$

76. $\displaystyle\int \sqrt{R}\,du = \frac{(2au+b)\sqrt{R}}{4a} - \frac{D}{8a}\int\frac{du}{\sqrt{R}}.$

77. $\displaystyle\int \frac{u\,du}{\sqrt{R}} = \frac{\sqrt{R}}{a} - \frac{b}{2a}\int\frac{du}{\sqrt{R}}.$

78. $\displaystyle\int \frac{du}{u\sqrt{R}} = -\frac{1}{\sqrt{c}}\ln\left|\frac{\sqrt{R}+\sqrt{c}}{u} + \frac{b}{2\sqrt{c}}\right| + C \qquad \text{if } c > 0$

79. $\displaystyle\int \frac{du}{u\sqrt{R}} = \frac{1}{\sqrt{-c}}\operatorname{Sin}^{-1}\left(\frac{bu+2c}{u\sqrt{D}}\right) + C \qquad \text{if } c < 0 \text{ and } D > 0.$

80. $\displaystyle\int \frac{du}{R\sqrt{R}} = \frac{-2(2au+b)}{D\sqrt{R}} + C.$

81. $\displaystyle\int \frac{u\,du}{R\sqrt{R}} = \frac{2(bu+2c)}{D\sqrt{R}} + C.$

82. $\displaystyle\int R^{3/2}\,du = \frac{2au+b}{8a}\left(R - \frac{3D}{8a}\right)R^{1/2} + \frac{3D^2}{128a^2}\int\frac{du}{\sqrt{R}}.$

Forms Involving Trigonometric Functions

83. $\displaystyle\int \sin^2 u\,du = \frac{u}{2} - \frac{\sin 2u}{4} + C.$

84. $\displaystyle\int \cos^2 u\,du = \frac{u}{2} + \frac{\sin 2u}{4} + C.$

TABLE II Continued

85. $\int \sin^3 u\, du = \frac{\cos^3 u}{3} - \cos u + C.$

86. $\int \cos^3 u\, du = \sin u - \frac{\sin^3 u}{3} + C.$

87. $\int \sin^2 au \cos^2 au\, du = \frac{u}{8} - \frac{1}{32a} \sin 4au + C.$

88. $\int \tan^2 u\, du = \tan u - u + C.$

89. $\int \cot^2 u\, du = -\cot u - u + C.$

90. $\int \sec^3 u\, du = \frac{1}{2} \sec u \tan u + \frac{1}{2} \ln|\sec u + \tan u| + C.$

91. $\int \csc^3 u\, du = -\frac{1}{2} \csc u \cot u + \frac{1}{2} \ln|\csc u - \cot u| + C.$

92. $\int u \sin u\, du = \sin u - u \cos u + C.$

93. $\int u \cos u\, du = \cos u + u \sin u + C.$

94. $\int u^2 \sin u\, du = 2u \sin u - (u^2 - 2) \cos u + C.$

95. $\int u^2 \cos u\, du = 2u \cos u + (u^2 - 2) \sin u + C.$

96. $\int \sin^n u\, du = -\frac{\sin^{n-1} u \cos u}{n} + \frac{n-1}{n} \int \sin^{n-2} u\, du.$

97. $\int \cos^n u\, du = \frac{\cos^{n-1} u \sin u}{n} + \frac{n-1}{n} \int \cos^{n-2} u\, du.$

98. $\int \tan^n u\, du = \frac{\tan^{n-1} u}{n-1} - \int \tan^{n-2} u\, du.$

99. $\int \cot^n u\, du = -\frac{\cot^{n-1} u}{n-1} - \int \cot^{n-2} u\, du.$

100. $\int \sec^n u\, du = \frac{\tan u \sec^{n-2} u}{n-1} + \frac{n-2}{n-1} \int \sec^{n-2} u\, \mathrm{du}.$

101. $\int \csc^n u\, du = -\frac{\cot u \csc^{n-2} u}{n-1} + \frac{n-2}{n-1} \int \csc^{n-2} u\, du.$

102.
$$\begin{aligned}
\int \cos^m u \sin^n u\, du &= \frac{\cos^{m-1} u \sin^{n+1} u}{m+n} + \frac{m-1}{m+n} \int \cos^{m-2} u \sin^n u\, du \\
&= -\frac{\sin^{n-1} u \cos^{m+1} u}{m+n} + \frac{n-1}{m+n} \int \cos^m u \sin^{n-2} u\, du \\
&= -\frac{\sin^{n+1} u \cos^{m+1} u}{m+1} + \frac{m+n+2}{m+1} \int \cos^{m+2} u \sin^n u\, du \\
&= \frac{\sin^{n+1} u \cos^{m+1} u}{n+1} + \frac{m+n+2}{n+1} \int \cos^m u \sin^{n+2} u\, du.
\end{aligned}$$

In 103 through 105 it is assumed that $a^2 \neq b^2$.

103. $\int \sin au \sin bu\, du = \frac{\sin (a-b)u}{2(a-b)} - \frac{\sin (a+b)u}{2(a+b)} + C.$

104. $\int \sin au \cos bu\, du = -\frac{\cos (a-b)u}{2(a-b)} - \frac{\cos (a+b)u}{2(a+b)} + C.$

105. $\int \cos au \cos bu\, du = \frac{\sin (a-b)u}{2(a-b)} + \frac{\sin (a+b)u}{2(a+b)} + C.$

106. $\int \frac{du}{a + b \cos u} = \frac{2}{\sqrt{a^2 - b^2}} \operatorname{Tan}^{-1} \frac{\sqrt{a^2 - b^2} \tan \frac{u}{2}}{a + b} + C \qquad a^2 > b^2.$

TABLE II Continued

107. $\int \frac{du}{a + b\cos u} = \frac{1}{\sqrt{b^2 - a^2}} \ln \left| \frac{a + b + \sqrt{b^2 - a^2}\tan\frac{u}{2}}{a + b - \sqrt{b^2 - a^2}\tan\frac{u}{2}} \right| + C \qquad a^2 < b^2.$

108. $\int \frac{du}{a + b\sin u} = \frac{2}{\sqrt{a^2 - b^2}} \operatorname{Tan}^{-1} \left| \frac{a\tan\frac{u}{2} + b}{\sqrt{a^2 - b^2}} \right| + C \qquad a^2 > b^2.$

109. $\int \frac{du}{a + b\sin u} = \frac{1}{\sqrt{b^2 - a^2}} \ln \left| \frac{a\tan\frac{u}{2} + b - \sqrt{b^2 - a^2}}{a\tan\frac{u}{2} + b + \sqrt{b^2 - a^2}} \right| + C \qquad a^2 < b^2.$

110. $\int \frac{du}{a^2\cos^2 u + b^2\sin^2 u} = \frac{1}{ab} \operatorname{Tan}^{-1} \left(\frac{b\tan u}{a} \right) + C.$

Forms Involving Exponential Functions

111. $\int ue^{au}\,du = \frac{e^{au}}{a^2}(au - 1) + C.$

112. $\int u^2 e^{au}\,du = \frac{e^{au}}{a^3}(a^2u^2 - 2au + 2) + C.$

113. $\int e^{au}\sin bu\,du = \frac{e^{au}}{a^2 + b^2}(a\sin bu - b\cos bu) + C.$

114. $\int e^{au}\cos bu\,du = \frac{e^{au}}{a^2 + b^2}(a\cos bu + b\sin bu) + C.$

Forms Involving Inverse Trigonometric Functions (Assuming a > 0)

115. $\int \operatorname{Sin}^{-1}\frac{u}{a}\,du = u\operatorname{Sin}^{-1}\frac{u}{a} + \sqrt{a^2 - u^2} + C.$

116. $\int \operatorname{Cos}^{-1}\frac{u}{a}\,du = u\operatorname{Cos}^{-1}\frac{u}{a} - \sqrt{a^2 - u^2} + C.$

117. $\int \operatorname{Tan}^{-1}\frac{u}{a}\,du = u\operatorname{Tan}^{-1}\frac{u}{a} - \frac{a}{2}\ln(a^2 + u^2) + C.$

118. $\int u\operatorname{Sin}^{-1}\frac{u}{a}\,du = \frac{1}{4}(2u^2 - a^2)\operatorname{Sin}^{-1}\frac{u}{a} + \frac{u}{4}\sqrt{a^2 - u^2} + C.$

119. $\int u\operatorname{Cos}^{-1}\frac{u}{a}\,du = \frac{1}{4}(2u^2 - a^2)\operatorname{Cos}^{-1}\frac{u}{a} - \frac{u}{4}\sqrt{a^2 - u^2} + C.$

120. $\int u\operatorname{Tan}^{-1}\frac{u}{a}\,du = \frac{1}{2}(u^2 + a^2)\operatorname{Tan}^{-1}\frac{u}{a} - \frac{au}{2} + C.$

Miscellaneous Transcendental Forms

121. $\int u^n\sin au\,du = -\frac{1}{a}u^n\cos au + \frac{n}{a}\int u^{n-1}\cos au\,du.$

122. $\int u^n\cos au\,du = \frac{1}{a}u^n\sin au - \frac{n}{a}\int u^{n-1}\sin au\,du.$

123. $\int u^n e^{au}\,du = \frac{u^n e^{au}}{a} - \frac{n}{a}\int u^{n-1}e^{au}\,du.$

124. $\int u^n\ln au\,du = u^{n+1}\left[\frac{\ln au}{n + 1} - \frac{1}{(n + 1)^2}\right] + C.$

TABLE II Continued

125. $\int u^n (\ln au)^m \, du = \frac{u^{n+1}}{n+1}(\ln au)^m - \frac{ma^{n+1}}{n+1}\int u^n (\ln au)^{m-1} \, du.$

126. $\int \sin (\ln u) \, du = \frac{u}{2}[\sin (\ln u) - \cos (\ln u)] + C.$

127. $\int \cos (\ln u) \, du = \frac{u}{2}[\sin (\ln u) + \cos (\ln u)] + C.$

Definite Integrals

128. $\int_0^\pi \sin^2 au \, du = \int_0^\pi \cos^2 au \, du = \frac{\pi}{2}$ (2a an integer).

129. $$\int_0^{\pi/2} \sin^n u \, du = \int_0^{\pi/2} \cos^n u \, du = \begin{cases} \dfrac{2 \cdot 4 \ldots (n-1)}{3 \cdot 5 \ldots n} & n \text{ an odd integer} > 1 \\ \dfrac{1 \cdot 3 \ldots (n-1)}{2 \cdot 4 \ldots n} \dfrac{\pi}{2} & n \text{ an even integer} > 0. \end{cases}$$

130. $$\int_0^{\pi/2} \sin^m u \cos^n u \, du = \begin{cases} \dfrac{2 \cdot 4 \ldots (n-1)}{(m+1)(m+3) \ldots (m+n)} & n \text{ an odd integer} > 1 \\ \dfrac{2 \cdot 4 \ldots (m-1)}{(n+1)(n+3) \ldots (n+m)} & m \text{ an odd integer} > 1 \\ \dfrac{[1 \cdot 3 \ldots (m-1)][1 \cdot 3 \ldots (n-1)]}{2 \cdot 4 \cdot 6 \ldots (m+n)} \dfrac{\pi}{2} & m \text{ and } n \text{ both even integers} > 0. \end{cases}$$

Answers to Problems and Odd-Numbered Exercises

Answers are given for all of the problems following illustrative examples and for odd-numbered exercises at the end of each section. For exercises containing lettered parts the answers are given for parts (a), (c), (e), etc. Answers to some of the exercises are intentionally omitted.

The figures which illustrate and solve certain problems and exercises are grouped on various pages of this section. They may not be drawn to scale but the basic shape is indicated.

Page 3, Problems 1.1

1. -8 sq mi, 2 s.

Page 3, Exercises 1.1

Group A

1. (a) 7. (c) -75. **2.** (a) 20. (c) 20. **3.** 6. **5.** No. As close as we desire. No. 2.

Group B

1. 22 lb/in.^{2} **2.** (a) 79.6 in.^{3} **3.** 20. **5.** There is none. Yes.

Page 5, Problems 1.2

1. $F = \{(-1, 5), (-1, -7), (-1, 0), (0, 5), (0, -7), (0, 0), (2, 5), (2, -7), (2, 0)\}$. **2.** $\{(-2, 12), (-1, 3), (0, 0), (1, 3), (2, 12)\}$. **3.** There is no number whose square is one and also when added to two is zero.

Page 6, Exercises 1.2

Group A

1. (a) $F = \{(-2, 4), (-2, -1), (3, 4), (3, -1), (0, 4), (0, -1)\}$. (c) $F = \{(a, s), (a, t), (a, n), (i, s), (i, t), (i, n)\}$. **2.** (a) $F = \{(-2, 4), (-1, 1), (0, 0), (1, 1), (2, 4)\}$. (c) $F = \{(\frac{1}{3}, 8), (1, 10)\}$. (e) $F = \{(\frac{1}{2}, 6), (\frac{1}{3}, 4), (\frac{1}{4}, 3)\}$. **3.** (a) $F = \{(-2, -2), (-1, -1), (0, 0), (1, 1), (2, 2)\}$. (c) $F = \{(-2, 16), (-1, 4), (0, 0), (1, 4), (2, 16)\}$.

Group B

1. $\text{QII} = \{(x, y) \mid x < 0, y > 0\}$, $\text{QIII} = \{(x, y) \mid x < 0, y < 0\}$, $\text{QIV} = \{(x, y) \mid x > 0, y < 0\}$. **2.** (a) Fig. E.1. (c) Fig. E.2. **3.** Fig. E.3. Yes.

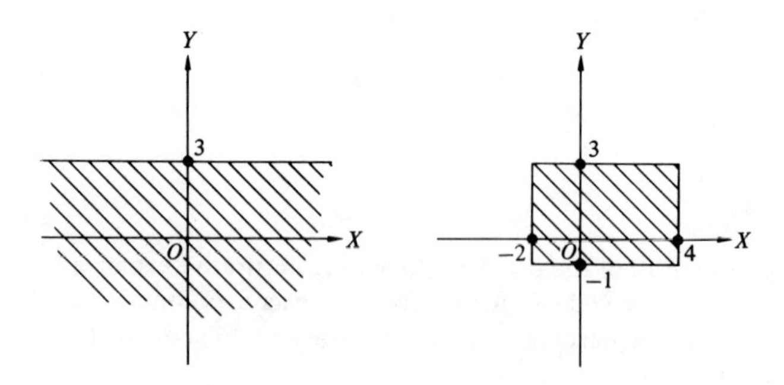

Figure E.1 **Figure E.2**

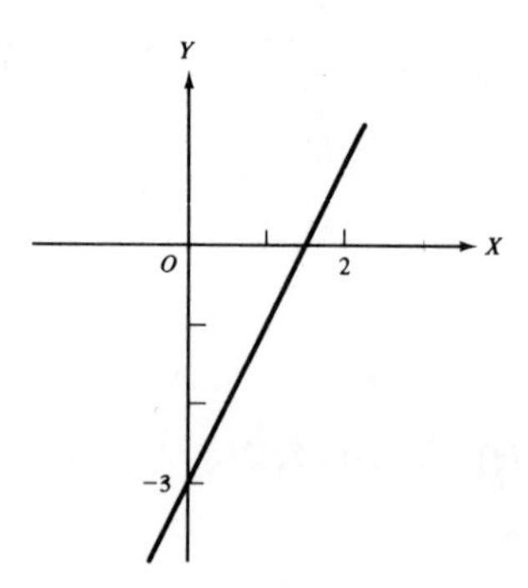

Figure E.3

Page 7, Problems 1.3

1. 9. **2.** (a) For an input of -5 the output of the f function is 2. (b) For an input of 0 the output of the G function is -6. (c) For an input of $x + h$ the output of the g function is $x^2 + 2hx + h^2$. (d) The output of the f function for an input of $(x + h)$.

3. 16; 1; h^2; $x^2 + 2hx + h^2$; $2hx + h^2$. **4.** -7; -4; 5. $\{(0, -7), (1, -4), (-2, 5)\}$. **5.** 7; 7; 7; 7; 7. The constant function. 0. **6.** 1; -7; h; $x + h$. Identity function. h. Both. **7.** 3; 0; 6. **8.** (a) $\pi r^2 h$. (b) $\pi r^2 h$. **9.** $\pm\sqrt{25 - x^2}$. No. Two values of y for one value of x. **10.** (a) Domain: $\{x \mid x \leqq 5\}$. Range: The set of all nonnegative numbers. (b) Domain: The set of all numbers except -5. Range: The set of all numbers except 1.

Page 11, Exercises 1.3

Group A

1. (a) For an input of -3 the output of the f function is 0. (c) For an input of x the output of the f function. **3.** $-9, 9, 3x + 3h, 3h$. **5.** (a) Yes. (c) No. (e) Yes. **6.** (a) x independent, y dependent. (c) s dependent, t independent. (e) y independent, x dependent. (g) x dependent, t independent. **7.** (a) D: all real numbers; R: all nonnegative real numbers. (c) D: $|x| \leqq 2$; R: $0 \leqq y \leqq 2$. (e) D: all x; R: $y \geqq 0$. (g) D: $r \geqq 0$; R: $V \geqq 0$. (i) D: $t \geqq 0$; R: $Q \geqq 0$.

Group B

1. (a) $f(t) = \frac{1}{2}gt^2$. (c) (i) $\frac{1}{2}g(t^2 + 2t\,\Delta t + \Delta t^2)$. (ii) $\frac{1}{2}g(2t\,\Delta t + \Delta t^2)$. **2.** (a) $f(-2) = 19$, $f(2) = 19$. **3.** (a) $f(2) = 4$, $f(-2) = -4$. **5.** (a) Odd. (c) Neither. **6.** (a) 12. (c) 3. **7.** $A = f(x) = x(20 - x)$. **9.** $C = kv^3$, \$84.38. **11.** $S = kwd^2$, 4800. **13.** $R = kl/r^2$, 40 Ω.

Group C

1. D: values of x common to both domains of f and g. R: product of ranges of f and g for the common domain. **3.** Odd (even); nothing.

Page 15, Problems 1.4

1. Fig. P.1. **2.** Fig. P.2. **3.** Fig. P.3. **4.** Fig. P.4. **5.** Fig. P.5. **6.** Fig. P.6.

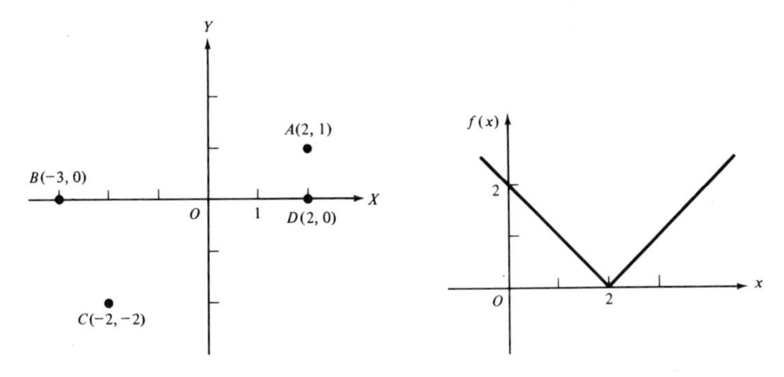

Figure P.1 **Figure P.2**

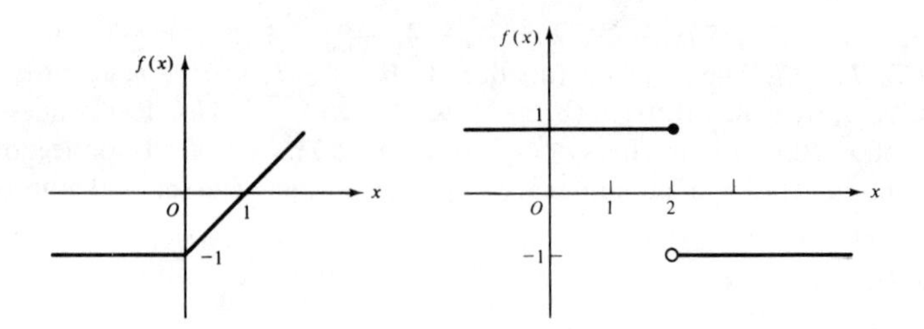

Figure P.3

Figure P.4

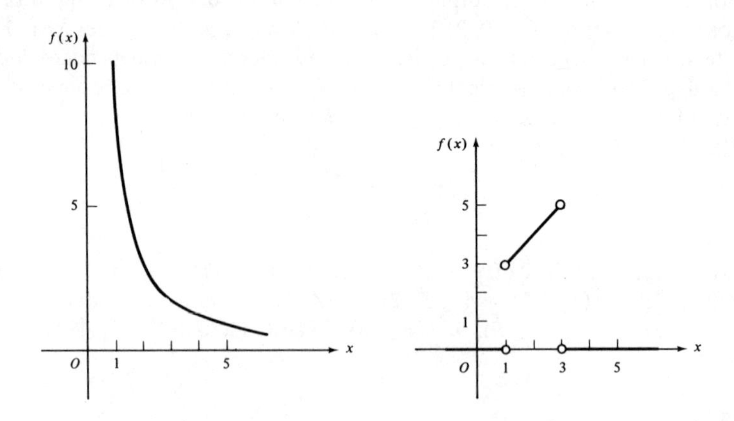

Figure P.5

Figure P.6

Page 21, Exercises 1.4

Group A

1. (a) $AB = 2$, $AC = 6$, $CB = -4$. **3.** Fig. E.4. **5.** (a) $y^{2/3}$. (c) $4 - y^{2/3}$. **7.** Fig. E.5. **9.** Fig. E.6. **11.** (a) Fig. E.7. (c) Fig. E.8. (e) Fig. E.9. (g) Fig. E.10. **12.** (a) $f(x) = u(x-2) - u(x-3)$. (c) $f(x) = 3[u(x-1) - u(x-2)]$. **13.** Fig. E.11.

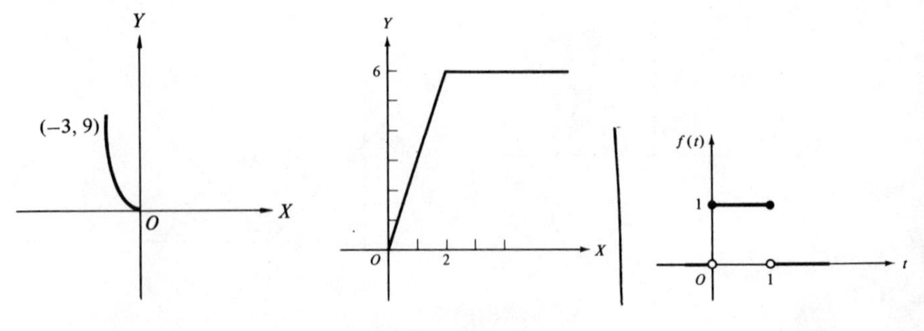

Figure E.4

Figure E.5

Figure E.6

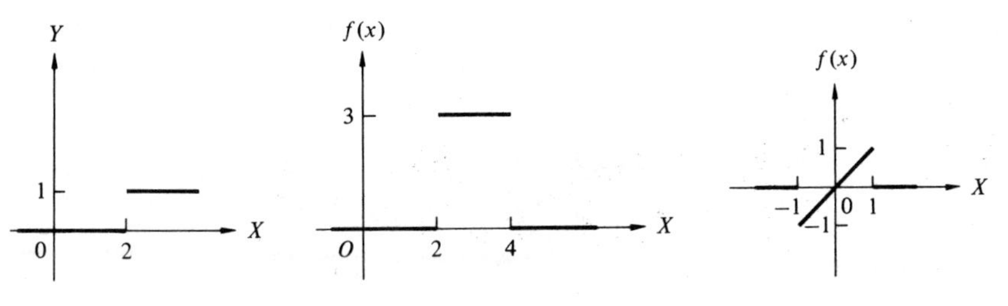

Figure E.7 Figure E.8 Figure E.9

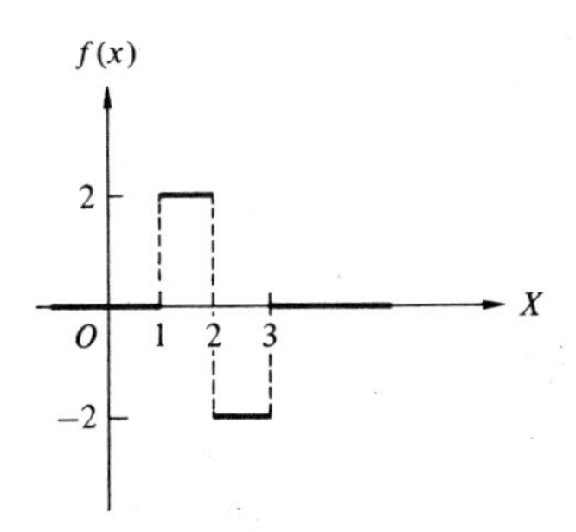

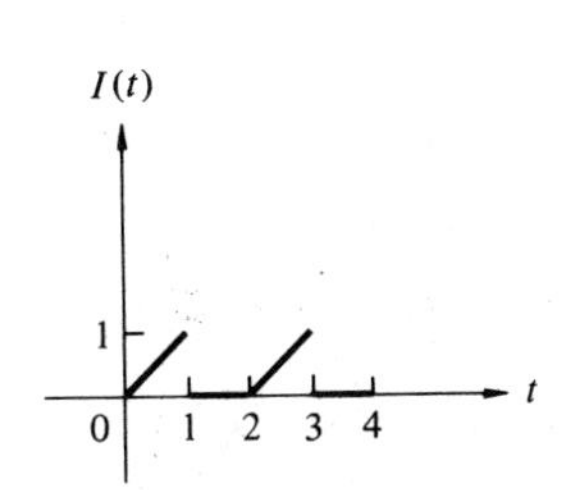

Figure E.10 Figure E.11

Group B

3. $I = f(R) = 12/R$. Fig. E.12. **5.** Fig. E.13. **7.** Fig. E.14. **9.** $f(x) = 2[u(x-1) - u(x-3)] - 2[u(x-3) - u(x-5)]$.

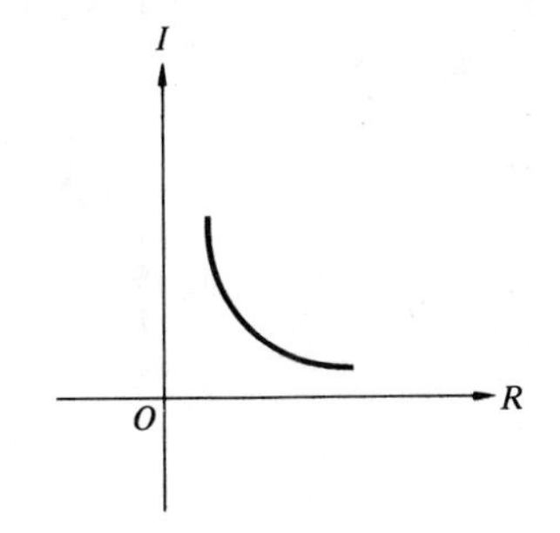

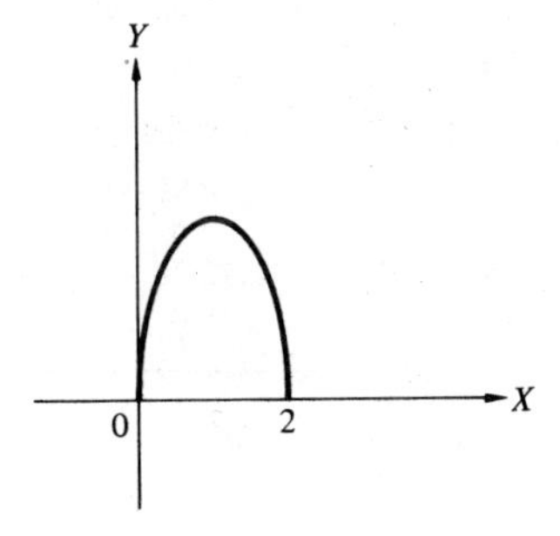

Figure E.12 Figure E.13

Page 25, Exercises 1.5

Group A

1. (a) 5. (c) 0. (e) Does not exist. (g) Indeterminate. **2.** (a) 3. (c) 1, 4. **3.** (a) 0. (c) $-\frac{3}{2} = -1.5$.

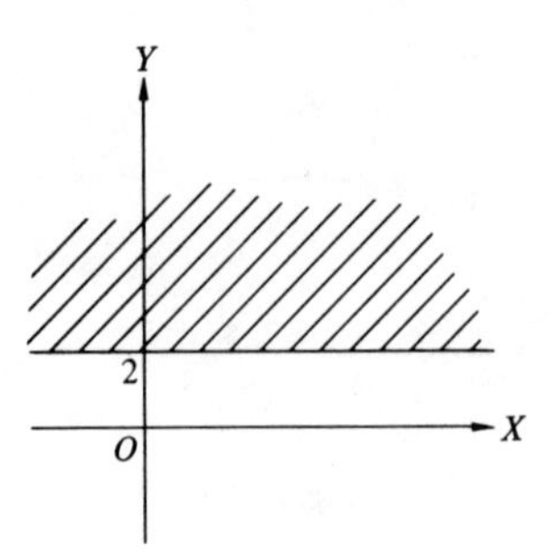

Figure E.14

Group B

1. (a) Becomes very large. (c) Approaches zero. (e) Becomes very large. (g) Approaches zero. (i) Approaches $\frac{2}{3} = 0.667$.

Page 28, Problems 1.6

1. 8. **2.** 0. **3.** Does not exist. **4.** 8. **5.** $\frac{1}{27} = 0.037$.

6.

x	x^2	$\frac{x^2 - 9}{x - 3}$	$\frac{1}{(x - 3)^2}$
2.0	4.0	5.0	1
2.5	6.25	5.5	4
2.9	8.41	5.9	100
2.99	8.9401	5.99	10,000
2.999	8.994001	5.999	1,000,000
$x \longrightarrow 3$ (a) $\lim_{x\to 3} x^2 = 9$.	(b) $\lim_{x\to 3} \frac{x^2 - 9}{x - 3} = 6$.	(c) $\lim_{x\to 3} \frac{1}{(x - 3)^2}$ does not exist.	
3.001	9.006001	6.001	1,000,000
3.01	9.0601	6.01	10,000
3.1	9.61	6.1	100
3.5	12.25	6.5	4
4.0	16.0	7.0	1

Page 31, Exercises 1.6

Group A

1. 2. **2.** (a) 4. (c) Does not exist. **3.** (a) 4. (c) Does not exist. (e) 6. (g) 3. **4.** (a) 0. (c) $2x$. (e) $6t$. (g) $-3/x^2$.

Group B

1. nu^{n-1}. **3.** 3.

Group C

1. Fig. E.15. **2.** Fig. E.16. (a) -2. (c) No. **3.** (a) Speed.

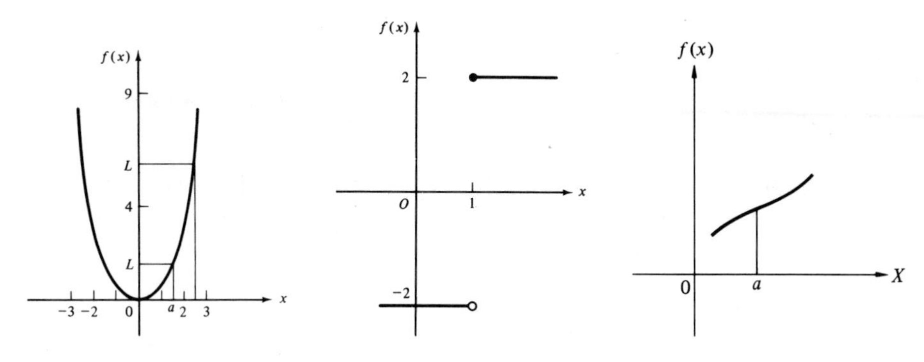

Figure E.15 **Figure E.16** **Figure E.17**

Page 33, Problems 1.7

1. (a) Not at $x = -7$. (b) Yes. (c) Not for (a); Yes for (b). **2.** $I = \begin{cases} 4, & 0 < t \leqq 2 \\ 0, & 2 < t \leqq 5. \end{cases}$ Yes, at $t = 2$. Different limits as $t \longrightarrow 2$ from above and below.

Page 34, Exercises 1.7

Group A

1. (a) Discontinuous at $x = 3$. (c) Continuous. (e) Discontinuous at $t = 2$. (g) Continuous. **2.** (a) Continuous for $-1 \leqq x \leqq 1$. (c) Discontinuous at $x = 0$. Different $f(x)$ as $x \longrightarrow 0^-$ and $x \longrightarrow 0^+$. (e) Discontinuous at $t = 0$. Different $g(t)$. (g) Discontinuous at $t = 1$. Division by zero. **3.** (a) Fig. E.17.

Group B

1. For continuity at a point $\lim_{x \to a^-} f(x) = \lim_{x \to a^+} f(x) = f(a)$.

Page 38, Problems 2.1

1. Fig. P.7.

t	0	1	2	3	4	5	6
s	0	-5	-8	-9	-8	-5	0
dist.	0	5	8	9	10	13	18

2. (a) -2 ft/s (to the left). (b) 1 ft/s (to the right). (c) 4 ft/s (to the right). **3.** (a) 0.4 ft/s. (b) 1 ft/s. (c) 0.25 ft/s. **4.** $2t - 6 + \Delta t$.

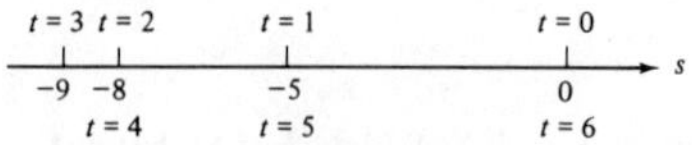

Figure P.7

Page 41, Exercises 2.1

Group A

1. (a) 0. (c) 0 ft/s. **2.** (a) $6x + 3\Delta x$; 27. (c) $2x - 2 + \Delta x$; 2.1. (e) $3x^2 + 3x(\Delta x) + (\Delta x)^2$, 19. (g) -5, -5. (i) $2x + 4 + \Delta x$, 5.

3.

t_1	Q_1	t_2	Q_2	Δt	ΔQ	$\frac{\Delta Q}{\Delta t}$
3.0	18.0	3.2	20.48	0.2	2.48	12.4
3.0	18.0	3.1	19.22	0.1	1.22	12.2
3.0	18.0	3.01	18.1202	0.01	0.1202	12.02
3.0	18.0	3.001	18.012002	0.001	0.12002	12.002
3.0	18.0	3.0001	18.00120012	0.0001	0.0012002	12.0002

$I \longrightarrow 12$

Group B

1. (a) 2. (c) 2. **2.** (a) $2t + \Delta t$; 7. (c) $2t + \Delta t$. **3.** 0, -10, -16, -18, -16, -10, 0, 14, 32. (c) 0. (e) 6. **5.** $-32t - 16\Delta t$. **7.** 25. **9.** $6t - 1 + 3\Delta t$. **11.** $4\pi r^2 + 4\pi r\Delta r + (\frac{4}{3})\pi(\Delta r)^2$.

Group C

1. Yes.

Page 43, Problems 2.2

1.

t_2	s_2	Δt	$\Delta s = s_2 - 4$	$\frac{\Delta s}{\Delta t}$
0.51	4.1616	0.01	0.1616	16.16
0.501	4.016016	0.001	0.016016	16.016
0.5001	4.00160016	0.0001	0.00160016	16.0016

As $t \longrightarrow 0.5$, $v = 16$ ft/s.

2. $v(0.5) = \lim_{t \to 0.5} 16(t + 0.5) = 16$ ft/s. **3.** 128 ft/s. **4.** (a) $a = \lim_{\Delta t \to 0} \frac{f(t + \Delta t) - f(t)}{\Delta t}$

(b) $I = \lim_{\Delta t \to 0} \dfrac{f(t + \Delta t) - f(t)}{\Delta t}$. (c) slope $= \lim_{\Delta x \to 0} \dfrac{f(x + \Delta x) - f(x)}{\Delta x}$ **5.** $3x^2 + 5$.
6. 24. **7.** 5, 29. At $t = 1$ the charge Q is changing 5 times as fast as t. At $t = 5$ the charge Q is changing 29 times as fast as t. **8.** (a) 0. (b) 0.

Page 49, Exercises 2.2

Group A

1. (a) $2x$. (c) $10t$. (e) 6. **2.** (a) $3x^2$. (c) $6t^2 - 2t$. (e) $3t^2 - 2t + 3$. **3.** (a) 3. (c) $-1/x^2$. (e) $1/2x^{1/2}$. (g) $-1/2x^{3/2}$. **5.** 4 A, 16 A, 58 A. **6.** (a) -1, 19, 49. (c) -3, 13, 37. **7.** -1 cm/s.

Group B

1.

t_2	s_2	Δt	$\Delta s = s_2 - 64$	$\dfrac{\Delta s}{\Delta t}$
1.9	57.76	-0.1	-6.24	62.4
1.99	63.3616	-0.01	-0.6384	63.84
1.999	63.936016	-0.001	-0.63984	63.984

As $t \longrightarrow 2^-$, av. vel. $\longrightarrow 64$.
3. (a) 0. (c) 0. **7.** 94.2 in.^{3}/in. **8.** (a) $3x^2$. (c) $\sin x$. **9.** (a) $\tan\theta$; $\sec^2\theta$. (c) x^n; nx^{n-1}.

Page 51, Problems 2.3

1. (a) $5x^4$. (b) $-12x^3$. (c) $42x^6$. (d) $7t$. **2.** (a) $2x^{-1/3}$. (b) $-6/t^4$. (c) $1/2\sqrt{t}$. **3.** (a) $18x^5 - 10x + 2$. (b) 5. (c) 12, does not exist, 12. (d) $15t^2 - 3 + 1/\sqrt{t} + 7/t^2$.

Page 53, Exercises 2.3

Group A

1. (a) 3. (c) $6x$. (e) $10t - 4$. (g) $32t + 32$. **2.** (a) $3x^2 + 4x^{-3}$. (c) $1/4x^{3/4}$. (e) $5 - 10/x^3$. **3.** (a) -7, -3, -1. (c) 0, 0, 0. (e) $-1/4$, does not exist, $-1/4$.

Group B

1. (a) 354. **2.** (a) 33. **3.** 4,256, falling. **5.** 3. **7.** 51.5. **9.** (a) 15.5. **11.** $4s^3 + 4s$. **13.** (a) $\frac{9}{5} = 1.8$. **14.** (a) 11. **15.** (a) 23. **17.** 3, 3, 3. **19.** 2. **21.** (a) $-15x^2$. **22.** (a) Fig. E. 18. **23.** (a) No.

Page 58, Problems 2.4

1. -3.5. **2.** $(-2, 2)$. **3.** 109.4°. **4.** $(0, -4)$.

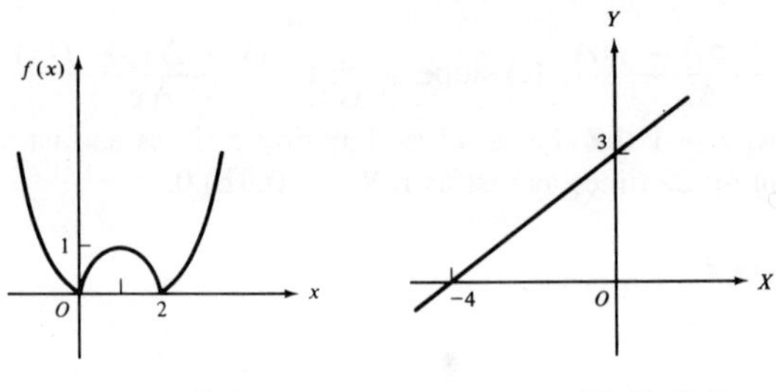

Figure E.18

Figure P.8

Page 62, Exercises 2.4

Group A

1. (a) 1. (c) 1. (e) Does not exist. (g) Does not exist. **2.** (a) (3, 2). (c) (3, 1) (e) (4, −4). (g) (−3, −5). **3.** (a) 1. (c) −1. (e) Does not exist. (g) $-\sqrt{3} = -1.7$. **4.** (a) $\frac{1}{3}$, −3. **5.** (a) −1.5. (c) Does not exist. **6.** (a) 63.4°. **7.** 45° or 135°.

Group B

1. (a) $m = -1$. **3.** $\pm\sqrt{2} = \pm 1.4$.

Page 63, Problems 2.5

1. $y = 7x - 19$. **2.** $y = 3x + 2$. **3.** 1.5, 3.5. **4.** A straight line parallel to the X-axis and 2 units below it. Zero. **5.** A straight line parallel to the Y-axis and 2.5 units to the right of it. Not defined. **6.** Fig. P.8. **7.** (a) 4. (b) −4. (c) 4. (d) The current is (a) increasing 4; (b) decreasing 4; (c) increasing 4 times as fast as the time.

Page 67, Exercises 2.5

Group A

1. (a) $3x - y - 8 = 0$. (c) $x + 2y + 14 = 0$. **2.** (a) 3, (0, 2). (c) $-\frac{2}{3} = -0.67$, (0, 0). **3.** (a) −1. (c) 2. (e) $-\frac{3}{2}$. **4.** (a) Parallel. (c) Neither. (e) Parallel. **5.** (a) Fig. E.19. (c) Fig. E.20. (e) Fig. E.21. (g) Fig. E.22. (i) Fig. E.23.

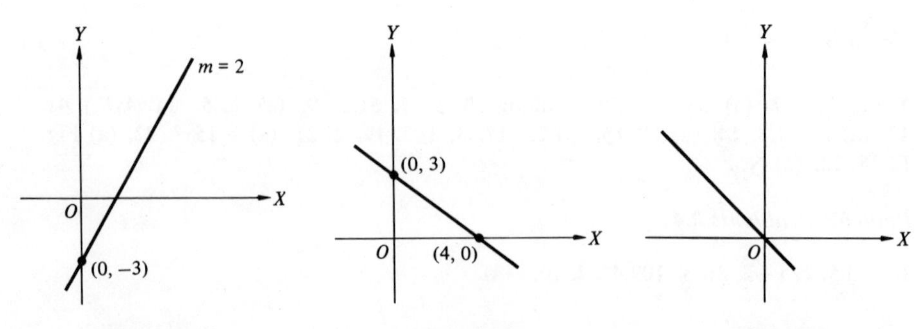

Figure E.19

Figure E.20

Figure E.21

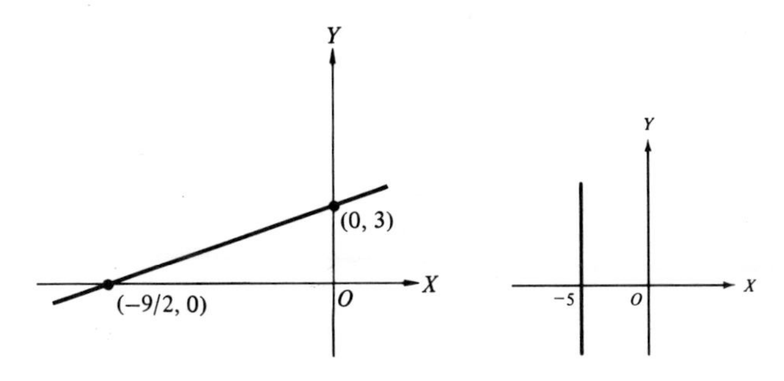

Figure E.22 Figure E.23

Group B

1. (a) $y = -3$. (c) $x = -4$. **2.** (a) $5x + y - 13 = 0$. (c) $2x + 5y - 10 = 0$. (e) $2x + 3y = 0$. **3.** 15.3°. **5.** $x - 1.73y + 10.66 = 0$. **7.** (a) 20. (c) −20. **8.** (a) 5. (c) 0.05 V.

Page 70, Problems 2.6

1. 9.49. **2.** (2.5, 2.5). **3.** 1.85.

Page 73, Exercises 2.6

Group A

1. (a) 5. (c) $\sqrt{65} = 8.06$. (e) $4\sqrt{5} = 8.94$. **2.** (a) (2, 3), (c) (9.71, 13). (e) (−4, 1.5). **3.** (a) 2. (c) 1.7. (e) 0. (g) $\frac{1}{2}\sqrt{2} = 0.707$.

Group B

1. (a) $m_{BA} = \frac{4}{3} = 1.33$; $m_{BC} = \frac{4}{15} = 0.267$; $m_{AC} = -\frac{4}{9} = -0.444$. (c) $\frac{4}{7} = 0.571$; 0; −4. **3.** 2.46.

Group C

1. (a) $y - y_0 = (B/A)(x - x_0)$. (c) $\dfrac{|Ax_0 + By_0 + C|}{\sqrt{A^2 + B^2}}$. Yes.

Page 76, Problems 2.7

1. 1, −1. **2.** (3, −9). **3.** $y = 8x - 16$. **4.** 4. **5.** $x + 2y - 5 = 0$. **6.** 347.5°, 63.4°.

Page 78, Exercises 2.7

Group A

1. (a) 6, −6. (c) 12, 0, 3, 12. **2.** (a) (0, 0). (c) $(-\frac{1}{3}, \frac{1}{3})$. **3.** (a) $x + y + 1 = 0$. (c) $3x - y$

$-2 = 0$. **4.** (a) $x - y - 3 = 0$. (c) $x + 3y - 4 = 0$. **5.** (a) (0, 0), 45°; (1, 1), 18.4°. (c) (0, 0), 0°. **6.** (a) Down, up. (c) Up, down.

Group B

3. $4x - y - 4 = 0$, $4x + y - 12 = 0$, 28.2°. **5.** 1.

Group C

3. (a) Yes. **4.** (a) 1 for $x > 0$; -1 for $x < 0$. (c) Yes.

Page 82, Problems 2.8

1. Inc. for $t < 1$ and $t > 3$. Dec. for $1 < t < 3$.

Page 82, Exercises 2.8

Group A

1. (a) Increase for $x > 1$; decrease for $x < 1$. (c) Increase for $x < -2$, $x > 1$; decrease for $-2 < x < 1$. **3.** Never. **5.** At $t = 2$. **7.** Increase for $t > 1$; decrease for $t < 1$. **9.** (a) Fig. E.24. (c) Decreases to zero.

Page 83, Problems 2.9

1. $15x^4 - 8x^3 + 2x$, $60x^3 - 24x^2 + 2$, 0, 408. **2.** Dec. for $x < 0.5$. Inc. for $x > 0.5$.

Page 84, Exercises 2.9

Group A

1. (a) 2. (c) $36x^2 - 12x$. **2.** (a) $-\frac{1}{4}x^{-3/2}$. (c) 0. **3.** (a) 142. (c) 64, 32. **4.** (a) Increase for $x > 1$; Decrease for $x < 1$. (c) Increase for $x < -1, x > 0$; decrease for $-1 < x < 0$. **5.** (a) 5.59. (c) Does not exist. (e) 5.59.

Group B

1. $v = 0, a = 32$; yes ($t < 2$), no. **3.** Increase $t < 0.48$, decrease $t < 0.48$. **5.** $x = \pm 1$. **7.** Yes, $t < \frac{1}{4}$. **8.** (a) 10. (c) 4. **9.** 0.

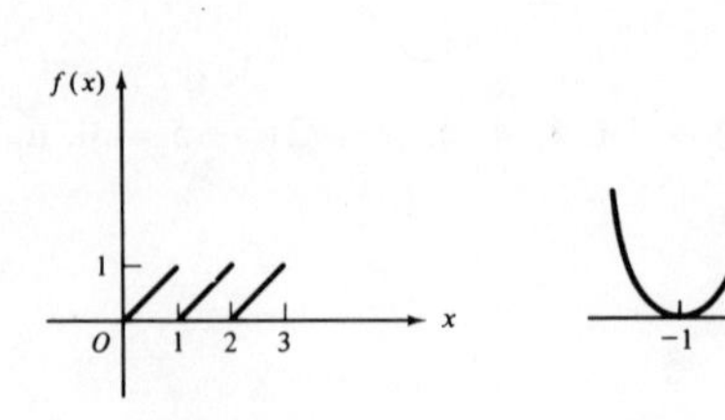

Figure E.24

Figure P.9

Page 86, Problems 2.10

1. $26(4x^3 - 5x^2 + 7)^{12}(12x^2 - 10x)$. **2.** $(9/2)(5t^2 - 7t)^{1/2}(10t - 7)$. **3.** (a) 0. (b) $13(6t^2 - 9)$. **4.** $(1/2)(x^4 + 16)^{-1/2}(4x^3) - 10(x^3 - x)^4(3x^2 - 1)$. **5.** 7.58. **6.** 0.16. **7.** $3000 \cos 60t$. **8.** $-36t^2 \sin 2t^3$. **9.** 21.3. **10.** 12.

Page 91, Exercises 2.10

Group A

1. (a) $(3u^2 - 1)5$. (c) $3(10x - 7)$. **2.** (a) $(2x/3)(x^2 + 25)^{-2/3}$. (c) $-2(2x^3 - x^{-1})^{-3} \cdot (6x^2 + x^{-2})$. **3.** (a) 219, 615. (c) -0.575. **4.** (a) 0. (c) 0. (e) $\pi^2 t^{2.14}$. (g) $5(2t - 3)^4(2) - 2(3t^2 - t)^{-1/2}(6t - 1)$. **5.** (a) $(3t - 1)(2t - 3) + (t^2 - 3t + 5)(3)$. (c) $(t^2 - 1)(2t - 2) + (t^2 - 2t + 1)(2t)$. **6.** (a) 15.6. (c) 16. **7.** (a) $-5/(2x - 1)^2$. (c) $-3/x^2$. (e) $-2/(x + 1)^3$. (g) $[6(5x + 7)(3x - 1) - 5(3x - 1)^2]/(5x + 7)^2$. **8.** (a) 3.5. (c) -20. **9.** (a) $15 \cos 5x$. (c) $-35 \sin 7x$. (e) $6x^2 \cos x^3$. (g) $-(x + 1) \sin (x + 1)^2$. (i) $6x \cdot \sin^2 2x \cos 2x + \sin^3 2x$. (k) $(\sin 2x - 2x \cos 2x)/\sin^2 2x$. **10.** (a) 5.73. (c) 0.459. (e) 2.76.

Group B

1. (a) $-1/x^2$. (c) Same. **2.** (a) $(-x + 1)/(x + 1)^3$. (c) Same. **3.** (a) $9x^2 - 22x + 8$. (c) Same. **5.** 95.5. **7.** $[v(du/dx) - u(dv/dx)]/v^2$. Yes. **9.** 0.077. **11.** $\sec^2 x$.

Page 96, Problems 3.1

1. Yes. **2.** (a) No. dy/dx is not defined at the origin. (b) Smooth for $x > 0$.

Page 97, Exercises 3.1

Group A

1. (a) Smooth. (c) Smooth. (e) Smooth. (g) Not smooth at (0, 0). **2.** (a) Continuous; not smooth at $x = 2$. (c) Continuous; not smooth at $t = 5$. (e) Not continuous at all integers except $x = 0$ and $x = 2$; not smooth at any integer.

Group B

1. All smooth.

Group C

1. Theorem 3.1.2.

Page 100, Problems 3.2

1. $(-1, 3)$ rel. max. $(1, -1)$ rel. min. **2.** No relative maxima or minima. **3.** Rel. max. $(0, 1)$, $(2\pi/3, 1)$, $(4\pi/3, 1)$ and $(2\pi, 1)$. Rel. min. $(\pi/3, -1)$, $(\pi, -1)$ and $(5\pi/3, -1)$. **4.** $(0, 1)$. **5.** No inflection point. **6.** Rel. max. $(0, 1)$. Rel min. $(-1, 0)$ and $(1, 0)$. Infl. pts. $(-0.58, 0.44)$ and $(0.58, 0.44)$. Fig. P.9.

Page 104, Exercises 3.2

Group A

1. (a) C.P. $(-1, 2)$, $(1, -2)$. Maximum $(-1, 2)$, minimum $(1, -2)$. (c) C.P. $(-1, 0)$; minimum $(-1, 0)$. (e) C.P. $(0, 2)$; minimum. $(0, 2)$. (g) No C.P. minimum $(0, 0)$. **2.** (a) C.P. $(0, 2)$, $(-2, 6)$; maximum $(-2, 6)$, minimum $(0, 2)$; I.P. $(-1, 4)$. (c) C.P. $(0, 4)$; I.P. $(0, 4)$; no maximum or minimum (e) C.P. $(0, 16)$; minimum $(0, 16)$; no I.P. (g) C.P. $(0, 1)$, $(-1, 3)$, $(1, -1)$; $(-1, 3)$ maximum, $(1, -1)$ minimum, I.P. $(0, 1)$, $(-0.71, 2.23)$, $(0.71, -0.23)$. **3.** (a) C.P. $(1, 0)$; No max. or min. I.P. $(1, 0)$. (c) C.P. $(0, 1)$; $(0, 1)$ max. I.P. $(-0.58, 0.75)$, $(0.58, 0.75)$. (e) C.P. $(0, 1)$, $(\pi, -1)$, $(2\pi, 1)$; max. $(0, 1)$, $(2\pi, 1)$; min. $(\pi, -1)$. I.P. $(\pi/2, 0)$, $(3\pi/2, 0)$. (g) C.P. $(0, 0)$ min, $(2.03, 1.82)$ max, $(4.91, -4.81)$ min; I.P. $(1.08, 0.95)$, $(3.64, -1.74)$.

Group B

3. Maximum (x_1, y_1), (x_5, y_5); minimum (x_3, y_3), (x_9, y_9); I.P. (x_2, y_2), (x_4, y_4), (x_6, y_6), (x_7, y_7), (x_8, y_8). **5.** Critical point at $t = 1$ sec. I.P. at $t = 0.58$ s. **7.** Minimum at $t = 0$. **9.** Maximum $t = 0.5$; minimum at $t = 0.8$. **11.** No. Not defined at $t = 2$.

Group C

3. 0. **5.** No. Not a function at b. **7.** (a) 1. (c) Fig. E.25. (e) No.

Page 109, Problems 3.3

1. Fig. P.10. **2.** Fig. P.11.

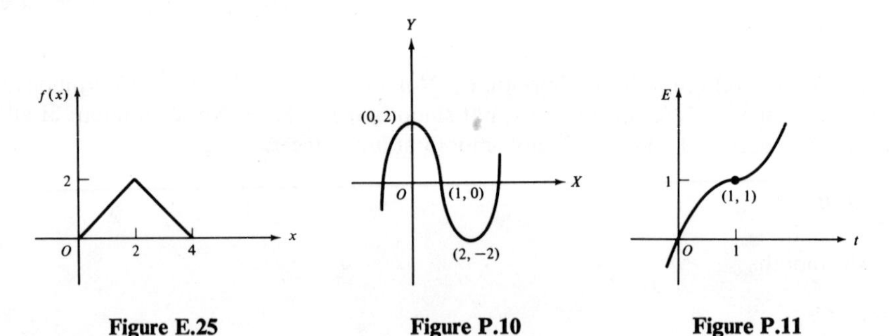

Figure E.25 Figure P.10 Figure P.11

Page 111, Exercises 3.3

Group A

1. (a) Fig. E.26. (c) Fig. E.27. (e) Fig. E.28. **2.** (a) Fig. E.29. (c) Fig. E.30. **3.** (a) Fig. E.31. (c) Fig. E.32. **4.** (a) Fig. E.33. (c) Fig. E.34. (e) Fig. E.35.

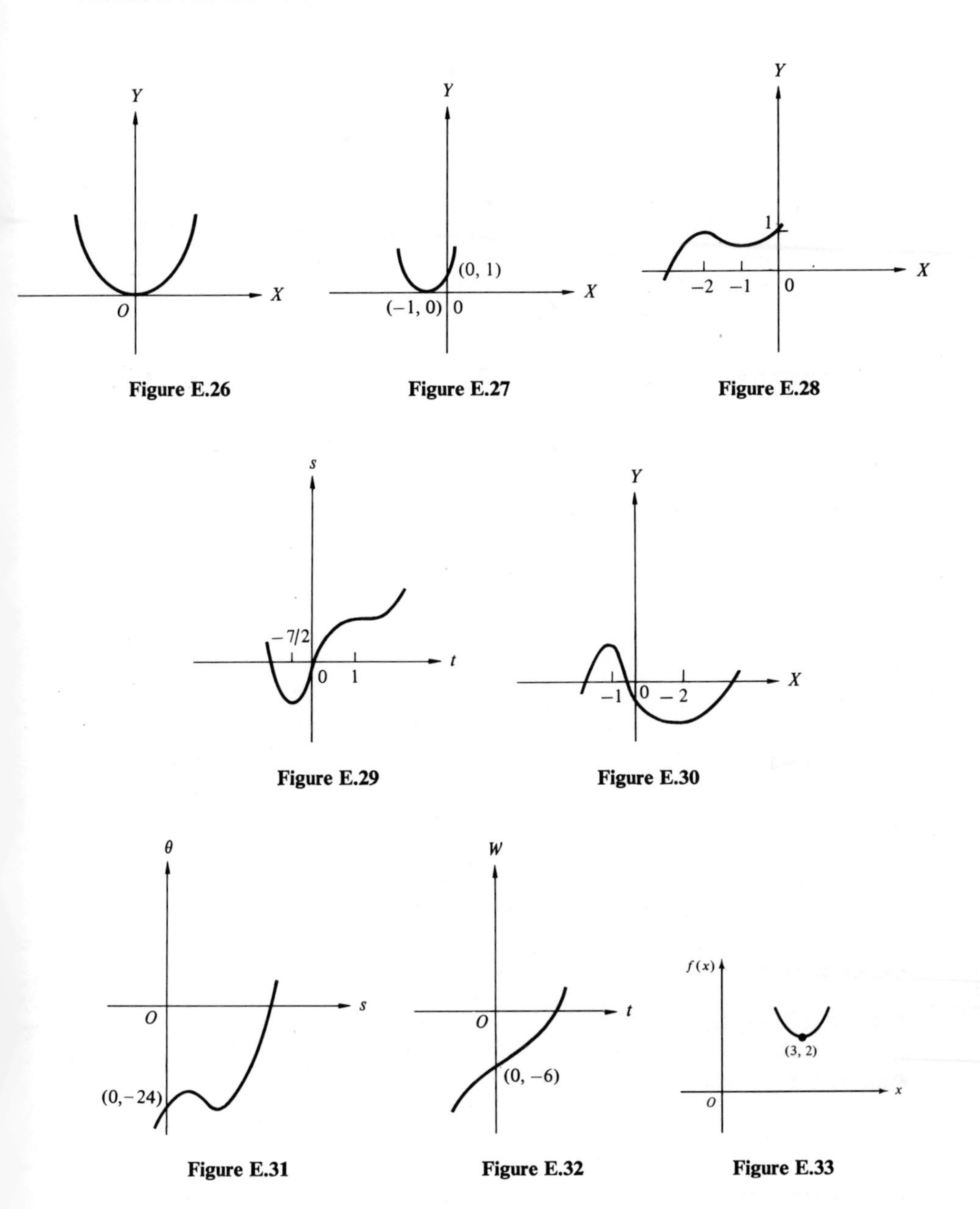

Figure E.26 Figure E.27 Figure E.28

Figure E.29 Figure E.30

Figure E.31 Figure E.32 Figure E.33

Group B

1. Fig. E.36. **3.** $V = 4x^3 - 32x^2 + 64x$. Fig. E.37. **5.** Fig. E.38. **7.** Fig. E.39. **9.** Same relative minimum. Both open up. Graph of $y = x^4$ flatter at the origin. **11.** Graph of $y = (x - 1)^4$ is one unit to the right of the graph of $y = x^4$.

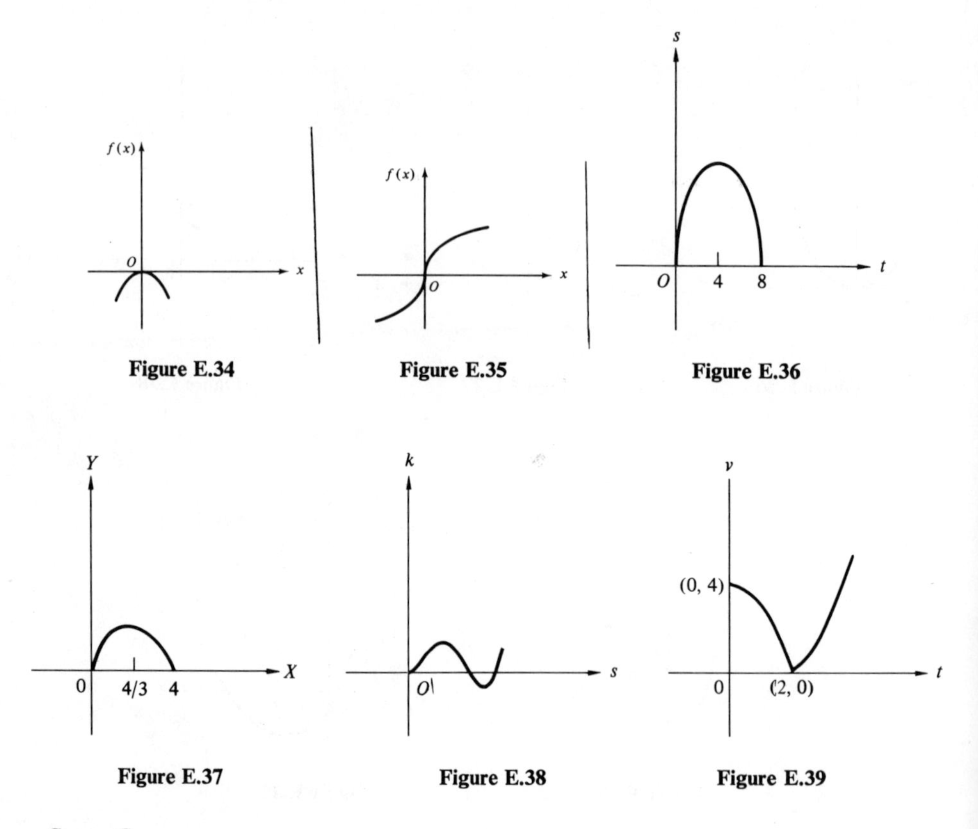

Figure E.34 Figure E.35 Figure E.36

Figure E.37 Figure E.38 Figure E.39

Group C

1. $c = -3a$, $b = 0$, $d = 4$, $a > 0$, for example $a = 1$, $c = -3$, $b = 0$, $d = 4$.
2. (e) The value of C shifts the graph vertically. Up for $C > 0$, down for $C < 0$.

Page 115, Exercises 3.4

1. $6 \times 6 \times 3$ cm. **2.** 696.5 cm^3.

Page 117, Exercises 3.4

Group A

1. (a) Relative minimum (0, 0); absolute minimum (0, 0); absolute maximum (1, 1), (−1, 1). (c) Absolute minimum at (0, 2); no absolute maximum, y not defined at $x = 2$. (e) No absolute maximum nor minimum; y not defined at end points. (g) Relative maximum (−1, 1); relative minimum $(-\frac{1}{3}, \frac{23}{27})$; no absolute maximum nor minimum, y not defined at end points. (i) No relative maximum nor minimum, absolute minimum at (1, 1); no absolute maximum; y not defined at $x = 0$. (k) No relative maximum nor minimum, no absolute maximum nor minimum.

Group B

1. 30,625 ft². **3.** 6″ × 6″. **5.** 15.2″ × 18.6″. **7.** 6, 18. **9.** 8 × 8. **11.** 210. **13.** $I = 4$. **17.** 4.09°C.

Group C

3. No, slope is never zero. Yes, at end points.

Page 119, Problems 3.5

1. 96 ft/s; $0 < t < 3$, s inc., $3 < t < 6$, s dec.; 144 ft; 128 ft, −32 ft/s, −32 ft/s². **2.** Fig. P.12.

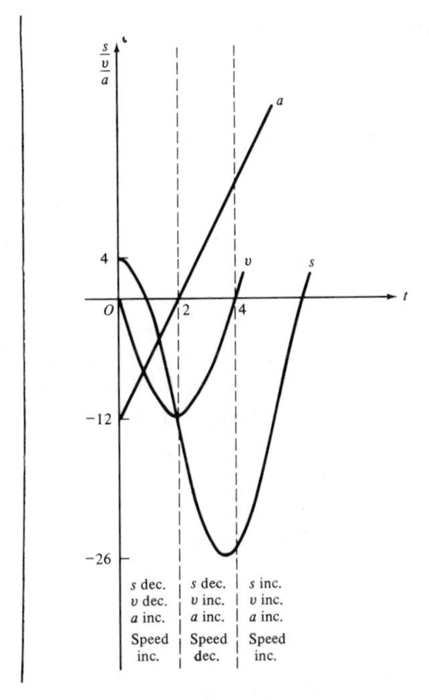

Figure P.12

Page 121, Exercises 3.5

Group A

1. (a) 9, 3.75; decreases 5.25. **2.** (a) s increases for $t < 19$, decreases for $t > 19$; v decreases for all t; speed decreases for $t < 19$; increases for $t > 19$; (c) s increases for $t > \frac{3}{2}$, decreases for $t > \frac{3}{2}$, v increases for all t, speed decreases for $t < \frac{3}{2}$, increases for $t > \frac{3}{2}$. (e) s increases for all t, v increases for $t > \frac{3}{2}$, decreases for $t < \frac{2}{3}$, speed increases for $t > \frac{2}{3}$, decreases for $t < \frac{2}{3}$.

Group B

1. Increase for $0 < t < 2$, decrease for $2 < t < 4$. **3.** Increase for $0 < t < 3$; decrease for $3 < t < 6$. **5.** s decrease for $0 < t < 0.07$; increase for $t > 0.07$. **6.** (a) 1.57. (c) 1.57, 3.

Page 125, Problems 3.6

1. 2.104.

Page 125, Exercises 3.6

Group A

1. 2.33. **3.** 0.33. **5.** 0.441. **6.** (a) 2.236. (c) 1.260.

Page 127, Problems 4.1

1. $\Delta y = (6x - 1)\,dx + 3(dx)^2$; $dy = (6x - 1)\,dx$.

Page 128, Exercises 4.1

Group A

1. (a) dx. (c) $(2x - 2)dx$. (e) $(4x^3 + 6x + 1)dx$. **2.** (a) 0, 0. (c) $3x^2dx$, $3x^2dx + 3x(dx)^2 + (dx)^3$. (e) $(4x^3 - 9x^2 + 4x)dx$, $(4x^3 - 9x^2 + 4x)dx + (6x^2 - 9x + 2)(dx)^2 + (4x - 3)(dx)^3 + (dx)^4$. **3.** (a) 0.2, 0.2. (c) 0.08, 0.0816. (e) 0, 0. **4.** (a) $dy = 3(3x^2 - 5x)^2(6x - 5)dx$. (c) $dv = [-19/(2t - 5)^2]dt$. (e) $-10 \sin 5x\, dx$.

Group B

1. 0.9616, 0.96. **3.** 8.937 ft^3, 8.1 ft^3. **5.** 0.03, 1.51. **7.** 0.1204, 0.12. **8.** (a) 0.35, −0.75. **9.** 0.024, 0.024001.

Group C

3. Yes. $y = ax + b$.

Page 130, Problems 4.2

1. 19 V; 0.0115 V. **2.** 3.019. **3.** (a) 0.2%. (b) 0.6%. (c) 0.4%.

Page 132, Exercises 4.2

Group A

1. (a) 3.6. (c) 12.6. (e) 0.015. **2.** (a) 0.45; 45%. (c) 1; 100%. (e) 0.0035; 0.35%. **3.** (a) 4.125. (c) 2.33. (e) 11.96. **4.** (a) $3x^2dx$. (c) $2\pi rh\,dr$, πr^2dh. **5.** (a) $3\,dx/x$. (c) dh/h.

Group B

1. 0.282 in.^{3} **3.** 1.39%. **5.** ± 0.22 in. **7.** 2, 0.188.

Page 134, Problems 4.3

1. (a) $x^3 + C$. (b) $(3/5)x^{5/3} + (3/2)x^2 + C$. **2.** (a) $-(5/3)\cos 3x + C$. (b) $-(1/3)\sin 3\pi t + C$.

Page 134, Exercises 4.3

Group A

1. (a) $x^2 + C$. (c) $x^4 + C$. (e) $x^{1/5} + x + C$. (g) $(1/36)x^6 + (1/25)x^5 - (1/16)x^4 + C$. (i) $1.76x^{1.7} + C$. **2.** (a) $-1/x + C$. (c) $-\frac{1}{2}x^{-2} + C$. (e) $-\frac{1}{2}x^{-2} - \frac{1}{3}x^{-3} + C$. (g) $2x^{1/2} + \frac{1}{5}x^5 + C$. **3.** (a) $(2/3)x^3 - (3/2)x^{-2} + C$. (c) $-t^{-1} - (1/3)t^3 + C$. **4.** (a) $-\cos 2t + C$. (c) $(5/3)\sin 3x + C$. (e) $-(11/6)\cos 60t + C$. (g) $(3/2)\sin 2x + (5/2)\cos 2x + C$.

Group B

1. (a) $x^3/3 + 3x^2/2 + 7x + C$. (c) $\frac{10}{152}x^{19/10} + 2x + C$. (e) $a_0x^{n+1}/(n+1) + C$. (g) $\dfrac{a_0x^{n+1}}{n+1} + \dfrac{a_1x^n}{n} + \ldots + a_nx + C$. **2.** (a) $\frac{4}{17}x^{17} + \frac{3}{16}x^{16} + \frac{3}{5}x^{10} + x^2 + 5x + C$. (c) $\frac{4}{101}x^{101} + \frac{3}{76}x^{76} + \frac{2}{51}x^{51} + \frac{1}{26}x^{26} + x + C$. (e) $-\frac{1}{3}x^{-3} + x^6 + C$. (g) $-\frac{1}{3}x^{-3} + \frac{1}{5}x^5 + C$.

Page 136, Problems 4.4

1. $Q = t^4 - (1/2)t^2 + C$. **2.** $y = 0.5x^4 + 2$. **3.** $v = gt + 5, s = (g/2)t^2 + 5t$. 101 ft/s. 159 ft.

Page 138, Exercises 4.4

Group A

1. (a) $x^4/4 + C$. (c) $(1/25)t^5 + (5/24)t^{6/5} + 2t + C$. **2.** (a) $y = x^4 + 16$. (c) $y = x^{1/3}$. (e) $y = 15x^{1/5} - 24$. **3.** (a) $y = 2\sin x + 1$. (c) $Q = -1.5\cos 2t + 1.5$.

Group B

1. $y = x^2 - 5$. **3.** 176 ft. **5.** 3.5 Ω. **7.** $\frac{3}{2}t^2 + 4t + \frac{3}{2}$. **9.** $\frac{1}{3}t^3 + t^2 + t + 5$. **11.** 56.

Group C

1. $y = \frac{1}{2}x^2 + C$. **3.** $y = \frac{1}{3}x^3 + x^5 + C$. **5.** (a) $x + y^2 = C$. (b) $x + y^2 - 1 = 0$.

Page 142, Problems 4.5

1. 30.

Page 142, Exercises 4.5

Group A

1. (a) 6. (c) 4.4. (e) 20.8. **2.** (a) 2.67. (c) 0.167. (e) 6.25. **3.** 1.

Page 145, Problems 5.2

1. 43.

Page 146, Exercises 5.2

Group A

1. (a) 41.88. (c) 0.5. (e) 0.427. **2.** (a) 41.7. (c) 0.5. (e) 0.4.

Group B

1. (a) 1.11. (c) 19.7. (e) 0.65. (g) 1.26.

Group C

1. $y = f(x)$ is a straight line. **7.** 11.087.

Page 150, Problems 5.3

1. 42. **2.** 21.08.

Page 152, Exercises 5.3

Group A

1. (a) 2.67. (c) 6. (e) 2. **2.** (a) 1.75. (c) 5.375. (e) 1.6. **3.** (a) 3.08. (c) 8. **4.** (a) 3. (c) 2.

Group B

1. (a) 4. (c) 12. (e) 225. **2.** (a) 10.67. (c) 21.3. **3.** (a) 64. (c) 180.

Group C

1. 0.68; 0.68. **3.** (a) 2.67. (c) 4.

Page 155, Problems 5.4

1. 80. **2.** $(1/2)x^6 + C$. **3.** -31.25.

Page 156, Exercises 5.4

Group A

1. (a) 9. (c) 4.67. (e) 11.25. (g) 4.4. (i) 366.2. **2.** (a) $x^3/3 + C$. (c) $t^5/5 + t^4/2 + t^2/2 + t/2 + C$. (e) $\frac{3}{5}y^{5/3} + \frac{1}{2}y^2 + 2y^{1/2} + C$. (g) $\frac{2}{3}x^{3/2} + (\sqrt{2}/2)x^2 + x + C$. (i) $\frac{1}{5}u^5 + \frac{1}{4}u^4 - \frac{1}{3}u^3 - \frac{1}{2}u^2 + u + C$. **3.** (a) 3. (c) 0. **4.** (a) $(-5/3)\cos 3x + C$. (c) $4\sin(x/2) + C$.

Group B

1. (a) 3.75. (c) 6.875. (e) 2.4. **2.** (a) 7.17. (c) 0.07. (e) 0.4. (g) -8.

Page 159, Problems 5.5

1. 0.

Page 159, Exercises 5.5

Group A

1. (a) 0.2. (c) 4. **2.** (a) 0.67. (c) 2.67. **3.** (a) 65. (c) 196. **4.** (a) 1.4. (c) 6.75.

Page 163, Problems 6.1

1. 4.5 sq units. **2.** 4.5 sq units.

Page 165, Exercises 6.1

Group A

1. (a) 21.3. (c) 20.8. **2.** (a) 32. (c) 4.5. **3.** (a) 4.5. (c) 1.3.

Group B

1. (a) 0.5. (c) 6.03. **2.** (a) 9. (c) 0.021. **3.** (a) 12.8. (c) 19.2. **5.** (a) 1. **7.** 0.414.

Group C

1. (a) AB $4x - 3y + 2 = 0$; BC $2x + y - 14 = 0$; AC $3x - y - 1 = 0$. (c) 5.

Page 168, Problems 6.2

1. (a) 12. (b) 7. **2.** 75, 45. **3.** (1.2, 0.7). **4.** (2.5, 1.5). **5.** (3, 3.6). **6.** (0.17, 2.4).

Page 173, Exercises 6.2

Group A

1. (a) 4; 0. (c) 54; 54. **2.** (a) 1.3ρ; 1.3ρ. (c) 0.53ρ; 0. **3.** (a) (0, 0.29). (c) (2, 2). **4.** (a) (0.67, 0.67). (c) (0, 0.4). **5.** $(b/3, h/3)$. **6.** (1.6, 2).

Group B

1. (a) (2.5, 5). (c) (1, −2). **2.** (a) (0.75, 1.6). (c) (−0.4, 0.86). **3.** (a) (2.4, 1.6). **5.** $m = \rho \int_0^{\pi/2} \cos x\, dx$; $M_Y = \rho \int_0^{\pi/2} x \cos x\, dx$; $M_X = (\rho/2) \int_0^{\pi/2} \cos^2 x\, dx$. **8.** (a) (2, 3.3). **9.** (a) $m = \rho \int_0^4 x^{3/2}\, dx$; $M_Y = \rho \int_0^4 x^{5/2}\, dx$; $M_X = \rho \int_0^4 (1/2)x^3 dx$. (c) $m = \rho \int_0^8 y^{2/3}\, dy$; $M_Y = \rho \int_0^8 (1/2)y^{4/3}\, dy$; $M_X = \rho \int_0^8 y^{5/3}\, dy$.

Group C

1. $((2d + a)/2,\ (2c + b)/2)$. **2.** (a) 7.23 in. from left edge, 8.85 in. above bottom. (c) 7.84 in. from left edge, 9.7 in. above bottom.

Page 176, Problems 6.3

1. (a) 64. (b) 21. **2.** 2.53; 1.45. **3.** 34.7; 2.08. **4.** 3.15ρ.

Page 179, Exercises 6.3

Group A

1. (a) $i_Y = 56$; $k_Y = 2$. $i_X = 56$; $k_X = 2$. (c) $i_Y = 204$; $k_Y = 2.75$. $i_X = 204$; $k_X = 2.75$. **3.** 156.25ρ. **5.** 0.167ρ. **7.** 273.07ρ. **9.** 64.8ρ. **11.** (a) 1.79. (c) 1.34.

Group B

1. 1.6ρ. **3.** 38.4ρ. **5.** $ml^2/3$; $l/\sqrt{3} = 0.58l$.

Group C

1. (a) 30. **3.** 1.1. **5.** (a) 0.305ρ. (c) 2.97ρ. (e) 1192ρ.

Page 182, Problems 6.4

1. 25.1. **2.** 0.52. **3.** 0.52. **4.** $i_{\text{axis}} = \rho\pi h r^2/10 = 3r^2 m/10$. **5.** 1.

Page 186, Exercises 6.4

Group A

1. 56.5. **3.** 152.7. **5.** 67.02. **7.** 0.419. **8.** (a) 8.4. (c) 1.57. **9.** 25.1. **11.** 0.94. **13.** (a) 2.7. **14.** (a) 72.

Group B

1. $\pi r^2 h$. **4.** (a) 381.2. **5.** (a) 52.1. **6.** (a) 40.2. **7.** 490. **8.** (a) (i) $\pi \int_0^2 x^6\, dx$.

(ii) $2\pi \int_0^8 (2y - y^{4/3})\,dy$. (c) (i) $\pi \int_0^8 y^{2/3}\,dy$. (ii) $2\pi \int_0^2 (8x - x^4)\,dx$. (e) (i) $\pi \int_0^8 (4y^{1/3} - y^{2/3})\,dy$. (ii) $2\pi \int_0^2 (16 - 8x - 2x^3 + x^4)\,dx$.

Page 189, Problems 6.5

1. 6 ft lb. **2.** −30,000 ft lb. **3.** 30,630 ft lb.

Page 191, Exercises 6.5

Group A

1. (a) 69.3. (c) 33.2. **3.** 81.7 ft lb. **5.** $0.75Q_1Q_2$.

Group B

1. 87.5 ft lb. **3.** 2.45×10^7 ft lb. **5.** −10.5. **7.** 396 ft lb.

Page 192, Problems 6.6

1. 4. **2.** 105 W. **3.** 3.46. **4.** 77.8.

Page 194, Exercises 6.6

Group A

1. (a) 6. (c) 6.75. **2.** (a) −3. (c) 9. **3.** 64. **5.** 12. **6.** (a) 6 V. (c) 7.64 V. **7.** (a) 4.16. (c) 0.18.

Group B

3. 27.6. **4.** (a) 4; 4.6. (c) 7; 8.3. **5.** (a) 1. (c) Same.

Page 197, Exercises 7.1

Group A

1. (a) Algebraic. (c) Not algebraic. (e) Not algebraic. **2.** (a) 13.4, (7.1.2, 7.1.1). (c) 13.5, (7.1.1, 7.1.4). (e) 3, (7.1.4).

Page 198, Problems 7.2

1. 0. **2.** $13(3x^2 - 10x)$. **3.** $4x - (1/3)x^{-2/3} + 4$.

Page 199, Exercises 7.2

Group A

1. (a) 0. (c) $3x^2 + 13 + 1/2\sqrt{x}$. (e) $20x^3 + 20x + \frac{15}{2}x^{1/2}$. (g) $\frac{3}{2}(-t^{-3/2} + t^{-1/2})$.

Group B

1. $s = 9188$ cm, $t = 2.5$ s. **3.** 87. **5.** $-1 < x < 1$, $x \neq 0$.

Group C

1. (a) $(d/dx)(u_1 + u_2 + \ldots + u_n) = (du_1/dx) + (du_2/dx) + \ldots + (du_n/dx)$.
2. (a) $c(du/dx) + c(dv/dx)$.

Page 201, Problems 7.3

1. (a) $(3x^2 - x)(12x^2 + 5) + (4x^3 + 5)(6x - 1)$. (b) $60x^5 - 16x^3 + 45x^2 - 10x$. Yes. **2.** $(t^3 - 2)(t^{2/3})(10t - 9) + (t^3 - 2)(5t^2 - 9)(2/3)(t^{-1/3}) + t^{2/3}(5t^2 - 9t)(3t^2)$.

Page 201, Exercises 7.3

Group A

1. (a) $14x^{13}$. (c) $\frac{7}{2}x^{5/2} + 6x^2 + \frac{3}{2}\sqrt{2x} + 2\sqrt{2}$. (e) $33y_{17/16}$. (g) $\frac{15}{2}t^{3/2} - 6\pi + t$ $2t^{-1/2}$. (i) $(t - 2)(t^2 + 3)(2) + (t - 2)(2t - 1)(2t) + (t^2 + 3)(2t - 1)$.

Group B

1. (a) $-(x^2 + 6)/x^4$. (c) $(-9 - 2t)/t^4$. (e) $4x^3 + 12x$. **3.** 87. **5.** 71.25.

Group C

2. (a) $10(5x + 3)$. (c) Yes. $an(ax + b)^{n-1}$.

Page 203, Problems 7.4

1. $(t^2 + 2)/5t^2$. **2.** $-15/(3t - 2)^2$.

Page 204, Exercises 7.4

Group A

1. (a) $2/(x + 1)^2$. (c) $(-4x + 21)/x^4$. (e) $-6/t^3$. (g) $-5/3t^{2/3}(t^{1/3} - 4)^2$. (i) $(-x^4 + 4x)/(x^3 + 2)^2$. (k) $\dfrac{-10x^2 - 3x + 3}{4x^{1/4}(2x^2 + 3x + 1)^2}$. (m) $\dfrac{18x^5 + x - 1}{x^{3/2}}$.

Group B

1. (a) $\dfrac{2x^2 - 14x + 20}{(2x - 7)^2}$. (c) $\dfrac{-3x^2 + 12x - 3}{x^2(2x - 1)^2}$. (e) $\dfrac{-\pi(9x - 4)}{(3x - 2)^2 x^3}$.
(g) $\dfrac{36t^2 + 18t - 12}{(t^2 - 1)^2(4t + 3)^2}$. **3.** Negative for all x, $x \neq 0$. Fig. E.40.

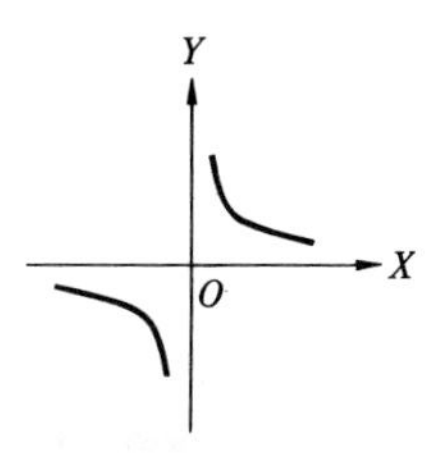

Figure E.40

Group C

2. (c) -3. (e) 1. (g) 0. **3.** $-1/2$.

Page 206, Problems 7.5

1. $4\pi(3t^4 - 2t^2)^2(12t^2 - 4t)$. **2.** $9(5x^3 - 3x)^8(15x^2 - 3)$. **3.** $x^2/(x^3 - 8)^{2/3}$.
4. $-105 \sin 5x$. **5.** $(3u^2 - 5)/8(2u - 1)^3$.

Page 208, Exercises 7.5

Group A

1. (a) $2(3x^2 - 7x)(6x - 7)$. (c) $[3(5x^2 + 4x + 7)^2 - 2(5x^2 + 4x + 7)](10x + 4)$.
2. (a) $3(x^2 + 2x + 5)^2(2x + 2)$. (c) $(x)(x^2 + 2)^{-1/2}$. (e) $2(x^{3/4} + x^{1/2} + x^{1/4}) \cdot (\frac{3}{4}x^{-1/4} + \frac{1}{2}x^{-1/2} + \frac{1}{4}x^{-3/4})$. (g) $(\sqrt{3} - 3t)/2\sqrt{t}$. **3.** (a) $\frac{3}{2}u$. (c) $(2u - 3)/7u^6$.

Group B

1. $\dfrac{4x - 5}{\sqrt{4x^2 - 10x + 4}}$. **3.** 628.3. **5.** 527.8. **7.** (a) $-15/(2x - 5)^2$. (c) $(-15x + 7)/(3x - 1)^3$. **8.** (a) $15 \cos^2 5x \sin 5x$. (c) $30x^2 \sin x^3 \cos x^3$. **9.** (a) -0.75.

Group C

2. (a) $dy/dx = (dy/du)(du/dv)(dv/dx)$. **5.** (a) $1/5x^4$. (c) $1/6$.

Page 210, Problems 7.6

1. $[(2t + 5)^3(1/2)(3t^2 - 16)^{-1/2}(6t) - (3t^2 - 16)^{1/2}(3)(2t + 5)^2(2)]/(2t + 5)^6$.
2. $4[(5v)(3v^2 - v)]^3[(5v)(6v - 1) + (3v^2 - v)(5)]$. **3.** $(1/2)[(2t + 3)/(3t^2 - 7)^{-1/2} \cdot \{[(3t^2 - 7)(2) - (2t + 3)(6t)]/(3t^2 - 7)^2\}$. **4.** $-4t\, dt/(16 - 4t^2)^{1/2}$.

Page 211, Exercises 7.6

Group A

1. (a) $(5x^2 + 2)/2\sqrt{x}$. (c) $(-5x^3 + 1)/2\sqrt{x}\,(x^3 + 1)^2$. (e) $80x^3(x^4 + 1)$.
2. (a) $9[(x^5 + 2x^3 + 3) + (8x^4 + 2x^2 + 1)]^8(5x^4 + 32x^3 + 6x^2 + 4x)$.

(c) $\dfrac{7x^{7/2} + 13x^{3/2} + 3x^{1/2}}{2(x^3 + 3x + 1)^{1/3}}$. (e) $(4x^3 - 3x^5)/(1 - x^2)^{3/2}$. (g) $9[(3x^2 - x)(2x)]^8(18x^2 - 4x)$. **3.** (a) $(2x/3)(x^2 + 1)^{-2/3}\,dx$. (c) $\dfrac{-7x^2 + 1}{4x^{3/4}(x^2 + 1)^2}\,dx$. (e) 0. **4.** (a) 79.47. **5.** (a) $30(3x^2 + 2)^3(27x^2 + 2)$. (c) $2(x^3 - 2x)(21x^2 - 6x - 6) + 2(3x^2 - 2)(7x^3 - 3x^2 - 6x + 2)$.

Group B

1. (5, 0), (0.71, 16.4). **3.** 7.07×3.54. **5.** $0.707c$. **7.** $r = 3$ in., $h = 6$ in. **9.** $r = 4$ in., $h = 4.5$ in. **11.** $\frac{2}{3}r_0$. **13.** $k/2$.

Group C

1. (a) $xy = c$. (c) $x^2y^2 = c$. (e) $x = cy$.

Page 215, Problems 7.7

1. Fig. P.13. **2.** (a) 0. (c) 0.5. **3.** Fig. P.14.

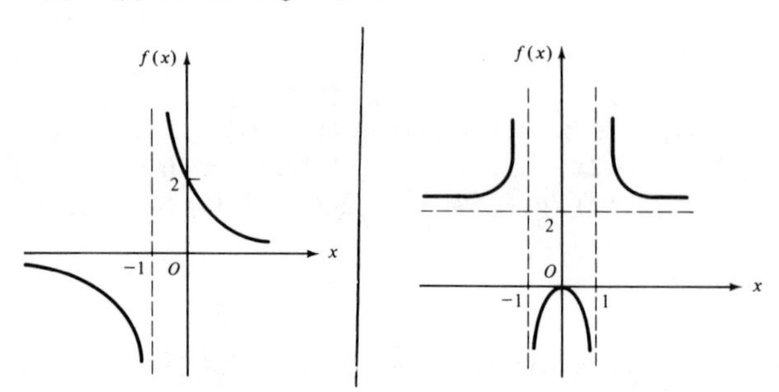

Figure P.13

Figure P.14

Page 218, Exercises 7.7

Group A

1. (a) $x \to 0^-$, $f(x) \to -\infty$; $x \to 0^+$, $f(x) \to \infty$; $x \to \infty$, $f(x) \to 0^+$; $x \to -\infty$, $f(x) \to 0^-$; $x = 0$, $f(x) = 0$. (c) $x \to -5^-$, $f(x) \to -\infty$; $x \to -5^+$, $f(x) \to \infty$; $x \to \infty$, $f(x) \to 0^+$; $x \to -\infty$, $f(x) \to 0^-$; $x = -5$, $f(x) = 0$. (e) $x \to 0^-$, $f(x) \to \infty$; $x \to 0^+$, $f(x) \to \infty$; $x \to \infty$, $f(x) \to 0^+$; $x \to -\infty$, $f(x) \to 0^+$; $x = 0$, $f(x) = 0$. (g) $x \to 0^-$, $f(x) \to -\infty$; $x \to 0^+$, $f(x) \to \infty$; $x \to \infty$, $f(x) \to 0^+$; $x \to -\infty$, $f(x) \to 0^-$; $x = 0$, $f(x) = 0$. (i) $x \to -2^-$, $f(x) \to \infty$; $x \to -2^+$, $f(x) \to -\infty$; $x \to 2^-$, $f(x) \to -\infty$; $x \to 2^+$, $f(x) \to \infty$; $x \to \infty$, $f(x) \to 0^+$; $x \to -\infty$, $f(x) \to 0^+$; $x = \pm 2$, $f(x) = 0$. (k) $x \to -2.3^-$, $f(x) \to \infty$; $x \to -2.3^+$, $f(x) \to -\infty$; $x \to 2.3^-$, $f(x) \to -\infty$; $x \to 2.3^+$, $f(x) \to \infty$; $x \to \infty$, $f(x) \to 0.33^+$; $x \to -\infty$, $f(x) \to 0.33^+$; $x = \pm 2.3$, $f(x) = 0.33$. **2.** (a) Fig. E.41. (c) Fig. E.42. (e) Fig. E.43. (g) Fig. E.44. (i) Fig. E.45. (j) Fig. E.46.

Group B

1. (a) Fig. E.47. (c) Fig. E.48. (e) Fig. E.49. **2.** (a) Fig. E.50.

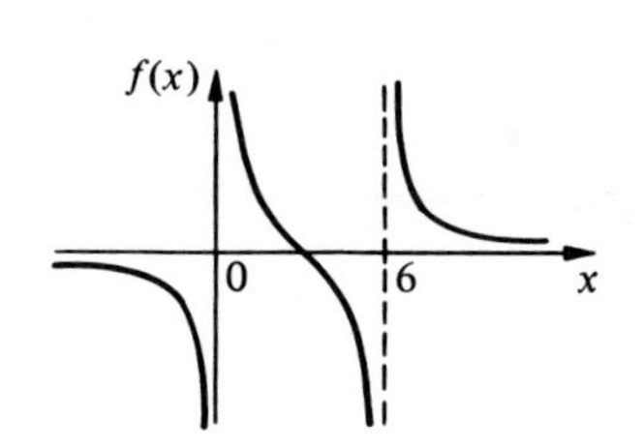

Figure E.41

Figure E.42

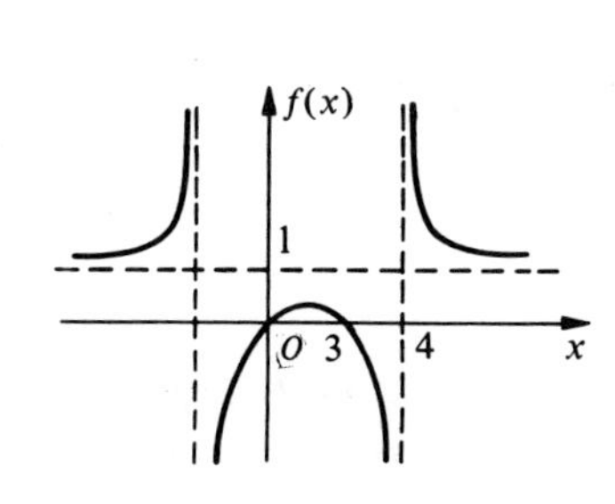

Figure E.43

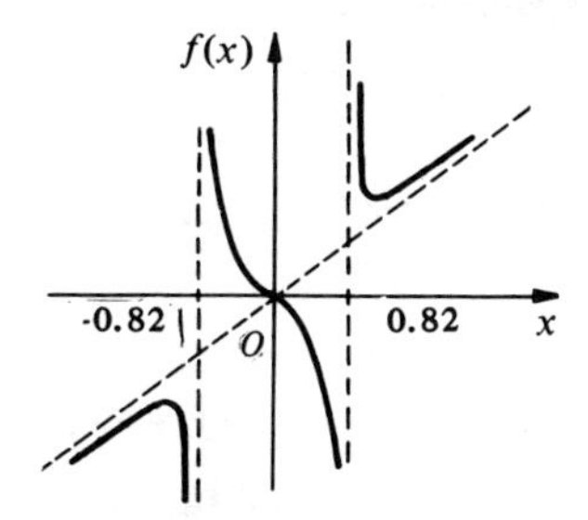

Figure E.44

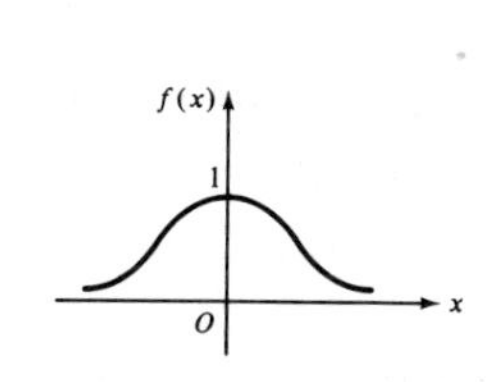

Figure E.45

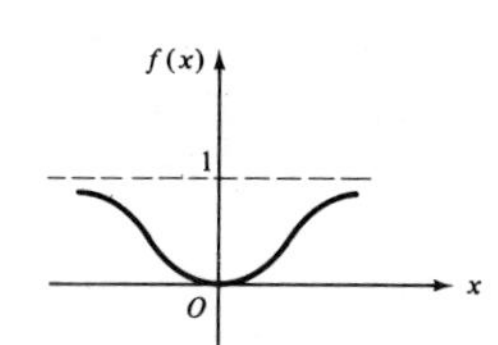

Figure E.46

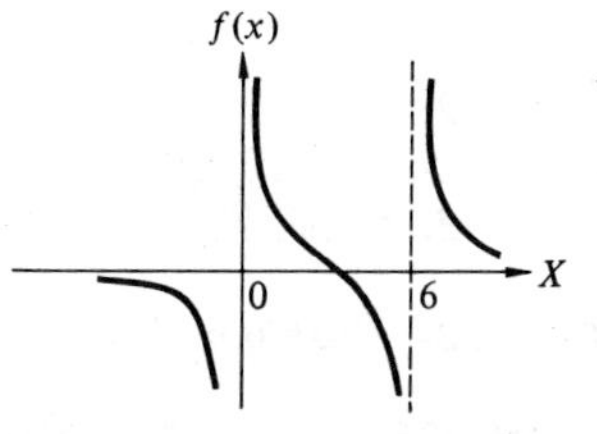

Figure E.47

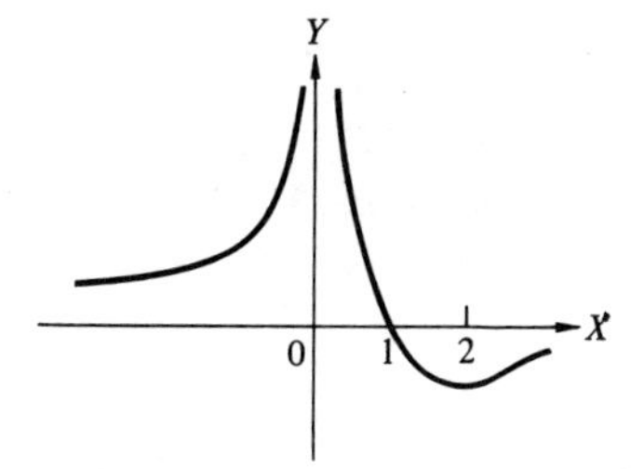

Figure E.48

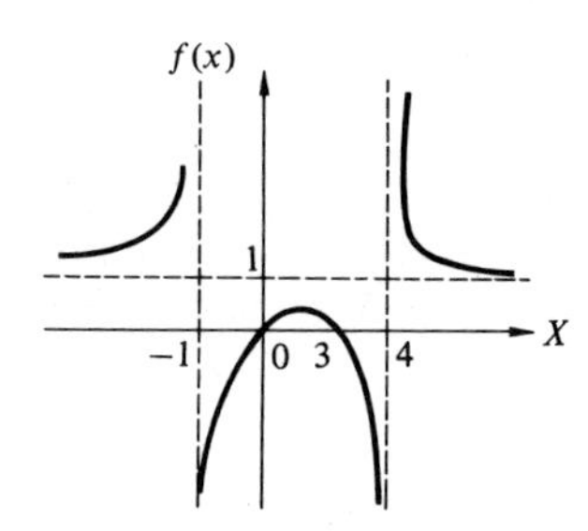

Figure E.49

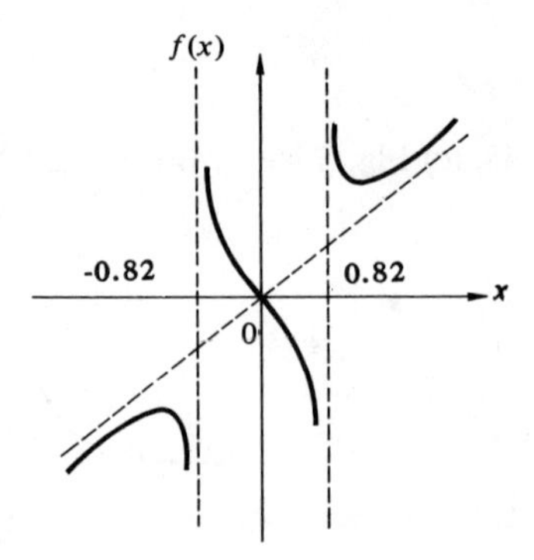

Figure E.50

Page 220, Problems 7.8

1. $3x^2/2y$. **2.** $(3x^2y - 2xy^3)/(3y^2 - x^3)$. **3.** $5x + 8y - 18 = 0$.

Page 221, Exercises 7.8

Group A

1. (a) $4/3y^2$. (c) $-x/y$. (e) $-(32x\sqrt{xy} + y)/(x - \sqrt{xy})$. **2.** (a) $2y^2/(y^2 + 2xy - x^2)$. (c) $\dfrac{4x\sqrt{1+x} - y}{2\sqrt{1+x}(\sqrt{1+x} + 2y)}$. (e) $[3(xy)^{2/3} - y]/[x + 2x^{2/3}y^{1/3}]$. **3.** (a) -0.5. (c) 0.

Group B

1. (a) $y = \pm\sqrt{4 - x^2}$. (c) $-x/\sqrt{4 - x^2}$, $x/\sqrt{4 - x^2}$. (e) $x + \sqrt{3}\,y = 4$, $x - \sqrt{15}\,y = 8$. **2.** (a) $y = \pm 3x^{1/2}$. (c) $(3/2)/x^{1/2}$, $-(3/2)/x^{1/2}$. **3.** $\sqrt{2}\,x - 6y = 9\sqrt{2}$. **4.** (a) $(y^2 + 2xy)(4y + 4x)/(x^2 + 2xy)^2 - 2x(y^2 + 2xy)^2/(x^2 + 2xy)^3 - 2y/(x^2 + 2xy)$. **5.** None. **7.** 1.05. **8.** (a) $(5xy^2 - x^3)/(9y^3 - 5x^2y)$. (c) (0, 0).

Page 224, Problems 7.9

1. $5dx/(25 - x^2)^{1/2}$. **2.** $\sqrt{100t^2 + 64}\,dt$. **3.** $5y^2 = 64x$. **4.** (a) 0.21. (b) -0.0087.

Page 226, Exercises 7.9

Group A

1. (a) $\sqrt{1 + 16x^6}\,dx$. (c) $\sqrt{1 + (1/4x)}\,dx$. (e) $\sqrt{1 + (4y^3 + 2y)^2}\,dy$.
2. (a) $\sqrt{1 + 4t^2}\,dt$. (c) $\sqrt{1 + 1/4t}\,dt$. (e) $\sqrt{4t^2 + 1}\,dt$. (g) dt. **3.** (a) $y = x^2$. (c) $y = x^{3/2}$. (e) $x^2 + y^2 = 1$. **4.** (a) 0.25; -0.03. (c) 0.35; -0.09. (e) -1; 2.8.

Group B

1. (a) $x = t$, $y = t^{1/2}$. (c) $x = 5\cos t$, $y = 5\sin t$. **2.** (a) $y = x^2$. (c) Yes. They are not unique.

Group C

1. $dS = 2\pi x^2\sqrt{1 + 4x^2}\,dx$. **3.** (a) $-1/4t^3$. (c) $-8(8 - t^2)^{-3/2}$.

Page 228, Problems 7.10

1. Fig. P.15. **2.** Fig. P.16. (3, 5); 5.8, 59°. **3.** Fig. P.17.

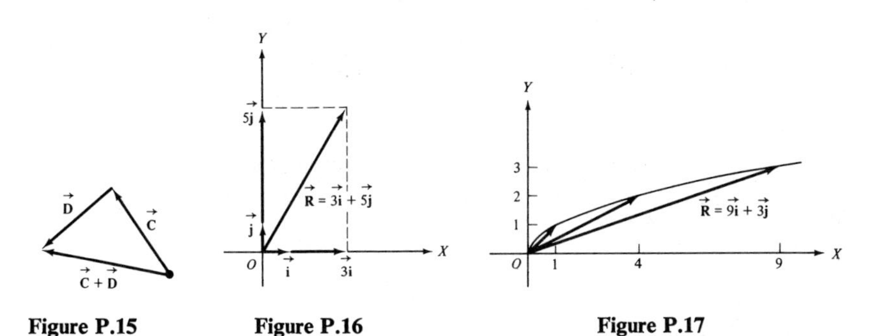

Figure P.15 Figure P.16 Figure P.17

Page 231, Exercises 7.10

Group A

1. (a) Fig. E.51. (2, 2); 2.8, 45°. (c) Fig. E.52. (−2, −3); 3.6, 236°. (e) Fig. E.53. (4, 0); 4, 0°. **2.** (a) $3\mathbf{i} + 2\mathbf{j}$; 3.6, 33.7°. (c) $-0.5\mathbf{i} - 0.4\mathbf{j}$; 0.64, 218.7°. (e) $-2\mathbf{i}$; 2, 180°. **3.** (a) $t\mathbf{i} + t^2\mathbf{j}$; Fig. E.54. (c) $t^3\mathbf{i} + t\mathbf{j}$; Fig. E.55.

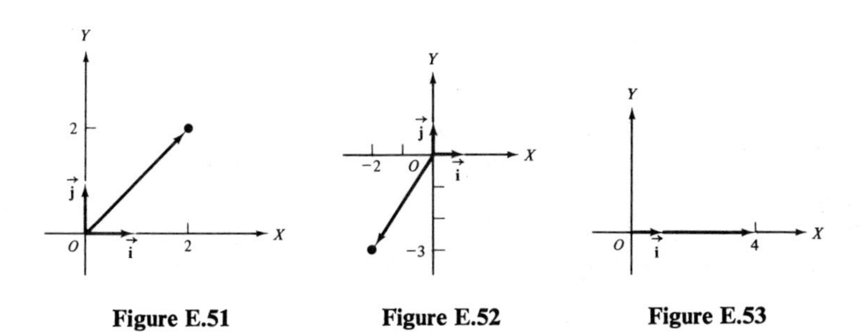

Figure E.51 Figure E.52 Figure E.53

Group B

2. (a) $3\mathbf{i} + 5\mathbf{j}$. (c) $-2\mathbf{i} + 6\mathbf{j}$. **4.** (a) $5.2\mathbf{i} + 3\mathbf{j}$. (c) $-4\mathbf{j}$. **6.** (a) $\mathbf{i}$. (c) $-2\mathbf{i} - 3\mathbf{j}$.

Page 234, Problems 7.11

1. $\mathbf{V} = 2\mathbf{i} + 3\mathbf{j} = 3.2\underline{/56.3^\circ}$ ft/s; $\mathbf{A} = 2\mathbf{i} + 6\mathbf{j} = 6.3\underline{/71.6^\circ}$ ft/s². $v = 27.7$ ft/s.
2. 6.1 m/s², 1.6 m/s².

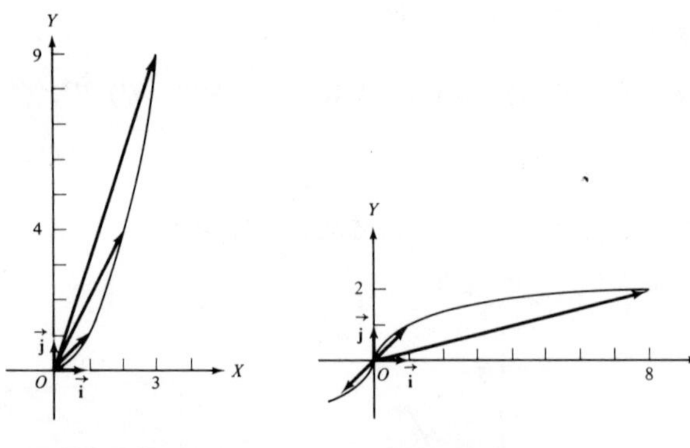

Figure E.54 Figure E.55

Page 236, Exercises 7.11

Group A

1. (a) 4.1. (c) 10.05. (e) 1.4. **2.** (a) $3\mathbf{i} + 2\mathbf{j} = 3.6\underline{/33.7^\circ}$. (c) $\mathbf{i} + \mathbf{j} = 1.4\underline{/45^\circ}$. **3.** 8 ft/s^2 towards the center. **5.** $0.89\mathbf{i} + 1.79\mathbf{j} = 2\underline{/63.6^\circ}$, $-0.32\mathbf{i} + 0.64\mathbf{j} = 0.72\underline{/116.6^\circ}$.

Group B

1. $4.01\underline{/-3.6^\circ}$. **3.** $17.9\underline{/26.5^\circ}$. **5.** $1 \leqq t \leqq 3$; not defined.

Page 237, Problems 7.12

1. 226.2 ft^3/s. **2.** 1.08 ft/min. **3.** 2.25 ft/s.

Page 239, Exercises 7.12

Group A

1. 6283 in.^{3}/s. **3.** 1 ft/s. **5.** -628 ft^3/min. **7.** 3.3 ft/s. **9.** 0.064 cm/s.

Group B

1. 25 ft lb/s. **3.** 0.0082 s/s. **5.** -0.08 ft/min. **7.** -0.096 atmospheres/min. **9.** -866 ft/s^2.

Page 242, Problems 7.13

1. $\frac{1}{3}(\frac{1}{3}x^3 - 5)^3 + C$. **2.** $\sqrt{x^2 - 6} + C$. **3.** -2600.

Page 242, Exercises 7.13

Group A

1. (a) $\frac{1}{3}(3x^2 + 2x + 1)^3 + C$. (c) $\sqrt{x^2 + 2} + C$. (e) $\frac{3}{4}(x^2 + 3)^{2/3} + C$. (g) $-\frac{1}{6}(16 - x^4)^{3/2} + C$. **2.** (a) 0.33. (c) 6556. (e) 10.7. **3.** (a) $\frac{1}{2}\sin^2 x + C$. (c) $-\frac{1}{4}\cos 2x + C$.

Page 246, Problems 8.2

1. 32. **2.** Fig. P.18.

Page 247, Exercises 8.2

Group A

1. (a) $(6)^2$. (c) $(\frac{1}{3})^3$. (e) $(4)^5$. **3.** Fig. E.56.

Group B

3. (a) The power of one is always one. **5.** $b^0 = 1$.

Page 249, Exercises 8.3

Group A

1. (a) e^5. (c) e^{t-u}. **3.** (a) Fig. E.57. (c) Fig. E.58. **4.** (a) exp $3x$. **5.** (a) exp $(4 + x)$.

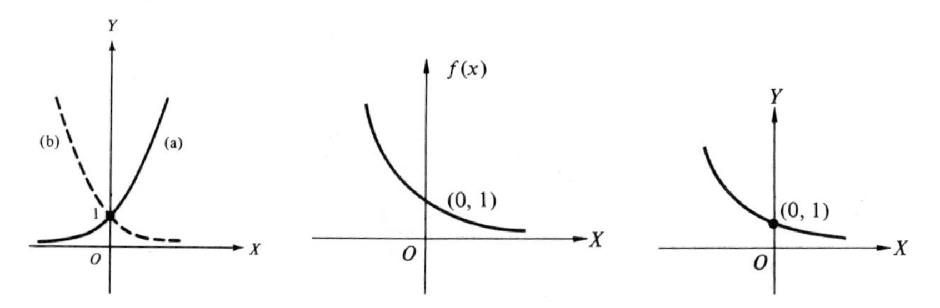

Figure P.18 Figure E.56 Figure E.57

Group B

3. $e^x(e^{\Delta x} - 1)/\Delta x$. No. **5.** $y = (H/2w) \cosh (wx/H)$. **6.** Fig. E.59.

Page 251, Problems 8.4

1. $(x - 2)/3$. **2.** No. For each value of $f(x)$ there is more than one value of x.

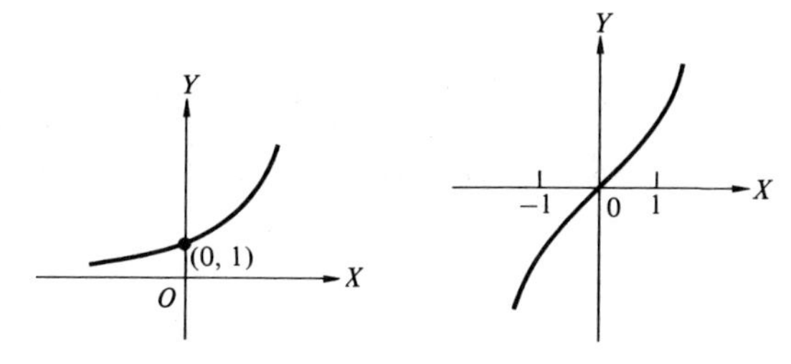

Figure E.58 Figure E.59

Page 251, Exercises 8.4

Group A

1. (a) Fig. E.60. $f^{-1}(x) = \frac{1}{6}(x - 2)$. (c) Fig. E.61. No inverse. (e) Fig. E.62. No inverse. **2.** (a) Fig. E.63. (c) Fig. E.64.

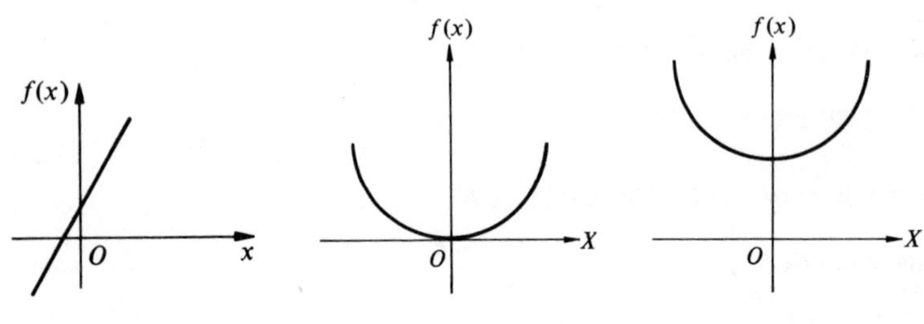

Figure E.60 **Figure E.61** **Figure E.62**

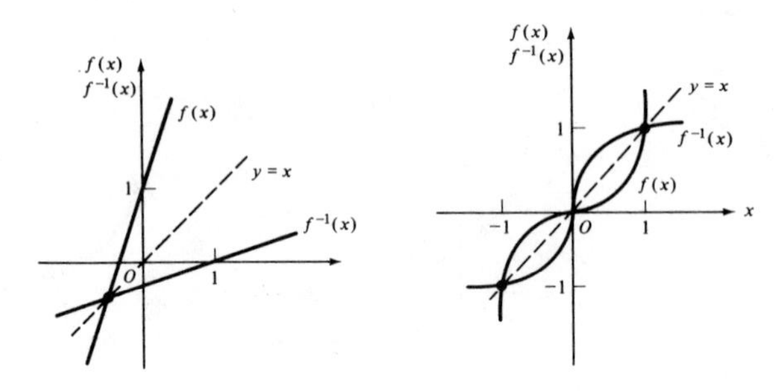

Figure E.63 **Figure E.64**

Page 252, Problems 8.5

1. (a) $5 = \log_2 32$. (b) $10^3 = 1000$. **2.** $x = 3.17$.

Page 254, Exercises 8.5

Group A

1. (a) $3 = \log_3 27$. (c) $0 = \log 1$. (e) $(32)^{1/5} = 2$. (g) $3 = \log_{1/2} (8)^{-1}$. (i) $3 = \ln 20.1$. **2.** (a) 10. (c) -4. **3.** (a) $f^{-1}(x) = \ln x$. (c) $f^{-1}(x) = 10^x$. **4.** (a) $\log 20$. (c) $\log_b 128$. (e) $\log 1250$. **5.** (a) 0.6931. (c) 0.3010. (e) 2.7081. **6.** (a) 1.467. **7.** (a) 2.3026. (c) -0.3067. (e) 1.65. (g) 2.535.

Group B

3. $\frac{1}{5} \ln 10 - \frac{1}{5} \ln (10 - I)$. **6.** (a) $-kt = \ln (1 - C/A_0)$.

Group C

4. (a) No. $\ln(MN) = \ln M + \ln N$. (c) No. $\ln N - \ln M = \ln(N/M)$.

Page 257, Problems 8.6

1. $6t^2/(2t^3 - 5)$. **2.** $(24t^2 + 8)/(t^3 + t) + 30/(5t - 7)$. **3.** $2.1715(6x^2 - 5)/(2x^3 - 5x)$. **4.** Fig. P.19. No critical points. No inflection point. Slope always positive and decreases as x increases. **5.** $\ln x$.

Page 258, Exercises 8.6

Group A

1. (a) $3/x$. (c) $4/t$. (e) $(3x^2 - 3)/(x^3 - 3x + 2)$. **2.** (a) $2(\log e)/x$. (c) $(5x^4 + 2)(\log e)/(x^5 + 2x + 1)$. **3.** (a) $dy = 3dx/x$. (c) $dy = 3dt/t$. **4.** (a) $1/x$. (c) $2/x$. (e) $2x \ln x^2$. **5.** $\log x$.

Group B

1. (a) $(2x + 2)/3(x^2 + 2x)$. (c) $(3x^2)/2(x^3 + 1) - (4x^3 + 5)/(x^4 + 5x)$. (e) $\dfrac{2(1 + x^2)\ln x - 2x^2(\ln x)^2}{x(1 + x^2)^2}$. **2.** (a) $[2(x + 1)/3(x^2 + 2x)] \log e$. (c) $[(7x^6 + 3)/2(x^7 + 3x) + 2] \log e - (x^2 \log e)/(x^3 + 2)$. **3.** e^x. **5.** $x^x[1 + \ln x]$. **6.** (a) $(2x)(y^3 + 3)/(x^2 + 2)(3y^2)$. **7.** (a) $(\ln 6x^2)/x$. **8.** (a) $1 + \ln 2x$. (c) $t + (2t - 1)\ln(t - 1)$. **9.** (a) $-1/x^2$. **10.** (a) Fig. E.65. (c) Fig. E.66. **13.** 0.5.

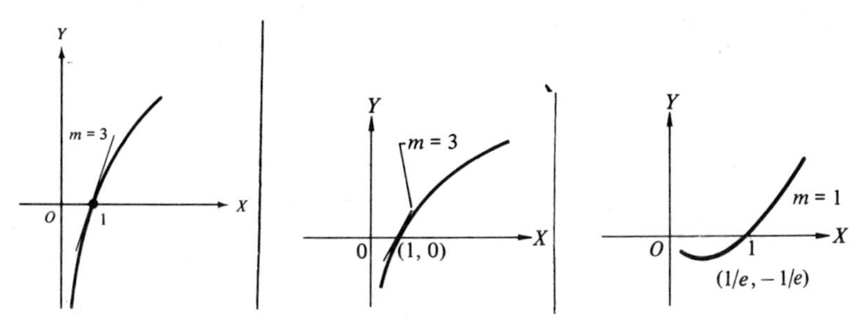

Figure P.19 Figure E.65 Figure E.66

Page 261, Problems 8.7

1. $6te^{t^2}$. **2.** $9.7t(5^{3t^2})$. **3.** $xe^{3x}(3x + 2)$. **4.** Fig. P.20.

Page 262, Exercises 8.7

Group A

1. (a) $2e^{2x}$. (c) $(2x + 2)e^{(x^2+2x+1)}$. (e) $(5x^4 + 8x^3 + 6x)e^{(x^5+2x^4+3x^2+1)}$. (g) $-6xe^{-3x^2}$. (i) 0. **2.** (a) $2e^{2x}$, $4e^{2x}$, $8e^{2x}$, $16e^{2x}$. **3.** (a) $3^x \ln 3$. (c) $2t(\ln 2)2^{t^2}$. **5.** Fig. E.67.

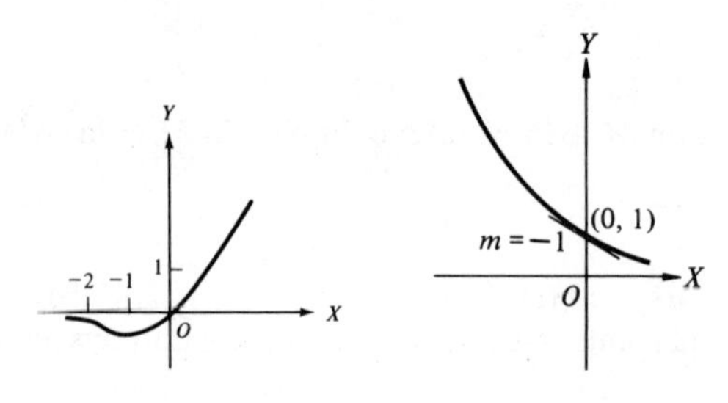

Figure P.20

Figure E.67

Group B

1. (a) $(x^2 + 2x)e^x$. (c) $(5x^2 - 2x + 5)e^{5x+2}/(x^2 + 1)^2$. (e) $[(6x^2 + 8x)/3(x^3 + 2x^2 + 1)^{1/3}]e^{(x^3+2x^2+1)^{2/3}}$. **3.** 2.27×10^{-4}. **5.** $x/(x^2 + 2)$. **7.** $-y/x$.

Page 264, Problems 8.8

1. $\ln t - 1/t + C$. **2.** 0.629. **3.** 2.34. **4.** 5.63. **5.** 1.3.

Page 266, Exercises 8.8

Group A

1. (a) $\frac{1}{4}\ln|4x + 1| + C$. (c) $2e^{x^3+6x} + C$. (e) $1/2a \ln|ax^2 + b| + C$. (g) $\frac{3}{4}(e^{2x} + 3)^{2/3} + C$. (i) $\frac{1}{4}\ln|e^{4x} + 1| + C$. **2.** (a) $\frac{1}{2}\ln 2 = 0.3466$. (c) $\ln\frac{5}{2} = 0.9163$. (e) $(e^{16} - 1)/4e^7 = 2026$. (g) $2(e - 1) = 3.44$. **3.** (a) 0.116. (c) $(ax + b)[\ln(ax + b) - 1]/a$.

Group B

1. (a) $-e^{-x} + C$. (c) $\frac{1}{2}\ln|e^{x^2} + 2| + C$. (e) $x^2/2 - 2x + 8\ln|x + 2| + C$. **2.** (a) $\frac{1}{2} + \ln 4 = 1.8863$. (c) $\frac{1}{2}(e - 1) = 0.8592$. (e) $3 + \ln 3 = 4.0986$. (g) $\frac{3}{2} + \ln\frac{3}{2} = 1.9055$. (i) $\frac{1}{2}\ln\frac{3}{4} = -0.1438$. **3.** (a) $(1 - e^{-2t})/4$.

Group C

2. (a) $5^x/\ln 5 + C$. (c) $(ae)^x/\ln(ae) + C$. (e) $2/\ln 3 = 1.82$.

Page 268, Exercises 8.9

Group A

1. (a) $2x - y - e = 0$, $x + 2y - 3e = 0$. (c) $(\ln 5)x - 5y + (1 + \ln 5) = 0$, $25x + 5(\ln 5)y - (\ln 5 - 25) = 0$. (e) $3x - y + 1 = 0$, $x + 3y - 3 = 0$. **2.** (a) Fig. E.68. (c) Fig. E.69. (e) Fig. E.70. **3.** Fig. E.71. $(-2, 0.54)$, $(0, 0)$. **5.** $(0.717, -0.123)$. **7.** $(1.82, 0.303)$. **9.** (a) 2.83. (c) 2.83. (e) 2.83. (g) $(e^2 + 1)^{3/2}/e^2 = 3.3$. **10.** (a) 0.693. (c) $(e^2 - 1)/2 = 3.2$.

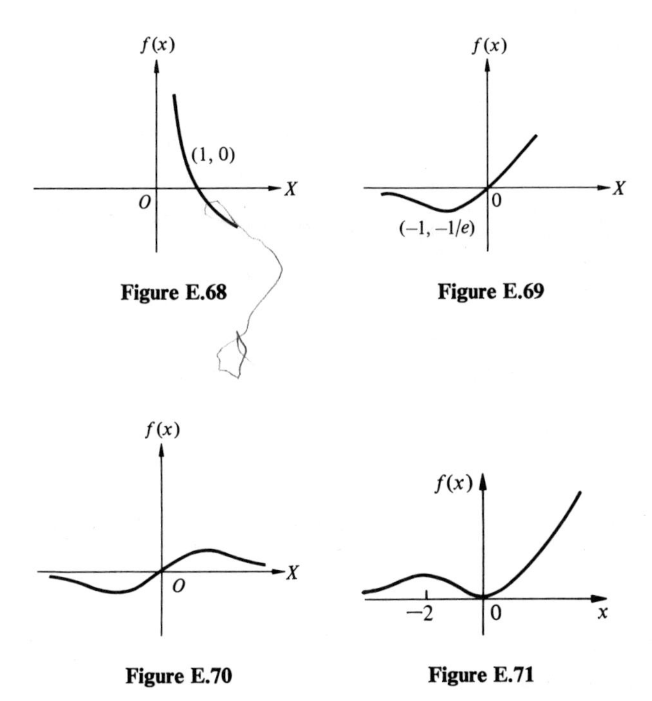

Figure E.68

Figure E.69

Figure E.70

Figure E.71

Group B

1. $v = 50(1 - e^{-16t/25})$. $v \longrightarrow 50$ ft/s. **3.** 0.2454, 0.424. **5.** (a) $\frac{1}{2}\ln\frac{5}{2} = 0.4582$. **7.** $nGT\ln(V_T/V_O)$. **9.** $s = e^{-t} + 2t - 1$. **10.** (a) 1.1. (c) 0.1. **11.** (a) 0.71. **12.** (c) 17.6. **13.** (a) 95,500 ft lb. **15.** 0.607.

Page 274, Exercises 9.2

Group A

1. Fig. E.72. **2.** (a) 1. (c) 0.

Group B

1. (0.707, 0.707); 0.707, 0.707, 1. **3.** 0.866, 0.5, 1.732. **5.** 0.

Group C

1. (a) $x^2 + y^2 = 1$; $\sin^2 u + \cos^2 u = 1$. (c) $1 + (x/y)^2 = 1/y^2$; $1 + \cot^2 u = \csc^2 u$. **3.** Fig. E.73. $|PM| = |P_rM_r|$, $|OM| = |OM_r|$ and $|T(u)| = |T(u_r)|$. **5.** (a) sin 1.34. (c) tan 0.46. (e) $-\sec 1.06$. (g) $-\sin 1.4$.

Page 279, Problems 9.3

1. -0.936.

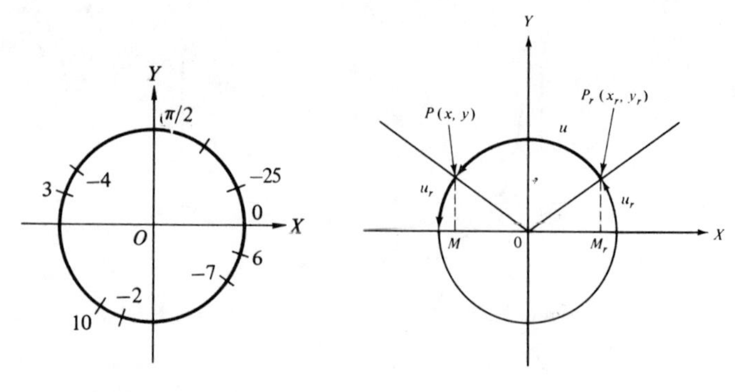

Figure E.72

Figure E.73

Page 279, Exercises 9.3

Group A

1. (a) (9.3.35, 36, 37). (c) (9.3.41, 42, 43). **2.** (a) 0.566. (c) −0.143. (e) −0.122.

Group B

3. (c) $13 \sin (t + \varphi)$ where $\sin \varphi = \frac{5}{13}$, $\cos \varphi = -\frac{12}{13}$.

Page 281, Problems 9.4

1. Fig. P.21. **2.** Fig. P.22.

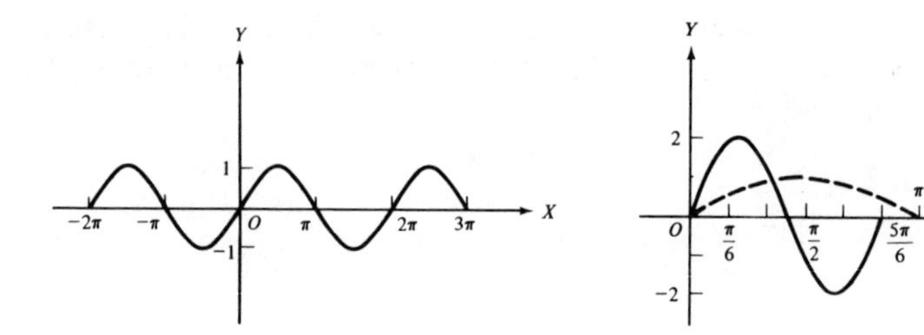

Figure P.21

Figure P.22

Page 282, Exercises 9.4

Group A

1. (a) Fig. E.74. (c) Fig. E.75. (e) Fig. E.76. (g) Fig. E.77.

Group B

1. (a) Fig. E.78. (c) Fig. E.79. $f(x) = \sin 2x$. **3.** Fig. E.80.

Page 284, Problems 9.5

1. (a) $6 \sec^2 2x$. (b) $72 \tan 4x \sec^2 4x$. **2.** (a) $-12 \csc^2 3x$. (b) $-28 \cot 7t \csc^2 7t$. **3.** (a) $14 \sec 2t \tan 2t$. (b) $24 \sec^2 4t \tan 4t$. **4.** (a) $-7 \sin 7x$. (b) $-30 \cos 5x \sin 5x$.

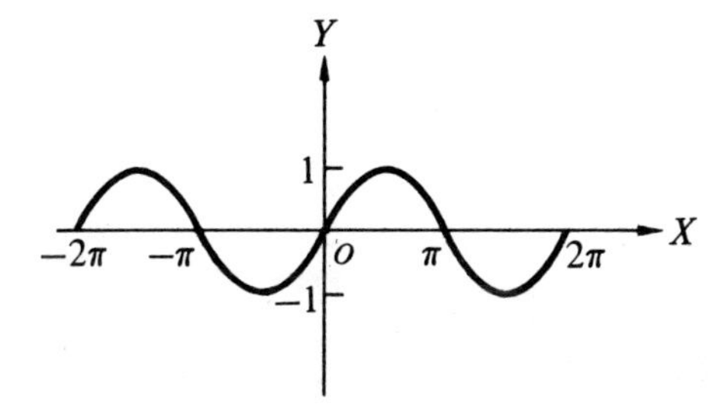

Figure E.74

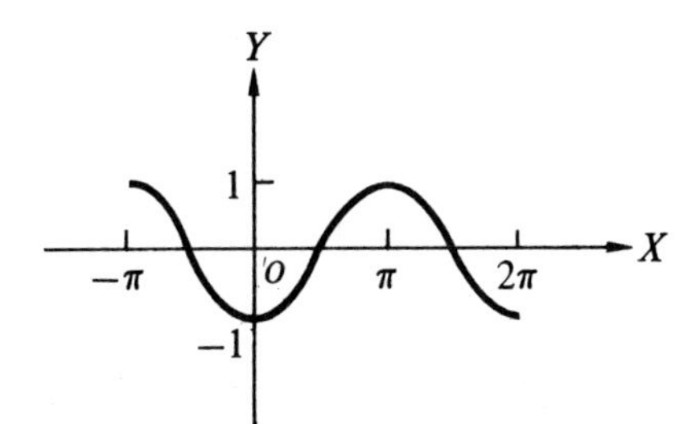

Figure E.75

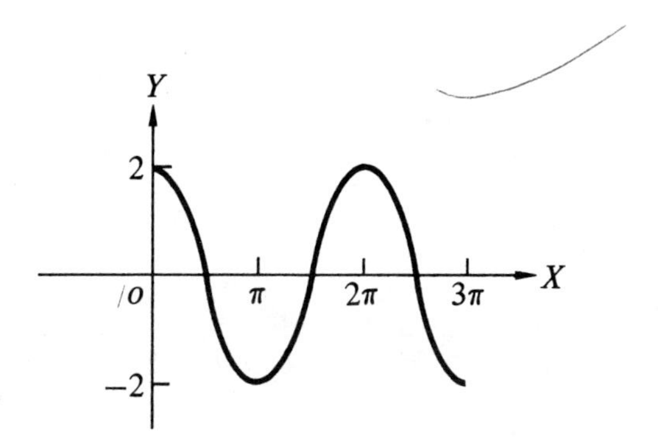

Figure E.76

5. (a) $-10 \cos 5t$. (b) $6 \sin^2 2t \cos 2t$. **6.** (a) $6 \csc 2x \cot 2x$. (b) $-6 \csc^2 3x \cot 3x$. **7.** $-4t^4 \sin t^2 + 6t^2 \cos t^2$. **8.** $24 \tan^3 5x \sec^4 2x + 30 \tan 5x \sec^6 2x$. **9.** $(-e^{-3t}/2) \cdot (7 \sin 5t + 11 \cos 5t) - (4/3) \cos 2t$. **10.** $[6x \sin (x^2 - y)]/[3 \sin (x^2 - y) - 1]$.

Page 291, Exercises 9.5

Group A

1. (a) $3 \cos 3x$. (c) $-5 \sin 5x$. (e) $3 \cos 3x - 2 \sin x$. (g) $2x + 3 \cos 3x$. (i) $7e^t - 550 \cos 110t$. **2.** (a) $6 \sec^2 2x$. (c) $-35 \sec 7x \tan 7x$. (e) $2t \sec^2 t^2$. (g) $-2 \csc^2 t$.

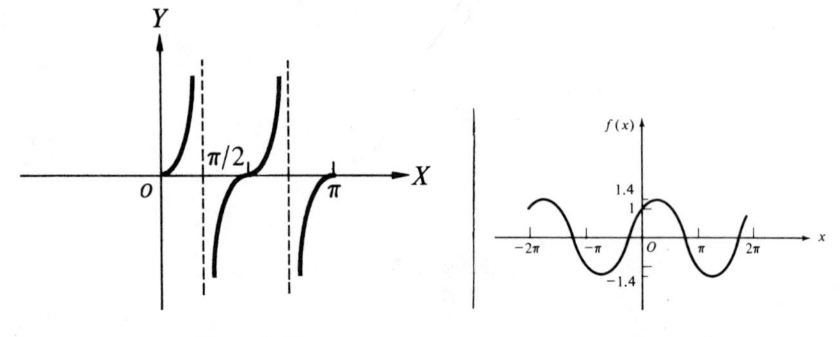

Figure E.77 Figure E.78

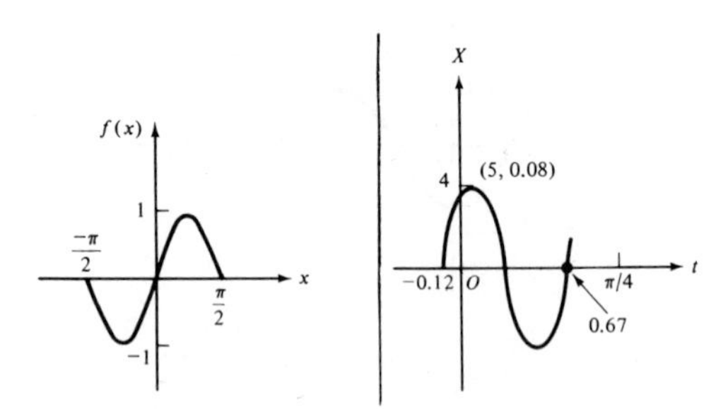

Figure E.79 Figure E.80

(i) $2(\sec 2x \tan 2x - \csc 2x \cot 2x)$. **3.** (a) $6x^2 \cos x^2 + 3 \sin x^2$. (c) $5(x^2 + 3) \cos 5x + 2x \sin 5x$. (e) $6 \sin 3x \cos 3x = 3 \sin 6x$. (g) $-6x^3 \sin x^3 + 2 \cos x^3$. (i) $-2(x^3 - 3x^2) \sin x^2 + (2x - 3) \cos x^2$. (k) $110e^{-3t} \cos 110t - 3e^{-3t} \sin 110t$. **4.** (a) $2 \cos 4x$. (c) $-\sin^2 (3x + 1) \sin 2x + 3 \cos^2 x \sin (6x + 2)$. **5.** (a) $\sec x(\sec^2 x + \tan^2 x)$. (c) $\tan x$. (e) $-e^{-\tan t} \sec^2 t$. (g) $\sec^2 3x(6x \tan 3x + 1)$. (i) $\sec x$. **6.** (a) $-12 \sin 2x$. (c) $-16 \sin 2x - 12 \cos 2x$. (e) $-9 \sin^2 x \cos 3x - 6 \sin 2x \sin 3x + 2 \cos 2x \cos 3x$. **7.** (a) $2 \sec^2 x \tan x$. (c) $18 \sec^2 3v(2 \tan^2 3v + \sec^2 3v)$. (e) $6 \sec^2 3t(3t \tan 3t + 1)$. **8.** (a) $6 \cos 2x\, dx$. (c) $(3x \cos 3x + \sin 3x)\, dx$. **9.** (a) $15(\sec^2 5x)dx$. (c) $x(\sec x)(x \tan x + 2)dx$.

Group B

1. 2.921. **2.** (a) $3 \cot 3x$. (c) $(x \cos x - \sin x)/x^2$. (e) $[(\cos \sqrt{t})/\sqrt{t} - \sin \sqrt{t}]/2$. (g) $-e^v \sin e^v$. **3.** (a) $x^{\tan x}[(\tan x)/x + (\sec^2 x) \ln x]$. (c) $(\cot x)^x[\ln \cot x - (x \csc^2 x/\cot x)]$. (e) $\tan (\theta/2) \sec^2 (\theta/2)$. **4.** (a) 1.540. (c) 1.830. **5.** (a) 25.8. (c) $-\pi$. (e) 0. **6.** (a) $(-x^2 \sin x - 2x \cos x + 2 \sin x)/x^3$. (c) $e^{\sin 2t}[4 \cos^2 2t - 4 \sin 2t]$. (e) $2e^x \cos x$. **7.** (a) $e^x(2 \sec^2 x \tan x + 2 \sec^2 x + \tan x)$. (c) $8 \csc^2 2x \cot 2x$. (e) $e^{ax}(2b^2 \csc^2 bx \cot bx - 2ab \csc^2 bx + a^2 \cot bx)$. **8.** (a) $-(\sin y + y \cos x)/(\sin x + x \cos y)$. (c) $-y/x$. **9.** (a) $2y/x^2$. **10.** (a) $\dfrac{\sec^2 (x - y)}{1 + \sec^2 (x - y)}$. (c) $\dfrac{y \sec^2 (x + y)}{1 - y \sec^2 (x + y)}$. (e) $\dfrac{-\csc^2 (x + y)}{[e^y + \csc^2 (x + y)]}$.

Page 295, Problems 9.6

1. Fig. P.23. **2.** (a) $t = 0, 3.1, 6.3, \cdots$, or $n\pi$, $n = 0, 1, 2, 3 \cdots$. (b) $t = 1.6, 4.7, \cdots$, or $(2n - 1)\pi/2$, $n = 1, 2, 3, \cdots$. $x = 0$. (c) $t = 0, 3.1, 6.3, \cdots$, or $n\pi$, $n = 0, 1, 2, \cdots$. $x = 4, -4, 4, \cdots$. **3.** $dv = 0.02$.

Page 297, Exercises 9.6

Group A

1. (a) 1, 0, −1, 0, 1. **5.** (a) 0.5151. (c) 1.07. (e) −0.896. **7.** $\sqrt{a^2 + b^2}$.

Group B

1. Fig. E.81. **2.** (a) 16 in./s. (c) −13.8 in./s. (e) $t = n\pi/8$, $n = 0, 1, 2, \cdots$. $x = 0$. **3.** 7.26. **5.** (a) (1) 8, 0, −32. (2) −8, 0, 32. (3) 0, 16, 0. (4) 7.27, −6.66. −29.09. (b) $\pi/12 = 0.262$. **7.** 45°. **9.** $\tan\theta = \mu$. **11.** (a) $ds = dt$. **12.** (a) $(-5 \sin t)\mathbf{i} + (5 \cos t)\mathbf{j}$. (c) Perpendicular.

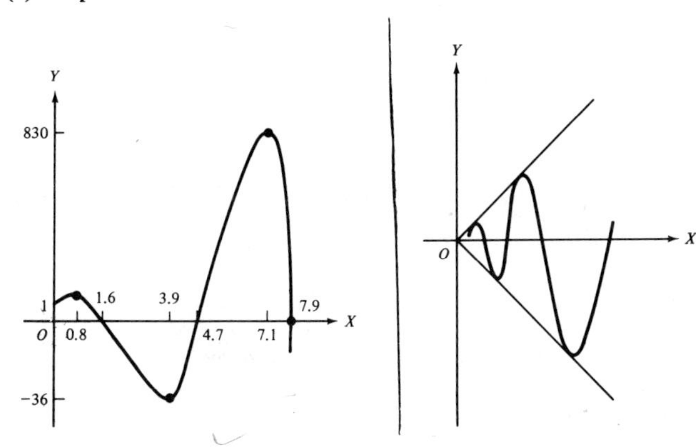

Figure P.23   Figure E.81

Page 300, Problems 9.7

1. $(1/8) \sin (8x + 5) + C$. **2.** $(1/7) \sec 7x + C$. **3.** $-\cos (\ln x) + C$. **4.** 0. **5.** $(-1/8)\cdot \cot^4 2x + C$.

Page 303, Exercises 9.7

Group A

1. (a) $-\frac{1}{2} \cos 2x + C$. (c) $\frac{1}{7} \tan 7x + C$. (e) $\frac{1}{2} \sec 2x + C$. (g) $-\frac{1}{3} \ln |\cos 3x| + C$. (i) $\frac{1}{5} \ln |\sec 5x + \tan 5x| + C$. **2.** (a) $\frac{1}{2} \sec 2x + C$. (c) $-\frac{1}{2} \cos e^{2x} + C$. (e) $\frac{1}{3} \ln |\sin 3x| + \frac{1}{3} \sin 3x + C$. (g) $-\frac{1}{4} \cos 2u + C$. **3.** (a) 1. (c) 0.949. (e) 0. **4.** (a) $\ln |\csc x - \cot x| + \cos x + C$. (c) $\tan x - x + C$.

Group B

1. (a) $\ln|\sin x + 4| + C$. (c) $-2\sqrt{\cos^2 \phi + 2} + C$. (e) $\frac{1}{5}(4 + \sec^2 x)^{5/2} + C$. (g) $2\sqrt{4 - \cot x} + C$. **2.** (a) 0.0606. (c) 0.693. (e) 0.2. **4.** (a) $\frac{1}{3}\sin^3 x + C$. (c) $\frac{1}{8}\sin^4 2x + C$. (e) $-\frac{1}{7}\cos^7 x + C$. **5.** (a) 0.5. (c) 0.5. Yes. **6.** (a) $(1/2)\tan^2 x + C$. (c) $(1/8)\cdot \tan^4 2x + C$. (e) $(-1/9)\cot^3 3x + C$. **7.** (a) $(1/4)\sec^4 x + C$. (c) $(-1/5)\csc^5 x + C$.

Group C

5. $\int \sin^m x[1 - \sin^2 x]^{(n-1)/2} \cos x\, dx$. **7.** $\int \cos^n x[1 - \cos^2 x]^{(m-1)/2} \sin x\, dx$.
8. (a) $x/8 - (1/32)\sin 4x + C$. (c) $(3/8)x - (1/4)\sin 2x + (1/32)\sin 4x + C$.
9. (a) $\tan x - x + C$. (c) $(1/3)\tan^3 x - \tan x + x + C$. (e) $(-1/3)\cot^3 x + \cot x + x + C$. **10.** (a) $(1/5)\sec^5 x - (1/3)\sec^3 x + C$. **11.** (a) $(-1/12)\cos 6x - (1/4)\cdot \cos 2x + C$. (c) $(-1/8)\sin 4x + (1/4)\sin 2x + C$.

Page 307, Problems 9.8

1. $7\ln|\sqrt{25 + x^2} + x| + C$. **2.** Yes. Avoids using a negative sign.
3. $(1/16)\sqrt{(4t^2 - 9)^3}(4t^2 + 6) + C$.

Page 309, Exercises 9.8

Group A

1. (a) $\ln|\sqrt{9 + x^2} + x| + C$. (c) 11.8. (e) $\ln|x + \sqrt{x^2 - 25}| + C$. (g) $\frac{1}{9}x/\sqrt{x^2 + 9} + C$. (i) $-\sqrt{x^2 + 1}/x + C$.

Group B

1. (a) 1.29. (c) 2.72. (e) $-\sqrt{4 - (x + 1)^2} + C$. (g) $0.84a^2$.

Page 311, Problems 9.9

1. −0.07. **2.** 2. Same. **3.** 19.7.

Page 312, Exercises 9.9

Group A

1. (a) 2. (c) 1.76. **3.** (a) 22.2. (c) 0.674. **5.** 0.4.

Group B

1. 12.6. **2.** (a) 12.3. (c) 0.023. **3.** $x = \ln|(t + \sqrt{t^2 - 25})/5| - \sqrt{t^2 - 25}/t$.
5. 76.4 volts. **7.** (a) 0.71. (c) 84.9. **8.** (a) 8.5. **9.** 3.14.

Page 317, Problems 10.1

1. $-\pi/6$. **2.** $\sqrt{3}/2 = 0.866$. **3.** $\pm\sqrt{1 - y^2}/y$, + for $y > 0$; − for $y < 0$. **4.** $(1 - x)/$

$(1 + x^2)$. **5.** 0.28. **6.** (a) $(2\pi/3) \pm 2n\pi$, and $(4\pi/3) \pm 2n\pi$, $n = 0, 1, 2, \cdots$. $u = 2\pi/3$. (b) No. It restricts the answer to only one number. **7.** Their domain is restricted.

Page 320, Exercises 10.1

Group A

1.

Function Value	Domain	Range
$y = \text{Sin } x$	$-90° \leqq x \leqq 90°$	$-1 \leqq y \leqq 1$
$y = \text{Cos } x$	$0 \leqq x \leqq 180°$	$-1 \leqq y \leqq 1$
$y = \text{Tan } x$	$-90° < x < 90°$	All real numbers.
$y = \text{Cot } x$	$0° < x < 180°$	All real numbers.
$y = \text{Sec } x$	$0° \leqq x < 90°$ or, $-180° \leqq x < -90°$	$y \geqq 1$, or $y \leqq -1$.
$y = \text{Csc } x$	$0° < x \leqq 90°$, or $-180° < x \leqq -90°$	$y \geqq 1$, or $y \leqq -1$.
$y = \text{Sin}^{-1} x$	$-1 \leqq x \leqq 1$	$-90° \leqq y \leqq 90°$
$y = \text{Cos}^{-1} x$	$-1 \leqq x \leqq 1$	$0° \leqq y \leqq 180°$
$y = \text{Tan}^{-1} x$	All real numbers.	$-90° < y < 90°$
$y = \text{Cot}^{-1} x$	All real numbers.	$0° < y < 180°$
$y = \text{Sec}^{-1} x$	$x \geqq 1$, or $x \leqq -1$.	$0° \leqq y < 90°$, or $-180° \leqq y < -90°$.
$y = \text{Csc}^{-1} x$	$x \geqq 1$, or $x \leqq -1$.	$0° < y \leqq 90°$, or $-180° < y \leqq -90$.

2. (a) $\pi/6$. (c) 0. (e) $-\pi/2$. (g) $-\pi/4$. (i) $-\pi/6$. (j) $3\pi/4$. **3.** (a) 0.85. (c) 1.94. (e) -0.95. **4.** (a) $\frac{5}{13}$. (c) $\frac{1}{2}$. (e) $\frac{24}{25}$. (g) $-\frac{4}{3}$. (i) $-\frac{7}{24}$. (k) $\frac{15}{8}$. **5.** (a) $2x$. (c) $y/\sqrt{1 - y^2}$. (e) $1 - 2u^2$. (g) $2v/(1 + v^2)$. (i) 0.25. (k) -0.25. (m) -0.11. **6.** (a) $\text{Tan}^{-1}(1/x)$. (c) $\text{Cos}^{-1}(1/x)$. (e) $\text{Sin}^{-1}(1/3x)$.

Group B

2. (a) Fig. E.82. (c) Fig. E.83. (e) Fig. E.84. **4.** (a) Fig. E.85. (c) Fig. E.86. (e) Fig. E.87.

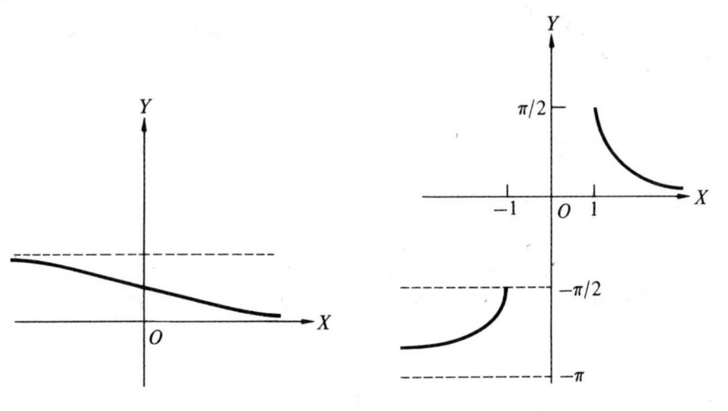

Figure E.82 Figure E.83

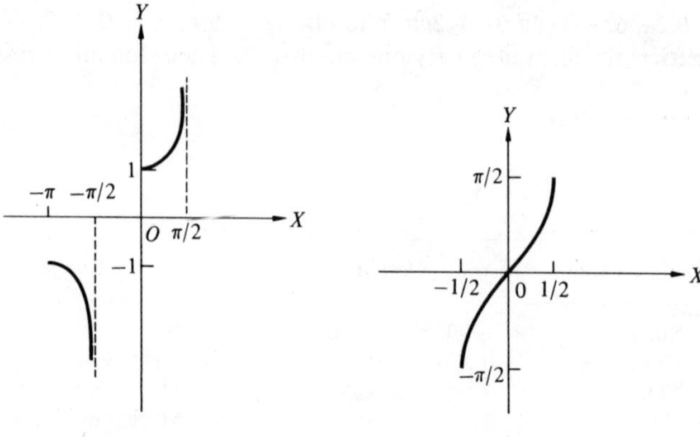

Figure E.84

Figure E.85

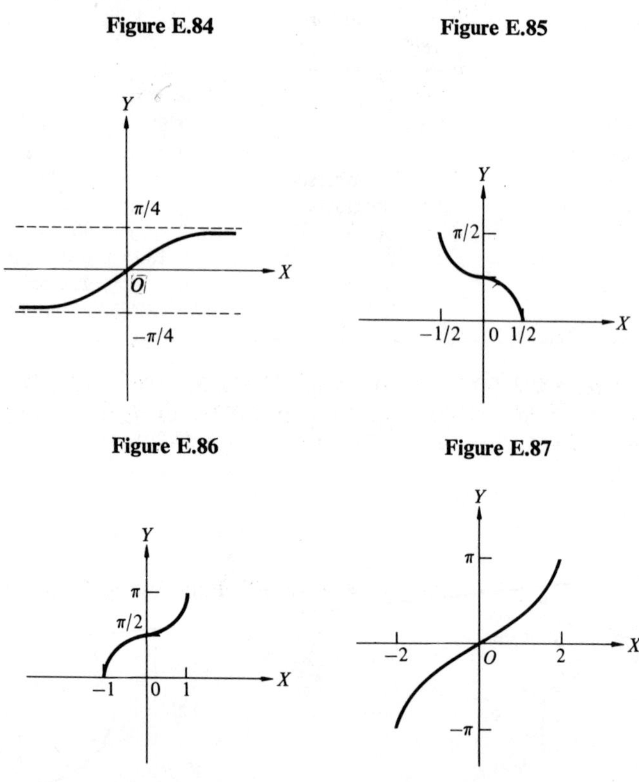

Figure E.86

Figure E.87

Figure E.88

Figure E.89

(g) Fig. E.88. (i) Fig. E.89. **5.** (a) Always. (c) Never true. (e) Always. **6.** (a) $\sin\theta = \frac{1}{2}$ has many solutions. $\theta = \operatorname{Sin}^{-1}(\frac{1}{2})$ has only one solution. **7.** $-\pi/2 \leqq \operatorname{Sin}^{-1}\theta \leqq \pi/2$.

Page 323, Problems 10.2

1. $5/\sqrt{1-25t^2}$. **2.** $2\tan 2x/\sqrt{1-(\ln\cos 2x)^2}$.

Page 324, Exercises 10.2

Group A

1. (a) $2/\sqrt{1-4x^2}$. (c) $-3/\sqrt{1-9x^2}$. (e) $2/(1+4x^2)$. **2.** (a) $3/\sqrt{3+6x-9x^2}$. (c) $2/(x^2+4)$.

Group B

1. (a) $2x/\sqrt{1-x^4}$. (c) $1/x(1+\ln^2 x)$. (e) $x/\sqrt{1-x^2}+\text{Sin}^{-1}\,x$. (g) $2/(1+4x^2)\cdot \text{Tan}^{-1}\,2x$. **2.** (a) 1.79. (c) -0.29. **3.** (a) 4.63.

Page 325, Problems 10.3

1. 0.93. **2.** $(1/24)\,\text{Tan}^{-1}\,(2x/3)+C$. **3.** 0.245.

Page 326, Exercises 10.3

Group A

1. (a) $\text{Sin}^{-1}\,x/3+C$. (c) $\frac{1}{2}\,\text{Tan}^{-1}\,x/2+C$. (e) $(1/\sqrt{2})\,\text{Sin}^{-1}\,\sqrt{\frac{2}{3}}\,x+C$. (g) $\frac{1}{3}\,\text{Tan}^{-1}\,(x+2)/3+C$. (i) $\frac{1}{4}\,\text{Tan}^{-1}\,(x+2)/4+C$. **2.** (a) 0.52. (c) 0.39. (e) 1.57.

Group B

1. (a) $\text{Sin}^{-1}\,(x+4)/5+C$. (c) $\dfrac{1}{\sqrt{2}}\,\text{Tan}^{-1}\,(x+2)/\sqrt{2}+C$. **2.** (a) 1.57. (c) 0.55.

Group C

3. (a) 0.52.

Page 328, Exercises 10.4

Group A

1. (a) 0.1. (c) 0.835. **3.** (a) $y=x$, $y=-x$. (c) $y=-1.15x+1.51$, $y=0.87x+2.52$.

Group B

1. (0.72, 0). **2.** (a) Fig. E.90. **3.** $\theta=\text{Tan}^{-1}\,bx/(b^2+2x^2)$, $x=0.71b$. **5.** 12 ft. **7.** 0.79. **9.** 0.016.

Page 332, Problems 11.2

1. $(9/14)(x-2)^{4/3}(2x+3)+C$. **2.** $(1/9)\ln|9x^2+16|-(5/12)\,\text{Tan}^{-1}\,(3x/4)+C$. **3.** $(1/4)\,\text{Tan}^{-1}\,[(t+2)/4]+C$. **4.** $(3/2)\ln|t^2+4t+20|-(7/4)\,\text{Tan}^{-1}\,[(t+2)/4]+C$.

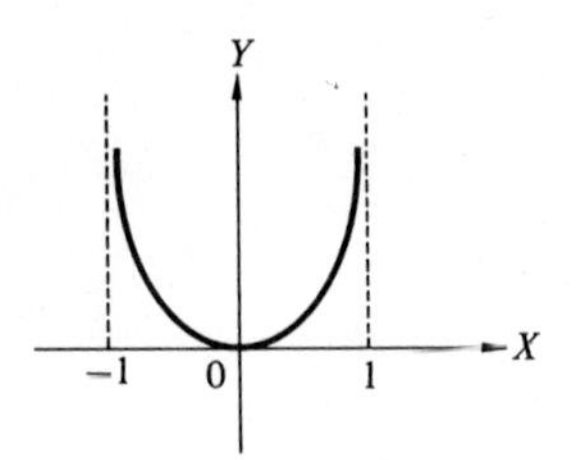

Figure E.90

Page 334, Exercises 11.2

Group A

1. (a) $\frac{1}{3}(x^2 + 2)^{3/2} + C$. (c) $\frac{1}{6}(x^4 - 16)^{3/2} + C$. (e) $-\frac{1}{3}(25 - x^2)^{3/2} + C$. (g) $-\frac{1}{3}(7 - x^2)^{3/2} + C$. (i) $\frac{1}{3}(25 + x^2)^{3/2} + C$. (k) $\sqrt{x^2 - 4} - 2\,\text{Tan}^{-1}\sqrt{x^2 - 4}/2 + C$. (m) $\frac{1}{9}(e^{3x} + 4)^3 + C$. (o) $\frac{1}{2}\ln|x^2 + 16| + \text{Tan}^{-1}x/4 + C$. (q) $\frac{1}{9}\ln|9x^2 + 25| - \frac{16}{15}\text{Tan}^{-1}3x/5 + C$. (s) $\frac{3}{2}\text{Tan}^{-1}(x - 2)/2 + C$. (u) $\text{Sin}^{-1}(x + 1)/5 + C$. (w) $\frac{1}{2}\ln|x^2 + 4x + 13| - \frac{2}{3}\text{Tan}^{-1}(x + 2)/3 + C$.

Group B

1. (a) $(x^2/4)\sqrt{25 - 16x^4} + \frac{25}{16}\text{Sin}^{-1}4x^2/5 + C$. (c) $\frac{1}{96}(25 + 16x^4)^{3/2} + C$. (e) $(2/\sqrt{3})\,\text{Tan}^{-1}(2x + 3)/\sqrt{3} + C$.

Group C

2. (a) $\frac{1}{10}\ln|(x - 5)/(x + 5)| + C$. (c) $\ln|x + \sqrt{x^2 + 9}| + C$. (e) $\ln|x + \sqrt{x^2 - 9}| + C$.

Page 336, Problems 11.3

1. $(x^2/4)(2\ln x - 1) + C$. **2.** -2. **3.** 1.91. **4.** $x\,\text{Cos}^{-1}x - \sqrt{1 - x^2} + C$.

Page 338, Exercises 11.3

Group A

1. (a) $(e^{2x}/4)(2x - 1) + C$. (c) $\frac{1}{4}\cos 2x + 2x\sin 2x + C$. (e) $(e^{-t}/2)(-\cos t + \sin t) + C$. (g) $x\,\text{Sin}^{-1}2x + \frac{1}{2}\sqrt{1 - 4x^2} + C$. (i) $x^3(\frac{1}{3}\ln x - \frac{1}{9}) + C$. **3.** $(e^3 - 4)/9e^3 = 0.089$.

Group B

1. (a) $e^t(t^2 - 2t + 2) + C$. (c) $[e^{ax}/(a^2 + b^2)](a\cos bx + b\sin bx) + C$. (e) $\frac{1}{2}\sec x\tan x + \frac{1}{2}\ln|\sec x + \tan x| + C$. **2.** (a) 9.87. (c) -0.19. (e) 3.04. **3.** 19.74. **5.** 0.174.

Group C

1. $(e^{3x}/27)(9x^2 - 6x + 2) + C$. **2.** (a) $2x \sin x - (x^2 - 2) \cos x + C$. (c) (1) $e^{2x}(\frac{1}{2}x^4 - x^3 + \frac{3}{2}x^2 - \frac{3}{2}x + \frac{3}{4}) + C$, (2) $x^5 \sin x + 5x^4 \cos x - 20x^3 \sin x - 60x^2 \cos x + 120x \sin x + 120 \cos x + C$.

Page 341, Problems 11.4

1. $\ln|(x-2)^2/(x+1)^3| + C$. **2.** $\text{Tan}^{-1}(x/2) - \ln|(x-1)^3| + C$.

Page 342, Exercises 11.4

Group A

1. (a) $\ln|(x+2)(x-1)^2| + C$. (c) $\ln|(x+1)^2/(x+2)| + C$. (e) $\ln|(x+1)/(x-1)(x-2)^2| + C$. (g) $\text{Tan}^{-1} x - \ln|x+1| + C$. (i) $\ln|(x-1)| + \frac{1}{8}\ln|4x^2+25| + \frac{3}{10}\text{Tan}^{-1} 2x/5 + C$.

Group B

1. (a) $\ln\left|\dfrac{(x+2)(x-2)^2}{(x+3)^2}\right| + C$. (c) $\ln\left|\dfrac{x^2+x+1}{x+1}\right| + C$. (e) $\ln\left|\dfrac{(x+1)^2}{(x+2)(x+3)}\right| + C$. **2.** (a) $\ln(\frac{81}{80}) = 0.0125$. (c) $\ln(\frac{9}{10}) = -0.1054$. (e) $\ln 3 + \pi/8 = 1.49$. **3.** $[1/k(a-b)]\ln|(ab-bx)/(ab-ax)|$.

Group C

1. (a) $\ln|(x-1)/(x-2)^2| - 1/(x-2) + C$.

Page 344, Problems 11.5

1. $(120x^2 + 240x + 400)(5-2x)^{3/2} + C$. **2.** $(5/4)x^2\sqrt{16x^4 - 9} - (45/16)\ln|4x^2 + \sqrt{16x^2 - 9}| + C$. **3.** 0.67. **4.** 0.089. **5.** 0.744.

Page 345, Exercises 11.5

Group A

1. (a) $-\frac{1}{4}\ln|(x-2)/(x+2)| + C$. (c) $\ln|x + \sqrt{x^2-9}| + C$. (e) $(x/2) - \frac{1}{16}\sin 8x + C$. (g) $\frac{1}{3}[\tan 3x - 3x] + C$. (i) $(x/\sqrt{16-4x^2}) - \frac{1}{2}\text{Sin}^{-1} x/2 + C$. **2.** (a) 1.15. (c) 5.06.

Group B

1. (a) $x/3\sqrt{3-4x^2} + C$. (c) $10e^{2x}(\frac{1}{2}x^4 - x^3 + \frac{3}{2}x^2 - \frac{3}{2}x + \frac{3}{4}) + C$. (e) $-5x/\sqrt{x^2+3} + 5\ln|x + \sqrt{x^2+3}| + C$. (g) $\frac{1}{8}\ln|4x^2+5x+1| - \frac{5}{24}\ln(2-4x)/(7+4x)| + C$. (i) $-\frac{1}{2}(\cos x) - \frac{1}{10}(\cos 5x) + C$. (k) $\frac{1}{27}[6x \sin 3x - (9x^2 - 2)\cos 3x] + C$. (m) 0.59. (o) 0.44.

Page 347, Problems 11.6

1. 0.019. **2.** 1.57. **3.** Does not exist. **4.** 1.05.

Page 349, Exercises 11.6

Group A

1. (a) 1. (c) $1/(p-1)$. (e) 1.57.

Group B

1. (a) Does not exist. (c) $1/(1-p)$. (e) 1.57. (g) 1.57. (i) 1. **3.** 3.14.

Group C

1. (a) $f(s) = 1/s^2$, defined for $s > 0$. **2.** Does not exist.

Page 355, Problems 12.2

1. (a) Symmetric with respect to the Y-axis. (b) Symmetric with respect to the X-axis. (c) Symmetric with respect to the origin.

Page 355, Exercises 12.2

Group A

1. Symmetric with respect to: (a) Both axes and origin. (c) X-axis. (e) Origin. (g) Both axes and origin. (i) No symmetry.

Group B

1. (a) Y-axis. (c) No symmetry

Group C

1. (a) Y-axis. (c) Both axes and origin. **2.** (a) Y-axis. (c) Both axes and origin. (e) X-axis.

Page 358, Problems 12.3

1. Circle, center at origin, radius of 3. **2.** Circle, center at $(-2, 1)$, radius is 3.87. **3.** $C(1, -2)$, $a = 3$. **4.** $x^2 + y^2 + 4x - 4y + 4 = 0$. **5.** 153.9. **6.** 37.7.

Page 362, Exercises 12.3

Group A

1. (a) Circle: $C(0, 0)$; $a = 2$. (c) Circle: $C(0, 0)$; $a = 2.31$. (e) Circle: $C(-3, 0)$; $a = 4$.

(g) A point at (0, 0). **2.** (a) $A = 1$, $B = 0$, $C = 1$, $D = E = 0$, $F = -4$. (c) $A = 3$, $B = 0$, $C = 3$, $D = E = 0$, $F = -16$. (e) $A = 1$, $B = 0$, $C = 1$, $D = 6$, $E = 0$, $F = -7$. (g) $A = 1$, $B = 0$, $C = 1$, $D = E = F = 0$. **3.** (a) (2, 3); 2.83. (c) (1, −2); 2. (e) (0, −2); 4.47. **4.** (a) $x^2 + y^2 = 4$. (c) $(x - \frac{1}{3})^2 + y^2 = 1$. (e) $(x + \frac{1}{4})^2 + (y + 3)^2 = \frac{1}{4}$. **5.** $(x - 1)^2 + (y - 1)^2 = 25$. **7.** $(x - 2)^2 + (y + 2)^2 = 4$.

Group B

1. $\frac{3}{4} = 0.75$. **3.** $\{(x, y) \mid (x - h)^2 + (y - k)^2 = a^2\}$. **5.** No. (a point); no. **7.** $x^2 + y^2 = a^2$. **9.** πa^2. **11.** 4.6. **13.** $\pi a^4/4$. **15.** 1.67 ft/s. **17.** $2\pi a$. **18.** (a) $4\pi a^2$. **19.** $V = L[(h - a)\sqrt{2ah - h^2} + a^2 \text{Sin}^{-1}(h - a)/a + \pi a^2/2]$; 3.4 ft; 4 gal.

Page 365, Problems 12.4

1. Parabola, $F(-3, 0)$, $V(0, 0)$, directrix $x = 3$, opens to left. **2.** Fig. P.24. $F(0.75, 0)$, f.w. = 3. **3.** Fig. P.25. $V(1, 2)$, $F(1, 4)$, f.w. = 8. **4.** $x^2 - 6x + 4y - 3 = 0$. **5.** $s = \int_0^3 \sqrt{1 + (3 - 2y)^2}\, dy$.

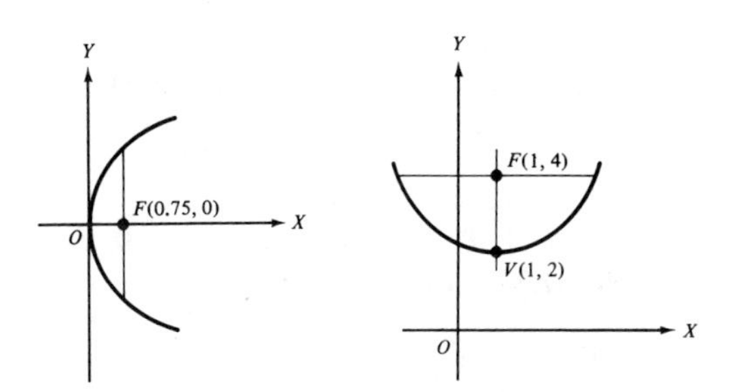

Figure P.24 **Figure P.25**

Page 370, Exercises 12.4

Group A

1. All parabolas. (a) $V(0, 0)$; $F(1, 0)$; f.w. = 4. (c) $V(0, 0)$; $F(-2, 0)$; f.w. = 8. (e) $V(0, 0)$; $F(0, 1)$; f.w. = 4. (g) $V(0, 0)$; $F(0, -\frac{1}{4})$; f.w. = 1. **2.** All parabolas. (a) $V(1,1)$; $F(1, 2)$; f.w. = 4. (c) $V(1, 1)$; $F(1, -1)$; f.w. = 8. (e) $V(1, 1)$; $F(\frac{5}{4}, 1)$; f.w. = 1. (g) $V(2, 4)$; $F(2, \frac{15}{4})$; f.w. = 1. **3.** (a) $(y + 2)^2 = 4(x - 1)$. (c) $(x - 2)^2 = 16(y + 1)$. **4.** (a) Maximum (4.5, 20.25). Fig. E.91. (c) Minimum (0, −4). Fig. E.92. (e) Minimum (0, 4). Fig. E.93. (g) Minimum (−1.5, −0.25). Fig. E.94. **5.** (a) 10.7. (c) 10.7. **7.** (0, −1.6).

Group B

1. (a) $y^2 = 4x$. **2.** (a) $x^2 = 4y$. **3.** Vertex, focus, focal width. **5.** $\{(x, y) \mid y^2 = 4px\}$. **7.** $2x^2 + y - 2 = 0$. **9.** 10.1. **11.** $s = 16t^2 + c_1 t + c_2$.

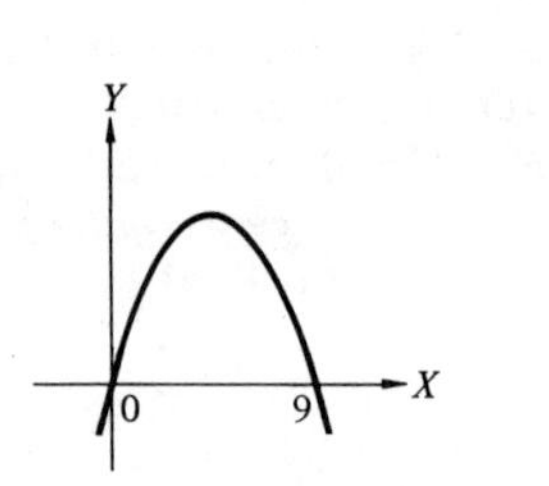

Figure E.91

Figure E.92

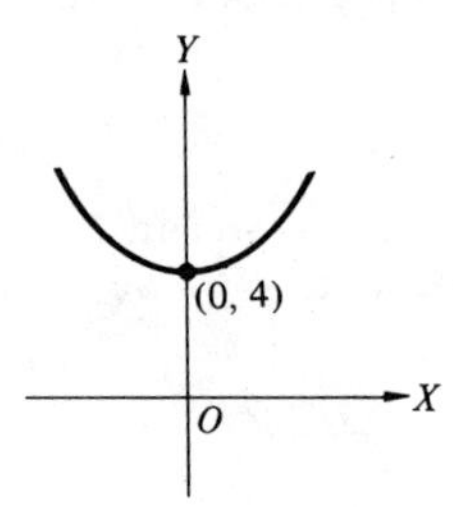

Figure E.93

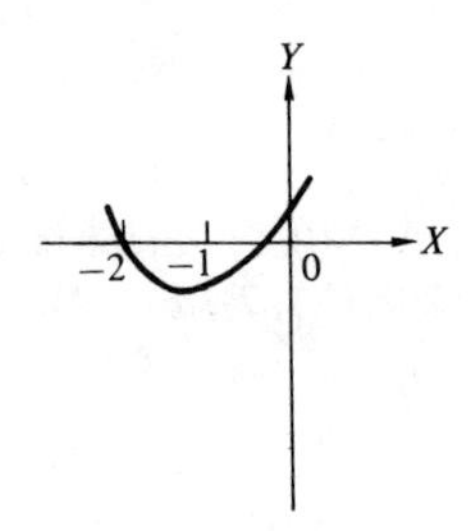

Figure E.94

Group C

1. $A = B = E = F = 0$, $C = 1$, $D = -8$. **3.** $y^2 + 8x = 0$. **7.** $3x^2 - 1000y = 0$ for origin at vertex. 211 ft. **9.** (a) 68. (c) 1.177. (e) 0.251.

Page 375, Problems 12.5

1. Fig. P.26. $C(0, 0)$, $F(\pm 1.7, 0)$, $V(\pm 2, 0)$. **2.** Fig. P.27. $V(0, \pm 5)$, $F(0, \pm 4.6)$. **3.** Fig. P.28. $C(-1, 2)$, $V(-1, 6)$, $V'(-1, -2)$, $F(-1, 5)$, $F'(-1, -1)$. **4.** $[(x + 1)^2/9] + [(y + 2)^2/5] = 1$.

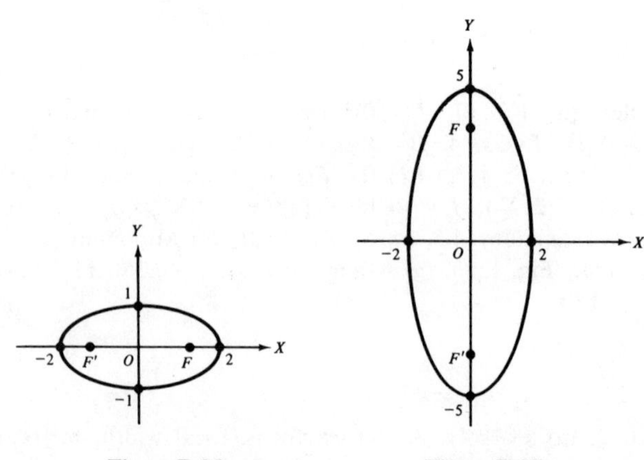

Figure P.26

Figure P.27

Page 379, Exercises 12.5

Group A

1. All are ellipses. (a) $C(0, 0)$; $V(\pm 5, 0)$; $F(\pm 3, 0)$. (c) $C(0, 0)$; $V(\pm 2, 0)$; $F(\pm 1, 0)$. (e) $C(0, 0)$; $V(0, \pm 4)$; $F(0, \pm 2.6)$. (g) $C(0, 0)$; $V(0, \pm 2)$; $F(0, \pm 1.7)$. **2.** All are ellipses. (a) $C(1, 2)$; $V(-4, 2)$, $(6, 2)$; $F(4, 2)$, $(-2, 2)$. (c) $C(-1, 2)$; $V(-1, 6)$, $(-1, -2)$; $F(-1, 5)$, $(-1, -1)$. **3.** (a) $(x^2/16) + (y^2/9) = 1$. (c) $(x^2/16) + (y^2/25) = 1$. (e) $[(x - 1)^2/25] + [(y - 2)^2/16] = 1$.

Group B

3. $A = b^2$, $B = 0$, $C = a^2$, $D = -2b^2h$, $E = -2a^2k$, $F = b^2h^2 + a^2k^2 - a^2b^2$. **5.** No. **7.** $y = \pm(b/a)\sqrt{a^2 - x^2}$. No. Two values for y. Yes. $|x| \leqq a$. **9.** $12\pi = 37.7$. **11.** $\pi a^3 b/4$. **13.** πab. Same. **15.** 1 unit. **17.** $7\sqrt{21}/16 = 2$.

Group C

1. $16x + 15y - 100 = 0$. $\alpha_1 = \alpha_2 = 65.7°$.

Page 383, Problems 12.6

1. Fig. P.29. $C(0, 0)$, $V(\pm 2, 0)$, $F(\pm 3.6, 0)$, $y = \pm 3x/2$. **2.** Fig. P.30. $C(1, -1)$, $V(1, 0.4)$, $V'(1, -2.4)$, $F(1, 1.2)$, $F'(1, -3.2)$.

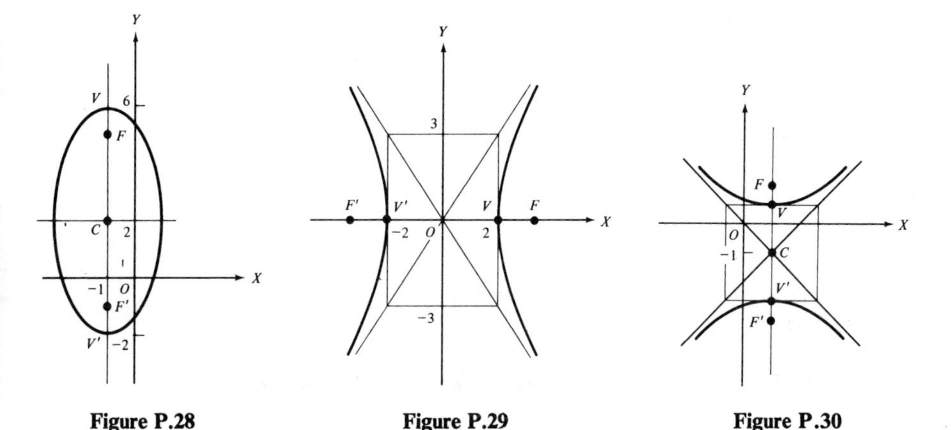

Figure P.28 Figure P.29 Figure P.30

Page 386, Exercises 12.6

Group A

1. All hyperbolas. (a) $C(0, 0)$; $V(\pm 4, 0)$, $F(\pm 5, 0)$; $y = \pm 3x/4$. (c) $C(0, 0)$; $V(\pm 1, 0)$; $F(\pm 1.4, 0)$; $y = \pm x$. (e) $C(0, 0)$; $V(0, \pm 4)$; $F(0, \pm 5)$; $y = \pm 4x/3$. (g) $C(0, 0)$; $V(0, \pm 5)$; $F(0, \pm 13)$; $y = \pm 5x/12$. **2.** All hyperbolas. (a) $C(1, 2)$; $V(3.4, 2)$, $(-1.4, 2)$;

$F(4.3, 2)$, $(-2.3, 2)$. (c) $C(1, -1)$; $V(1, 3)$, $(1, -5)$; $F(1, 4)$, $(1, -6)$. **3.** (a) $(x^2/16) - (y^2/9) = 1$. (c) $(x^2/16) - (y^2/9) = 1$. (e) $[(y + 1)^2/4] - [9(x + 2)^2/16] = 1$. **5.** 5.2. **7.** 4.2.

Group B

1. Yes, if $c = d = e \neq 0$. **5.** $A = b^2$, $B = 0$, $C = -a^2$, $D = -2b^2h$, $E = 2a^2k$, $F = b^2h^2 - a^2k^2 - a^2b^2$. $A = -a^2$, $B = 0$, $C = b^2$, $D = 2a^2h$, $E = -2b^2k$, $F = b^2k^2 - a^2h^2 - a^2b^2$. **7.** $y = \pm(b/a)\sqrt{x^2 - a^2}$. No. Two values for y. $|x| \geqq a$. **9.** (a) Hyperbola. (c) An asymptote, $S_t = 0$. **11.** 18.5.

Group C

1. $(b/a)(x - \sqrt{x^2 - a^2})$. **5.** (a) See Fig. 12.6.5.

Page 389, Problems 12.7

1. $(-1, -1.5)$. **2.** $[(x')^2/4] - [(y')^2/9] = 1$. $(-3, 2)$.

Page 390, Exercises 12.7

Group A

1. (a) $(-4, 9)$. (c) $(-2, 4)$. (e) $(-2, 10)$. **2.** (a) $x'^2 + y'^2 = 4$. (c) $x'^2 = 4y'$. **3.** (a) $(-2, 3)$. (c) $(1, -2)$.

Group B

1. No.

Group C

1. Fig. E.95. $x'^2/2 - y'^2/2 = 1$.

Page 393, Problems 13.1

1. Fig. P.31. **2.** $(-2.6, -1.5)$. **3.** $(5, 233.1°)$. **4.** $r = 8 \sin\theta/\cos^2\theta$.

Page 395, Exercises 13.1

Group A

1. Fig. E.96. **3.** (a) $(-2.5, 4.3)$. (c) $(0.25, -0.43)$. (e) $(0, -6)$. (g) $(-6, 0)$. (i) $(2.3, -1.93)$. **4.** (a) $(13, 67.4°)$. (c) $(2.2, 153.4°)$. (e) $(6, \pi/2)$. **5.** $(2.8, \pi/4)$. (c) $(1, -\pi/2)$. (e) $(2, \pi)$. **6.** (a) $r = 4$. (c) $r = 4\cos\theta/\sin^2\theta$. (e) $r^2 = 4\sec 2\theta$. **7.** (a) $x^2 + y^2 = 81$. (c) $y = x$. (e) $3x^2 + y^2 = 4$. (g) $(x^2 + y^2)^2 = 4(x^2 - y^2)$.

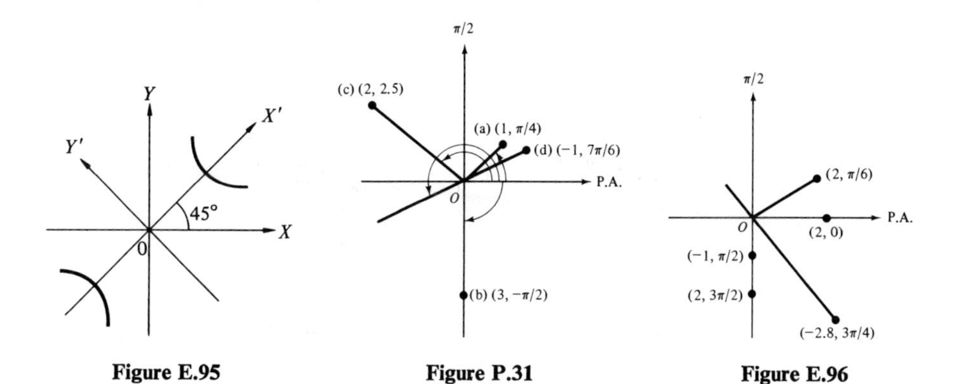

Figure E.95 **Figure P.31** **Figure E.96**

Page 397, Problems 13.2

1. Fig. P.32. $x^2 + y^2 - 2y = 0$. **2.** Fig. P.33. $x^2 + y^2 = 25$. **3.** Fig. P.34. $y = -0.58x$. **4.** Fig. P.35. $x^4 + y^4 - 4x^3 - 5x^2 - 9y^2 + 2x^2y^2 - 4xy^2 = 0$. **5.** Fig. P.36. $x^4 + y^4 + 4y^3 - x^2 + 3y^2 + 2x^2y^2 + 4x^2y = 0$. **6.** Fig. P.37. $(x^2 + y^2)^7 = 16(x^3 - 3y^2x)^2$. **7.** Fig. P.38. $(x^2 + y^2)^2 = 18xy$. **8.** Fig. P.39. **9.** Fig. P.40.

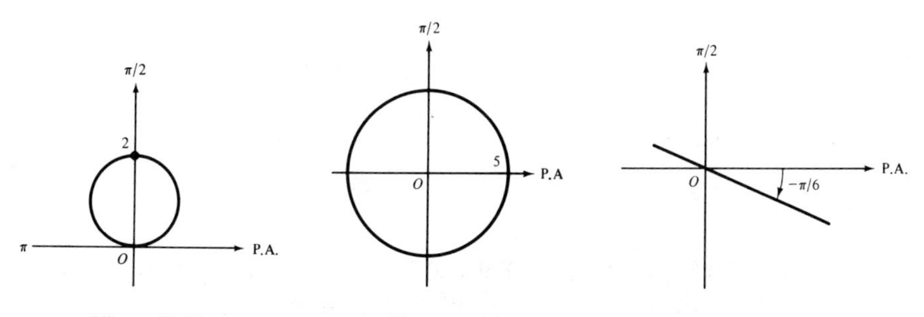

Figure P.32 **Figure P.33** **Figure P.34**

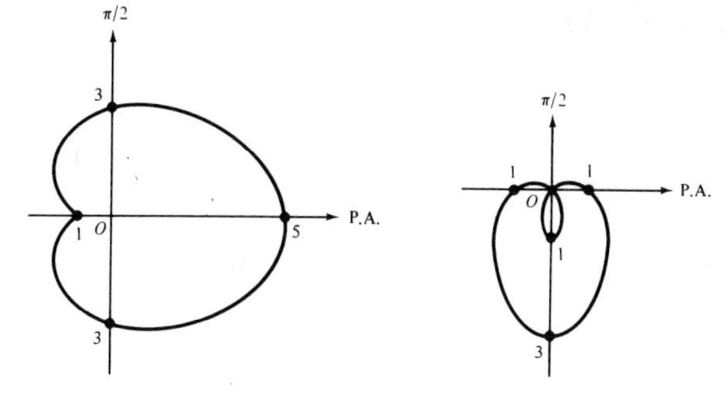

Figure P.35 **Figure P.36**

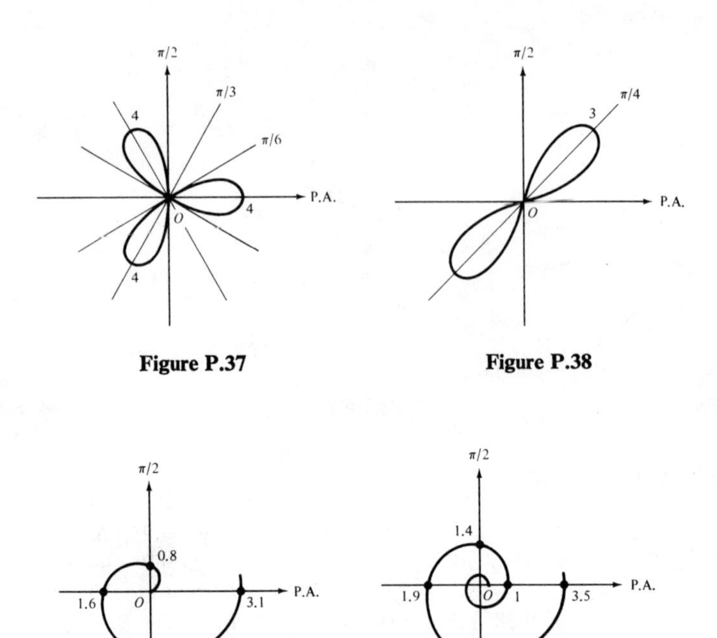

Figure P.37

Figure P.38

Figure P.39

Figure P.40

Page 402, Exercises 13.2

Group A

1. (a) Fig. E.97. (c) Fig. E.98. (e) Fig. E.99. (g) Fig. E.100. (i) Fig. E.101. (k) Fig. E.102. (m) Fig. E.103. (o) Fig. E.104. (q) Fig. E.105. (s) Fig. E.106.

Group B

1. It can. **3.** Polar for (a), (c), (e).

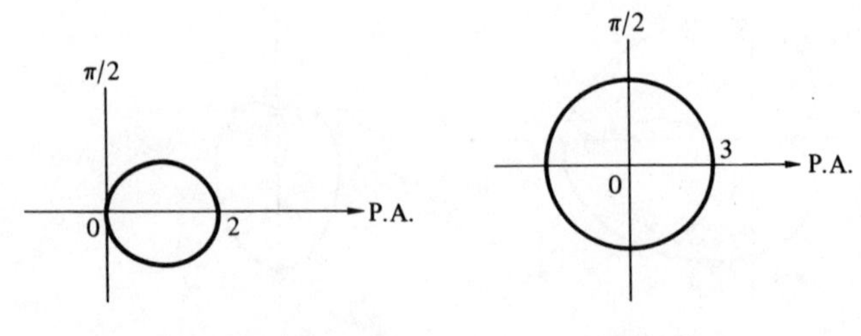

Figure E.97

Figure E.98

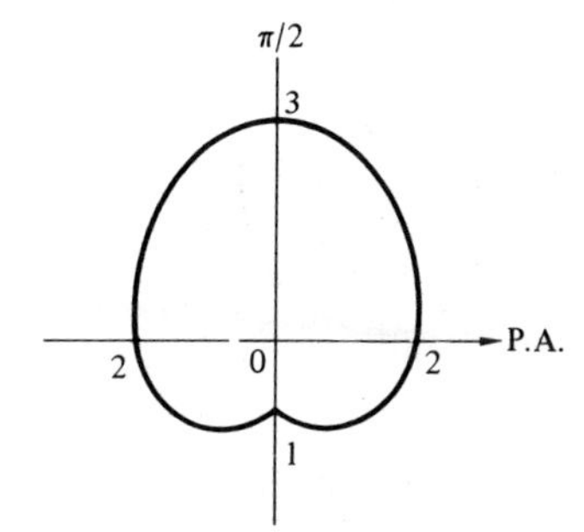

Figure E.99

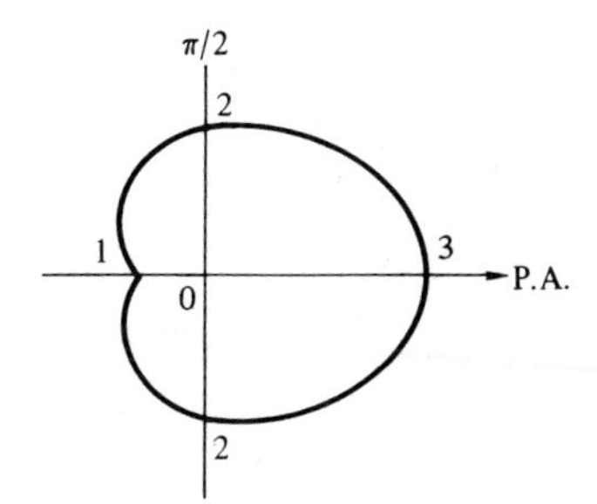

Figure E.100

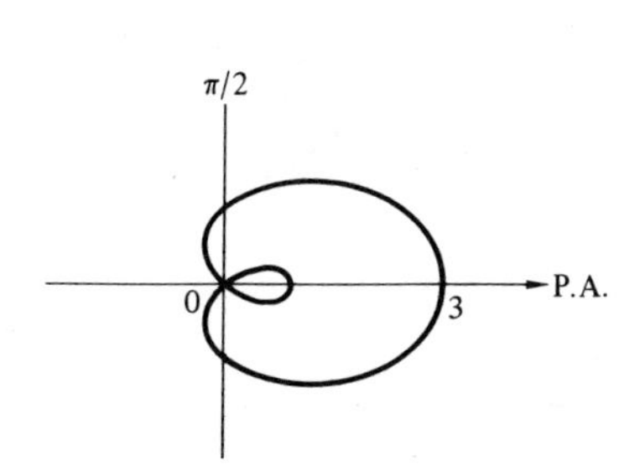

Figure E.101

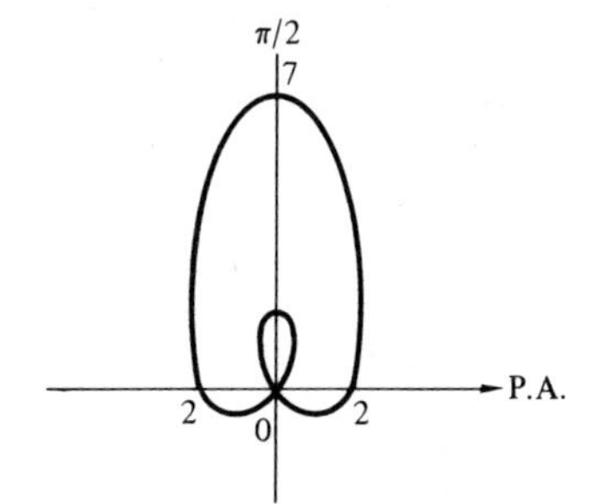

Figure E.102

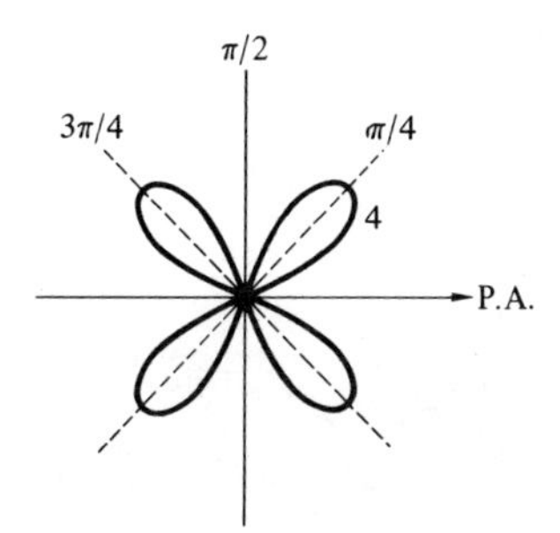

Figure E.103

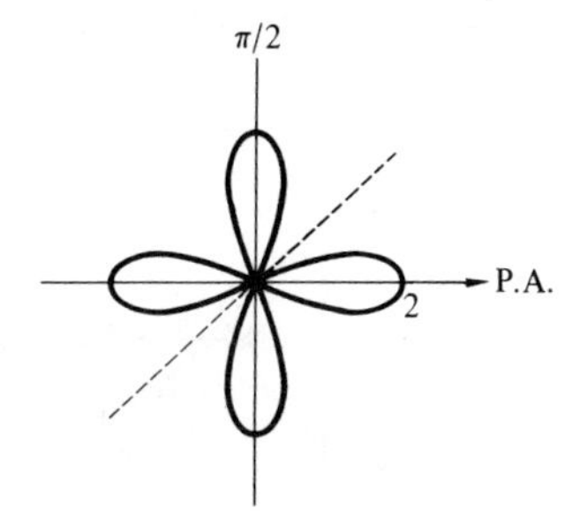

Figure E.104

Page 403, Problems 13.3

1. $r = -2 \csc \theta$. **2.** Hyperbola.

Page 404, Exercises 13.3

<u>*Group A*</u>

1. (a) $r = -4 \sec \theta$. (c) $\theta = \pi/6$. (e) $r^2 - 16r \sin \theta + 48 = 0$. **2.** (a) $e = 1$, $p = 2$,

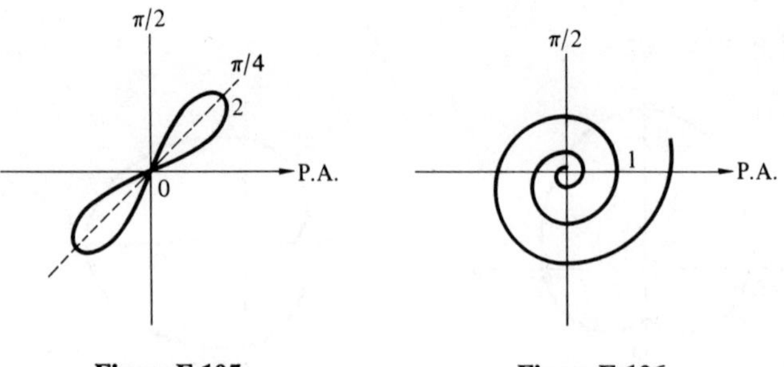

Figure E.105 Figure E.106

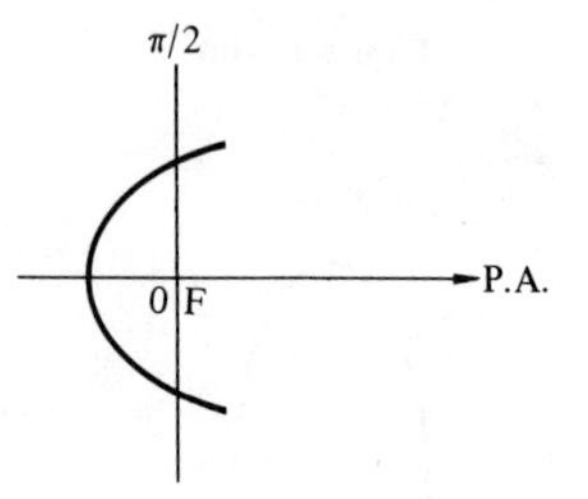

Figure E.107

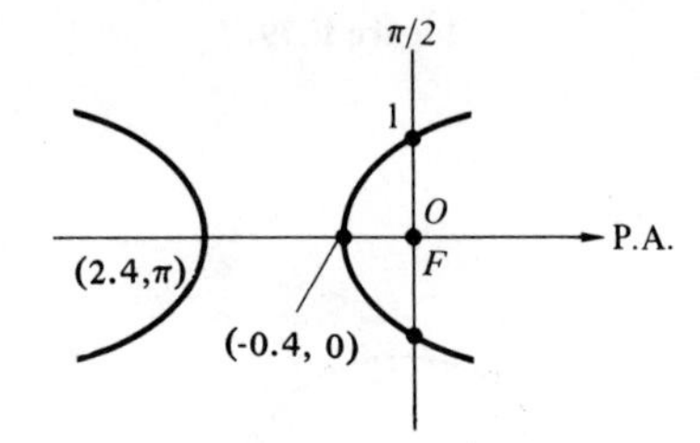

Figure E.108

parabola. (c) $e = \frac{3}{2}$, $p = 1$, hyperbola. **3.** (a) Parabola. Fig. E.107. (c) Hyperbola. Fig. E.108.

Group B

1. (a) Fig. E.109. (c) Fig. E.110. (e) Fig. E.111.

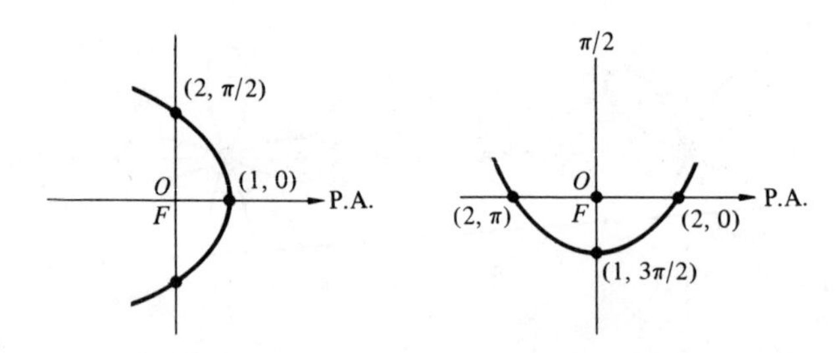

Figure E.109 Figure E.110

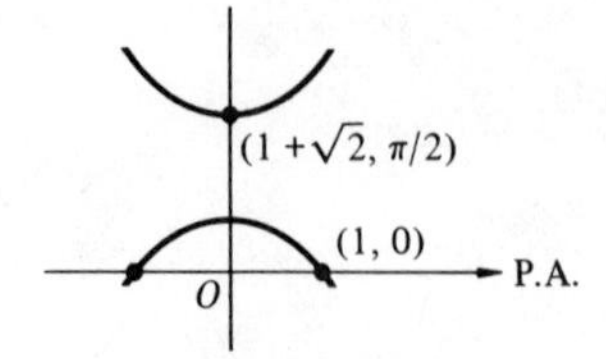

Figure E.111

Page 407, Problems 13.4

1. 4.4 units/s.

Page 407, Exercises 13.4

Group A

1. 8 units/s. **3.** 3 rad/min ccw. **5.** (a) $1/\sqrt{2 + 2\cos\theta}$. (b) 0.5 rad/s. ccw. (c) 0.71 rad/s. ccw. **7.** 6, 1.7, 6.23 units/s.

Page 409, Problems 13.5

1. $11\pi = 34.6$. **2.** $s = \int_{-\pi/2}^{\pi/2} \sqrt{13 + 12\sin\theta}\, d\theta$.

Page 409, Exercises 13.5

Group A

1. (a) 3.14. (c) 18.8. (e) 14.1. (g) 0.54. **2.** (a) $\int_0^{\pi/2} \sqrt{25 + 24\cos\theta}\, d\theta$. (c) $2\sqrt{2} \int_0^{3\pi/5} (1 + \cos\theta)^{-3/2}\, d\theta$. (e) $2\sqrt{2} \int_0^{2\pi} \sqrt{1 + \cos\theta}\, d\theta$.

Group B

1. 1.4. **3.** 0.79. **5.** 11.3. **7.** 34.6.

Group C

1. (a) 7.65. (c) $1.92a^2$. (e) 7.71.

Page 413, Problems 14.1

1. $-c_1\cos x - c_2\sin x + c_1\cos x + c_2\sin x = 0$. **2.** One. **3.** $y = (3 + 2t)e^{-t}$.

Page 415, Exercises 14.1

Group A

1. (a) First order, first degree. (c) First order, second degree. **6.** (a) 2. (c) 1. **7.** (a) $y = 5e^t$. (c) $y = 3\sin t$. (e) $y = \cos t + 2\sin t$. (g) $Q = 10 - 5e^{-t}(2\cos 2t + \sin 2t)$. **8.** (a) Two. **9.** $-1, 2$.

Page 417, Problems 14.2

1. $yx = c$. **2.** $\frac{1}{5}\,\mathrm{Tan}^{-1}(y/5) - \frac{1}{2}\ln(x^2 + 1) = -\pi/20$.

Page 417, Exercises 14.2

Group A

1. (a) $x = cy^3$. (c) $(y - 2) = cx^3$. (e) $\frac{1}{3}\text{Tan}^{-1} y/3 = \text{Sin}^{-1} x/3 + C$. (g) $I = 2 \sin t - 3 \cos t + C$. (i) $I = (e^{-t}/2)(-\cos t + \sin t) + C$. **2.** (a) $\ln (y^2 - 8) = \frac{1}{2}\text{Tan}^{-1}(x/4) - (\pi/8)$. (c) $xy = \frac{1}{3}(e^{3x} - 1)$. (e) $x^2 y = \ln x + 2$.

Group B

1. (a) $(y^3 + 1)^2(x^2 + 1)^3 = C$. (c) $2x^2y^2 = y^2 + C$. (e) $y + 3 = C \cos x$. (g) $\sec x + \tan y = C$.

Group C

1. (a) $y = \frac{1}{6}x^3 + c_1 x + c_2$. (c) $Q = e^{-t} + c_1 t + c_2$. (e) $y = \frac{1}{12}x^4 - \frac{1}{3}x^3 + \frac{3}{2}x^2 + 1$. **2.** (a) $x = 3yx^2 + cy$. (c) $x^3y^2 = c$. (e) $y^2 = cx^4$.

Page 420, Problems 14.3

1. $y = x^3/8 + cx^{-5}$. **2.** 9.3A.

Page 421, Exercises 14.3

Group A

1. (a) $y = \frac{1}{5}x^2 + cx^{-3}$. (c) $y = -3 + cx$. (e) $y = \frac{1}{2}e^x + ce^{-x}$. (g) $y = x \cos x + c \cos x$. **2.** (a) $y = x^3 - x^2 + x$. (c) $I = \frac{1}{200}(\sin 60t - 3 \cos 60t - \frac{3}{200}e^{-20t})$. -0.0096. (e) $Q = \frac{4}{15}(1 - e^{-75t/8})$.

Group B

1. (a) $y \sin x = \ln \sec x + C$. (c) $y = e^{-x^2}(x + C)$. (e) $y = \frac{1}{4}(x^4 - 1) + Ce^{-x^4}$.

Group C

1. (a) $x = 1 + Ce^{-y}$. (c) $xy = \sin y - y \cos y + C$.

Page 423, Problems 14.4

1. $y = x^2 + 3$. **2.** $y = \ln (1/\sqrt{x}) + C$. **3.** $2y^2 + x^2 - 41 = 0$. **4.** $I = \frac{20}{41}(\frac{5}{2} \sin 2t - 2 \cos 2t) + \frac{40}{41}e^{-(5/2)t}$; -0.279A. **5.** $4B_o$. **6.** 15.7 ft/s.

Page 428, Exercises 14.4

Group A

1. $y^2 = Cx$. **3.** $y^2 = 2x + C$; $y^2 = 2x + 16$. **5.** $I = \frac{1}{2}(1 - e^{-20t})$. **6.** (a) $I = \frac{270}{481}e^{-20t}$

$+ \frac{30}{481}(20 \sin 9t - 9 \cos 9t)$. **7.** $Q = \frac{1}{5}(1 - e^{-10t})$. **8.** (a) $Q = \frac{2}{9}(e^{-t} - e^{-10t})$. **9.** 78.5 min. **11.** (a) 57.8°C. (b) 15.5 min. **13.** 6 s. **15.** (a) 20 ft/s. (b) $s = 20t + 100e^{-t/5} - 100$. $v = 20(1 - e^{-t/5})$.

Group B

1. $A = 1000e^{-0.346t}$. **3.** $v = 50 - 10e^{-gt/50}$ ft/s down. $s = 50t + 500/ge^{-gt/50} - 500/g$, from position where chute opened; 50 ft/s down. **5.** $Q = 10 - 9(1 + 0.1t)^{-100}$. $I = 90(1 + 0.1t)^{-101}$. **7.** (b) 347 m/s. **9.** 33.9 g.

Page 431, Problems 14.5

1. $y = c_1e^{x} + c_2e^{-3x/2}$. **2.** $Q = (c_1 + c_2)e^{t/2}$. **3.** $y = e^{-3x}(c_1 \cos 2x + c_2 \sin 2x)$. **4.** $x = \frac{3}{5} \sin 5t$.

Page 434, Exercises 14.5

Group A

1. (a) $y = c_1e^{2x} + c_2e^{3x}$. (c) $y = c_1e^{x} + c_2$. (e) $y = (c_1 + c_2x)e^{-3x}$. (g) $y = e^{-2x}(c_1 \cos 3x + c_2 \sin 3x)$. (i) $y = e^{-x/2}(c_1 \cos \frac{1}{3}x + c_2 \sin \frac{1}{3}x)$. (k) $y = c_1 \cos 5x + c_2 \sin 5x$. **2.** (a) $y = c_1e^{x} + c_2e^{-2x}$. (c) $y = (c_1 + c_2x)e^{2x}$. (e) $x = c_1 \cos 4t + c_2 \sin 4t$. (g) $y = e^{2x}(c_1 \cos 3x + c_2 \sin 3x)$. (i) $Q = e^{-6t}(c_1 \cos 8t + c_2 \sin 8t)$. **3.** (a) $y = 2e^{-x} - e^{-2x}$. (c) $x = \frac{5}{7}e^{3t} + \frac{2}{7}e^{-4t}$. (e) $y = 4.6e^{-0.41x} - 1.6e^{-2.41x}$. (g) $y = 2e^{-t} \cos t$.

Group B

1. $Q = Q_0e^{-Rt/2L}\left[\cos \omega t + \frac{R}{2L\omega} \sin \omega t\right]$, $I = -Q_o\left[\frac{R^2 + 4L^2\omega^2}{4L^2\omega}\right]e^{-Rt/2L} \sin \omega t$,
where $\omega = \sqrt{\frac{1}{LC} - \frac{R^2}{4L^2}}$.

Page 437, Problems 14.6

1. $y = c_1e^{-x} + c_2e^{2x} - 2x + 1$. **2.** $y = c_1 \cos 3x + c_2 \sin 3x - \frac{1}{8} \sin 5x$. **3.** $Q = 2e^{-t/2}[\cos (\sqrt{3}/2)t + (1/\sqrt{3}) \sin (\sqrt{3}/2)t] - 2 \cos t$.

Page 439, Exercises 14.6

Group A

1. (a) $y = c_1e^{2x} + c_2e^{-3x} - 4$. (c) $y = e^{-x}(c_1 \cos 2x + c_2 \sin 2x) + \frac{1}{5} \sin x - 10 \cos x$. (e) $y = c_1e^{3t} + c_2e^{-5t} - \frac{1}{15}t^2 - \frac{34}{225}t - \frac{98}{3375} - \frac{1}{16}e^{-t}$. (g) $x = c_1 \cos 3t + c_2 \sin 3t + \frac{16}{9}t^2 - \frac{32}{81} + \frac{8}{5} \cos 2t$. (i) $y = c_1 + c_2e^{-25t} - \frac{1}{100}e^{-5t}$. **2.** (a) $x = \frac{9}{5}e^{2t} + \frac{6}{5}e^{-3t} - 2$. (c) $x = \frac{3}{5}e^{t} - \frac{2}{5}e^{-t} - \frac{1}{5} \cos 2t$. (e) $x = \frac{5}{2} \cos t - \frac{1}{2} \sin t + t^2 - 2 - \frac{1}{2}e^{-t}$.

Group B

1. (a) $x_c = c_1e^{3t} + c_2e^{-5t}$. **3.** (a) $x = c_1e^{-3t} + c_2e^{2t} + \frac{1}{5}te^{2t}$. (c) $(c_1 + c_2t)e^{-t} + \frac{1}{2}t^2e^{-t}$.

Page 441, Problems 14.7

1. $x = \frac{5}{4}\sin 4t$; 1.2 ft; 3.3 ft/s. **2.** $Q = e^{-3t}[\frac{1}{3}\cos 4t + \frac{1}{4}\sin 4t] - \frac{1}{3}\cos 5t$. $I = e^{-3t}[-\frac{25}{12}\sin 4t + \frac{1}{3}\sin 5t]$. -0.24 A. $I_{ss} = \frac{1}{3}\sin 5t$.

Page 444, Exercises 14.7

Group A

1. $x = \frac{1}{4}\cos 16t, v = -4\sin 16t$. **3.** $x = (e^{-8t}/2)(\cos 8t + \sin 8t)$, damped oscillatory. **5.** $Q = -4e^{-5t} + e^{-20t} + 3$, 3 is steady state: $I = 20e^{-5t} - 20e^{-20t}$, no steady state. **7.** $Q = e^{-2t}(\cos t + 2\sin t) + 5$; $I = -5e^{-2t}\sin t$.

Group B

1. (a) $x = \frac{25}{3}e^{-t} - \frac{1}{3}e^{-7t} + 6\sin t - 8\cos t$. (b) $v = -2.34$ ft/s. (d) 6 ft/s. **3.** (a) $Q = 4t\sin 2t$; as t inc., Q inc.

Page 446, Problems 14.8

1. $5/s$. **2.** $5/s^2$. **3.** $s/(s^2 + a^2)$. **4.** $5/s - 3s/(s^2 + 4)$. **5.** $s\mathcal{L}\{F(t)\} + 2$. **6.** $\cos 3t$. **7.** $(3/\sqrt{7})\sin\sqrt{7}\,t$. **8.** $e^{2t}(\cos 3t + \frac{1}{3}\sin 3t)$. **9.** $3e^{-2t} - 5e^t$.

Page 450, Exercises 14.8

Group A

1. (a) $(2/s^3) + (3/s^2) + (1/s)$. (c) $(120/s^6) + [1/(s-2)^2]$. (e) $2/(s-3) - 4/(s+3)$. (g) $[1/2(s-2)] - (1/2s)$. (i) $[54/(s^2+9)^2]$. **2.** (a) 1. (c) $2\cos 4t$. (e) $e^{-4t} + t^3e^{3t}$. (g) $e^{2t}\sin\sqrt{6}\,t$. (i) $(\cos 2t - \cos 3t)/5$. **3.** (a) $\mathcal{L}\{y\} = \dfrac{s+4}{s^2+2s+3}$. (c) $\mathcal{L}\{y\} = \dfrac{2}{s^3(2s+1)(s+1)} + \dfrac{3}{s(2s+1)(s+1)} + \dfrac{2s}{(2s+1)(s+1)} + \dfrac{3}{(2s+1)(s+1)}$. (e) $\mathcal{L}\{y\} = \dfrac{s}{(s^2+8s+2)(s^2+4)} + \dfrac{2s}{s^2+8s+2} + \dfrac{17}{s^2+8s+2}$.

Group B

1. (a) $\frac{1}{2}(e^{2t} - 1)$. (c) $e^{2t} - \cos 2t$. (e) $e^{-t}(3\cos 2t - 2\sin 2t)$. (g) $-e^{-t} - 12e^t + 15e^{2t}$. (i) $2e^{2t}(\cos 4t - \frac{3}{4}\sin 4t)$.

Group C

1. (b) $\frac{1}{2}(1 - \cos 2t)$. **3.** $4[e^t - t - 1]$.

Page 452, Problems 14.9

1. $x = \frac{5}{16}(1 - \cos 4t)$. **2.** $x = te^{3t}(\frac{5}{2}t + 4)$. **3.** 0.11.

Page 454, Exercises 14.9

Group A

1. (a) $x = \frac{5}{4}\cos 8t$. (c) $x = e^{-2t}(2t^2 + 2t - 1)$. (e) $x = \frac{1}{2}t^4e^{-3t}$. (g) $x = e^{-5t} - e^{3t} + 8\sin t$. (i) $x = \frac{1}{2}e^{-t}\sin 2t$.

Group B

1. (a) $I = e^{-4t}(\frac{1}{5}t + 3)$. (c) $Q = \frac{1}{6}(5\sin 2t - \sin 10t)$, $I = \frac{10}{6}(\cos 2t - \cos 10t)$. (e) $x = e^{-16t}(\frac{1}{2} + 18t)$, $v = 10e^{-16t} - 288te^{-16t}$.

Group C

1. (c) $Q = \sin 3t$.

Page 458, Problems 14.10

1. 1.68.

Page 459, Exercises 14.10

Group A

1. (a) 0.49. (c) 1.03. (e) 3.1. (g) 1.51. (i) 0.51.

Group B

1. 1.34 A.

Page 462, Problems 15.1

1. (a) -10. (b) 27. **2.** Fig. P.41. **3.** 8.2.

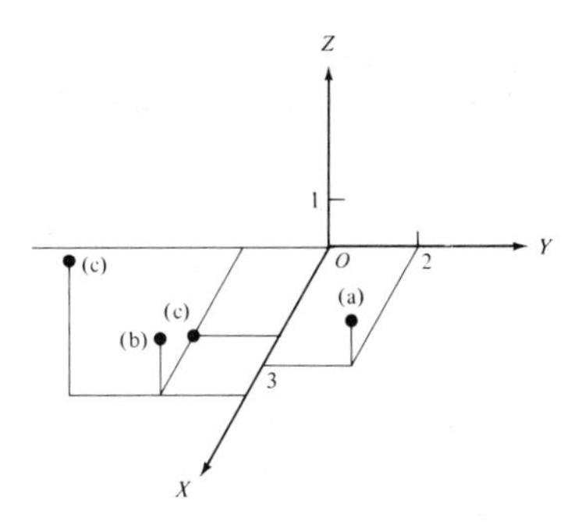

Figure P.41

Page 465, Exercises 15.1

Group A

1. (a) $\{-1, 1, 0, 1\}$. (c) $\{60, 3, 0\}$. **2.** (a) Yes. (c) No, two values for z. (e) Yes. (g) Yes. **3.** Fig. E.112. **4.** (a) 3. (c) 3.

Page 466, Problems 15.2

1. $x = 0$, $y + 3z - 6 = 0$; $y = 0$, $2x + 3z - 6 = 0$; $z = 0$, $2x + y - 6 = 0$. Fig. P.42. **2.** Fig. P.43. **3.** Sphere: $(2, 0, -3)$, 3. **4.** Fig. P.44.

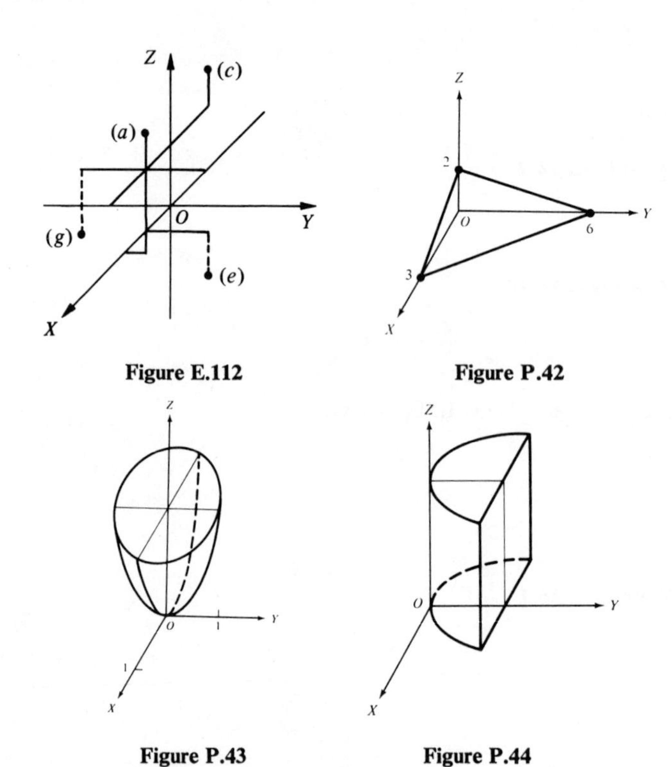

Figure E.112

Figure P.42

Figure P.43

Figure P.44

Page 471, Exercises 15.2

Group A

1. (a) Fig. E.113. (c) Fig. E.114. (e) Fig. E.115. **2.** (a) Fig. E.116. (c) Fig. E.117. **3.** (a) Fig. E.118. (c) Fig. E.119. (e) Fig. E.120. (g) Fig. E.121. **4.** (a) Fig. E.122. (c) Fig. E.123. (e) Fig. E.124. (g) Fig. E.125. (i) Fig. E.126. **5.** (a) Fig. E.127. (c) Fig. E.128. (e) Fig. E.129. **6.** (a) Fig. E.130. (c) Fig. E.131. (e) Fig. E.132.

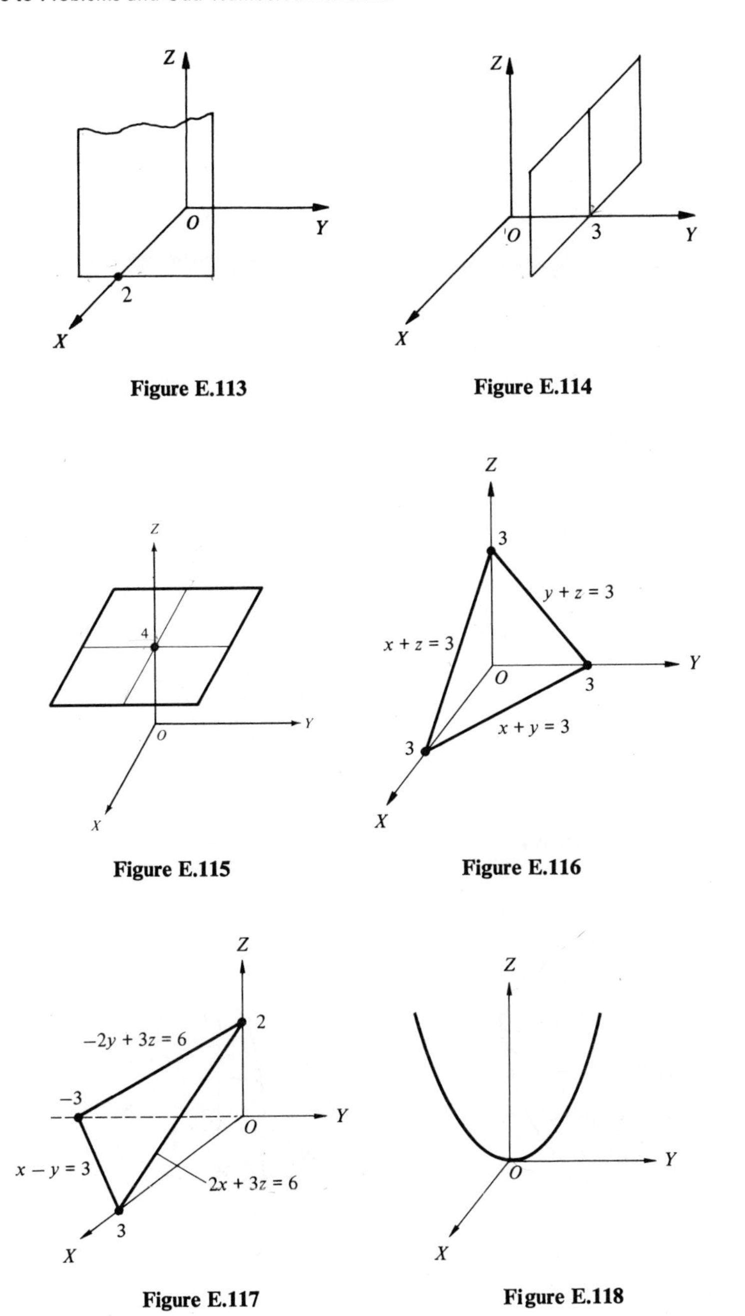

Page 472, Problems 15.3

1. Fig. P.45. Circles parallel to the XY-plane. Parabolas parallel to the XZ- and YZ-planes.

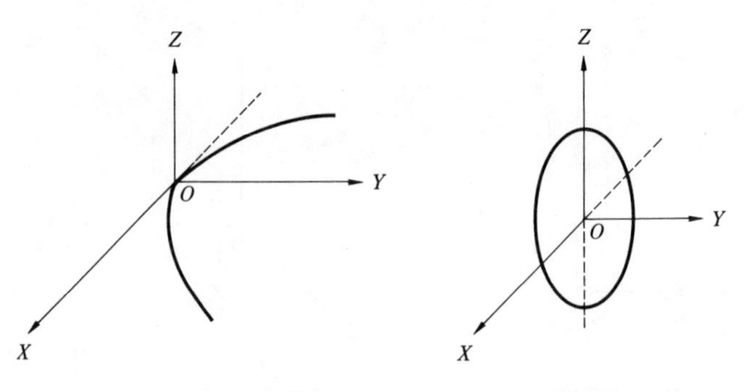

Figure E.119 **Figure E.120**

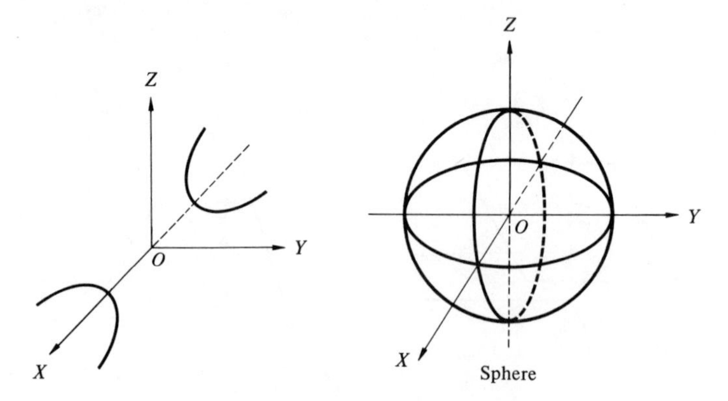

Figure E.121 **Figure E.122**

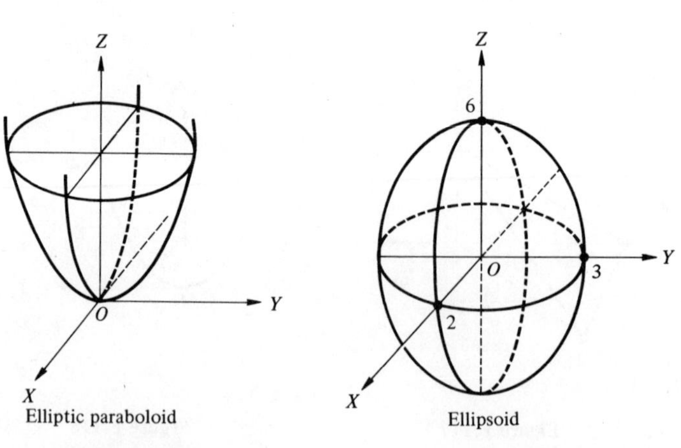

Figure E.123 **Figure E.124**

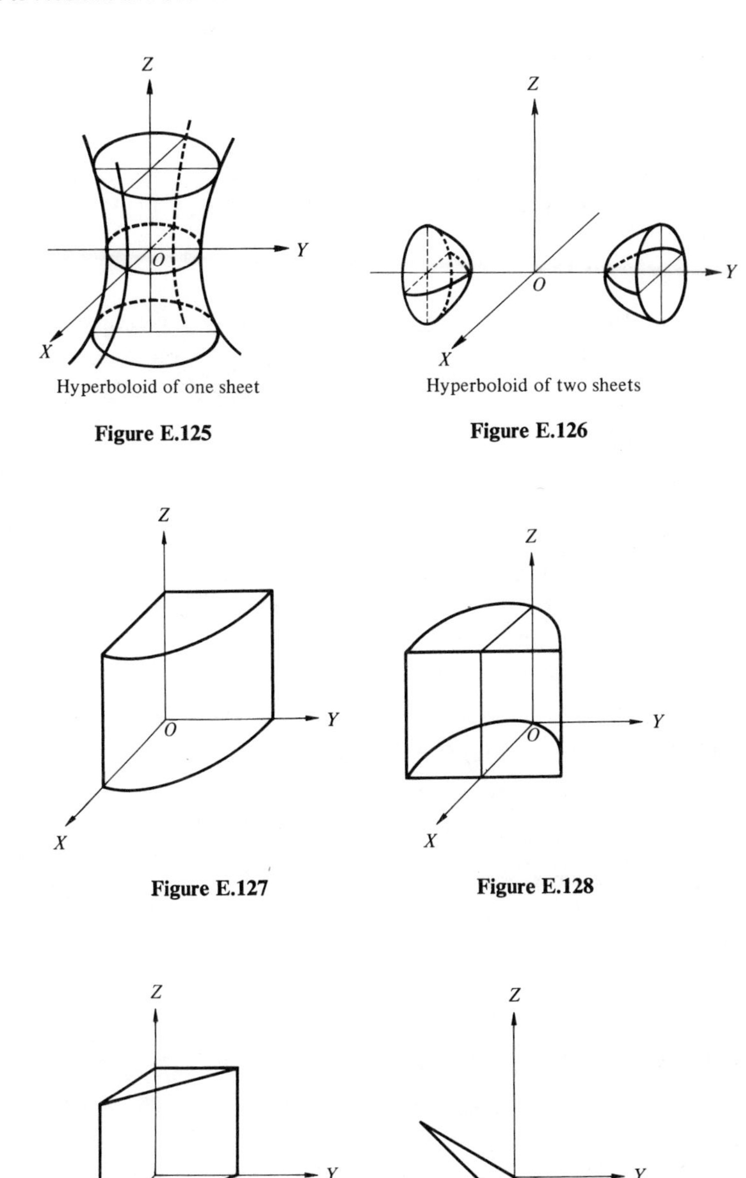

Figure E.125

Figure E.126

Figure E.127

Figure E.128

Figure E.129

Figure E.130

Page 473, Exercises 15.3

Group A

1. (a) Fig. E.133. (c) Fig. E.134. **2.** (a) Fig. E.135. (c) Fig. E.136. (e) Fig. E.137. **3.** (a) Fig. E.138. (c) Fig. E.139. **4.** (a) Fig. E.140. (c) Fig. E.141.

Page 474, Problems 15.4

1. $\partial z/\partial x = 12x^3y^2 - y^3, \partial z/\partial y = 6x^4y - 3y^2x$. **2.** $\partial z/\partial x = 6x^2y^2u^{1/2}v^{-2/3}, \partial z/\partial y = 4x^3yu^{1/2}v^{-2/3}$, $\partial z/\partial u = x^3y^2u^{-1/2}v^{-2/3}$, $\partial z/\partial v = (-4/3)x^3y^2u^{1/2}v^{-5/3}$. **3.** 0.36. **4.** $\partial z/\partial x = -2$.

Page 476, Exercises 15.4

Group A

1. (a) $\partial z/\partial x = 4x^3 + 3x^2y + 2xy^2$, $\partial z/\partial y = x^3 + 2x^2y - 3y^2$. (c) $\partial z/\partial x = 2x/(x^2 + y^2)$, $\partial z/\partial y = 2y/(x^2 + y^2)$. (e) $\partial z/\partial x = y/(x^2 + y^2)$, $\partial z/\partial y = -x/(x^2 + y^2)$. **3.** (a) -0.5, -3.32. (c) 2.5, 5.

Group B

1. (a) 83.8 in.^{3}/in. **3.** (a) 51.3. (c) 69.1. **5.** $V/k, P/k, -kT/V^2$.

Group C

1. (a) 2, 2, 0. (c) $-e^y \sin x$, $e^y \sin x$, $e^y \cos x$.

Page 478, Problems 15.5

1. 2.33 psi. **2.** 22.

Page 479, Exercises 15.5

Group A

1. $dP = 2IRdI + I^2dR$. **3.** (a) $dw = 2xdx + 2ydy$. (b) $dx = -r \sin\theta\, d\theta + \cos\theta\, dr$, $dy = r\cos\theta\, d\theta + \sin\theta\, dr$. (c) $2rdr$. **4.** (a) Minimum (2, −1, −9). (c) Minimum (6, −4, −2). (e) Maximum (−0.48, −0.93, −4.26). **5.** (a) −0.17. (c) 0.044.

Group B

1. (a) $dx = \sin\phi\cos\theta\, dr + r\cos\phi\cos\theta\, d\phi - r\sin\phi\sin\theta\, d\theta$, $dy = \sin\phi\sin\theta\, dr + r\cos\phi\sin\theta\, d\phi + r\sin\phi\cos\theta\, d\theta$, $dz = \cos\phi\, dr - r\sin\phi\, d\phi$. (b) $2rdr$. **3.** 4330.08 in^3. **5.** −40 A. **7.** $x = 6$, $y = 6$, $z = 12$.

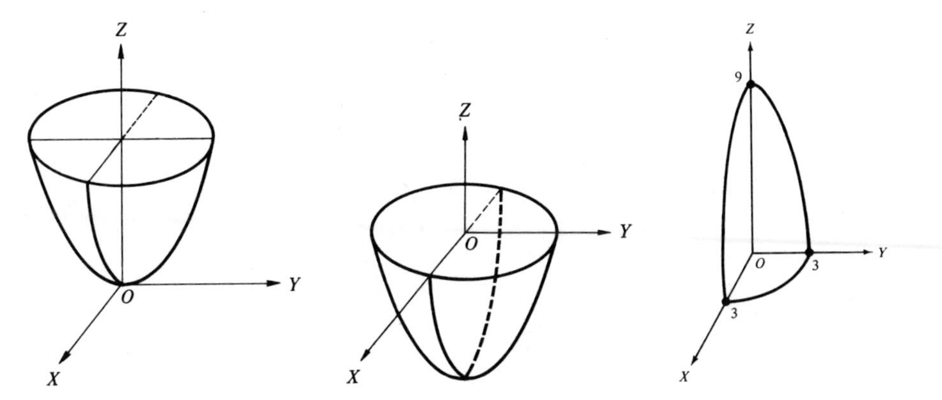

Figure E.131 Figure E.132 Figure P.45

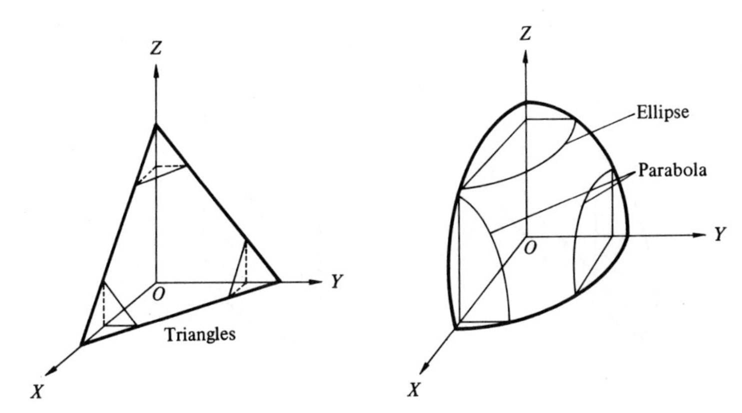

Figure E.133 Figure E.134

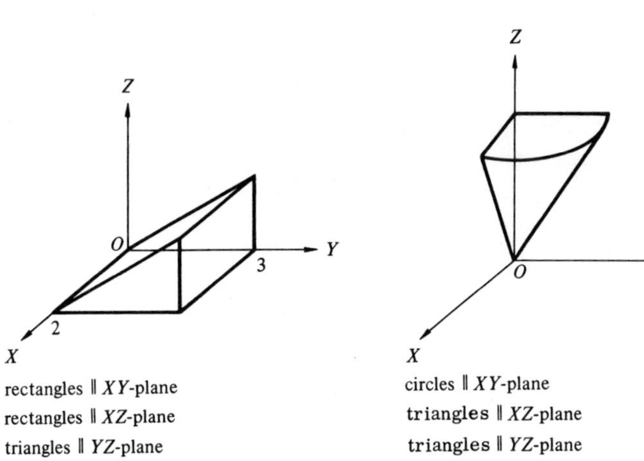

Figure E.135 Figure E.136

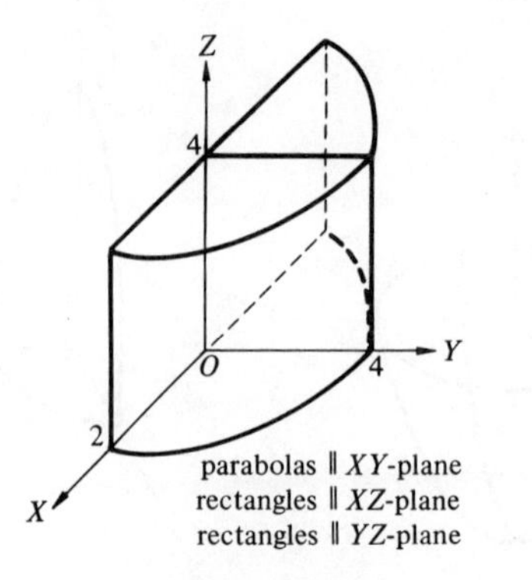

Figure E.137

Z
O
1
Y
1
X

Figure E.138

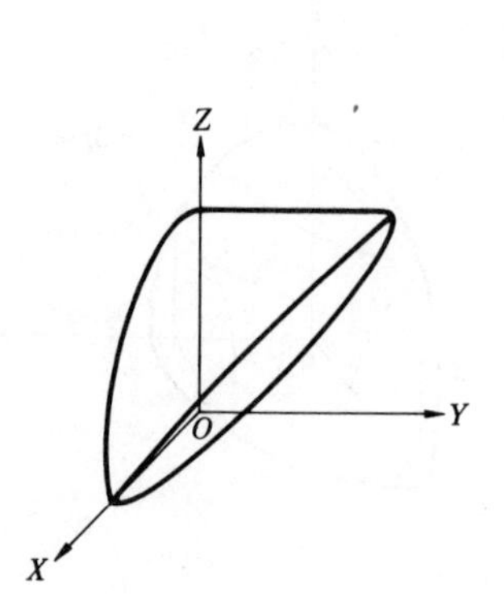

Figure E.139

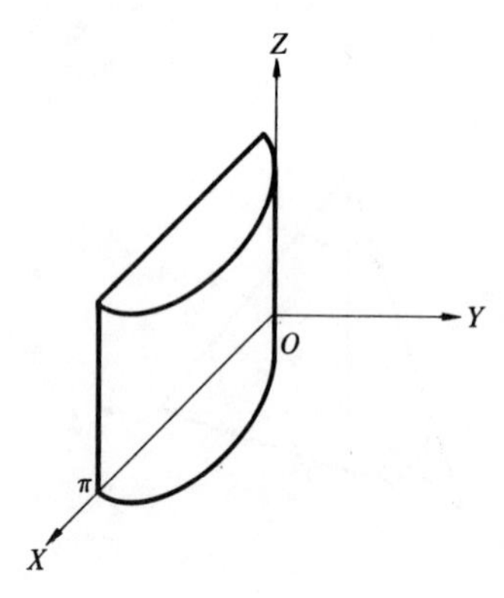

Figure E.140

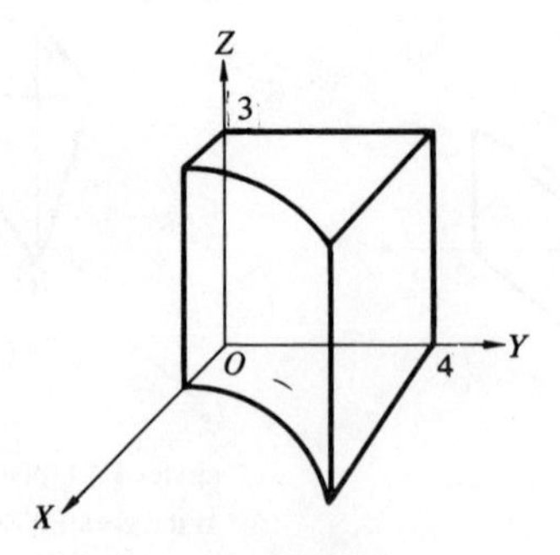

Figure E.141

Group C

1. (a) $y = -\frac{19}{26}x + \frac{30}{26}$. (c) $y = \frac{3}{2}x + \frac{1}{6}$.

Page 482, Problems 16.1

1. 30. **2.** 0.53. **3.** 0.81. **4.** -0.33.

Page 484, Exercises 16.1

Group A

1. (a) 0.75. (c) 3.38. **2.** (a) 1. (c) 0. **3.** (a) 0.79. (c) -0.22.

Group B

1. (a) 0.33. (c) 0.067. **2.** (a) 0.41. (c) 18.6. **3.** (a) 0.26. (c) 33.5.

Group C

1. (a) -1.125. (c) -0.53.

Page 487, Problems 16.2

1. 0.17. **2.** 4.5. **3.** 0.79.

Page 489, Exercises 16.2

Group A

1. (a) 0.17. (c) 0.1. (e) 9. (g) 6.75. (i) 0.39. (k) 2.46.

Group B

1. 0.33. Same. **3.** (a) 36. (c) 0.083.

Page 492, Problems 16.3

1. $(\pi/4, \pi/8) = (0.79, 0.39)$. **2.** $(2/3\pi, 1/2) = (0.21, 0.5)$. **3.** $i_Y = 604.2$, $i_X = 156.2$, $i_0 = 760.4$. **4.** (a) $i_L = 65$. (b) $i_L = 275$.

Page 496, Exercises 16.3

Group A

1. (a) (2, 1.6). (c) (2.4, 0). (e) (2, 5.6). (g) (0.083, 0.083). (i) $i_Y = a^3b\pi/4$, $i_X = ab^3\pi/4$, $i_o = (ab\pi/4)(a^2 + b^2)$. (k) $\pi a^4/4$. (m) $\pi a^4/2$. (o) 244. (q) 2.36. (s) 2.45. (u) (0.83, 0).

Group B

3. 852 in^4.

Page 500, Problems 16.4

1. $27\pi = 84.8$. **2.** $49\pi/8 = 19.2$.

Page 501, Exercises 16.4

Group A

1. $\pi/6 = 0.52$. **3.** $8\pi = 25.1$. **5.** $48\pi = 150.8$. **7.** $\pi/32 = 0.098$. **9.** $81\pi/32 = 7.95$. **11.** $32\pi = 100.5$.

Group B

1. 0.17.

Page 504, Problems 17.1

1. 3. **2.** 1/2, 2/9, 3/28, 4/65; $n/(n^3 + 1)$; $(n - 1)/[(n - 1)^3 + 1]$; $(n + 1)/[(n + 1)^3 + 1]$. **3.** $1/(n + 3)(n + 1)$.

Page 506, Exercises 17.1

Group A

1. (a) Finite. (c) Infinite. **2.** (a) 1.5. (c) Does not exist. **3.** (a) $1, \frac{1}{2}, \frac{1}{3}, \frac{1}{4}$. (c) $-1, \frac{1}{4}, -\frac{1}{9}, \frac{1}{16}$. (e) $1, \frac{1}{2}, \frac{1}{6}, \frac{1}{24}$. (g) $x, -x^3/6, x^5/120, -x^7/5040$. (i) $1, x, x^2/2, x^3/6$. **4.** (a) $(n + 2)\cdot(n + 1)$. (c) $1/(3n + 1)$.

Group B

1. (a) $1/(n - 1)$. (c) $1/n(n^2 + 3n + 3)$. **4.** (a) $\frac{1}{2}, \frac{1}{6}, \frac{1}{12}, \frac{1}{20}, \frac{1}{30}$. (c) 1, 4, 9, 16, 25. **5.** (a) 1, 3, 5, 7. (c) $3, \frac{9}{2}, \frac{9}{2}, \frac{27}{8}$. **6.** (a) $3! + 6! + 9! + 12! + 15!$. (c) No.

Group C

3. (a) Converge. (c) Converge.

Page 508, Problems 17.2

1. (a) $\lim_{n\to\infty} (3/5n) = 0$. (b) $\lim_{n\to\infty} [3n/(5n + 1)] = 3/5 \neq 0$. **2.** Div. **3.** Conv. **4.** Conv. **5.** Div. **6.** $-1 < x \leqq 1$.

Page 512, Exercises 17.2

Group A

1. (a) Further test. (c) Diverges. (e) Further test. **2.** (a) Converges. (c) Diverges. (e) Diverges. (g) Diverges. (i) Converges. **3.** (a) Converges. (c) Converges. (e) Diverges. **4.** (a) Converges. (c) Diverges. (e) Converges. (g) Test fails. **5.** (a) $-1 \leqq x \leqq 1$. (c) $-1 < x < 1$. (e) All x. (g) All x.

Group B

2. (a) $-2 < x \leqq 0$. (c) All x. **3.** No.

Group C

1. (a) Yes. (c) Nothing. **2.** (a) Conv. (c) Conv.

Page 515, Problems 17.3

1. $1 - x^2/2! + x^4/4!$; $(-1)^{n+1}x^{2n-2}/(2n-2)!$; $\sum_{n=1}^{\infty}(-1)^{n+1}x^{2n-2}/(2n-2)!$. **2.** $1 + x + x^2 + x^3 + \cdots$; Yes.

Page 516, Exercises 17.3

Group A

1. (a) $e^x = 1 + x + (x^2/2!) + (x^3/3!) + \cdots$. (c) $\sin 2x = 2x - (8x^3/3!) + (32x^5/5!) + \cdots$. (e) $\ln(1 - x) = -x - x^2/2 - x^3/3 + \cdots$.

Group B

1. (a) $\tan x = x + (x^3/3) + \frac{2}{15}x^5 + \cdots$. (c) $\mathrm{Tan}^{-1}\, x = x - (x^3/3) + (x^5/5) + \cdots$. (e) $\sin(x + \pi/4) = (1/\sqrt{2})[1 + x - x^2/2 - x^3/6 + \cdots]$. (g) $\ln \cos x = -(x^2/2) - (x^4/12) - (x^6/45) - \cdots$.

Group C

1. (a) $y = 1 + x + (x^2/2!) + (x^3/3!) + \cdots = e^x$. (c) $y = 1 + (x^2/2!) + (x^3/3!) + (x^4/4!) + \cdots = e^x - x$. **3.** $(x^3/3) - [x^7/(7)3!] + [x^{11}/(11)5!] - [x^{15}/(15)7!] + [x^{19}/(19)9!]$.

Page 518, Problems 17.4

1. $\frac{\sqrt{3}}{2} + \frac{1}{2}\left(x - \frac{\pi}{3}\right) - \frac{\sqrt{3}}{8}\left(x - \frac{\pi}{3}\right)^2 + \cdots$. Near $x = \frac{\pi}{3} = 1.05$. **2.** $\ln 2 + (1/2)(x - 1) - (1/8)(x - 1)^2 + \cdots$. Near $x = 1$.

Page 519, Exercises 17.4

Group A

1. (a) $e^x = e^2[1 + (x - 2) + (x - 2)^2/2! + \cdots]$. (c) $\ln x = (x - 1) - \dfrac{(x-1)^2}{2!} + \dfrac{2(x-1)^3}{3!} - \dfrac{6(x-1)^4}{4!} + \cdots$. (e) $\sqrt{x} = 1 + \frac{1}{2}(x - 1) - \frac{1}{8}(x - 1)^2 + \frac{1}{16}(x - 1)^3 + \cdots$. (g) $\sec x = 2 + 2\sqrt{3}(x - \pi/3) + 7(x - \pi/3)^2 + \cdots$. (i) $\sin x = \sqrt{3}/2 - \frac{1}{2}(x - 2\pi/3) - (\sqrt{3}/4)(x - 2\pi/3)^3 + \cdots$.

Group B

1. (a) $\mathrm{Tan}^{-1}\, x = \pi/4 + \frac{1}{2}(x - 1) - \frac{1}{4}(x - 1)^2 + \frac{1}{12}(x - 1)^3 + \cdots$. (c) $\sqrt{1 + x^2} = \sqrt{2} + (1/\sqrt{2})(x - 1) + (1/4\sqrt{2})(x - 1)^2 + \cdots$. (e) $(1/\sqrt{x^2 - 1}) = (1/\sqrt{3}) - (2/3\sqrt{3})(x - 2) + (1/2\sqrt{3})(x - 2)^2 + \cdots$.

Group C

1. $y = 1 + (x - 1) - (x - 1)^2 + (x - 1)^3 + \cdots$.

Page 520, Problems 17.5

1. 1.2214. **2.** 0.7885. **3.** 2.0655. **4.** $x - (2/3)x^{3/2} + (1/4)x^2 - (1/15)x^{5/2} + \cdots + C$. **5.** 0.00133.

Page 522, Exercises 17.5

Group A

1. 0.1736. **3.** 3.1, one thousand. **5.** 0.1823. **7.** Less than 0.000021. **9.** (a) 1.809. (c) 0.602.

Group B

1. $1 + (x^2/2!) + (x^4/4!) + (x^6/6!) + \cdots$. **3.** $x + (x^3/3) + \frac{2}{15}x^5 + \cdots$.

Page 524, Problems 18.1

1. 3. **2.** Fig. P.46. **3.** Fig. P.47. 1.5, 0.

Page 526, Exercises 18.1

Group A

1. (a) Fig. E.142. $f(\frac{1}{2}) = 1$, $f(3.2) = 0$, $f(-1) = 1$. (c) Fig. E.143. $f(\pi/2) = 1$, $f(3\pi/2) = -1$, $f(5\pi/2) = 1$. (e) Fig. E.144. $f(1) = 1$, $f(2) =$ not defined, $f(3) = 1$.

Group B

1. (a) Fig. E.145. (c) Fig. E.146. (e) Fig. E.147. (g) Fig. E.148.

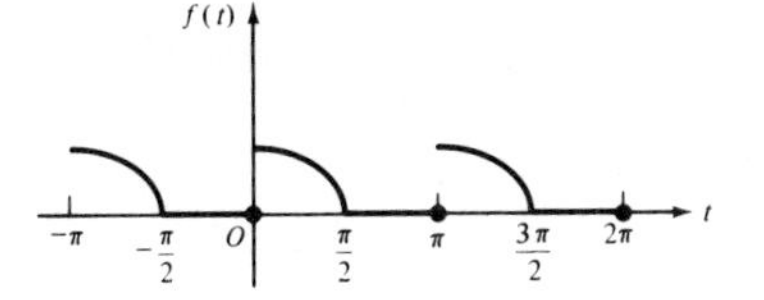

Figure P.46

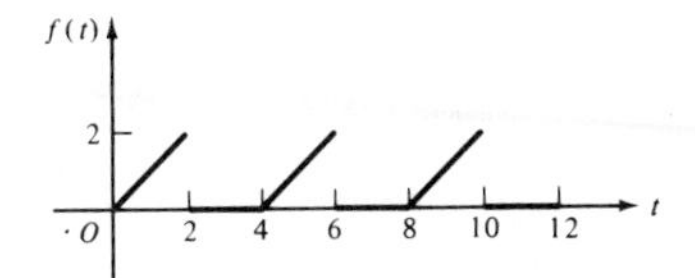

Figure P.47

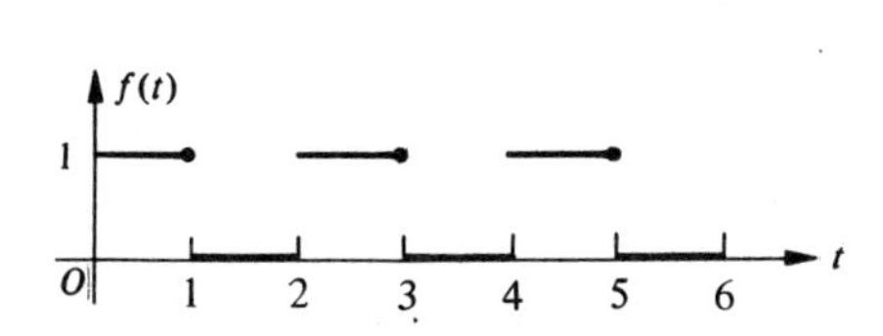

Figure E.142

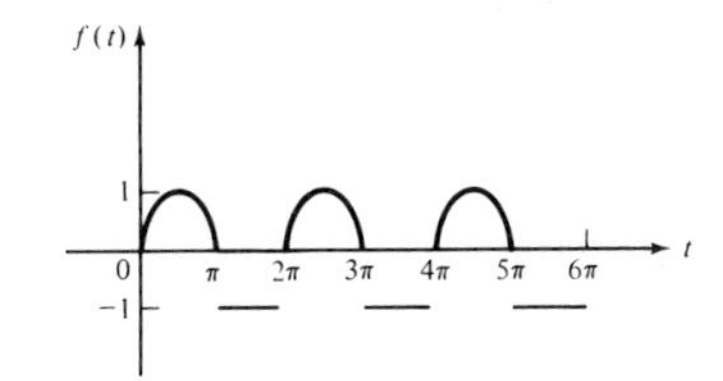

Figure E.143

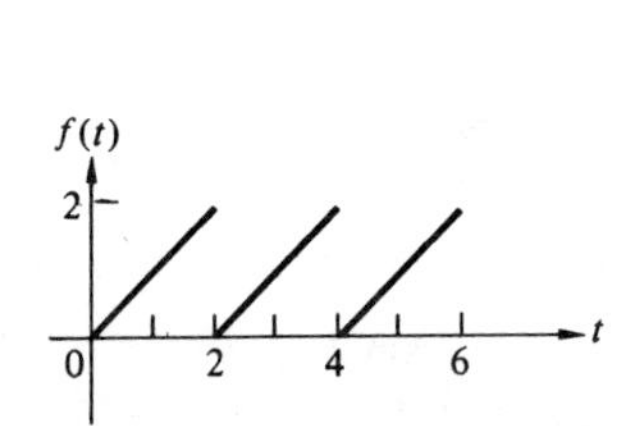

Figure E.144

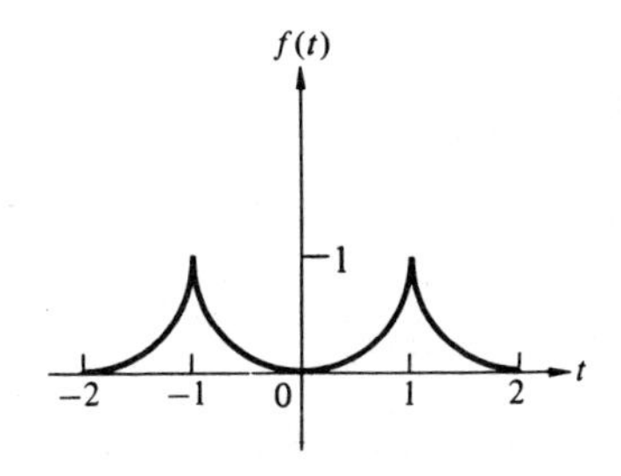

Figure E.145

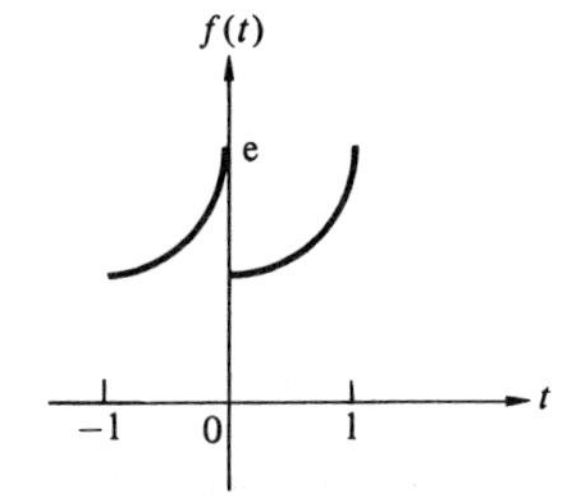

Figure E.146

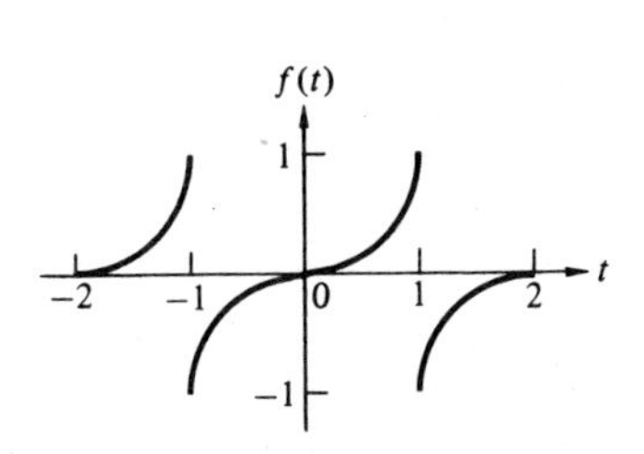

Figure E.147

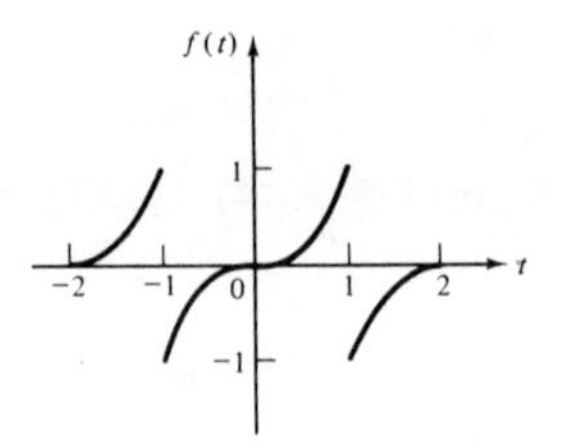

Figure E.148

Page 531, Problems 18.2

1. $f(t) = \frac{4}{\pi}\sin\left(\frac{\pi}{2}t\right) + \frac{4}{3\pi}\sin\left(\frac{3\pi}{2}t\right) + \frac{4}{5\pi}\sin\left(\frac{5\pi}{2}t\right) + \frac{4}{7\pi}\sin\left(\frac{7\pi}{2}t\right) + \cdots$. **2.** $f(t) = \frac{12}{\pi} - \frac{6}{\pi}\cos 4t + \frac{3}{4\pi}\cos 6t - \frac{6}{5\pi}\cos 8t + \cdots + 6\sin 2t + \cdots$.

Page 534, Exercises 18.2

Group A

1. (a) $\frac{3}{2} + \cos(\pi t/3) + \frac{1}{2}\cos(2\pi t/3) + \cdots + \sin(\pi t/3) + \frac{1}{4}\sin(2\pi t/3) + \cdots$. (c) $\frac{1}{2} - (4/\pi)[\cos(2\pi t/5) + \frac{1}{3}\cos(6\pi t/5) + \frac{1}{5}\cos(10\pi t/5) + \cdots]$. **2.** (a) $f(t) = \frac{1}{2} + (2/\pi)(\sin \pi t + \frac{1}{3}\sin 3\pi t + \cdots)$. (c) $f(t) = 1 + (4/\pi)(\sin t + \frac{1}{3}\sin 3t + \frac{1}{5}\sin 5t + \cdots)$. (e) $f(t) = \frac{1}{4} - (2/\pi^2)(\cos \pi t + \frac{1}{9}\cos 3\pi t + \frac{1}{25}\cos 5\pi t + \cdots) + (1/\pi)(\sin \pi t - \frac{1}{2}\sin 2\pi t + \frac{1}{3}\sin 3\pi t + \cdots)$.

Group B

1. (a) $f(t) = (e-1) + 2(e-1)\left(\frac{\cos 2\pi t}{1+4\pi^2} + \frac{\cos 4\pi t}{1+16\pi^2} + \cdots\right) - 4\pi(e-1)\cdot \left(\frac{\sin 2\pi t}{1+4\pi^2} + \frac{\sin 4\pi t}{1+16\pi^2} + \cdots\right)$. (c) $f(t) = (\pi/4) - (2/\pi)(\cos t + \frac{1}{9}\cos 3t + \cdots) + (\sin t - \frac{1}{2}\sin 2t + \cdots)$. (e) $f(t) = (1/\pi)(\frac{8}{3}\sin t - \frac{16}{15}\sin 2t + \frac{24}{35}\sin 3t + \cdots)$. (g) $f(t) = -\frac{1}{3} + (4/\pi^2)(\cos \pi t - \frac{1}{4}\cos 2\pi t + \frac{1}{9}\cos 3\pi t + \cdots) + (2/\pi)(\sin \pi t - \frac{1}{2}\sin 2\pi t + \frac{1}{3}\sin 3\pi t + \cdots)$.

Page 536, Problems 18.3

1. $f(t) = \cdots - \frac{4i}{\pi}e^{-3i\pi t} - \frac{12i}{\pi}e^{-i\pi t} + 6 + \frac{12i}{\pi}e^{i\pi t} + \frac{4i}{\pi}e^{3i\pi t} + \cdots$.

Page 538, Exercises 18.3

Group A

1. (a) $f(1) = 0$, $f(2) = \frac{1}{2}$, $f(3) = \frac{1}{2}$. (c) $f(-2) = 4$, $f(-1) = 1$, $f(0) = 0$, $f(2) = 4$, $f(3) = 1$. **4.** (a) $f(t) = \cdots + ie^{-\pi i t/2} + (\pi/2) - ie^{\pi i t/2} + \cdots$. (c) $f(t) = \cdots (1/8\pi)\cdot(-\sqrt{3} + 3i)e^{-4\pi i t/3} + (1/4\pi)(\sqrt{3} + 3i)e^{-2\pi i t/3} + \frac{1}{3} + (1/4\pi)(\sqrt{3} - 3i)e^{2\pi i t/3} + (1/8\pi)(-\sqrt{3} - 3i)e^{4\pi i t/3} + \cdots$.

Group B

1. (a) $f(t) = (-i/4\pi)e^{-4\pi it} - (i/2\pi)e^{-2\pi it} + \frac{1}{2} + (i/2\pi)e^{2\pi it} + (i/4\pi)e^{4\pi it} + \cdots$.
(c) $f(t) = \cdots + [(i/4\pi) + (1/8\pi^2)]e^{-4\pi it} + [(i/2\pi) + (1/2\pi^2)]e^{-2\pi it} + \frac{1}{3} + [(-i/2\pi) + (1/2\pi^2)]e^{2\pi it} + [(-i/4\pi) + (1/8\pi^2)]e^{-4\pi it} + \cdots$.

Group C

1. $E(t) = \sum_{n=-\infty}^{n=\infty} c_n e^{200\pi it}$, where $c_n = -12i/n\pi$, n odd. $c_0 = 6$. $z = 150 + i[2n\pi - (5000/n\pi)]$, $I(t) = E(t)/z$.

Page 544, Exercises A.2

1. $y = 2.67x - 1.13$. **3.** $y = 0.99x^2 + 2.98$. **5.** $I = 5.05e^{-0.5t} + 1.96$.

Index

A

D

E

F

G

K

L

M

N

O

P

Q

R

S

T

U

V

W

Z